# Understanding Petroleum Reservoirs: Towards an Integrated Reservoir Engineering and Geochemical Approach

**Special Publication reviewing procedures**

The Society makes every effort to ensure that the scientific and production quality of its books matches that of its journals. Since 1997, all book proposals have been refereed by specialist reviewers as well as by the Society's Books Editorial Committee. If the referees identify weaknesses in the proposal, these must be addressed before the proposal is accepted.

Once the book is accepted, the Society has a team of Book Editors (listed above) who ensure that the volume editors follow strict guidelines on refereeing and quality control. We insist that individual papers can only be accepted after satisfactory review by two independent referees. The questions on the review forms are similar to those for *Journal of the Geological Society*. The referees' forms and comments must be available to the Society's Book Editors on request.

Although many of the books result from meetings, the editors are expected to commission papers that were not presented at the meeting to ensure that the book provides a balanced coverage of the subject. Being accepted for presentation at the meeting does not guarantee inclusion in the book.

Geological Society Special Publications are included in the ISI Index of Scientific Book Contents, but they do not have an impact factor, the latter being applicable only to journals.

More information about submitting a proposal and producing a Special Publication can be found on the Society's web site: www.geolsoc.org.uk.

It is recommended that reference to all or part of this book should be made in one of the following ways:

CUBITT, J. M., ENGLAND, W. A. & LARTER, S. (eds) 2004. *Understanding Petroleum Reservoirs: towards an Integrated Reservoir Engineering and Geochemical Approach*. Geological Society, London, Special Publications, **237**.

MEULBROEK, P., CATHLES, L. & GODDARD, W. A. 2004. HCToolkit/EOSInterface: an open source, multi-platform phase equilibria framework for exploring phase behaviour of complex mixtures. *In*: CUBITT, J. M., ENGLAND, W. A. & LARTER, S. (eds) 2004. *Understanding Petroleum Reservoirs: towards an Integrated Reservoir Engineering and Geochemical Approach*. Geological Society, London, Special Publications, **237**, 89–98.

GEOLOGICAL SOCIETY SPECIAL PUBLICATION NO. 237

# Understanding Petroleum Reservoirs: towards an Integrated Reservoir Engineering and Geochemical Approach

EDITED BY

J. M. CUBITT
ES Information & Consultancy Ltd, Wrexham, UK

W. A. ENGLAND
BP, Sunbury-on-Thames, UK

and

S. R. LARTER
University of Calgary, Canada

2004

Published by

The Geological Society

London

# THE GEOLOGICAL SOCIETY

The Geological Society of London (GSL) was founded in 1807. It is the oldest national geological society in the world and the largest in Europe. It was incorporated under Royal Charter in 1825 and is Registered Charity 210161.

The Society is the UK national learned and professional society for geology with a worldwide Fellowship (FGS) of 9000. The Society has the power to confer Chartered status on suitably qualified Fellows, and about 2000 of the Fellowship carry the title (CGeol). Chartered Geologists may also obtain the equivalent European title, European Geologist (EurGeol). One fifth of the Society's fellowship resides outside the UK. To find out more about the Society, log on to www.geolsoc.org.uk.

**The Geological Society Publishing House** (Bath, UK) produces the Society's international journals and books, and acts as European distributor for selected publications of the American Association of Petroleum Geologists (AAPG), the American Geological Institute (AGI), the Indonesian Petroleum Association (IPA), the Geological Society of America (GSA), the Society for Sedimentary Geology (SEPM) and the Geologists' Association (GA). Joint marketing agreements ensure that GSL Fellows may purchase these societies' publications at a discount. The Society's online bookshop (accessible from www.geolsoc.org.uk) offers secure book purchasing with your credit or debit card.

To find out about joining the Society and benefiting from substantial discounts on publications of GSL and other societies worldwide, consult www.geolsoc.org.uk, or contact the Fellowship Department at: The Geological Society, Burlington House, Piccadilly, London W1J 0BG: Tel. +44 (0)20 7434 9944; Fax +44 (0)20 7439 8975; E-mail: enquiries@geolsoc.org.uk.

For information about the Society's meetings, consult *Events* on www.geolsoc.org.uk. To find out more about the Society's Corporate Affiliates Scheme, write to enquiries@geolsoc.org.uk.

Published by The Geological Society from:
The Geological Society Publishing House
Unit 7, Brassmill Enterprise Centre
Brassmill Lane
Bath BA1 3JN, UK

(*Orders*: Tel. +44 (0)1225 445046
Fax +44 (0)1225 442836)
Online bookshop: http://bookshop.geolsoc.org.uk



**British Library Cataloguing in Publication Data**
A catalogue record for this book is available from the British Library.

ISBN 1-86239-168-8

Typeset by J W Arrowsmith Ltd, Bristol, UK
Printed by Alden Press, Osney Mead, Oxford OX2 0EF, UK

**Distributors**

*USA*
AAPG Bookstore
PO Box 979
Tulsa
OK 74101-0979
USA
*Orders*: Tel. +1 918 584-2555
Fax +1 918 560-2652
E-mail bookstore@aapg.org

*India*
Affiliated East-West Press Private Ltd
Marketing Division,
G-1/16 Ansari Road, Daryaganj,
New Delhi 110 002
India
*Orders*: Tel. +91 112 327-9113, +91 112 326-4180
Fax +91 112 326-0538
E-mail affiliat@vsnl.com

*Japan*
Kanda Book Trading Co.
Cityhouse Tama 204
Tsurumaki 1-3-10
Tama-shi
Tokyo 206-0034
Japan
*Orders*: Tel. +81 (0)423 57-7650
Fax +81 (0)423 57-7651
E-mail geokanda@ma.kcom.ne.jp

# Contents

# Introduction

WILLIAM A. ENGLAND[1] & STEVE LARTER[2*]

[1]*BP Exploration, Chertsey Road, Sunbury-on-Thames, Middlesex TW16 7LN, UK*
[2]*NRG Petroleum Group, School of Civil Engineering and Geosciences, University of Newcastle-upon-Tyne, UK*
**Present address: Department of Geology and Geophysics, University of Calgary, Calgary, Canada*

Since the first Geological Society Special Publication on reservoir geochemistry (Cubitt & England 1995), there has been a steady increase in the number of publications and conferences on this topic. Reservoir geochemistry is now a routine tool of the petroleum industry with new companies now providing dedicated reservoir geochemical services. With 95% of future world oil production likely to come from already discovered accumulations this steady increase in research and application of reservoir geochemistry is welcome. Despite this, there remain many challenges and great areas of ignorance and missed opportunities. This volume therefore provides a summary of the current status of reservoir geochemistry and the challenges that remain.

Reservoir geochemistry studies the compositional variations of petroleum reservoir fluids (waters, oils and gases) at a variety of spatial and temporal scales. These studies reveal information about petroleum basin development, details of reservoir filling and leaking, and about petroleum mixing and alteration (Larter & Aplin 1995). This information is of interest from both the academic and applied points of view, and provides insights not available from other methods. Reservoir geochemistry is concerned with the 3D compositional variations in reservoir fluids that are commonly observed in petroleum reservoirs as free fluids or as fluid inclusions in mineral cements. It strives to understand the origin of these heterogeneities and to apply them to enhance the understanding of the formation of petroleum accumulations and exploration, appraisal, development and production strategies. Key components of the approach are (1) the analytical methods for determination of fluid compositional heterogeneity, (2) the physical, chemical and numerical models for interpreting the compositional differences in terms of basin history and reservoir connectivity and (3) the case studies of application. By way of introduction, we attempt to summarize these main areas, which provide the basic currency of development within the field of reservoir geochemistry.

## New approaches, analytical and geochemical developments

The development of analytical methods and theoretical and practical methodologies for the determination and interpretation of fluid compositional variation in petroleum reservoirs has been one of the foundation stones of reservoir geochemistry and new developments continue to appear.

**Thompson** presents a study of the molecular distributions of petroleum fluids, as revealed by standard PVT data. The method relies on the observation that carbon number profiles may be fitted by *two* exponential decline curves ($C_3$–$C_5$ and $C_{10+}$). Using PVT data principally from western Canada, Thompson proposes that changes to the $C_3$–$C_5$ and $C_{10+}$ slopes can be related to maturation and phase separation. The addition of allochthonous gas can also be recognized. Given the ubiquity of PVT data, these techniques will prove useful additions to the interpretive armoury.

**Wilhelms & Larter** review progress in understanding the quantitative evolution of molecular marker compositions in reservoired petroleum. They suggest that all oils are mixtures, and that use of many of the high molecular weight biological marker geochemical proxies is compromised by orders of magnitude concentration variations of these components, within even related oil charges or even within single reservoirs. They identify the less exotic non-biomarker molecules as best for assessing reservoir continuity and charge mixing, and they suggest that the concept of maturity needs revising given this complex scenario. They consider that the apparent ubiquity of petroleum mixing, whilst a complication, may herald a new era of petroleum

*From*: CUBITT, J. M., ENGLAND, W. A. & LARTER, S. (eds) 2004. *Understanding Petroleum Reservoirs: towards an Integrated Reservoir Engineering and Geochemical Approach*. Geological Society, London, Special Publications, **237**, 1–5. 0305-8719/04/$15.00 

geochemistry, with great potential for improved resolution of charging processes.

**Philp *et al.*** summarize their decade of work on high molecular weight hydrocarbons (HMWHC) in oils, emphasising the analytical issues associated with separation of pure wax and asphaltene fractions (which frequently contaminate one another). Improved understanding of the variability in composition of the fraction may ultimately aid in prediction and avoidance of flow assurance problems associated with wax deposition. These HMWHC compounds, with more than 40 carbon atoms, are now known to be a complex mixture of up to eight homologous series of alkanes and alkyl aromatic hydrocarbons. They are found in oils sourced from essentially all source rocks facies, not just source rocks containing higher plant debris.

**Holba *et al.*** have studied the impact of petroleum biodegradation on reservoir fluid properties and well productivity in the Kuparuk River field, Alaska. Very significantly, they show that the major factor affecting the hydrocarbon degradation process is anaerobic biodegradation by dissimilatory iron reduction that selectively removes long chain alkyl benzenes and alkyl toluenes. The classical first signatures of oil biodegradation (removal of *n*-alkanes) are not seen in this system, but nevertheless the process severely reduces oil quality and flow properties deteriorate. This study is a good example of how studies focused on practical production objectives can also substantially improve our knowledge of basic processes, such as biodegradation, that are still poorly understood.

**Meulbroek *et al.*** have created an innovative computational framework to estimate phase equilibria, equation of state properties and compositionally dependent viscosity for the modelling of reservoir processes. This toolkit has been designed with the reservoir geochemist in mind and therefore is created as a set of modules that can be utilized within standard spreadsheet applications.

## Physical, chemical and numerical models

The interpretation of chemical compositional data from petroleum systems in the context of quantitative models was one of the major advances in petroleum geoscience in the 1980s and 1990s. Great strides were made in basin modelling and in understanding the timescales of petroleum system processes such as petroleum mixing within reservoirs. Despite this, we lack effective models or model calibrations in many areas of petroleum geoscience. We consider effective models of petroleum charging and mixing in reservoirs crucial to useful interpretation of reservoir geochemical data, yet utilitarian models capable of integrating compositional and reservoir property data in an effective manner still elude us. This volume contains several new and welcome developments in this area.

**Smalley *et al.*** studied the rates at which pressure differences, contact differences, density differences and chemical compositional differences equilibrate in a simplified reservoir model. They supply analytical solutions, supported by numerical simulations. These solutions should prove useful for screening studies, by comparing the various mixing timescales with the time available since the reservoir was filled. This is a much quicker process than building and running a full-scale reservoir model.

**Stainforth** presents new models for reservoir filling and mixing. He presents data to support the idea that, in many cases, petroleum does not mix at all during reservoir filling. As new petroleum enters the trap, it fills from the crest of the structure, forcing previously emplaced petroleum downwards. This is a result of the general decrease in fluid density with maturity. This model predicts that the shapes of saturation pressure versus depth curves are related to trap geometry (depth versus volume curves) as well as source rock kitchen parameters. Field data are presented to support this model.

**Skeie *et al.*** use the new approach of kinetics of reservoired oil asphaltene thermal degradation and basin modelling to try to better define the source kitchen maturity and extent in inadequately sampled source kitchens. They use the approach to re-evaluate the evolution of the Tampen Spur petroleum systems, with particular emphasis on the charging of the Snorre Field. They suggest that the Møre Basin was not a major source of petroleum to the Snorre Field, with the bulk of the charge coming from the east. Though still controversial, this represents an ambitious attempt to try to deal with our inability to adequately characterize source kitchens.

**Di Primio & Skeie** use compositional data from closed system pyrolysis experiments to construct a compositional kinetic model of petroleum generation from the Draupne Formation. Although this, in itself, is not new, they developed the model in a format compatible with the data resolution used in PVT models that are familiar to reservoir engineers. Most importantly, they tuned their PVT models to the observed properties of actual petroleum accumulations observed in a well-defined portion of the North Sea. This represents a significant advance in our ability to make direct comparisons between model predicted and actual compositions and phase behaviour of reservoired petroleums.

## Case studies – water and petroleum fluid compositional variations

In recent years, interest is increasingly moving from the techniques employed in reservoir geochemistry to the interpretation of results in a wider geological context. In particular, there is increased emphasis on understanding the processes responsible for fluid variations (e.g. gravitational segregation, filling direction, biodegradation) and relating these to the rates of the processes that lead to reductions in fluid heterogeneity (diffusion, thermal and density-driven convection).

**Lines & Auld** present a detailed analysis of the complex fluid distribution found among the fields of the J-Block, UKCS, involving several reservoir intervals. In an integrated study, the authors identify differences in maturity and source input using high resolution GC fingerprinting. Pressure data and strontium isotope analyses were used to investigate reservoir connectivity/compartmentalization. 2D basin modelling was employed to estimate the timing and extent of petroleum expulsion and migration. These results were integrated to produce a charge model to explain the differences observed, enabling the authors to assess the likely impact on field development.

**Wavrek & Mosca** report a study of the giant oil fields of the Val D'Agri region of Italy. They identified gravitationally induced compositional grading as the dominant cause of variation between samples in this complex area. An extensive analytical programme was undertaken in order to establish the key molecular fractions and molecular signatures that best characterize the alteration signature. They included PVT data, conventional geochemical measurements and measurements on the organic sulphur fraction.

**Mueller & Scholz** describe an application of reservoir geochemistry to the Voelkersen dry gas field in Germany. Due to the well depths (>4700 m), it is important to identify large connected compartments for economic production. Condensate and gas isotope variations were found to be valuable in establishing compartment boundaries. The geochemically-derived compartments correlated well with those based on conventional characterization methods. Further refinements were suggested based on the reservoir geochemical study. Practical sampling issues were also discussed, as samples at the well head are impossible to acquire due to gas pressure and temperature.

**Weissenburger & Borbas** describe a study on the fluid properties and compartmentalization in the Magnolia deep-water field from the Gulf of Mexico. They highlight observations that suggest that mixing is incomplete within what are expected to be well-connected compartments. For example, fine-scale MDT sampling across one reservoir shows a gradient in methane carbon isotope of 2.5 parts per thousand per 100 feet, despite the apparent absence of pressure discontinuities. Given that Miocene, Pliocene and Pleistocene reservoirs were deposited about 7.0 to 1.4 Ma ago, it is clear that this is a very recent hydrocarbon-bearing province. Presently occurring filling and leakage, coupled with low reservoir temperatures, are also factors making complete compositional homogenization less likely.

**Taylor *et al.*** describe how a petrographic and fluid inclusion study, coupled with quartz precipitation kinetic modelling, of a deep gas reservoir in the Norphlet trend in Mobile Bay reveals a very complex charge and leakage history of oil, prior to a late final gas charge. Reservoir connectivity and performance is influenced by quartz cementation, with preservation of porosity and formation of tight zones being controlled by the extent of chlorite coating of quartz grains controlling quartz diagenetic deposition rates.

**Dutton & Flanders** examined the diagenesis and reservoir quality of the East Ford Field, Texas, which is undergoing a carbon dioxide flood. The compartmentalization of the arkosic sandstone reservoir is chiefly controlled by authigenic calcite layers associated with the tops and bottoms of turbidite sandstone units. Gases produced from zones below low permeability calcite cemented sandstone intervals in new infill wells have high carbon dioxide contents. These, together with geophysical log interpretations, indicate that the calcite-cemented zones are laterally continuous and act as vertical barriers in the reservoir.

**McCartney *et al.*** analysed a very significant data set of formation water composition and formation pressure data to determine whether regional barriers to fluid flow affect formation water compositions, evolution of overpressure distribution and changing hydrocarbon migration pathways through time. They suggest the Jurassic and Triassic aquifers of the Central North Sea have essentially behaved as a large composite aquifer, which has many implications for the deep prospectivity of the CNS.

**Karlsen *et al.*** investigated petroleum migration and overpressure evolution in the Haltenbanken area of the Norwegian offshore. Using petroleum geochemical characterization of current petroleums from reservoir core extracts and of palaeo-petroleum trapped in fluid inclusions, they demonstrate a very complex evolution of petroleum in the reservoirs. Both currently filled and currently dry structures in the Smørbukk field area were initially charged with black oil some 50–70 Ma ago, the currently dry structures then being filled with gas condensate prior to seal failure and leakage of the petroleum from the reservoir. The leakage events are interpreted as being caused by overpressure development related to diagenesis in the Smørbukk fault zone.

Finally, **Rushing *et al.*** have created a mass balance approach that directly integrates dynamic elements of the total hydrocarbon system with all static reservoir engineering, geochemical and petrophysical components. This prospect evaluation technique provides an independent estimate of hydrocarbon resource-in-place for comparison with that computed by the reservoir engineering approach.

## Future directions in reservoir geochemistry

We have come a long way since the first Special Publication on reservoir geochemistry (Cubitt & England 1995). Despite this, however, little progress has been made in some key areas. The coupling of basin models, reservoir filling/mixing models and detailed chemical compositions of petroleum has hardly started. In principle, numerical basin models coupled to reservoir models could address many questions. Given the large number of possible processes that need to be modelled, it is expected that progress along these lines will be difficult. Nevertheless, work along these lines is inevitable and urgently needed. **Stainforth** has shown that considering the petroleum system (source rock/migration/trapping system) as one entity has important impacts. It is not clear yet how widely applicable Stainforth's observations will turn out to be, but he has clearly demonstrated the importance of linking reservoir geochemistry to models of source kitchen maturation and reservoir filling.

There still appears to be more work to be done on understanding vertical gradients in fluid composition. Stainforth's work suggests that many vertical gradients may be due to filling and do not represent true vertical thermodynamic equilibrium of thermal and gravitational forces. Recent work by Ratulowski *et al.* (2003) has demonstrated the use of a laboratory centrifuge to simulate reservoir-scale isothermal gravity segregation. Their work suggests that there is a very valuable place for carefully designed laboratory work in reservoir geochemistry to help understand field data. So far no laboratory work has been reported on combined thermo-gravitational equilibrium, despite some theoretical work.

The lack of equilibration noted by **Weissenburger & Borbas** in the young deep water Gulf of Mexico petroleum system also point to the importance of fully understanding the timescales of the creation and elimination of fluid heterogeneity. The expense of deep-water exploration and appraisal means that reservoir geochemical deductions on compartment presence will become ever more important. Due to the very recent timing of trap formation and filling, together with active biodegradation and leakage, it is clear that our interpretative methods developed in less dynamic basins will not be directly applicable to many deep-water basins.

The increasingly obvious dynamic nature of most petroleum systems, often involving synchronous charging from variably mature source rocks, leakage/spilling and alteration of petroleums in traps, poses great challenges for petroleum geochemists. Petroleum geochemists have traditionally used approaches and parameters best suited to describe petroleums as single entities with simple origins in terms of source or maturity-weighted averages. Our analytical approaches have improved enormously in the time since the mid 1980s, when the vast bulk of the current petroleum geochemical source and maturity parameter genre was developed. We feel there is now an enormous potential for a new wave of theoretical and practical development in petroleum geochemistry focused on using component concentrations and a more evolutionary chemical model of reservoired petroleum. This would be a significant advance on the 'parameter ratio' types of information that were developed in the past to answer rather different questions. Such approaches, if coupled directly to developments in integrative utilitarian reservoir charge/mixing/production numerical models, may make the 21st century the century of reservoir geochemistry.

As in most scientific fields, the more we know, the more questions are posed. Reservoir geochemistry is no exception, and it is clear that there remains much work to be done in this fascinating field.

We acknowledge the support and encouragement of the Geological Society, and in particular John Cubitt, over a number of years in promoting conferences and publications in the area of Reservoir Geochemistry.

## References

CUBITT, J. M. & ENGLAND, W. A. (eds) 1995. *The Geochemistry of Reservoirs.* Geological Society, London, Special Publications, **86**.

LARTER, S. R. & APLIN, A. C. 1995. Reservoir geochemistry: methods, applications and opportunities. *In*: CUBITT, J. M & ENGLAND, W. A. (eds) *The Geochemistry of Reservoirs.* Geological Society, London, Special Publications, **86**, 5–32.

RATULOWSKI, J., FUEX, A. N, WESTRICH, J. T. & SIELER, J. J. 2003. *Theoretical and Experimental Investigations of Isothermal Compositional Grading.* SPE Reservoir Evaluation and Engineering, paper **84777**, June 2003.

# Interpretation of charging phenomena based on reservoir fluid (PVT) data

KEITH F. M. THOMPSON

*Petroleum/Geochemical Data, P.O. Box 671264, Dallas, TX 75367-1264, USA (e-mail: keithfmt@earthlink.net)*

**Abstract:** Molar concentration profiles of reservoir fluids reveal accumulation history and alteration. Slope Factors (SF) define rates of exponential decrease in concentration of light *n*-alkanes ($C_3$–$nC_5$) and liquid pseudo-components ($P_{10+}$) with increasing carbon number. Most petroleum fluids are substantially characterized by SF($C_3$–$nC_5$) and SF($P_{10+}$), the former invariably being the greater. Specific paired values of SF($C_3$–$nC_5$) and SF($P_{10+}$) are process-diagnostic. In oils, maturation, gas injection, evaporative fractionation and migration depletion involving loss of gas, are recognizable. SF data are interpreted in the light of PVT analyses representing oils (principally from western Canada) and gas-condensates (from numerous basins), also asphaltene pyrolysis experiments and equation of state calculations. Covariant increase of SF($C_3$–$nC_5$) and SF($P_{10+}$) during maturation is demonstrated, but correlation is frequently destroyed by modification of the light ends by the admixture of allochthonous gas, increasing only SF($C_3$–$nC_5$). Secondary gas-enrichment is a requisite process for the generation of gas-condensates by evaporative fractionation. Compositional criteria for the recognition of enrichment are provided for the first time, particularly attainment of a value of SF($C_3$–$nC_5$) exceeding 1.69 (a tentative limit). Available data indicate that the process has occurred in a large proportion of oil accumulations, ranging from 23% of 30 reservoirs in the Jurassic Smackover Formation in Alabama, to 78% of 36 in the northern North Sea.

The objectives of this study are to illustrate, in terms of commercial PVT data, and Slope Factors (SF) summarizing the compositional information therein, three fundamental processes affecting the make-up of petroleums: the maturation of oil, its modification by the admixture of gas and gas-condensate, and the generation of gas-condensates by evaporative fractionation as a consequence of gas addition.

Oils and gases have been extensively studied separately, but less frequently in their natural associations. Pioneering work was carried out by Zhuze & Yushkevitz (1959), Zhuze *et al.* (1962, 1963) and Zhuze & Ushakova (1981). Geochemical evaluations of vapour–liquid equilibria and PVT data include Price *et al.* (1983), Thompson (1987, 1988, 1992, 2002), England and co-workers (1989, 1991, 2002), Larter & Mills (1991), Knudsen & Meisingset (1991), Dzou & Hughes (1993), Curiale & Bromley (1996), Meulbroek *et al.* (1998), di Primio (1998, 2002), Khavari-Khorasani *et al.* (1998), van Graas *et al.* (2000), Meulbroek (2002) and Losh *et al.* (2002). Meulbroek, as well as di Primio, are contributors to this volume. The present work employs PVT reports which provide analyses of reservoir fluids from methane through $C_{30+}$ for oils in western Canada, but only $C_1$–$C_{7+}$ data for other regions. Analyses in PVT reports represent the fluid as present in the reservoir, without loss of either gaseous or liquid components. Up to 41 individual compound and pseudo-component concentrations may be reported (where a pseudo-component comprises a summed suite of compounds, ideally all of the same carbon number, further defined below). It is stressed that PVT gas analytical data differ substantially in composition from those representing surface samples of petroleum gas from production equipment such as separators etc. Separator gases are strongly and variably fractionated in comparison with reservoir fluids, with losses, versus methane, of light and liquid components. Stock tank oils are similarly fractionated, but without substantial effect beyond approximately $nC_5$ if the sample is stored well.

## Slope Factors

The physical properties reported in PVT analyses, such as gas–oil ratio (GOR), saturation pressure, viscosity, shrinkage factors and data concerning phase changes during simulated production, are expressions of the chemical make up of the reservoir fluid. Before physical properties can be understood and modelled, the chemical nature of a reservoir fluid must be rationalized. This requires a means of reducing the number of compositional variables. An applicable technique involves the determination of Slope

*From*: CUBITT, J. M., ENGLAND, W. A. & LARTER, S. (eds) 2004. *Understanding Petroleum Reservoirs: towards an Integrated Reservoir Engineering and Geochemical Approach.* Geological Society, London, Special Publications, **237**, 7–26. 0305-8719/04/$15.00 

Factors (Kissin 1987; Thompson 2002). The approach relies upon the fact that pseudo-component concentrations form exponential progressions by carbon number, a fact discovered by Lohrenz & Bray (1964), applied by Pedersen *et al.* (1992), and widely utilized in petroleum engineering calculations. Investigations of the exponential distribution of polymer degradation products by molecular weight are traceable to Kuhn (1930) and Flory (1936). Exponential progressions of $C_{7+}$ *n*-alkane concentrations in oils were reported by Kissin (1987) and of gaseous and light *n*-alkanes by the writer (Thompson 2002). The latter reference discusses the definition, calculation, theory and mechanism of generation of petroleum slopes and Slope Factors.

The slope of the exponential equation fitted to petroleum molar concentration data over a range of carbon numbers is transformed into a Slope Factor. The model is

$$y = A\,e^{(-an)} \qquad (1)$$

where $y$ equals the molar percentage of a given component, $A$ is a normalization factor, $a$ is the slope, and $n$ is the carbon number of the component. Slope Factors are defined thus:

$$\mathrm{SF} = e^{a}. \qquad (2)$$

SF equals the constant ratio expected to be observed between adjacent petroleum components, regarded here as a sequence which increases with decreasing carbon number. The exponent, $a$, is thus considered positive, in keeping with the mechanism of creation of the sequence, the progressive fragmentation of long-chain precursor molecules and the accumulation of short chains.

Two Slope Factors can be defined in PVT analytical data, identified here as $SF(C_3–nC_5)$ and $SF(P_{10+})$, the first representing light alkanes, the second, liquid pseudo-components. Employing gas-chromatographic analysis, data are summed in PVT reports so that, for example, pseudo-component $P_{10}$ includes all compounds eluting after $nC_9$, up to, and including, $nC_{10}$, resolved and unresolved, integrated from an 'empty column' baseline. The series $P_7–P_{30}$ conforms to an exponential sequence, occasionally with slight excess in the $C_6–C_9$ region believed to be attributable to liquid components accompanying secondarily added gas, as discussed later. It was found (Thompson 2002) that the normal-alkanes exhibit two exponential series, the first from propane to $nC_6$, the second from $nC_7$ to $nC_{30}$ and higher, the former invariably steeper than the latter, the slope break occurring at $nC_6$ in asphaltene pyrolysates. Unfortunately, PVT analyses report individual *n*-alkanes only

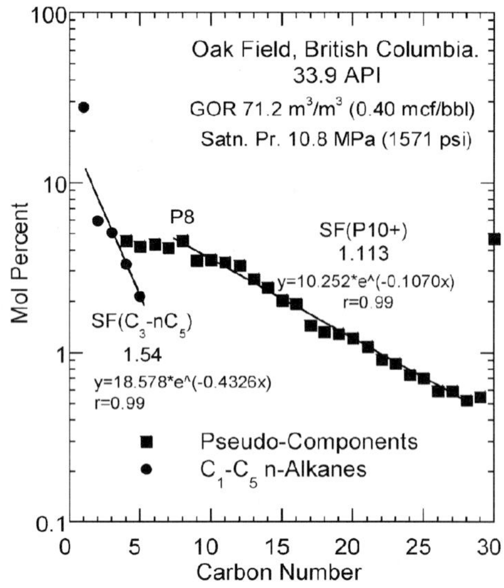


**Fig. 1.** Molar concentration profile of low maturity oil, Oak field, British Columbia, illustrating assessment of $SF(C_3–nC_5)$ and $SF(P_{10+})$.

through pentane, providing pseudo-component data at higher carbon numbers. In petroleums, methane and ethane are variable in concentration, conforming poorly to the exponential progression which, using the available data, comprises propane, *n*-butane and *n*-pentane, yielding $SF(C_3–nC_5)$. In almost all petroleums, methane is present in excess, while ethane is deficient, as seen in Figure 1. Liquid *n*-alkanes comprise the greater part of the pseudo-components, therefore the slope break in PVT data manifests itself between $SF(C_3–nC_5)$ and $SF(P_{10+})$. The cause of the break in slope is discussed in Thompson (2002).

Employing two variables, $SF(C_3–nC_5)$ and $SF(P_{10+})$, a major data reduction is brought about and the composition of a petroleum is substantially quantified. The two parameters are co-variant as generated in oils although secondary processes commonly degrade the relationship. Investigation of one such process, the injection of gas into oil, is a major topic herein. It was postulated (Thompson 1987) that the majority of gas-condensates are generated by evaporative fractionation subsequent to the injection of migratory gas into oil. The process was originally recognized on the basis of changes in gasoline composition involving the enrichment of aromatic and naphthenic hydrocarbons in the residual oil and, ultimately, a similar expression in daughter gas-condensates. In 1987, the only available evidence of migratory, allochthonous, gas was the occurrence of isotopically light, biogenic, methane in certain oils. Slope Factor evidence presented here is indicative of the commonplace occurrence of the injection of gas and condensate into oil accumulations in western Canada and in other basins, leading to partial or complete vaporization and the generation of further gas-condensates enriched in liquid fraction components. This evaporative, fractionating, process appears to largely control SF($P_{10+}$) in gas-condensates, subject to the effects of confining pressure and temperature. With few possible exceptions, all of the gas-condensates evaluated here are evaporative, their Slope Factors having no relationship to the maturity of the liquid components.

## Data provenance

Reservoir fluid analyses employed in this study originate in commercial petroleum engineering laboratories and have been made available by the Alberta Energy and Utilities Board, the Province of British Columbia Ministry of Energy and Mines, and the Canada–Newfoundland Offshore Petroleum Board, representing petroleums found in their respective areas of administration. Limited representation of other areas is also provided. Evaluation of the range of commonly encountered Slope Factors has involved the collection of PVT studies of oils and gas-condensates which provide analytical data from $C_1$ to $C_{30+}$. This protocol has been followed in Canada since approximately 1980. Here, 125 such are employed, largely from the Western Canada Sedimentary Basin, the suite which was utilized in Thompson (2002). A suite of gas-condensate data is newly added, providing 48 similarly detailed analyses including virtually all available cases from Alberta, also representatives from the Gulf of Mexico region, Columbia, the North Sea, offshore Newfoundland, and Nigeria.

To assist in defining the Slope Factors of an unaltered sequence of oils of increasing maturity, pyrolysates of a petroleum asphaltene were employed (data of Thompson 2002). Aliquots of a purified asphaltene from a mature, clastic-sourced, oil were heated in sealed quartz capillaries for a period of 48 h at temperatures of 320, 340, 350 and 360 °C (MSSV method of Horsfield *et al.* 1989). The two lower temperatures yielded petroleum-like fluids of very low maturity, less that that of a 30° API oil, while the 350 and 360 °C pyrolyses attained the maturity level of mid-range of oils and were strikingly oil-like in their associated values of light end and liquid component Slope Factors, as evidenced in Table 1.

**Table 1.** *Slope Factors representing asphaltene pyrolysates, and oils of Rainbow and adjacent fields, northwest Alberta*

| Case | Location | SF($C_2$–$nC_6$) | SF($C_3$–$nC_5$) | SF($P_{10+}$) | Properties of oils | | |
|---|---|---|---|---|---|---|---|
| | | | | | API | GOR ($m^3/m^3$) | Bubble point (MPa) |
| Asphaltene 48/320 °C | | 1.12 | 1.16 | 1.02 | | | |
| Asphaltene 48/340 °C #1 | | 1.32 | 2.00 | 1.04 | | | |
| Asphaltene 48/340 °C #2 | | 1.46 | 1.72 | 1.07 | | | |
| Asphaltene 48/350 °C | | 1.56 | 1.69 | 1.12 | | | |
| Asphaltene 48/360 °C | | 1.54 | 1.91 | 1.16 | | | |
| 701, Rainbow S. | 107-10* | | 1.69 | 1.16 | 41.6 | 73.30 | 13.47 |
| 702, Rainbow | 108-9 | | 1.52 | 1.11 | 35.4 | 66.31 | 8.14 |
| 703, Rainbow | 108-10 | | 2.06† | 1.12 | 38.7 | 89.66 | 9.43 |
| 704, Rainbow | 109-6 | | 1.53 | 1.12 | 40 | – | 15.38 |
| 705, Rainbow | 109-8 | | 1.56 | 1.12 | 36.4 | 73.65 | 8.66 |
| 706, Rainbow | 108-6 | | 1.57 | 1.13 | 41.4 | 131.1 | 14.84 |
| 707, Rainbow | 108-7 | | 1.50 | 1.15 | 45 | 186.95 | 16.93 |
| 708, Rainbow | 110-7 | | 1.79† | 1.10 | 34.4 | 78.59 | 12.4 |
| 709, Rainbow S. | 107-9 | | – | 1.16 | 44.8 | 109.31 | 16.84 |
| 710, Rainbow S. | 107-9 | | – | 1.18 | 44.1 | 104.40 | 16.46 |
| 711, Rainbow S. | 107-10 | | 1.57 | 1.14 | 41.5 | – | 13.96 |
| 712, Rainbow | 111-6 | | 1.53 | – | – | – | 6.9 |
| 713, Virgo | 114-6 | | 1.53 | 1.12 | 38.4 | 117.15 | 14.58 |
| 714, Amber | 114-8 | | – | 1.13 | 41.6 | 88.84 | 15.65 |
| 715, Virgo | 115-5 | | 1.57 | 1.11 | 29.9 | 37.13 | 5.25 |

* Township and Range, Alberta W6.
† Elevated values attributed to gas enrichment.

The present study employs values of SF($C_3$–$nC_5$) and SF($P_{10+}$) determined from fitted equations as shown in Figure 1. The molar profile of a representative oil is depicted, in this instance an oil of low maturity assessed on the basis of SF($P_{10+}$) and API gravity (see below), postulated to be unaltered from its condition upon generation. The low saturation pressure, 10.8 MPa, is noteworthy, indicating undersaturation upon generation and, minimally, during its initial migration. The presence of 'excess' methane, compared with the projection of the light end exponential series, is incompatible with free radical cracking of linear precursors (which results in a substantial deficiency of methane) but is unexplained, as is the 'deficiency' of ethane. These features are observed in the majority of oils. It is evident, and not uncommon, that the $P_{10+}$ data exhibit minor departures from an exponential progression (equation obtained employing the application 'KaleidaGraph') due to analytical shortcomings, yet a very high correlation coefficient is returned. This indicates the need for careful appraisal of computed results. However, precision can be extremely high, as illustrated above by the study of oils representing the Rainbow field, Alberta. The determination of SF($P_{10+}$) has an advantage over most indices of maturity in representing an average over numerous large peak area measurements.

## Slope Factor relationships in oils

Given a large set of PVT analyses, in effect randomly selected, a primary task is to distinguish paired values of SF($C_3$–$nC_5$) and SF($P_{10+}$) which reflect only progressive maturation. Only then can other relationships be assigned to the effects of secondary alteration. Two approaches were employed to distinguish a maturation sequence: (1) appraisal of the Slope Factors in a suite of oils known to be co-genetic and to represent a maturation sequence on the basis of published geochemical data, and (2) comparison of oil data with that representing asphaltene pyrolysates, where cracking conditions unequivocally define maturity.

### *An oil maturity sequence*

The configuration of the Western Canada Sedimentary Basin, with highly mature source rocks in a folded foredeep, facilitating migration onto a wide foreland homocline (Michael & Bachu 2001), suggests that a large sample of oils should exhibit a wide range of maturities. A suite meeting the first criterion above (geochemical appraisal of a co-genetic maturation sequence) is provided by the oils of the Middle Devonian Muskeg and Keg River Formations in northwestern Alberta analysed by Li *et al.* (1999), produced principally from the Rainbow field, also from the smaller Zama, Virgo, Amber, Shekilie, Nipisi and Mitsue fields. Lower Keg River Formation carbonates, deposited in a hypersaline environment, are the postulated source of the high-sulphur (0.4–1.1%) oils. Various geochemical maturity parameters were published, including terpane-based and methylphenanthrene isomeric ratios, exhibiting progressive increase in the suite of oils, described as ranging from low to high API gravity. The Rainbow oils examined here represent wells other than those sampled by Li *et al.* (1999).

PVT analyses representing 15 oils from the Rainbow field and adjacent Muskeg–Keg River fields were examined, as detailed in Table 1. As shown in Figure 2 and equation (3), API gravity is correlated with SF($P_{10+}$), confirming the role of SF($P_{10+}$) as an index of maturity (Kissin 1987).

$$\mathrm{SF}(P_{10+}) = 0.96 + 0.0044.\mathrm{API}, \quad r = 0.83. \quad (3)$$

The relationships of SF($C_3$–$nC_5$) and SF($P_{10+}$) in Rainbow region oils and in the pyrolysates are examined in a subsequent section.

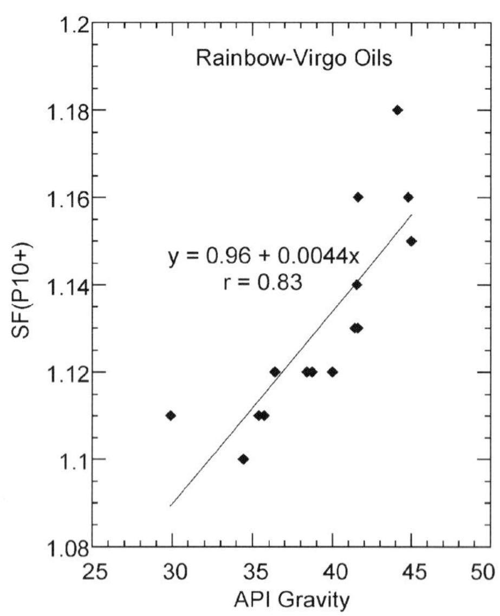


**Fig. 2.** Relationship of SF($P_{10+}$) and API gravity in oils of increasing maturity, Rainbow field area, Middle Devonian, northwestern Alberta, Canada.

In the 340, 350 and 360 °C asphaltene pyrolysis experiments, values of SF($C_3$–$nC_5$) substantially greater than those of SF($C_2$–$nC_6$) reflect the observed 'excesses' of propane (above the $C_2$–$nC_6$ exponential progression), as discussed and illustrated below. Non-identical results in the two 340 °C cases are indicative of deficient experimental control.

## *Comparison of oil and pyrolysate data*

Table 1 and Figure 3 illustrate the degree of similarity which can be adduced between asphaltene pyrolysate data (350 °C/48 h) and oils selected on the basis of simultaneous near-matches of the two Slope Factors, given the following caveat. The best-developed light end exponential series in the asphaltene pyrolysates involves the series ethane through $nC_6$ (Thompson 2002). Figure 3 shows that the pyrolysis data exhibit an oil-like excess of methane but, as opposed to the case in oils, ethane conforms relatively well to an exponential progression involving $nC_6$, $nC_5$ and $nC_4$, that is, propane is present in 'excess' in the pyrolysates. Relatively elevated levels of propane result in high values of SF($C_3$–$nC_5$). These features represent a systematic difference between oils and the pyrolysates, in addition to the differences exhibited by the liquid components described earlier

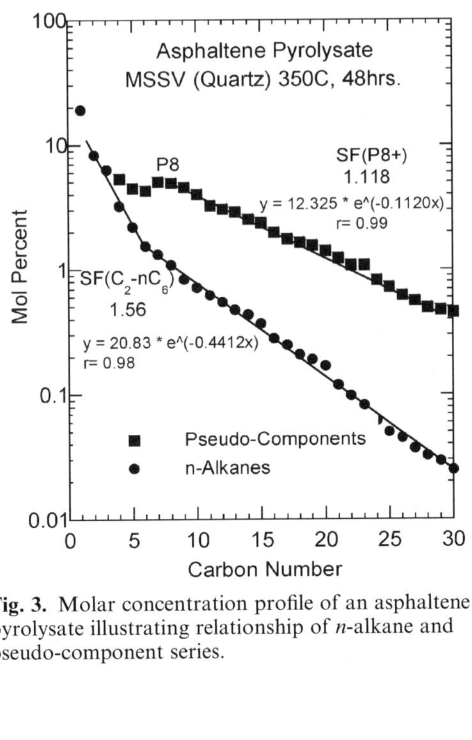


**Fig. 3.** Molar concentration profile of an asphaltene pyrolysate illustrating relationship of $n$-alkane and pseudo-component series.

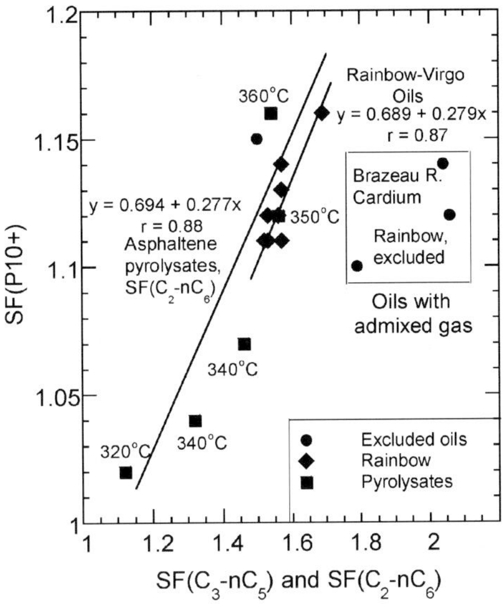


**Fig. 4.** Relationship of light end and liquid component Slope Factors in asphaltene pyrolysates and in a series of oils of increasing maturity.

(Thompson 2002). In the pyrolysates, the slope break in the $n$-alkane series occurs at $nC_6$, indicating that $nC_6$ should be included in the light end exponential progression. The absence of $nC_{6+}$ $n$-alkane data in PVT analyses precludes the measurement of SF($C_2$–$nC_6$). Thus, the best-developed light end exponential series in oils, SF($C_3$–$nC_5$), is compared here with that best developed in the pyrolysates, SF($C_2$–$nC_6$).

SF($C_2$–$nC_6$) and SF($P_{10+}$) increase simultaneously in the pyrolysates, as shown in Figure 4, according to the relationship

$$SF(P_{10+}) = 0.694 + 0.277.SF(C_2\text{–}nC_6),\quad r = 0.88. \qquad (4)$$

Of the 11 paired values of SF($C_3$–$nC_5$) and SF($P_{10+}$) representing Rainbow oils (Table 1), eight exhibit substantial concordance with the trend of the pyrolysis data. Two Rainbow oils, cases 703 and 708, are excluded in the comparison, having relatively elevated values of SF($C_3$–$nC_5$) attributed to the admixture of gas, as discussed in the following section. Excluding a third case (707, Table 1) which might also be regarded as an outlier, results in the regression of equation (5), almost identical with equation (4), although a degree of fortuitousness is involved:

$$SF(P_{10+}) = 0.689 + 0.279.SF(C_3\text{–}nC_5),\quad r = 0.87. \qquad (5)$$

## Gas injection into oil

### *Anomalous associated values of SF($C_3$–$nC_5$) and SF($P_{10+}$)*

The above comparisons indicate that, during maturation, SF($C_3$–$nC_5$) increases in value more rapidly than does SF($P_{10+}$) by a factor of 3.6 in both synthetic and natural series of oils, in a fashion similar to that represented by equations (4) and (5). Examination (in a subsequent section) of all of the available oil data which yield paired values of SF($C_3$–$nC_5$) and SF($P_{10+}$) shows that this relationship is far from representative. Overall, in oils, there is no correlation. SF($C_3$–$nC_5$) spans a wide range of values, from 1.16 to 2.73, versus a limited range of values of SF($P_{10+}$), 1.09 to 1.25. This is attributed to the commonplace occurrence of secondary alteration, in particular, the addition of gas possessing elevated values of SF($C_3$–$nC_5$), an interpretation favoured by consideration of the overall ratios of oil to gas in petroleum accumulations in the Western Canada Sedimentary Basin. Table 2 summarizes initial volumes in-place of oil, and of gas, as published by the Alberta Energy and Utilities Board.

Only the Middle Devonian is principally oil-productive, as confirmed by the following criteria. A sample of 141 PVT analyses representing oil accumulations in western Canada possesses a mean GOR of 105 $m^3/m^3$ (0.59 mcf/bbl). Only 11% exceed 178 $m^3/m^3$ (1.0 mcf/bbl). In general, only volatile oils rarely exhibit GORs greater than 500 $m^3/m^3$, above which level complete vaporization commonly occurs. Most stratigraphic units must, therefore, principally produce gas-condensate.

In the suite of Rainbow oils the highest value SF($C_3$–$nC_5$) occurring in a case where SF($C_3$–$nC_5$) and SF($P_{10+}$) are correlated is 1.69. Values of SF($C_3$–$nC_5$) greater than 1.69 are attributed to admixed gas. SF($C_3$–$nC_5$) values in the range 2.0 to 3.5 predominate in gas-condensates in western Canada, representing suitable candidate fluids for admixture. The value 1.69 is proposed as a working upper limit. Future studies which facilitate the certain recognition of unaltered oils of higher maturity will probably raise the proposed limiting value. The Robinson Creek Smackover oil of Alabama (SF($C_3$–$nC_5$) 1.76), discussed in the section entitled 'Actual oils and gas-condensates', is possibly such a case. The converse problem, of decrease in SF($C_3$–$nC_5$) due to alteration, for example by gas depletion during migration, also remains to be detailed.

In the past it was believed that gas-condensates are invariably of elevated maturity. Immature gas-condensates were described by Garrigues *et al.* (1988). Thompson (1987, 1988) suggested that evaporative fractionation can be invoked in all cases. Early, and continuing, migration of oil of increasing maturity into structures on basin flanks may be, and commonly is, followed by the migration of highly mature gas along numerous of the same migration paths, as documented by James (1990) in western Canada. The same phenomena, involving the entry of gas-condensates bearing highly mature gases into early-emplaced oils, are documented in South Pass 61 field, coastal Louisiana (Holba *et al.* 1996). In general, gas–oil interactions result in gas-enrichment of normal oils, and are postulated to frequently result in gasification, that is, the generation of gas-condensates by evaporative fractionation. Such gas-condensates comprise liquids of various maturities vaporized from oils, combined with gas and light ends which are, if only in part, of high maturity. The high maturity of the added gas may be masked by mixing with indigenous, oil-dissolved, gas. The concepts of gas injection presented here parallel those developed by Gussow (1954), but stress gas–oil interaction, rather than displacement of oil by gas.

At the moment of phase separation, oil and derived gas-condensate are saturated, the saturation pressures of both being equal to the confining pressure. Basin evolution, for example

**Table 2.** *Initial volumes of petroleum in place in Alberta by stratigraphic interval (excluding tar sands)*

| Stratigraphic interval | Oil ($10^6$ $m^3$) | Gas ($10^9$ $m^3$) | GOR ($m^3/m^3$) | GOR (mcf/bbl) |
|---|---|---|---|---|
| Upper Cretaceous | 2160 | 1318 | 610.2 | 3.426 |
| Lower Cretaceous | 1074 | 2629 | 2247.9 | 13.743 |
| Jurassic | 108 | 147 | 1361.1 | 7.642 |
| Permo-Triassic | 337 | 229 | 679.5 | 3.815 |
| Mississippian | 605 | 1214 | 2006.6 | 11.266 |
| Upper Devonian | 2461 | 1489 | 605 | 3.397 |
| Middle Devonian | 958 | 113 | 118 | 0.662 |

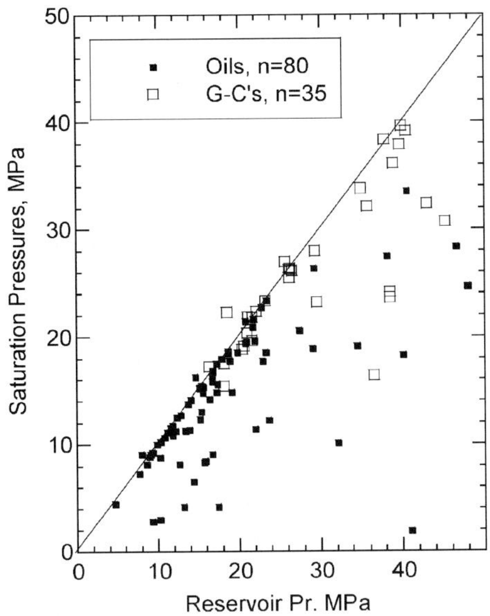


**Fig. 5.** Relationships of saturation pressure and reservoir pressure in oil accumulations (western Canada) and gas-condensates (numerous basins).

subsidence and burial or faulting after fluid emplacement, can result in conditions of undersaturation (Kuo 1997). Despite uplift in western Canada, a substantial proportion of oil and gas accumulations are undersaturated, as illustrated in Figure 5. Among 80 oil accumulations with available data, reservoir pressure exceeds bubble point by more than 10% in 35 cases. Reservoir pressure exceeds dew point by more than 10% in 13 of 36 gas-condensate reservoirs. Reservoir pressure in excess of bubble point may also reflect the relatively low saturation pressures of all oils as generated (suggested by the data of Table 1), followed by migration at depth under pressures greater than those required to maintain saturation. Equality of saturation and reservoir pressure suggests migration under conditions of migration depletion, and reservoir charging which is recent in terms of the timing of basin burial. Michael & Bachu (2001) evaluate fluid and pressure distributions in the uplifted Alberta basin, attributing certain instances of sediment overpressure to gas generation in the deep basin.

## *Evidence of gas admixture in oil*

Oil produced from the Brazeau River Cardium (Cretaceous) pool, detailed in Table 3, contributes unique information which is suggested to be most readily explicable in terms of the introduction of allochthonous, light, gas-condensate at elevated pressure. The Slope Factors of this oil are plotted in Figure 4, in addition to those of the two Rainbow field oils which are suggested to be gas-enriched. Rainbow field is almost entirely oil productive, but the extreme southwestern area possesses limited numbers of gas-condensate wells, evidencing the local availability of gas.

The Brazeau River Cardium oil accumulation is underlain, some 770 m deeper, by the Brazeau River Nisku (Upper Devonian) gas-condensate field. The latter exhibits abnormal reservoir pressures, with gradients up to 15.1 MPa/km (0.67 psi/ft). The Cardium reservoir pressure is 29.2 MPa (4235 psi), the bubble point of the oil 26.37 MPa (3825 psi). Reservoir pressures of representative Nisku gas accumulations are of the order of 34.5–48.3 MPa (5000–7000 psi). The hypothesis is proposed that Nisku gas migrated into a Cretaceous oil reservoir above, raising the reservoir pressure, adding light ends, raising API gravity and the value of SF($C_3$–$nC_5$), but not adding sufficient $P_{10+}$ material to significantly alter SF($P_{10+}$).

The Brazeau River Cardium oil is regarded as Cretaceous. The Pembina Cardium oil, in a field contiguous to the northeast, is assigned to a Cretaceous source, the Second White Speckled Shale, in the genetic classification of Li *et al.* (2001). Several oil accumulations occur in the

**Table 3.** *Comparison of PVT characteristics of oil in the Brazeau River Cardium field with those of oils in other Cretaceous reservoirs*

| Characteristic | Brazeau River Cardium oil | Cretaceous oils ($n$ = 30) |
|---|---|---|
| Pressure Grad. (MPa/km) | 12.27 | 8.58 |
| Pressure Grad. (psi/ft) | 0.55 | 0.38 |
| GOR ($m^3/m^3$) | 290.3 | 99.7 |
| GOR (mcf/bbl) | 1.63 | 0.56 |
| API gravity | 41.6 | 33.2 |
| SF($C_3$–$nC_5$) | 2.04 | 1.69 |
| SF($P_{10+}$) | 1.14 | 1.13 |

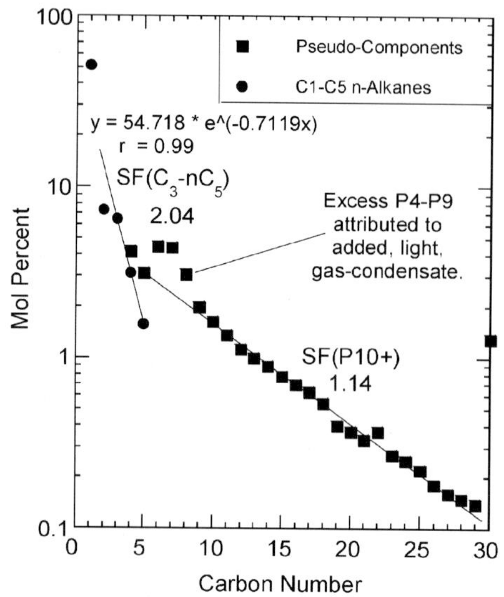


**Fig. 6.** Molar concentration profile of the Cardium Formation oil, Brazeau River field, west-central Alberta, postulated to exhibit addition of gas-condensate.

Brazeau River Nisku field, but exhibit a mean reservoir pressure of 33.9 MPa (4912 psi), and a normal pressure gradient (10.62 MPa/km), believed to be incapable of resulting in the observed Cardium values by vertical migration of Devonian oil. The sample mean SF($C_3$–$nC_5$) in Nisku gas-condensate reservoirs is 2.16. The value of SF($C_3$–$nC_5$) in the Brazeau River Cardium oil allows that a substantial volume of Nisku gas-condensate is admixed. Instances of elevation of reservoir pressure accompanying gas migration were described by Mumme & Ferrell (1979).

The molar profile of the Brazeau River Cardium oil (Fig. 6) exhibits a conspicuous excess of pseudo-components $P_4$–$P_9$, compared with predicted concentrations based on the projection of the $P_{29}$–$P_{10}$ exponential trend. $P_9$, $P_8$, possibly also $P_7$, are expected to lie on this projection in normal oils (Fig. 1). It is suggested that the anomaly reflects the addition not only of gaseous and light alkanes, but of an associated, light, gas-condensate, containing an insignificant quantity of $P_{10+}$. Patterns with excess material similar to that shown in Figure 6 are almost invariably associated with the occurrence of SF($C_3$–$nC_5$) greater than 1.69 (termed an 'elevated' value) as indicated by the following. Examining the reservoir fluid molar profiles of 125 oils from the Western Canada Sedimentary Basin, 32 oils are biodegraded, gas-depleted, or both, preventing evaluation of their histories. Excess pseudo-components $P_4$–$P_9$ occur in 29 of the remaining 93 cases. Only five of these fail to possess, in addition, elevated values of SF($C_3$–$nC_5$) which would be expected. Of the five, two exhibit evidence of gas depletion obscuring the question of their having possessed elevated values of SF($C_3$–$nC_5$) in the past. In its initial expression, migration depletion lowers concentrations of $C_1$–$C_3$, tending to decrease SF($C_3$–$nC_5$). Elevated values of SF($C_3$–$nC_5$) without excess $P_4$–$P_9$ occur in a further 26 cases, attributable to the addition of relatively dry gas. Elevated values of SF($C_3$–$nC_5$) occur, therefore, in 50 cases, 53.7% of the set which can be evaluated. Normal profiles occur in 38 cases.

Biodegradation is expressed in molar profiles as a 'valley', initially centred on $P_{10}$, which progressively widens and deepens. Two cusps develop in the profile, one at each end of the depleted range. The cusp at the lower end, commonly at $P_8$, remains when the front end has been markedly depleted, commonly by degasification in addition to biodegradation. The residual $P_8$ cusp serves to distinguish front end depletion due to biodegradation from that due to evaporative fractionation. The latter process leaves a smooth front end profile as illustrated later.

On the basis of the tentative limit for the recognition of gas-enrichment, a value of SF($C_3$–$nC_5$) exceeding 1.69, the process was sought in other basins and regions employing PVT data reported only to $P_{7+}$. Gas enrichment has occurred in 64% of 33 oil accumulations in Indonesia, in 54% of 54 oil accumulations in the Gulf of Mexico Tertiary–Quaternary section, in 23% of 30 accumulations in the Jurassic Smackover Formation in the eastern Gulf of Mexico region, in 78% of 36 oil accumulations in the northern North Sea, compared with 53.7% of oil accumulations evaluated above in the Western Canada Sedimentary Basin.

Figure 7 represents an example of the extreme differences in the relationships of SF($C_3$–$nC_5$) and SF($P_{10+}$). Both are controlled by maturity in certain instances, recalling the correlation of SF($C_3$–$nC_5$) with SF($P_{10+}$) in the Rainbow oils and in pyrolysates. In the general case, however, plotting all of the available data, SF($P_{10+}$) and API gravity exhibit a limited correlation (correlation coefficient 0.39, significant at 1% level, 121 degrees of freedom) while the relationship of SF($C_3$–$nC_5$) and API gravity is random. It is suggested that the correlation of SF($C_3$–$nC_5$) with API gravity which would be predictable in unaltered oils is destroyed in a large fraction of oils by modification of the light ends by the admixture of allochthonous gas.

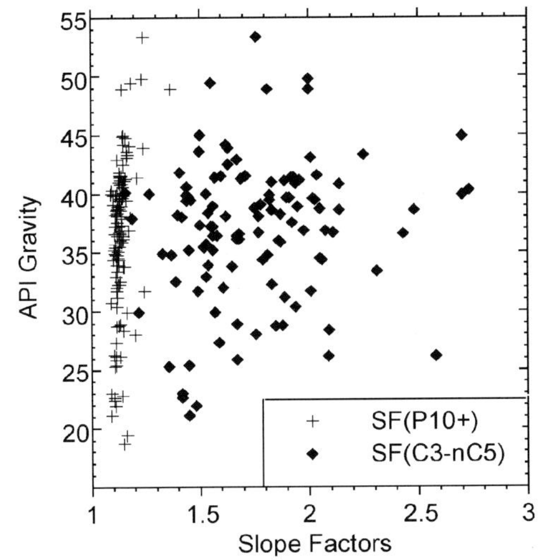


**Fig. 7.** Contrasting relationships of $SF(P_{10+})$ and $SF(C_3\text{–}nC_5)$ with API gravity in oils from western Canada. $SF(C_3\text{–}nC_5)$ is postulated to be frequently perturbed by the addition of gas to oil.

The occurrence of gas–oil mixing usually results also in elevated values of GOR, as in the instance of the Brazeau River Cardium oil. Figure 8 presents the relationship of GOR and $SF(C_3\text{–}nC_5)$ in the western Canadian oil data, defining a weak correlation. Oils exhibiting values of $SF(C_3\text{–}nC_5)$ greater than 1.69 generally possess 'elevated' GORs (greater than the principal mode, 60–80 $m^3/m^3$), compatible with a concept of gas enrichment.

## *Elevated hydrogen sulphide concentrations*

Employing available detailed PVT data, mole percent hydrogen sulphide is plotted against $SF(C_3\text{–}nC_5)$ in Figure 9 for both oils and gas-condensates. Only cases with greater than an arbitrary level of 2% $H_2S$ are shown, to emphasize the following observation. Mole percent $H_2S$ and $SF(C_3\text{–}nC_5)$ exhibit an inverse correlation in gas-condensates, but a positive correlation in $H_2S$-bearing oils. If thermochemical sulphate reduction (TSR) involving the oxidation of light hydrocarbons (Krouse *et al.* 1988) is invoked as the sole cause of elevated $H_2S$ (and $CO_2$) content, and the process is operative in both types of fluid, the effect on $SF(C_3\text{–}nC_5)$ would be expected to operate in the same sense in both. Krouse *et al.* (1988) showed that the carbon isotopic ratios of methane, ethane and propane are modified by TSR, all becoming isotopically heavier due to selective oxidation. There are significant losses of these compounds, also of *n*-butane. The phenomena are best developed in deep, fold belt gas accumulations, reservoired in carbonates associated with

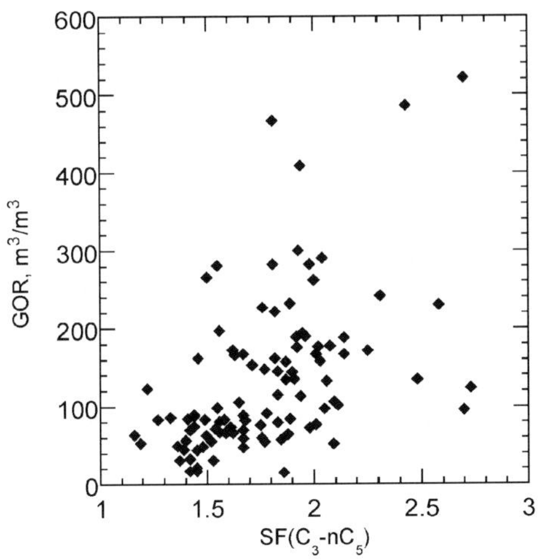


**Fig. 8.** Relationship of gas–oil ratios and values of $SF(C_3\text{–}nC_5)$ in 108 PVT analyses of oils, principally from the Western Canada Sedimentary Basin. Values of $SF(C_3\text{–}nC_5)$ in excess of 1.69, associated with GORs greater than approximately 70 $m^3/m^3$, are attributed to gas-enrichment.

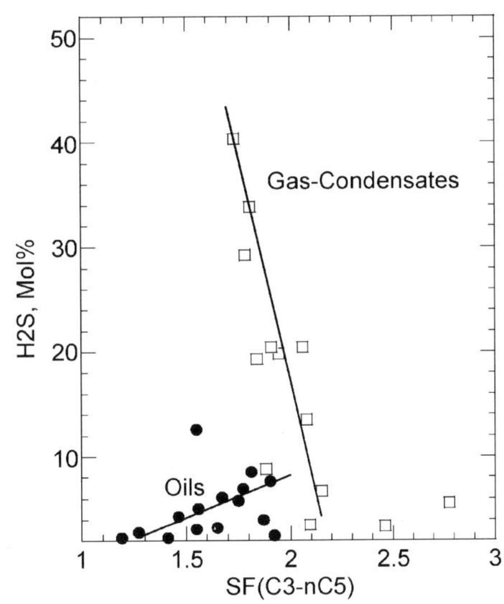


**Fig. 9.** Decreasing values of $SF(C_3\text{–}nC_5)$ accompanied by increasing levels of $H_2S$ in gas-condensates are attributed to thermochemical sulphate reduction. Concomitant increase in $SF(C_3\text{–}nC_5)$ and $H_2S$ in oils is postulated to be due to admixture of $H_2S$-bearing gas-condensate.

evaporites. A decreasing degree of oxidation of hydrocarbons with increasing molecular weight lowers SF($C_3$–$nC_5$). If this is an original characteristic of gas reservoirs exhibiting TSR, what phenomenon could reverse the expression in oil accumulations? The hypothesis is suggested that in the oils illustrated in Figure 9, increasing SF($C_3$–$nC_5$) is due principally to increasing admixture of $H_2S$-rich gas-condensates, TSR not having occurred to an appreciable extent in the oil accumulations. Such gas-condensates continue to exhibit elevated values of SF($C_3$–$nC_5$), despite decrease attributable to TSR.

## Comparison of oils and gas-condensates

Figure 10 presents all of the available Slope Factor data, representing Canadian oils, and gas-condensates from numerous basins. Values of SF($C_3$–$nC_5$) in oils extend to substantially higher levels than those illustrated in Figure 4 and that encountered in the Brazeau River Cardium oil. The higher-valued cases do not exhibit significantly higher values of SF($P_{10+}$) and therefore are not in accord with the requirements of a process of maturation. Figure 10 includes the Rainbow suite and asphaltene pyrolysate values defining the simultaneous increase of SF($P_{10+}$) and SF($C_3$–$nC_5$) attributed to maturation. Elevated values of SF($C_3$–$nC_5$) in gas-condensates are attributable either to advanced thermal cracking (to levels higher than those reached in the asphaltene pyrolysates described here), to evaporative fractionation at relatively low pressures, or to migration depletion, as investigated below. In gas-condensates these processes increase both Slope Factors.

The gas-condensate data of Figure 10 occupy a realm of Slope Factor values almost entirely separate from that of oils. This is largely due to the total dichotomy in physical–chemical factors controlling slope in oils and gases. Both Slope Factors in oils are initially controlled by reaction kinetics and cracking mechanisms in liquid systems, while slopes in gases (generally gas-condensates because of their ability to precipitate liquid) are largely controlled by the fugacities of the components, in turn dependent upon system composition, pressure and temperature. Fugacity control is operative during thermodynamic equilibration between liquid and vapour at the time of separation. Taking an extreme view, all gas-condensates may ultimately be regarded as

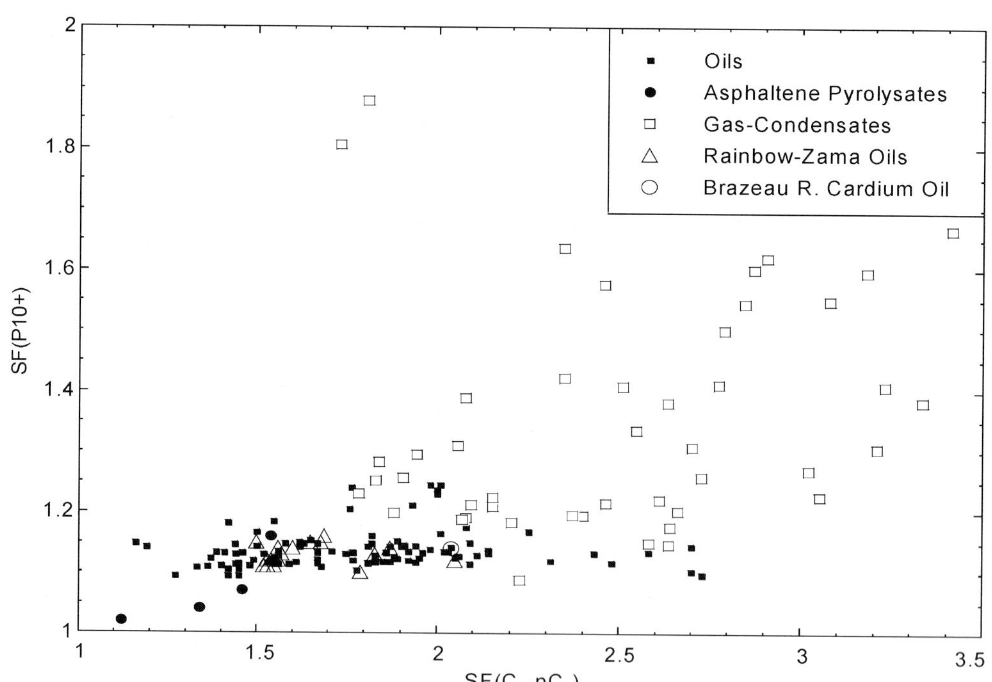


**Fig. 10.** Values of SF($C_3$–$nC_5$) and SF($P_{10+}$) in the oils and gas-condensates of this study. Sets by fluid type are almost disjoint because of different principal controlling factors: cracking in liquids, fugacity in gases. Rainbow–Zama fields present a maturation sequence of oils.

products of evaporative fractionation, even if the vapour–liquid separation step takes place in a highly mature source rock or oil accumulation undergoing cracking to gas. The degree of fractionation may be small, as under conditions close to the critical point of the system. The usual locale of evaporative fractionation is an oil reservoir, phase separation occurring either upon the introduction of sufficient gas or upon pressure reduction due possibly to uplift, or, as proposed earlier, to the opening of a fault migration pathway (Thompson 1987).

Both oils and gases are subject to a special case of evaporative fractionation, termed here migration depletion, upon depressurization below saturation pressure accompanying movement along an ascending pathway. Gas-condensates selectively precipitate heavier, lower fugacity, components, increasing both SF($C_3$–$nC_5$) and SF($P_{10+}$), maintaining dew point at pathway pressure. Thus, as shown below, dew point is correlated with SF($P_{10+}$) in gas-condensates, while a subordinate role is played by system composition. Oils selectively lose methane and ethane upon migration, with smaller losses of propane and heavier components in the general case, with decrease in SF($C_3$–$nC_5$), but without effect upon SF($P_{10+}$) (Thompson 2002). Again, saturation pressure, here the bubble point, is maintained at ambient pressure by the losses.

## *Theoretical oils and gas-condensates*

The phenomena which determine SF($C_3$–$nC_5$) and SF($P_{10+}$) in gas-condensates are illustrated, firstly, by computation using PVT software, secondly, empirically, by examination of the values occurring in oils and their cap gases, and thirdly, by examination of those occurring in oils and gas-condensates closely associated in stacked reservoirs in a single structure.

WinProp is a PVT program (Computer Modelling Group, Ltd., Calgary) which employs a cubic equation of state and calculates multiphase equilibria and fluid properties. WinProp sub-routines were employed to calculate $C_1$ to $C_{30}$ compositions, GORs, specific gravities, and other characteristics in problems of the following type. A parent oil was chosen from the available data, possessing values of 1.438 and 1.146 for SF($C_3$–$nC_5$) and SF($P_{10+}$), respectively. The oil is of intermediate maturity, 40.6 API gravity, GOR 87.6 $m^3/m^3$ (0.492 mcf/bbl), exhibiting a bubble point of 12.48 MPa (1810 psi) at reservoir temperature, 62.2 °C: a representative normal oil from Boundary Lake field, Halfway Formation (Triassic), British Columbia, believed to be virtually unaltered from its condition upon generation. In an initial test, using the given $C_1$ to $C_{30}$ compositional data, WinProp calculated a bubble point of 12.74 MPa, and a GOR of 92.6 $m^3/m^3$.

Three series of calculations (Experiments 1 to 3 below) were carried out to simulate the effects on the oil of the conceptual addition of gas, and subsequent modification of the mixture by evaporative fractionation. Gas addition was simulated using the 'recombination' routine of WinProp, intended to determine reservoir fluid composition from separator gas and stock tank oil analyses. Here the tank oil is replaced by the full reservoir fluid analysis of the Boundary Lake oil. The separator gas is replaced by a hypothetical reservoir fluid representing the added gas. Post-injection oils were made up to a GOR of 178.11 $m^3/m^3$ at 13.79 MPa (1.0 mcf/bbl, 2000 psi) at the temperature of the experiment, compositions which would yield substantially higher GORs employing a conceptual separator operated at normal pressure and temperature. Experimental results in terms of Slope Factors are illustrated in Figure 11.

In Experiment 1 the added gas was 'wet', comprising 75 mol% methane, 12% ethane, components thereafter decreasing at SF($C_2$–$nC_5$) of 2.00, i.e. propane was 6%, and so on. This procedure ignored the 'deficiency' of ethane seen

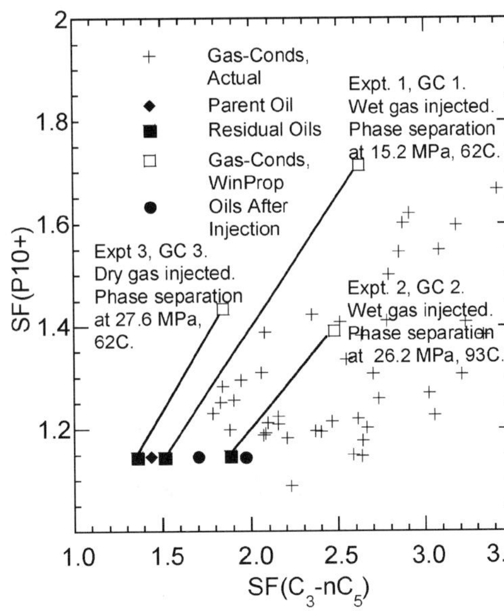


**Fig. 11.** Slope Factors based on compositions generated using the PVT program 'WinProp' (CMG Ltd., Calgary), illustrating gas-enriched residual oils, derived gas-condensates, residua and parent.

**Table 4.** *Properties of parent oil, added gas and post-injection oil, Experiment 1*

| Property | Parent oil | Input gas | Post-injection oil |
|---|---|---|---|
| SF($C_3$–$nC_5$) | 1.438 | 2.000 | 1.705 |
| SF($P_{10+}$) | 1.146 | – | 1.146 |
| Molecular weight | 100.32 | 22.78 | 62.81 |
| Mole volume (l/mol) | 0.125 | 0.146 | 0.135 |
| Density (kg/m$^3$) | 805.11 | 156.57 | 466.35 |
| Composition (vol%) | | | 47.77% oil, 52.23% input gas |
| Composition (mol%) | | | 51.63% oil, 48.37% input gas |

in nature, and generated a post-injection oil with a bubble point of 27.87 MPa (4041.5 psi) at the chosen temperature of 62.2 °C. Further properties are given in Table 4. The product is comparable with the gas-enriched oils recognized above, SF($C_3$–$nC_5$) having been raised from 1.438 to 1.705, while SF($P_{10+}$) was unaltered. An oil undergoing gas mixing, or injection, will remain single phase as long as the increasing bubble point remains lower than the confining pressure. Gas-condensate can form by evaporative fractionation should gas injection raise the bubble point above reservoir pressure, or upon sufficient pressure reduction.

In Experiment 2 the same parent oil was employed, mixed with a hypothetical wet, but highly mature, gas with 70 mol% methane, 13% ethane, SF($C_2$–$nC_5$) of 3.00, resulting in $nC_5$ of 0.48%. $P_7$ was set at 2.16%, other pseudo-components decreasing at a Slope Factor of 1.9, contributing material to the mixture up to $P_{15}$. The post-injection oil had a GOR of 344.1 m$^3$/m$^3$ (1.93/2 mcf/bbl), a bubble point of 32.10 MPa (4656.2 psi), calculated at a temperature of 93.3 °C, that of the subsequent separation. SF($P_{10+}$) was 1.149, and SF($C_3$–$nC_5$) 1.972, intermediate between the parent oil, 1.438, and the input gas, 3.00.

In Experiment 3 the same parent oil was employed but mixed with dry gas comprising 85% methane and 15% ethane, obviously incapable of raising SF($C_3$–$nC_5$). The mixture possessed a GOR of 214.8 m$^3$/m$^3$ (1.206 mcf/bbl) and a bubble point of 37.93 MPa at 62.2 °C.

In the second part of each experiment the 'Phase Separation' routine of WinProp was used, generally employed to simulate the effect of a pressure reduction step with progression from single to two phase conditions, as in a gas separator. Here the conceptual separator represents a postulated second reservoir at lower pressure than the first, connected by a fault pathway. The separation pressure is the saturation pressure common to the residual oil and the gas-condensate. The experiments are designed to test whether these concepts generate gas-condensates having realistic Slope Factors, GORs and dew points. Details concerning phase separation products are given in Table 5.

**Table 5.** *Slope Factor and phase separation data for Experiments 1, 2 and 3*

| Experiment and fluid | Dew point (MPa) | Bubble point (MPa) | Temperature (°C) | GOR (m$^3$/m$^3$) | SF($C_3$–$nC_5$) | SF($P_{10+}$) |
|---|---|---|---|---|---|---|
| Experiment 1 | | | | | | |
| Parent | | 12.48 | 62.2 | 87.6 | 1.438 | 1.146 |
| Oil after injection | | 27.87 | 62.2 | – | 1.705 | 1.146 |
| Post-separation residual oil | | 15.16 | 62.2 | 195.2 | 1.513 | 1.145 |
| Gas-condensate | 15.16 | | 62.2 | 89 573.0 | 2.623 | 1.713 |
| Experiment 2 | | | | | | |
| Parent | | 12.48 | 62.2 | 87.6 | 1.438 | 1.146 |
| Oil after injection | | 32.1 | 93.3 | 344.1 | 1.972 | 1.149 |
| Post-separation residual oil | | 26.2 | 93.3 | 242.4 | 1.887 | 1.147 |
| Gas-condensate | 26.2 | | 93.3 | 5 572.8 | 2.477 | 1.391 |
| Experiment 3 | | | | | | |
| Parent | | 12.48 | 62.2 | 87.6 | 1.438 | 1.146 |
| Oil after injection | | 37.93 | 62.2 | 214.8 | 1.438 | 1.146 |
| Post-separation residual oil | | 27.57 | 62.2 | 340.9 | 1.358 | 1.144 |
| Gas-condensate | 27.57 | | 62.2 | 8 552.6 | 1.845 | 1.433 |

The data of Table 5 are plotted in Figure 11, which also illustrates the distribution of gas-condensate Slope Factors in nature, employing the PVT data of this study. The slope characteristics of the computed gas-condensate compositions are comparable with those of natural occurrence, though with a slightly lower suite of values of SF($C_3$–$nC_5$). Differences could be eliminated by tuning the calculations. The experiments illustrate the increase in SF($C_3$–$nC_5$) in 'gas-enriched' oils with little or no change in SF($P_{10+}$), and the occurrence of elevated saturation pressures and GORs. Again, SF($P_{10+}$) is not significantly altered in the residual oils, i.e. the 'gas-enriched' oils after phase separation and the generation of daughter gas-condensates. The latter, however, exhibit large increases in both SF($C_3$–$nC_5$) and SF($P_{10+}$). A subsequent illustration shows that the vectors joining parent and daughter in terms of SF($C_3$–$nC_5$) and SF($P_{10+}$) are similar to those joining actual parent oil and daughter gas-condensate pairs.

## *Actual oils and gas-condensates*

Figure 12 illustrates oils and gas-condensates which occur either as oil and gas-cap, or together in the same structure in close vertical association. Details are given in Table 6. The oil and gas-condensate pairs represent widely separated

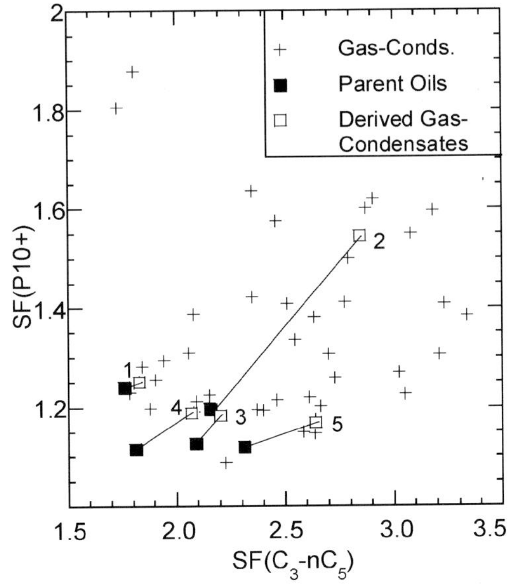


**Fig. 12.** Values of SF($C_3$–$nC_5$) and SF($P_{10+}$) in oils and closely related gas-condensates (oil and gas cap or same structure).

basins. They involve oils which are simultaneously rich in gas and possess elevated values of SF($C_3$–$nC_5$), as required by the theory developed above. All cases appear to have involved gas-rich-condensate addition to oil, followed by evaporative fractionation. It is suggested that there is adequate reason to postulate that the majority of gas-condensates studied here are generated in this fashion. The maturity of the gasified oil has no bearing upon the gasification process, nor upon the Slope Factors of the gas-condensate produced. It may be noted anecdotally that, in examining Slope Factor data in the Western Canada Basin, it has been found that, if a given formation in a field is predominantly gas-productive, and hosts both wells producing oil, and others producing gas-condensate, the majority of the oils, if not all, are gas-enriched.

The Smackover Formation oil and gas-condensate represented by Robinson Creek and Hatter's Pond fields in Alabama, in the eastern US Gulf Coast region, are of unusual interest. Slope Factors and other characteristics are given in Table 6. They are cogenetic fluids, relatively similar in saturation pressures, though substantially separated geographically. They are suggested to represent any high maturity pair of closely associated oil and gas-condensate fluids. Reservoir temperature is 166 °C at Hatter's Pond, a temperature which has persisted since, minimally, the Miocene. The calculated equivalent vitrinite reflectance based on methylphenanthrene isomer concentrations is 1.99% (present author's unpublished data). Robinson Creek oil is at 127 °C, but has an API gravity of 43.4, a characteristic comparable with that of the highest maturity and highest slope oil encountered in Rainbow field (Alberta). The latter is of 47.1 API and SF($P_{10+}$) equals 1.18, but SF($C_3$–$nC_5$) could not be evaluated because of gas depletion. It is suggested that Hatter's Pond gas-condensate possesses characteristics appropriate to thermal cracking. It is visualized that a highly mature oil undergoing *in-situ* cracking is brought first to the condition of a volatile oil (one of elevated GOR and under conditions close to its critical point), then completely or partially vaporized, the added gaseous components being autochthonous. Robinson Creek has volatile oil characteristics, though now cooled by migration. It is noteworthy that the paired slope values of Hatter's Pond and Robinson Creek are closely similar despite a fractionation step, as would be the case for oil and gas-condensate close to critical conditions for the system. At the critical point, a single phase fluid exists which, upon minor changes in pressure and temperature, transforms into either a

**Table 6.** *Characteristics of closely associated oil and gas-condensate pairs*

| Case, Figure 12 | Fluid | Field, location | Formation, age | Gas-condensate association | Sat'n Pr (MPa) | GOR ($m^3/m^3$) | API | SF($C_3$–$nC_5$) | SF($P_{10+}$) |
|---|---|---|---|---|---|---|---|---|---|
| 1 | Oil | Robinson Creek, Alabama | Smackover, Jur. | | 24.65 | 226.2 | 43.4 | 1.76 | 1.24 |
| | G-C | Hatter's Pond, Alabama | Smackover, Jur. | Co-genetic | 22.37 | 340.2 | 49.4 | 1.83 | 1.25 |
| 2 | Oil | Hoffman, S. Texas | Eocene | | 26.35 | 162.1 | – | 2.16 | 1.20 |
| | G-C | Hoffman, S. Texas | Eocene | Gas-cap | 26.35 | 6255.2 | 64.8 | 2.85 | 1.54 |
| 3 | Oil | Brazeau River, Alberta | Nisku, Dev. | | 20.44 | 202.2 | – | 2.09 | 1.13 |
| | G-C | Brazeau River, Alberta | Nisku, Dev. | Co-genetic | 32.41 | 576.9 | 46.4 | 2.20 | 1.18 |
| 4 | Oil | Hibernia, Offshore Newfoundland | Cretaceous | | 40.53 | 466.7 | 34.8 | 1.81 | 1.12 |
| | G-C | Hibernia, Offshore Newfoundland | Cretaceous | Same well | 38.33 | 1223.6 | 49.2 | 2.07 | 1.19 |
| 5 | Oil | Cusiana, Columbia | Cretaceous | | 39.34 | 486.2 | 36.6 | 2.43 | 1.13 |
| | G-C | Cusiana, Columbia | Cretaceous | Same well | 39.16 | 1241.4 | 42.1 | 2.64 | 1.17 |

gas-condensate with traces of precipitated liquid, or into an oil with minor exsolved gas.

The oils of Table 6 would be termed 'gas-enriched'. It is noteworthy that vectors joining oils and related gas-condensates in Figure 11, and related fluid pairs in Figure 12, are similar in orientation and in the values which they join. This argues in favour of a necessary initial mixing of oil and gas, raising SF($C_3$–$nC_5$) in the parent liquid prior to an evaporative fractionating step.

## Relationship of Slope Factors and saturation pressure

### *SF($P_{10+}$) in gas-condensates*

The principal control of SF($P_{10+}$) in gas-condensates is dew point. The available detailed gas-condensate data shows that these variables are significantly correlated (Fig. 13), as suggested by Thompson (2002) on the basis of limited data. Correlation is due to control of SF($P_{10+}$) by fugacity. Decreasing pressure results in a continually decreasing capability of the suite of solvent components, principally methane, ethane, propane and butane, to retain heavier components in solution. The principles involved are those of the process of constant volume depletion, during which fluid pressure is progressively lowered below the original dew point, with assessment of the percentage of dissolved liquid precipitated at each intermediate pressure, each a new

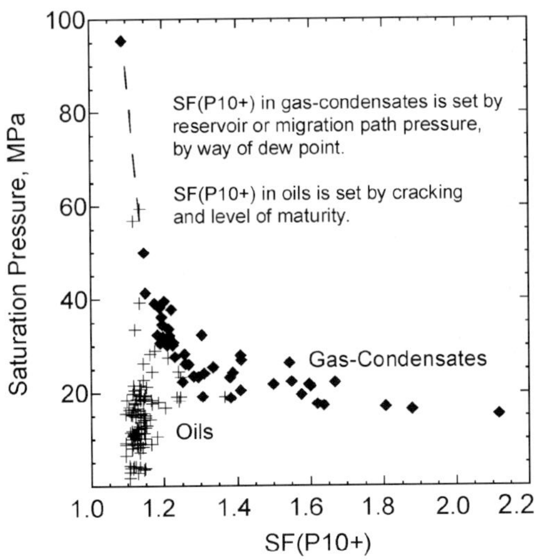


**Fig. 13.** Correlation of saturation pressure (dew point) of gas-condensates with SF($P_{10+}$) and absence of correlation with bubble point in oils.

dew point. Liquid precipitation is accompanied by increase in SF($P_{10+}$) which is correlated with increasing API gravity and decreasing mean molecular weight of the condensate liquid.

Figure 13 also shows that bubble point in oils is not correlated with SF($P_{10+}$). The linear correlation coefficient is 0.27. On the other hand, as saturation pressure is related to the gas and light end content of oils, bubble point is significantly correlated with SF($C_3$–$nC_5$) ($r = 0.49$, employing a linear equation, illustrated in Figure 14, $n = 109$). This suggests the further significance of gas enrichment and its recognition, not merely as a harbinger of gasification and gas-condensate occurrence, but as a control of bubble point and the pressure conditions under which gas released from oil in the reservoir during production will begin to adversely affect reservoir permeability.

In Figure 13, an unusual gas-condensate is represented at a dew point of 95.43 MPa and SF($P_{10+}$) of 1.09, occurring in a significantly overpressured reservoir at a depth of 5890 m in the offshore Gulf of Mexico. The fluid is liquid-rich, GOR 659 $m^3/m^3$, carries an unusually large proportion of $P_{7+}$ components (10.17 mol%), and has a stock tank liquid gravity of 28.9° API. It is of interest because of its unusually low value of SF($P_{10+}$), the lowest in the entire suite of oils and gas-condensates reviewed here, as shown in Figure 10 at a value of SF($C_3$–$nC_5$) of 2.23. The cause of the low value of SF($P_{10+}$) is uncertain. Hypothetical possibilities include low maturity (the author's unpublished data show that an oil in the same well exhibits a methylphenanthrene maturity of Ro 0.87%, and unequilibrated steranes) or simply evaporative fractionation at high pressure. In the latter instance, on the basis of the collected gas-condensate data (Fig. 13), the observed value of SF($P_{10+}$) could reasonably be deemed appropriate to the elevated saturation pressure. Gasification would result in the parent oil being depleted across the entire measured carbon number range, except that of the added gas. Neither parent nor daughter would exhibit front end depletion, or 'roll over' (see below) as observed in evaporative fractionation at lower pressures, in the region of 30–50 MPa. Furthermore, it appears that the value of SF($P_{10+}$) in the gas-condensate could be lower than that of the parent oil.

### *SF($C_3$–$nC_5$) in oils and gas-condensates*

Further comparisons of natural and theoretical, computed, gas-condensates (Table 5) are

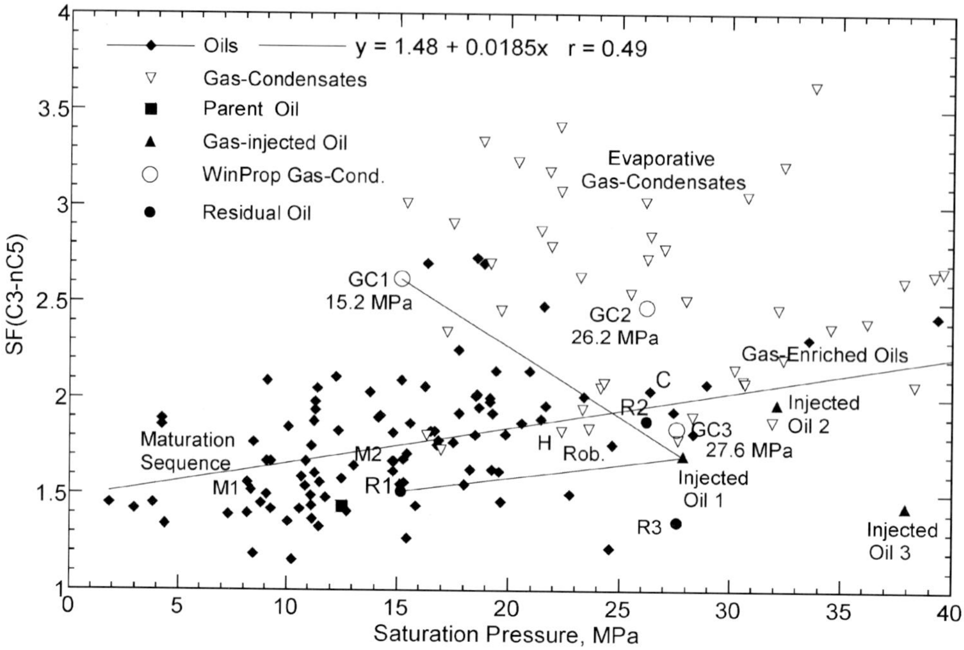


**Fig. 14.** SF($C_3$–$nC_5$) versus saturation pressures for the oils and gas-condensates of this study. Three data regions are distinguishable, that of the Maturation Sequence of oils (M1, M2), that of Gas-Enriched Oils, including Brazeau River Cardium (C), and that of Evaporative Gas-Condensates. An area between the first two regions also hosts data representing volatile oils, e.g. Robinson Creek (Rob.), often of high maturity, i.e. oils close to their critical points, and similar gas-condensates, some believed to be of thermal origin, e.g. Hatter's Pond (H).

illustrated in Figure 14 which plots all of the SF($C_3$–$nC_5$) data of this study against saturation pressure. The three theoretical oils generated by gas enrichment (Oils after injection, Experiments 1, 2 and 3) attain calculated saturation pressures greater than that of the Brazeau River Cardium oil (C, Fig. 14), the archetypal case of gas-enrichment, several also having similar values of SF($C_3$–$nC_5$). The theoretical gas-condensates GC1, GC2 and GC3 exhibit SF($C_3$–$nC_5$) values in inverse proportion to pressure (dew point). The properties of GC1 and GC2 allow the conclusion that natural gas-condensates with similarly associated values of slope and saturation pressure are attributable to evaporative fractionation. The points H and Rob represent the properties of Hatter's Pond gas-condensate and Robinson Creek oil (Table 6), discussed above. The points M1 and M2 define the maturation trend, as so far documented, defined in the Rainbow field area.

Figure 14 facilitates the recognition of three, broad, realms of associated values of SF($C_3$–$nC_5$) and saturation pressures: those of the Maturation Sequence of oils, those of Gas-Enriched Oils, and of Evaporative Gas-Condensates. It is suggested that the first two regions embrace between them the domain of thermal gas-condensates, insofar as they can be documented by maturity, and of volatile oils often close to their critical points. It is noted in addition that, as shown in Figure 13, oils having saturation pressures (bubble points) greater than 20 MPa also possess elevated values of SF($P_{10+}$), an index of maturity.

## Repeated evaporative fractionation

The majority of western Canadian oils examined possess $P_8$ as the dominant pseudo-component, and do not exhibit, to a significant degree, the phenomenon termed here 'front end roll over' (Thompson 1987) whereby the dominant liquid pseudo-component progressively migrates from $P_8$ to higher carbon numbers. Many western Canadian oils are gas-enriched and are considered to be parent to gas-condensates. Most are believed, however, to represent a single, or few, gasification steps, or limited occurrence of evaporative fractionation.

Repeated evaporative fractionation steps were envisioned by Thompson (1987). Tertiary age reservoirs in the Gulf Coast region of the United States, onshore and on the continental shelf, have experienced substantial depletion with multiple or continuous additions of gas to oil, followed by evaporative fractionating and the migration of gas-condensates. Comparison with the Western Canada Sedimentary Basin, itself predominantly a gas basin, may be made on the basis of overall gas/oil ratios. Losh (2002) reports a net, original, in-place gas/oil ratio of 6477 $m^3/m^3$ in the Tiger Shoal region (some 500 $km^2$) on the inner shelf of western Louisiana, a value almost three times greater than the maximum encountered in any stratigraphic interval in western Canada. The concept of continuous gas-washing in the area was modelled and detailed by Meulbroek *et al.* (1998) and a regional appraisal of the process on the continental shelf, offshore Louisiana, was given by Losh *et al.* (2002). Curiale and Bromley (1996) describe evaporative fractionation in Vermilion 14 field, offshore Louisiana, and Holba *et al.* (1996) in South Pass 61.

A simulation of repeated evaporative fractionation steps was generated using WinProp. The parent oil was of a hypothetical composition with 53.1% methane, a molar $C_1/C_2$ ratio of 4.62, SF($C_3$–$nC_5$) equal to 1.80, $P_7$ equal to 2.08%, and SF($P_{10+}$) of 1.16, detailed to $P_{30}$. The injected gas was also hypothetical with 75% methane, $C_1/C_2$ of 6.15, SF($C_3$–$nC_5$) equal to 4.48, (extremely high and mature), terminating at $nC_5$. Six gas addition steps were simulated at 44.8 MPa (6500 psi) and 121 °C, the added gas comprising 67.3 volume percent of the system in each instance. Each addition was followed by a phase separation step at 27.6 MPa and 121 °C. Figures 15 and 16 depict, respectively, developmental stages of the six progressively depleted oils and their accompanying gas-condensates.

In Figure 15, the residual oils show light end depletion, particularly strong at $P_6$ and higher, beyond the compositional range of the added gas, resulting in 'front end roll-over'. The dominant pseudo-component recedes from $P_{10}$ to $P_{17}$. The light ends steepen, SF($C_3$–$nC_5$) changing from 1.90 in the first residual, R1, to 3.11 in R6. As depletion removes progressively heavier liquid components the compositional segment exhibiting exponential distribution is progressively shortened, from approximately $P_{20}$ to $P_{30}$ in R1, to approximately $P_{23}$ to $P_{30}$ in R6. However, the SF values in these segments remain little altered, changing from 1.14 to 1.139.

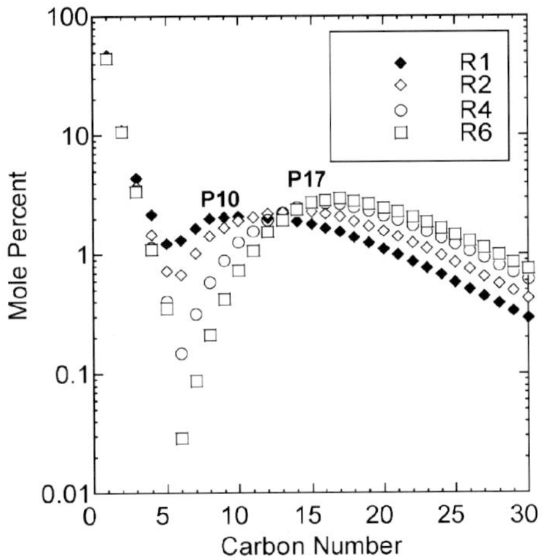


**Fig. 15.** Calculated (WinProp) compositions of residual oils subsequent to six successive stages of evaporative fractionation, showing characteristic regression of the modal pseudo-component and increasing front end roll-over accompanying depletion. Demonstrates increasing SF($C_3$–$nC_5$) with progressive gas injection.

Figure 16 shows progressive modifications of the derived gas-condensates, G1 to G6. Each reflects compositional features of its parent oil. Thus, front end roll-over develops in the gas-condensates as the parent oil is stripped, but the modal pseudo-component does not regress

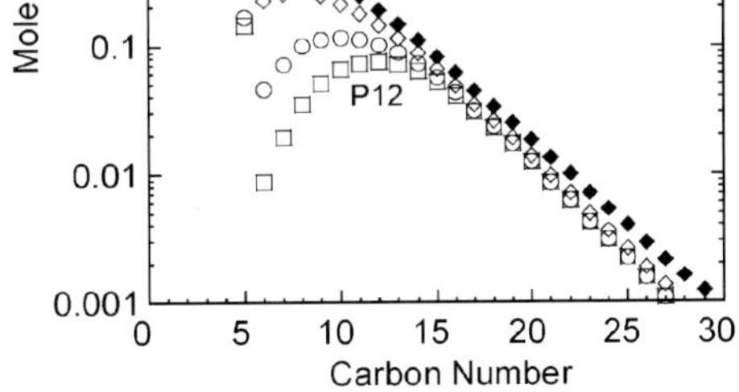


**Fig. 16.** Calculated (WinProp) compositions of gas-condensates derived from the oils of Figure 15, generated in six stages of evaporative fractionation. Illustrates regression of the modal pseudo-component and increasing front end roll-over inherited from residual oil.

beyond $P_{12}$. SF($C_3$–$nC_5$) changes from 2.49 in G1 to 4.38 in G6, asymptotically approaching that of the stripping gas. The value of SF($P_{15+}$) changes remarkably little, from 1.346 in G1 to 1.352 in G6.

## Summary

It is proposed that various associations of values of SF($C_3$–$nC_5$) and SF($P_{10+}$) are diagnostic of petroleum characteristics and history. Certain paired values characterize unaltered oils and define their level of maturation, both parameters increasing simultaneously. Unaltered oils, i.e. single phase fluids comprising liquid and gaseous components generated simultaneously in the source and never phase-separated, are recognized in terms of their Slope Factors. Two examples which span the presently-definable range of maturities in terms of Slope Factors possess values of SF($C_3$–$nC_5$) from 1.50 to 1.69, of SF($P_{10+}$) between 1.11 and 1.18, API gravity from 34° to 44° and GOR from 71 to 105 $m^3/m^3$. Other oils have paired Slope Factors with an equal range of values of SF($P_{10+}$), but with elevated values of SF($C_3$–$nC_5$). It is postulated that these cases represent oils to which allochthonous gas of a higher maturity than that of the indigenous gas has been added. This phenomenon has not previously been widely recognized for lack of appropriate tools, though it has occasionally been documented by gas isotopic studies.

Employing detailed PVT analyses of some 125 oils principally from the Western Canada Sedimentary Basin, and 47 gas-condensates largely from the same region but including examples from five other basins, it is shown that paired values of SF($C_3$–$nC_5$) and SF($P_{10+}$) representing oils and gas-condensates occupy nearly separate domains in a plot of SF($P_{10+}$) versus SF($C_3$–$nC_5$). The parameters are controlled by wholly distinct processes, in unaltered oils by cracking and kinetics in the source rock, in gas-condensates by fugacities which depend upon system composition, pressure and temperature. Elevated correlation coefficients linking dew point (the pressure at which a gas-condensate is finally equilibrated with a separate liquid phase) and SF($P_{10+}$) show that temperature and composition play a subsidiary role. In gas-condensates, SF($P_{10+}$) is correlated with API gravity, mean molecular weight, viscosity and other characteristics of the liquid fraction, without regard to the maturity of the parent oil.

Paired values of SF($C_3$–$nC_5$) and saturation pressure facilitate the recognition of three, broad, data realms: that of the Maturation Sequence of oils, of Gas-Enriched Oils, and of Evaporative Gas-Condensates. The first two regions are suggested to embrace between them the domains of thermal gas-condensates, insofar as they can be documented by maturity, of volatile oils often close to their critical points, and of oils deemed to be of high maturity on the basis of their saturation pressure and SF($P_{10+}$).

Gratitude is expressed to staff members and the managements of Conoco Inc, Saga Petroleum, Statoil a.s., Texaco Inc and Unocal Corporation for support of the studies involved and for permission to publish. Thanks are extended to Peter Kelley, Texaco Inc, for information enhancing the capabilities of WinProp, a software product of the Computer Modeling Group, Calgary, Alberta, also to R. di Primio, M. A. Bastow and J. Cubitt for valuable reviews.

## References

Curiale, J. A. & Bromley, B. W. 1996. Migration of petroleum into Vermilion 14 field, Gulf Coast, U.S.A.—molecular evidence. *Organic Geochemistry*, **24**, 563–579.

di Primio, R., Dieckmann, V. & Mills, N. 1998. PVT and phase behavior analysis in petroleum exploration. *Organic Geochemistry*, **29**, 207–222.

di Primio, R. 2002. Unraveling secondary migration effects through the regional evaluation of PVT data: a case study from Quadrant 25, NOCS. *Organic Geochemistry*, **33**, 643–654.

Dzou, L. I. P. & Hughes, W. B. 1993. Geochemistry of oils and condensates, K Field, offshore Taiwan: a case history in migration fractionation. *Organic Geochemistry*, **20**, 437–462.

England, W. A. & Mackenzie, A. S. 1989. Some aspects of the organic geochemistry of petroleum fluids. *Geologisch Rundschau*, **78**, 274–288.

England, W. A., Andrews, S. R. & Ruckheim, J. 1991. Experimental simulation of gas-phase migration: results and interpretations. *In*: Manning, D. (ed.) *Organic Geochemistry Advances and Applications in Energy and the Natural Environment*. Manchester University Press, 146–150.

England, W. A. 2002. Empirical correlations to predict gas/gas-condensate phase behavior in sedimentary basins. *Organic Geochemistry*, **33**, 665–673.

Flory, P. J. 1936. Molecular size distribution in linear condensation polymers. *Journal of the American Chemical Society*, **57**, 1877–1885.

Garrigues, P., Connan, J., Bellocq, J., Parlanti, E. & Ewald, M. 1988. Relationship between rank and distribution of methyl aromatic hydrocarbons for condensates of different origins. *In*: Mattavelli, L. & Novelli, L. (eds) Advances in Organic Geochemistry 1987. *Organic Geochemistry*, **13**, 1115–1121.

GUSSOW, W. C. 1954. Differential trapping of hydrocarbons. *Bulletin of the American Association of Petroleum Geologists*, **38**, 816–853.

HOLBA, A. G., DZOU, L. I. P., HICKEY, J. J., FRANKS, S. G., MAY, S. J. & LENNEY, T. 1996. Reservoir geochemistry of South Pass 61 Field, Gulf of Mexico: compositional heterogeneities reflecting filling history and biodegradation. *Organic Geochemistry*, **24**, 1179–1198.

HORSFIELD, B., DISKO, U. & LEISTNER, F. 1989. The micro-scale simulation of maturation: outline of a new technique and its potential applications. *Geologisch Rundschau*, **78**, 361–374.

JAMES, A. T. 1990. Correlation of reservoired gases using the carbon isotopic compositions of the wet gas components. *Bulletin of the American Association of Petroleum Geologists*, **74**, 1441–1458.

KHAVARI-KHORASANI, G., DOLSON J. C. & MICHELSEN, J. K. 1998. The factors controlling the abundance of heavy versus light oils, as constrained by data from the Gulf of Suez. Part 1. The effect of expelled petroleum composition, PVT properties and petroleum system geometry. *Organic Geochemistry*, **29**, 255–282.

KISSIN, Y. V. 1987. Origin of *n*-alkanes in petroleum crudes. *Geochimica et Cosmochimica Acta*, **51**, 2445–2458.

KNUDSEN, K. & MEISINGSET, K. K. 1991. Evaporative fractionation effects for the oils in the Gullfaks South area. *In*: MANNING, D. (ed.) *Organic Geochemistry Advances and Applications in Energy and the Natural Environment*. Manchester University Press, 162–165.

KROUSE, H. R., VIAU, C. A., ELIUK, L. S., UEDA, A. & HALAS, S. 1988. Chemical and isotopic evidence of thermochemical sulphate reduction by light hydrocarbon gases in deep carbonate reservoirs. *Nature*, **333**, 415–419.

KUHN, W. 1930. Über die Kinetik des Abbaues hochmolekularer. *Berichte der Deutschen Chemischen Gesellschaft, Abteilung B*, **63**, 1503–1509.

KUO, L.-C. 1997. Gas exsolution during fluid migration and its relation to overpressure and petroleum accumulation. *Marine and Petroleum Geology*, **14**, 224–229.

LARTER, S. & MILLS, N. 1991. Phase-controlled molecular fractionations in migrating petroleum charges. *In*: ENGLAND W. A. & FLEET, A. J. (eds) *Petroleum Migration*. Geological Society, London, Special Publications, **59**, 137–147.

LI, M., FOWLER, M. G., OBERMAJER, M., STASIUK, L. D. & SNOWDON, L. R. 1999. Geochemical characterization of Middle Devonian oils in NW Alberta, Canada: possible source and maturity effect on pyrrolic nitrogen compounds. *Organic Geochemistry*, **30**, 1039–1058.

LI, M., HUANG, Y., OBERMAJER, M., JIANG, C., SNOWDON, L. R. & FOWLER, M. G. 2001. Hydrogen isotopic compositions of individual alkanes as a new approach to petroleum correlation: case study from the Western Canada Sedimentary Basin. *Organic Geochemistry*, **32**, 1387–1400.

LOHRENZ, J. & BRAY, B. G. 1964. The calculation of bubble points of reservoir fluids from their compositions. Society of Petroleum Engineers Paper **792**, Rocky Mountain S.P.E. Meeting, Casper, Wyoming, May 1964.

LOSH, S., CATHLES, L. & MUEULBROEK, P. 2002. Gas-washing of oil along a regional transect, offshore Louisana. *Organic Geochemistry*, **33**, 655–664

MEULBROEK, P. 2002. Equations of state in exploration. *Organic Geochemistry*, **33**, 613–634.

MEULBROEK, P., CATHLES, L. & WHELAN, J. 1998. Phase fractionation at South Eugene Island Block 330. *Organic Geochemistry*, **29**, 223–240.

MICHAEL, K. & BACHU, S. 2001. Fluids and pressure distributions in the foreland-basin succession in the west-central part of the Alberta basin, Canada: Evidence for permeability barriers and hydrocarbon generation and migration. *Bulletin of the American Association of Petroleum Geologists*, **85**, 1231–1252.

MUMME, S. T. & FERRELL, R. E. 1979. Geopressure in the Houma and Hollywood fields, Louisiana. *Transactions of the Gulf Coast Association of Geological Societies*, **29**, 321–327.

PEDERSEN, K. S., BLILIE, A. L. & MEISINGSET, K. K. 1992. PVT Calculations on petroleum reservoir fluids using measured and estimated compositional data for the plus fraction. *Industrial and Engineering Chemical Research*, **31**, 1378–1384.

PRICE, L. C., WENGER, L. M., GING, T. & BLOUNT, C. W. 1983. Solubility of crude oil in methane as a function of pressure and temperature. *Organic Geochemistry*, **4**, 201–221.

THOMPSON, K. F. M. 1987. Fractionated aromatic petroleums and the generation of gas-condensates. *Organic Geochemistry*, **11**, 573–590.

THOMPSON, K. F. M. 1988. Gas-condensate migration and oil fractionation in deltaic systems. *Marine and Petroleum Geology*, **5**, 237–246.

THOMPSON, K. F. M. 1992. Contrasting characteristics attributed to migration observed in petroleums reservoired in clastic and carbonate sequences in the Gulf of Mexico Region. *In*: ENGLAND, W. A. & FLEET, A. J. (eds) *Petroleum Migration*. Geological Society, London, Special Publications, **59**, 191–205.

THOMPSON, K. F. M. 2002. Compositional regularities common to petroleum reservoir fluids and pyrolysates of asphaltene and kerogens. *Organic Geochemistry*, **33**, 829–841.

VAN GRAAS, G. W., GILJE, A. E., ISOM, T. P. & TAU, L. A. 2000. The effects of phase fractionation on the composition of oils, condensates, and gases. *Organic Geochemistry*, **31**, 1419–1439.

ZHUZE, T. P. & USHAKOVA, G. S. 1981. Dependency of the phase behavior of petroleum–natural gas systems upon their composition at high pressures. *Zeitschrift für Angewandte Chemie*, **27**, 37–49.

ZHUZE, T. P. & YUSHKEVICH, G. N. 1959. Solubility of oil and its heavy fractions in compressed gases. *Trudy Institut Nefti Akademia Nauk SSSR*, **13**, 262–274.

ZHUZE, T. P., USHAKOVA, G. S. & YUSHKEVICH, G. N. 1962. The influence of high pressures and temperatures on the content and properties of condensate in the gas phase of gas–oil deposits. *Geochemistry*, **8**, 797–806.

ZHUZE, T. P., YUSHKEVICH, G. N., USHAKOVA, G. S. & TUMAREV, K. D. 1963. Use of phase composition data in the system oil-gas at high pressures for ascertaining the genesis of some pools. *Petroleum Geology*, **7**, 186–191.

# Shaken but not always stirred. Impact of petroleum charge mixing on reservoir geochemistry

ARND WILHELMS[1†] & STEVE LARTER[2*]

[1]*Norsk Hydro, O&E Exploration, 0240 Oslo, Norway*
[2]*NRG Petroleum Group, School of Civil Engineering and Geosciences, University of Newcastle upon Tyne, UK*
[*]*Present address: Department of Geology and Geophysics, University of Calgary, Calgary, Canada*
[†]*Present address: Norsk Hydro, Technology, 5020 Bergen, Norway*

**Abstract:** Essentially all petroleums are mixtures with different components charged from source rocks at different temperatures. This heterogeneous charge is the basis for compositional differences in reservoirs that are the basic elements of reservoir geochemical approaches. Because many classical petroleum geochemical tracers of source facies and maturity, such as the cyclic biomarker hydrocarbons, show several orders of magnitude variation in concentration in petroleum systems these compounds do not reliably track facies or maturity signals in mixed oil situations. Light hydrocarbon and aromatic hydrocarbon parameters are more reliable in this sense but, as mixtures are the norm, the concept of the maturity of oils needs revising. We suggest an alternative approach is needed which tracks the maturity/petroleum mass fraction relationships for reservoired oils (mass fraction maturity) and allows the bracketing of source kitchen maturity. We strongly advise against using compound ratios in reservoir geochemical studies without having knowledge of the compounds concentration range variations within the petroleum system being studied.

Petroleum generation is the rate-determining step in reservoir charging (England *et al.* 1987). Oil and associated solution gas is generated from oil prone source rocks over a temperature range of approximately 100–150 °C in most sedimentary basins with oil cracking to gas in source rocks at temperatures from 150 to 200 °C (Pepper & Corvi 1995). With typical geological heating rates of 1–10 °C/Ma at any location within a petroleum system, this means that source rocks are charging traps for timescales from a few million to a few tens of millions of years. During this period, the expelled and trapped petroleum shows a progressive evolution of both bulk (e.g. saturated and aromatic hydrocarbon contents, gas/oil ratios) and molecular (biomarker and non-biomarker) maturity parameters. It is this evolution in composition of a charged petroleum through time from even a single source rock, coupled with often inefficient mixing in the trap, that results in the compositional differences often seen across petroleum reservoirs (England *et al.* 1987; Larter & Aplin 1995) that reservoir geochemists have exploited with such great success. Even in basins with a single uniform source rock, petroleum reservoirs integrate an evolving petroleum composition with time. Thus, nearly all reservoired oils being mixtures of compositionally different petroleum derived from the source rock(s) at different temperatures. In many basins, multiple source systems charge traps producing mixed oils and thus we can distinguish two end-member mixing regimes:

(1) mixing in traps of petroleums expelled from similar organic facies source rocks at different maturities;
(2) mixing in traps of petroleums from two or more different source rocks.

Petroleum mixing has been a self-evident property of petroleum systems for many decades. Clear evidence of mixing of fresh and degraded oils can be seen in biodegraded oil populations (e.g. Volkman 1984; Cassani & Eglinton 1986; Barnard & Bastow 1991; Horstad & Larter 1997; Wilhelms *et al.* 2001) and, as inferred above, in the mixed petroleum compositions seen in most petroleum reservoirs, with many examples of multiple source charges to petroleum traps being reported. Koopmans *et al.* (2002) showed that mixing of oils in reservoirs containing biodegraded petroleum is the primary control on oil physical properties such as viscosity. Migration contamination of petroleum by lower maturity markers picked up during secondary migration is another end member example of petroleum mixing in nature.

Given the premise that petroleum mixing in traps is common (indeed, we would say inevitable, pervasive and universal in all oil charges), it is

*From*: CUBITT, J. M., ENGLAND, W. A. & LARTER, S. (eds) 2004. *Understanding Petroleum Reservoirs: towards an Integrated Reservoir Engineering and Geochemical Appraoch*. Geological Society, London, Special Publications, **237**, 27–35. 0305-8719/04/$15.00 

surprising that petroleum geochemists have continued to use the relative abundances of, in particular, saturated hydrocarbon and other biomarkers to a large degree in correlations and maturity/facies assessment of reservoired oils and gas/condensates. This is despite evidence that the concentrations of these compounds vary greatly in petroleum sourced from different source rock types and vary greatly for single facies source rocks of differing maturities. This is particularly seen with biomarkers that decrease in concentration with increasing oil maturity. Most biomarkers, including isoprenoid alkanes, are generated from kerogens early in the oil window (before 0.8 Ro%) and are then diluted by later generated biomarker depleted petroleum (Larter *et al.* 1979; Rullkötter *et al.* 1984; Øygard *et al.* 1988; Larter 1990; Bishop & Abbott 1993; Li *et al.* 1994 1995; Jiang & Li 2002).

Li *et al.* (1995) suggested that in the Liaohe Basin, China, mixing of low maturity petroleums carrying high concentrations of biomarkers with more abundant but more mature equivalent oils (assessed using aromatic hydrocarbon maturity parameters, resulted in so-called mixed maturity oils that had low apparent maturities based on alkane biomarker ratios and which had been interpreted previously as 'immature oils'. They suggested that many apparently low maturity petroleums were in fact mixtures of oils with a wide range of maturity. Jiang & Li (2002) extended this concept to an examination of the Williston Basin petroleum systems, showing convincingly that while many of the oils have alkane biomarker distributions characteristic of the Madison Group source rock facies, in fact aromatic hydrocarbon distributions in the oils indicate most oils in the basin are Bakken formation sourced. This was due to the order of magnitude differences in concentrations of commonly used biomarker compounds between the two oil systems, the less dominant source (Madison group) having much greater biomarker concentrations than the dominant Bakken source rocks.

So should we stick with steranes and is there hope for hopanes? Yes, but not in the way we have been using them. In this paper, we describe the results of quantitative analysis of key petroleum system molecular markers of all compound types from oil suites around the world and suggest an alternative approach to reinventing a more utilitarian kind of geochemistry with recommendations for how petroleum geochemistry should now develop. A revolution in geochemistry is needed if the grand goals of being able to determine the quantitative contributions to a reservoir petroleum of component oils charged from source rocks at different temperatures and the residence times of these oils in petroleum traps is to be realized. These parameters, unlike existing approaches, which provide only non-weighted average maturities and no trap-residence time information, would allow source kitchen and reservoir charging scenarios to be accurately determined and thus actually constrain basin modelling efforts. In this paper, we attempt to take the first steps towards that goal.

## Samples and methods

The data presented come from analysis of 351 petroleum fluids from drill stem test (DST), repeat formation test (RFT) and production tests from basins in the North Sea, China, West Africa, USA and Germany. All data are quantitative measurements of the concentrations of various saturated and aromatic hydrocarbons in 'dead' (separator degassed) oils and condensates from these petroleum systems. Concentrations are reported relative to the dead fluid, as would normally be the case.

Samples were analysed according to The Norwegian Industry Guide to Organic Geochemical Analysis (Weiss *et al.* 2000).

## Results and interpretation

Figure 1 shows the absolute concentrations in dead oils and condensates of several major geochemically usable compound types in 351 petroleums covering the full range of maturities found in petroleum system fluids. Samples suspected of biodegradation were filtered out based on light hydrocarbon and *n*-alkane isoprenoid parameters.

Whereas the concentrations of *n*-alkanes and pristane and phytane typically vary by an order of magnitude, aromatic hydrocarbons and light hydrocarbons in most oils vary approximately over two orders of magnitude in absolute concentration. These variations are trivial compared with the range of variation seen in absolute concentrations shown by biomarker or diamondoid hydrocarbons in the sample set. Biomarkers vary by nearly 5 orders of magnitude with diamondoids varying by up to 3 orders of magnitude in absolute concentration. Samples with biomarker concentrations lower than −3.0 log wt% are normally not interpreted due their low concentrations and provide a lower boundary. Hence, interpretable biomarkers span approximately 3 orders of magnitude in concentration range.

The maturity of any large oil data set will vary substantially and Figure 2 shows the range of

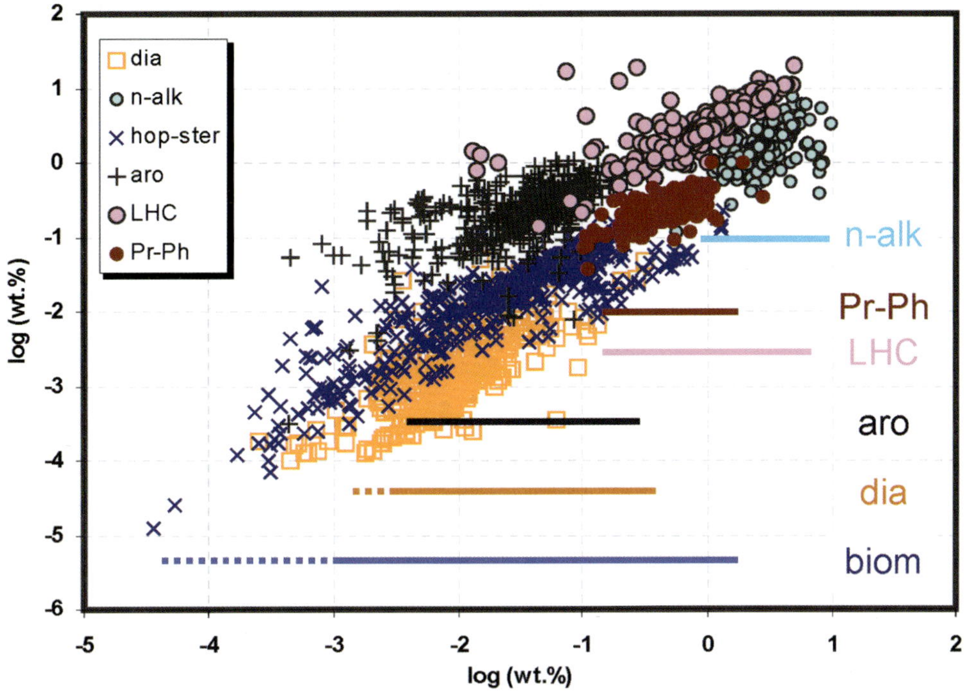


**Fig. 1.** Variation in concentrations for major and minor components of geochemical interest in crude oils (plotted as log of wt% of oil). The bars represent the orders of magnitude concentration variation for the following compound types (*Y* axis variable first; *X* axis second): diamandoid hydrocarbons (dia): sum of dimethyl diamantanes versus sum of adamantanes. *n*-alkanes (*n*-alk): sum of $n$-$C_{15}$ to $n$-$C_{19}$ versus $n$-$C_{20}$ to $n$-$C_{24}$. Hopanes and steranes (hop-ster): sum of (Ts+Tm, 29Ts, $28\alpha\beta H$, $29\alpha\beta H$, $30\alpha\beta H$ and $C_{31}$ to $C_{35}$ S&R hopanes) versus sum of ($C_{27}$ to $C_{30}$ $\beta\beta$R&S steranes plus $C_{27}$ S&R diasteranes). Aromatic hydrocarbons (aro): sum of methylphenanthrenes versus sum of $C_2$ napthalenes. Light hydrocarbons (LHC): sum of $C_7$ compounds (excluding $n$-$C_7$) versus $n$-$C_7$. Pristane-phytane (Pr-Ph): pristane versus phytane.

maturities present in the data set studied, assessed here by the light hydrocarbon parameters defined by Thompson (1983).

Figure 3 shows the effect of maturity (assessed here as Ro% equivalent by the light hydrocarbon parameters defined by Schaeffer & Littke 1988) on the biomarker concentrations in these fluids, indicated here by the hopanes. The figure shows how, with increased fluid maturity, the absolute concentrations of hopanes drop over 3 orders of magnitude through the oil window. Thus, hopane distributions from low maturity fluids in a mixture will invariably dominate in mixed oils sourced even from a single source rock.

When oils with different concentrations of a petroleum marker compound are mixed in different decimal fractions, the actual concentration of a compound in the mixture is given by:

$$\text{concentration}_{\text{mix}} = \text{fraction}_{\text{oil1}} \times \text{concentration}_{\text{oil1}} + (1 - \text{fraction}_{\text{oil1}}) \times \text{concentration}_{\text{oil2}}. \quad (1)$$

Where the concentration of specific components in each oil are vastly different, binary mixtures of the two oils will have compound distributions biased towards the oil with largest concentrations of the compounds. By taking end member oil concentrations of several components, classical biomarker or other compound ratios can be calculated for any mixtures of end-member oils in any proportions. If we take a single petroleum system, we can examine the effects of these types of concentration variations. Figure 4 shows an effective geochemical facies screen based on typical molecular parameters, which defines two end member oil types in the system, oil A and oil B. Figure 5 shows that the biomarker concentrations in the end member oils vary by orders of magnitude. Even oils with a substantial fraction of oil B mixed in will look like oil A in terms of sterane and hopane signatures and concentrations.

In contrast, Figure 6 also shows cross plots of normal and isoprenoid alkane and tetracyclic alkane parameters for the end member oils and

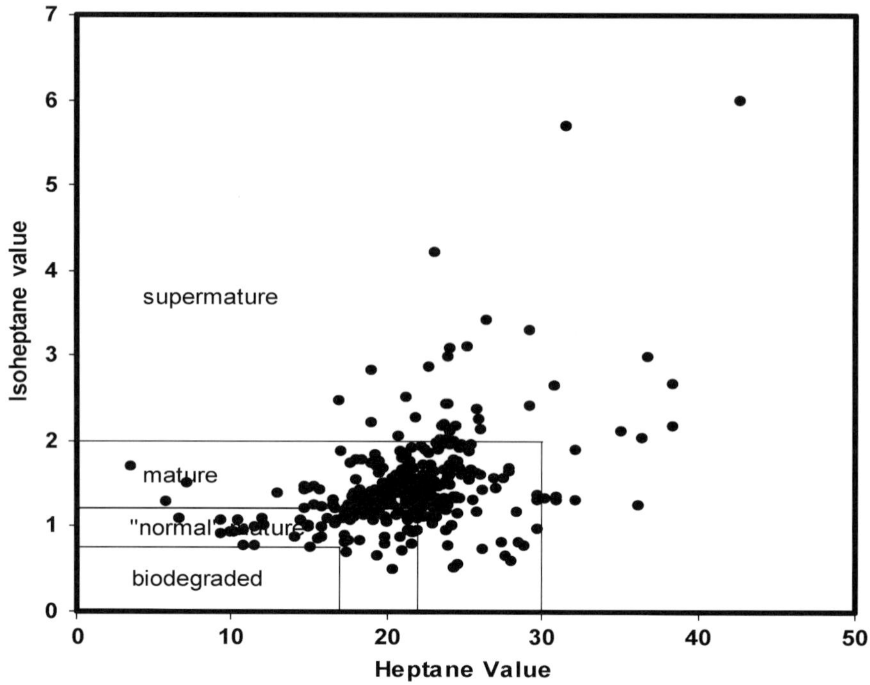


**Fig. 2.** Variation in maturity of the sample set studied based on light hydrocarbon maturity parameters (based on Thompson 1983).

calculated binary mixing ratios calculated using the end member oil concentrations and equation (1) to determine the concentrations of each component in the mixed oils. Whereas the acyclic alkane parameters (compounds abundant and subequal in concentration in most oils) readily resolve the effects of mixing differences, the biomarker parameters (low abundance and very unequal concentrations in the end member oils) show an oil A signature even when the mixture contains 90% oil B. Such problems, with an inability to resolve mixtures of oils with

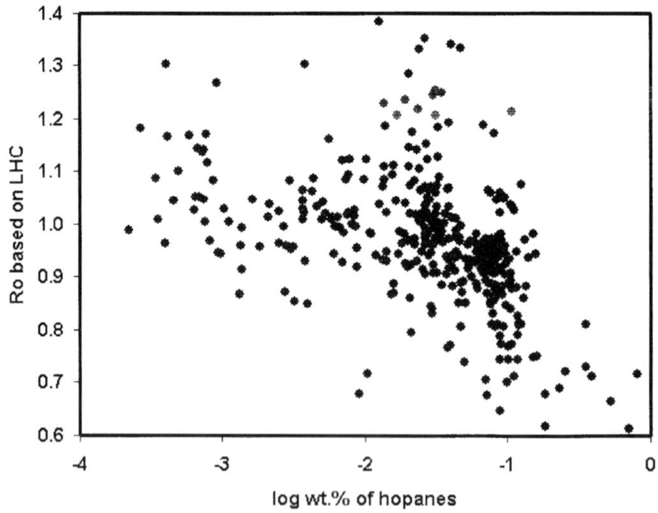


**Fig. 3.** Variation in concentration of hopanes in the sample set with maturity based on light hydrocarbons (LHC).

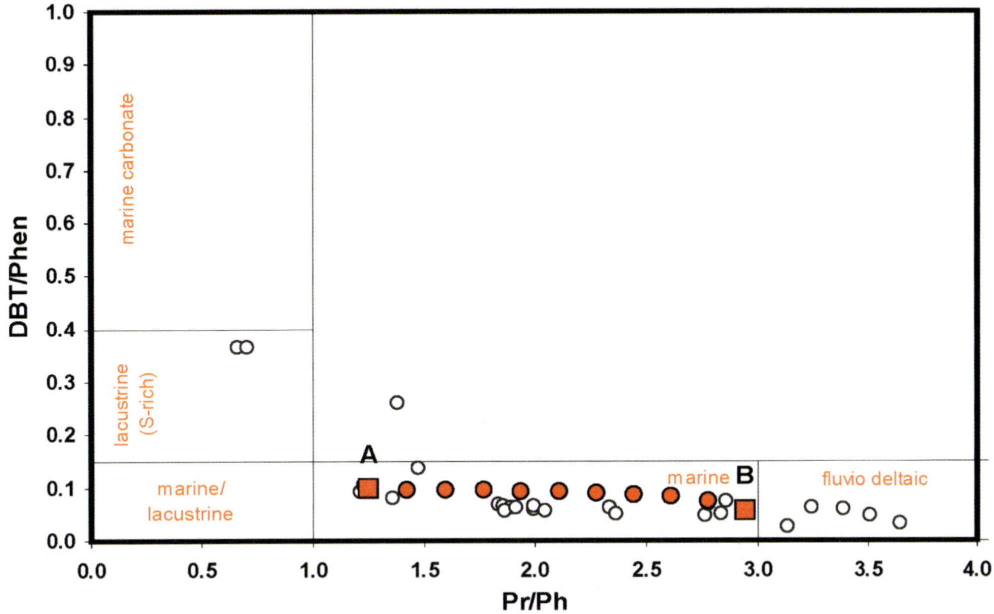


**Fig. 4.** Definition of oil families in a major petroleum basin based on relatively abundant petroleum components defined by parameters proposed by (Hughes *et al.* 1995). Two typical oils (A, B) are indicated and considered further below. Open circles are actual oil data, red squares are selected end member oils, red circles represent calculated mixtures of oil A and B in 10% increments.

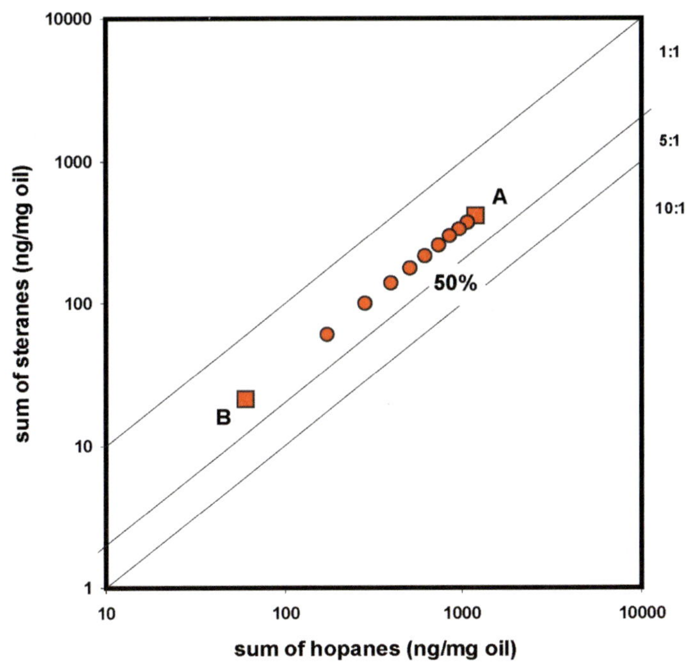


**Fig. 5.** Sterane and hopane concentrations (as defined in Fig. 1) in end member oils A and B, and calculated mixtures of these oils. Red squares are selected end member oils; red circles represent calculated mixtures of oils A and B in 10% increments.

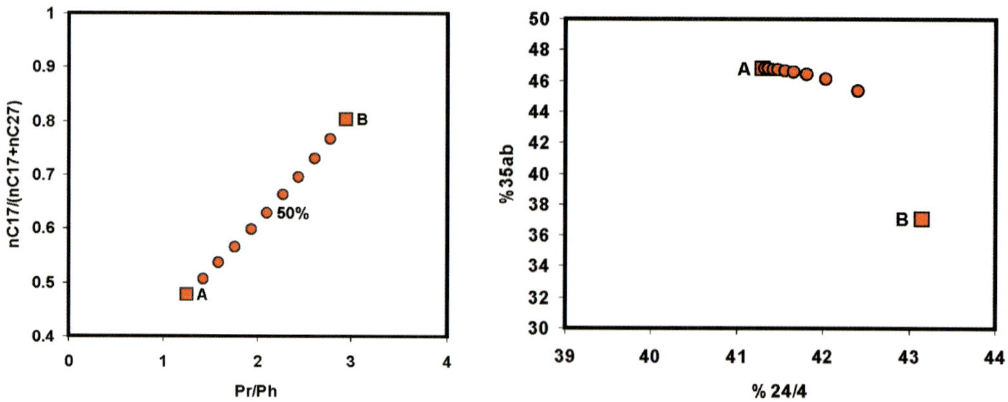


**Fig. 6.** Acyclic alkane, hopane and tetracyclic alkane cross plots of end member oils A and B and mixtures of each. Red squares are selected end member oils; red circles represent calculated mixtures of oils A and B in 10% increments. Definition of ratios: %35ab = 100 × S&R $C_{35}$ homohopane/(S&R $C_{34}$ homohopane + S&R $C_{35}$ homohopane); %24/4 = 100 × $C_{24}$ tetracyclic terpane/($C_{24}$ tetracyclic terpane + $C_{23}$ tricyclic and $C_{25}$ tricyclic terpane). Note that for the acyclic alkanes which have similar concentration ranges the mixing curve is essentially linear, whereas for the cyclic biomarkers with very different end member concentrations, the mixing curve is distinctively nonlinear. In this illustrative case of course: the actual variations in %24/4 are not sensibly interpretable.

biomarker ratios is common in our experience, and suggests, as Li *et al.* (1995) and Jiang & Li (2002) have concluded, that interpretations of facies and maturities of mixtures of oils with biomarker parameters is a risky business without a detailed quantitative study of component concentrations in end member oils. This simple analysis also suggests that oil fingerprints based on the major resolvable hydrocarbon components of gas chromatograms (abundant alkanes and aromatic hydrocarbons), which will have similar order of magnitude concentrations in most oils, are the most reliable means of detecting compositional differences in reservoir geochemical settings in many petroleum systems.

This linear mixing behaviour for pristane and phytane also explains why parameters based on these compounds are usually very good correlation parameters and facies indicators even in mixed oil situations. As biomarker concentrations also vary greatly with maturity for even single facies oil charges, the implications of oil mixing for maturity assessments using biomarkers are also severe. We examine now more generally the concept of oil maturity in general.

## Concepts of maturity

The concept of oil maturity is in need of revision. Although we talk of oils have a maturity, this is clearly an absurd concept, as all oils will be mixtures even if from a single source rock. Various fractions of the oil will have been generated at different temperatures both from a single point in a source rock and at various times across the hundreds of square kilometres typifying most petroleum source kitchens. Thus different components of oil have in essence different maturities and what is needed is a deeper concept of petroleum maturity that describes the contributions different fractions of petroleum generated at different source temperatures make to a reservoired oil or gas. We suggest an alternative approach is needed which tracks the maturity/petroleum mass fraction relationships for individual components or fractions of reservoired oils (mass fraction maturity) and allows the bracketing of source kitchen maturity.

We have defined in-house calibrations of source rock maturity from various petroleum systems, assessed as source rock vitrinite reflectance equivalent as a proxy for maximum source rock temperature. The molecular parameter versus Ro% calibrations referred to below are either based on published source rock parameters we have validated to our satisfaction (e.g. Schaefer & Littke 1988) or from in-house work on source rock maturity sequences. Inevitably, as all oils are mixtures we cannot accurately define point maturities for reservoired oils but by judicious choice of calibrant source-rock samples we feel the calibrations referred to are as robust as is possible in dealing with spatially and temporally complex entities such as source basins. The oil maturities we report are those interpreted classically based on correlations between measured oil molecular ratios and the

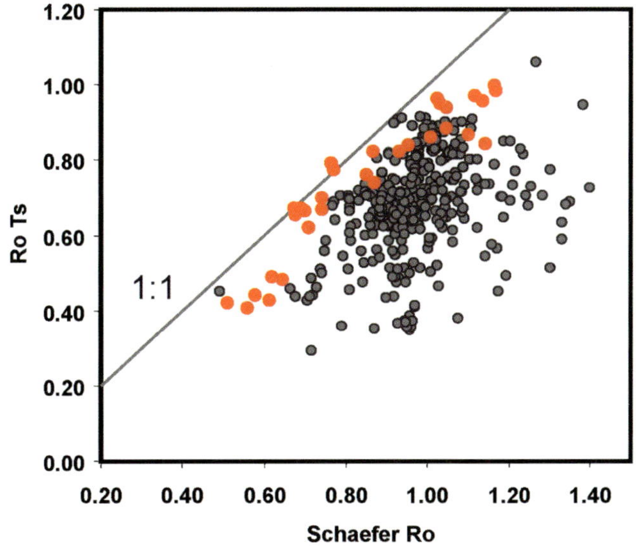


**Fig. 7.** Cross plots of calculated vitrinite reflectance equivalent biomarker maturity (Ro Ts) and light hydrocarbon maturity (Schaefer Ro) effectively indicates that almost all oils are mixtures in the true sense with a range of maturities of charge being present together in the trap. Only the flagged oils (red dots) may represent single maturity point charges, but even here that is doubtful. A mass fraction maturity concept better describes the maturity of an oil.

source rock calibration sets to derive a single maturity estimate for each compound class.

Comparison of typical biomarker based source maturity estimates versus more quantitatively representative light hydrocarbon based maturity assessments are shown in Figure 7. This figure shows that for virtually all reservoired oils there is a substantial mismatch between biomarker and light hydrocarbon maturities, the latter being more representative of the bulk of the oil reflecting, of course, their much greater concentration. Thus we can safely conclude essentially all oils are mixtures and the maturity of an oil cannot be described by a single value for the whole material.

As can be seen in Figure 8, different petroleum components measure processes across different source rock temperature and maturity ranges with biomarkers being most active and abundant at low source maturity and light hydrocarbons and aromatic hydrocarbons being more dominant later in the oil window. However, even the more reliable light hydrocarbon and aromatic hydrocarbon parameters still represent the behaviour of mixtures.

Figure 8 shows the apparent distribution of conventional 'net' oil maturities seen with three different compound classes in a suite of oils. As can be seen the various fractions track different charge maturity ranges and this approach can thus readily detect oil mixing. The range of fraction-specific maturities of an oil would effectively bracket the source kitchen maturity range for a reservoired oil but does not provide any relationship between the mass fraction of total reservoired oil charged at each temperature increment from the basin. The maturity of a petroleum component could more effectively be specified by a statistical summary of the source temperature range for that component in the given oil. The maturity distribution of oil would then represent the mass fraction versus source temperature at expulsion profile for all the components in that reservoired fluid—a mass fraction maturity. Currently our approaches do not allow us to effectively produce maturity distribution profiles for reservoired petroleums. Ideally, a mass fraction maturity relating the distribution of source temperature versus total mass of petroleum component reservoired needs to be defined—a mass fraction maturity. Such mass fraction maturities would allow much greater resolution of charge histories as they would provide a vehicle for distinguishing, in a quantitative way, the impact of many geological scenarios for a reservoir or prospect suite such as early charge loss, or the late charge of high maturity petroleums from scenarios where a complete source basin charge is captured. Ability to perform these tasks

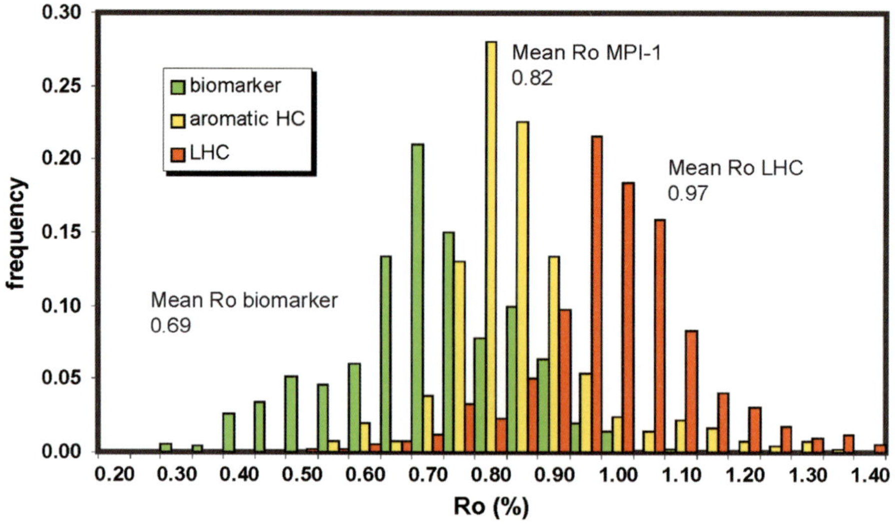


**Fig. 8.** The distribution of fraction-specific 'maturity' for the oils and condensates in the data set. Mean maturities for the data set based on representative biomarker, aromatic hydrocarbon and light hydrocarbon parameters are indicated, based on parameters proposed by Radke (1988) and Schaefer & Littke (1988). Biomarker maturities are based on proprietary source rock correlations. The cyclic biomarkers are generated earlier than the aromatic hydrocarbons with the quantitatively most abundant light hydrocarbons being generated, on average at higher maturity than the more structurally exotic molecules.

would transform the use of molecular geochemistry in exploration and provide some constraint on basin models, which are currently constrained largely by petroleum location and gross petroleum properties. Critically, a mass fraction maturity approach allows explorationists to estimate from analysis of suites of reservoired oils the fraction of total source basin charge reservoired in a basin, something we cannot currently do with any accuracy.

## Conclusions

Essentially all petroleums are mixtures with different components charged from the same source rock(s) at different temperatures. Because many classical petroleum geochemical tracers of source facies and maturity such as cyclic biomarkers show several orders of magnitude variation in petroleum systems they do not reliably track facies or maturity signals in mixed oil situations. Light hydrocarbon and aromatic hydrocarbon parameters are more reliable in this sense but, as mixtures are the norm, the concept of the maturity of oil needs revising. We suggest an alternative approach is needed which tracks the maturity/petroleum mass fraction relationships for a reservoired oil (mass fraction maturity) and allows the bracketing of source kitchen maturity. Ideally, a mass fraction maturity relating the distribution of source temperature versus total mass of petroleum component reservoired needs to be defined—a mass fraction maturity. Such mass fraction maturities would allow much greater resolution of charge histories as they would provide a vehicle for distinguishing, in a quantitative way, the impact of many geological scenarios for a reservoir or prospect suite such as early charge loss, or the late charge of high maturity petroleums from scenarios where a complete source basin charge is captured. Critically, a mass fraction maturity approach allows explorationists to estimate from analysis of suites of reservoired oils the fraction of total source basin charge reservoired in a basin, something we cannot currently do with any accuracy.

Reservoir geochemical protocols must address the real variations in absolute concentrations of the various commonly used geochemical tracers as lack of variation in some biomarker parameters across oilfields in reservoir geochemical studies where oil admixing has occurred is frequently not apparent using simple peak ratio protocols.

The project is a collaboration under the Bhudda II project. We thank Norsk Hydro for support and permission to publish. We acknowledge the help provided in review by Colin Snape and an anonymous reviewer.

## References


BARNARD, P. C. & BASTOW, M. A. 1991. *Petroleum Generation, Migration, Alteration, Entrapment and Mixing in the Central and Northern North Sea.* Geological Society, London, 167–190.

BISHOP, A. N. & ABBOTT, G. D. 1993. The interrelationship of biological marker maturity parameters and molecular yields during contact metamorphism. *Geochimica et Cosmochimica Acta*, **57**, 3661–3668.

CASSANI, F. & EGLINTON, G. 1986. Organic geochemistry of Venezuelan extra-heavy oils 1. Pyrolysis of asphaltenes: A technique for the correlation and maturity evaluation of crude oils. *Chemical Geology*, **56**, 167–183.

ENGLAND, W. A., MACKENZIE, A. S., MANN, D. M. & QUIGLEY, T. M. 1987. The movement and entrapment of petroleum fluids in the subsurface. *Journal of the Geological Society London*, **144**, 327–347.

HORSTAD, I. & LARTER, S. R. 1997. Petroleum migration, alteration, and remigration within Troll Field, Norwegian North Sea. *American Association of Petroleum Geologists Bulletin*, **81**, 222–248.

HUGHES, W. B., HOLBA, A. G. & DZOU, L. I. P. 1995. The ratios of dibenzothiophene to phenathrene and pristane to phytane as indicators of depositional environment and lithology of petroleum source rocks. *Geochimica et Cosmochimica Acta*, **59**, 3581–3598.

JIANG, C. & LI, M. 2002. Bakken/Madison petroleum systems in the Canadian Williston Basin. Part 3: geochemical evidence for significant Bakken-derived oils in Madison Group reservoirs. *Organic Geochemistry*, **33**, 761–787.

KOOPMANS, M. P., LARTER, S. R., ZHANG, C., MEI, B., WU, T. & CHEN, Y. 2002. Biodegradation and mixing of crude oils in Eocene Es3 reservoirs of the Liaohe basin, northeastern China. *American Association of Petroleum Geologists Bulletin*, **86**, 1833–1843.

LARTER, S. R. 1990. The molecular characterization of kerogen. Applications to primary and secondary migration studies and to maturation modelling. *Review of Palaeobotany and Palynology*, **65**, 379–391.

LARTER, S. R. & APLIN, A. C. 1995. Reservoir Geochemistry: Methods, Applications and Opportunities. *In*: CUBITT, J. M. & ENGLAND, W. A. (eds) *The Geochemistry of Reservoirs.* Geological Society, London, Special Publications, **86**, 5–32.

LARTER, S. R., SOLLI, H., DOUGLAS, A. G., DE LANGE, F. & DE LEEUW, J. W. 1979. The occurrence and significance of prist-1-ene in kerogen pyrolysates. *Nature*, **279**, 405–408.

LI, M., LARTER, S. R. & MEI, B. 1994. Early generation of 20S-5a(H),14a(H),17a(H)- and 5a(H),14b(H), 17b(H)-steranes in low salinity lacustrine shales. *Organic Geochemistry*, **22**, 893–902.

LI, M., LARTER, S., MEI, B. & WU, T. 1995. Mixed maturity signatures in oils and source rocks of the Liaohe Basin. *In*: GRIMALT, J. & DORRANSORRO, C. (eds) *Organic Geochemistry, Developments and Applications to Energy, Climate, Environment and Human History*, 421–423, AIGOH, Donostia, San Sebastian.

ØYGARD, K., LARTER, S. R. & SENFTLE, J. 1988. The Control of Maturity and Kerogen Type on Analytical Pyrolysis Data. *Organic Geochemistry*, **13**, 1153–1162.

PEPPER, A. S. & CORVI, P. J. 1995. Simple kinetic models of petroleum formation. Part I: oil and gas generation from kerogen. *Marine and Petroleum Geology*, **12**, 291–319.

RADKE, M. 1988. Application of aromatic compounds as maturity indicators in source rocks and crude oils. *Marine and Petroleum Geology*, **5**, 224–236.

RULLKÖTTER, J., MACKENZIE, A. S., WELTE, D. H., LEYTHAEUSER, D. & RADKE, M. 1984. Quantitative gas chromatography–mass spectrometry analysis of geological samples. *Organic Geochemistry*, **6**, 817–827.

SCHAEFER, R. G. & LITTKE, R. 1988. *Maturity-Related Changes in the Low Molecular-Weight Hydrocarbon Fraction of Toarcian Shales.* Pergamon Press, Oxford, 887–892.

THOMPSON, K. F. M. 1983. Classification and thermal history of petroleum based on light hydrocarbons. *Geochimica et Cosmochimica Acta*, **47**, 303–316.

VOLKMAN, J. K. 1984, Biodegradation of aromatic hydrocarbons in crude oils from the Barrow Subbasin of Western Australia. *Organic Geochemistry*, **6**, 619–632.

WEISS, H. M., WILHELMS, A., MILLS, N., SCOTCHMER, J., HALL, P. B., LIND, K. & BREKKE, T. 2000. *NIGOGA—The Norwegian Industry Guide to Organic Geochemcial Analysis* [online]. Edition 4.0. Published by Norsk Hydro, Statoil, Geolab Nor, SINTEF Petroleum Research and the Norwegian Petroleum Directorate. www.npd.no

WILHELMS, A., LARTER, S. R., HEAD, I., FARRIMOND, P., DI PRIMIO, R. & ZWACH, C. 2001. Biodegradation of oil in uplifted basins prevented by deep-burial sterilization. *Nature*, **411**, 1034–1037.

# An overview of developments related to the characterization and significance of high molecular weight paraffins/hydrocarbons ($>C_{40}$) in crude oils

R. PAUL PHILP*, MICHAEL HSIEH & FAZEELAT TAHIRA

*School of Geology and Geophysics, University of Oklahoma, Norman, OK 73019, USA*
**(Fax: 001-405-325-3140; e-mail: pphilp@ou.edu)*

**Abstract:** Over the past 10 years, investigations into the characteristics of the high molecular weight hydrocarbon (HMWHC) fraction in crude oils and, to a lesser extent, source rock extracts have continued to reveal novel information concerning the distribution of hydrocarbons $>C_{40}$. The major impetus for this work has come from the fact that HMWHCs can cause significant production problems in certain geographical regions and particularly deepwater frontier areas. Since these HMWHCs appear to be ubiquitous in crude oils, the primary questions that need to be addressed are: what are these compounds, where do they come from, and how do they affect physical properties of oils? Here, we review our work over the past decade and discuss the significance of these results and their potential application to reservoir and production problems involving paraffins and asphaltenes.

It was commonly believed for many years that only oils derived from source rocks containing higher plant source material would have a high paraffin content. However, it is now abundantly clear that oils derived from lacustrine and marine source rocks also contain relatively high concentrations of higher molecular weight hydrocarbons. In addition to developing methods for the qualitative and quantitative separation of HMWHCs from asphaltenes, progress has been made in identifying individual components of the high molecular weight fraction. This fraction is not a simple mixture of *n*-alkanes but a complex mixture of seven or eight homologous hydrocarbon series, each with significantly different physical properties. A knowledge of these structures is important in predicting crude oil properties such as cloud point and pour point. Series identified to date include alkylcyclopentanes, alkylcyclohexanes, alkylbenzenes and various branched hydrocarbons.

In summary, since the 1970s most of the geochemical research emphasis has been placed on compounds below $C_{40}$. Whilst compounds above $C_{40}$ may not have the same degree of structural specificity as the traditional biomarkers, the amount of information available from these compounds could be extremely beneficial in the long term, particularly for reservoir characterization and production purposes and other problems involving high molecular weight hydrocarbons.

High molecular weight hydrocarbons (HMWHCs) in crude oils have had an undistinguished history over the years, apart from a few landmark papers such as the classic paper by Hedberg (1968) who proposed that waxes in crude oils reflected terrigenous organic source material. Furthermore the oils were typically restricted to non-marine stratigraphic sequences deposited during the Devonian to Pliocene periods. Gas chromatography capillary columns capable of withstanding high temperatures (Lipsky & Duffy 1986*a*, *b*) advanced our ability to characterize HMWHCs ($C_{40+}$) in crude oils and more recently in source rock extracts (Mueller & Philp 1998). Improvements in analytical technology, such as high temperature gas chromatography (HTGC) and methods to isolate and concentrate HMWHCs has resulted in observations dispelling earlier beliefs that HMWHCs are restricted to higher-plant-derived organic matter (Hedberg 1968; Kinghorn 1983). A multitude of evidence is now available which demonstrates the ubiquity of HMWHCs in crude oils and source rocks originating from terrigenous, marine and lacustrine source materials, varying in age and maturity (Moldowan *et al.* 1985; Tegelaar *et al.* 1989; Carlson *et al* 1993; Heath *et al.* 1995; Philp *et al.* 1995; Hsieh 1999; Hsieh & Philp 2001).

The occurrence of HMWHCs in crude oils has typically been determined by simulated distillation but the poor chromatographic resolution associated with this approach has been a limiting factor in understanding the complexity of the HMWHC fraction (Trestianu *et al.* 1985; Thomson & Rynaski 1992; Neer & Deo 1995; Durand *et al.* 1998). Routine analyses of waxy crude oils by HTGC have revealed that HMWHCs are composed of complex mixtures of alkyl-cycloalkanes, methylbranched alkanes, and alkyl-aromatic hydrocarbons, and may represent a significant fraction of the whole oil (Hsieh & Philp 2001).

*From*: CUBITT, J. M., ENGLAND, W. A. & LARTER, S. (eds) 2004. *Understanding Petroleum Reservoirs: towards an Integrated Reservoir Engineering and Geochemical Approach.* Geological Society, London, Special Publications, **237**, 37–51. 0305-8719/04/$15.00 

For example, Burger *et al.* (1981) characterized an oil sample composed of 14% wax, 58% of which included hydrocarbons in the $C_{46+}$ range; and Barker (1995) noted that microcrystalline waxes in a refined petroleum sample was composed of 20-40% *n*-alkanes, 15-40% *iso*-alkanes and ~35% cycloalkanes. The HMWHCs in the $C_{40+}$ region are of particular interest since they are thought to be the primary culprit in wax deposition problems. Case studies have demonstrated that a minimal microcrystalline wax content of 2% in an oil can result in wax deposition problems (Holder & Winkler 1965; Tuttle 1983; Ajienka & Ikoku 1990).

This paper will attempt to highlight and summarize the developments that have taken place in the characterization of HMWHCs. Data acquired over the past decade will be used and individual samples identified where appropriate and necessary. For the most part the samples are used simply to illustrate a particular development and will not be described in great detail. Similarly the analytical methods are somewhat generic at this stage but are summarized briefly below for convenience. Detailed descriptions of analytical procedures and conditions are described in Hsieh (1999), Thanh *et al.* (1999), Hsieh *et al.* (2000) and Hsieh & Philp (2001).

## Experimental

### *Wax/asphaltene isolation*

A major problem associated with the characterization of the HMWHCs is the quantitative and qualitative separation of waxes from asphaltenes. Thanh *et al.* (1999) developed a method, subsequently modified (Hsieh *et al.* 2000), which involved the adsorption of the crude oil on alumina, followed by extraction of the alumina to remove the maltenes, including the HMWHCs, leaving the asphaltenes adsorbed to the alumina. The maltenes were subsequently separated into wax and non-wax fractions via acetone precipitation (Burger *et al.* 1981) and the asphaltenes were recovered by secondary extraction of the alumina. Characterization of the respective fractions demonstrated that the waxes were asphaltene-free and the asphaltenes were wax-free.

### *High temperature gas chromatography*

HTGC is now a routine analytical tool and high temperature columns, of varying quality, are available from a number of suppliers. If necessary these columns can be temperature programmed to an upper limit of approximately 480 °C, permitting separation of hydrocarbons up to approximately $C_{120}$ (Hsieh & Philp 2001). Most analyses only require upper temperatures of approximately 380 °C because the majority of oil samples collected at the well-head typically do not contain hydrocarbons above $C_{80}$. The solution of HMWHCs must be warmed and thoroughly shaken prior to injection with a heated syringe. Failure to do this will result in non-reproducible analyses.

### *Gas chromatography–mass spectrometry*

In an effort to identify as many components in the wax fraction as possible, short (15 m) HTGC columns interfaced to the mass spectrometer are used. This procedure satisfactorily separates many of the major series of compounds and permits compounds up to about $C_{50}$ to elute directly into the mass spectrometer. Many of the compounds described below have been identified on the basis of mass spectral interpretation or comparison with other samples where their identities have been confirmed (Hsieh & Philp 2001). Confirmation of the components tentatively identified ultimately requires co-injection with authentic standards.

## Discussion

### *Review of advances in the analysis of HMWHCs*

There are numerous papers and references in the literature to early studies of waxes from a wide variety of sources containing compounds, which at the time of discovery, could be thought of as HMWHCs. For example, in 1916, Tsujimoto first identified squalene ($C_{30}H_{50}$) in shark liver oil, and Chapman (1917) isolated a similar compound from Portuguese shark liver oil which he called spinacene and which was probably squalene. As time has passed and our analytical capabilities have improved, the known carbon number distribution of compounds in the HMWHC fraction has moved to higher and higher carbon numbers.

In the mid-1980s, a great deal of emphasis was placed on the discovery of new and useful sterane and terpenoid biomarkers in crude oils and source rocks, and the majority of compounds discovered were in the carbon number range below $C_{35}$ (Peters & Moldowan 1993). There were a couple of exceptions, including hopanes extending to $C_{40}$ in extracts from the Thornton bitumen (Rullkötter & Philp 1981). Around the same time, HTGC was starting to emerge as a useful tool for extending hydrocarbon analyses into higher carbon number ranges (Lipsky & Duffy 1986*a*, *b*). The use of HTGC demonstrated that previously it was the analytical limitations that were preventing advances into the higher carbon number ranges and not the absence of the higher carbon numbered compounds. Perusal of the geochemical literature reveals that this has been a common factor – every advance in analytical capabilities has lead to a significant advance in the upper carbon number limit of biomarkers identified in oils and source rocks.

Oklahoma, where many of our studies were conducted, is a state with abundant oil wells, and wells with abundant wax or paraffin deposition problems despite the fact that most oil is produced from marine source rocks (in particular the Woodford Shale). In our early work several waxes responsible for blocking oil wells in various locations of the Oklahoma oilfields were examined by HTGC, and it was noted that the wax deposits were typically dominated by hydrocarbons maximizing around $C_{40}$–$C_{50}$ (del Rio & Philp 1992). However, oils produced from the same well did not show any anomalous concentrations of HMWHCs when analysed by conventional or HTGC. A method was necessary to concentrate the waxes from the oils, bearing in mind that oil samples collected from the well-head will have already lost a large proportion of the wax component through deposition in the well bore. Furthermore, a significant proportion of the HMWHCs may also have precipitated in the reservoir if there was a significant temperature gradient between source rock and reservoir.

A wax concentration method was developed which incorporated the UOP method (i.e. UOP Method 46-64) modified by Burger *et al.* (1981). However, this led to the discovery of another problem of major significance in accurately characterizing an oil's wax and asphaltene content. Typically, after precipitation from a crude oil little else is done with the asphaltene fraction apart from quantification. During our method development, asphaltene fractions precipitated in the traditional manner with pentane were analysed by HTGC and observed to contain significant proportions of HMWHCs ($>C_{40}$) that are also insoluble in pentane (Figure 1). A similar observation has been reported by Teixeira & Gonçalves (2000), who demonstrated the tendency of long-chain alkanes to coprecipitate with asphaltenes in low-boiling-point solvents.

The discovery of HMWHCs in the asphaltene fraction clearly necessitated the development of a new method that would eliminate the presence of HMWHCs from the asphaltene fractions. In 1999, Thanh *et al.* successfully developed a method involving adsorption of the oil on alumina, and extraction with *iso*-octane to recover the maltene fraction followed by extraction with dichloromethane/methanol to recover the asphaltenes from the alumina. This method produced an asphaltene fraction devoid of HMWHCs and a maltene fraction devoid of asphaltenes, permitting a far more accurate indication of the gross compositional characteristics of the oil. Furthermore, asphaltene-free waxes could be isolated from the maltenes and, if required, further divided into macro- and micro-crystalline fractions providing additional and useful information of interest to the production engineers.

## *Isolation and characterization of HMWHCs*

Following the work described above, the situation was such that a reliable and reproducible method was now in place for the isolation and characterization of waxes. Additional papers were being published on the topic of HMWHCs. Heath *et al.* (1997) noted the resistance to biodegradation of HMWHCs beyond $C_{40}$. This was an extremely important observation since residues previously thought to be devoid of *n*-alkanes when analysed by conventional GC, effectively concentrated hydrocarbons above $C_{40}$ following extensive biodegradation. Hsieh (1999) observed a similar effect after concentrating waxes from a heavily biodegraded oil (sourced from higher plant material) that appeared to be devoid of *n*-alkanes, with a relatively high abundance of pristane and phytane. The isolated wax contained an homologous series of well-preserved *n*-alkanes ($C_{20}$–$C_{35}$) along with a complex mixture of long-chain methylbranched alkanes and alkylcycloalkanes ($C_{25}$–$C_{60}$; Figure 2).

Carlson *et al.* (1993) noted variations in co-eluting monocyclic/acyclic hydrocarbons with the nature of the depositional environment in which the original source material was deposited. The homologous series of compounds, subsequently tentatively identified as alkylcyclopentanes, was found to extend into the $C_{60}$ region of crude oils. Carlson *et al.* (1993) observed that in marine oils this series of compounds had a distinct odd/even predominance pattern (Figure 3a), whereas oils from saline lacustrine environments demonstrate a high even/odd

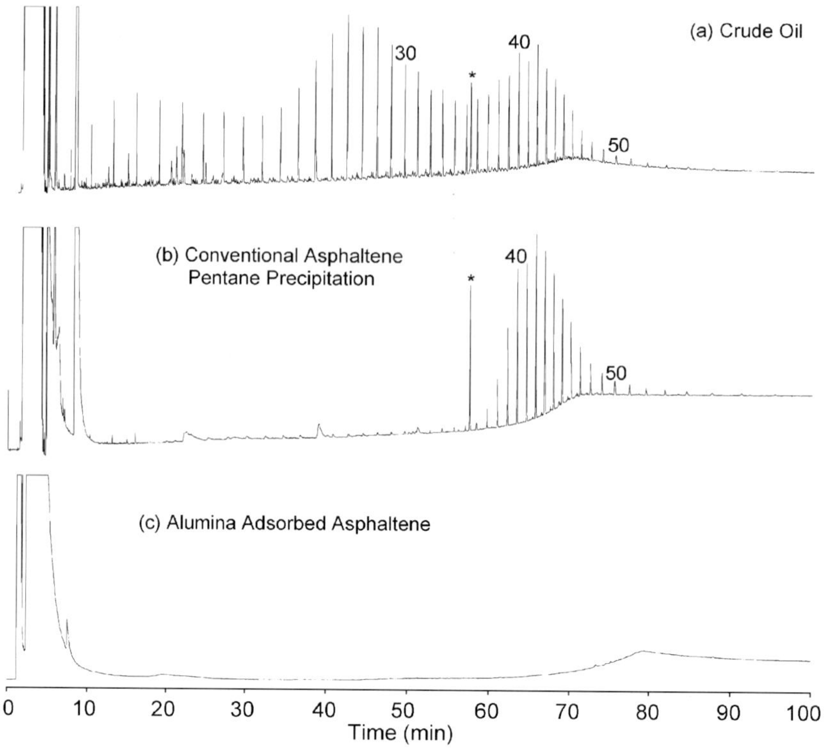


**Fig. 1.** Asphaltene fractions precipitated in the traditional manner with pentane and analysed by HTGC often contain significant proportions of HMWHCs (>$C_{40}$) which are also insoluble in pentane, as illustrated in this example.

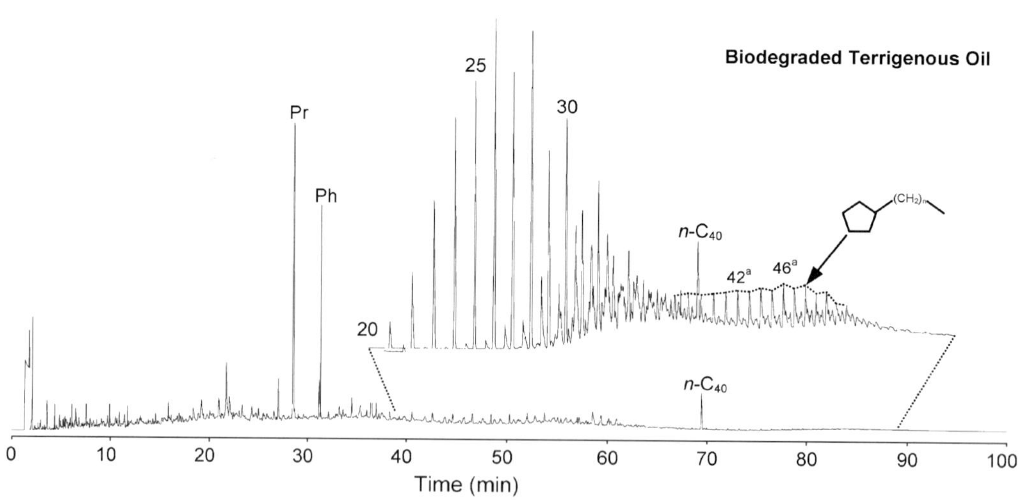


**Fig. 2.** Extensively biodegraded oils may not appear to contain any *n*-hydrocarbons, but by applying the wax concentration process homologous series of well preserved *n*-alkanes ($nC_{20}$–$nC_{35}$) and a complex mixture of long-chain methylbranched- and alkylcyclo-alkanes ($nC_{25}$–$nC_{60}$) could be isolated from this sample (Hsieh 1999).

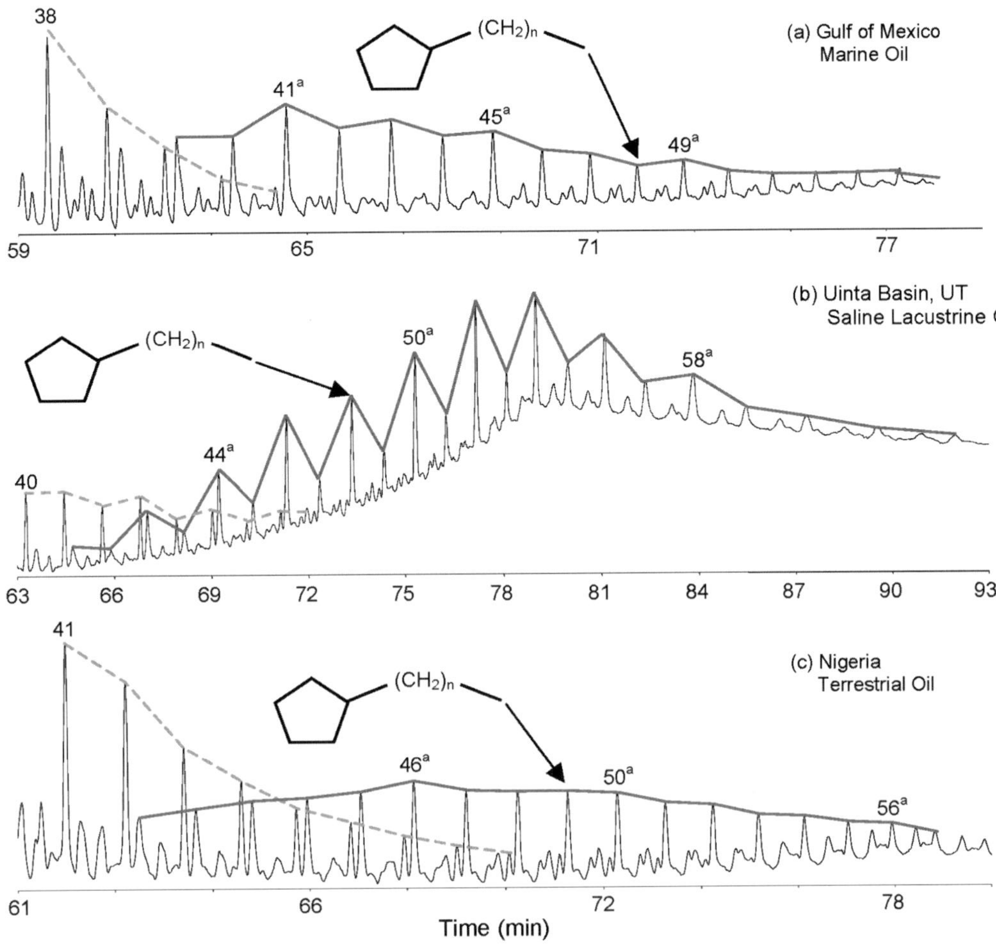


**Fig. 3.** (**a**) Oils derived from marine source rocks are characterized by a distinct odd/even predominance pattern for the alkylcyclopentanes, (**b**) oils from saline lacustrine environments have a high even/odd predominance pattern, and (**c**) freshwater lacustrine oils a low even/odd or no clear predominance pattern.

predominance pattern (Figure 3b), and freshwater lacustrine oils were characterized by a low even/odd or no clear predominance (Figure 3c). They identified these compounds as co-eluting monocyclic/acyclic hydrocarbons, and more recent work has shown the presence of both alkylcyclohexanes and alkylcyclopentanes (Hsieh *et al.* 2000). While it is not suggested that these distributions will replace the utility of the more conventional biomarker parameters, they may be useful when conventional biomarker distributions provide ambiguous or misleading results. Additional papers discussed the relationship between pour points of crude oils and their wax content (Carlson *et al.* 1994). The isotopic compositions of the HMWHCs were found to be similar to those of the LMWHCs, suggesting that these hydrocarbons were probably derived from the same or similar sources (Carlson *et al.* 1997).

### *Preferential precipitation of waxes during secondary migration*

Carbon number distribution and molecular structures of waxes vary significantly in oils, affecting solubility properties and behaviour of crude oils. An in-depth understanding of wax composition and distribution would greatly benefit models for predicting wax deposition potential and allow improvements in the development of wax remediation procedures used during oil production. Our studies have demonstrated that the variability of carbon number distribution of waxes is not restricted to differences in geographic location or source material. As part of a study to investigate changes in wax content and composition resulting from migration, a number of oils from the Rimbey–Meadowbrook reef trend in east-central Alberta are being characterized for their wax content. These oils sourced from the Duvernay Formation have been previously investigated by Li *et al.* (1998) and divided into two sub-families. Our preliminary study has simply investigated wax content of both sub-families with increasing migration distance; future studies will investigate whether or not any changes in the wax content or distribution can be related to these sub-families. A review of these preliminary results suggests that as the oils migrate updip the wax content decreases and the carbon number distribution moves towards lower carbon number distribution maxima indicating deposition of the higher molecular weight hydrocarbons (Figure 4). If this is indeed correct and it can be demonstrated that wax deposition is taking place with increasing migration distance, this opens the possibility that such an observation may ultimately be developed into a migration distance parameter.

## *Occurrence of HMWHCs in source rocks*

Recent work has demonstrated the presence of HMWHCs in source rock extracts (Tuo & Philp 2003). These compounds may have been overlooked in the past through the inefficient extraction processes and not using HTGC to analyse the extracts. To illustrate the presence of these compounds in rock extracts, HTGC chromatograms from the extracts of source rocks from basins in China are shown in Figure 5, indicating the presence of HMWHCs extending to $C_{60+}$ in these particular samples. The *n*-alkanes and alkylcyclopentanes are clearly evident, although the methylalkanes are present in very low concentrations. A more detailed discussion of these results has been published elsewhere (Tuo & Philp 2003). Distribution of

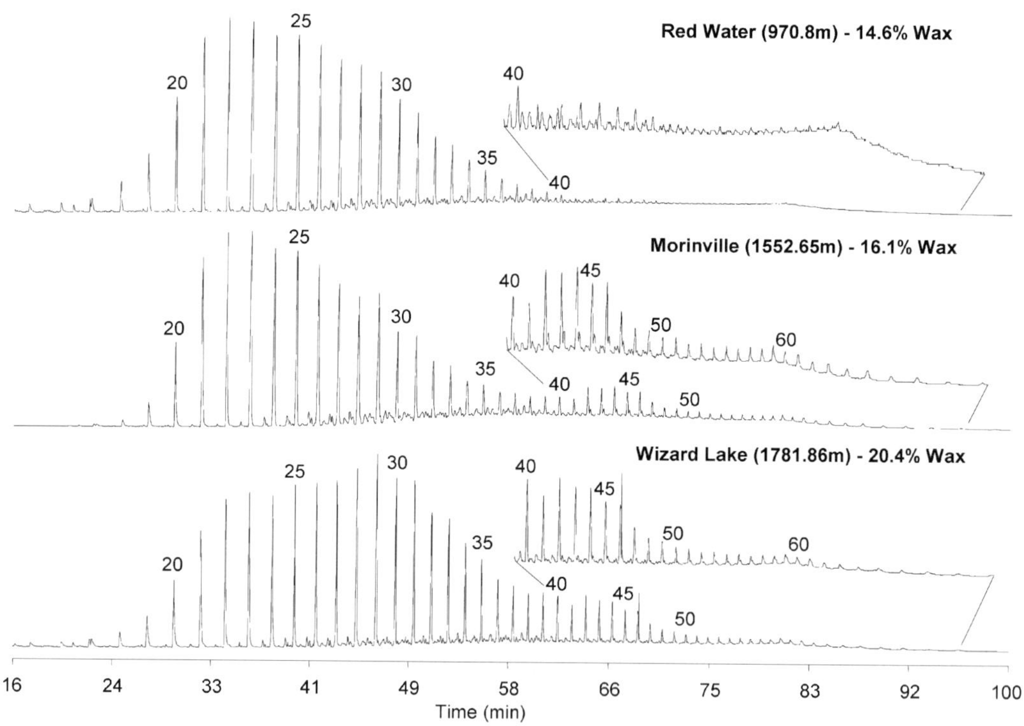


**Fig. 4.** A number of oils from the Rimbey–Meadowbrook reef trend in east-central Alberta have been characterized for wax content. HTGC chromatograms for three of these oils sourced from the Duvernay Formation are shown in this figure. Note that as these oils migrate to the shallower reservoirs, the wax content decreases and the carbon number distribution moves to lower carbon numbers as a result of depositions of the higher carbon numbered components.

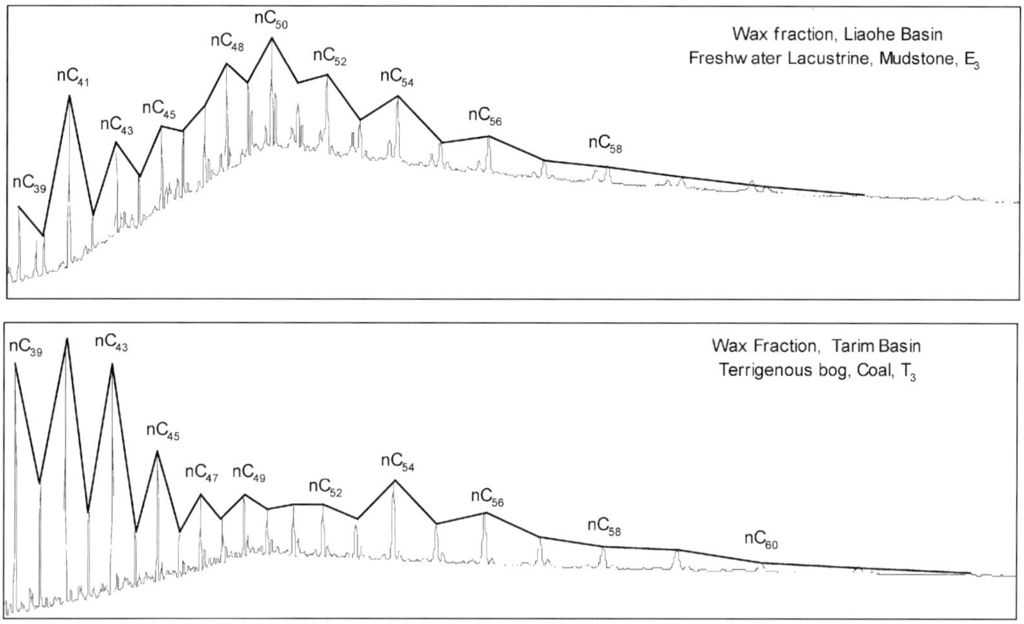


**Fig. 5.** Source rock extracts have now been shown to possess HMWHCs with distributions similar to those observed in crude oils, as illustrated by these examples from the Tarim and Liahoe Basins, China. The chromatograms shown here represent the concentrated wax fractions isolated from two samples to illustrate both the presence of the HMWHCs in the source rock extracts and variations with their depositional environments.

these HMWHC compounds may be used to provide information on the depositional environment and to possibly undertake oil/source rock correlations. As mentioned above, a knowledge of the wax content of the source rocks and how this changes in the oils will assist in the development of migration parameters based on changes in wax content. A major advantage of using wax content as a migration tool is the fact that changes are based primarily on the temperature gradient and as a result are less affected by mineral interactions and other processes which have led to the demise of other proposed migration parameters.

## *Origin and mechanisms for the formation of HMWHCs*

As work on lipid chemistry has developed, a tremendous number of HMWHCs of one type or another have been found in high abundance in the lipids of various plants, animals, microorganisms and insect waxes (Nelson & Blomquist 1995). High molecular weight esters are components of cuticular waxes of higher plants ($C_{36}$–$C_{52}$; Cranwell & Volkman 1981). Lipids and insoluble fractions from several freshwater and marine algae have been suggested as possible precursors for HMWHCs (de Leeuw & Largeau 1993; Gelin *et al.* 1999), and more recently high molecular weight lipids ($C_{80+}$) have been detected in freshwater microalgae (Allard *et al.* 2002). A variety of bacteria and blue-green algae possess the capability to synthesize lipids with varying carbon number ranges. Other suggested sources for HMWHCs include indigenous components of living organisms (e.g. cuticular waxes; blue-green algae or bacteria associated with hypersaline, carbonate environments), fatty acids or wax esters (Dembicki *et al.* 1976; Tissot & Welte 1984; Kissin 1990). Cyclopropyl acids are common in bacterial and algal mats and are reported precursors of monomethylalkanes (MMAs) in sediments (Cardoso *et al.* 1978; Makushina *et al.* 1978).

Many of the hydrocarbons in insect lipids are also of rather high molecular weight and it is not unusual to find reports of lipids containing hydrocarbons up to $C_{150}$, not necessarily as homologous series but simply as individual compounds. The surface lipids of the field cricket (*Nemobius fasciatus*) are 100% hydrocarbons (Nelson & Blomquist 1995). Long-chain branched alkanes have been reported in the grasshopper (*Schistocerca vaga*), where the major series include monomethyl alkanes ($C_{33}$–$C_{55}$), dimethyl alkanes ($C_{35}$–$C_{55}$) and trimethylalkanes ($C_{38}H_{78}$) (Nelson 1978). This is not the place to review insect lipid geochemistry, and for a more detailed coverage the interested reader is referred to the monograph by Nelson & Blomquist *et al.* (1995). Suffice it to say insects should be considered very carefully as possible sources for at least some of the HMWHCs discussed in this paper. Kenig *et al.* (1995) suggested that some of the HMWHCs in Abu Dhabi algal mats may have been derived from insects feeding on the algal mats.

HMWHCs may also be produced through various mechanisms such as oligomerization of low molecular weight fatty acids (del Rio & Philp 1992), decarboxylation of fatty acids, or decarbonylation of aldehydes (Bianchi 1995). Possible precursors of HMWHCs in oils might include long-chain fatty acids and alcohols (extending as high as $C_{80}$) and long-chain dicarboxylic acids (as high as $C_{120}$) which have been reported in the outer cell walls of freshwater and marine microalgae (Allard *et al.* 2002). Killops *et al.* (2000) recently proposed that long-chain *n*-alkanes in coals may be formed by thermal decarboxylation of cutan and cutin derivatives at early stages of maturation. Many of the above suggestions were made before the identity of the major homologous series of compounds had been determined. Recent work, described below, has led to additional information on the possible origins for these compounds. It should also be mentioned that future isotopic studies of individual compounds may shed further light on this topic, particularly as our capability to use combined HTGC combined with isotope ratio mass spectrometry permits us to obtain isotope ratios on individual compounds in the higher molecular weight region above $C_{40}$, as shown in the early work of Carlson *et al.* (1997).

## *New developments and identification of homologous series of HMWHCs*

As noted above, Carlson *et al.* (1993) initially identified the presence of monocyclic/acyclic branched alkanes in the high molecular weight fractions. More recently Hsieh *et al.* (2000) noted the presence of alkylaromatic compounds in this fraction along with a wide variety of branched hydrocarbons, alkylcyclohexanes and alkylcyclopentanes. However, the need to document the identities of all the compounds completely cannot be over-emphasized. Physical properties of the branched hydrocarbons, particularly melting points, vary significantly with degree of branching (Table 1). Therefore, if the individual compounds can be identified it will greatly improve any modelling efforts designed to predict the anticipated extent of wax deposition and assist in predicting temperatures at which wax deposition may be expected to occur. Carlson *et al.* (1993) noted that pour point varies with wax content, but if the individual series of compounds are identified it may be possible to refine this relationship even further. García *et al.* (2000) have also demonstrated that iso- and cyclo-alkane concentrations of greater than 40 wt% can hinder wax crystal arrangement and increase the cloud point of an oil. Most thermodynamic models consider the wax to be composed entirely of *n*-alkanes, which is not correct. More than 50% of wax is a very complex mixture of branched and cyclic compounds composed of at least seven to eight homologous series of hydrocarbons in differing proportions, all of which need to be considered when modelling properties of the oils.

**Table 1.** *Melting points of some individual compounds to illustrate variations resulting from branching positions in the linear chain*

| Hydrocarbon | Melting point (°C) |
|---|---|
| $C_{24}H_{50}$ | |
| *n*-Tetracosane | 51.5 |
| 2-Methyltricosane | 42.0 |
| 2,2-Dimethyldocosane | 34.6 |
| 5-*n*-Butyleicosane | 8.0 |
| $C_{26}H_{54}$ | |
| *n*-Hexacosane | 56.4 |
| 5-*n*-Butyldocosane | 20.8 |
| 7-*n*-Butyldocosane | 3.2 |
| 11-*n*-Butyldocosane | 0.0 |

As a result of improved molecular sieving, and analysis of sieved fractions by GCMS using short GC columns, a number of additional homologous series of branched hydrocarbons have been identified in the HMWHC fractions. A total ion current chromatogram for a branched/cyclic HMWHC fraction is shown in Figure 6, and illustrates the complexity of the fraction. Interpretation of the mass spectra for a number of these components reveals that the major components are homologous series of monomethylalkanes, *n*-alkylcyclohexanes, methyl-*n*-alkylcyclohexanes, *n*-alkylcyclopentanes and *n*-alkylbenzenes, *n*-alkyltoluenes and *n*-alkylxylenes. Six homologous series of monomethylalkanes have been tentatively identified in these samples from the interpretation of mass spectral data and literature data (Klomp 1986; Fowler & Douglas 1987; Summons 1987; Shiea *et al.* 1990;

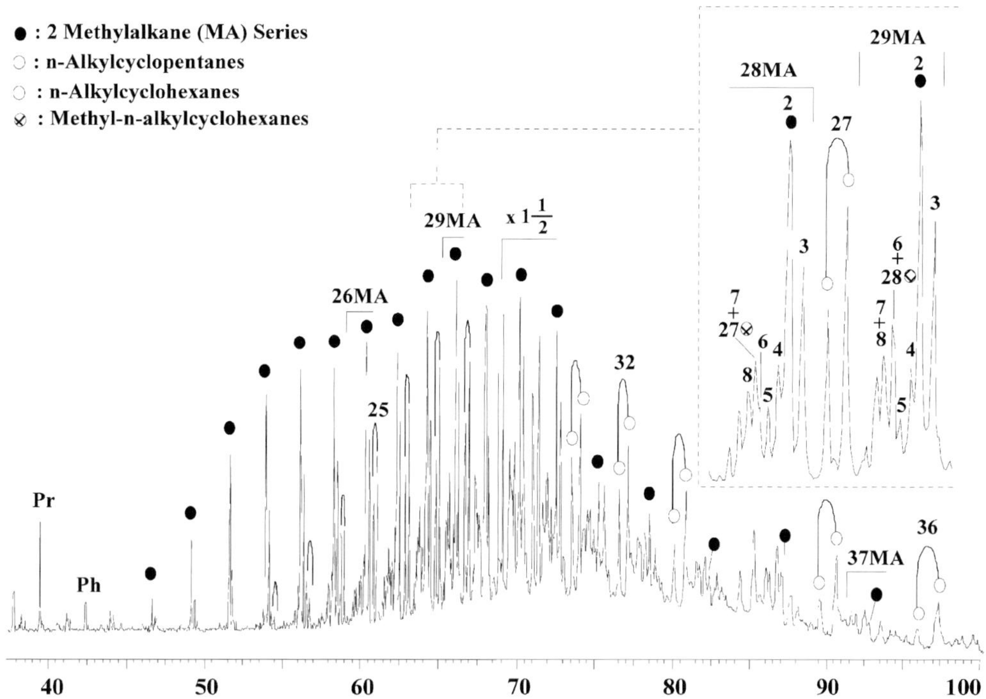


**Fig. 6.** The complexity of the branched/cyclic HMWHC fraction is illustrated in this total ion current chromatogram. The major components in this fraction are homologous series of monomethylalkanes, *n*-alkylcyclohexanes, methyl-*n*-alkylcyclohexanes, *n*-alkylcyclopentanes and *n*-alkylbenzenes, *n*-alkyltoluenes and *n*-alkylxylenes as summarized in Table 1.

Kenig *et al.* 1995) and the series of branched hydrocarbons extends to at least $C_{60+}$. The 2-, 3-, 4-, 5-, 6- and 7-methylalkane series (MMA) have been tentatively identified through the presence of characteristic ions at *m/z* 56, 70, 84, 98 and 112, respectively, which arise from cleavage of the carbon–carbon bond at the branching position (Figure 7). The homologous series of 2-methylalkanes is the most abundant series of MMAs in these branched/cyclic HMWHC fractions. A more detailed report on the identification will be published elsewhere after identities have been confirmed through co-injection of authentic standards.

In a recent study, Lehmler *et al.* (2002) have also recognized the difficulties in identifying branched alkanes $>C_{40}$, but successfully synthesized nine mid-chain branched alkanes (carbon chain between $C_{38}$–$C_{53}$), which will be used to assist in the identification of the homologous series in the HMWHC fractions. The presence of the *n*-alkylcyclopentanes, *n*-alkylcyclohexanes and methyl-*n*-alkylcyclohexanes was tentatively confirmed through the monitoring of the ions at *m/z* 68, 82 and 96 (Figure 8), respectively. The complexity of methyl-*n*-alkylcyclohexanes, as illustrated in the *m/z* 96 chromatogram, results from the existence of several isomers. The major series is identified as the 1-methyl-4-alkylcyclohexanes based on mass spectral data and comparison with literature data (Fowler *et al.* 1986; Dong *et al.* 1993; Hsieh *et al.* 2000).

Several homologous series of *n*-alkylmonoaromatic compounds including *n*-alkylbenzenes, *n*-alkyltoluenes and *n*-alkylxylenes have been tentatively identified in the HMW aromatic hydrocarbon fraction (Figure 9). The complexity of the mixture results, in part, from the presence of the *o*-, *m*- and *p*-isomers, as illustrated by the complexity of the *m/z* 105/106 and 119/120 chromatograms (Dong *et al.* 1993). It appears that these homologous series of aromatic hydrocarbons extend beyond $C_{50+}$, although these identifications remain to be confirmed by co-injection with authentic standards.

## *Future work and scope of research*

With the identification of these additional homologous series of compounds in both the saturate and aromatic fractions, a number of wax

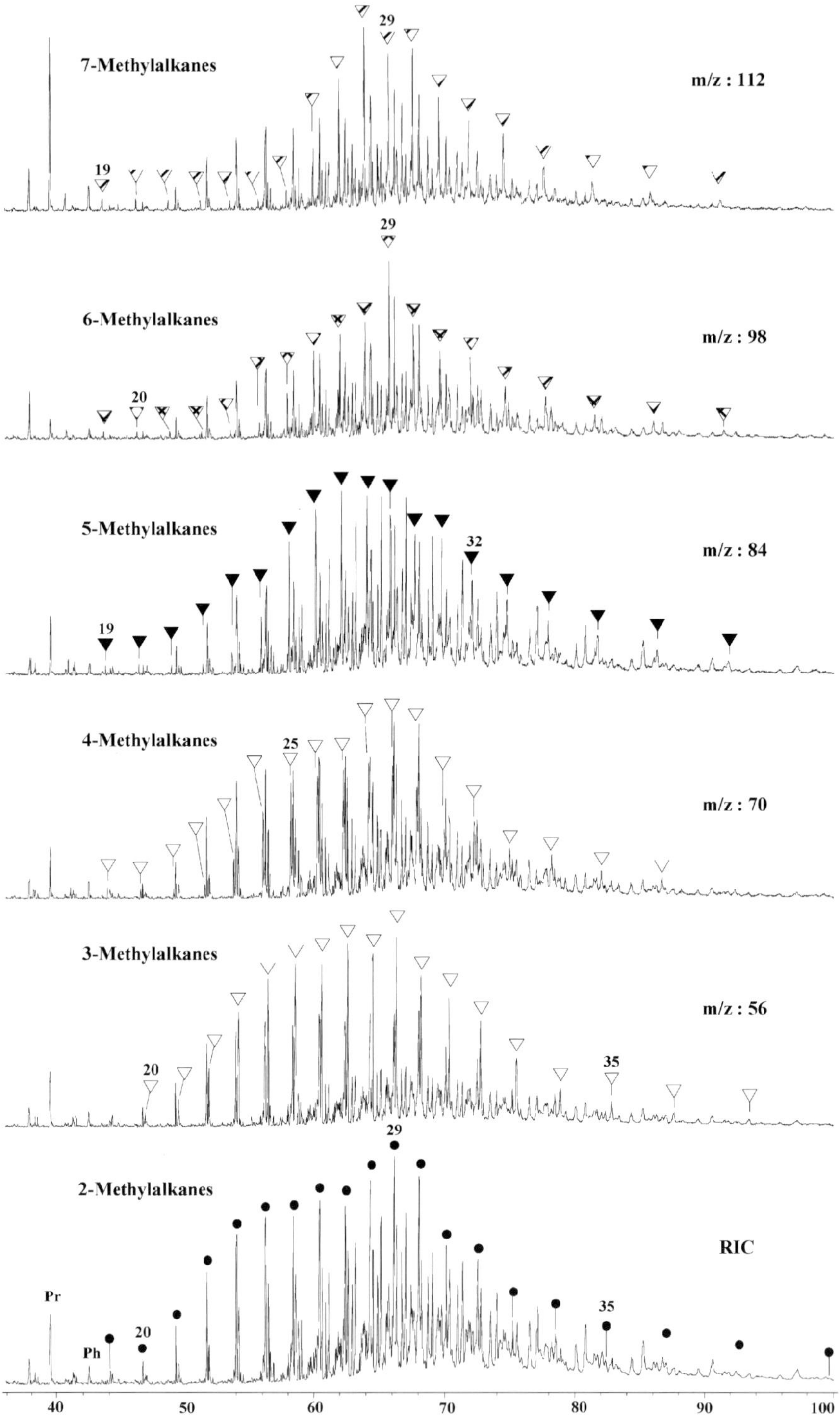


**Fig. 7.** Six homologous series of monomethylalkanes have been tentatively identified in the HMWHC fractions. Single ion chromatograms of the ions at *m/z* 56, 70, 84, 98 and 112 are used to illustrate the presence of these branched alkanes with the methyl group at carbon numbers 2, 3, 4, 5, 6 and 7 respectively.

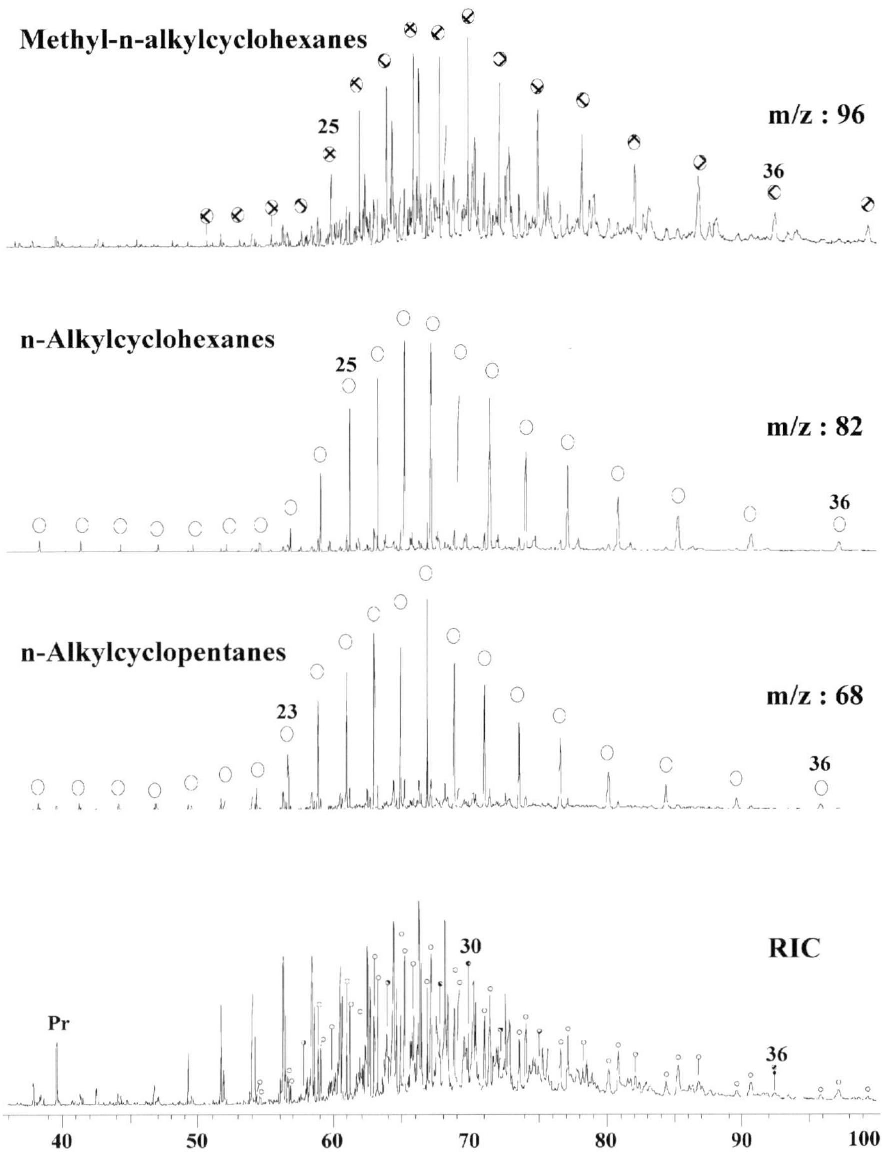


**Fig. 8.** *n*-Alkylcyclopentanes, *n*-alkylcyclohexanes and methyl-*n*-alkylcyclohexanes can be readily detected in the HMWHC fractions through the use of single monitoring of the ions at *m/z* 68, 82 and 96, respectively, as illustrated above.

concentrates isolated from oils sourced from different types of environments are currently being re-examined to obtain additional information on the variability of the methylalkanes with depositional environments. Paraffin deposits in pipelines and production facilities will also be re-examined to determine whether there is a trend in the deposition of one or more particular series of branched hydrocarbons relative to another. While this work will not necessarily prevent paraffin deposition, a better understanding of the origin and occurrence of the HMW compounds in crude oils (structures, physical/chemical properties and origin) could reduce clean-up and remedial expenses, lower operation costs and investigative/research expenses, and improve production (Tuttle 1983; Escobedo & Mansoori 1992; Leontaritis 1996). It should

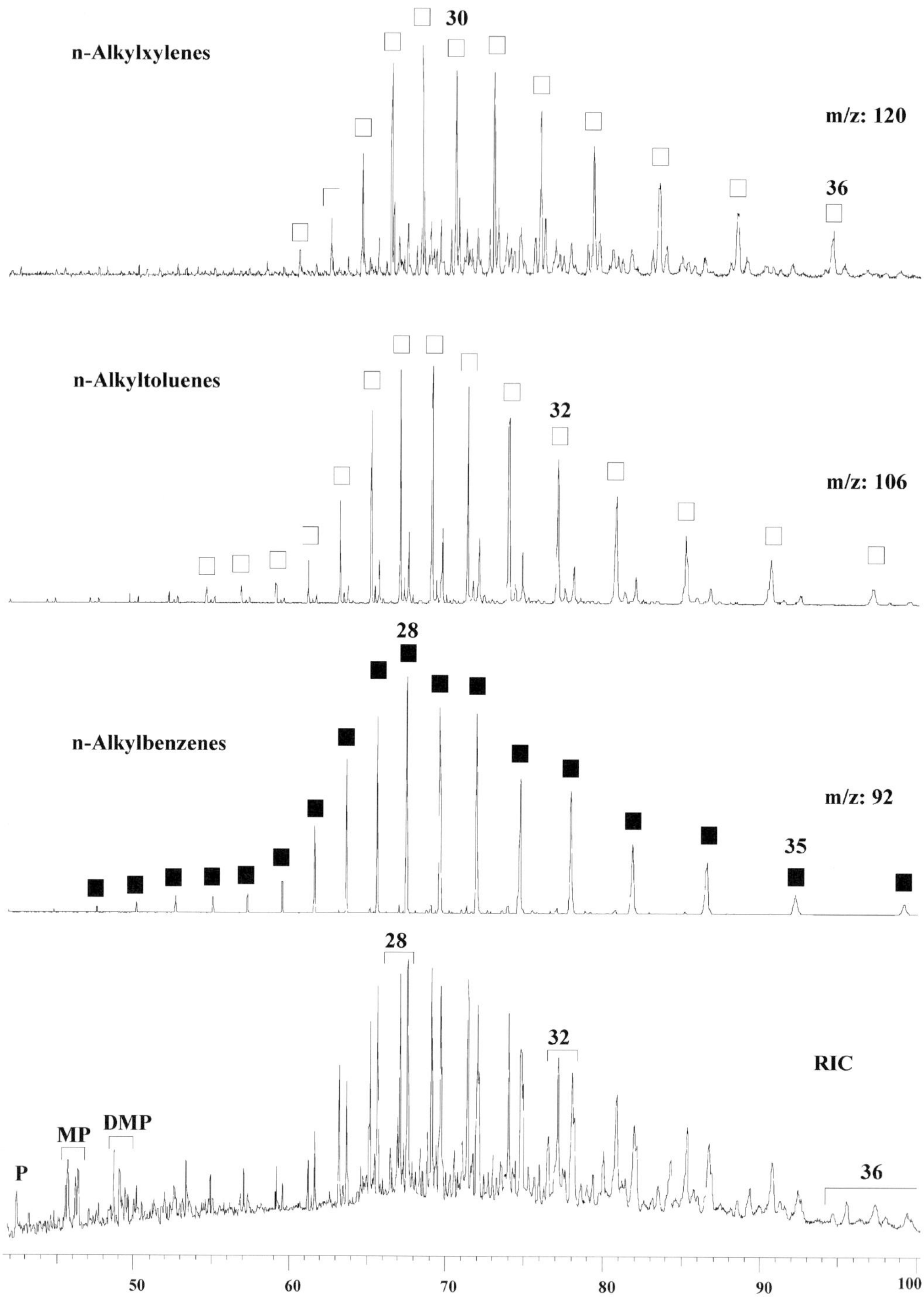


**Fig. 9.** *n*-Alkylbenzenes, *n*-alkyltoluenes and *n*-alkylxylenes can be readily detected in the HMWHC fractions through the use of single monitoring of the ions at *m*/*z* 92, 106 and 120, respectively, as illustrated above.

also permit one to more efficiently develop wax inhibitors and solvent systems necessary to remove or reduce paraffin deposits if the deposit to be removed is more accurately characterized before attempting any remediation. Delineation of the structure and distribution of the HMW branched alkanes ($C_{40+}$) will enable us to better understand the behaviour of waxy oils, develop/implement more accurate models to predict wax precipitation, and develop preventative and remedial measures before precipitation and deposition become a major issue.

## Summary

The occurrence and distribution of HMWHCs in crude oils is ubiquitous and varies significantly, with the depositional environment from which the oil is sourced (e.g. marine, freshwater or lacustrine environments). Analytical limitations have been our primary hindrance in characterizing HMWHCs, and have been overcome by the introduction of HTGC, development of a method to qualitatively and quantitatively separate the two primary classes of heavy organics (i.e. asphaltenes and waxes), and more recently have improved the ability to molecular sieve HMWHCs. Each of these advances has greatly enhanced our understanding of the composition of waxes in crude oils, and taken us a step closer to solving the questions surrounding waxy crude oils.

The distribution of specific homologous series of HMWHCs can reveal the nature of the source of crude oils. Carlson *et al.* (1993) identified a series of co-eluting monocyclic/acyclic hydrocarbons ranging between $C_{40}$ and $C_{60}$ which appeared to demonstrate predominance patterns which could be correlated to depositional environments. Hsieh *et al.* (2000) tentatively identified these compounds as alkylcyclopentanes, and in their evaluation of a large collection of oils observed that the alkylcyclopentanes carbon preference should be restricted to the $C_{40}$–$C_{46}$ range, for predicting depositional environments.

Long-chain alkylcyclopentanes, *n*-alkanes and monomethylalkanes have been isolated from source rock extracts, with HMWHCs extending well into the $C_{60+}$ range. These compounds have been overlooked in the past, due to limitations in the extraction process and because extracts were not analysed by HTGC. Since HMWHCs are present in crude oils, their presence in source rocks should not be surprising. Our extraction of source rocks at higher temperatures, along with wax isolation, has demonstrated that these compounds are more prevalent than previously thought. More in-depth studies of these compounds in source rocks may enlighten our understanding of the origins of HMWHCs, and possibly be applied to oil/source rock correlation, or used in the development of an oil migration tool. One example has also been provided to illustrate how the wax content of an oil changes quantitatively and qualitatively as the oil migrates from deeper hotter regions to shallower cooler reservoirs. If such a change could be calibrated this could lead to the development of an additional migration distance indicator.

To date, we have identified homologous series of monomethylalkanes, *n*-alkylcyclohexanes, methyl-*n*-alkylcyclohexanes, *n*-alkylcyclopentanes, *n*-alkylbenzenes, *n*-alkyltoluenes and *n*-alkylxylenes, in waxy crude oils. Each of these compounds has unique physical properties, and will affect the overall behaviour of an oil (e.g. melting point properties and wax crystal arrangement). Identification of these compounds will improve any efforts to model or predict the wax deposition potential of oils. As we come to know the exact structure of constituents in the HMWHC fraction, we will be able to gain better insights into the origin and mechanisms for the formation of HMWHCs in crude oils and source rocks.

We wish to thank Dr. Maowen Li of the Geological Survey of Canada for supplying the samples of the Rimbey–Meadowbrook oils from the Western Canadian Sedimentary Basin. We also appreciate the comments from the referees, Dr. Abid Bhullar and Dr. Joe Curiale, which were of great assistance in improving the overall quality of this paper.

## References

Ajienka, J. A. & Ikoku, C. U. 1990. Waxy crude oil handling in Nigeria: practices, problems, and prospects. *Energy Sources*, **12**, 463–478.

Allard, B., Rager, M. N. & Templier, J. 2002. Occurrence of high molecular weight lipids ($C_{80+}$) in the trilaminar outer walls of some freshwater microalgae. A reappraisal of algaenan structure. *Organic Geochemistry*, **33**, 789–801.

Barker, A. 1995. The chromatographic analysis of refined and synthetic waxes. *In*: Adlard, E. R. (ed.) *Chromatography in the Petroleum Industry*. Journal of Chromatography Library Series, **56**, 55–93.

Bianchi, G. 1995. Plant waxes. *In:* Hamilton, R. J. (ed.) *Waxes: Chemistry, Molecular Biology and Functions*. The Oily Press, Scotland, 175–222.

Burger, E. D., Perkins, T. K. & Striegler, J. H. 1981. Studies of wax deposition in Trans Alaska Pipeline. *Journal of Petroleum Technology*, **33**, 1076–1086.

CARDOSO, J., WATTS, C. D., MAXWELL, J. R., GOODFELLOW, R., EGLINTON, G. & GOLUBIC, S. 1978. A biochemical study of Abu Dhabi Algal mats: A simplified ecosystem. *Chemical Geology*, **23**, 273–291.

CARLSON, R. M. K. *ET AL.* 1993. High temperature gas chromatography of high-wax oils. *Indonesian Petroleum Association, 22nd Annual Convention Proceedings*. Jakarta, Indonesia, 483–507.

CARLSON, R. M. K., JACOBSON, S. R., MOLDOWAN, J. M. & CHAN, E. I. 1994. Potential applications of high temperature gas chromatography to Middle Eastern petroleum exploration and production. *The Middle East Petroleum Geoscience Conference*, Geo '94. 25–27 April, Bahrain, 258–267.

CARLSON, R. M. K., DIAS, R. F. & SCHOELL, M. 1997. Origins of high molecular weight alkanes $>C_{40}$ in waxes from natural crude oils and bitumens based on carbon isotopic evidence. *In*: *Abstracts of 18th International Meeting on Organic Geochemistry*, 22–26 September. Maastricht, The Netherlands, 399–400.

CHAPMAN, A. C. 1917. Spinacene: A new hydrocarbon from certain fish liver oils. *Journal of the Chemical Society*, **111**, 56–69.

CRANWELL, P. A. & VOLKMAN, J. K. 1981. Alkyl and steryl esters in a Recent lacustrine sediment. *Chemical Geology*, **32**, 29–43.

DEMBICKI, H. J., MEINSCHEIN, W. G. & HATTIN, D. E. 1976. Possible ecological and environmental significance of the predominance of even-carbon number $C_{20}$-$C_{30}$ *n*-alkanes. *Geochimica et Cosmochimica Acta*, **40**, 203–208.

DONG, J.-Z., VORKINK, W. P. & LEE, M. L. 1993. Origin of long chain alkylcyclohexanes and alkylbenzenes in a coal-bed wax. *Geochimica et Cosmochimica Acta*, **57**, 837–849.

DURAND, J. P., BRÉ, A., BÉBOULÉNE, J. J., DUCROZET, A. & CARBONNEAUX, S. 1998. Simulated distillation methods for petroleum fractions with minimal residue in the boiling range of 35–700 °C. *Journal of Chromatographic Science*, **36**, 431–434.

ESCOBEDO, J. & MANSOORI, G. A. 1992. Heavy organic deposition and plugging of wells (Analysis of Mexico's Experience). *Society of Petroleum Engineers, paper number* **SPE 23696**, 349–362.

FOWLER, M. G. & DOUGLAS, A. G. 1987. Saturated hydrocarbon biomarkers in oils of Late Precambrian age from Eastern Siberia. *Organic Geochemistry*, **11**, 201–213.

FOWLER, M. G., ABOLINS, P. & DOUGLAS, A. G. 1986. Monocyclic alkanes in Ordovician organic matter. *Organic Geochemistry*, **10**, 815–823.

GARCÍA, M. C., CARBOGNANI, L., OREA, M. & URBINA, A. 2000. The influence of alkane class-types on crude oil wax crystallization and inhibitors efficiency. *Journal of Petroleum Science and Engineering*, **25**, 99–105.

GELIN, F., VOLKMAN, J. K., LARGEAU, C., DERENNE, S., SINNINGHE DAMSTÉ, J. S. & DE LEEUW, J. W. 1999. Distribution of aliphatic, nonhydrolyzable biopolymers in marine microalgae. *Organic Geochemistry*, **30**, 147–159.

HEATH, D., MOFFATT, R. L. & ROWLAND, S. 1995. Quantification of the $C_{30+}$ fraction of North Sea gas condensates by high temperature capillary gas chromatography. *Analytical Proceedings*, **32**, 485–487.

HEATH, D. J., LEWIS, C. A. & ROWLAND, S. J. 1997. The use of high temperature gas chromatography to study the biodegradation of high molecular weight hydrocarbons. *Organic Geochemistry*, **26**, 769–785.

HEDBERG, H. D. 1968. Significance of high-wax oils with respect to genesis of petroleum. *American Association of Petroleum Geologists Bulletin*, **52**, 736–750.

HOLDER, G. A. & WINKLER, J. 1965. Wax crystallization from distillate fuels, Parts 1, 2, and 3. *Journal of the Institute of Petroleum*, **51**, 228–252.

HSIEH, M. 1999. Characterization of waxes in high pour-point crude oils. MS thesis, University of Oklahoma.

HSIEH, M. & PHILP, R. P. 2001. Ubiquitous occurrence of high molecular weight hydrocarbons in crude oils. *Organic Geochemistry*, **32**, 955–966.

HSIEH, M., PHILP, R. P. & DEL RIO, J. C. 2000. Characterization of high molecular weight biomarkers in crude oils. *Organic Geochemistry*, **31**, 1581–1588.

KENIG, F., SINNINGHE DAMSTÉ, J. S., KOCK-VAN DALEN, A. C., RIJPSTRA, W. I. C., HUC, A. Y. & DE LEEUW, J. W. 1995. Occurrence and origin of mono-, di-, and trimethylalkanes in modern and Holocene cyanobacterial mats from Abu Dhabi, United Arab Emirates. *Geochimica et Cosmochimica Acta*, **59**, 2999–3015.

KILLOPS, S. D., CARLSON, R. M. K. & PETERS, K. E. 2000. High-temperature GC evidence for the early formation of $C_{40+}$ *n*-alkanes in coals. *Organic Geochemistry*, **31**, 589–597.

KINGHORN, R. R. F. 1983. *An Introduction to the Physics and Chemistry of Petroleum*. Wiley, Chichester.

KISSIN, Y. V. 1990. Catagenesis of light cycloalkanes in petroleum. *Organic Geochemistry*, **15**, 575–594.

KLOMP, U. C. 1986. The chemical structure of pronounced series of iso-alkanes in South Oman crudes. *Organic Geochemistry*, **10**, 807–814.

DE LEEUW, J. W. & LARGEAU, C. 1993. A review of macromolecular organic compounds that comprise living organism and their role in kerogen, coal and petroleum formation. *In*: ENGEL, M. H. & MACKO, S. (eds) *Organic Geochemistry*. Plenum Press, New York.

LEHMLER, H. J., BERGOSH, R. G., MEIER, M. S. & CARLSON, R. M. K. 2002. A novel synthesis of branched high molecular weight ($C_{40+}$) long-chain alkanes. *Bioscience, Biotechnology and Biochemistry*, **66**, 523–531.

LEONTARITIS, K. J. 1996. Offshore asphaltene and wax deposition: problems/solutions. *World Oil*, **217**, 57–63.

LI, M., YAO, H., FOWLER, M.G. & STASIUK, L.D. 1998. Geochemical constraints on models for secondary petroleum migration along the Upper Devonian Rimbey–Meadowbrook reef trend in central Alberta. *Organic Geochemistry*, **29**, 163–182.

LIPSKY, S. R. & DUFFY, M. L. 1986a. High temperature gas chromatography: the development of new aluminum clad flexible fused silica glass capillary columns coated with thermostable nonpolar phases: Part 1. *Journal of High Resolution Chromatography*, **9**, 376–382.

LIPSKY, S. R. & DUFFY, M. L. 1986b. High temperature gas chromatography: the development of new aluminum clad flexible fused silica glass capillary columns coated with thermostable nonpolar phases: Part 2. *Journal of High Resolution Chromatography*, **9**, 725–730.

MAKUSHINA, V. M., AREFEV, O. A., ZABRODINA, M. N. & PETROV, A. 1978. New relic alkanes of petroleum. *Neftekhimiya*, **18**, 847–854 (in Russian).

MOLDOWAN, J. M., SEIFERT, W. K. & GALLEGOS, E. J. 1985. Relationship between petroleum composition and depositional environment. *American Association of Petroleum Geologists Bulletin*, **69**, 1255–1268.

MUELLER, E. & PHILP, R. P. 1998. Extraction of high molecular weight hydrocarbons from source rocks: an example from the Green River Formation, Uinta Basin, Utah. *Organic Geochemistry*, **28**, 625–631.

NEER, L. A. & DEO, M. M. 1995. Simulated distillation of oils with a wide carbon number distribution. *Journal of Chromatographic Science*, **33**, 133–138.

NELSON, D. R. 1978. Long-chain methyl-branched hydrocarbons: occurrence, biosynthesis, and function. *Advances in Insect Physiology*, **13**, 1–33.

NELSON, D. R. & BLOMQUIST, G. J. 1995. Insect waxes. *In*: HAMILTON, R. J. (ed.) *Waxes: Chemistry, Molecular Biology and Functions*. The Oily Press, Scotland, 1–90.

PETERS, K. E. & MOLDOWAN, J. M. 1993. *The Biomarker Guide: Interpreting Molecular Fossils in Petroleum and Ancient Sediments*. Prentice-Hall, Englewood Cliffs, NJ.

PHILP, R. P., BISHOP, A. N., DEL RIO, J. C. & ALLEN, J. 1995. Characterization of high molecular weight hydrocarbons ($>C_{40}$) in oils and reservoir rocks. *In*: CUBITT, J. M. & ENGLAND, W. A. (eds) *The Geochemistry of Reservoir Rocks*. Geological Society, London, Special Publications, **86**, 71–85.

DEL RIO, J. C. & PHILP, R. P. 1992. High molecular weight hydrocarbons: A new frontier in organic geochemistry. *Trends in Analytical Chemistry*, **11**, 187–193.

RULLKÖTTER, J. & PHILP, R. P. 1981. Extended hopanes up to $C_{40}$ in Thornton bitumen. *Nature*, **292**, 616–618.

SHIEA, J., BRASSELL, S. C. & WARD, D. M. 1990. Mid-chain branched mono- and dimethyl alkanes in hot spring cyanobacterial mats: A direct biogenic source for branched alkanes in ancient sediments. *Organic Geochemistry*, **15**, 223–231.

SUMMONS, R. E. 1987. Branched alkanes from ancient and modern sediments: Isomer discrimination by GC/MS with multiple reaction monitoring. *Organic Geochemistry*, **11**, 281–289.

TEGELAAR, E. W., MATTHEZING, R. M., JANSEN, J. B. H., HORSFIELD, B. & DE LEEUW, J. W. 1989. Possible origin of n-alkanes in high-wax crude oils. *Nature*, **342**, 529–531.

TEIXEIRA, M. A. G. & GONÇALVES, M. L. A. 2000. Evaluation of paraffinic material in asphaltenes isolated by precipitation with light alkanes. *Petroleum Science and Technology*, **18**, 273–286.

THANH, N. X., HSIEH, M. & PHILP, R. P. 1999. Waxes and asphaltenes in crude oils. *Organic Geochemistry*, **30**, 119–132.

THOMSOM, J. S. & RYNASKI, A. F. 1992. Simulated distillation of wax samples using supercritical fluid and high temperature gas chromatography. *Journal of High Resolution Chromatography*, **15**, 227–234.

TISSOT, B. P. & WELTE, D. H. 1984. *Petroleum Formation and Occurrence*. Springer, New York.

TRESTIANU, S., ZILIOLI, G., SIRONI, A., SARAVALLE, C. & MUNARI, F. 1985. Automatic simulated distillation of heavy petroleum fractions up to 800 °C TBP by capillary gas chromatography, Part I: Possibilities and limits of the method. *Journal of High Resolution Chromatography & Chromatography Communications*, **8**, 771–781.

TSUJIMOTO, M. 1916. A highly unsaturated hydrocarbon in shark liver oil. *Journal of Industrial and Engineering Chemistry*, **8**, 889–896.

TUO, J. & PHILP, R. P. 2003. Occurrence and distribution of high molecular weight hydrocarbons in selected non-marine source rocks from Liaohe, Qaidam and Tarim Basins, China. *Organic Geochemistry*, in press.

TUTTLE, R. N. 1983. High pour-point and asphaltic crude oils and condensates. *Journal of Petroleum Technology*, **35**, 1192–1197.

# Effects and impact of early-stage anaerobic biodegradation on Kuparuk River Field, Alaska

ALBERT G. HOLBA[1], LISA WRIGHT[2], RICK LEVINSON[2], BRADLEY HUIZINGA[1] & MARK SCHEIHING[2]

[1]*ConocoPhillips, Integrated Geological Analysis, 600 North Dairy Ashford Road, Houston, TX 77079, USA*
*(e-mail: Albert.G.Holba@conocophillips.com)*
[2]*ConocoPhillips Alaska Inc., 700 G Street, Anchorage, AK 99510, USA*

**Abstract:** Anaerobic processes have only recently been recognized as an important mechanism in the biodegradation of crude oils. They are normally invoked to explain extensively biodegraded oils with little or no possibility of contact by oxygenated waters from an active aquifer. This work with Kuparuk Field indicates that early stages of anaerobic biodegradation can be subtle and easily missed, yet have economic impact.

Kuparuk River Field, North Slope of Alaska, comprises two reservoir intervals: vertically stratified and imbricated lower shoreface sandstones (A sands), and overlying shallow marine sandstones with complex permeability structure (C sands). The vertical and lateral distribution of viscous oil (less than 20° API) shows a strong relationship to structure and faulting in the Kuparuk Field. Multiple mechanisms for the origin of tars and viscous oils can be proposed, including early aerobic biodegradation, anaerobic biodegradation, inorganic oxidation and gas deasphalting. This geochemical study, integrated with stratigraphic, structural and production data, was undertaken to help understand the origin and distribution of tar and viscous oil in the field.

Obvious depletion of *n*-alkanes and other paraffins, classically regarded as indicative of early biodegradation, is not observed in examined samples. However, Kuparuk viscous oils show slight to extreme selective depletion in long-chain alkyl aromatic (LCAA) hydrocarbon families (e.g. alkylbenzenes and alkyltoluenes). This is interpreted as indicative of an early stage of anaerobic microbial degradation that likely destabilized the oil to promote subsequent precipitation of asphaltenes as tar.

Depletions in LCAAs in core samples in the field are linked to decreased hydrocarbon/nonhydrocarbon ratio and to an increase in the high molecular weight ($>C_{50+}$) components of Rock-Eval 6 pyrolysates. Using a calibration curve constructed from oil Rock-Eval 6 pyrolysis, the API gravity of core oil plus bitumen can be estimated. Tar-plugged formations with depleted LCAAs have estimated API gravities <8°. Portions of the Kuparuk reservoir with higher iron content tend to show greater depletions in LCAA. Anaerobic biodegradation is likely mediated by dissimilatory iron-reducing bacteria. Biodegradation likely destabilizes the oil with respect to asphaltene precipitation such that later arrival of petroleum leads to tar in the reservoir. Increased tar and depleted LCAAs correspond to intervals with lower productivity indices, thus indicating a significant impact on petroleum producibility.

The ability of bacteria to biodegrade petroleum has been long recognized (ZoBell 1945; Hunt 1953). Geoscientists have recognized that both aerobic and anaerobic microorganisms biodegrade oil in the subsurface (Bailey *et al.* 1973; Milner *et al.* 1977), and that reservoir souring, $H_2S$ generation, is caused by sulphate-reducing bacteria (Bastin 1926). However, a correlation of fresh meteoric water with extent of biodegradation, and laboratory experiments with aerobic and anaerobic microorganisms led to acceptance that biodegradation of petroleum was caused by aerobic microorganisms (Winters & Williams 1969; Bailey *et al.* 1973; Harwood 1973; ZoBell 1973). Aerobic microbial activity has remained the standard explanation for biodegraded oils in the petroleum industry. However, recent work has indicated that anaerobic biodegradation is an important, if not dominant, mode of biodegradation in many petroleum accumulations (Connan *et al.* 1994; Holba *et al.* 1996; Larter *et al.* 2003).

The presence of tar in the Kuparuk River Formation, Kuparuk River Field, North Slope, Alaska (Fig. 1), was first observed in 1971 in a cored well, West Sak 1. A number of subsequent wells have encountered viscous oils and occasionally tar staining. Most of these wells are located at the periphery of the field, and generally downdip on the east and south flanks of the field.

*From*: CUBITT, J. M., ENGLAND, W. A. & LARTER, S. (eds) 2004. *Understanding Petroleum Reservoirs: towards an Integrated Reservoir Engineering and Geochemical Approach*. Geological Society, London, Special Publications, **237**, 53–88. 0305-8719/04/$15.00 

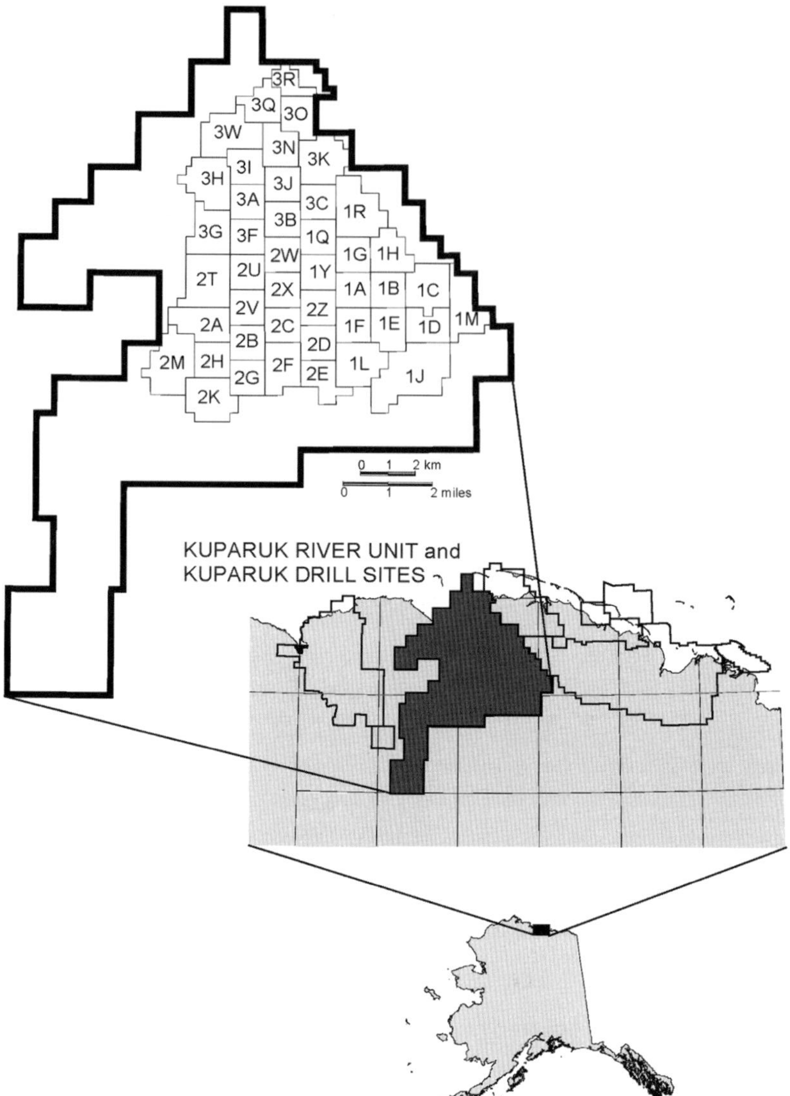


**Fig. 1.** General location map of Kuparuk River Unit shows drill site identifications. Inset maps show Kuparuk River Unit relative to other North Slope Units and the location of the Kuparuk River Field area on a map of Alaska.

However, viscous oil is found within grabens near the crest of the structure. A variety of mechanisms have been advanced to explain the presence of this tar/viscous oil. These include: (1) early aerobic biodegradation of oil when the field was shallow; (2) anaerobic biodegradation of oil after filling of the current accumulation; (3) inorganic oxidation by one of the inorganic mineral phases present as early diagenetic cements; and/or (4) gas deasphalting late in the field's history.

This study was undertaken in an attempt to improve our ability to predict the distribution of viscous oils and tar within the field as part of an overall effort to identify at-risk reserves and improve recovery.

## Geological setting and petroleum production

The Kuparuk River Field, located on the North Slope of Alaska (Fig. 1), is a mature giant field

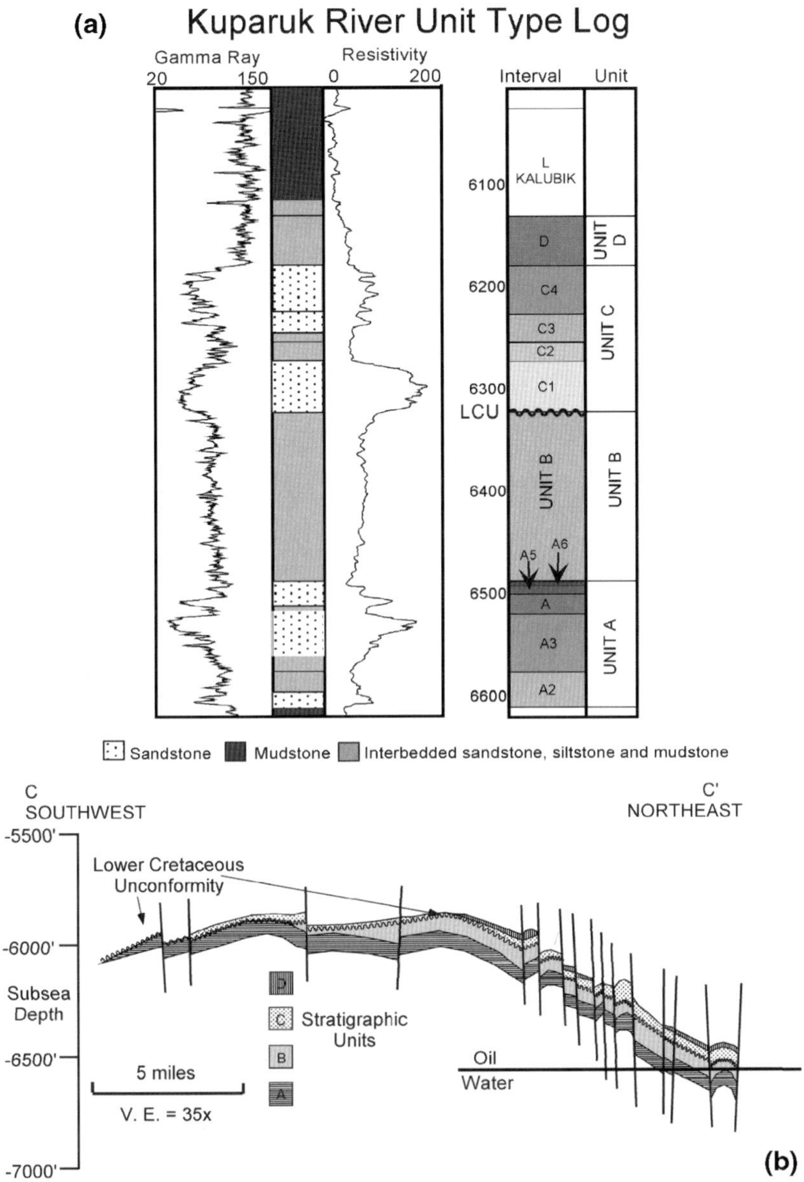


**Fig. 2.** Kuparuk River stratigraphy: (**a**) unit type log and stratigraphic diagram showing Lower Cretaceous intervals and formation names, (**b**) cross-section across Kuparuk Field.

that is productive from two major sandstone intervals of the Lower Cretaceous Kuparuk River Formation, the A sands and overlying C sands (Fig. 2). The two sand units are separated by the intervening Unit B shales, siltstones and sandstones that range from 180 feet thick on the east side of the field to completely truncated by the Lower Cretaceous Unconformity (LCU) on the west side. The field is intensely faulted (Fig. 3). Faulting is normal and throws range from subseismic to several hundred feet. Gross interval thickness of the Kuparuk River Formation is about 300 feet. Average API oil gravity is 24°, and oil viscosity averages 2.5 cP. Present day asphaltene content of produced oils ranges from 2 to 22%. Viscous oil (i.e. less than 20° API) occurs around the field periphery, near the oil/water contact(s) and in grabens across the structural axis of the field. Asphaltene-rich petroleum charge is dominantly sourced by the calcareous facies of the Triassic

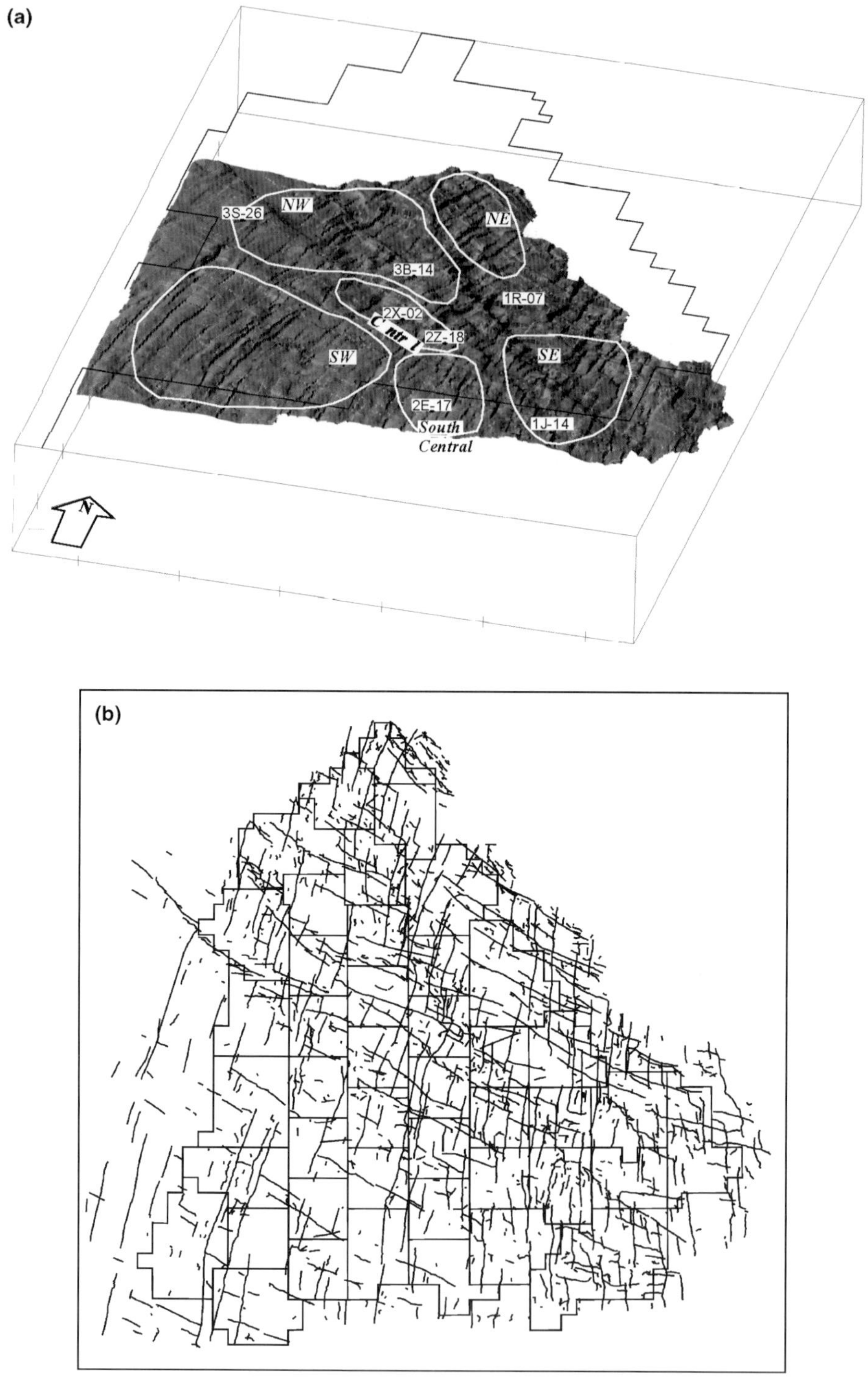


**Fig. 3.** Kuparuk River Field fault distribution: Note, large older NW–SW trending faults bounding larger grabens, and the more numerous small, later NNE–SSW faults. (**a**) Kuparuk River Field structural map at the base of the Kuparuk River A Formation showing areas used in text and tables. (**b**) Two populations of faults intersecting top of the Kuparuk River Formation.

Shublik formation (Masterson 2001; Masterson *et al.* 2001). A younger reservoir that overlies the Kuparuk River Field, the West Sock Formation, contains an accumulation of moderately biodegraded viscous oil (Masterson *et al.* 2001).

### *A sand reservoir geology*

The Kuparuk A formation is characterized as very fine-grained, vertically stratified, imbricated sandstones with relatively low permeability (horizontal air permeabilities on the order of tens to hundreds of millidarcies). They are interpreted as sand and mud deposits from storm-driven waves and currents deposited in a lower shoreface setting. The Kuparuk A sands consist of six separate sandstone bodies (Fig. 2), A1–A6 (base-to-top), that are interpreted to represent the erosional remnants of shorefaces. These remnant shorefaces are oriented generally SW–NE and step basinward, i.e. prograde, to the SE (Masterson & Paris 1987). Ankerite (ferroan dolomite) cement locally occludes porosity, particularly in units A3 and A4.

### *C sand reservoir geology*

The depositional and diagenetic history of the C sands is more complex than that of the A sands. Deposition of the C sands post-dates a major period of uplift and incision by the LCU associated with early rifting in the North Slope area. This rifting created local highs to the east (Prudhoe High) and west (Colville High) that fed sediment into NW-SE fault grabens in a tectonic saddle between the highs. Four subintervals of the C sand are commonly recognized, the basal C1 to uppermost C4 sand (Fig. 2). The C1 and C4 sands are the principal reservoir quality sands and responsible for the highest production rates observed in the field. They are characterized as poorly sorted, medium-grained, often glauconitic sands with local thicks controlled by growth in NW–SE-oriented, fault-bounded grabens. The C2 and C3 sub-intervals have a more blanket-like character and are interpreted to have been deposited in a lower shoreface setting. Glauconite is locally abundant. The presence of siderite (iron carbonate) locally renders sandstone non-pay. Massive siderite is most common at the top of the C4 sub-interval.

### *Structure and faulting*

The structure of the Kuparuk field is defined by a gentle anticlinal structure with a crest at 5600 feet subsea and flanks below 7000 feet subsea (Krantz & Longden 2002). The axis of the structure trends NW–SE (Fig. 3a). The trapping mechanism is a combination of stratigraphic pinchout of the C sands and erosional truncation of Units A and B by the regional LCU to the south and west, and structural dip closure to the north and east (Masterson & Paris 1987). At present, about 5200 faults have been mapped in the field, ranging from several hundred foot throws of regional extent across the field to faults a few hundred feet long with throws less than 15 feet (Fig. 3b). Two major fault trends are evident: (1) a NW–SE normal fault system began forming during C sand deposition, and (2) a later NNE–SSW normal fault system cutting the earlier syndepositional episode. The somewhat different expression of these fault styles between units A and C is likely due to differences in mechanical properties between the two units (Krantz & Longden 2002). Faulting exerts a significant impact on reservoir compartmentalization and sweep efficiency.

### *Impact of viscous oil distribution on recovery*

Although much of the production is commingled A and C sand oils, analysis of end member A and C sand oils reveals A sand oils tend to be lower average gravity than C sand oils (Fig. 4). Distribution of low gravity (<20° API) oil also appears to differ between Kuparuk A and C sand units.

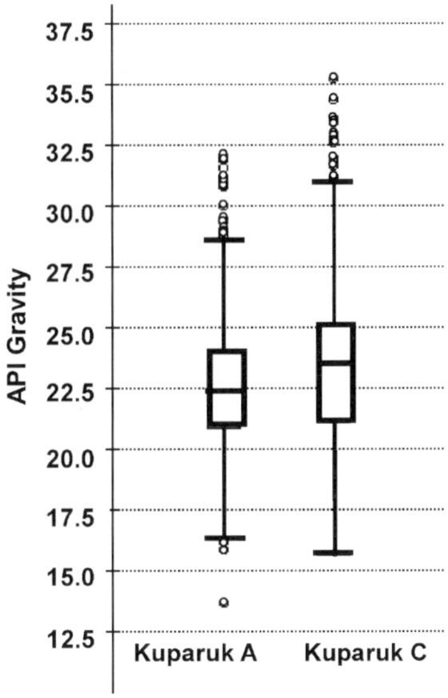


**Fig. 4.** Box plot distribution of Kuparuk A and Kuparuk C crude oil API gravity.

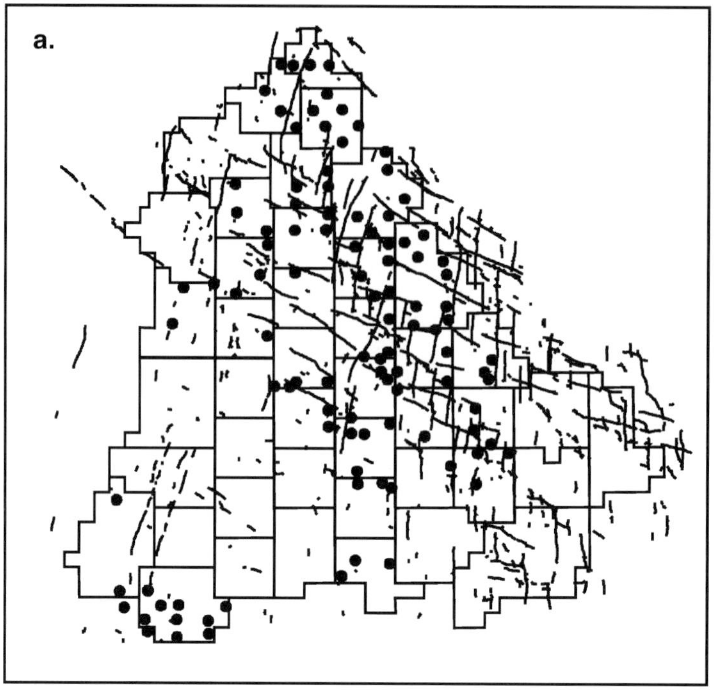


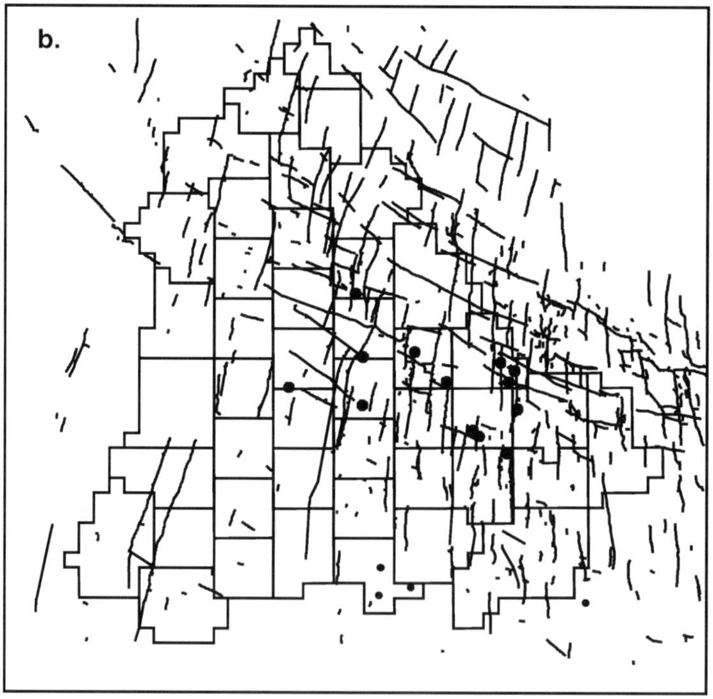


**Fig. 5.** Locations of oil samples with <20° API superimposed on map of base of Kuparuk formation faults with >50 feet of vertical displacement. (**a**) Kuparuk A sandstone, (**b**) Kuparuk C sandstone.

Locally, the locations of low oil gravity in Kuparuk A and C sands are independent of each other (Fig. 5). Low gravity A sand oils tend to occur in wells that are proximate to larger scale faults, in wells that are close to the oil/water contact, or in wells located in the central graben structures of the field. Lower gravity C sand oils are recovered from tar zones around the edges of the field, particularly from areas of greater permeability, areas of higher fault density and magnitude, and near the oil/water contact.

The occurrence of low gravity (higher viscosity) oils can have a locally negative impact on field production rate. The Productivity Index (PI) is defined as barrels of liquid produced per day divided by the pressure difference between the average reservoir pressure and the sand face pressure in a well. The preponderance of data indicates that oil gravity and viscosity are very important determining factors for PI, although PI is also affected by other factors (i.e. permeability, thickness or multiphase flow). Maps of the PI in producer polygons in the Kuparuk reservoir (Fig. 6) show a correlation between lower than average PI and the occurrence of oil of $<20^{\circ}$ API (Fig. 5). Within the field, tar is sufficiently extensive to prevent oil flow in only a few wells. Well 2E-17 did not flow despite presence of high quality C sand as indicated on wireline logs. 2E-17 was cored and showed extensive tar filling most pore space. A recent well on the field periphery, 1J-14, also encountered extensive tar in Kuparuk C. Other Kuparuk intervals under production have had occurrences of tar shows, i.e. 1D-30. Geochemical characteristics of these wells will be examined in greater detail later, resulting in key observations concerning the causes for heavier oils and tars in the field.

## Samples and methods

Analysed gases and oils are bottom-hole samples unless otherwise indicated in the text or figures. Bottom-hole samples (BHS) and cores generally predate enhanced oil recovery activity (1979–1984), thus represent original field conditions. Exceptions include multiple dynamic formation test (MDT) oils, repeat formation tester (RFT) oils, sidewall cores (SWC), or separator oil samples from recent wells usually at the periphery of the field.

Gas composition and isotopic ratios (relative to PDB) were measured by Isotech Laboratories, Champaign, IL, using their normal laboratory methods. Gas sample identification, gas carbon and deuterium isotopic data are provided in Table 1. All liquid sample preparation and chromatographic analyses were performed at Baseline/DGSI Inc., The Woodlands, TX. Electron microprobe analysis was performed by Ben Powell, Jim Hickey and Gary Lovell at the Phillips Petroleum Company facility, Bartlesville, Oklahoma.

Whole oil gas chromatography (WGC) was performed on an HP 5890 gas chromatograph equipped with an FID detector (350 °C), an HP 7673 autosampler and a 60 m DB-1 column. Samples were injected using split injection at 275 °C using constant flow helium carrier gas. The column oven programme initial temperature was 30 °C with a 3 °C/min ramp to a final temperature of 320 °C (total run time = 121.66 min). Internal standard (*trans*-2-heptene) at 0.035 mg/10.0 mg oil was used for quantification with data acquisition and integration on HP ChemStation software. Oil sample identification and whole oil GC ratios (Thompson 1987 [ratios B, F and H]; Mango 1994 [ratios P2 and P3]) are provided in Table 2.

Rock samples were extracted using methylene chloride and a Soxhlet apparatus and the resulting extract was further fractionated by a semi-quantitative SARA separation. After asphaltenes were removed from the concentrated extract by precipitation with excess pentane, the pentane soluble portion of the sample was separated by medium-pressure liquid chromatography (MPLC) using a deactivated silica gel precolumn and an activated silica gel main column by eluting the saturate and aromatic hydrocarbon fractions from the activated silica column with hexane in the forward and backflush modes respectively. Polar nonhydrocarbons were backflushed from the precolumn with methylene chloride-methanol. Carbon isotopes were done on a subset of hydrocarbon fractions at Coastal Sciences Labs, Austin, TX (Table 3).

Gas chromatography/mass spectrometry (GC/MS) of aromatic hydrocarbon fractions was done on an HP 6890 gas chromatograph equipped with an HP 7683 autosampler, a 60 m DB-1 column connected to an HP 5973 mass selective detector operated in the scan mode from 50 to 450 amu. On-column sample injection was done at programmed temperature at 3 °C above column oven temperature with constant flow helium carrier gas. The column oven programme initial temperature was 70 °C programmed to 315 °C (total run time 131.33 min). Internal standard (*ortho*-terphenyl) at 300 ppm was used for quantification with data acquisition and integration on HP ChemStation software. Ratios may be found in Table 4.

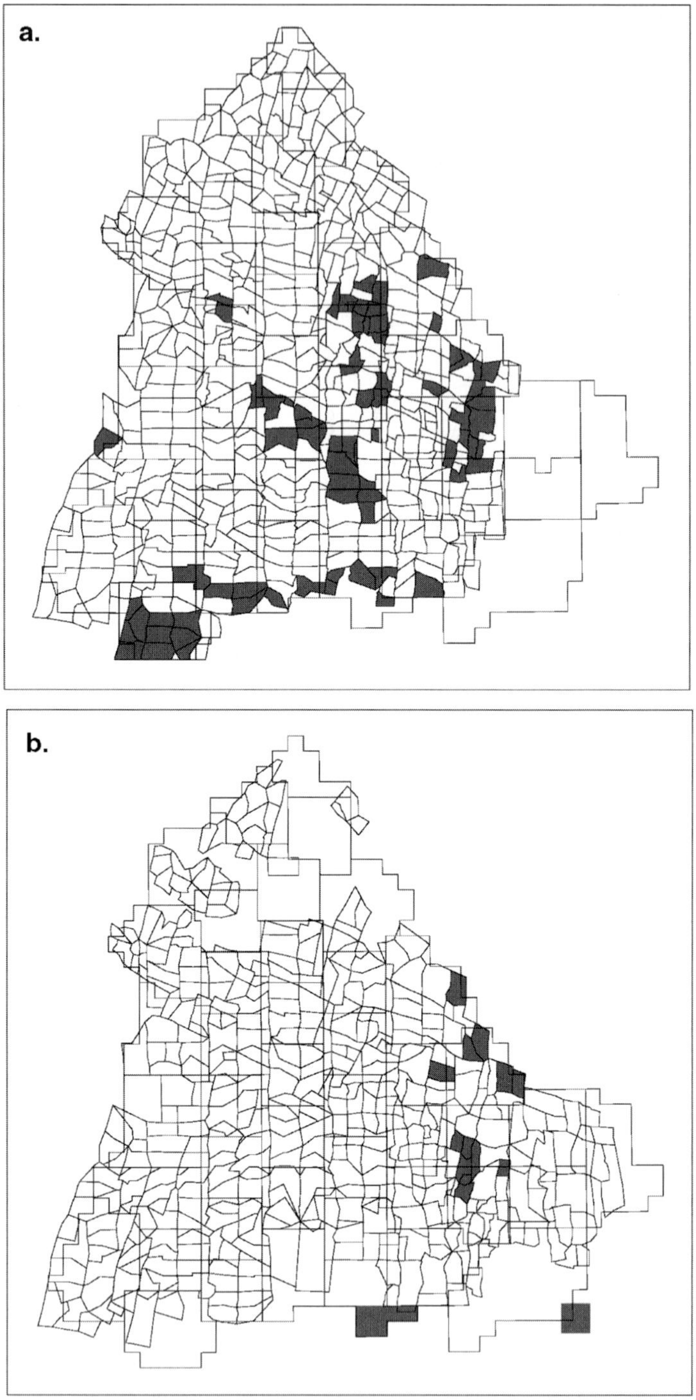


**Fig. 6.** Shaded polygons for (**a**) Kuparuk River A Formation or (**b**) Kuparuk River C Formation denote the occurrence of heavy oil and disappointing petroleum producibility as defined by the producibility index (see text).

**Table 1.** *Gas isotope data*

| Gas sample ID | Well | Fm | Type | Area family | $\delta^{13}C\ CO_2$ | $\delta^{13}C\ C_1$ | $\delta^{13}C\ C_2$ | $\delta^{13}C\ C_3$ | $\delta^{13}C\ iC_4$ | $\delta^{13}C\ nC_4$ | $\delta D\ C_1$ |
|---|---|---|---|---|---|---|---|---|---|---|---|
| 01G00023 | 2A-2 | A | BHS | SW | −10.67 | −44.77 | −31.14 | −29.91 | −29.53 | −29.95 | −200.7 |
| 01G00024 | 2A-2 | C | BHS | SW | −12.29 | −44.83 | −31.35 | −29.97 | −29.52 | −29.98 | −202.4 |
| 01G00025 | 2B-5 | A | BHS | SW | −11.56 | −45.73 | −31.80 | −30.38 | −29.96 | −30.21 | −208.8 |
| 01G00027 | 2D-3 | A | BHS | SC | −11.09 | −46.63 | −32.74 | −31.11 | −30.71 | −30.96 | −213.7 |
| 01G00028 | 2E-10 | A | BHS | SC | −11.94 | −47.59 | −32.97 | −30.90 | −30.62 | −30.84 | −217.9 |
| 01G00029 | 2F-9 | A | BHS | SW | −11.42 | −46.21 | −32.18 | −30.64 | −30.28 | −30.47 | −208.6 |
| 01G00030 | 2G-10 | A | BHS | SW | −11.67 | −45.71 | −31.82 | −30.63 | −30.09 | −30.48 | −208.5 |
| 01G00031 | 2H-8 | A | BHS | SW | −11.12 | −45.06 | −31.31 | −30.12 | −29.84 | −30.11 | −206.3 |
| 01G00032 | 2T-10 | A | BHS | SW | −12.75 | −44.21 | −30.90 | −29.49 | −28.90 | −29.61 | −196.0 |
| 01G00034 | 2V-3 | A | BHS | SW | −11.62 | −45.67 | −31.85 | −30.26 | −29.84 | −30.27 | −206.9 |
| 01G00033 | 2U-16 | A | BHS | C | −10.35 | −45.14 | −32.46 | −31.27 | −30.92 | −31.42 | −200.9 |
| 01G00037 | 2X-7 | C | BHS | C | −11.20 | −45.05 | −32.44 | −31.13 | −30.69 | −31.21 | −202.3 |
| 01G00022 | 1Q-12 | A | BHS | NW | −11.06 | −42.89 | −32.95 | −32.25 | −31.23 | −31.89 | −190.9 |
| 01G00035 | 2W-12 | A | BHS | NW | −12.27 | −42.97 | −32.81 | −32.01 | −31.26 | −31.73 | −191.4 |
| 01G00036 | 2W-12 | C | BHS | NW | −12.71 | −42.88 | −32.76 | −31.99 | −31.11 | −31.82 | −192.3 |
| 01G00039 | 3A-16 | A | BHS | NW | −12.02 | −41.12 | −32.74 | −32.37 | −31.17 | −31.94 | −180.3 |
| 01G00040 | 3B-3 | A | BHS | NW | −11.58 | −41.31 | −32.71 | −32.33 | −31.32 | −32.07 | −179.2 |
| 01G00044 | 3J-8 | A | BHS | NW | −12.36 | −41.50 | −32.57 | −32.27 | −31.12 | −31.92 | −180.9 |
| 01G00045 | 3K-11 | A | BHS | NW | −12.05 | −42.34 | −31.95 | −32.12 | −31.18 | −31.72 | −181.4 |
| 01G00046 | 3M-19 | ? | BHS | NW | −12.42 | −40.79 | −32.53 | −32.13 | −30.94 | −31.66 | −178.7 |
| 01G00042 | 3H-18 | ? | BHS | NW | −15.72 | −40.67 | −33.94 | −33.14 | −31.78 | −32.49 | −175.8 |
| 01G00015 | 3S-26 | C | Test | NW | −15.78 | −41.80 | −34.44 | −33.16 | −32.07 | −32.49 | −183.7 |
| 01G00041 | 3C-6 | A | BHS | NE | | −39.94 | −35.03 | | | | −180.6 |
| 01G00047 | 3N-13 | A | BHS | NE | −11.81 | −41.39 | −32.27 | −32.39 | −31.32 | −31.83 | −180.1 |
| 01G00048 | 3O-7 | ? | BHS | NE | −11.01 | −42.03 | −31.78 | −31.91 | −31.16 | −31.29 | −180.3 |
| 01G00052 | 1J-14 | A | MDT | SE | −19.64 | −44.05 | −31.55 | −31.37 | −31.51 | −31.95 | −193.7 |
| 01G00018 | 1D-4 | C | BHS | SE | −7.03 | −41.11 | −30.94 | −31.27 | −30.89 | −31.21 | −193.3 |
| 01G00019 | 1E-10 | ? | BHS | SE | −11.11 | −47.48 | −32.86 | −30.80 | −30.58 | −30.82 | −213.4 |
| 01G00020 | 1E-11 | A | BHS | SE | −8.66 | −44.27 | −32.23 | −31.51 | −31.02 | −31.34 | −198.2 |
| 01G00021 | 1F-16 | West Sak | BHS | | | −48.13 | −27.13 | −27.17 | −28.90 | −29.13 | −306.7 |
| 01G00038 | 2Z-4 | West Sak | BHS | | | −48.14 | −27.10 | −27.17 | −28.91 | −29.13 | −308.3 |
| 95G00029 | Kalubik-1 | Kuparuk. | RFT | | −18.86 | −41.14 | −34.73 | −33.56 | | | −178.8 |

**Table 2.** *Whole oil GC ratios*

| Oil Sample ID | Well | Fm | Sample type | Area family | Pr/Ph | $nC_{17}$ /Pr | $nC_6$/ $nC_{16}$ | B | F | H | P2 | P3 |
|---|---|---|---|---|---|---|---|---|---|---|---|---|
| 01X07854 | 2A-2 | A | BHS | SW | 1.51 | 1.49 | 1.09 | 1.34 | 0.41 | 16.2 | 10.0 | 3.22 |
| 01X07855 | 2A-2 | C | BHS | SW | 1.44 | 1.53 | 1.29 | 1.23 | 0.44 | 16.6 | 10.6 | 3.56 |
| 01X07893 | 2A-16 | A + C | prod | SW | 1.43 | 1.51 | 1.09 | 1.26 | 0.43 | 16.2 | 10.5 | 3.64 |
| 01X07856 | 2B-5 | A | BHS | SW | 1.45 | 1.55 | 1.28 | 1.22 | 0.44 | 16.6 | 10.6 | 3.53 |
| 01X07859 | 2E-10 | A | BHS | SC | 1.46 | 1.53 | 1.58 | 1.25 | 0.51 | 17.8 | 11.6 | 4.01 |
| 01X07860 | 2F-9 | A | BHS | SW | 1.48 | 1.53 | 1.33 | 1.19 | 0.45 | 16.8 | 10.8 | 3.55 |
| 01X07861 | 2G-10 | A | BHS | SW | 1.46 | 1.55 | 1.30 | 1.25 | 0.42 | 16.3 | 10.2 | 3.56 |
| 01X07862 | 2H-8 | A | BHS | SW | 1.44 | 1.54 | 1.25 | 1.27 | 0.43 | 16.4 | 10.4 | 3.63 |
| 01X07863 | 2T-10 | A | BHS | SW | 1.47 | 1.54 | 1.16 | 1.23 | 0.43 | 16.5 | 10.4 | 3.44 |
| 01X07866 | 2V-3 | A | BHS | SW | 1.61 | 1.58 | 1.57 | 1.12 | 0.47 | 17.2 | 11.3 | 3.74 |
| | | | | | | | | | | | | |
| 01X07865 | 2U-16 | A | BHS | C | 1.37 | 1.49 | 1.57 | 0.85 | 0.56 | 18.7 | 12.9 | 4.36 |
| 01X07864 | 2U-10 | ? | sep. liq | C | 1.35 | 1.52 | 1.57 | 0.87 | 0.56 | 18.9 | 12.7 | 4.38 |
| 01X07894 | 2U-10 | A + C | prod | C | 1.39 | 1.51 | 1.83 | 0.83 | 0.60 | 19.6 | 13.4 | 4.47 |
| 01X07869 | 2X-7 | C | BHS | C | 1.33 | 1.49 | 1.80 | 0.88 | 0.55 | 18.5 | 13.0 | 4.48 |
| | | | | | | | | | | | | |
| 01X07853 | 1Q-12 | A | BHS | NW | 1.10 | 1.56 | 1.44 | 0.74 | 0.58 | 18.8 | 13.8 | 5.03 |
| 01X07867 | 2W-12 | A | BHS | NW | 1.21 | 1.50 | 1.49 | 0.75 | 0.59 | 19.0 | 13.8 | 4.74 |
| 01X07868 | 2W-12 | C | BHS | NW | 1.23 | 1.45 | 1.58 | 0.73 | 0.60 | 19.2 | 14.0 | 4.94 |
| 01X07895 | 2W-12 | A + C | prod | NW | 1.22 | 1.45 | 1.34 | 0.70 | 0.62 | 20.1 | 14.1 | 4.80 |
| 01X07871 | 3A-16 | A | BHS | NW | 1.19 | 1.40 | 1.52 | 0.71 | 0.59 | 18.8 | 14.3 | 5.17 |
| 01X07872 | 3B-3 | A | BHS | NW | 1.20 | 1.46 | 1.24 | 0.74 | 0.59 | 19.1 | 13.6 | 4.83 |
| 01X07874 | 3F-14 | ? | sep. liq | NW | 1.19 | 1.47 | 1.49 | 0.67 | 0.61 | 19.6 | 14.3 | 5.01 |
| 01X07884 | 3G-25X | C4 + A3 + A2 | prod | NW | 1.16 | 1.56 | 1.28 | 0.70 | 0.60 | 19.7 | 13.8 | 4.75 |
| 01X07883 | 3H-20 | A | prod | NW | 1.18 | 1.43 | 1.37 | 0.66 | 0.60 | 19.1 | 14.4 | 5.10 |
| 01X07877 | 3J-8 | A | BHS | NW | 1.16 | 1.44 | 1.50 | 0.66 | 0.59 | 18.9 | 14.5 | 5.20 |
| 01X07878 | 3K-11 | A | BHS | NW | 1.14 | 1.52 | 1.23 | 0.78 | 0.52 | 17.6 | 13.7 | 5.28 |
| 01X07879 | 3M-19 | ? | BHS | NW | 1.20 | 1.46 | 1.43 | 0.68 | 0.60 | 19.3 | 14.5 | 5.18 |
| 01X07875 | 3H-18 | ? | BHS | NW | 1.27 | 1.51 | 1.53 | 0.63 | 0.64 | 20.0 | 14.5 | 5.04 |
| 01X07842 | 3S-26 | C | Test | NW | 1.28 | 1.49 | 1.60 | 0.62 | 0.64 | 20.4 | 14.4 | 4.66 |
| | | | | | | | | | | | | |
| 01X07873 | 3C-6 | A | BHS | NE | 1.15 | 1.49 | 1.41 | 0.79 | 0.53 | 17.7 | 13.8 | 5.28 |
| 01X07896 | 3C-15 | A + C | prod | NE | 1.16 | 1.44 | 1.19 | 0.75 | 0.55 | 18.2 | 13.7 | 4.97 |
| 01X07880 | 3N-13 | A | BHS | NE | 1.11 | 1.62 | 2.04 | 0.79 | 0.54 | 17.8 | 13.8 | 5.62 |
| 01X07881 | 3O-7 | ? | BHS | NE | 1.18 | 1.62 | 1.93 | 0.72 | 0.56 | 18.0 | 14.4 | 5.96 |
| 01X07882 | Oliktok Pt. 2 | ? | sep. liq | NE | 1.18 | 1.61 | 1.28 | 0.71 | 0.56 | 18.4 | 14.1 | 5.41 |

| | | | | | | | | | | | | |
|---|---|---|---|---|---|---|---|---|---|---|---|---|
| 01X07887 | 1B-03 | C | prod | SE | 1.42 | 1.52 | 0.44 | 0.97 | 0.48 | 18.0 | 11.5 | 3.95 |
| 01X07888 | 1C-04 | C | prod | SE | 1.16 | 1.48 | 0.55 | 0.95 | 0.48 | 17.4 | 12.3 | 4.41 |
| 01X07849 | 1D-4 | C | BHS | SE | 1.45 | 1.48 | 1.39 | 0.79 | 0.53 | 17.7 | 13.5 | 5.06 |
| 01X07889 | 1D-08 | C | prod | SE | 1.24 | 1.48 | 0.87 | 0.94 | 0.49 | 17.1 | 12.5 | 4.48 |
| 00X07836 | 1D-30 | C3-C1 | prod | SE | 1.25 | 1.47 | 1.88 | 0.72 | 0.56 | 18.1 | 14.4 | 5.31 |
| 01X07890 | 1E-07 | A + C | prod | SE | 1.23 | 1.49 | 0.70 | 0.94 | 0.49 | 17.6 | 12.2 | 4.35 |
| 01X07850 | 1E-10 | ? | BHS | SE | 1.63 | 1.55 | 1.35 | 1.21 | 0.50 | 17.8 | 11.6 | 3.90 |
| 01X07851 | 1E-11 | A | BHS | SE | 1.29 | 1.54 | 1.58 | 0.85 | 0.54 | 18.0 | 13.0 | 4.86 |
| 01X07891 | 1E-36 | C | prod | SE | 1.36 | 1.46 | 1.04 | 0.92 | 0.51 | 18.1 | 12.3 | 4.13 |
| 01X07892 | 1H-21 | C | prod | SE | 1.33 | 1.54 | 0.63 | 0.87 | 0.47 | 17.2 | 12.4 | 4.48 |
| 00X07843 | 1J-14 | A | MDT | SE | 1.39 | 1.46 | 0.97 | 0.82 | 0.54 | 18.3 | 12.9 | 4.66 |
| 01R00048 | 1J-14 | C | SWC | SE | 1.58 | 1.50 | | | | | | |
| 01R00049 | 1J-14 | C | SWC | SE | 1.52 | 3.45 | | | | | | |
| 01R00050 | 1J-14 | C | SWC | SE | 1.52 | 1.42 | | | | | | |
| 01X07852 | 1F-16 | W.Sak | BHS | | bd | bd | bd | 1.41 | 0.34 | 11.2 | 11.6 | 8.82 |
| 01X07870 | 2Z-4 | W.Sak | BHS | | bd | bd | bd | 1.41 | 0.34 | 11.2 | 11.4 | 8.83 |
| 93X03259 | Kalubik-1 | Kuparuk | RFT | | 1.35 | 1.62 | 1.38 | 0.76 | 0.54 | 17.9 | 13.6 | 4.74 |

**Table 3.** *Selected carbon isotopic data for hydrocarbon fractions*

| Lab number | Well | Formation | Sample type | Depth TVD ss | Carbon isotopes | | Comments |
|---|---|---|---|---|---|---|---|
| | | | | | Saturate | Aromatic | |
| 92R03712 | PBU U-13 | Shublik A | core | | −30.3 | −29.6 | Source rock |
| 92R03716 | PBU U-13 | Shublik A | core | | −30.5 | −29.1 | Source rock |
| 92R03719 | PBU U-13 | Shublik C | core | | −29.6 | −28.7 | Source rock |
| 01R00048 | 1J-14 | Kuparuk C | SWC | 6351 | −30.3 | −29.8 | Tar |
| 01R00049 | 1J-14 | Kuparuk C | SWC | 6373 | **−29.6** | **−29.1** | Tar, most depleted LCAA |
| 01R00050 | 1J-14 | Kuparuk C | SWC | 6389 | −30.2 | −29.7 | Tar |
| 01X07843 | 1J-14 | Kuparuk A | MDT oil | 6455 | −30.3 | −30.0 | |
| 90X02558 | 1D-04 | Kuparuk C | BHS | | −30.1 | −29.7 | |
| 00X07836 | 1D-30 | Kuparuk C | Prod. oil | | −30.3 | −29.9 | Produced oil below tar zone |
| 90X02566 | 1E-11 | Kuparuk A | BHS | | −30.1 | −29.7 | |
| 90X02571 | 1Q-12 | Kuparuk A | BHS | | −30.1 | −29.7 | |
| 90X02565 | 2A-02 | Kuparuk C | BHS | | −30.1 | −29.5 | |
| 90X02574 | 2A-02 | Kuparuk A | BHS | | −29.9 | −29.4 | |
| 90X02562 | 2B-05 | Kuparuk A | BHS | | −29.9 | −29.5 | |
| 90X02563 | 2F-09 | Kuparuk A | BHS | | −30.0 | −29.5 | |
| 90X02564 | 2G-10 | Kuparuk A | BHS | | −30.1 | −29.5 | |
| 90X02577 | 2T-10 | Kuparuk A | BHS | | −30.0 | −29.5 | |
| 90X02569 | 2U-16 | Kuparuk A | BHS | | −30.2 | −29.6 | |
| 90X02560 | 2V-03 | Kuparuk A | BHS | | −29.9 | −29.4 | |
| 90X02567 | 2W-12 | Kuparuk A | BHS | | −30.0 | −29.6 | |
| 90X02568 | 2W-12 | Kuparuk C | BHS | | −30.0 | −29.5 | |
| 90X02559 | 2X-07 | Kuparuk C | BHS | | −30.0 | −29.6 | |
| 90X02556 | 3A-16 | Kuparuk A | BHS | | −30.1 | −29.5 | |
| 90X02570 | 3B-03 | Kuparuk A | BHS | | −29.9 | −29.5 | |
| 90X02561 | 3C-06 | Kuparuk A | BHS | | −30.0 | −29.6 | |
| 90X02557 | 3N-13 | Kuparuk A | BHS | | −30.2 | −29.8 | |

Rock-Eval 6 (manufactured by Vinci Technologies) analyses (Lafargue *et al.* 1997) were performed at Baseline/DGSI using approximately 80 mg rock/analysis. Pyrolysis conditions used an initial condition of 180 °C for 10 min followed by a temperature ramp of 10 °C/min to 650 °C with a final 1 min hold time. Oxidation conditions used an initial temperature of 400 °C for 3 min with a 25 °C/min ramp to 450 °C (5 min hold) and then a ramp at 25 °C/min to 750 °C. Data are typically acquired for an additional 3 min after the end of pyrolysis and oxidation programmes. However, Kuparuk River formation rock samples frequently contain considerable siderite, and a specimen of siderite was pyrolysed on the instrument to determine the effect of siderite breakdown on the Rock-Eval 6 results. We stopped integration of the $CO_2$/CO peak area in the pyrolysis step at 350 °C for all Kuparuk River Formation samples. No correction was made in the oxidation step, but the results suggest the residual carbon, Rc, is likely affected by mineral decomposition unrelated to organic content of the rock and not applicable to API gravity prediction for this study. Rock-Eval data may be found in Table 5.

## Results

Geochemical results are organized into four sections.

(1) Gas geochemistry and WGC light hydrocarbon peak ratios were examined to determine mega-compartments within the Kuparuk River Field, to establish processes that affect the compartments and to establish basic geochemical background patterns.
(2) GC/MS of aromatic fractions from core extracts of specific zones within the Kuparuk River Formation were analysed to determine the degree of depletion of long-chain alkyl aromatics. The core samples analysed were chosen based on available cores for their relationship to production issues (tar, oil quality and production rates).
(3) Rock-Eval 6 pyrolysis was utilized to determine the relative amount of heavy asphaltic material in the same cores analysed by aromatic hydrocarbon GC/MS.
(4) The inorganic geochemistry of the Kuparuk River Formation was examined using the electron microprobe.

**Table 4.** *Aromatic hydrocarbon GC/MS ratios*

| Sample ID | Well | Fm | Sample type | Field area | TVD ss | LCAAR-1 | LCAAR-2 | LCAAR-3 | DBT /Phe | Rc |
|---|---|---|---|---|---|---|---|---|---|---|
| 02R00034 | 3B-14 | C4 | Core | NW | 5911.5 | 0.78 | 0.093 | 0.24 | 0.46 | 0.82 |
| 02R00035 | 3B-14 | C4 | Core | NW | 5923 | 0.76 | 0.099 | 0.27 | 0.45 | 0.82 |
| 02R00036 | 3B-14 | C1 | Core | NW | 5946.9 | 0.77 | 0.088 | 0.26 | 0.36 | 0.81 |
| 02R00037 | 3B-14 | A3 | Core | NW | 6088.7 | 0.71 | 0.107 | 0.21 | 0.65 | 0.86 |
| 02R00038 | 3B-14 | A3/A2 | Core | NW | 6119.2 | 0.69 | 0.092 | 0.18 | 0.65 | 0.85 |
| 01X07842 | 3S-26 | C | Test oil | NW | | 0.72 | 0.093 | 0.17 | 0.66 | 0.85 |
| 02R00032 | 3A-09 | A3 | Core | NW | 6022.3 | 0.74 | 0.125 | 0.26 | 0.56 | 0.87 |
| 02R00033 | 3A-09 | A2 | Core | NW | 6048.2 | 0.75 | 0.125 | 0.26 | 0.55 | 0.85 |
| 02R00039 | 3M-16 | A2 | Core | NW | 6170.1 | 0.74 | 0.130 | 0.22 | 0.60 | 0.87 |
| 02R00040 | 3M-16 | A1 | Core | NW | 6197.5 | 0.71 | 0.114 | 0.20 | 0.66 | 0.86 |
| 02R00022 | 2X-02 | C4 | Core | C | 5753.7 | 0.74 | 0.107 | 0.25 | 0.60 | 0.81 |
| 02R00023 | 2X-02 | C2 | Core | C | 5789.8 | 0.73 | 0.100 | 0.24 | 0.50 | 0.83 |
| 02R00024 | 2X-02 | A4 | Core | C | 5957 | 0.66 | 0.083 | 0.20 | 0.57 | 0.84 |
| 02R00025 | 2X-02 | A3 | Core | C | 5987.8 | 0.62 | 0.065 | 0.15 | 0.62 | 0.82 |
| 02R00026 | 2X-02 | A3 | Core | C | 6000 | 0.62 | 0.062 | 0.16 | 0.60 | 0.81 |
| 02R00027 | 2X-02 | A2 | Core | C | 6052.7 | 0.67 | 0.082 | 0.18 | 0.59 | 0.81 |
| 02R00028 | 2Z-18 | C3 | Core | C | 5803.8 | 0.74 | 0.093 | 0.25 | 0.42 | 0.83 |
| 02R00029 | 2Z-18 | A4 | Core | C | 5986.9 | 0.67 | 0.084 | 0.23 | 0.51 | 0.87 |
| 02R00030 | 2Z-18 | A3 | Core | C | 6032 | 0.63 | 0.073 | 0.17 | 0.58 | 0.84 |
| 02R00031 | 2Z-18 | A2 | Core | C | 6065.3 | 0.57 | 0.043 | 0.16 | 0.60 | 0.84 |
| 02R00021 | 2K-11 | A4 | Core | SW | 6104.1 | 0.68 | 0.083 | 0.23 | 0.47 | 0.87 |
| 02R00070 | 2E-17 | C4 | Core | SC | 5918 | 0.59 | 0.041 | 0.15 | 0.40 | 0.82 |
| 02R00071 | 2E-17 | C4 | Core | SC | 5930 | 0.66 | 0.043 | 0.19 | 0.32 | 0.81 |
| 02R00072 | 2E-17 | C4 | Core | SC | 5935.7 | 0.58 | 0.038 | 0.15 | 0.46 | 0.80 |
| 02R00073 | 2E-17 | C4 | Core | SC | 5942 | 0.64 | 0.043 | 0.18 | 0.36 | 0.81 |
| 02R00020 | 2E-10 | A5 | Core | SC | 5914.5 | 0.68 | 0.082 | 0.22 | 0.50 | 0.85 |
| 00X07843 | 1J-14 | A | MDT oil | SE | 6455 | 0.52 | 0.054 | 0.23 | 0.46 | 0.84 |
| 01R00048 | 1J-14 | C | SWC | SE | 6351 | 0.48 | 0.038 | 0.18 | 0.48 | 0.76 |
| 01R00049 | 1J-14 | C | SWC | SE | 6373 | 0 | 0 | 0 | 0.51 | 0.73 |
| 01R00050 | 1J-14 | C | SWC | SE | 6389 | 0.39 | 0.024 | 0.13 | 0.48 | 0.77 |
| 02R00015 | 1E-14 | C4/C3 | Core | SE | 6102.5 | 0.68 | 0.091 | 0.21 | 0.53 | 0.87 |
| 02R00016 | 1E-14 | A5 | Core | SE | 6258.1 | 0.64 | 0.085 | 0.19 | 0.58 | 0.86 |
| 02R00014 | 1D-08 | A5 | Core | SE | 6538 | 0.64 | 0.080 | 0.20 | 0.46 | 0.88 |
| 00X07836 | 1D-30 | C3-C1 | Prod. Oil | SE | | 0.40 | 0.033 | 0.15 | 0.63 | 0.82 |
| 00R00553 | 1D-30 | C4 | Cuttings | SE | | 0.32 | 0.017 | 0.07 | 0.48 | 0.77 |
| 00R00554 | 1D-30 | C4 | Cuttings | SE | | 0.35 | 0.022 | 0.12 | 0.54 | 0.83 |
| 00R00555 | 1D-30 | C3/2 | Cuttings | SE | | 0.39 | 0.029 | 0.14 | 0.43 | 0.84 |
| 00R00556 | 1D-30 | C1 | Cuttings | SE | | 0.44 | 0.041 | 0.18 | 0.45 | 0.83 |
| 02R00017 | 1R-07 | C1 | Core | NE | 6376 | 0.69 | 0.099 | 0.19 | 0.63 | 0.86 |
| 02R00018 | 1R-07 | A3 | Core | NE | 6543.1 | 0.62 | 0.077 | 0.15 | 0.64 | 0.85 |
| 02R00019 | 1R-07 | A2 | Core | NE | 6614.3 | 0.61 | 0.071 | 0.16 | 0.56 | 0.84 |

**Table 5.** *Rock-Eval 6 data.*

| Rock sample ID | Well | Fm | Sample type | Field area | Depth TVD ss | R1 | R2 | R3 | RC | Y | API calc. |
|---|---|---|---|---|---|---|---|---|---|---|---|
| 02R00034 | 3B-14 | C4 | Core | NW | 5911.5 | 5.88 | 13.68 | 4.88 | 0.78 | 0.800 | 24.0 |
| 02R00035 | 3B-14 | C4 | Core | NW | 5923 | 3.41 | 7.67 | 4.71 | 0.61 | 0.702 | 16.1 |
| 02R00036 | 3B-14 | C1 | Core | NW | 5946.9 | 1.78 | 4.44 | 2.28 | 0.57 | 0.732 | 18.1 |
| 02R00037 | 3B-14 | A3 | Core | NW | 6088.7 | 11.81 | 24.47 | 9.88 | 0.90 | 0.786 | 22.6 |
| 02R00038 | 3B-14 | A3/A2 | Core | NW | 6119.2 | 10.04 | 19.11 | 9.48 | 0.88 | 0.755 | 19.9 |
| 02R00032 | 3A-09 | A3 | Core | NW | 6022.3 | 7.60 | 22.85 | 10.78 | 0.89 | 0.739 | 18.6 |
| 02R00033 | 3A-09 | A2 | Core | NW | 6048.2 | 9.40 | 22.63 | 10.32 | 0.97 | 0.756 | 20.0 |
| 02R00039 | 3M-16 | A2 | Core | NW | 6170.1 | 5.68 | 14.12 | 5.57 | 2.88 | 0.780 | 22.1 |
| 02R00040 | 3M-16 | A1 | Core | NW | 6197.5 | 11.35 | 27.41 | 10.61 | 1.11 | 0.785 | 22.5 |
| 01R00120 | 3H-09 | C4 | Core | NW | 5868.42 | 2.03 | 4.73 | 9.32 | 1.98 | 0.420 | 5.1 |
| 01R00121 | 3H-09 | C4 | Core | NW | 5872.4 | 1.06 | 4.05 | 2.60 | 0.45 | 0.663 | 13.7 |
| 01R00122 | 3H-09 | C4 | Core | NW | 5880.36 | 1.45 | 4.65 | 2.37 | 0.58 | 0.720 | 17.3 |
| 01R00123 | 3H-09 | C4 | Core | NW | 5882.85 | 1.78 | 6.41 | 3.40 | 0.76 | 0.707 | 16.4 |
| 01R00124 | 3H-09 | C4 | Core | NW | 5883.85 | 1.86 | 6.83 | 3.73 | 0.81 | 0.700 | 15.9 |
| 01R00125 | 3H-09 | C4 | Core | NW | 5886.83 | 1.56 | 9.11 | 4.70 | 0.57 | 0.694 | 15.6 |
| 01R00126 | 3H-09 | C4 | Core | NW | 5895.69 | 2.29 | 9.02 | 4.67 | 0.58 | 0.708 | 16.5 |
| 01R00127 | 3H-09 | C4 | Core | NW | 5899.48 | 2.74 | 10.89 | 5.38 | 0.63 | 0.717 | 17.1 |
| 01R00128 | 3H-09 | C4 | Core | NW | 5906.25 | 2.27 | 8.11 | 4.56 | 0.62 | 0.695 | 15.6 |
| 01R00129 | 3H-09 | C4 | Core | NW | 5912.72 | 2.68 | 12.07 | 7.87 | 0.78 | 0.652 | 13.1 |
| 01R00130 | 3H-09 | C4 | Core | NW | 5918.09 | 1.46 | 6.90 | 4.55 | 0.55 | 0.648 | 12.9 |
| 01R00131 | 3H-09 | C4 | Core | NW | 5920.18 | 0.49 | 2.94 | 1.58 | 0.40 | 0.685 | 15.0 |
| 02R00022 | 2X-02 | C4 | Core | C | 5753.7 | 6.96 | 8.97 | 3.43 | 0.60 | 0.823 | 26.3 |
| 02R00023 | 2X-02 | C2 | Core | C | 5789.8 | 5.93 | 10.46 | 6.24 | 0.72 | 0.724 | 17.6 |
| 02R00024 | 2X-02 | A4 | Core | C | 5957 | 9.76 | 19.10 | 12.92 | 1.66 | 0.691 | 15.4 |
| 02R00025 | 2X-02 | A3 | Core | C | 5987.8 | 9.66 | 21.04 | 23.90 | 3.65 | 0.562 | 9.1 |
| 02R00026 | 2X-02 | A3 | Core | C | 6000 | 8.09 | 14.85 | 15.30 | 14.99 | 0.600 | 10.6 |
| 02R00027 | 2X-02 | A2 | Core | C | 6052.7 | 6.92 | 13.03 | 11.96 | 8.85 | 0.625 | 11.8 |
| 02R00028 | 2Z-18 | C3 | Core | C | 5803.8 | 5.88 | 10.68 | 4.36 | 0.58 | 0.792 | 23.1 |
| 02R00029 | 2Z-18 | A4 | Core | C | 5986.9 | 6.07 | 19.57 | 11.86 | 9.74 | 0.684 | 14.9 |
| 02R00030 | 2Z-18 | A3 | Core | C | 6032 | 6.43 | 13.36 | 10.38 | 1.26 | 0.656 | 13.3 |
| 02R00031 | 2Z-18 | A2 | Core | C | 6065.3 | 7.78 | 14.70 | 9.78 | 1.17 | 0.697 | 15.7 |
| 01R00103 | 2U-16 | C4 | Core | C | 5682.62 | 0.31 | 1.78 | 0.53 | 0.21 | 0.798 | 23.7 |
| 01R00104 | 2U-16 | C4 | Core | C | 5690.31 | 1.53 | 10.53 | 6.00 | 0.53 | 0.668 | 14.0 |
| 01R00105 | 2U-16 | C4 | Core | C | 5699.91 | 0.41 | 2.12 | 0.59 | 0.24 | 0.811 | 25.0 |
| 01R00106 | 2U-16 | C4 | Core | C | 5707.9 | 3.10 | 15.40 | 6.61 | 0.46 | 0.737 | 18.5 |
| 01R00107 | 2U-16 | C4 | Core | C | 5713.3 | 3.36 | 13.24 | 6.46 | 0.46 | 0.720 | 17.3 |
| 01R00108 | 2U-16 | C4 | Core | C | 5720.7 | 1.87 | 8.57 | 4.37 | 0.35 | 0.705 | 16.3 |
| 01R00109 | 2U-16 | C1/2/3 | Core | C | 5724 | 2.62 | 16.75 | 9.38 | 0.50 | 0.674 | 14.3 |

| | | | | | | | | | | | |
|---|---|---|---|---|---|---|---|---|---|---|---|
| 01R00110 | 2U-16 | C1/2/3 | Core | C | 5725.89 | 2.07 | 9.89 | 5.18 | 0.38 | 0.698 | 15.8 |
| 01R00111 | 2U-16 | C1/2/3 | Core | C | 5731.39 | 2.18 | 13.24 | 7.90 | 0.59 | 0.661 | 13.6 |
| 01R00112 | 2U-16 | C1/2/3 | Core | C | 5734.09 | 1.77 | 7.45 | 4.43 | 0.38 | 0.675 | 14.4 |
| 01R00113 | 2U-16 | A3 | Core | C | 5800.56 | 4.20 | 19.50 | 9.97 | 0.45 | 0.704 | 16.2 |
| 01R00114 | 2U-16 | A3 | Core | C | 5813.25 | 2.56 | 16.71 | 8.98 | 0.36 | 0.682 | 14.8 |
| 01R00115 | 2U-16 | A3 | Core | C | 5825.25 | 3.11 | 17.47 | 9.79 | 0.51 | 0.678 | 14.6 |
| 01R00116 | 2U-16 | A3 | Core | C | 5837.24 | 2.16 | 14.62 | 9.58 | 0.46 | 0.637 | 12.3 |
| 01R00117 | 2U-16 | A2 | Core | C | 5859.93 | 1.93 | 10.65 | 7.75 | 0.42 | 0.619 | 11.5 |
| 01R00118 | 2U-16 | A2 | Core | C | 5875.82 | 4.46 | 19.39 | 10.72 | 0.00 | 0.690 | 15.3 |
| 01R00119 | 2U-16 | A2 | Core | C | 5899.91 | 2.48 | 16.22 | 11.58 | 0.83 | 0.618 | 11.4 |
| 02R00021 | 2K-11 | A4 | Core | SW | 6104.1 | 6.74 | 16.64 | 8.19 | 0.62 | 0.741 | 18.8 |
| 01R00093 | 2E-10 | C4 | Core | SC | 5869.44 | 1.61 | 6.99 | 3.83 | 0.37 | 0.692 | 15.4 |
| 01R00094 | 2E-10 | C4 | Core | SC | 5871.24 | 1.18 | 4.19 | 3.26 | 0.51 | 0.622 | 11.6 |
| 01R00095 | 2E-10 | A5 | Core | SC | 5909.35 | 2.58 | 12.24 | 9.34 | 0.58 | 0.613 | 11.2 |
| 01R00096 | 2E-10 | A5 | Core | SC | 5913.34 | 3.79 | 11.27 | 6.37 | 0.39 | 0.703 | 16.1 |
| 02R00020 | 2E-10 | A5 | Core | SC | 5914.5 | 4.26 | 7.85 | 4.06 | 0.58 | 0.749 | 19.5 |
| 01R00097 | 2E-10 | A5 | Core | SC | 5920.52 | 3.75 | 14.06 | 9.65 | 0.56 | 0.649 | 12.9 |
| 01R00098 | 2E-10 | A5 | Core | SC | 5922.32 | 2.40 | 9.40 | 8.12 | 0.61 | 0.592 | 10.3 |
| 01R00099 | 2E-10 | A5 | Core | SC | 5924.41 | 3.71 | 14.56 | 10.64 | 0.66 | 0.632 | 12.1 |
| 01R00100 | 2E-10 | A4 | Core | SC | 5946.56 | 2.53 | 9.47 | 5.41 | 0.34 | 0.689 | 15.3 |
| 01R00101 | 2E-10 | A4 | Core | SC | 5965.01 | 1.25 | 8.12 | 5.44 | 0.27 | 0.633 | 12.1 |
| 01R00102 | 2E-10 | A4 | Core | SC | 5981.87 | 1.62 | 8.64 | 7.26 | 0.36 | 0.586 | 10.0 |
| 02R00070 | 2E-17 | C4 | Core | SC | 5918 | 0.51 | 1.70 | 3.65 | 0.60 | 0.377 | 4.3 |
| 02R00071 | 2E-17 | C4 | Core | SC | 5930 | 1.29 | 3.73 | 4.78 | 1.07 | 0.512 | 7.4 |
| 02R00072 | 2E-17 | C4 | Core | SC | 5935.7 | 4.62 | 13.78 | 18.86 | 2.27 | 0.494 | 6.9 |
| 02R00073 | 2E-17 | C4 | Core | SC | 5942 | 2.15 | 6.29 | 9.45 | 1.45 | 0.472 | 6.3 |
| 02R00015 | 1E-14 | C4/C3 | Core | SE | 6102.5 | 5.25 | 11.53 | 5.37 | 0.54 | 0.758 | 20.1 |
| 01R00073 | 1E-14 | C3 | Rock | SE | 6108.6 | 5.09 | 12.73 | 5.35 | 0.75 | 0.769 | 21.1 |
| 01R00074 | 1E-14 | C3 | Rock | SE | 6109.3 | 4.60 | 12.32 | 6.36 | 0.66 | 0.727 | 17.8 |
| 02R00016 | 1E-14 | A5 | Core | SE | 6258.1 | 6.06 | 13.06 | 6.89 | 0.80 | 0.735 | 18.4 |
| 01R00075 | 1E-14 | A5 | Rock | SE | 6261.09 | 5.60 | 16.68 | 8.51 | 1.04 | 0.724 | 17.6 |
| 01R00076 | 1E-14 | A5 | Rock | SE | 6262.08 | 6.16 | 15.55 | 8.41 | 0.90 | 0.721 | 17.3 |
| 01R00071 | 1D-08 | A5 | Core | SE | 6537.8 | 2.54 | 5.53 | 2.76 | 0.52 | 0.745 | 19.2 |
| 01R00072 | 1D-08 | A5 | Core | SE | 6538.3 | 2.76 | 5.88 | 2.53 | 0.66 | 0.774 | 21.5 |
| 01R00077 | 1E-30 | A5 | Rock | SE | 6236.23 | 4.73 | 17.16 | 10.10 | 1.30 | 0.684 | 15.0 |
| 01R00078 | 1E-30 | A5 | Rock | SE | 6237.16 | 5.18 | 15.27 | 8.38 | 1.09 | 0.709 | 16.6 |
| 01R00079 | 1F-05 | A4 | Rock | SE | 6092.51 | 5.71 | 13.90 | 6.45 | 2.23 | 0.752 | 19.7 |
| 01R00080 | 1F-05 | A4 | Rock | SE | 6093.47 | 4.90 | 15.76 | 7.27 | 4.51 | 0.740 | 18.7 |
| 02R00014 | 1D-08 | A5 | Core | SE | 6538 | 5.11 | 11.85 | 6.47 | 0.90 | 0.724 | 17.6 |
| 02R00017 | 1R-07 | C1 | Core | NE | 6376 | 3.69 | 6.02 | 2.45 | 0.31 | 0.799 | 23.8 |
| 02R00018 | 1R-07 | A3 | Core | NE | 6543.1 | 6.81 | 9.54 | 4.08 | 0.60 | 0.800 | 24.0 |
| 02R00019 | 1R-07 | A2 | Core | NE | 6614.3 | 7.73 | 16.03 | 8.27 | 0.85 | 0.742 | 18.9 |

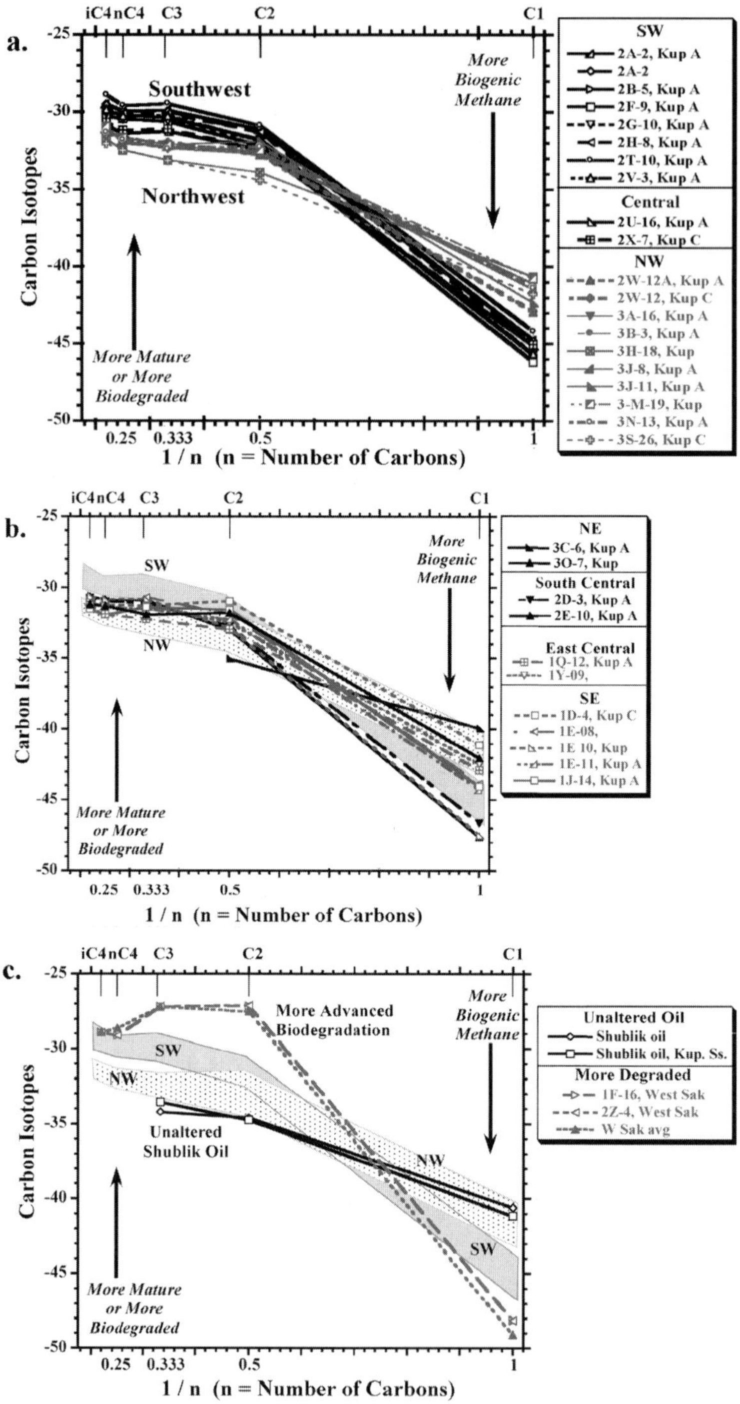


**Fig. 7.** Gas line plot (after Chung *et al.* 1988) for Kuparuk River Field: (**a**) gas data from northwestern (NW) and southwestern (SW) parts the field cluster by geographic area, (**b**) gas data from northeast, southeast, and central graben areas and (**c**) gas data from more biodegraded West Sak gases and pristine Shublik derived gas from an independent accumulation outside the field area. Backgrounds in Figures 7b and 7c show the range of NW (stippled) and SW (shaded) gas data.

*Field compartmentalization*

Gas data from the northwestern (NW) and southwestern (SW) parts of the field tend to separately cluster together with distinct carbon isotopic profiles (Fig. 7a), thus representing mega-compartments within the field. Data for two wells from the far western edge of field (3H-18 and 3S-26) diverge, having unusually light ethane, propane and butane isotopic values. Gas data from the eastern portion of Kuparuk River field [designated in Table 1 as northeast (NE), southeast (SE) and south-central (SC)] is much more scattered (Fig. 7b). For comparison, a gas line plot of data from a smaller independent petroleum accumulation NW of Kuparuk Field, Kalubik-1 (Kuparuk Fm.), charged by a similar Shublik source facies, but with no indication of biodegradation, was plotted with the NW and SW Kuparuk gas types stippled/shaded in the background (Fig. 7c). Also on Figure 7c are gas data from the shallow West Sak field illustrating gas in a more advanced stage of biodegradation (Masterson 2001; Masterson *et al.* 2001).

The more biodegraded West Sak gases have the greatest biogenic methane admixed with thermogenic methane. Biogenic methane may accumulate in a structure before thermogenic charge or co-migrate with thermogenic gas into the field, but it has also been observed in fields with significant petroleum biodegradation (Sweeney & Taylor 1999; Pallasser 2000; Boreham *et al.* 2001). Methanogens could be utilizing by-products of petroleum biodegradation produced by other anaerobes or possibly be participating in the biodegradation directly (Zengler *et al.* 1999). $C_1$ and $C_2$ gases from the eastern areas (Fig. 7b) are highly variable, with some methane isotopes from the eastern half of the field approaching values found in West Sak (Fig. 7c), indicating higher biogenic methane content. In contrast, NW and other eastern gases have methane isotopes consistent with unaltered Shublik gas with little biogenic component (Fig. 7c). SW gas has intermediate biogenic methane contents.

Heavier (more positive) carbon isotope ratios for $C_2$–$C_4$ gases can indicate either increasing maturity gas charge or increasing degree of biodegradation. $C_2$–$C_4$ gases show increasing values in order from Kalubik-1, NW, SW and West Sak gases. $C_3$ and $C_4$ carbon isotope values for eastern gases tend to cluster, overlapping both NW and SW data clusters. Ethane is among the more resistant gases to biodegradation (Boreham *et al.* 2001), thus the relatively heavy ethane isotopes in some gases likely represents more mature gas charge in addition to the effects of mild biodegradation as indicated by the biogenic methane content.

Kuparuk C is in contact with Kuparuk A in the western half of the field (Fig. 2), and thus, within each western area, communication between Kuparuk A and C is likely extensive. Since the eastern half of the field is closer to the oil/water contact, then the scattered nature of the eastern gas isotope data could be due to processes such as biodegradation, or different petroleum charges to sands that are better separated by silts and shales (Fig. 2; also see Masterson & Eggert 1992, figure 10).

Whole oil/gas chromatographic (WGC) light hydrocarbon (LHC) ratios also tend to show clustering by geographic area (Fig. 8). The B and F ratios in Figure 8 are frequently related to petroleum alteration, but may also be affected by source charge (after Thompson 1987). Figure 8 coherently groups LHC data from SW and NW oils as well as from the eastern half of the field (NE and SE). Eastern oil WGC data again tend to be intermediate to data between NW and SW samples, like eastern $C_2$–$C_4$ gas data. Oils from one of the central grabens (Fig. 3a) from 2U and 2X wells tend to cluster with the eastern oils. Plots of other WGC ratios (see Table 2) show similar clustering, but data from the eastern oils tend to have numerous rogue values, which plot independent of the main cluster. Eastern oils tend to have the lowest *n*-alkane/isoprenoid ratios, suggesting biodegradation or lower thermal maturity.

The petroleum filling history of the field is likely complex, with possible multiple episodes or combinations of filling, leakage and degradation. Shifts in the $C_2$–$C_4$ gases could be from maturity or a combination of maturity and biodegradation. The range in methane isotopic values is more strongly suggestive of an anaerobic biodegradation process contributing biogenic methane. Regardless of cause, northwestern and southwestern portions of the field have consistent, but different, geochemical signatures within their areas in both the gas and oil data. The downdip eastern and southeastern portions of the field are more diverse in terms of geochemical attributes and are, in general, regionally less distinct. This is likely to be, in part, due to better separation and isolation of sands (C, A3, A2 etc.) by intervening shales, but may also reflect locally greater effects of the oil degradation or diverse charge. The original Kuparuk field at discovery was highly compartmentalized into mega-compartments in the SW, NW and within different intervals in the eastern part of the field. Subsequent water-flooding and EOR

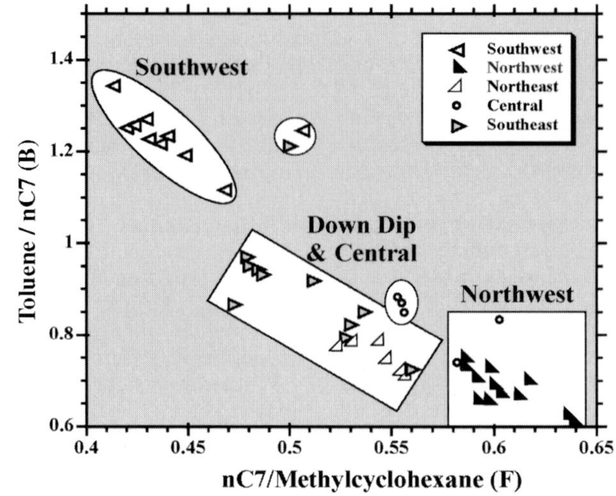


**Fig. 8.** Plot of light hydrocarbon (LHC) whole oil GC ratios for Kuparuk oils (after Thompson 1987). Northwest, southwest and downdip areas from eastern and southeastern parts of the field cluster.

processes have changed the pressure and saturation distribution across the field.

## *Molecular patterns of anaerobic degradation*

A review of WGC data for 38 oil samples (from this study and an archive of older data) was done to search for indications of biodegradation. Generally, analyses of whole oils and saturate hydrocarbon GCs in samples from heavily tar-stained core extracts as well as the better producing zones showed none of the usual signs of microbial biodegradation; i.e. no obvious depletions in *n*-alkanes relative to branched and cyclic hydrocarbons (Fig. 9).

Total ion currents of full scan GC/MS analysis of aromatic fractions for a Shublik source rock sample and an extract of an oil-stained core from a good producing portion of the Kuparuk River field (well 3B-14) are shown in Figure 10. The source rock sample provides a view of the aromatic fraction characteristics of the Shublik charge to Kuparuk River Field. The 3B-14 core has depletions in the naphthalenes and other more volatile low molecular weight aromatics. Considering that many of the cores available for this study were stored for long periods of time (up to 20 years), then evaporative loss of light ends during storage and handling is likely and cannot be attributed to a subsurface process. Note that the dibenzothiophenes and phenanthrenes (Fig. 10) remain strong in the 3B-14 core.

Long-chain alkyl aromatics or LCAAs (e.g. long-chain alkyl benzenes and toluenes, from aromatic hydrocarbon GC/MS $m/z = 92$ and 106 respectively) are readily sensitive to biodegradation (Grimalt *et al.* 1991; Blanc & Connan 1994), but are present in high concentrations in the Shublik source rock and in Shublik-derived oils in good producing portions of the field (Fig. 11). LCAA $<C_{17}$ are commonly depleted in core extracts in this study, probably from evaporative loss during long term storage (Fig. 11c,d; see Fig. 10b). However, examination of aromatic hydrocarbon GC/MS from Kuparuk C SWC extracts of a recent well, 1J-14, from the east boundary of the field (Fig. 12), revealed unusual depletions of LCAAs relative to normal LCAA patterns in Shublik-derived oils (Fig. 11). 1J-14 SWC extracts were taken from the Kuparuk C sand as attempts at an MDT failed to produce petroleum flow. The middle of the Kuparuk C (6373 ft) had the most severe depletion of LCAAs (Fig. 12b, f) with almost no survival of LCAA constituents other than a few ortho-alkyl toluene isomers. The background peaks marked with asterisks are typically small and barely notable (Fig. 11), but are the most prominent peaks in the mass chromatograms of 1J-14 (Fig. 12). The elution time of the background peaks and their mass spectra ($M+ = 212$) indicates that they are minor fragments of dimethyldibenzothiophenes. The deepest (6389 ft) and shallowest (6351 ft) SWCs from 1J-14 have some lower molecular weight LCAAs ($C_{12}$–$C_{18}$) but are severely depleted in the heavier homologues. An MDT oil from the Kuparuk A sand is also depleted in the higher molecular weight LCAAs (Fig. 12d,h). Such

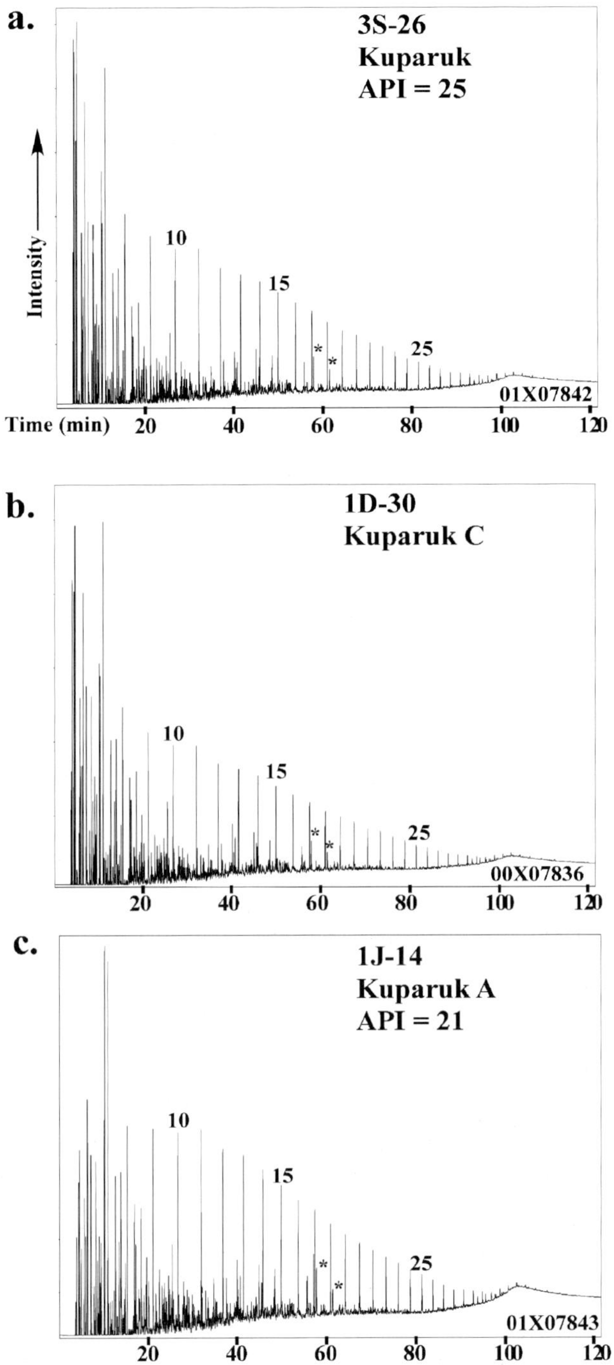


**Fig. 9.** Whole oil gas chromatograms of oils from diverse parts of the field show little indication of alteration or biodegradation: (**a**) high API MDT from Kuparuk C northwest area, (**b**) production oil Kuparuk C3-C1 from southeastern area below tar stained Kuparuk C4 and (**c**) MDT from Kuparuk A southeastern boundary of the field in most downdip location.

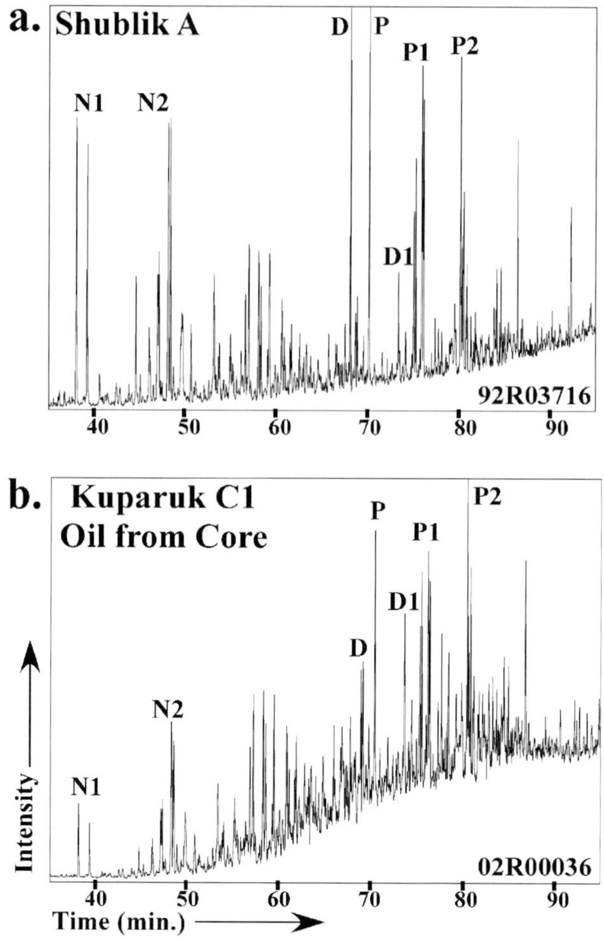


**Fig. 10.** GCMS total ion current (TIC) for aromatic hydrocarbon fractions are dominated by two and three ring aromatic compounds: (**a**) Shublik A source rock extract (Prudhoe Bay Field) and (**b**) Kuparuk River Field oil-stained core, 3B-14 well, after long-term storage. Aromatic family labels are: N1, methylnaphthalenes; N2, dimethylnaphthalenes; D, dibenzothiophene; D1, methyldibenzothiophenes; P, phenanthrene; P1, methylphenanthrenes; P2, dimethylphenanthrenes.

depletions of LCAA have been observed previously (Blanc & Connan 1994) and attributed to 'incipient' biodegradation.

The thorough and selective removal of LCAAs without a similar severe removal of saturate or other aromatic species is most likely caused by the earliest stage biodegradation. Oxidation of aromatic hydrocarbons in the laboratory under mild conditions is readily performed on compounds with benzylic hydrogens (Morrison & Boyd 1966); thus, the selective early removal of LCAAs is not surprising.

Biodegradation, particularly under anaerobic conditions, of petroleum is usually accompanied by a shift to more positive values in carbon isotopes (Sofer 1984; Wilkes *et al.* 2000). Coincident with the most severe removal of LCAA (Fig. 12b,f) is a slight shift to heavier carbon isotopic ratios for both saturate and aromatic hydrocarbon fractions compared with other Kuparuk C SWCs from 1J-14 (22 ft above and 16 ft below). Comparison of isotope data from 1J-14 SWC with produced oils from around the field (Table 3) indicates that the 6373 ft SWC was slightly heavier than typical Kuparuk oils. The isotope values for the 1J-14 SWC extracts are well within the range observed for Shublik source rock extracts (Table 3). Examination of maturity parameters, e.g. MPI-1, or an estimated vitrinite reflectance of hydrocarbon generation, Rc, calculated from MPI-1 (Radke 1987) in Table 4, shows that the tar-plugged core or

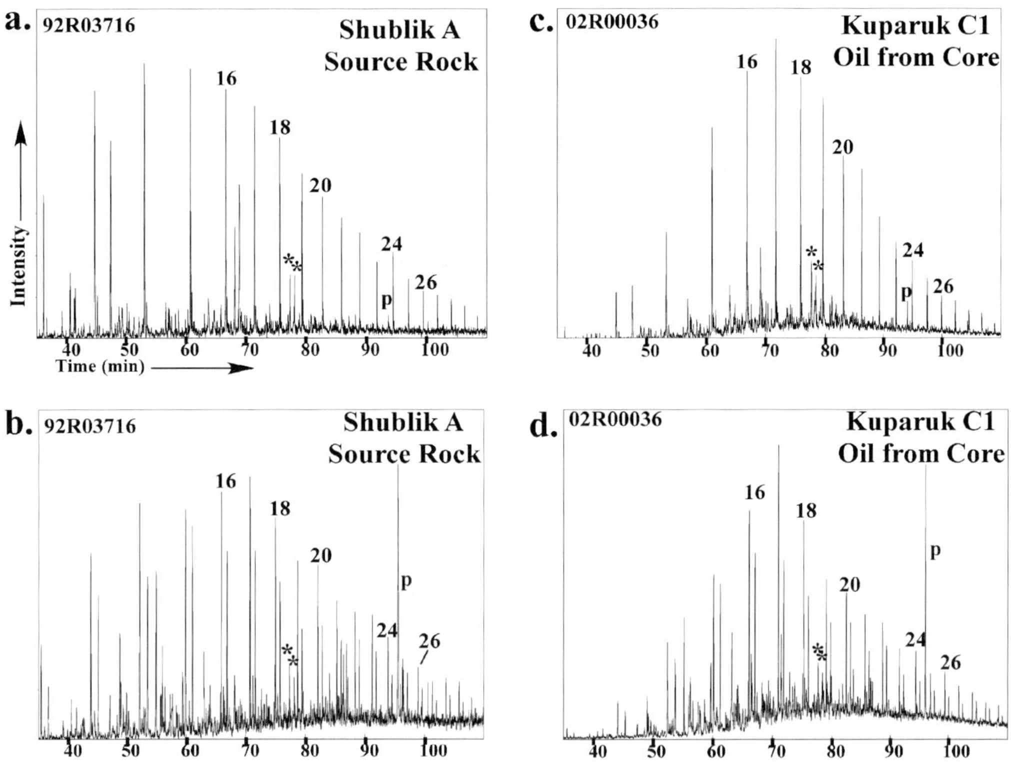


**Fig. 11.** Long chain alkyl aromatic normal patterns have a full homologous series. (**a**) Shublik A source rock Prudhoe Bay Field long chain alkyl benzenes. (**b**) Shublik A long chain alkyl toluenes. (**c**) Kuparuk River Field oil stained core 3B-14 well long chain alkyl benzenes. (**d**) 3B-14 long chain alkyl toluenes. Numbers designate carbon numbers; p is phytanyl-substituted aromatic; asterisk designates background fragment peaks. Note, toluenes have *ortho*, *meta* and *para* isomers for each carbon number.

SWC extracts such as those from 1J-14 or 2E-17 tend to have the lowest maturities. Lower maturity alone corresponds to lighter rather than heavier carbon isotopic values, thus suggesting that these subtle shifts could be due to the degree of biodegradation. For the middle 1J-14 SWC the shift in isotopes for both saturate and aromatic hydrocarbons also suggests that degradation has been active with other molecular types than the LCAAs.

To capture the depletion of LCAAs in a numerical form, a series of ratios was constructed comparing them with other more degradation-resistant components within the oil or extract (Table 4). Ratios were constructed using the peak areas of $C_{19+}$ alkylbenzenes or alkyltoluenes with the areas of the background peaks in the 92 or 106 mass chromatograms (Figs 11 & 12). Examination of the total ion current (TIC) for GC/MS of aromatic hydrocarbons for 1J-14 SWC extracts (Fig. 13) generally do not show evidence for water washing (Palmer 1993). The TIC for 1J-14 samples and in cores with no tar, i.e. less probable biodegradation (Fig. 10), suggested that the phenanthrene and dibenzothiophene families of aromatic hydrocarbons may be more resistant to the biodegradation process at the early stages observed in Kuparuk River field as previously observed (Connan 1984) although they are not immune to degradation during more advanced stages of degradation (Larter *et al.* this volume). Ratios were also constructed for LCAAs with the sums of phenanthrene plus its methyl homologues ($m/z = 178$, 192, 206) and dibenzothiophene and its methyl homologues ($m/z = 184$, 198). After testing various combinations, values for potentially useful ratios are provided in Table 4 (see Appendix for ratio definitions).

Figure 14 provides depth profiles of three LCAA ratios constructed from data for selected wells with problematic or problem-free production (see Fig. 3a for locations): Figures 14a–c all show that the heavily tar-plugged 2E-17 well

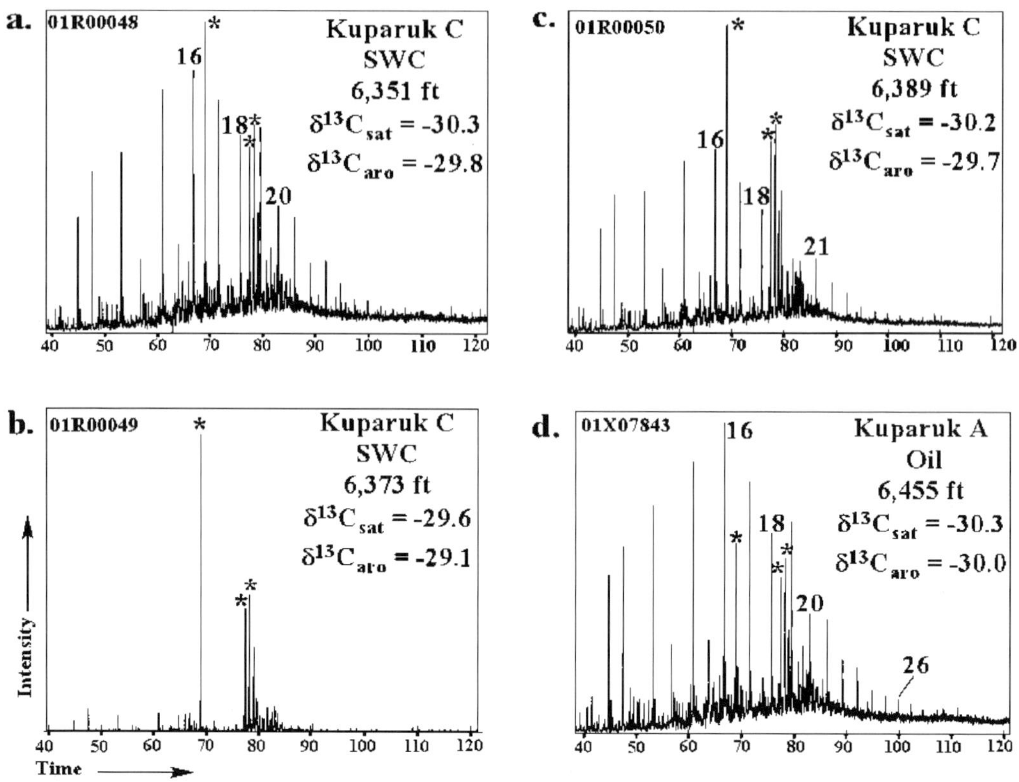


**Fig. 12.** Long chain alkyl aromatics for the most downdip well, 1J-14. Kuparuk River C Formation did not flow when MDTs were attempted. Long chain alkyl benzenes ($m/z = 92$) show variable depletions: (**a**) SWC-43, (**b**) SWC-31, (**c**) SWC-18 and (**d**) MDT oil.

has the greatest depletion of LCAAs. The 3B-14 core extracts generally have the least depletion in LCAA, except that the Kuparuk A sand extracts can show depletions in some ratios, suggesting that some biodegradation may have also occurred in 3B-14 Kuparuk A sands too.

Wells in the central graben of the field (2X-02 and 2Z-18) have good production from the Kuparuk C and disappointing production or no production from the Kuparuk A. The Kuparuk C in the two graben wells (2X-02 and 2Z-18) generally show low if any depletions in LCAAs, but the data from cores from the relatively poorer producing Kuparuk A sand intervals reveal significant depletions in LCAAs, almost as severe as in the 2E-17 tar-plugged well. Figure 14d shows depth profiles for LCAAR-2 in two downdip wells (1J-14 and 1R-07). 1R-07 is from the east downdip flank of the field in which the Kuparuk A tends to have more viscous oil than the Kuparuk C (Fig. 5) and LCAAR-2 shows more depletion of LCAAs in Kuparuk A relative to Kuparuk C. The key observation is that depletion of LCAAs relative to more resistant aromatic species appears to parallel oil quality and production characteristics.

### *Asphaltic bitumen content by Rock-Eval 6*

Pyrolysis methods may be used to characterize the API gravity of heavy oils and tar deposits (Trabelsi *et al.* 1994; Lafargue *et al.* 1997). Rock-Eval 6 pyrolysis affords greater resolution by molecular size than previous Rock-Eval methods. Previous Rock-Eval methods produced two pyrolysis peaks measured by an FID detector: S1, containing free, distillable, hydrocarbons that are eluted or 'flashed' at 300 °C for 3 min, and S2, that contains organic matter that results from thermal breakdown of high molecular weight organics (kerogens, bitumens or asphalts) from 300 to 600 °C. The Rock-Eval 6 provides user-defined oven starting temperatures, allowing the S1/S2 peaks to be further resolved into a third peak. These 'new' peaks, also measured

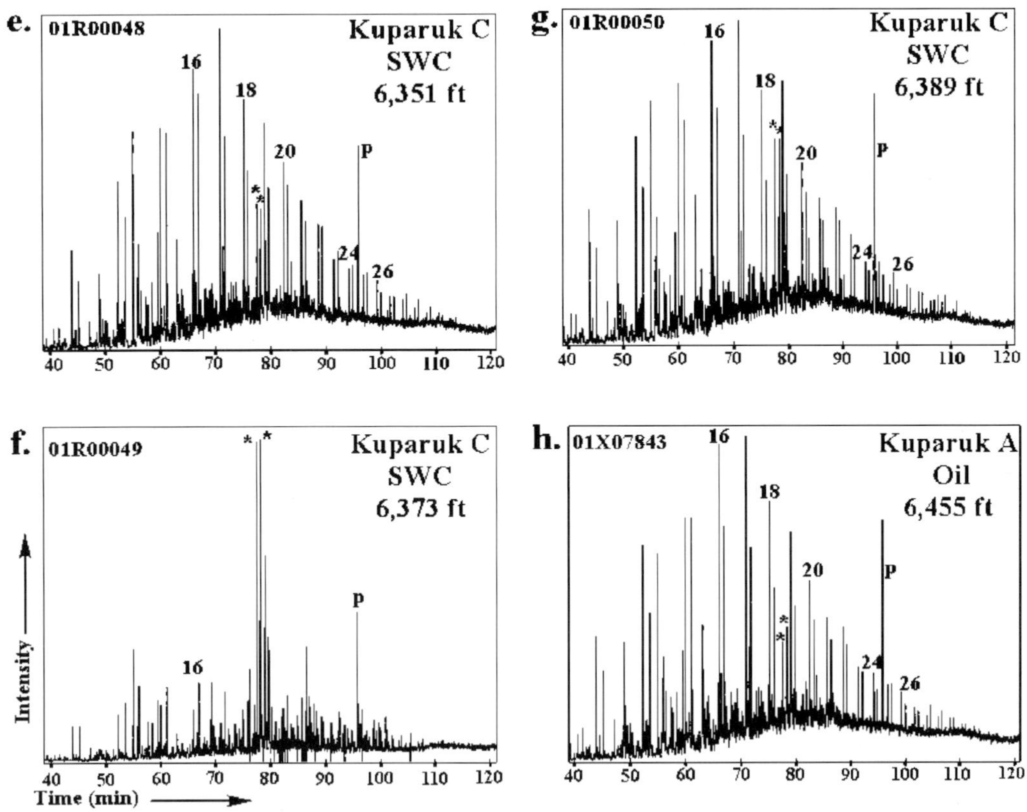


**Fig. 12.** Long chain alkyl toluenes ($m/z = 106$) also show variable depletions. (**e**) SWC-43, (**f**) SWC-31, (**g**) SWC-18 and (**h**) MDT oil. Depths reflect measured values rather than TVD. Asterisks designate normal background peaks from a minor fragment and 'P' designates a phytanyl-substituted aromatic.

using an FID detector, are referred to as R1, R2 and R3 (Fig. 15). For pyrolysis of oils or reservoir rocks, R1 represents the volatile distillable organics ($C_1$–$C_{24}$) that elute at 180 °C when the oven temperature is held for 10 min. R2 represents less volatile (distillable) organics ($C_{24}$–$C_{50}$) that elute following the 180 °C hold up to ~370 °C; and R3 represents less volatile organics and organics cracked from solid-phase organics ($C_{50+}$) that elute at temperatures $\gtrsim$370 °C. Figure 15 shows pyrograms for two samples analysed previously for LCAA content (Fig. 14a–c); cores from Kuparuk C4 and A2 from Central Graben well 2X-02. Kuparuk A sands in the area have poorer production characteristics than Kuparuk C sands. The pyrograms show that the Kuparuk A2 core sample has greater amounts of heavier ($C_{50+}$) organic matter likely from asphaltic tar staining (peak R3).

Dow *et al.* (2000) recently proposed a parameter (*X* coefficient) that relates oil API gravity to a ratio of $(R1 + R2)/(R1 + R2 + Ra)$ where $Ra = R3 + (10Rc)/0.9$. The RC (residual carbon) term represents heavy bitumens or recycled kerogens not directly volatilized by pyrolysis, but that could be oxidized to $CO_2$ or CO in the separate oxidation step (Lafargue *et al.* 1997). However, the Kuparuk formation often contains cements, siderite among others, which decompose to $CO_2$ or CO at temperatures reached in the oxidation step of Rock-Eval 6 analysis. Unacceptable variability in RC was observed in pyrolysis of Kuparuk samples, possibly because of decomposition of carbonate minerals. Therefore, a *Y* coefficient was adopted to correlate Rock-Eval 6 pyrolysis results to petroleum density, where $Y = (R1 + R2)/(R1 + R2 + R3)$.

Oil samples from Kuparuk River Field and biodegraded West Sak Field were pyrolysed in order to construct a calibration curve of Rock-Eval 6 *Y* coefficient to measured oil API gravity over the range 10.1° to 33.1° (Fig. 16a). An exponential function was used to fit the data

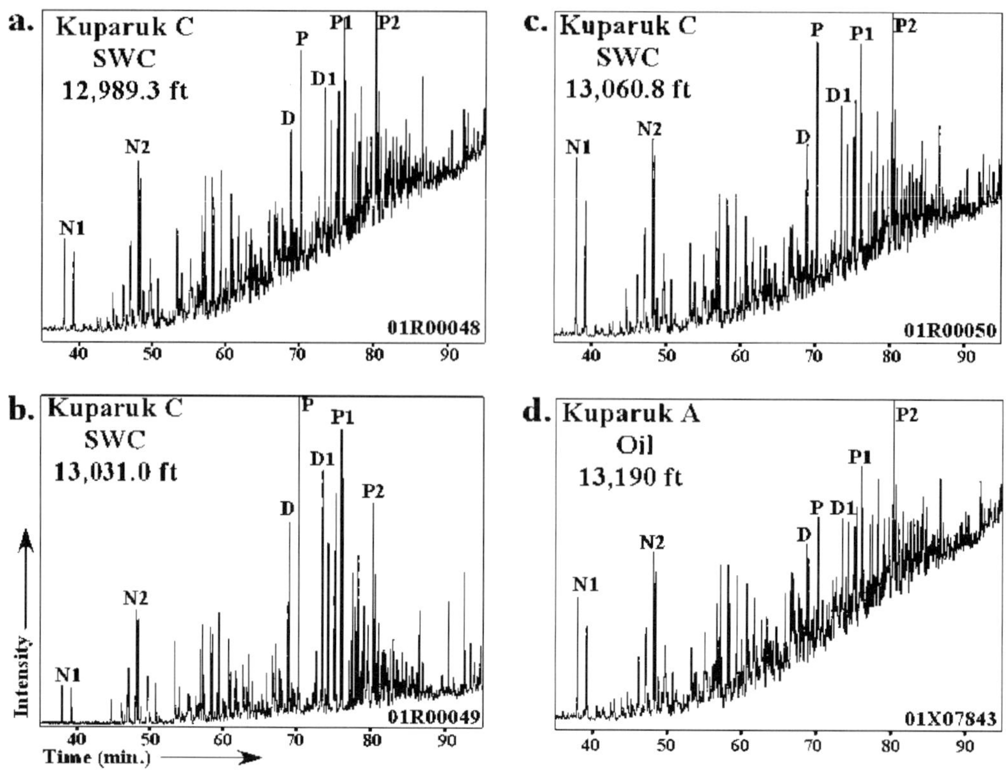


**Fig. 13.** GCMS total ion current (TIC) for aromatic hydrocarbon fractions from 1J-14 well have low amounts of low molecular weight two and three ring aromatics like other Kuparuk sandstone core extracts: (**a**) SWC-43, (**b**) SWC-31, (**c**) SWC-18 and (**d**) MDT oil.

giving a linear correlation coefficient, $R$, of 0.905 (Fig. 16a). A predicted API gravity for petroleum in core samples was calculated using the exponential fit equation from Figure 16a (see Table 5). Comparison of the predicted API gravity to the hydrocarbon/nonhydrocarbon ratio from the semi-quantitative fractionation of core extracts (Fig. 16b) generally shows increasing hydrocarbon content correlates with increasing predicted API gravity. However, many of the core samples pyrolysed contained visible staining of immobile tars. Pyrolysis will cause breakdown of not only the asphaltenes in producible oil but also the immobile asphaltic tar observed in cores. Thus, it is inevitable that the predicted API gravity will reflect a combination of producible oil and immobile oil staining for some Kuparuk samples and may be lower than the value of the producible oil. Figure 16c plots the predicted API gravity from Rock-Eval 6 with API gravities measured on oil extracted from core samples shortly after coring or from production oils. Note that data from the northwestern area typically has an excellent correspondence, but that many of the areas shown earlier to have had anaerobic depletion of LCAAs frequently have predicted API gravities lower than that measured on oil extracted from core. The LCAA depletion suggests that where anaerobic biodegradation has occurred that excess asphaltic bitumen stains the reservoir sands.

A plot of the predicted API gravity with depth (Fig. 17a) for the four wells previously showing variable LCAA depletions (Fig. 14) reveals a pattern similar to that observed in the LCAA plots. The heavily tar stained core from 2E-17 with the greatest LCAA depletions and no petroleum flow had predicted API gravities from 4 to 7.5°. The 3B-14 cores from good productive intervals, especially Kuparuk C, had the highest predicted gravities, but the predicted value is generally lower than that of the produced oils. Volatile lighter hydrocarbons were likely lost from the core over 20 years of storage as noted earlier (Figs 10 & 11). The more productive Kuparuk C formations from the Central graben wells had predicted API gravities in the same range as the 3B-14 cores. The Kuparuk A cores

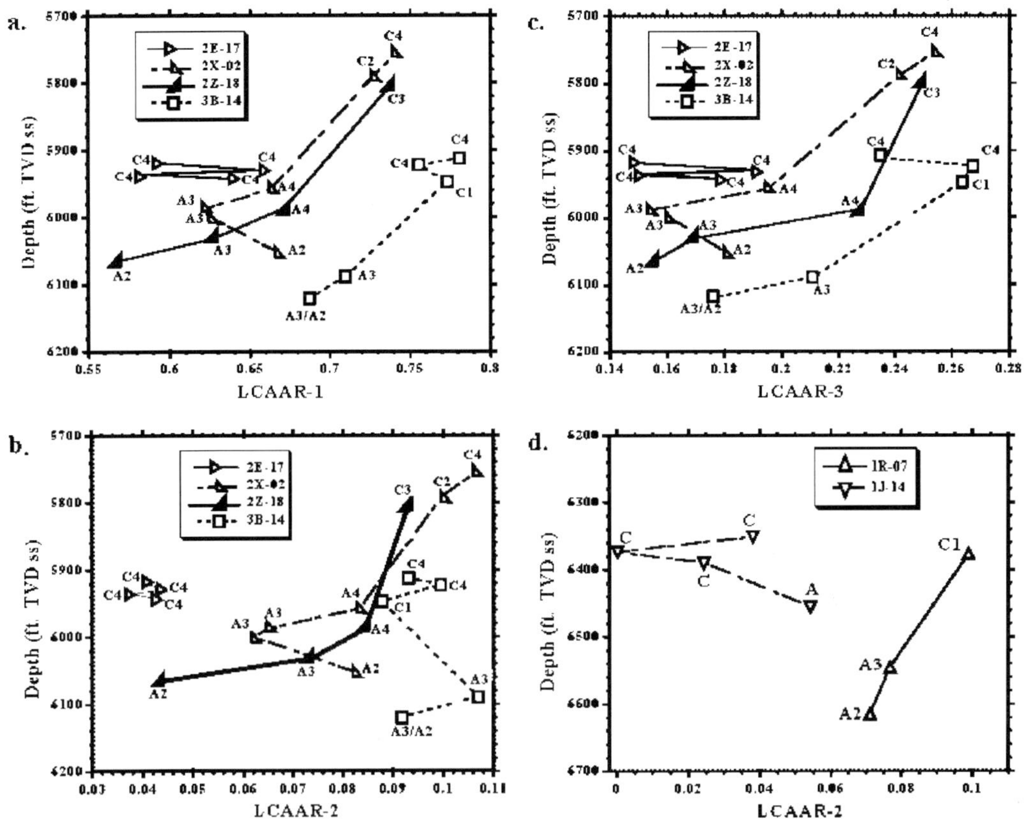


**Fig. 14.** Plot of LCAA ratios with depth (true vertical sub-sea) for Kuparuk River formation core extracts from selected wells: 2E-17 heavily tar stained Kuparuk C4; central graben wells 2X-02 and 2Z-18 with good producing Kuparuk C and poor producing Kuparuk A; and 3B-14 good producing well; or downdip wells 1J-14 and 1R-07 for different LCAA ratios: (**a**) LCAAR-1, (**b**) LCAAR-2, (**c**) LCAAR-3 and (**d**) LCAAR-2. LCAA ratios are defined in the appendix.

from the central graben wells have lower predicted API gravities corresponding to the greater depletion of LCAAs (Fig. 14). Plots of LCAA depletion ratios (Table 4) with predicted Rock-Eval 6 API gravity (Table 5) indicate linear relationships; API (RE-6 predicted, Table 5) versus LCAA ratios (Appendix and Table 4) gave linear correlation coefficients of LCAA-1 $R = 0.596$, LCAA-2 $R = 0.765$ and LCAA-3 $R = 0.500$.

Examination of Rock-Eval 6 pyrolysis data over closer spacing through a series of Kuparuk C, A3 and A2 sands for two western wells 2U-16 and 3H-09 (Fig. 17b) revealed two patterns. First, there appear to be zones of lower API asphaltics towards the tops of formations, often in better quality reservoir sandstone. Second, predicted API gravities tend to decrease with depth within each formation (C4, A3, A2 etc.) as would be expected for gravity segregation, but over a narrower depth range than is expected for the process of gravity segregation. Asphaltene contents for oils within the field fall into two categories: those that have a linear relationship with true vertical depth (TVD) and those with excess asphaltic tar.

Thus, Rock-Eval 6 pyrolysis data can be used to predict API gravity of producible oil where the rock is not heavily tar-filled. In cases where tar occurs, Rock-Eval 6 reveals information about the asphaltic bitumen that degrades production characteristics in the analysed portion of the reservoir.

## *Reservoir diagenesis and inorganic chemistry of Kuparuk Formation*

There appears to be an association between the most extreme biodegradation, indicated by the

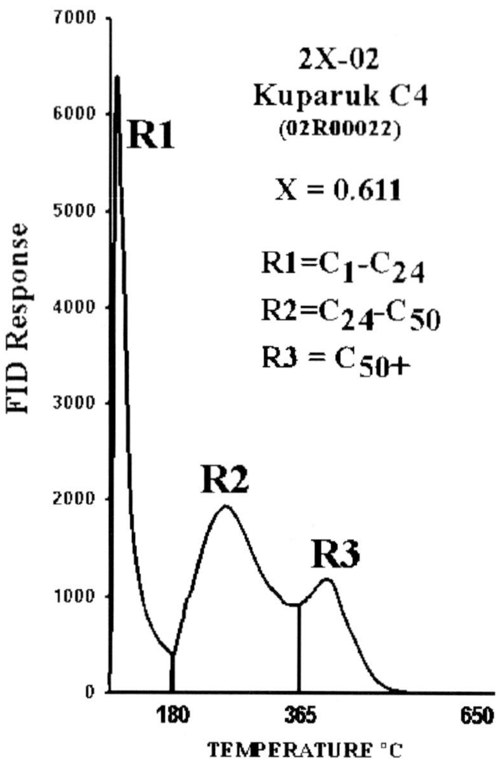


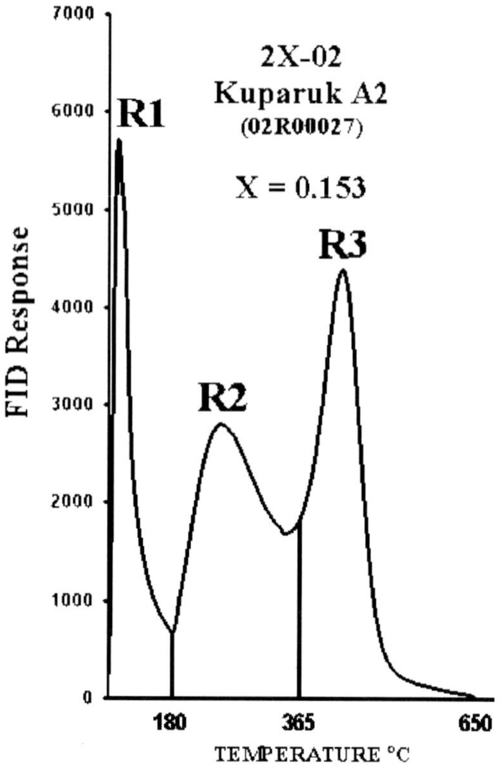


**Fig. 15.** Rock-Eval 6 pyrograms of cores from 2X-02 well: (**a**) Kuparuk C4 with good production characteristics and (**b**) Kuparuk A2 with disappointing production characteristics.

presence of tar, and the occurrence of iron enrichment in glauconite and shale pellets in the Kuparuk C reservoir. The exact mineral species and oxidation state of this iron enrichment is unknown at this time, but is possibly in the form of iron oxide or iron hydroxide. Black, hard, rounded grains have been observed in core and sidewall cores from 2E-17 and 1J-14 respectively, two wells that have tar filling pore space in core samples. These grains are opaque in thin-sections, which suggests a concentration of metals (likely in a combined form). In an effort to learn more about these grains, 12 Kuparuk C sand samples from four peripheral Kuparuk River Unit wells with known tar presence underwent microprobe analysis. Backscatter electron images (BSE) and energy dispersive spectrometer (EDS) spectra show that most of these dense dark rounded grains are altered glauconite pellets.

Glauconite is a mica-like hydrated potassium and iron (ferrous and ferric) aluminosilicate mineral often found in shallow marine settings, but it is also a term used to describe pellet shaped morphologies of heterogeneous composition (Bentor & Kastner 1965; Fairbridge & Bourgeois 1978; Kazalov 1982). Glauconization is the alteration of clay pellets, commonly faecal in origin, to the mineral glauconite. In this process, the pellets become enriched in iron and potassium, turn green in colour, and pellets may exhibit surficial creases or 'valleys'. This likely requires marine water and reducing conditions. Glauconite interaction with oxygenated waters during early diagenesis produces a colour change from green to brown or orange. The FeO/OH enrichment of Kuparuk glauconite observed through microprobe analysis is substantially over and above that produced by normal or oxidized glauconization. EDS spectra from normal green glauconite, from brown oxidized glauconite and from black FeO/OH-enriched glauconite in these samples show a progressive increase in iron and oxygen content (Fig. 18). The degree of Fe enrichment is variable and it commonly occurs as amorphous forms or filaments on a 1–2μm scale (Figs 19 & 20). It is unknown why some glauconites are

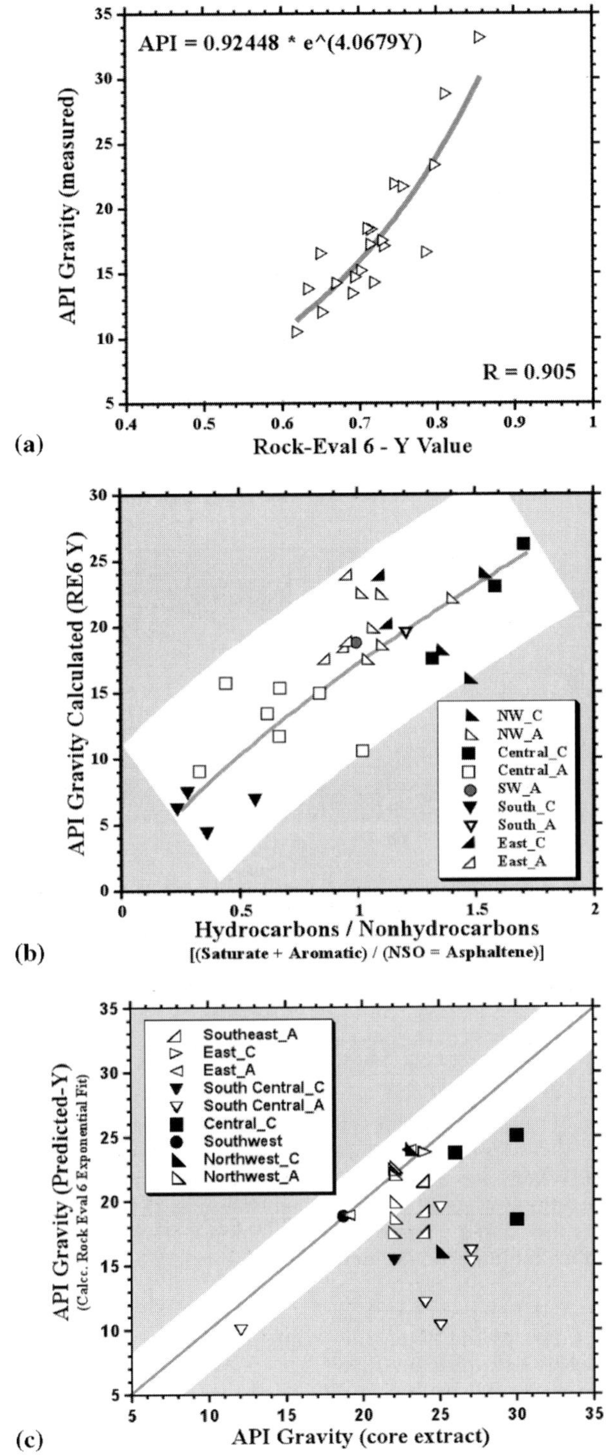


**Fig. 16.** Rock-Eval 6 calibrations: (**a**) calibration curve relating Rock-Eval 6 Y factor to oil API gravity and (**b**) predicted API for core samples from Rock-Eval 6 Y as a function of extract hydrocarbon/nonhydrocarbon ratio. (**c**) Plot of predicted API for core samples compared with API gravity measured from core extracts.

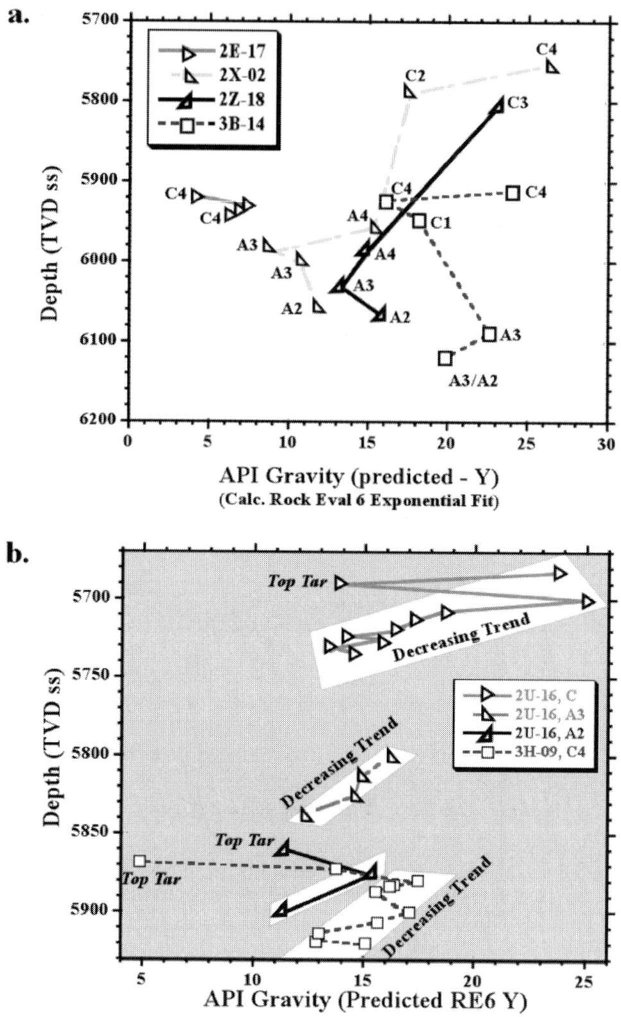


**Fig. 17.** Plot of Rock-Eval 6 with depth (true vertical subsea) for Kuparuk River formation core extracts: (**a**) for 2E-17, 2X-02, 2Z-18 and 3B-14 (see Fig. 14a–c) and (**b**) with higher density sampling showing gravity segregation-like trends for 2U-16 and 3H-09 wells.

FeO/OH-enriched while others are not. Many FeO/OH-enriched glauconites have a rim of illite (Fig. 21). Other associated diagenetic products include scattered kaolinite books and quartz rosettes (Fig. 21).

Tar fills intergranular and rim pore space in these samples and also saturates the illitic rims of these grains. Tar is easily identified on EDS spectra by a large carbon peak and small oxygen and sulphur peaks (Fig. 21). Tar is distinguished from liquid hydrocarbons in pore spaces because it is immobile despite washing from core acquisition, slabbing, plugging, and hexane extraction prior to preparation of polished petrographic thin sections. Tar also commonly exhibits shrinkage fractures (Fig. 21). The associated FeO/OH enrichment found in the same samples with tar may indicate a common biogenic origin.

## Discussion

### *Biodegradation*

The epitome of biodegradation in Kuparuk Field is the most heavily tar filled SWC from 1J-14 well. The severe but selective depletion of LCAA hydrocarbons, despite a lack of obvious

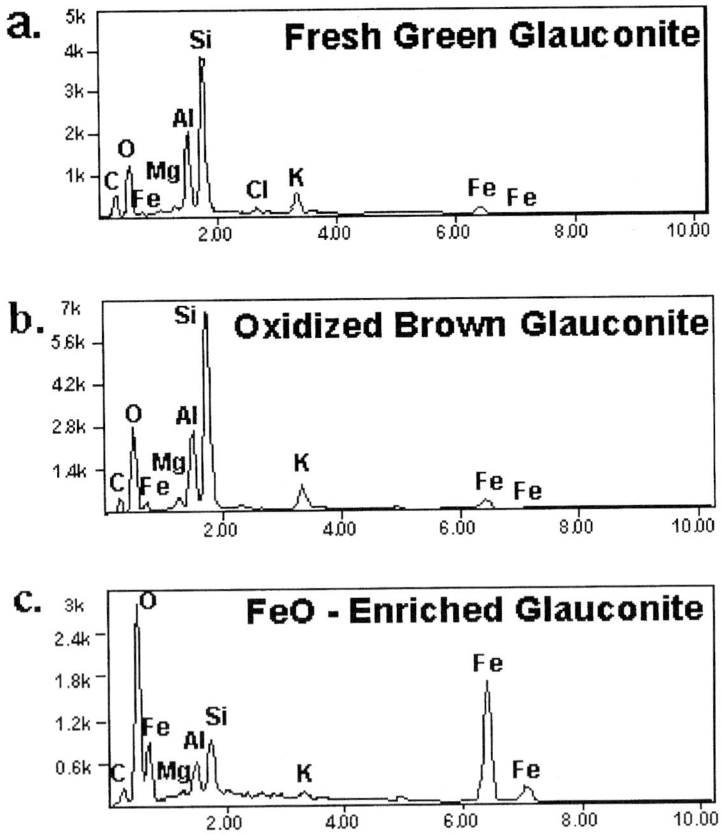


**Fig. 18.** EDS spectra of (**a**) un-oxidized green glauconite 1L-07, (**b**) oxidized brown glauconite, 2E-17 and (**c**) Fe-enriched glauconite, 1J-14. Brown glauconite has a higher oxygen peak than green glauconite, but otherwise these glauconites are very similar. The Fe-enriched glauconite has substantially stronger iron and oxygen peaks, indicating presence of FeO/OH.

alteration of general saturate and aromatic hydrocarbons, is suggestive of a microbial biodegradation process. The LCAAs are attacked preferentially because activation of the benzylic position by the aromatic ring makes oxidation much easier. The complete removal of alkylbenzenes was previously observed in a 'heavy oil' from the Pécorade field (Aquitaine basin) as 'incipient biodegradation' (Blanc & Connan 1994). The heavy oil contained undegraded *n*-alkanes, naphthalenes, benzothiophenes and dibenzothiophenes. Lighter produced oils from the Pécorade field had a normal suite of alkylbenzenes.

Locally, over a short depth interval in 1J-14 SWCs, the concomitant shift in isotopes and LCAA removal supports a biodegradation effect. The most tar-filled core or SWC extracts tend to also have the lowest values for maturity parameters, MPI-1 or estimated vitrinite reflectance equivalent Rc (Radke 1987). The observed maturity parameters have the opposite trend to the isotope data in that the most tar-rich and least mature samples are isotopically heaviest. This supports a biodegradation process as being related to tar occurrence. Lower maturity parameters also suggests that emplacement of the tar was earlier in the filling history of the field restricting continued charge of oil where tar staining was heaviest. The heaviest tar occurrence also tends to correspond to the highest iron contents (Figs 18–21).

### *Anaerobic dissimilatory ferric iron-reducing microbes*

The depletion in LCAA and deposition of tar has taken place in the downdip periphery as well as in deeper Kuparuk A sands in the central grabens of the present day accumulation (Fig. 5) suggesting that biodegradation took place when the field

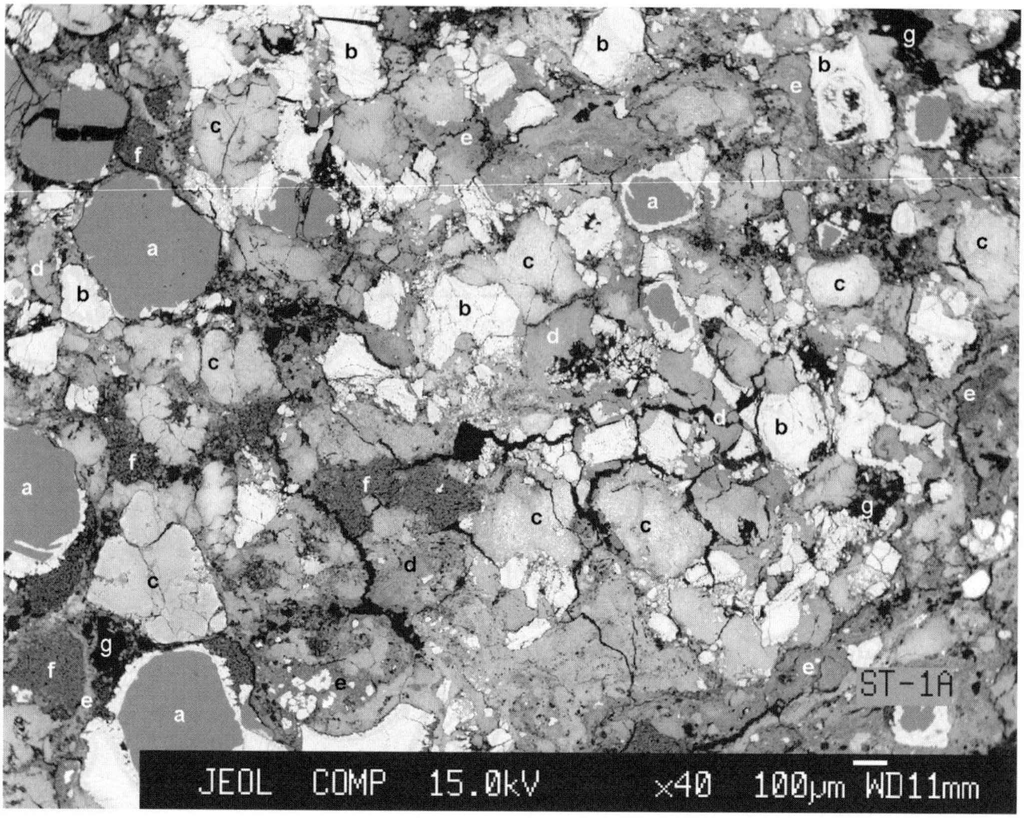


**Fig. 19.** Low magnification back-scatter electron (BSE) image of Kuparuk C sandstone from 1J-14 showing framework and diagenetic mineralogies. BSE images are sensitive to the average elemental composition of the mineral phase under examination. The lighter the grey tone on the image, the greater the atomic mass of the material being examined. Darker toned features have lower atomic masses. Scale bar is in the lower right corner is 100 μm. Grain compaction and fracturing is due to percussion acquisition of this sidewall core sample. (**a**) Quartz: dark grey framework grain, flat texture, commonly serves as nucleation site for bright angular siderite. (**b**) Siderite: bright white (Fe rich) to light grey (Mg rich) angular, diagenetic $FeCO_3$ cement. Siderite commonly exhibits oscillatory banding between the two chemical end members, which reflect changing water chemistry. (**c**) FeO/OH enriched glauconite: variable light grey, common amorphous texture. Lighter toned than normal glauconite due to greater atomic mass of iron. (**d**) Normal fresh or oxidized glauconite: medium to dark grey, common amorphous texture. (**e**) Illite clay: medium to dark grey, laminar, commonly deformed, often forms thin rims on glauconite grains. (**f**) Kaolinite clay: dark grey, occurs as small stacks or platelets or 'books'. (**g**) Tar: black amorphous, fills intergranular pore space and saturates illite and kaolinite fragments and rims.

was charged and near its present-day dimensions. Thus, there is little likelihood of an aerobic process. Recent work indicates that subsurface biodegradation of petroleum hydrocarbons is most likely anaerobic in nature (Larter *et al.* this volume).

To date, most studies of anaerobic biodegradation of hydrocarbons in the subsurface have concentrated upon individual bacteria and terminal electron acceptors (Widdel & Rabus 2001; Magot *et al.* 2000), but subsurface anaerobic biodegradation of hydrocarbons is often done by consortia of organisms (Zwolinski *et al.* 2000), e.g. archaea with sulphate-reducing bacteria (Boetius *et al.* 2000). Anaerobic microbes can biodegrade hydrocarbons using several possible electron acceptors: nitrate, sulphate or high valence metals, e.g. Fe(III) or Mn(IV) (Stetter *et al.* 1993; Rabus *et al.* 1999; Magot *et al.* 2000; Spormann & Widdel 2000; Zwolinski *et al.* 2000; Widdel & Rabus 2001). Anaerobic hydrocarbon degradation is often associated with methanogenesis (Zengler *et al.* 1999; Sweeney & Taylor 1999; Pallasser 2000;

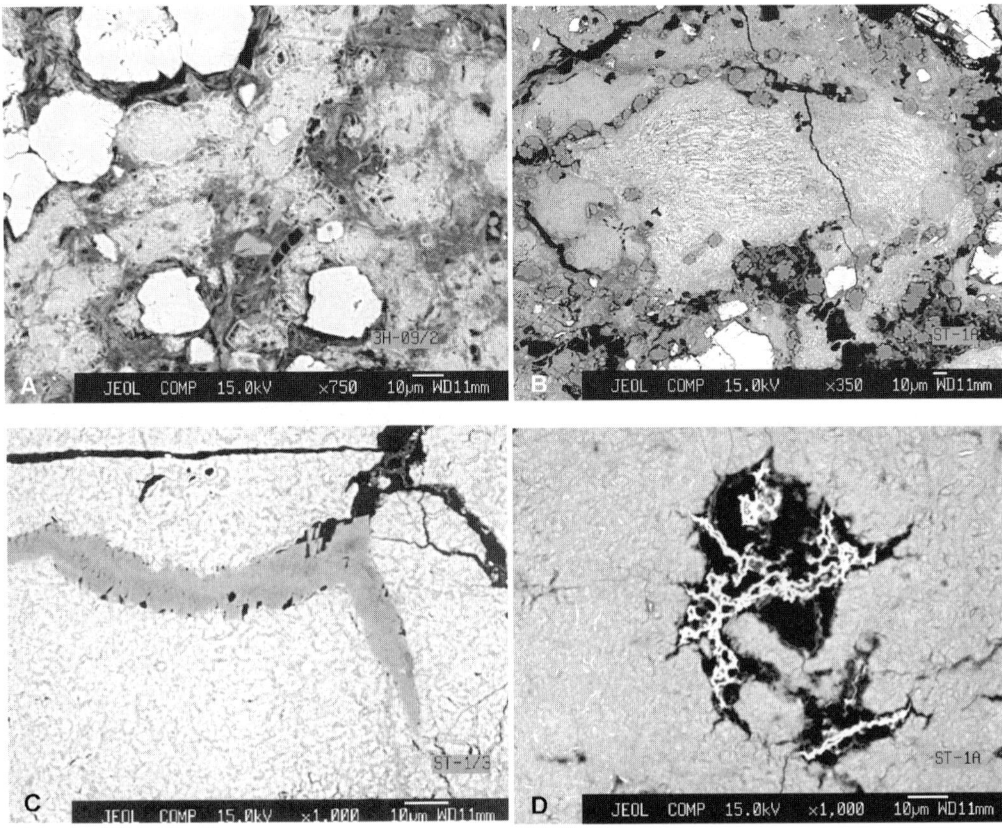


**Fig. 20.** BSE images showing different morphologies of FeO/OH in enriched glauconite and shale. **(a)** Amorphous to filament-like globules of medium-grey FeO/OH in shale clast from 3H-09. Solid bright angular features are secondary siderite crystals. **(b)** Lamellae/stringers of FeO/OH in deformed glauconite from 1J-14. Medium grey quartz rosettes and small dark grey kaolinite books surround the 'bison-like' deformed grain. **(c)** Glauconite heavily replaced by amorphous FeO/OH from 1J-14. Fracture is partially filled with zoned grey siderite. **(d)** Boxwork filaments of FeO/OH in tar-filled dissolution pores in glauconite from 1J-14. Glauconite groundmass also contains small worm-like filaments of FeO/OH.

Boreham *et al.* 2001). Sulphate-reducing bacteria have been most commonly studied because of the impact on reservoir souring and are known to degrade hydrocarbons in the laboratory (Rueter *et al.* 1994; Wilkes *et al.* 2000; Elshahed *et al.* 2001) and in subsurface petroleum reservoirs (Holba *et al.* 1996; Zengler *et al.* 1999; Widdel & Rabus 2001). However, for Kuparuk River formation waters, a replenishable source of sulphate or nitrate is not readily available. 'Fe(III) is the most abundant terminal electron acceptor in most anaerobic aquifers' (Lovley 1991; Zwolinski *et al.* 2000), thus a dissimilatory iron-reducing bacterial pathway is a candidate for anaerobic degradation of hydrocarbons in Kuparuk River field. Dissimilatory iron-reducing bacteria (DIRB) anaerobically oxidize hydrocarbons (Lovley & Longergan 1990; Lovley 1991, 1993; Anderson *et al.* 1998) converting ferric iron to magnetite, siderite or 'green rust' depending upon pH, Eh and environmental conditions (Fredrickson *et al.* 1998). Not all ferric iron in the rock or sediment is available to microbes (Lovley & Phillips 1987). The limited electron microprobe analyses indicate that the most severe depletions in LCAAs in Kuparuk and the most tar-filled zones tend to also have the greatest excess of iron (Figs 18–21). Specifically, the 1J-14 well samples had the greatest iron enrichments, supporting a link of LCAA depletion and DIRB.

Long-chain alkylbenzenes are a preferred and early target of petroleum biodegradation in the subsurface (Grimalt *et al.* 1991; Blanc & Connan 1994), and specifically for anaerobic degradation of $C_1$–$C_5$ alkylbenzenes in the

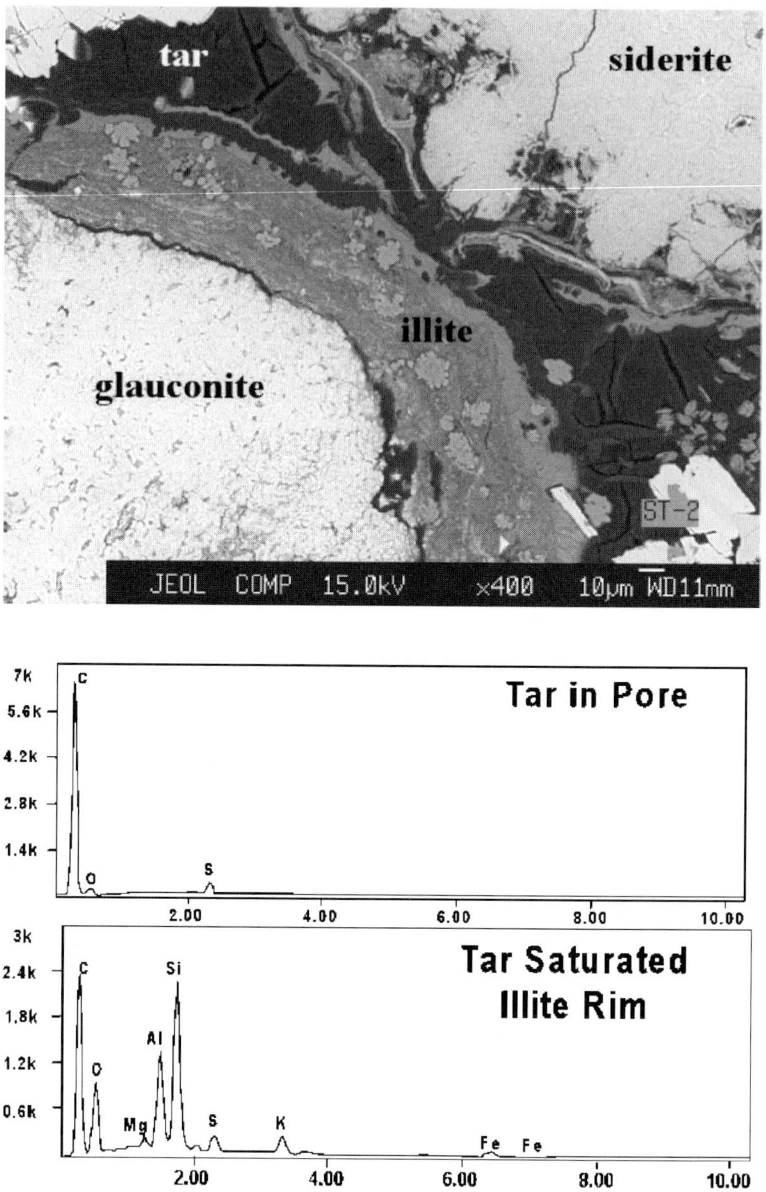


**Fig. 21.** High magnification BSE image of bright amorphous FeO/OH enriched glauconite pellets with grey illite rims. The rims contain quartz rosettes and kaolinite books, and the rims are saturated with tar. Intergranular pore space contains tar that exhibits shrinkage fractures. Light grey angular zoned siderite crystals and grey kaolinite books are also present in the intergranular pores. EDS spectra (top) shows typical signature for tar with large carbon peak and small oxygen and sulphur peaks. EDS spectra (bottom) shows the signature for an illite rim, which is saturated with tar. It contains the same carbon, oxygen and sulphur peaks, plus large silicon, aluminium peaks and smaller potassium, magnesium and iron peaks.

laboratory or shallow contaminated aquifers (Rabus *et al.* 1999; Harms *et al.* 1999). Anaerobic biodegradation may also alter the polar (resin) and asphaltene fractions, as well as the hydrocarbon fractions (Thorn & Aiken 1998). The nonhydrocarbon components of Kuparuk accumulation may also have been altered during the early stages of biodegradation, possibly affecting asphaltene solubility, but this is beyond the scope of this study. Interestingly, DIRB are known to utilize soluble humic substances in shallow terrestrial or aquatic environments as electron shuttles between electron acceptors (i.e. ferric iron) in forms that are generally not available to the bacteria (Lovley *et al.* 1996). The possibility of petroleum nonhydrocarbons as electron shuttles in the biodegradation of petroleum in the reservoir remains uninvestigated.

Many biodegraded petroleum accumulations have elevated content of isotopically light biogenic methane. Biogenic methane can be picked up by petroleum migration through immature biogenic-methane rich formations during migration to the final trap (e.g. Weissenburger & Borbas (this volume)). However, migration of petroleum from deep Shublik sources through immature source formations is not likely in this case. Certain downdip area wells (1D, 2D and 2E), and southwestern-central graben (2A–2X) wells tend to have the greatest biogenic methane contents (Table 1). The probable source of this biogenic methane is an association of methanogens in a consortium during petroleum biodegradation.

## *Thermal maturity and timing*

Examination of core extract maturities by Rc (or estimated vitrinite reflectance, calculated from MPI-1, Radke 1987) reveals that samples from the periphery of the field tend to indicate the highest maturities (e.g. 1R-07, 1D-08, 1E-14, 2K-11, 3M-16 and 3S-26; Rc=0.86-0.88 in Table 4). Wells in the middle of the field (e.g. 2X-02, 2Z-18 or 3B-14) tend to have intermediate–lower maturities suggesting that either they had not received more mature charge due to local viscous oil creating a barrier to mixing of early and late charge, or to their physical location. Heavily tar-filled samples from 1J-14, 1D-30 and 2E-17 tend to have the greatest depletions in LCAAs and the lowest maturities measured by Rc in the field (1J-14, 1D-30 and 2E-17; Rc=0.77–0.83, Table 4). As noted previously, the lowest maturities and a slight heavy shift in isotopes (Table 3) tend to correspond to the greatest depletion in LCAAs. A possible explanation of these observations is that subtle removal of LCAAs by biodegradation was related to tar precipitation, which may locally restrict further charge of petroleum. Highest tar precipitation freezes the maturity of the hydrocarbons at the time of asphaltene precipitation at some point after the degradation of the oil.

## *Asphaltene—tar precipitation*

LCAAs have a polarizable aromatic moiety and a long hydrophobic alkyl sidechain. Thus, they interact with both the paraffinic saturate hydrocarbons and the more aromatic polar parts of crude oil helping to stabilize the mixture. Likewise, the polar fraction of the oil bridges the solubility of the asphaltenes and the remainder of the oil. The removal of LCAA destabilizes the oil relative to the solubility of asphaltenes in the bulk oil. The overall extent of anaerobic biodegradation is subtle and seems insufficient to cause the elevated tar precipitation and viscous oil by itself. The final 'trigger' for precipitation of asphaltenes to form tar or viscous oil is most likely subsequent migration of gas and more mature petroleum, or a drop in reservoir pressure to around the bubble point during one of the faulting/fracturing episodes (Fig. 3).

Higher gas maturity, as indicated by carbon isotopes of ethane through butanes for SW gases (Fig. 7a), coincides with extensive tar and reduced productivity index (Fig. 6a) in Kuparuk A sands in the southwestern part of the field. Gas migrating up the deep faults bounding the central grabens could likewise have interacted with LCAA-depleted and destabilized oil in the deeper Kuparuk A of 2X and 2Z wells, promoting asphaltic precipitation (Fig. 17a). The calculated Rc representing oil range hydrocarbons for 2Z-18 shows subtly higher values in the deeper Kuparuk A than the C. The Kuparuk A in the central graben has precipitated tar corresponding to decreased productivity index (Fig. 6a).

If the tar was a residue from a very early aerobic biodegradation, early in the field's charge history when the field was shallow, then a more localized and restricted pattern of tar occurrence would be present only at the palaeo-high of the field. Instead, one observes tar across the present-day perimeter of the field. Also, aerobic degradation would likely have been more severe, destroying more than the LCAAs, and later charge would have replaced all hydrocarbons including the LCAAs. Alternatively, if an inorganic oxidation process were involved, it would require a mineralogical oxidant strong enough to remove an LCAA in a water-wet system. No evidence has been documented for such an inorganic oxidation process on the volume necessary without microbial mediation.

Dissimilatory ferric iron reducing-microbes require a continued supply of ferric iron and other nutrients to degrade a significant volume of petroleum. Thus, the original emplacement of extensive ferric iron in a form accessible to microbes during early diagenesis of the Kuparuk sands or the resupply of ferric iron from migration of formation waters is required to drive anaerobic biodegradation to a significant extent. The diagenetic emplacement of ferric iron or the later migration of brines is likely controlled to some extent by depositional setting, structure, and faulting. Structural and local mineralogical elements likely control location and severity of biodegradation as well as subsequent asphaltene, tar, deposition. For example, the middle SWC from 1J-14 has the most extensive removal of LCAAs (Fig. 14d), and is severely plugged with tar to the point that no later recharge of petroleum contributed more LCAAs and froze the maturity indicating ratios in the phenanthrenes (MPI) at a lower maturity.

## Conclusions

(1) Compartmentalization studies suggest that several mega-compartments existed at the time of field discovery, e.g. NW and SW parts of the field.
(2) Long chain alkyl aromatic (LCAA) hydrocarbons have been selectively depleted. Core and sidewall core samples from the most severely tar plugged sands (2E-17 and 1J-14) had the most extensive removal of LCAAs. All analysed incidences of tar or viscous oil corresponded to depleted LCAA.
(3) Selective removal of LCAAs is most likely from an anaerobic biodegradation process—possibly from dissimilatory ferric iron-reducing bacteria (DIRB) or a consortium involving methanogens. Samples with greatest depletion of LCAA and most pervasive tar tend to have the highest available iron-rich mineralogies.
(4) Increased tar or the incidence of viscous oil corresponds (**a**) to increased non-hydrocarbon content of petroleum (lower hydrocarbon/nonhydrocarbon ratio), (**b**) increased high molecular weight organic matter observed in Rock-Eval 6 pyrograms and (**c**) an increased pore-filling petroleum density (as evidenced by lower predicted API gravity from Rock-Eval 6.
(5) The actual precipitation of asphaltic tars may be related directly to anaerobic biodegradation processes, but it is more likely to result from a subsequent process such as gas or more mature petroleum charge or from pressure reduction through faults during tectonic events after destabilization of the petroleum by anaerobic biodegradation.
(6) Recognition of the anaerobic biodegradation process and its drivers is the first step in development of predictive models to understand tar distribution and location and has important consequences on the economics of field development.

The authors thank Susan Reynolds, David Shafer, Russ Bone, Dennis Wegener, James Hickey, Ben Powell, Rick Tharp, Susan Singletary Gordon and Leslie Deal for their assistance in data acquisition or interpretation. The authors especially thank Paul Walker for assistance with Rock-Eval 6 calibration and analysis. The manuscript was improved by reviews by S. Larter and K. Weissenburger.

## Appendix. LCAA ratio definitions.

| LCAA ratio | Numerator | Denominator |
|---|---|---|
| LCAAR-1 | $\sum C_{19}$–$C_{22}$ alkyl benzenes ($m/z = 92$) | Methyldibenzothiophene ($m/z = 92$) + $C_{19}$–$C_{22}$ AB ($m/z = 92$) |
| LCAAR-2 | $\sum C_{19}$–$C_{26}$ alkyl benzenes ($m/z = 92$) | Phenanthrene ($m/z = 178$) + $\sum$ Methylphenanthrenes ($m/z = 192$) + $C_{19}$–$C_{22}$ AB ($m/z = 92$) |
| LCAAR-3 | $\sum C_{19}$–$C_{26}$ alkyl toluenes ($m/z = 106$) | DBT ($m/z = 184$) + $\sum$ Methyl-DBTs ($m/z = 198$) + $C_{19}$–$C_{26}$ AT ($m/z = 106$) |

## References

ANDERSON, R. T., ROONEY-VARGA, J. N., GAW, C. V. & LOVLEY, D. E. 1998. Anaerobic benzene oxidation in the Fe(III) reduction zone of petroleum-contaminated aquifers. *Environmental Science & Technology*, **32**, 1222–1229.

BAILEY, N. J. L., KROUSE, H. R., EVANS, C. R. & ROGERS, M. A. 1973. Alteration of crude oil by waters and bacteria—evidence from geochemical and isotope studies. *American Association of Petroleum Geologists Bulletin*, **57**, 1276–1290.

BASTIN, E. S. 1926. The problem of the natural reduction of sulphates. *American Association of Petroleum Geologists Bulletin*, **10**, 1270–1299.

BENTOR, Y. K. & KASTNER, M. 1965. Notes on the mineralogy and origin of glauconite. *Journal of Sedimentary Petrology*, **35**(1), 155–166.

BLANC, P. & CONNAN, J. 1994. Preservation, degradation, and destruction of trapped oil. *In*: MAGOON, L. B. & DOW, W. G. (eds) *The Petroleum System—From Source to Trap*. American Association of Petroleum Geologists Memoir, **60**, 237–247.

BOETIUS, A. *ET AL*. 2000. A marine microbial consortium apparently mediating anaerobic oxidation of methane. *Nature*, **407**, 623–626.

BOREHAM, C. J., HOPE, J. M. & HARTUNG-KAGI, B. 2001. Understanding source, distribution, and preservation of Australian natural gas: A geochemical perspective. *APEA Journal*, 523–547.

CHUNG, H. M., GORMLY, J. R. & SQUIRES, R. M. 1988. Origin of gaseous hydrocarbons in subsurface environments: theoretical considerations of carbon isotope distribution. *Chemical Geology*, **71**, 97–103.

CONNAN, J. 1984. Biodegradation of crude oils in reservoirs. *In*: BROOKS, J. & WELTE, D. (eds) *Advances in Petroleum Geochemistry*, volume 1. Academic Press, London, 299–335.

CONNAN, J., DESSORT, D. & MAGOT, M. 1994. Advances in our knowledge of alteration of hydrocarbons in reservoirs. Conference abstract, p 11.

DOW, W. G., ALLEN, J. R. & KUHNEL, C. J. 2000. Determination of API gravity from very small samples of oils, tar mats, and solid bitumens with the Rock-Eval 6 instrument. *In*: TRINDADE, L. A. F., MACEDO, A. C. & BARBANTI, S. M. (eds) *New Perspectives on Organic Geochemistry for the Third Millennium*. Proceedings, 7th Latin American Congress on Organic Geochemistry. 22–26 October 2000, Foz do Iguacu, Brazil, ALAGO, 232–236.

ELSHAHED, M. S., GIEG, L. M., MCINERNEY, M. J. & SUFLITA, J. M. 2001. Signature metabolites attesting to the in situ attenuation of alkylbenzenes in anaerobic environments. *Environmental Science & Technology*, **35**, 682–689.

FAIRBRIDGE, R. W. & BOURGEOIS, J. 1978. Glauconite. *In*: *The Encyclopedia of Sedimentology*. Dowden, Hutchinson & Ross Inc., Stroudsburg, PA, 364–366.

FREDRICKSON, J. K., ZACHARA, J. M., KENNEDY, D. W., DONG, H., ONSTOTT, T. C., HINMAN, N. W. & LI, S. 1998. Biogenic iron mineralization accompanying the dissimilatory reduction of hydrous ferric oxide by a groundwater bacterium. *Geochimica et Cosmochimica Acta*, **62**(19/20), 3239–3257.

GRIMALT, J. O., GRIFOLL, M., SOLANAS, A. M. & ALBAIGÉS, J. 1991. Microbial degradation of marine evaporitic crude oils. *Geochimica et Cosmochimica Acta*, **55**, 1903–1913.

HARMS, G., ZENGLER, K., RABUS, R., AECKERSBERG, F. A., MINZ, D., ROSSELLO-MORA, R. & WIDDEL, F. 1999. Anaerobic oxidation of *o*-xylene, *m*-xylene, and homologous alkylbenzenes by new types of sulphate-reducing bacteria. *Applied & Environmental Microbiology*, **65**(3), 999–1004.

HARWOOD, R. J. 1973. Biodegradation of oil. *In*: *The Geology of Fluids and Organic Matter in Sediments*. National Conference on Earth Science, 30 April–4 May 1973, Banff, Alberta, Canada. Edmonton: University of Alberta, 149–156.

HOLBA, A. G., DZOU, L. I. P., HICKEY, J. J., FRANKS, S. G., MAY, S. J. & LENNY, T. 1996. Reservoir geochemistry of South Pass 61 Field, Gulf of Mexico: Compositional heterogeneities reflecting filling history and biodegradation. *Organic Geochemistry*, **24**, 1179–1198.

HUNT, J. M. 1953. Composition of Crude oil and its relation to stratigraphy in Wyoming. *American Association Petroleum Geologists Bulletin*, **37**, 1837–1872.

KAZALOV, G. A. 1982. Glauconites as geochemical indicators of facies conditions of sediment genesis. *Geochemistry International*, **19**, 1–14.

KRANTZ, R. & LONGDEN, M. R. 2002. Fault characterization and reservoir development at Kuparuk River Field, part 2: Understanding fault zones and predicting fault seal. *American Association of Petroleum Geologists Pacific Section Meeting*, Anchorage, Alaska, 85.

LAFARGUE, E., ESPITALIÉ, J., MARQUIS, F. & PILLOT, D. 1997. Rock-Eval 6 applications in hydrocarbon exploration, production and in soil contamination. *Revue de l'Institut Français du Pétrole*, **53**(4), 421–437.

LARTER, S. *ET AL*. 2003. The controls on the composition of biodegraded oils in the deep subsurface-part 1: biodegradation rates in petroleum reservoirs. *Organic Geochemistry*, **34**, 601–613.

LOVLEY, D. R. 1991. Dissimilatory Fe(III) and Mn(IV) reduction. *Microbiology Review*, **55**, 259–287.

LOVLEY, D. R. 1993, Dissimilatory metal reduction. *Annual Review of Microbiology*, **47**, 263–290.

LOVLEY, D. R. & LONGERGAN, D. J. 1990. Anaerobic oxidation of toluene, phenol, and *p*-cresol by the dissimilatory iron-reducing organism, GS-15. *Applied & Environmental Microbiology*, **56**(6), 1858–1864.

LOVLEY, D. R. & PHILLIPS, E. J. P. 1987. Rapid assay for microbially reducible ferric iron in aquatic sediments. *Applied & Environmental Microbiology*, **53**(7), 1536–1540.

LOVLEY, D. R., COATES, J. D., BLUNT-HARRIS, E. L., PHILLIPS, E. J. P. & WOODWARD, J. C. 1996. Humic substances as electron acceptors for microbial respiration. *Nature*, **382**, 445–448.

MAGOT, M., OLLIVIER, B. & PATEL, B. K. C. 2000. Microbiology of petroleum reservoirs. *Antonie van Leeuwenhoek*, **77**, 103–116.

MANGO, F. 1994. The origin of light hydrocarbons in petroleum: Ring preference in the closure of carbocyclic rings. *Geochimica et Cosmochimica Acta*, **58**(2), 895–901.

MASTERSON, W. D. 2001. Petroleum filling history of central Alaskan North Slope Fields. PhD Dissertation, University of Texas at Dallas.

MASTERSON, W. D. & EGGERT, J. T. 1992. Kuparuk River Field–North Slope, Alaska. *In*: FOSTER, N. H. & BEAUMONT, E. A. (eds) *Stratigraphic Traps III*. AAPG Treatise of Petroleum Geology, Atlas of Oil and Gas Fields, 257–284, Tulsa, Oklahoma.

MASTERSON, W. D. & PARIS, C. E. 1987. Depositional history and reservoir description of the Kuparuk River Formation, North Slope, Alaska. *In*: TAILLEUR, I. & WEIMER, P. (eds) *Alaskan North Slope Geology*. Pacific Section SEPM & The Alaska Geological Society, Bakersfield & Anchorage, 95–107.

MASTERSON, W. D., DZOU, L. I. P., HOLBA, A. G., FINCANNON, A. L. & ELLIS, L. 2001. Evidence for biodegradation and evaporative fractionation in West Sak, Kuparuk and Prudhoe Bay field areas, North Slope, Alaska. *Organic Geochemistry*, **32**, 411–441.

MILNER, C. W. D., ROGERS, M. A. & EVANS, C. R. 1977. Petroleum transformations in reservoirs. *Journal Geochemical Exploration*, **7**, 101–153.

MORRISON, R. T. & BOYD, R. N. 1966. *Organic Chemistry*. Allyn and Bacon, Boston, 382–383.

PALLASSER, R. J. 2000. Recognising biodegradation in gas/oil accumulations through the $\delta^{13}C$ compositions of gas components. *Organic Geochemistry*, **31**, 1363–1373.

PALMER, S. E. 1993. Effect of biodegradation and water washing on crude oil composition. *In*: ENGEL M. H. & MACKO S. A. (eds) *Organic Geochemistry Principles and Applications*. Plenum Press, New York, 511–533.

RABUS, R., WILKES, H., SCHRAMM, A., HARMS, G., BEHRENDS, A., AMANN, R. & WIDDEL, F. 1999. Anaerobic utilization of alkylbenzenes and *n*-alkanes from crude oil in an enrichment culture of denitrifying bacteria affiliating with the β-subclass of *Proteobacteria*. *Environmental Microbiology*, **1**(2), 145–157.

RADKE, M. 1987. Organic geochemistry of aromatic hydrocarbons. *In*: BROOKS, J. & WELTE, D. (eds) *Advances in Petroleum Geochemistry*, Volume 2. Academic Press, London, 141–207.

RUETER, P., RABUS, R., WILKES, H., AECKERSBERG, F., RAINEY, F. A., JANNASCH, H. W. & WIDDEL, F. 1994. Anaerobic oxidation of hydrocarbons in crude oil by new types of sulphate-reducing bacteria. *Nature*, **372**, 455–458.

SOFER, Z. 1984. Stable carbon isotope compositions of crude oils: Application to source depositional environments and petroleum alteration. *American Association of Petroleum Geologists Bulletin*, **68**(1), 31–49.

SPORMANN, A. M. & WIDDEL, F. 2000. Metabolism of alklybenzenes, alkanes, and other hydrocarbons in anaerobic bacteria. *Biodegradation*, **11**, 85–105.

STETTER, K. O. *ET AL.* 1993. Hyperthermophilic archaea are thriving in deep North Sea and Alaskan oil reservoirs. *Nature*, **365**, 743–745.

SWEENEY, R. E. & TAYLOR, P. 1999. Biogenic methane derived from the biodegradation of petroleum under environmental conditions and in oil and gas reservoirs. *AAPG Hedberg Research Conference, Natural Gas Formation and Occurrence*. 6–10 June 1999, Durango, Colorado.

THOMPSON, K. F. M. 1987. Fractionated aromatic petroleums and the generation of gas-condensates. *Organic Geochemistry*, **11**, 573–590.

THORN, K. A. & AIKEN, G. R. 1998. Biodegradation of crude oil into nonvolatile organic acids in a contaminated aquifer near Bemidji, Minnesota. *Organic Geochemistry*, **29**(4), 909–931.

TRABELSI, K., ESPITALIÉ, J. & HUC, A. Y. 1994. Characterisation of extra heavy oils and tar deposits by modified pyrolysis. *In: European Symposium on Heavy Oil Technologies in a Wider Europe, Proceedings*. 7 and 8 June 1994, 30–40.

WEISSENBURGER, K. & BORBAS, T. 2004. Fluid properties, phase and compartmentalization predictions using geochemical data—Magnolia Field case study, deepwater, Gulf of Mexico, USA. *In*: CUBITT, J. M., ENGLAND, W. A. & LARTER, S. (eds) 2004. *Understanding Petroleum Reservoirs: towards an Integrated Reservoir Engineering and Geochemical Approach*. Geological Society, London, Special Publications, **237**, 231–256.

WIDDEL, F. & RABUS, R. 2001. Anaerobic biodegradation of saturated and aromatic hydrocarbons. *Biotechnology*, **12**, 259–276.

WILKES, H., BOREHAM, C., HARMS, G., ZENGLER, K. & RABUS, R. 2000. Anaerobic degradation and carbon isotope fractionation of alkylbenzenes in crude oil by sulphate-reducing bacteria. *Organic Geochemistry*, **31**, 101–115.

WINTERS, J. C. & WILLIAMS, J. A. 1969. Microbiological alteration of crude oil in the reservoir. *In*: *Proceedings of Symposium on Petroleum Transformation in Geologic Environments*. American Chemical Society, Division of Petroleum Chemistry, 7–12 September, New York City, Paper TETR 86: E22–31.

ZENGLER, K., RICHNOW, H. H., ROSSELLO-MORA, R., MICHAELLS, W. & WIDDEL, F. 1999. Methane formation from long-chain alkanes by anaerobic microorganisms. *Nature*, **401**, 266–269.

ZOBELL, C. E. 1945. The role of bacteria in the formation and transformation of petroleum hydrocarbons. *Science*, **102**, 364–369.

ZOBELL, C. E. 1973. Microbial degradation of oil: present status, problems, and perspectives. *In*: AHEARN, D. G. & MEYERS, S. P. (eds) *Microbial degradation of oil Pollutants Workshop*. Center for Wetland Resources, Louisiana State University Baton Rouge, LA, 3–16.

ZWOLINSKI, M. D., HARRIS, R. F. & HICKEY, W. J. 2000. Microbial consortia involved in the anaerobic degradation of hydrocarbons. *Biodegradation*, **11**, 141–158.

# HCToolkit/EOS interface: an open source, multi-platform phase equilibria framework for exploring phase behaviour of complex mixtures


PETER MEULBROEK,[1] LARRY CATHLES[2] & WILLIAM A. GODDARD[3]

[1] *Mc 139-74, Beckman Institute, California Institute of Technology, Pasadena, CA 91125, USA (e-mail: meulbroek@wag.caltech.edu)*

[2] *Department of Geological Sciences, Cornell University, Ithaca, NY 14853, USA (e-mail: cathles@geology.cornell.edu)*

[3] *Mc 139-74, Beckman Institute, California Institute of Technology, Pasadena, CA 91125, USA (e-mail: wag@wag.caltech.edu)*



**Abstract:** We present a computational framework that can be used to estimate phase equilibria, equation of state properties and composition-dependent viscosity, aimed at the geochemical community for modelling reservoir processes, and at the chemical community to quickly estimate continuum properties of known mixtures. The framework presented is an extensible, component-based set of modules that can be used in calculating phase properties (volumes and densities, compositions and conditions of phase separation, Henry's Law constants and viscosities) at a wide range of pressures, temperatures and starting compositions. There are two pieces to the model: the HCToolkit (a set of Perl modules that act as computational engines) and the EOSInterface (an ActiveX wrapper for the HCToolkit) which allow the models in the HCToolkit to be usable in Microsoft Office programs. An application hosted in Microsoft Excel is included within the distribution.

The EOS models are adaptable for mixtures of arbitrary complexity, with number and types of components only limited by hardware. Model runs with mixtures of over 100 components have been tested, and are perfectly feasible. Implemented within the software are four equations of state; additional equations of state can easily be added. Also implemented are $G^E$ models and mixture viscosities. Finally, the EOS models provide liquid–vapour flash calculations, liquid–liquid flash (designed for petroleum–aqueous solutions) and the generation of phase diagrams.


Fluid-phase equilibria/equation of state (EOS) models are powerful tools used to predict phase properties of mixtures where gathering experimental data is difficult, expensive or impractical. These models have seen widespread application in both chemical engineering and petroleum engineering. Within the petroleum engineering community, EOS models are ubiquitously used to predict the phase behaviour and PVT properties of reservoir petroleum fluids during production. They are primarily used in reservoir simulation, testing the efficacy of different exploitation strategies. Increasingly, rapid phase equilibria calculations are becoming a useful tool of the reservoir geochemist for such tasks as production allocation, reservoir continuity (Meulbroek 2002), common source evaluation and determining fluid history (di Primio 2002; Losh *et al.* 2002). These calculations typically rely on a cubic equation of state, such as Soave–Redlich–Kwong (SRK) driving a fluid-phase equilibria model. Though current commercial software implementations of these models exist, no general, flexible model has been released to the community. We present here the HCToolkit, a flexible, extensible open source framework for performing general FPE calculations. Implemented within the software are four equations of state: Van der Waals, Mathias (Mathias 1983), Aasberg-Petersen (Aasberg-Petersen & Stenby 1991) and the PSRK (Holderbaum & Gmehling 1991). Additional equations of state can easily be added, as detailed in this work. The EOS models are adaptable for mixtures of arbitrary complexity, with number and types of components only limited by hardware. Model runs with mixtures of over 100 components are routinely executed. Also implemented are the $G^E$ models UNIFAC (Fredenslund *et al.* 1975) and modified UNIFAC (Gmehling *et al.* 1993) models. Viscosities are calculated using the Chung *et al.* method (Chung *et al.* 1988). Finally, the EOS models provide liquid–vapour flash calculations, liquid–liquid flash (designed for petroleum–aqueous solutions), and the generation of phase diagrams (both *P-T* and *P-X*).

As an extension to the HCToolkit, we also release the EOSInterface, an MS Excel-hosted extension of the HCToolkit that allows the user access to all calculation abilities from within the Excel spreadsheet.


*From*: CUBITT, J. M., ENGLAND, W. A. & LARTER, S. (eds) 2004. *Understanding Petroleum Reservoirs: towards an Integrated Reservoir Engineering and Geochemical Approach.* Geological Society, London, Special Publications, **237**, 89–98. 0305-8719/04/$15.00

## Problem area

The design principles that resulted in the EOS-Interface are aimed at developing a flexible framework for continuum fluid calculations for the problem solver and developer. The goal was to develop a framework that could be used for any phase equilibrium problem including sub-surface and industrial processes. To provide this flexibility, the framework is equation-agnostic; that is, any calculation that the framework performs can be done with any of the embedded equations of state, and additional equation of state models can be easily added to the code base. Early on, we realized that the rapid development environment provided by a scripting language would speed code production and provide built-in extensibility. We chose Perl—the most popular multi-platform scripting language available—after considering and rejecting Visual Basic (limited to the Microsoft Windows platforms), Python (much smaller developer base) and Java (higher development overhead). Perl is an open source dynamic scripting language with good debugging and profiling capabilities, and an extensive library of freely-available extensions (the CPAN library). If execution speed becomes a bottleneck in any application, developers can also transparently link compiled code within a Perl code.

In order for the code to be applicable to mixtures consisting of a wide and expandable number of compounds, we decided to store all data in a relational database and to provide a variety of access methods to that database. In fact, any database that can be accessed by Perl-DBI or ODBC methods (which include all major commercial and open source products) can be used to store compound data. Since many of these databases are network-accessible, the HCToolkit framework can be driven by a common institution-wide (or multi-institution) database to ensure consistency of properties and mixtures in a calculation. We also store mixture information in the database to ease collaboration in studying specific compositions. Hence, all researchers in a specific institution can connect to the same database, and all can use consistent fluid compositions (e.g. fluid '72') in numeric simulations.

## Open source

The framework described here is open source (released under the GNU-GPL licence, http://www.gnu.org/licenses/gpl.html), a licensing method that has received a huge amount of publicity. Essentially, 'open source' means that the framework is released with source code, as a means of encouraging users to become active development partners (http://www.opensource.org/). All fair uses of the code including further distribution are allowed; however, any changes that a developer makes to the code must also be made available, in the spirit of collaboration. This is remarkably different from existing commercial licences, which forbid distribution and form a barrier to shared development.

In contrast to conventional wisdom, open source codes offer considerable advantages to software developers. If the goal is rapid framework improvement and extension, then an open source product advances much more quickly than a commercial framework, as all users of the framework are potentially developers. Open source codes also provide a guarantee to the user that the framework will not 'disappear'. Since the source code of a framework is available, if a development effort stalls at one location, another location can restart the development at a later date. Users can also customize the framework to their own needs and share those customizations.

## Abilities of the framework

The HCToolkit provides implementations for four equations of state—the original Van der Waals model; the Mathias (Mathias 1983) model, which improves predictions for polar mixtures; the Aasberg-Petersen model (Aasberg-Petersen & Stenby 1991) which improves predictions around the critical point; and the predictive RKS model (Holderbaum & Gmehling 1991), which utilizes the UNIFAC model to eliminate the need for empirical fitting parameters and thus allows predictions of phase behaviour without additional PVT (pressure, temperature, composition) data. These models can each be used to predict molar volume, density and fugacity for mixtures at arbitrary PVT conditions. Also included are several excess Gibbs free energy models, which predict the Gibbs free energy of mixing and activity coefficients for arbitrary mixtures. The two models implemented are the original UNIFAC model (Fredenslund *et al.* 1975) and the modified UNIFAC model (Gmehling *et al.* 1993).

The HCToolkit also presents several algorithms for fluid-phase equilibria calculations. They include the Rijkers & Heidemann (1986) model, an unpublished liquid–liquid algorithm and the Bunz *et al.* (1991) three-phase model. The HCToolkit also includes the Chung *et al.*

model for predicting viscosity of both liquid and vapour mixtures.

*Architecture*

The HCToolkit is a set of Perl modules that implement four equations of state, two flash algorithms and a multi-component, multi-phase, temperature- and pressure-dependent viscosity prediction. The modules have been successfully run on MS Windows, Mac OSX and Redhat Linux. These modules can be called from another Perl code, or (via an ActiveX interface) from a front-end written in either Visual Basic or a VBA application such as Excel.

In open source software, every user is a potential developer. Hence, technical knowledge of program architecture becomes relevant to the individual user. The HCToolkit design utilizes 'plugin' architecture; wherever possible, families of related modules are written to instance abstract interfaces. The interfaces are: Phase::EquationofState, Phase::Database, Phase::Flash and Phase::Unifac, as shown in Figure 1.

An example of the application of abstract interfaces can be found in the Flash module. Figure 2 shows the class layout for the Flash 'engine', an interface to a flash calculation. The Flash engine takes two defined 'phases' (equation of state objects) plus a database that provides properties for the components of the phase,

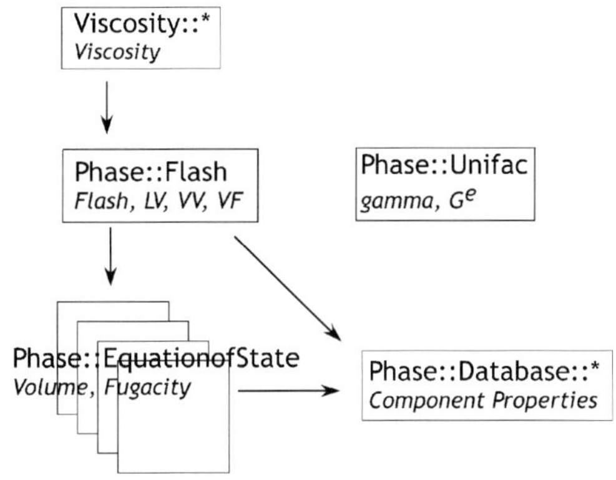


**Fig. 1.** Layout of the class structures in the HCToolkit.

and calculates the equilibrium distribution of components among the phases. The equation of state used to calculate properties for each phase is specified at run-time via a 'Phase::Equationof-State' object, and can even be different for different phases. All the EOS modules present the same interface to other modules by using abstract interfaces, allowing easy substitution of models for quick comparisons of calculation results. The Flash engine only expects the objects passed to it to conform to a generic interface (the 'Phase::EquationofState::Generic' interface, to be precise); any EOS object that is derived from that interface can be used in the flash calculation.

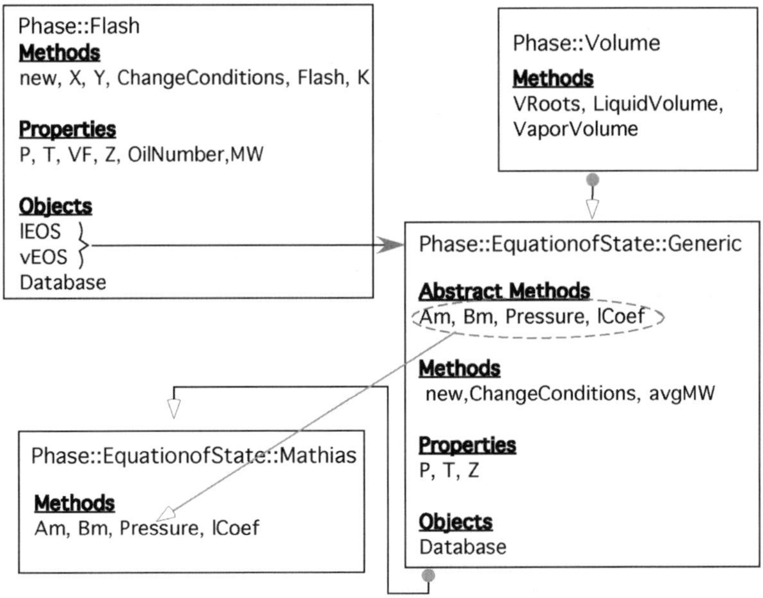


**Fig. 2.** Phase::Flash class layout.

Adding new equations of state to the framework is an easy and straightforward process. Two things tend to define most modern cubic equations of state: the mixing rules used to estimate incompressible volume ($b$) and attraction parameter ($a$), and the form of the attractive (second) term of the equation (see equation (1)). In order to add an additional EOS module, the user needs to supply an object-oriented Perl module that

(1) inherits from Phase::EquationofState::Generic;
(2) provides three functions: Am (the mixing rule for the attractive term coefficient from an equation of state), Bm (the mixing rule for the repulsive term coefficient from an equation of state) and Pressure (the pressure-explicit format for an equation of state);
(3) implements some functionality to return volume as a function of input variables composition, pressure and temperature (which is dependent on the form of the EOS).

For cubic equations of state, a pre-defined methodology allows easy implementation of (3), above. Most cubic equations of state are of the form

$$P = \frac{RT}{v - b} - \frac{a}{v^2 + uvb + wb^2} \qquad (1)$$

where $P$ is pressure, $R$ is the universal gas constant, $v$ is molar volume, *lb* is the incompressible volume, $a$ is the Van der Waals attractive parameter, and *lu* and *lw* are constants that define the equation of state (Poling 2000). For example, for the Van der Waals equation of state, $u = w = 0$, for the RKS equation of state, $u = 1$, $w = 0$. Equation (1) can be inverted to produce $v$ as a dependant variable. An analytical solution to the general relationship $v = f(P, T, a, b, u, w)$ that returns one or three real roots is possible, and implemented within the HCToolkit framework in the module "Phase::-Volume". In order for a new cubic equation of state module to predict molar volume using these functions, it merely needs to implement a function called 'lCoef' in the new EOS module that returns a list of the appropriate $u$, $w$, $P$, $T$, $a_{\mathrm{mix}}$ and $b_{\mathrm{mix}}$ values. This function is called by "Phase::Volume" in its general solution of the relationship $f$, above.

## *The EOSInterface Worksheet*

EOSInterface, an ActiveX/Excel interface for the HCToolkit, has a similar design philosophy to that of the HCToolkit. The EOSInterface Worksheet is a Microsoft Excel spreadsheet designed to interface with and form a front-end to the HCToolkit. The HCToolkit is completely functional without the EOSInterface: the EOSInterface sits on top of the HCToolkit, and exports most of the HCToolkit functionality in a more user-friendly environment. It is designed for the practical thermodynamicist who is content with simple flash calculations, and who is more comfortable with using an Excel spreadsheet than writing code in Perl. The EOSInterface interacts with the HCToolkit via an ActiveX gateway, compiled using ActiveState's PerlCtrl program. Since it is built on top of the HCToolkit, the EOSInterface provides the ability to do flash calculations using the four equations of state mentioned above. Figure 3 shows the object layout of the EOSInterface. Note that command-line scripting (via VBS) can also call the ActiveX object. This object shares with the HCToolkit the ability to interface with a large (>2500 species) database of compounds, including hydrocarbons, fluorocarbons and non-hydrocarbon species of interest to the geochemical community

## *PSRK equation of state*

Of particular interest to the users of the HCToolkit and EOSInterface is the implementation of the PSRK EOS (Holderbaum & Gmehling 1991). In traditional equations of state that are used in petroleum engineering, models take as input pure species critical properties and empirical binary adjustment parameters (the '*kij*' parameters) that adjust for non-ideal interactions between species. These *kij* parameters are typically derived from PVT experiments, and have both temperature and composition dependence, which make them impractical for complex systems or systems with unusual components. However, the PSRK EOS uses a new mixing rule that depends on a group-additivity $G^E$ model (UNIFAC) to estimate non-idealities, instead of using system/component-specific empirical data. UNIFAC considers molecules to be sets of interacting functional groups, the interactions of which are well known from a large number of studies (Fredenslund *et al.* 1975). To calibrate the PSRK EOS for any system only requires knowledge of molecular structure of all species to functional group resolution. Hence, this can be applied to complex systems with species whose interactions are not known. In the context of reservoir geochemistry, this means that, e.g. large biomarkers can be modelled as easily as well-known species such

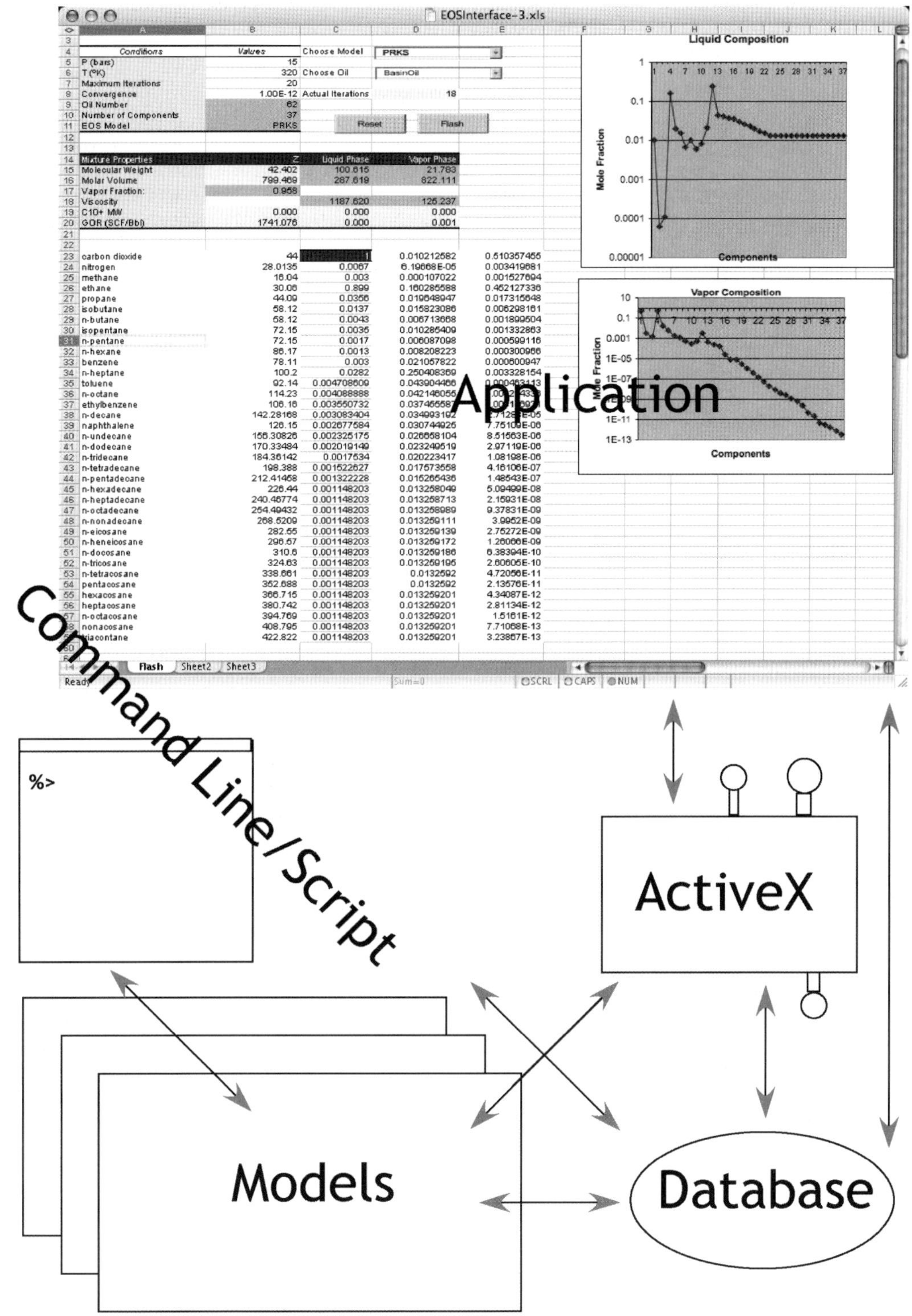


**Fig. 3.** Layout of the parts of the EOSInterface.

```
# initialize the database
$db = Phase::Database::PostgreSQL->new(
      UserName => 'guest',
      Password => 'guest',
);

# load a fluid
$db->LoadOilByNumber(75);

# initialize the flash obj.
my $flash = Phase::Flash->new(
      P => $P, T => $T,
      Z => Standard::List->new($x, 1-$x),
      Database=>$db,
      lEOS => Phase::EquationofState::PSRK->new( ),
      vEOS => Phase::EquationofState::PSRK->new( ),
);

$flash->ChangeConditions(P => $p);

# do the flash calculations
$flash->Flash(MaxIterations => 20, Precision => 1e-9);
```

**Fig. 4.** Example HCToolkit code.

as *n*-alkanes. The model is also adaptable to pseudo-components, if the composition of the pseudo-component is known to the level of functional groups (i.e., fraction methyl groups, fraction aromatic carbon etc.).

### *The Perl scripting front-end*

One of the most powerful aspects of the HCToolkit is its extensibility in Perl. This allows rapid development of new applications for specific tasks. For example, an included extension quickly calculates and plots a pressure–temperature phase diagram at user-specified conditions (PT-PhaseDiagram.pl, found in the /apps subdirectory of the distribution tree). In fact, a scripting application that does some sort of equilibrium calculation needs only do four things: create a database connection, use that connection to load a pre-defined fluid, initialize the flash object and do the flash calculation. These steps are shown in Figure 4.

## Calculation examples

The implementation of the EOS components allows very tight control over the accuracy of calculations. As proof of concept, two examples of calculations are included. The first is a prediction of vapour–liquid equilibrium in a binary $CO_2$–cyclopentane system. This system is somewhat non-ideal, due to the quadrupole movement of $CO_2$. The predicted phase envelope is shown in Figure 5.

Figure 5 shows a (*P*, *X*) phase diagram for $CO_2$–cyclopentane at a constant temperature. The calculation shows predicted isofractions compared with experimental data (Keshtkar *et al.* 1997). Overall fits are very good. The actual algorithm chosen to predict the phase envelope is actually quite naïve; a grid-search calculation was performed over (composition, *P*) space using the example code 'apps/PX-PhaseDiagram.pl'. Flash calculations were performed on each grid node of a $50 \times 50$ node grid, and the location of the phase envelope was interpolated. Though much more sophisticated methods are available (various bubble/dew point methods and phase envelope 'walking' methods) from the literature, the point of showing the calculations is that a quick-and-dirty solution for questions of interest that produces very good answers can be coded within the HCToolkit, and calculated fairly rapidly (under 5 min on an Apple Powerbook G4 laptop). Furthermore, these more sophisticated methods can also be implemented within the HCToolkit, if desired.

The second example shows the solubility of methane in water. In general, cubic equations of state do quite poorly with aqueous solutions, since most are tuned to non-polar compositions such as hydrocarbon mixtures. The HCToolkit allows one to quickly compare the abilities of

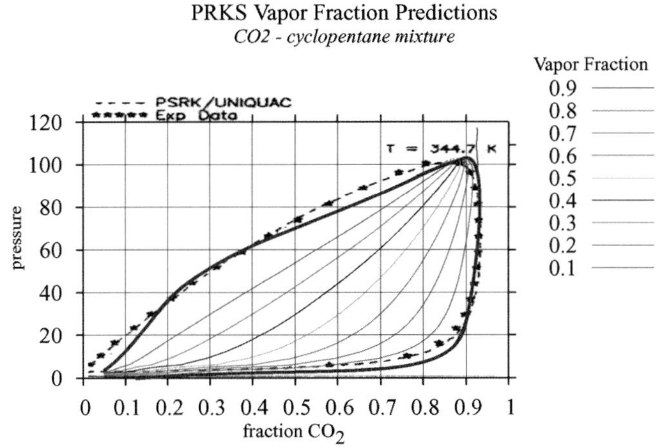


**Fig. 5.** Vapour–liquid equilibrium predictions using the HCToolkit. This figure shows the experimental data from (Keshtkar *et al.* 1997), an accurate binary calculation (UNIQUAC), the phase envelope (in bold) and vapour phase isofractions (shaded).

different equations of state. Figure 6 shows the prediction of methane solubility in water for two cubic equations of state—the Aasberg-Petersen & Stenby (1991) and the Mathias (1983)—compared with experimental data. The experimental data are taken from Li *et al.* (1997). At all temperatures, the Mathias EOS does a better job of predicting the solubility of methane in water (i.e. the amount of methane, by mole fraction, that can be dissolved in water as a function of pressure and temperature). This is not surprising, as the APS EOS is tuned for hydrocarbons and the Mathias EOS has a polar correction term for water mixtures. Unfortunately, the Mathias EOS only works well for the lowest temperature studied and works increasingly badly for higher temperatures and higher concentrations of $CH_4$.

## Field example

Unlike a monolithic code, the HCToolkit is designed to ease the development of tools to

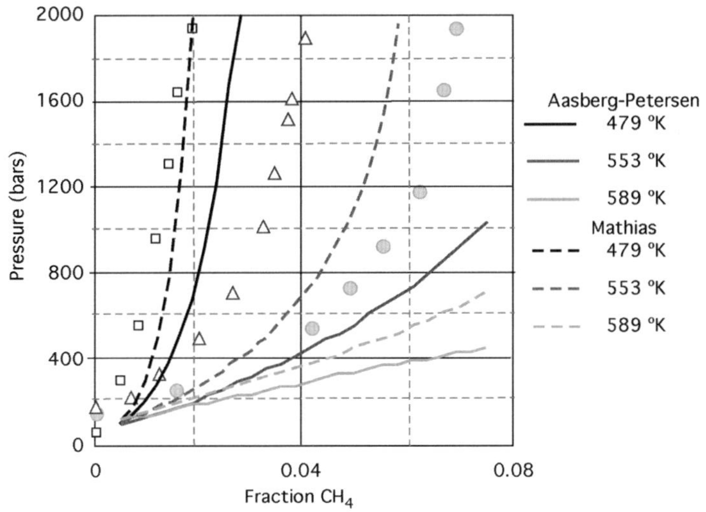


**Fig. 6.** Predictions of methane solubility in water. Shown is the bubble point pressure as a function of mixture composition (in mole fraction methane) at three different temperatures for two equations of state compared with experimental data.

solve phase-related problems in chemistry and geochemistry. HCToolkit comes with several example programs that are general, but can be used to address field-related questions. One such script is PT-PhaseDiagram.pl (found in the /apps directory), a script that calculates the phase distribution of a given mixture. These examples are not meant to be extensive; the toolkit is designed for the developer/user, giving the ability to quickly *develop* scripts to handle difficult problems.

In petroleum engineering and exploration geochemistry, equations of state are applied to 'PVT'-type compositions, which consist of several real components and a collection of 'pseudo-components' that are mixtures of undifferentiated compounds. Actually doing phase calculations with pseudo-components is only accurate for closed systems such as a reservoir (Meulbroek 2003), since pseudo-components are typically constructed of components that might have a similar molecular weight, but different partial pressures (fugacities). To better predict the complex partitioning and fractionation possible in an open system such as a migration conduit or a producing reservoir, one would prefer to use a more detailed specification, such as a GC analysis, that allows a more chemically-sensible lumping scheme. There are many problems with this approach, however. GC analyses are performed on 'dead' oils, where a large fraction of volatiles have separated and been lost from the fluid whose composition one is measuring. These compounds are precisely

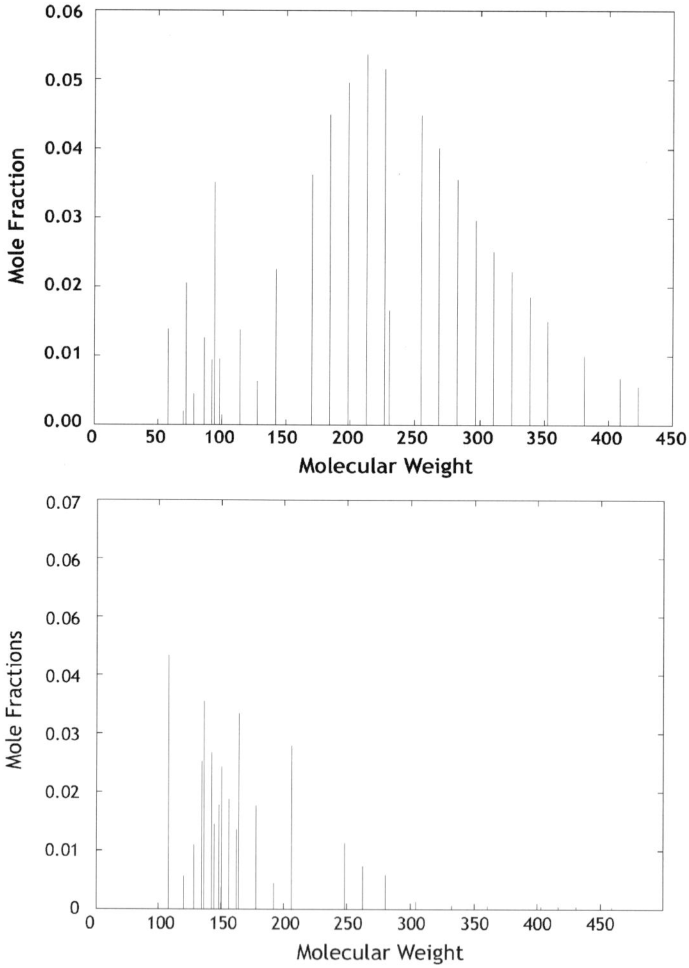


**Fig. 7.** Dead oil compositions, Vermillion area. The left figure shows a typical light oil (API gravity, 53°), while the right figure shows a condensate (original GOR, 15 K bbl/scf).

those that strongly influence phase behaviour, so estimating their loss is critical. Also, a great deal of oil never makes it through the GC column, including polar and asphaltene fractions. This is of lesser importance for vapour–liquid equilibrium, as these compounds have less of an effect on VLE (though a greater effect on aqueous–hydrocarbon equilibria). Hence, calculation of 'reincarnated' phase diagrams (using compositions where the volatile gasses are numerically reintroduced) is necessary. As an example of using GC for PVT calculations, several examples (one condensate, one light oil) are presented.

The examples presented here are dead oil compositions gleaned from GC data for several oils from the UNOCAL Vermillion area of offshore Louisiana, USA (Curiale & Bromley 1996*a*, *b*). The pseudo-GCs are presented in Figure 7. If we examine the composition of several of the fluids from that field, each shows the typical 'dead' oil profile of massive light-end loss (e.g. no methane, minimal ethane). In actuality, a great deal of methane and other light gases are associated with the original pre-separator fluids. For example, several PVT analyses from the area (unpublished UNOCAL data) shows a typical methane fraction of 0.55 for oils, 0.93 for a condensate. In order to use FPE models to analyse the fluid, an estimate of the live oil composition needs to be made. As an example of the profound phase behaviour effect of degassing, Figure 8 shows calculated phase envelopes for the oil from Figure 7a, where increasing amounts of methane have been numerically added. Each envelope (contiguous colour) is labelled with the mole percent methane present in the mixture. Note that the bottom phase

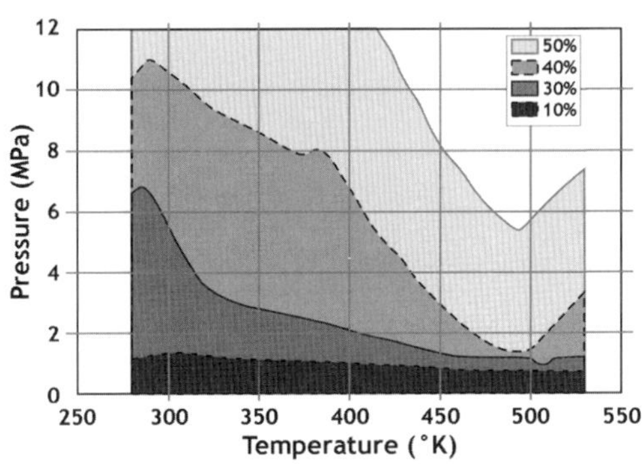


**Fig. 8.** Calculated phase envelopes for Figure 7 gas with added methane. Labelled are the fraction methane of the mixture whose phase envelope is shown. Note that a 'typical' reservoir has ($P = 30.3$ MPa, $T = 370.9$ K).

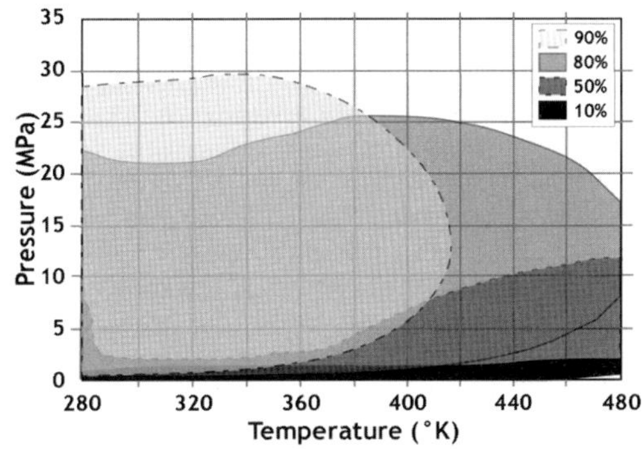


**Fig. 9.** Calculated phase envelopes for condensate with increasing amounts of added methane. Note that a 'typical' resevoir has ($P = 30.3$ MPa, $T = 370.9$ K).

boundary for each phase envelope lie very close to the 0 MPa pressure axis. By adding increasing amounts of methane (from 10 to 50% by mole), the maximum bubble point pressure moves from approximately 1.5 MPa to off the chart (over 12 MPa). A similar simulation can be performed for the condensate shown in Figure 7b, and is presented in Figure 9. Here, the phase diagram reaches much higher pressure (very close to reservoir pressures).

## Getting the code

The code for both the HCToolkit and the EOSInterface is available for download from http://ruby.wag.caltech.edu/EOSInterface. As the code is open source, all source code is published at this site, and users are encouraged to download and submit modifications to the code. An MS Windows installer program is available for the EOSInterface, making installation trivial. The EOSInterface program only works on the MS Windows platform. MS Office (MS Excel and MS Access) is required for standard installations. The HCToolkit works on most platforms and requires Perl (http://www.Perl.com) to be installed.

## Conclusions

Presented here is a new open source software framework for doing phase equilibria calculations. The framework consists of two components: the HCToolkit (a set of Perl classes that implement EOS calculations and flash calculations) and the EOSInterface (an ActiveX/MS Excel front end to the HCToolkit). Several calculations are also shown that demonstrate the flexibility of the programs presented here. A (composition, pressure) phase diagram for the system ($CO_2$–cyclopentane) demonstrates the usefulness of the PSRK EOS (Holderbaum & Gmehling 1991). A phase boundary calculation for the methane–water system shows the ability of the framework for comparing different equations of state in challenging calculations. The applicability to hydrocarbons is shown by phase diagrams generated from GC analyses. The framework is flexible, extensible and available for download at http://ruby.wag.caltech.edu/

The authors gratefully acknowledge the support of the US DOE (contract number DE-AC26-99BC15217) and MSC industrial partners.

## References

AASBERG-PETERSEN, K. & STENBY, E. 1991. Prediction of thermodynamic properties of oil and gas condensate mixtures. *Industrial & Engineering Chemistry Research*, **30**(1), 248–254.

BUNZ, A. P., DOHRN, R. & PRAUSNITZ, J. M. 1991. 3-Phase flash calculations for multicomponent systems. *Computers & Chemical Engineering*, **15**(1), 47–51.

CHUNG, T. H., AJLAN, M., LEE, L. L. & STARLING, K. E. 1988. Generalized multiparameter correlation for nonpolar and polar fluid transport-properties. *Industrial & Engineering Chemistry Research*, **27**(4), 671–679.

CURIALE, J. A. & BROMLEY, B. W. 1996a. Migration induced compositional changes in oils and condensates of a single field. *Organic Geochemistry*, **24**(12), 1097–1113.

CURIALE, J. A. & BROMLEY, B. W. 1996b. Migration of petroleum into Vermillion 14 field, Gulf Coast, USA—Molecular evidence. *Organic Geochemistry*, **24**(5), 563–579.

DI PRIMIO, R. 2002. Unraveling secondary migration effects through the regional evaluation of PVT data: a case study from Quadrant 25, NOCS. *Organic Geochemistry*, **33**(6), 643–653.

FREDENSLUND, A., JONES, R. L. & PRAUSNITZ, J. M. 1975. Group-contribution estimation of activity-coefficients in nonideal liquid-mixtures. *American Institute of Chemical Engineering Journal*, **21**(6), 1086–1099.

GMEHLING, J., LI, J. D. & SCHILLER, M. 1993. A modified Unifac model. 2. Present parameter matrix and results for different thermodynamic properties. *Industrial & Engineering Chemistry Research*, **32**(1), 178–193.

HOLDERBAUM, T. & GMEHLING, J. 1991. PSRK—a group contribution equation of state based on Unifac. *Fluid Phase Equilibria*, **70**(2/3), 251–265.

KESHTKAR, A., JALALI, F. & MOSHFEGHIAN, M. 1997. Evaluation of vapor–liquid equilibrium of $CO_2$ binary systems using UNIQUAC-based Huron-Vidal mixing rules. *Fluid Phase Equilibria*, **140**(1/2), 107–128.

LI, J. D., VANDERBEKEN, I., YE, S. Y., CARRIER, H. & XANS, P. 1997. Prediction of the solubility and gas-liquid equilibria for gas–water and light hydrocarbon–water systems at high temperatures and pressures with a group contribution equation of state. *Fluid Phase Equilibria*, **131**(1/2), 107–118.

LOSH, S., CATHLES, L. & MEULBROEK, P. 2002. Gas washing of oil along a regional transect, offshore Louisiana. *In*: MEULBROEK, P. & MACLEOD, G. (eds) *Organic Geochemistry*. Pergamon, Oxford & New York, **33**, 655–663.

MATHIAS, P. M. 1983. A versatile phase-equilibrium equation of state. *Industrial & Engineering Chemistry Process Design and Development*, **22**(3), 385–391.

MEULBROEK, P. 2002. Equations of state in exploration. *In*: MEULBROEK, P. & MACLEOD, G. (eds) *Organic Geochemistry*. Pergamon, Oxford & New York, **33**, 613–634.

POLING, B. E. 2000. *Properties of Gases and Liquids*. McGraw-Hill, New York.

RIJKERS, M. P. W. & HEIDEMANN, R. A. 1986. *Convergence behavior of single-stage flash calculations*. American Chemistry Society Symposium Series, **300**, 476–493.

# Rates of reservoir fluid mixing: implications for interpretation of fluid data

CRAIG SMALLEY[1], WILLIAM A. ENGLAND[1], ANN MUGGERIDGE[2], YAFES ABACIOGLU[3] & STEPHEN CAWLEY[4]

[1]*BP, Chertsey Road, Sunbury, Middlesex TW16 7LN, UK (e-mail: smalleypc@bp.com)*
[2]*Department of Earth Science & Engineering, Imperial College, London SW7 2AZ, UK*
[3]*BP 200 Westlake Park Bvd, Houston, Texas 77079, USA*
[4]*BP, Dyce, Aberdeen AB21 7PB, UK*

**Abstract:** This paper highlights the benefits of using knowledge of the rates of fluid mixing in the interpretation of reservoir fluid data. Comparison of the time it would take for a fluid difference to mix with the actual time available for mixing to occur allows two significant advances over a purely statistical analysis of reservoir fluid data: (1) differentiation of a step in fluid properties, indicative of a barrier to fluid communication, from a gradient indicative of incomplete mixing; and (2) quantitative estimation of the *degree* of compartmentalization that can readily be adapted into models for prediction of reservoir production performance.

We review the existing equations that estimate the mixing times for three main types of variation in fluid properties (fluid contacts, fluid density and fluid chemistry). In addition, a new relationship for fluid pressure mixing is presented. In each case the relationships were validated by comparison with numerical simulation. The different fluid mixing processes were compared by applying the equations to a range of simple fluid scenarios in one simple reservoir description. This shows that mixing times for fluid mixing processes are diffusion > fluid density > fluid contacts > fluid pressure. For each scenario, the processes were analysed in terms of the volume of fluid that must move in order to bring the system to equilibrium and the drive for fluid mixing (pressure difference × permeability/viscosity). Perhaps surprisingly, there is an excellent linear relation between fluid mixing times (a) calculated from the mixing equations and (b) estimated from volume/drive. This indicates that fluid volumes and mixing drive are the main controls on fluid mixing times. This can be used to derive simple interpretation guidelines to estimate mixing rates even in the absence of quantitative modelling. A simple field case study demonstrates how this understanding of fluid mixing times can add value to the interpretation of reservoir fluid data.

Understanding the rate at which fluids move and mix in the subsurface is crucial for correctly interpreting reservoir geochemical data, and is particularly important for extracting business value from such data. One of the main ways in which reservoir fluid data (here including fluid pressure, fluid physical properties, fluid chemical properties and fluid contacts) are applied is in the detection of reservoir compartmentalization, something that is key for optimizing reservoir development.

The presence of different fluid properties in different parts of a reservoir may result from compartmentalization, reflecting the inability of fluid variations to mix over geological timescales due to the presence of flow barriers.

In the past, most emphasis has been put on determining, statistically (e.g. cluster analysis: Stølum & Smalley 1992; Stølum *et al.* 1993)) or visually (e.g. Star diagrams: Kaufman *et al.* 1990; Hwang & Baskin 1994; Hwang *et al.* 1994), whether significant fluid variations exist. If 'significant' variations were detected in a reservoir, then this was taken as an indication of the presence of compartments. However, fluids take time to mix. Even if there is a significant fluid variation detected in a reservoir, it is then necessary to distinguish between (a) a variation that has simply not yet had time to mix, i.e. points on a continuous gradient, and (b) a variation that has had time to mix, but has not been able to, due to the presence of barriers to mixing. In this paper, we concentrate on the rates of fluid mixing, to provide ways of distinguishing these two scenarios.

The majority of past studies of reservoir compartmentalization have not considered mixing rates: this often resulted in a 'yes it is' or 'no it is not' interpretation of compartmentalization. However, in nature most scenarios are not simply black or white. The question should thus not be whether or not the reservoir is com-

*From*: CUBITT, J. M., ENGLAND, W. A. & LARTER, S. (eds) 2004. *Understanding Petroleum Reservoirs: towards an Integrated Reservoir Engineering and Geochemical Approach.* Geological Society, London, Special Publications, **237**, 99–113. 0305-8719/04/$15.00 

partmentalized, but *how* compartmentalized it is—what degree of communication exists between the compartments? Ideally, geological timescale reservoir simulation, either using a 3D reservoir simulation model adapted to deal with geological time scales (England *et al.* 1995; Muggeridge *et al.* 2004*a*, *b*) or a 3D basin model with a refined grid over the reservoir (Swarbrick *et al.* 2000), could co-simulate all the fluid data using a range of reservoir descriptions to determine the degree of compartmentalization. Another approach would be to set up scaled physical models of the reservoir in the laboratory, and study these to infer the degree of compartmentalization (Westrich *et al.* 1999). However, both these approaches are very time-consuming, requiring bespoke modification of simulation codes and/or workflows, or extensive laboratory facilities. Nevertheless, even in the absence of these approaches, quantitative or semi-quantitative knowledge of natural mixing rates would enable degree of compartmentalization to be assessed by determining the strength of a barrier to mixing; i.e. the retarding effect of the barrier to mixing, compared with the rate of mixing that would be expected in the absence of a barrier. There is also the intriguing possibility of being able to integrate data on a range of fluid properties, each of which mix at different rates, to set up a set of simultaneous equations that, when solved, would provide a much more accurate quantification of degree of compartmentalization than could not be possible from one data type alone.

In this paper, we examine mixing rates for the four main types of reservoir fluid data: fluid pressure, fluid contacts, fluid density and fluid chemistry. We bring together previously published mixing relations with a newly derived relation for pressure-driven flow, all validated against simulation models. By considering some simple mixing scenarios, we also examine the controls on mixing rates, to derive some simple interpretation guidelines for such data.

## Fluid variations: background

### *Pressure*

Fluid pressure is one of the most common parameters measured in oil and gas reservoirs. Highly precise measurements—to the sub-psi (pounds per square inch) level—are commonly obtained using wireline tools such as the Repeat Formation Tester (RFT) or Modular Formation Dynamics Tester (MDT), both of which can measure pressure at numerous depths within the reservoir. The quality of pressure data depends on the nature of the pressure gauge. Older strain gauges are accurate to ~0.1% and precise to ~0.1–0.5 psi, depending on conditions. Modern quartz gauges are accurate to a few psi, while precision may reach 0.005 psi or better. Permanent downhole gauges provide highly precise measurements from one point over a period of time. Less precise pressure estimations come from indirect methods, such as monitoring of mud weight during drilling.

Fluid pressures in general increase with depth due to the increasing overlying fluid column and rock overburden. The critical factor that can be used for assessing reservoir compartmentalization is thus not the absolute pressure, but the degree of pressure disequilibrium (overpressure) in different parts of the reservoir, i.e. the difference between the absolute pressure and the hydrostatic pressure gradient. With high-precision pressure measurements, even very small variations in fluid overpressure may be significant for compartment detection.

A number of mechanisms (Neuzil 1995) can generate pressure disequilibrium. The mechanisms that might cause pressure differences between different reservoir compartments include differing rates of burial, leading to different degrees of compaction disequilibrium, or different degrees of connectedness to regional overpressured or underpressured areas of either natural or man-made origin (e.g. depletion from nearby oil or gas production). Overpressure differences can be caused by petroleum generation or possibly by diagenetic reactions that cause volume changes or chemical compaction; however, it is not likely that these would cause differences within a reservoir.

Pressure variations will reduce with time by Darcy flow from the more overpressured area to the less overpressured area until the pressures have equilibrated and there is no longer any pressure gradient to drive fluid flow.

### *Density*

Fluid density is detected directly by obtaining fluid samples (from RFT/MDT sampling, or from a flowing well) and measuring their properties (e.g. density, usually expressed as an API gravity number) in the laboratory. The need for sample material limits the spatial coverage of density samples. Density can sometimes be estimated indirectly from the chemical composition of oil extracted from core material, which greatly increases the frequency of sampling possible within a well (Smalley *et al.* 1996, 1997). On a larger scale, density can be estimated from vertical pressure gradients, though less precisely.

Spatial fluid density variations are frequently inherited from the filling history of the reservoir. The initial fluids expelled from a source rock are relatively dense liquids. As a source rock becomes more thermally mature, it expels progressively lighter fluids and eventually gases. When such fluids fill a reservoir, and fill and spill from compartment to compartment within a reservoir, each part of the reservoir can end up with different proportions of fluids of different maturity and density. Field observations show that the segment of the reservoir closest to the source kitchen has often received the latest, lowest density charge. Those areas farthest away from the source kitchen may contain earlier denser fluids that have filled and spilled to their current location.

Other density variations can be caused by in-reservoir reactions, such as biodegradation or water washing. Parts of the reservoir that are in optimal conditions (temperature, good communication with oxygenated aquifer) for microbial activity may, for example, become more biodegraded and thus denser.

Adjacent areas with different fluid densities will mix by density-driven overturning, unless flow barriers inhibit such mixing (England *et al.* 1995). It is important to note here that the product of density-driven mixing may not be a homogeneous fluid density, but rather a gravitationally stable density gradient in which the lighter fluids overlay the denser ones. Density gradients are seen in many oil fields (e.g. Burgan, Kuwait: Kaufman *et al.* 1998; Gulf of Suez: Khavari-Khorasani *et al.* 1998). The key indicator for reservoir compartmentalization would be where the fluid densities are not gravity-stable (e.g. where different fluid densities exist at a common depth) and where it can be demonstrated that sufficient time has elapsed for density-driven mixing to have occurred in a non-compartmentalized system.

### *Contacts*

Fluid contacts (oil–water contacts (OWC), gas–water contacts (GWC) or gas–oil contacts (GOC)) are present in virtually all reservoirs. OWCs and GOCs are usually detected through resistivity logging, petroleum-saturated rocks usually being much more resistive than water-filled rocks. Much attention is placed routinely on the identification of contacts, because this affects the estimated volume of oil or gas contained within a field. However, production wells are often situated so as avoid intersecting contacts, limiting data availability.

Variations in fluid contacts may be inherited from the filling history, when different parts of the reservoir are filled from different directions, or by fill and spill pathways (e.g. England *et al.* 1987) or result from pressure communication with other, nearby, producing fields (see Coutts 1999, for example). In rare cases, they may result from natural movement of water in the aquifer (Winterhalder & Hann 1991). If the reservoir (and aquifer) is isolated from other fields but there is unrestricted communication within a reservoir, with time the oil or gas would move around until all fluid contacts had equilibrated to the same depth. The different densities of gas, oil and water would drive this mixing, in the same way as the fluid densities mentioned previously. The preservation of different distinct contact depths can be used as an indicator of lack of communication, whereas as a gradient in contact depth across the reservoir may simply indicate pressure communication with other aquifer or reservoir systems.

### *Chemistry*

Oil or water chemistry can be determined from fluid samples or from core extracts. Oil or gas variations may be inherited from the filling history, or be modified by in-reservoir processes such as described above for density. Water variations relate to either water–rock interaction (e.g. dissolution of salt or feldspars), or to regional flow patterns (e.g. influx of meteoric water).

Much effort has been put into finding parameters that can be measured in fluids that will reveal differences that relate to compartmentalization. Oil is a compositionally complex fluid, and there is a wide variety of parameters that could be measured. Generally, previous studies have used one of two approaches:

(1) a statistical approach, where for example peaks on GC (gas chromatograph) traces are compared from sample to sample, without necessarily knowing what compounds the peaks represent—so-called GC-fingerprinting (Kaufman *et al.* 1990; Smalley & Hale 1996);
(2) use of specific natural tracer compounds, such as oil 'biomarkers' that relate to maturity, source type, isotope signatures of dissolved gas or other compounds within oil, or the amount of dissolved gases in oil (England *et al.* 1987; Edman & Burk 1999).

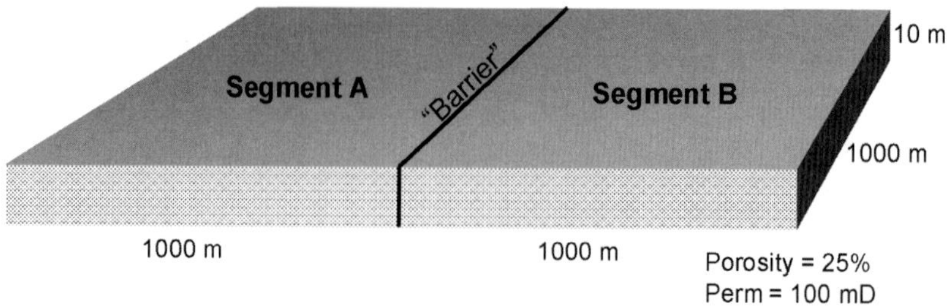


**Fig. 1.** Simple reservoir description used in analysis of fluid mixing. The porosity and permeability are homogeneously distributed.

In formation waters, perhaps the most useful natural tracer for compartment detection is the $^{87}Sr/^{86}Sr$ isotopic ratio. This ratio is easily affected by water–rock interaction (by dissolution of minerals with different $^{87}Sr/^{86}Sr$ compositions), such that there is a naturally wide range of potential water $^{87}Sr/^{86}Sr$ compositions. However, the $^{87}Sr/^{86}Sr$ ratio is not affected by dilution or precipitation. Hence, it is possible to measure the $^{87}Sr/^{86}Sr$ value of formation water using an aqueous extract from a core sample, even if the core has been allowed to dry out partially or completely. This Sr residual salt analysis (RSA) technique, originally developed at the Institute for Energy Technology in Norway in the mid 1980s, is now a standard means of tracking heterogeneities of formation water composition (e.g. Smalley *et al.* 1995; Mearns & McBride 2001).

If the only differences in fluids in a reservoir are chemical differences, then the only driving force for mixing is the chemical gradient that drives diffusional mixing. However, in many cases chemical differences will be accompanied by other factors, such as density differences. In such cases, diffusional mixing will occur in addition to the density-driven mixing, and the equilibrium state may not be a constant composition, but a gravitationally stable gradient (Whitson 1994; Schulte 1980; Sage & Lacey 1938; Muskat 1930).

## Controls on fluid mixing: some simple mixing scenarios

### *Base model*

The aim of this section is to achieve a better understanding of the controls on fluid mixing rates, in terms of volumes of material moved and the driving forces behind such movement. To do this, we take a simple reservoir geometry, and look at some simple end-member mixing scenarios involving differences in fluid pressure, density and fluid contacts.

A single base reservoir geometry is used for all of the subsequent calculations; see Figure 1. The reservoir consists of two adjacent segments, separated by a notional 'barrier.' Each reservoir segment is 1000 m × 1000 m areally, and 10 m thick. The key reservoir parameters and symbols are listed in Table 1. All equations use SI units but the table also lists some parameter values in field units, as these may be more familiar to some readers.

In the following scenarios, fluid differences are assumed between the two segments, and controls on mixing are considered.

### *Fluid pressures*

In this scenario, the initial pressures were assumed to be 4100 psi in one segment, and 4000 psi in the other. If this initial 100 psi pressure difference is to mix, such that the pressure in both segments equilibrates to 4050 psi, then material must move from the higher pressure segment into the lower pressure segment. The volume moved is

$$\text{volume moved} = 0.5(VC\,\Delta P) \qquad (1)$$

where the volume of the reservoir $V = L^2 H\phi$. The key to symbols is given in Table 1. How much material must move depends on rock and fluid compressibility ($C_{\text{total}} = C_{\text{fluid}} + C_{\text{rock}}$). In this simple set of scenarios, the rock compressibility is held constant, so fluid compressibility is the only parameter to vary in equation (1); we will thus ignore $C_{\text{rock}}$ for the present and assume $C_{\text{total}} = C_{\text{fluid}}$. In a completely water-filled system, then only 750 m$^3$ (0.02% of the reservoir fluid volume) of water needs to move from the higher-pressure segment to the lower-pressure segment in order for the pressure to equilibrate (Fig. 2). This is because water is so incompressible. Oil, being more compressible

**Table 1.** *Base model assumptions*

| Symbol | Property | Value/units | SI equivalent (value/units) |
|---|---|---|---|
| $L$ | Segment length | 1000 m | |
| $L$ | Segment width | 1000 m | |
| $H$ | Reservoir thickness | 10 m | |
| $H$ | Height of fluid contact | 1, 2, 3, 5 or 10 m | |
| $\phi$ | Porosity | 0.25 | |
| $k$ | Permeability | 100 mD | $10^{-13}$ m$^2$ |
| $C_r$ | Rock compressibility | $5 \times 10^{-6}$ psi$^{-1}$ | $7.2 \times 10^{-10}$ Pa$^{-1}$ |
| $C_w$ | Water compressibility | $3 \times 10^{-6}$ psi$^{-1}$ | $4.3 \times 10^{-10}$ Pa$^{-1}$ |
| $C_o$ | Oil compressibility | $1.5 \times 10^{-5}$ psi$^{-1}$ | $2.2 \times 10^{-9}$ Pa$^{-1}$ |
| $C_g$ | Gas compressibility | $2 \times 10^{-4}$ psi$^{-1}$ | $2.9 \times 10^{-8}$ Pa$^{-1}$ |
| $C_{vo}$ | Viscous oil compressibility | $1 \times 10^{-5}$ psi$^{-1}$ | $1.4 \times 10^{-9}$ Pa$^{-1}$ |
| $D_{sr}$ | Diffusion coefficient for Sr in water | m$^2$ s$^{-1}$ | |
| $D_{CH_4}$ | Diffusion coefficient for $CH_4$ in gas | m$^2$ s | |
| $D_{C_{12}}$ | Diffusion coefficient for $C_{12}$ in oil | m$^2$ s | |
| $D_{C_{200}}$ | Diffusion coefficient for $C_{200}$ in a viscous oil | m$^2$ s | |
| $G$ | Acceleration due to gravity | 9.81 m s$^{-2}$ | |
| $\mu_w$ | Water viscosity | 1 cP | $10^{-3}$ N s m$^{-2}$ |
| $\mu_o$ | Oil viscosity | 1 cP | $10^{-3}$ N s m$^{-2}$ |
| $\mu_g$ | Gas viscosity | 0.03 cP | $3 \times 10^{-5}$ N s m$^{-2}$ |
| $\mu_{vo}$ | Viscous oil viscosity | 100 cP | 0.1 N s m$^{-2}$ |
| $P_i$ | Initial pressure | 4000 psi | $28 \times 10^6$ Pa |
| $\rho_w$ | Water density | 1 g cm$^{-3}$ | 1000 kg m$^{-3}$ |
| $\rho_o$ | Oil density | 30 API | 876 kg m$^{-3}$ |
| $\rho_g$ | Gas density | 70 API | 685 kg m$^{-3}$ |
| $\rho_{vo}$ | Viscous oil density | 15 API | 966 kg m$^{-3}$ |

(Table 1), will require more fluid to move to effect equilibration. For a fairly low compressibility heavy oil this would be 2500 m$^3$ (0.05%), while a more typical 30° API oil would require 3750 m$^3$ (0.08%) to move. This analysis does not strictly apply to gas, except for small changes in pressure: its expansion is related to pressure through the non-Ideal Gas Law. However, using equation (1) as an approximation shows that, being far more compressible, a far greater volume of material would have to move (62 500 m$^3$, or 1.25%) in the case of gas, when compared with oil or water.

The instantaneous initial drive that causes fluid pressure to begin to equilibrate can be expressed by

$$\text{drive} = ak(\Delta P/\mu) \quad (2)$$

where $a$ is the cross-sectional area divided by the distance between the two reservoir compartments. It is set to 1 in these relative rate estimates, as we are not discussing different reservoir

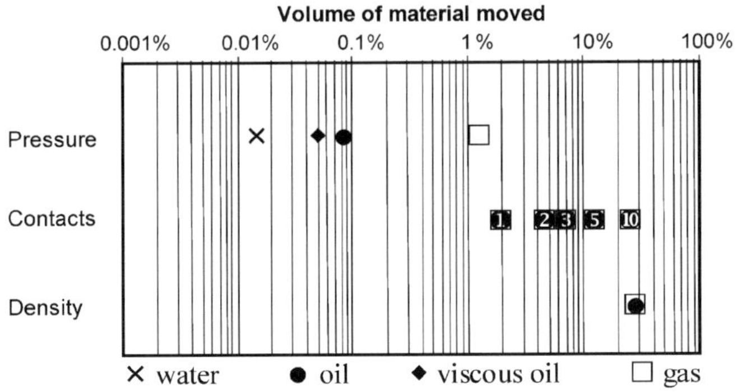


**Fig. 2.** Volume of fluid that must move in order for the system (Fig. 1) to reach equilibrium, for a fluid pressure difference of 100 psi, fluid contact differences of 1, 2, 3, 5 and 10 m, and density differences equivalent to 10 API gravity units.

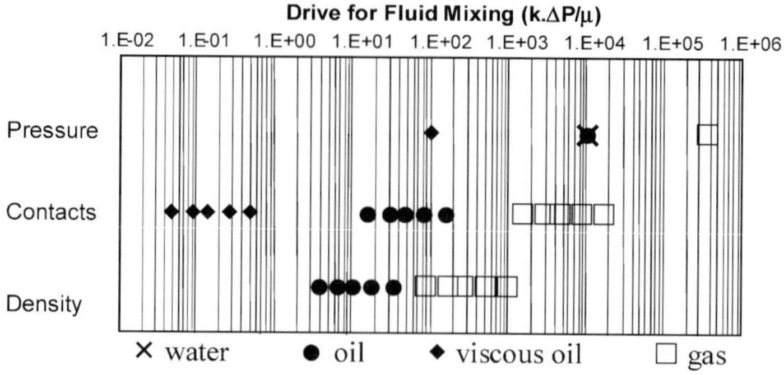


**Fig. 3.** Drive for fluid mixing $[\Delta P(k/\mu)]$ for the scenarios described in the caption to Figure 2.

geometries. Drive is equivalent to the instantaneous initial volumetric rate of fluid mixing. In the estimation of relative mixing times, drive can be used as a proxy for mixing rate.

For the present pressure mixing scenarios, the fluid viscosity ($\mu$) is the only parameter to vary in equation (2). The estimates of drive are illustrated in Figure 3. Gas, having by far the lowest viscosity, will have a rate about 30 times faster than both oil and water, which in turn will have a rate about 100 times greater than viscous oil ($\mu = 100\,\mathrm{cP}$).

A relative measure of overall fluid mixing time can be derived simply by dividing the volumes moved by drive:

$$\text{relative mixing time} \cong \text{volume/drive}. \qquad (3).$$

Obviously, this ignores the fact that, as more material moves, the driving force will decrease and hence the rate too. However, this will be true for all the scenarios; using the initial instantaneous rate (the maximum rate) here will underpredict the real time needed for equilibration, but it should correctly predict the *relative* times taken to equilibrate in the different scenarios we are considering.

This simple estimation of mixing times would suggest that the fastest fluid to mix would be water (Fig. 4), due to the small volume that is moved. Gas would take about twice as long: the ~60 times larger volume involved is mostly offset by the ~30 times faster mixing rate. Oil would take about twice as long as the gas, and viscous oil over 60 times longer than that, due to the very slow rates.

The mixing times will be revisited later and compared with quantitative modelling results.

### *Fluid contacts*

The equilibration of fluid contacts through time can be addressed using a similar approach to that used above. Using the same base reservoir model (Fig. 1), we assumed that Segment 1 is completely full of oil, viscous oil or gas, while Segment 2 has an oil–water contact or gas–water contact that is 1, 2, 3, 5 or 10 m above

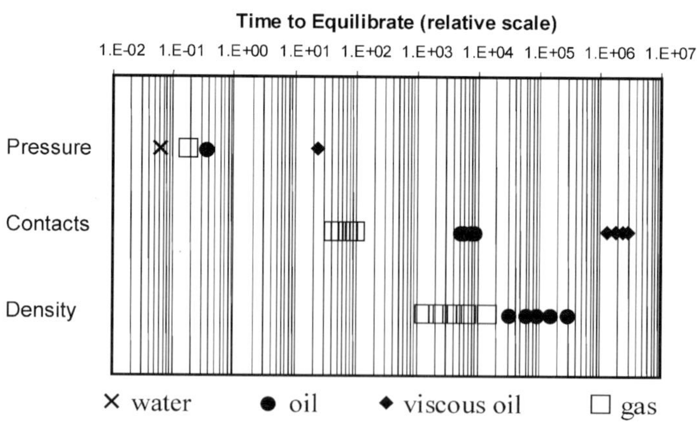


**Fig. 4.** Relative fluid mixing times for the scenarios described in the caption to Figure 2, estimated by dividing the volume change (Fig. 2) by the mixing drive (Fig. 3).

the base of Segment 2. Consider a 10 m initial difference in OWC. When this has equilibrated, the OWC in Segment 1 will have moved upwards by 5 m, while that in Segment 2 will have moved downwards by 5 m. It follows that the volume of oil or gas that must move from Segment 1 to Segment 2 is equal to half the initial contact height difference multiplied by the area of Segment 1 and its porosity (this will of course be balanced by the same volume of water moving in the opposite direction, but is ignored here as it will not affect the equilibration rate or time taken). Because this is a simple end-member scenario we have assumed that the fluids are completely segregated so there is no transition zone about the contact and neglected the effects of fluid compressibility (so as to differentiate this scenario completely from the fluid pressure scenario described above). With these assumptions, the volume of fluid that needs to move for equilibration to occur is dependent only on the initial contact height difference and is independent of fluid type (Fig. 2).

The drive causing fluid mixing to occur can be described by equation (2), where the $\Delta P$ is calculated from $\Delta\rho gh$, i.e.

$$\text{drive} = ak(\Delta\rho gh/\mu) \qquad (4)$$

where $\Delta\rho$ is the density difference between the fluids, $g$ is the acceleration due to gravity and $h$ is the initial fluid contact height difference. The results are shown in Figure 3. For any one fluid, the rates are higher the greater the initial contact height difference. For similar contact heights, the rates increase with decreasing viscosity and increasing $\Delta\rho$.

The relative mixing times for fluid contacts, calculated from equation (3), are illustrated in Figure 4. This shows that the initial fluid contact height difference only has a secondary impact on mixing time; the main control is the fluid type (oil, water or gas), with equilibration times for gas–water contacts estimated to be about 2 orders of magnitude shorter than oil–water contacts, which in turn are about 2 orders of magnitude shorter than viscous oil–water contacts. Fluid viscosities and density differences are the overriding controls here. All of the estimated relative mixing times are longer than for the simple pressure equilibration scenario (Fig. 4), by about 2 orders of magnitude for gas, up to about 5 orders of magnitude for viscous oil.

### *Density differences*

Fluid density differences can be treated very similarly to fluid contact differences, which are really just a special case of density differences where the fluids happen to be immiscible. Two cases were considered: (1) Segment 1 (Fig. 1) filled with an oil with an API gravity of 30°, and Segment 2 filled with oil 1, 2, 3, 5 and 10 API units higher (less dense); (2) Segment 1 filled with a very light oil/gas of 60° API, again with Segment 2 filled with oils with 1, 2, 3, 5 and 10 API units higher.

The volume of denser fluid that must move from Segment 1 into Segment 2 in order for the fluid densities to equilibrate is half of the volume of Segment 1, i.e. 25% of the total volume. This would be balanced by a similar amount of less dense fluid moving in the opposite direction. In this simple end-member scenario, where compressibility is ignored, the volume moved is 25% irrespective of fluid composition (Fig. 2). The volumes are much greater than involved with the pure pressure-driven scenario (Fig. 2) and right at the top end of volumes for contact-driven mixing.

The rate of fluid mixing, estimated using equation (4), is greater for gas than for oil due to its lower viscosity. The rates are about an order of magnitude less than for contact-driven mixing and about 2 orders of magnitude less than pressure-driven mixing of fluids of similar composition (Fig. 3). When it comes to estimating mixing times, using equation (3), mixing times are very long, about 1–2 orders of magnitude longer than contact-driven mixing and about 5 orders of magnitude longer than the pressure-driven scenario (Fig. 4). This is due to the combination of large volumes and slow rates involved with purely density-driven mixing (Figs 2 & 3). Figure 4 provides a summary of the relative mixing time expected in the different end-member scenarios considered.

### *Concentration differences*

Fluid compositional differences that are purely chemical in nature (i.e. not accompanied by differences in pressure or density) will mix through diffusion. In such a scenario, the volume of material moved can be expressed as

$$\text{volume moved} = 0.5\,\frac{M\,\Delta C}{\rho}\,V \qquad (5)$$

where $M$ is the molar mass ($\text{kg mol}^{-1}$) of the component diffusing, $\Delta C$ is the concentration difference ($\text{mol m}^{-3}$) and $\rho$ is the density of that component; $V$ is the total reservoir volume.

The drive for concentration differences is controlled by the effective diffusion coefficient $D$ ($\text{m}^2\,\text{s}^{-1}$) and the concentration difference $\Delta C$

(mol m$^{-3}$). *Assuming binary diffusion* it is given by

$$\text{drive} = a\frac{M}{\rho}D\Delta C. \tag{6}$$

The volumes moved and drives are not shown on Figures 2 and 3 because they are much smaller than for any of the other mechanisms. For example a 100 mg/l variation in Sr content in the formation water would require a volume change of 0.0004 of one percent of the reservoir volume (assuming a connate water saturation of 20%) while the drive would be 13 orders of magnitude less than the rate resulting from the movements of contacts in a viscous oil–water system. Thus, the overall mixing time is more than 8 orders of magnitude slower than the slowest mixing time calculated for the other mechanisms (the movement of contacts for a viscous oil–water system).

## Quantitative fluid mixing models

In this section, we consider some new and previously published quantitative models for fluid mixing, and compare these with the qualitative analysis above. For a more detailed treatment, see Muggeridge *et al.* 2004*a*, *b*.

### *Fluid pressures*

Figure 5 illustrates an impermeable but discontinuous shale separating an upper sand from an underlying constant pressure aquifer. The sands are penetrated by a vertical well, which is a distance $L$ (m) from the gap in the shale. The upper sand has a thickness $H$ (m), the shale has a thickness $h$ (m) and the width of the gap is $l$

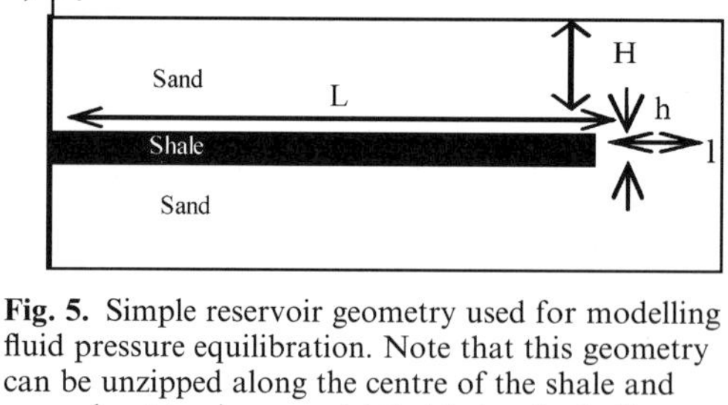


**Fig. 5.** Simple reservoir geometry used for modelling fluid pressure equilibration. Note that this geometry can be unzipped along the centre of the shale and opened out to give a model that is topologically equivalent to a cross section through the base reservoir description shown in Figure 1. Symbols described in Table 1, apart from $h$ (shale thickness) and $l$ (shale gap length).

(m). The upper sand has a permeability $k$ and the gap has a permeability $k_{gap}$; the shale is impermeable. The initial pressure difference between the upper and lower sands is $\Delta P$.

If the lower sand is assumed to have a constant pressure (i.e. it is an infinitely large aquifer), and the upper sand changes pressure until it equilibrates to the lower sand pressure, the time for the pressure in the upper sand to equilibrate with lower sand is given by

$$T = -0.09\frac{1}{\alpha_1^2}\frac{L^2\mu C\phi}{k}\ln\left\{\frac{\delta P}{\Delta P}\frac{R(R+1)+\alpha_1^2}{2R\sec\alpha_1}\right\} \tag{7}$$

where $T$ is in seconds, $\delta P$ is the pressure resolution of the measurement (i.e. the residual pressure difference at which the system is said to be equilibrated), $\alpha_1$ is the first root of the Transcendental Function

$$\alpha\tan\alpha = R$$

and $R = k_{gap}Ll/k(hH)$ ($R$ is the ratio of flow rate entering the sand to the flow rate along the sand towards the observation well). This solution is derived from that given by Carslaw and Jaeger (1959), for large times.

The topology of this system (Fig. 5) is identical to that of the two-reservoir segment base model (Fig. 1)—imagine the shale unzipped along its middle and opened out. The upper sand becomes Segment 1, the lower sand is Segment 2, and the gap at the edge of the shale becomes the interface between the two segments.

The analytical relation in equation (7) was tested against reservoir simulation results by recreating the system in Figure 5 in a commercial numerical reservoir simulator (VIP), and running a range of scenarios. In each scenario, the simulator was run until the pressures had equilibrated. The simulation results were then compared with the equilibration time calculated from equation (7). Numerous sensitivities were simulated: Figure 6 shows the results of four sensitivities to shale length. The results show that the analytical relation in equation (7) very closely approximates the equilibration time simulated using the VIP simulator. This lends strong support to the use of equation (7) for estimating pressure-driven fluid mixing rates and vice versa.

The equilibration times for fluid pressure differences in the base model illustrated in Figure 1 were calculated using equation (7) and the input parameters shown in Table 1. The fastest equilibration occurred in water, with the 100 psi difference being reduced to 1 psi in only 26 days (Fig. 8). Gas took 65 days

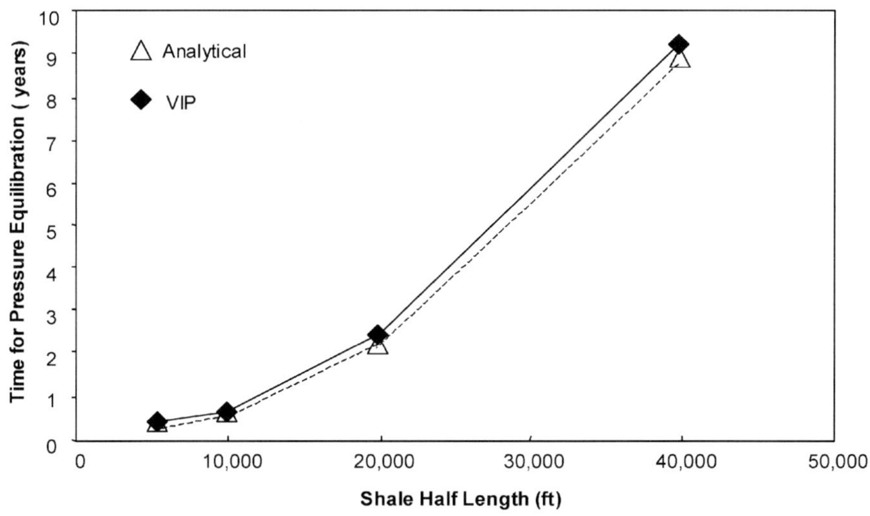


**Fig. 6.** Comparison of mixing times generated by the analytical relation in equation (7) and a numerical reservoir simulation model (VIP) of the same system.

and oil tool 130 days, while viscous oil took the longest at 23.7 years.

An alternative assumption is that *both* the upper and lower sand have variable pressure, and that with time the pressure in both sands will change until the pressures equilibrate to an intermediate pressure, lower than the initial pressure in the lower sand, but higher than the initial pressure in the upper sand. In this case, the mixing time is given by the relation

$$T = \left[\ln\left(\frac{2}{\pi}\right) - \ln\left(\frac{\delta P}{\Delta P}\right)\right]\frac{4L^2\phi C\mu}{k\pi^2}. \quad (8)$$

Pressure equilibration is in fact a diffusion process and can be treated by the same equations as those for diffusion or heat conduction, after making due allowance for the difference constants involved. This equation is derived from the expression given on page 101 of Carslaw and Jaeger (1959). Further simplification to give order of magnitude estimates leads to

$$T = \frac{L^2\phi C\mu}{k}. \quad (9)$$

The equilibration times for fluid differences in the base model illustrated in Figure 1 were calculated using equation (9) using the input parameters shown in Table 1. The fastest equilibration occurred in water, with the initial 100 psi difference being reduced to 1 psi in only 29 days. Oil took 147 days, whilst viscous oil took the longest at 27 years. Gas is not strictly amenable to this analysis unless pressure differences are relatively small, due to the nonlinear dependence of compressibility on pressure; however, if we assume the compressibility given in Table 1 we find it takes 74 days. These mixing times are very similar to (within ~15% of) those calculated from the constant-pressure lower sand model represented by equation (7). Hence either equation could be used here without substantially affecting the conclusions.

### *Fluid density*

An analytical relation for estimating density-driven fluid overturning was dealt with in detail by England *et al.* (1995). The relation they used was (symbols as in Table 1)

$$t = \frac{25}{4}\frac{L^2\mu\phi}{kgH\Delta\rho} \quad (10)$$

This relationship was derived from earlier work by Gardner *et al.* (1962), which treated the system as an initially vertical interface between two fluids of different densities (Fig. 7). With time, the linear interface rotates gradually towards the horizontal, with the less dense fluid over-riding the denser fluid (Fig. 7). The position of the interface is characterized by its gradient. England *et al.* (1995) considered a gradient of 0.01 as being essentially flat, i.e. the point where the density differences could be said to have equilibrated. They substituted this gradient into the relation of Gardner *et al.* (1962) to derive the mixing time in the form of equation (10).

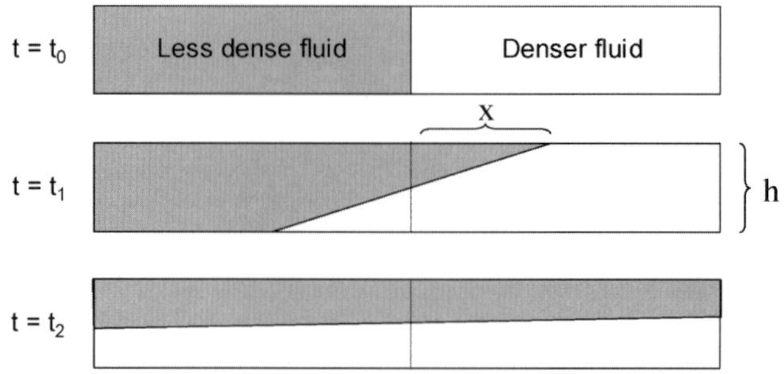


**Fig. 7.** Model for mixing of fluids by density-driven overturning. As the interface rotates through time, its progress can be represented by the parameter $2x/h$.

England *et al.* (1995) validated this relation by comparing mixing times derived from equation (10) with mixing times generated by a numerical compositional simulator. The results agreed very closely, validating the analytical relation.

When equation (10) is applied to the base model (Fig. 1) with 1–10° API gravity fluid density differences between the segments, the resulting mixing times are dependent on the fluid type and the initial density difference (Fig. 8). Gas would mix in about 20 000 to 200 000 years, while oil would take 400 000 to 4 million years. This time scale is over 5 orders of magnitude longer than pressure-driven mixing (Fig. 8).

## *Fluid contacts*

The model for density-driven fluid overturning discussed above (Fig. 7), and described in equation (10), is strictly designed to represent overturn and mixing of *miscible* fluids, such as oils of different density. However, reservoir simulation of such systems (England *et al.* 1995) shows that the actual amount of mixing that occurs at the fluid interface due to the density overturn is relatively small, limited to a narrow zone either side of the interface. More complete dispersion around the interface would result from additional diffusive mixing, not from the density-driven overturn process itself. It follows that the same relation (equation (10)) would closely approximate the process of density-driven overturning of *immiscible* fluids, such as oil and water or gas and water, if the reservoir thickness $H$ is replaced by the difference in contact heights $h$.

Using the density-overturning equation to model fluid contacts generates reasonable equilibration times, of about 1000 years for gas–water contacts, about 100 000 years for oil–water contacts, and about 10 million years for viscous oil–water contacts (Fig. 8). These timescales are about an order of magnitude faster than density differences in similar fluids, but still several orders of magnitude slower than pressure-driven mixing.

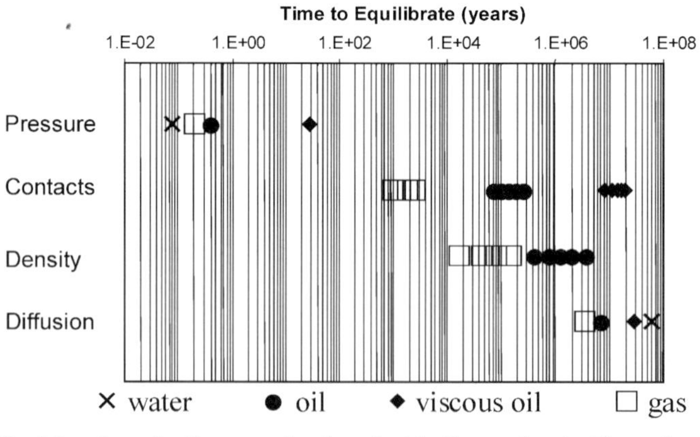


**Fig. 8.** Absolute fluid mixing times for the scenarios described in the caption to Figure 2, estimated using mixing equations. Compare with the relative fluid mixing times estimated in Figure 4.

### *Diffusional mixing*

Mixing of chemical differences in fluids through diffusion has been dealt with in some detail previously, for example, England *et al.* (1987) for oil and gas, and Smalley *et al.* (1995) for water. A general equation that gives an order of magnitude estimate of diffusional mixing times is

$$t(\mathrm{s}) = 0.1\,\frac{L^2}{D_{\mathrm{eff}}} \qquad (11)$$

where $L$ is the length (m) scale over which mixing occurs and $D_{\mathrm{eff}}$ ($\mathrm{m^2/s}$) is the effective diffusion coefficient:

$$D_{\mathrm{eff}} = \frac{D_{\mathrm{fluid}}\theta}{R}. \qquad (12)$$

In this equation, $\theta$ is the tortuosity of the porous medium (between 0 and 1; 0.5 would be typical for a consolidated sandstone), $D_{\mathrm{fluid}}$ is the diffusion coefficient in pure fluid and $R$ is a retardation factor, defined as

$$R = 1 + \frac{100-\phi}{\phi}\rho K_{\mathrm{D}} \qquad (13)$$

where $\phi$ is the porosity (%), $\rho$ (here estimated at 2.65 $\mathrm{g/cm^3}$) is the density of the solid framework of the rock and $K_{\mathrm{D}}$ is the distribution coefficient for sorption of the diffusing species onto the rock. Equation (11) has been applied to the mixing of Sr isotopes in water and validated against a 1D numerical model, with satisfactory results (Smalley *et al.* 1995).

The diffusional mixing model was applied to the base reservoir description shown in Figure 1, using the assumption that the diffusion distance ($L$) was 1000 m (i.e. the distance between the centre of Segment 1 and the centre of Segment 2). Four scenarios were modelled: (1) mixing of Sr isotopes in water, (2) mixing of methane (e.g. gas isotopes) in gas, (3) mixing of $C_{12}$ molecules (e.g. GC fingerprints) and (4) mixing of $C_{200}$ compounds in a viscous oil.

These calculations show that diffusional mixing is much the slowest of all the modelled processes (Fig. 8). Methane in gas and $C_{12}$ components in oil have the shortest diffusional mixing times, between one and ten million years, while $C_{200}$ components in viscous oil and strontium isotopes in water would take ten to a hundred million years to mix diffusionally between the two reservoir segments. In these simple calculations, pure fluids were assumed. In a more rigorous treatment, fluid saturations would have to be taken into account, including for example the increased tortuosity of the water phase with high oil or gas saturations (Worden *et al.* 1998). Wettability would also have to be taken into account; the present calculations assume a water-wet system, where the diffusion of solutes in the water phase is retarded by sorption onto mineral surfaces ($K_{\mathrm{D}}$ assumed to be 5 in equation (13)), while components of oil and gas are not significantly slowed ($K_{\mathrm{D}}$ assumed to be 0). Clearly, other wettability scenarios (oil-wet or mixed wettability) could significantly affect the model results.

## **Controls on fluid mixing time**

Using a simple reservoir base model (Fig. 1) we have derived estimates of fluid mixing times for fluid pressures, fluid contact and fluid density differences based on two completely separate approaches:

(1) We derived estimates of *relative* mixing times by considering the volume of fluid that is needed to move in each scenario in order to reach equilibration and the relative mixing drive (an instantaneous initial volumetric rate) as estimated from the permeability, viscosity and pressure difference or diffusion coefficient depending upon the mixing mechanism.
(2) Estimates of *absolute* mixing times were derived by applying validated analytical mixing equations to the same reservoir models.

Comparison of the relative mixing times (Fig. 4) and the absolute mixing time estimates (Fig. 8) show a surprising degree of similarity. Although the values are somewhat different, the ranking of the results is almost identical. Both approaches show pressure differences to be the fastest to equilibrate, with water being fastest of all, followed by gas, oil and viscous oil. Gas contacts and gas density differences are the next fastest, followed by oil contacts and density, and then viscous oil contacts. Diffusion is the slowest process.

Figure 9 shows a plot of relative mixing time versus absolute mixing times. The overall dataset (spanning pressure, contact and density mixing) defines a clear linear trend. A linear regression of the logs of the data (Fig. 9) has a good quality of fit ($R^2 = 0.97$), and an equation of

$$T_{\mathrm{abs}}\,(\mathrm{years}) = T_{\mathrm{rel}}^{1.22}. \qquad (14)$$

Using this equation to convert the estimates of relative mixing time (Fig. 4) into absolute mixing time provides a close match to the calculated mixing times in Figure 8. The results

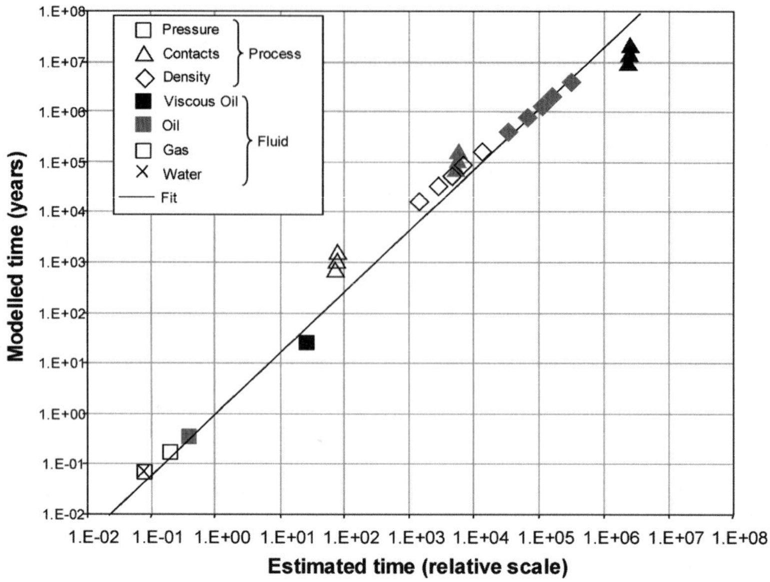


**Fig. 9.** Comparison of relative fluid mixing times (Fig. 4) and absolute fluid mixing times (Fig. 8). The linear fit has the equation $y = x^{1.22}$.

can be translated into a nomogram for rapid estimation of mixing times (Fig. 10). The implication of this is that the fluid mixing times are controlled principally by the factors included in the relative mixing time estimates: volumes moved ($\Delta V$), and drive ($\Delta P \times k/\mu$). In real-life scenarios, both vertical and horizontal permeabilities would have to be taken into account.

## Implications for interpretation of reservoir fluid data

It is now simple to estimate the fluid mixing times for fluid pressure differences, fluid contact differences, fluid density differences and fluid geochemical differences. These mixing times can be calculated for most reservoirs scenarios using the equations discussed herein (equations

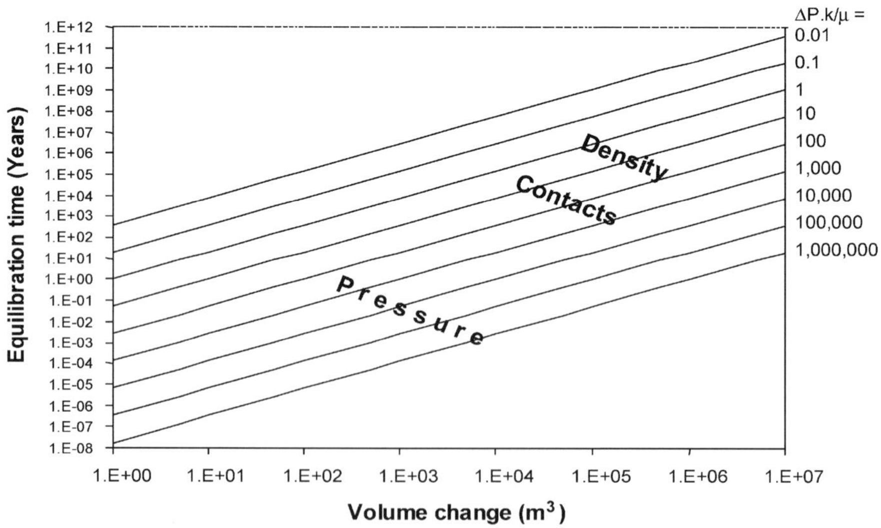


**Fig. 10.** Nomogram for calculating mixing times from knowledge of volume change and drive [$\Delta P(k/\mu)$].

(9)–(11)). If a quicker analysis is required, or is prompted by the poor state of knowledge of the reservoir, then mixing time can be estimated to within an order of magnitude using equations (1)–(3) and (14), which require only knowledge (or an estimate) of the reservoir fluid volume, permeability, fluid viscosity and pressure difference (which can be measured or calculated). Instead of using equation (14) to estimate the mixing time, the output from equations (1)–(3) can be placed onto Figure 10 and the mixing time simply read off.

This knowledge of mixing times can be used to distinguish between gradients and steps in fluid data. If the time available for mixing can be measured (for example using a basin model) or assumed, then this can be compared with the calculated mixing time for the fluid differences observed. If the time available is less than the mixing time, then it is likely that the fluid differences could represent points on a mixing gradient. If, however, the mixing time is much less than the time available, this is strong evidence of a step in fluid properties, reflecting reservoir compartmentalization.

If more than one type of data is available, this process can be used for each data type. If the data reflect processes with different rates, then the simultaneous interpretation of the full dataset should further constrain the degree of reservoir compartmentalization. If a field is known, or can reasonably assumed, to be compartmentalized then the quantitative equations presented here, or the semi-quantitative estimates, can be used to constrain the degree of compartmentalization. This can be done by reducing the permeability of the model until the fluid difference is preserved for the right amount of time. This will give an upper limit of the permeability of the reservoir barrier. In the case of pressure differences, this can be achieved more accurately using equation (7), in which the permeability of the interface between the compartments can be specified separately from the reservoir permeability.

The application of mixing times to the interpretation of reservoir fluid data can be illustrated with reference to pressures from the well characterized Ross Field, UKCS. (For background to the Ross Field, see Smalley & Hale (1996). All data referred to here are from this source). The Ross Field has a network of faults that are likely to be barriers to fluid communication (Fig. 11). Pressure data for the western part of the field shows that there is a 50 psi pressure difference between wells 13/28a-3 and 13/28a-7, and the two wells in the adjacent fault segment (13/28a-2 and 13/28a-5). A simple calculation of how rapidly these pressures would mix, using equation (9), indicated that this would occur within 5 years. Based on basin modelling work, it is unlikely that the pressure difference has arisen so recently. If it was assumed that the pressure difference had been preserved for, say, 1 million years, the calculations show that the

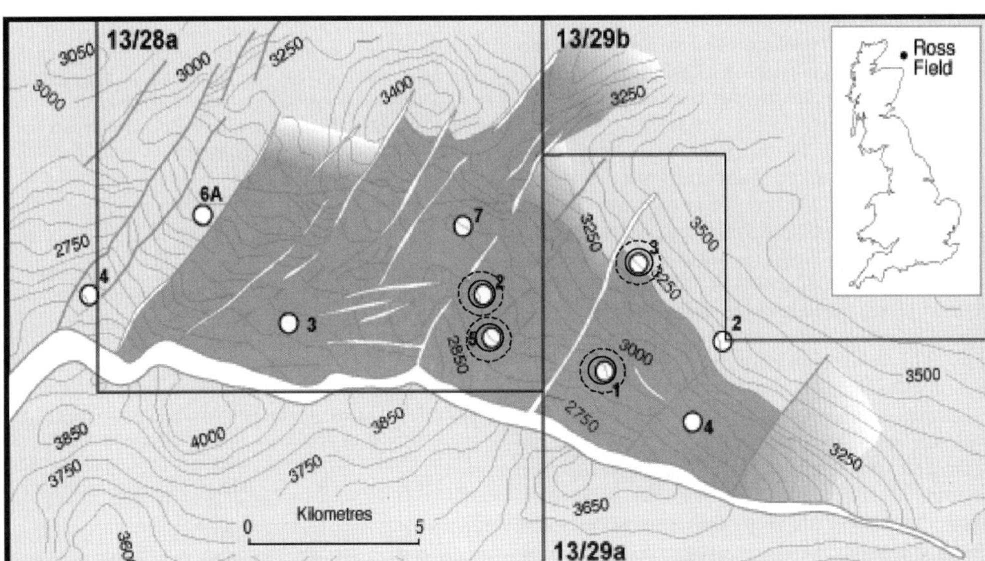


**Fig. 11.** Map of the Ross oil field. Inset shows location. The solid circles around four wells shows the distance diffused by Sr in 40 Ma; the dashed circles are the equivalent for biomarkers in oil.

permeability of the interface between the two segments would have to be about $1 \times 10^{-7}$ mD. Ten million years would demand a permeability of a little less than $1 \times 10^{-8}$ mD. This illustrates how knowledge of fluid mixing times can be used to constrain the *degree* of reservoir compartmentalization, something that would be very valuable for prediction of reservoir production performance. Such estimates of barrier permeability could easily be incorporated into reservoir simulation models as transmissibility multipliers.

In the central part of the Ross field, different $^{87}Sr/^{86}Sr$ RSA values of 0.7148 in well 13/28a-2 and 0.7158 in well 13/28a-5 are observed. Is this difference significant, indicating lack of communication between these two wells? The most likely mechanism for Sr isotope differences to mix is via diffusion. The amount of time available for diffusional mixing is not known. Basin modelling suggested that the reservoir filled with oil about 40 million years ago. If the Sr heterogeneity relates to the filling process, then the maximum amount of time available for mixing would be 40 m.y. Using the diffusion model (equation (11)), and neglecting the increased tortuosity due to the oil saturation, the maximum diffusion distance for Sr would be about 300 m, represented by solid circles around the well symbols on Figure 11. Clearly, there is no way that diffusion alone could mix the Sr between these wells in a sensible time frame. Thus the Sr differences indicate incomplete mixing, not compartmentalization.

Oil biomarker compositions were measured in samples from three wells in the central part of the Ross Field (Fig. 11). Wells 13/28-2 and 13/28a-5 had similar compositions, but the oil in 13/29a-3 has a different composition, related to a higher thermal maturity. If diffusion was the only mechanism leading to fluid mixing, then in 40 Ma the oil would be able to mix over a distance of about 520 m, illustrated by dashed circles around the wells in Figure 11. It is thus possible that the biomarker differences were inherited from the reservoir filling history, and subsequently have not had time to mix completely. This would indicate that the oil chemistry could not be interpreted to indicate compartmentalization. However, it is possible that the biomarker maturity parameters reflect oils of different density (unfortunately, density was not measured in oils from the crucial location). Indeed, there are density differences between the oils in wells 13/29a-1 and 13.29a-3. The density-mixing model (equation (10)) would indicate that density overturn would mix oil densities over the whole central part of the field in less than 3 Ma. Preservation of density differences (or compositional differences that could mix by density-driven overturn) would be good indicators of compartmentalization. This would support the structural evidence for sealing faults between these wells.

## Conclusions

(1) Models have been devised to represent fluid mixing related to pressure differences, contact differences, density differences and chemical differences. The models have all been validated against numerical models.
(2) The mixing times can be explained in terms of (a) the volume of fluid moved, and (b) the driving force for mixing, related to $\Delta P(k/\mu)$.
(3) Some simple examples from a North Sea oil field show the value of the fluid mixing models for distinguishing those fluid variations that are significant, and could reflect compartmentalization, from incomplete mixing that has no significance for fluid communication.

BP is thanked for funding the pressure equilibration modelling and for permission to publish.

## References

CARSLAW, H. S. & JAEGER, J. C. 1959. *Conduction of Heat in Solids*, 2nd edition. Oxford Publications.

COUTTS, S. D. 1999. Aquifer behaviour during Brent depressurization and the impact on neighboring fields. *Society of Petroleum Engineers, Reservoir Engineering and Evaluation*, **2**, 53–61.

EDMAN, J. D. & BURK, M. K. 1999. Geochemistry in an integrated study of reservoir compartmentalization at Ewing Bank 873, offshore Gulf of Mexico. *Society of Petroleum Engineers, Reservoir Engineering and Evaluation*, **2**, 520–526.

ENGLAND, W. A., MACKENZIE, A. S., MANN, D. M. & QUIGLEY, T. M. 1987. The movement and entrapment of petroleum fluids in the subsurface. *Journal of the Geological Society, London*, **144**, 327–347.

ENGLAND, W. A., MUGGERIDGE, A. H., CLIFFORD, P. J. & TANG, Z. 1995. Modelling density-driven mixing rates in petroleum reservoirs on geological time scales, with application to the detection of barriers in the Forties Field (UKCS). *In*: CUBITT, J. M. & ENGLAND, W. A. (eds) *The Geochemistry of Reservoirs*. Geological Society, London, Special Publications, **86**, 185–201.

GARDNER, G. H. F., DOWNIE, J. & KENDALL, H. A. 1962. Gravity segregation of miscible fluids in linear models. *Society of Petroleum Engineers Journal*, **95**, 95–104.

HWANG, R. J. & BASKIN, D. K. 1994. Reservoir connectivity and oil homogeneity in a large-scale reservoir. *Geoscience Geo94*, **2**, 529–541.

HWANG, R. J., AHMED, A. S. & MOLDOWAN, J. M. 1994. Oil composition variation and reservoir continuity: Unity Field, Sudan. *Organic Geochemistry*, **21**, 171–188.

KAUFMAN, R. L., AHMED, A. S. & ELSINGER, R. L. 1990. Gas chromatography as a development and production tool for fingerprinting oils from individual reservoirs: applications in the Gulf of Mexico. *In*: *9th Research Conference Proceedings*. Gulf Coast Section of the Society of Economic Paleontologists and Mineralogists, Earth Enterprises, Austin, Texas, USA, 263–282.

KAUFMAN, R. J., KABIR, C. S., ABDUL-RAHMAN, B., QUTTAINAH, R., DASHTI, H., PEDERSON, J. M. & MOON, M. S. 1998. Characterizing the Greater Burgan field using geochemical and other field data. Society of Petroleum Engineers Paper **49216**, SPE Annual Technical Conference and Exhibition, New Orleans, Louisiana, 27–30 September 1998.

KHAVARI-KHORASANI, G., MICHELSEN, J. K. & DOLSON, J. C. 1998. The factors controlling the abundance and migration of heavy vs. light oils, as constrained by data from the Gulf of Suez. Part II. The significance of reservoir mass transport processes. *Organic Geochemistry*, **29**, 283–300.

MEARNS, E. W. & MCBRIDE, J. J. 2001. Strontium isotope analysis can help define compartmentalization. *Oil & Gas Journal*, **99**, 70–75.

MUGGERIDGE, A. H., ABACIOGLU, Y., ENGLAND, W. A. & SMALLEY, P. A. 2004a. The rate of pressure dissipation from abnormally pressured compartments. *American Association of Petroleum Geologists Bulletin* (in press).

MUGGERIDGE, A. H., ABACIOGLU, Y., ENGLAND, W. A. & SMALLEY, P. A. 2004b. The dissipation of anomalous pressures in the subsurface. *Journal of Geophysical Research* (in press).

MUSKAT, M. 1930. Distribution of non-reacting fluids in the gravitational field. *Physical Review*, **35**, 1384–1393.

NEUZIL, C. E. 1995. Abnormal pressures as hydrodynamic phenomena. *American Journal of Science*, **295**, 742–786

SAGE, B. H. & LACEY, W. N. 1938. Gravitational concentration gradients in static columns of hydrocarbon fluids. *Transactions of the American Institute of Mechanical Engineers*, **132**, 120–131.

SCHULTE, A. M. 1980. Compositional variations within a hydrocarbon column due to gravity. Society of Petroleum Engineers Paper **9235**, presented at the 1980 SPE Annual Technical Conference and Exhibition, Dallas, 21–24 September.

SMALLEY, P. C. & HALE N. A. 1996. Early identification of reservoir compartmentalization by combining a range of conventional and novel data types. *Society of Petroleum Engineers, Formation Evaluation*, **September 1996**, 163–169.

SMALLEY, P. C., DODD, T. A., STOCKDEN, I. L., RAHEIM, A. & MEARNS, E. W. 1995. Compositional heterogeneities in oilfield formation waters: identifying them, using them. *In*: CUBITT, J. M. & ENGLAND, W. A. (eds) *The Geochemistry of Reservoirs*. Geological Society, London, Special Publications, **86**, 59–69.

SMALLEY, P. C., GOODWIN, N. S., DILLON, J. F., BIDINGER, C. R. & DROZD, R. J. 1996. New tools target oil quality sweetspots in viscous oil accumulations. *In*: *Proceedings, Volume W: Formation Evaluation and Reservoir Geology*. Society of Petroleum Engineers, Annual Technical Conference, Denver, 911–917.

SMALLEY, P. C., GOODWIN, N. S., DILLON, J. F., BIDINGER, C. R. & DROZD, R. J. 1997. New tools target oil-quality sweetspots in viscous oil accumulations. *Society of Petroleum Engineers, Reservoir Engineering*, **August 1997**, 157–161.

STØLUM, H. H. & SMALLEY, P. C. 1992. A deterministic method for assessing reservoir communication based on strontium fingerprinting. *In*: *European Petroleum Conference 1992, Proceedings*, Volume 1. Society of Petroleum Engineers, 451–464.

STØLUM, H. H., SMALLEY, P. C. & HANKEN, N. M. 1993. Prediction of large-scale communication in the Smørbukk fields from strontium fingerprinting. *In*: PARKER, J. R. (ed.) *Petroleum Geology of Northwest Europe*. Proceedings of the 4th Conference, 1421–1432.

SWARBRICK, R. E. *ET AL*. 2000. Integrated study of the Judy Field Block 30/7a—an overpressured Central North Sea oil/gas field. *Marine and Petroleum Geology*, **17**, 993–1010.

WESTRICH, J. T., FUEX, A. M., O'NEAL, P. M. & HALPERN, H. I. 1999. Evaluating reservoir architecture in the northern Gulf of Mexico with oil and gas chemistry. *Society of Petroleum Engineers, Reservoir Evaluation & Engineering*, **199**(2), 514–519.

WHITSON, C. 1994. Compositional gradients in oil reservoirs. *In*: University of Tulsa Centennial Petroleum Engineering Symposium. Society of Petroleum Engineers, Tulsa, OK, 443–459.

WINTERHALDER, J. R. & HANN, J. H. 1991. Wanaea/Cossack dynamic aquifer model: multidisciplinary approach. *In*: Asia–Pacific Conference, Proceedings. Society of Petroleum Engineers, 615–628.

WORDEN, R. H., OXTOBY, N. H. & SMALLEY, P. C. 1998. Can oil emplacement prevent quartz cementation in sandstones? *Petroleum Geoscience*, **4**, 129–138.

# New insights into reservoir filling and mixing processes

JOHN G. STAINFORTH

*Shell International E & P, 200 N. Dairy Ashford, Houston, TX 77079, USA*
*(e-mail: John.Stainforth@shell.com)*

**Abstract:** A common assumption is that hydrocarbon charge homogenizes with the petroleum already in a trap: thus, compositional gradients reflect only subsequent segregation of the petroleum under the combined influences of gravity, temperature and diffusion. Since such homogenization would entail an unfavourable generation of potential energy, a more plausible hypothesis is that hydrocarbons stack into traps roughly in the density sequences in which they arrive. The commonly accepted model suggests that homogeneous and gravitationally unstable petroleum columns move towards graded equilibrium ones, whereas the hypothesis preferred here implies that they gradually diffuse towards equilibrium from the opposite direction, i.e. poorly mixed initial states.

According to this hypothesis, trends of oil GOR and bubble point ($P_b$) are controlled by either (a) the charge GOR or (b) the evolving PT conditions at the GOC, depending on the gassiness of the charge. By contrast, API gravity trends mainly reflect the integrated maturity histories of source rock kitchens during the trap filling regardless of the charge gassiness. Because trap and kitchen histories are usually (very) different, a wide spectrum of API and GOR (and $P_b$) combinations is possible, from minor API gradients coupled with large GOR gradients and *vice versa*.

Charge modelling has moved beyond simply predicting the presence or absence of hydrocarbons to the prediction of hydrocarbon type (oil versus gas), and more recently to the prediction of hydrocarbon qualities. Amongst these qualities, the API gravity and GOR of oil, and the CGR of gas have very strong economic impacts, especially in the deepwater. API gravity, for example, affects the economics of an oilfield in at least three major ways: via (1) the recovery factors, (2) the production rates, and (3) the oil value at the refinery. Often these properties vary widely from pool to pool, and even in a given pool. Yet these variations are usually poorly understood in advance of production.

As our charge modelling has advanced, it has become clear that reliable predictions of oil and gas properties require not only an excellent charge model but also a realistic model of trap filling, spilling and leaking and of the major processes taking place in each reservoir. Thus, we have moved from simple basin models to ones that are integrated with trap filling models.

This paper considers two end-member hypotheses for trap filling: one in which the petroleum charge mixes perfectly in the trap as it fills, and another in which it does not mix at all. Clearly, the real world lies variously between these two extremes. Next some model results are compared with field data. Such comparisons suggest that the model that assumes poor mixing more closely approximates reality than the other.

## API gravity and GOR of oils

In the absence of reliable forward modelling methods, predictions of API gravity have often had to rely on poor to very poor correlations with variables such as present-day reservoir depth (e.g. Fig. 1a), or temperature (cf. Yu *et al.* 2002). Sometimes, a single API gravity value is assumed for an entire pool, based on limited data from one or two exploration wells, even though it is well known that APIs can vary greatly within a single pool (England *et al.* 1987).

Likewise, the solution GOR of oil correlates poorly with present-day depth (e.g. Fig. 1b). Often, it too is estimated using simple black oil correlations that assume saturation of oil with respect to gas (e.g. Beggs 1987; Glaso 1980), or some fixed degree of under-saturation. However, regional oil data usually reveal a wide range of saturations that seem to defy analysis (e.g. Fig. 2). Another approach has been to use empirical correlations between the solution GOR and the API gravity of the oil. These often correlate well in a given petroleum pool (Fig. 3a), but regionally there is usually large scatter (Fig. 3b), and globally there appears to be no single, consistent relationship at all (see Fig. 9). These limitations severely reduce the reliability of these empirical predictions of GOR.

*From*: CUBITT, J. M., ENGLAND, W. A. & LARTER, S. (eds) 2004. *Understanding Petroleum Reservoirs: towards an Integrated Reservoir Engineering and Geochemical Approach*. Geological Society, London, Special Publications, **237**, 115–132. 0305-8719/04/$15.00 

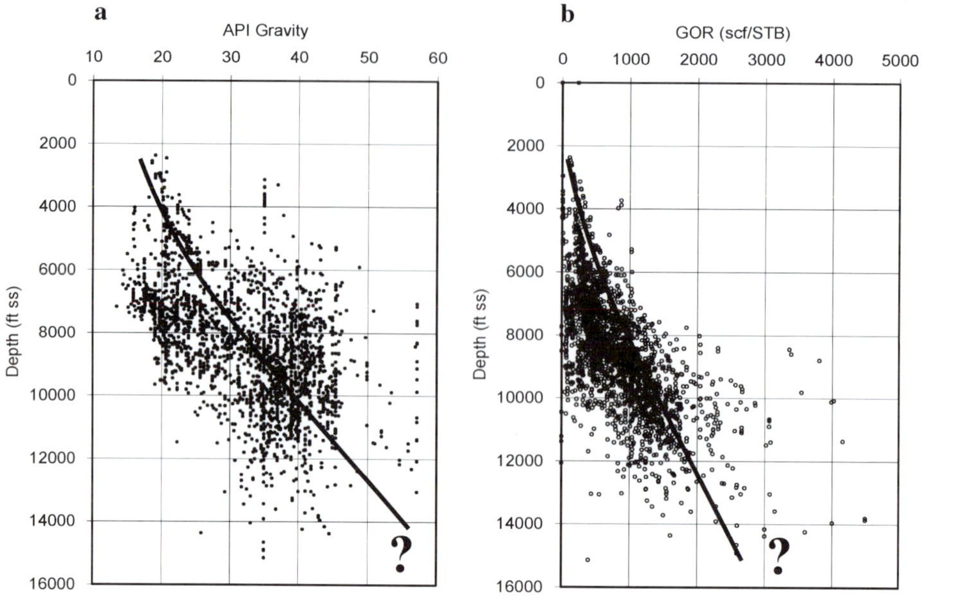


**Fig. 1.** Trends of (**a**) API gravity and (**b**) solution GOR of oil samples with depth, in a West African Tertiary delta. Note the poor correlations with depth.

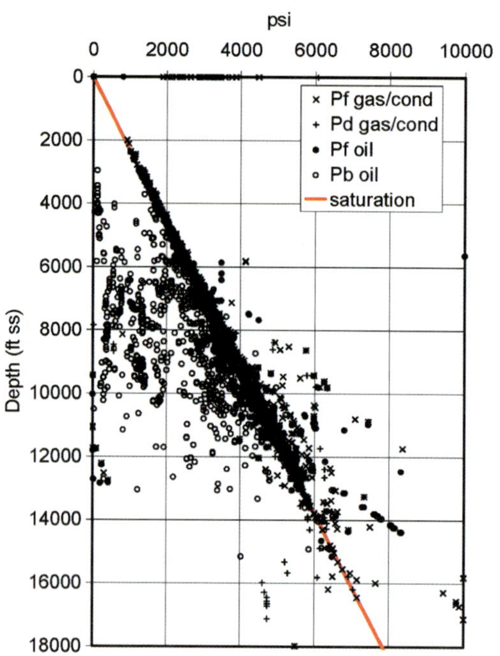


**Fig. 2.** Trends of fluid pressures versus depth in a West African Tertiary Delta: fluid pressures ($P_f$), saturation pressures (bubble points of oils, $P_b$, and dew points of gas-condensates, $P_d$). Note the wide scatter around the line of saturation for hydrostatic reservoirs and the extreme under-saturation of some of the oils.

## Saturation pressure versus depth

The saturation pressures (bubble and dew points) of petroleum liquid and vapour phases are closely related to the GOR of the liquid and the CGR of the vapour and strongly influence the production behaviour of these phases. Yet geologists and petroleum engineers have often assumed very simple values for these saturation pressures, for example:

(a) values constant throughout the liquid and vapour phases (Fig. 4a), or
(b) values equal to the reservoir pressures, i.e. the liquid and vapour phases are saturated with respect to each other at all depths in the reservoir (Fig. 4b).

Unfortunately, detailed PVT data often depart radically from such simple assumptions. Usually, the saturation pressure (dew and bubble point) trends are not constant or linear with respect to depth, but curved (e.g. Grant 1959; Havlena 1968; Nutakki *et al.* 1996; Araque & Auxiette 2002), and only converge on the reservoir pressures near the gas–oil contact (GOC) separating the two phases (Fig. 4c,d). This suggests that the two phases may only be in local thermodynamic equilibrium near the GOC. Often, the degree of under-saturation of the liquid phase decreases so rapidly with depth that the bubble points actually decrease with depth in an absolute sense (Fig. 4d).

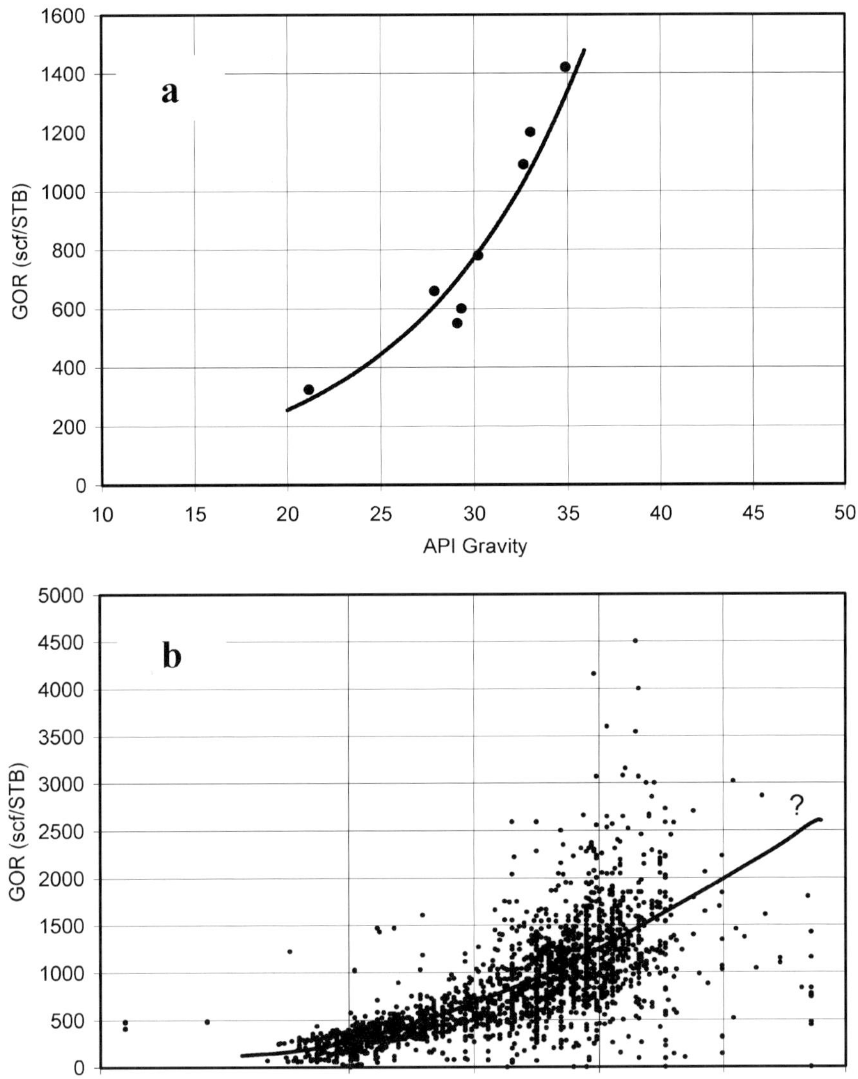


**Fig. 3.** Solution GOR and API gravity of oils from a West African Tertiary Delta: (**a**) in a single reservoir, showing good correlation, and (**b**) in the whole delta, showing poor correlation.

## Compositional gradients

Reservoir geochemistry often begins with the premise that petroleum in a given pool is so well mixed that the composition is more or less homogeneous. Separate compartments are invoked whenever the compositions at different locations differ by more than accepted amounts, especially if the reservoir pressures also imply a lack of communication.

However, many authors in the past half century have described strong compositional gradients in pools that are apparently in pressure communication throughout. This phenomenon is often ascribed to gravitational segregation acting on large or heavy oil columns (e.g. Grant 1959; Clinton & Ruiz 1968; Wadman *et al.* 1979; Schulte 1980; Creek & Schrader 1985; Ratulowski *et al.* 2000; Araque & Auxiette 2002). Yet global data (e.g. Figs 5–7) show that

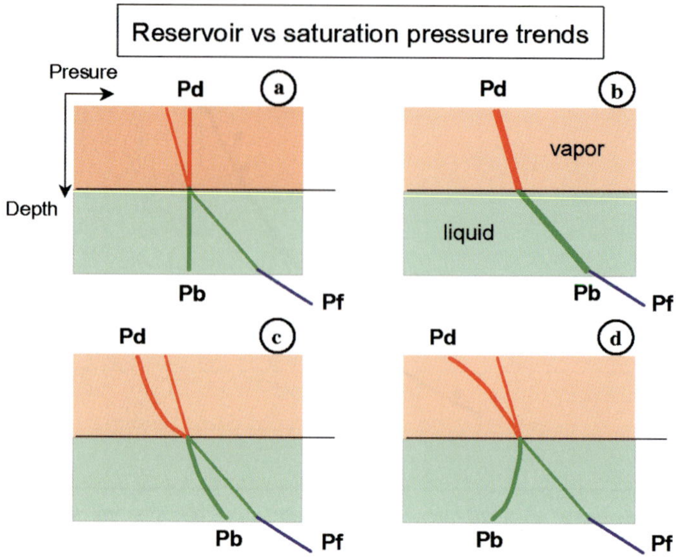


**Fig. 4.** Fluid pressure ($P_f$) and saturation pressure trends versus depth in reservoirs having both vapour and liquid columns. (**a**) Assumed constant dew point, $P_d$, and bubble point, $P_b$; (**b**) assumed linear $P_b$ and $P_d$ trends equal to $P_f$ at all depths; (**c**) observed concave up trends of $P_b$ and $P_d$; (**d**) observed concave down $P_b$ and $P_d$ trends (note absolute decrease in $P_b$ with depth).

these gradients can be strong in both large and small columns, and in heavy oils, light oils and gas-condensates alike (e.g. Sage & Lacey 1939; Espach & Fry 1951; Hunt 1979; Tissot & Welte 1979; Bath *et al.* 1980; Hirschberg 1984; Metcalfe *et al.* 1985; Nutakki *et al.* 1996). These compositional gradients manifest themselves in all the gross properties of the oils, such as the API gravity (Fig. 5), the solution GOR (Fig. 6) and the bubble point (Fig. 7), which closely follows the GOR. These compositional variations are not necessarily 'more or less linear with depth unless the reservoir fluids are near critical', as predicted by gravitational segregation theory (Hoier & Whitson 2001): in fact, they are often markedly curved.

It is important to note that the depth axis in these plots is not generally the depth (below the top of the oil column) in a single well but in many wells penetrating a given reservoir. Thus,

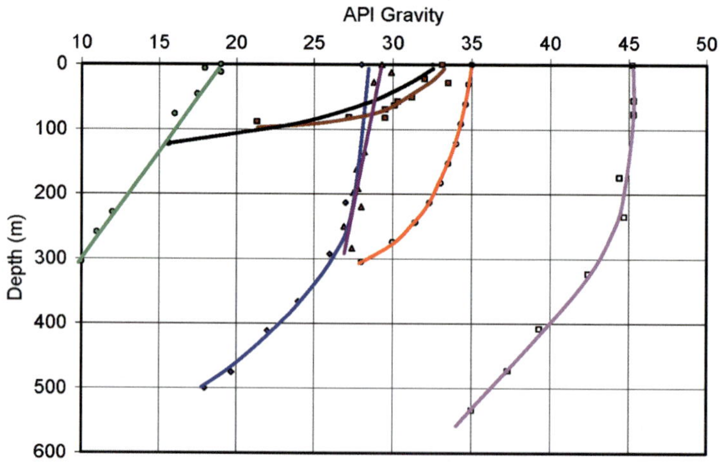


**Fig. 5.** Trends of API gravity of oils with depth below top of oil column in single reservoirs with multiple wells, from various fields around the world. Note strong, often curved, trends regardless of whether oils are heavy or light.

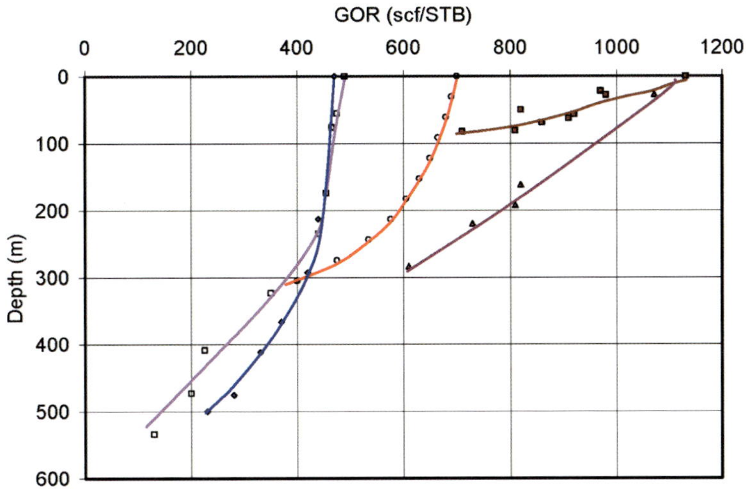


**Fig. 6.** Trends of GOR of oils with depth below top of oil column in single reservoirs with multiple wells, from various fields around the world. Again, note the strong and often curved gradients regardless of the gas content of the oils.

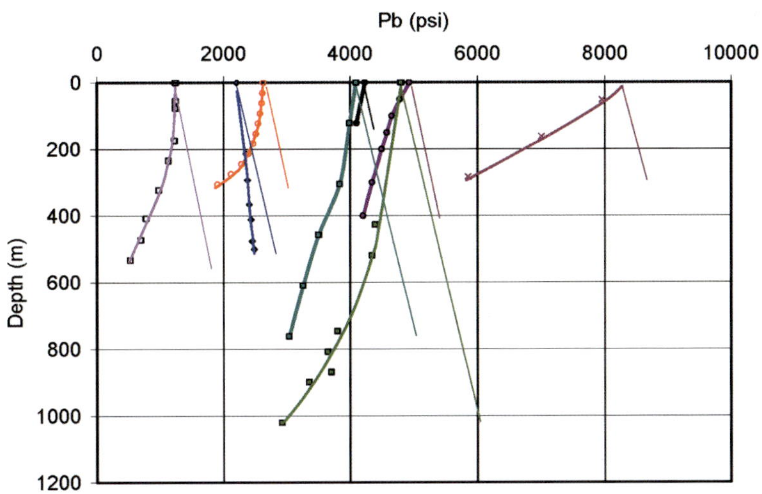


**Fig. 7.** Trends of bubble point ($P_b$) of oils with depth below top of oil column in single reservoirs with multiple wells, from various fields around the world. These correlate with the GOR trends. The thin lines are approximate hydrostatic gradients for reference only (they should not be taken literally to imply that the reservoirs are necessarily saturated at the top of the oil columns).

these plots really show the lateral changes in composition in the reservoir plotted against the vertical component of the sampling positions. Since the lateral distances between the wells are often quite large, this usually introduces some ambiguity into the interpretation of the plots. Do they show compositional gradients in single pools, or compositional changes across a number of barriers/baffles in a large field (e.g. Carrigan *et al.* 1998), or a combination of the two?

The compositional variations in the bubble point (or GOR) and API gravity of oils are usually well correlated in single oil pools (Figs 8 & 9). So we might expect these correlations to be controlled by the type or maturity of the petroleum charge, and for common global patterns to emerge. However, this does not appear to be the case. Correlations from different basins and geological settings are distributed widely in bubble point (or GOR) versus API gravity space and there appear to be no reliable

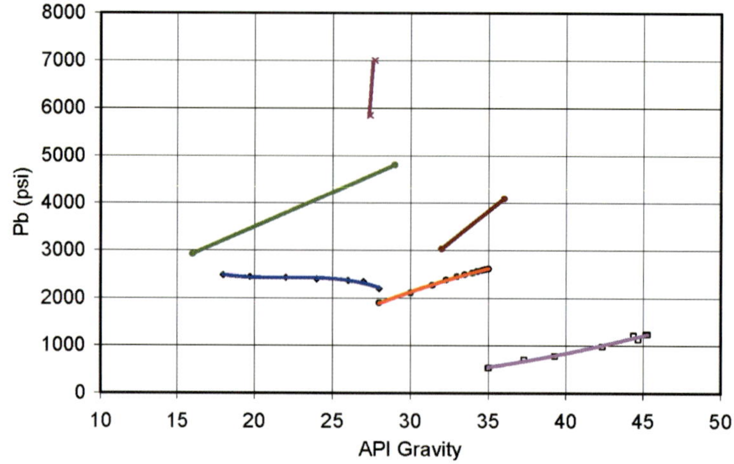


**Fig. 8.** Bubble point ($P_b$) versus API gravity of oils in single reservoirs from six fields around the world. Note the excellent correlations but their wide distribution in $P_b$-API space.

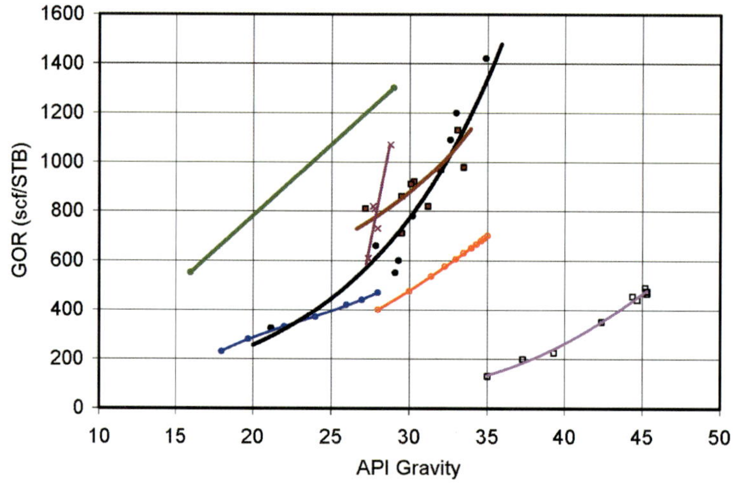


**Fig. 9.** GOR versus API gravity of oils in single reservoirs from seven fields around the world. Note the excellent correlations but their wide distribution in GOR-API space.

rules for light mature oils versus heavy immature oils, for example (Fig. 8). Instead, large gradients in API gravity can be associated with relatively mild gradients in bubble point and GOR, and *vice versa*, regardless of whether the oils are heavy or light (Figs 8 & 9).

## Models of trap filling

One popular model of trap filling is that reservoirs are well mixed, even where the composition of the incoming hydrocarbon charge must have changed greatly during the filling of the trap. This would demand that the incoming charge mixes so well with all the previously trapped petroleum that the resultant composition is more or less homogeneous and reflects the average charge during filling (Fig. 10a). Where compositional gradients are shown to exist (Fig. 10b), these have usually been ascribed to gravitational segregation in which the effects of gravity and temperature are balanced by diffusion, much like the Earth's upper atmosphere (e.g. Hoier & Whitson 2000). Theoretical studies suggest that the thermal (Soret) effect can either enhance or weaken the gravitational effect (Firoozabadi *et al.* 1996; Firoozabadi 1999),

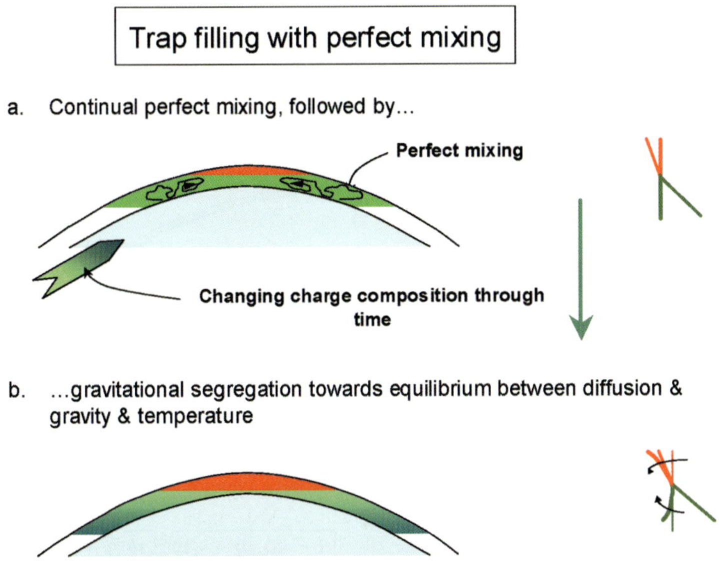


**Fig. 10.** Models of trap filling. Perfect mixing. In this widely accepted model, (**a**) a charge stream with a continuously changing composition mixes with the existing petroleum in the trap to produce homogeneous liquid and/or vapour phases, and (**b**) given enough geological time, these gravitationally segregate towards compositional gradients in which diffusion is in equilibrium with the gravity and temperature fields. The small inset figures show bubble and dew point and reservoir pressure curves in PT-space, as in Figure 4.

although some workers claim that the Soret effect is usually mitigating (e.g. Hoier & Whitson 2001). Theoretically, a counteracting Soret effect could be sufficiently strong to render a density profile unstable, leading to convection of the whole column and obliteration of the compositional gradient.

However, current thermodynamic theories of compositional equilibrium under the combined influence of gravity and temperature fields do not adequately explain the large compositional gradients that are often encountered, except at conditions close to critical (Schulte 1980; Holt *et al.* 1983; Creek & Schrader 1985; England *et al.* 1987; Nutakki *et al.* 1996). It is now quite common for the phenomenon of strong compositional grading to be associated with 'near-critical' fluids, but the definition of 'near-critical' fluids is rather broad and hazy! Another problem with these theories is that they often do not predict the shape of these compositional depth trends at all well. In fact, Hoier & Whitson (2001) doubt that most petroleum fields satisfy the fundamental assumptions in these models, especially that of zero mass flux (i.e. stationary state equilibrium).

A key factor in trap filling is the general inadequacy of diffusive mixing, as emphasized by England *et al.* (1987), England & Mackenzie (1989) and many subsequent authors. Many traps have aspect ratios like those shown in Figure 11a, with lateral dimensions that are an order of magnitude greater than the vertical, which is in turn an order of magnitude greater than the reservoir thickness. Mixing in this common type of trap frequently has to occur over large lateral distances that are (at least) an order of magnitude greater than characteristic diffusion distances. (For the sake of simplicity in this discussion, important differences in the diffusion coefficients of the different species in the oil are ignored. For example, the diffusivities of the gas molecules dissolved in the oil may be an order of magnitude greater than those of asphaltenes and resins, so that it should be easier for the gases dissolved in oils, and therefore the GOR, to equilibrate with the temperature and gravity fields, than the large and heavy oil compounds that have a dominating effect on API gravity.)

The inadequacy of lateral diffusive mixing is particularly acute in thin reservoirs that have filled in only the past few million years, which is often the case in Tertiary deltas, for example. By contrast, petroleum may mix effectively by vertical diffusion in reservoirs with large gross thicknesses that have had relatively long filling times (Fig. 11b). An example of the latter configuration is the Fahud Field in Oman, where oil analyses show very similar molecular

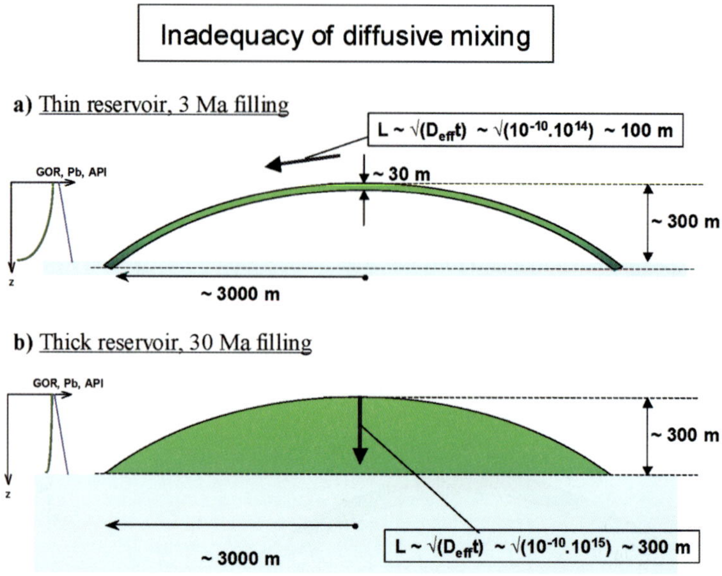


**Fig. 11.** The inadequacy of diffusive mixing, even after geological time periods. (**a**) Very poor mixing in a thin reservoir because the diffusive length scale $L$ is small relative to the lateral extent of the reservoir. ($D_{\text{eff}}$, effective diffusion coefficient, $m^2/s$; $t$, time period, s). (**b**) Moderate diffusive mixing in a thick gross reservoir. The small inset figures show generalized API gravity and/or GOR, and bubble point and reservoir pressure depth trends as in Figures 5–7.

compositions throughout the field, indicating a high degree of mixing on both vertical and lateral scales (Nederlof *et al.* 1995). The uniform oil composition indicates that the Natih and Huqf oils have completely commingled during the Tertiary and that no significant permeability barriers exist within the field, at least on this geologic time scale.

The inadequacy of diffusion begs the question of whether more efficient, advective mechanisms (such as convection) might be able to mix the petroleum in fields (e.g. Montel & Gouel 1985; Jacqmin 1990; Westrich *et al.* 1996, 1999; Ratulowski *et al.* 2000). Indeed, the inadequacy of diffusive mixing actually *promotes* local convection because it preserves the buoyancy associated with an influx of lightening petroleum. However, paradoxically the way in which this probably occurs in traps ensures that whole traps will *not* often be mixed, as explained now. Consider streams of charge entering a trap along preferential pathways resulting from geometrical focusing superimposed on the permeability architecture (Fig. 12). Commonly, but not necessarily, hydrocarbon charge becomes progressively lighter (less dense) with time as a source rock kitchen matures. Because diffusive mixing is slow, this lightening charge stream creates local fluid buoyancy that promotes gravitational instability. This causes the lighter charge to advect through the heavier trapped petroleum, towards the top of the trap, sufficiently fast that there is only limited mixing with the ambient petroleum (Fig. 12a). This, in turn, gradually forces the underlying petroleum downwards in the trap. The incremental charge may separate into vapour and liquid phases, given favourable PT conditions, in which case a vapour column can grow in addition to a liquid column. However, the details of individual cases are complicated by competition between the filling and various trap-leaking mechanisms, many of which favour gas over oil.

Less often, the hydrocarbon charge entering a trap becomes heavier with time. In this case, the heavier petroleum merely stacks onto the base of the existing petroleum column, again with limited diffusive mixing with the overlying petroleum. Thus, whether the column builds from the top or the base, or from intermediate positions, the petroleum tends to stack into the trap in a sequence of continually increasing density downwards. (Often, though, the charge enters a trap mainly from one side which, combined with the common inadequacy of mixing, can lead to lopsided compositional distributions in reservoirs.) Diffusive mixing attempts to move the heterogeneous mix towards one that is in equilibrium with the gravity and temperature fields

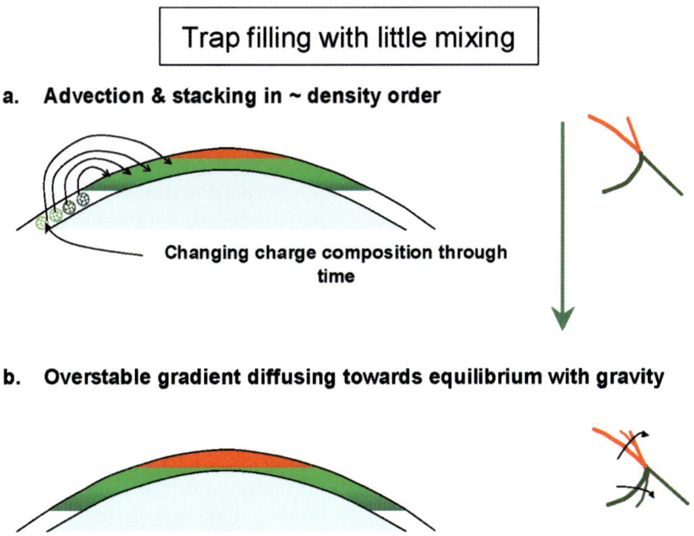


**Fig. 12.** Models of trap filling. Very poor mixing. In this model, in contrast to Figure 10, (**a**) a charge stream with a continuously changing composition stacks into the reservoir in approximate density order. Lighter petroleum than pre-existing petroleum advects to the top of the trap (or GOC), whereas heavier petroleum stacks onto the bottom of the column, or spills. (**b**) Given enough geological time, these highly graded columns diffuse towards less graded columns that are in equilibrium with the gravity and temperature fields.

(Fig. 12b). However, in the common cases where diffusive mixing is limited, the API gravity profile mainly reflects the changes in charge composition during trap filling rather than equilibrium with the gravity and temperature fields. This filling model is similar to that favoured by Mearns & McBride (1999) to explain 'fossilized' strontium isotope ratios ($^{87}Sr/^{86}Sr$) in formation waters in hydrocarbon columns.

This charging and trap-filling scenario explains the large compositional gradients in API gravity often observed. Charge modelling indicates that the range of composition of petroleum entering a trap is usually much larger than the modest range expected from gravitational segregation acting upon a homogenized fluid. According to the proposed model, the compositional profile changes very gradually (by diffusion) from a strongly graded and poorly mixed one towards a highly mixed and weakly graded one. It gradually moves from gravitational over-stability (Fig. 12a)—a state in which the gradient of density is greater than that in equilibrium with diffusion—towards a marginal, equilibrium stability (Fig. 12b). This is opposite to the implicit assumption that a homogeneous and unstable petroleum column (Fig. 10a) segregates towards a marginally stable equilibrium condition (Fig. 10b). That hypothesis involves an implausible move over an unnecessary potential energy hill: from a heterogeneous charge influx, through a supposedly homogeneous and gravitationally unstable intermediate state, and back into a heterogeneous stable mixture.

## *Testing the trap-filling model*

The two end-member models of trap filling, the first assuming perfect mixing in the trap and the second no mixing, were tested against data from a variety of deepwater fields worldwide. These models represent extreme versions of the two alternative hypotheses outlined above. Gravitational and thermal segregation (Soret effect) are not modelled. The computer code that we used to test these models also includes a number of different spilling and leaking mechanisms (hydraulic failure, diffusive seepage etc.), and a model for *in situ* biodegradation of oil controlled by temperature and residence time in the reservoirs. The specifics of these models are proprietary.

Comparing model results with field data is a long and convoluted process, complicated by non-unique results, in that one can obtain roughly the right results for myriad wrong reasons. To reduce this risk, the model is calibrated through several distinct steps:

(a) calibration of the thermal and maturity model;

(b) calibration of the predicted total petroleum volumes to field reserves estimates;
(c) ditto, for individual reservoir volumes.

Only then can we sensibly compare the predicted petroleum properties and their variations with the actual data from individual reservoirs.

## *Depth trends of API gravity*

The perfect mixing model frequently provides a poor fit to the distribution of petroleum properties in vertically stacked reservoirs and fails to account for the compositional grading within the individual reservoirs of the stack. Conversely, the second end-member model often fits the observed data surprisingly well, considering that it implies no mixing at all. This supports the contention that the petroleum in many, if not all, fields is poorly mixed. We show two examples here.

The first example is a West African field comprising a number of stacked Tertiary reservoirs. Figure 13 compares the vertical distribution of API gravity, predicted with the non-mixing model, with the field data. (The model reservoirs are a simplified representation of the reservoirs in this field in that sands that are in vertical communication are rolled into single gross reservoirs. This means that we are only looking to see whether the model reproduces the essential characteristics of the data rather than the details.) The earliest charge first enters the deepest of these reservoirs (red), which fills from the top down. This is because the oil charge becomes progressively lighter with time: the light charge works its way through the heavier surrounding oil to the top to the trap, forcing the rest of the petroleum downwards. Eventually, the petroleum in this deepest reservoir reaches the spill-point or a vertical leak point and the spilt oil moves up into the next shallower reservoir (blue), and so on. (The modelled filling is actually somewhat more complicated than this in that the reservoirs have gas caps during part or all of their filling histories and gas also leaks preferentially from the top of the traps.) The model predicts fairly strong gradients in the individual reservoirs, as a result of the lack of mixing, like those observed. The vertical distribution of the API gravity in this example is a common one in stacked reservoirs in Tertiary deltas. Models with efficient mixing of the petroleum in each reservoir do not produce these strong gradients.

These gradients in oil density largely reflect the evolving maturity of the source rock kitchen during trap filling. In this case, our dynamic model of biodegradation predicts only a minor effect on the oil gravity in the shallowest reservoir. Generally, we find that in stacked reservoirs

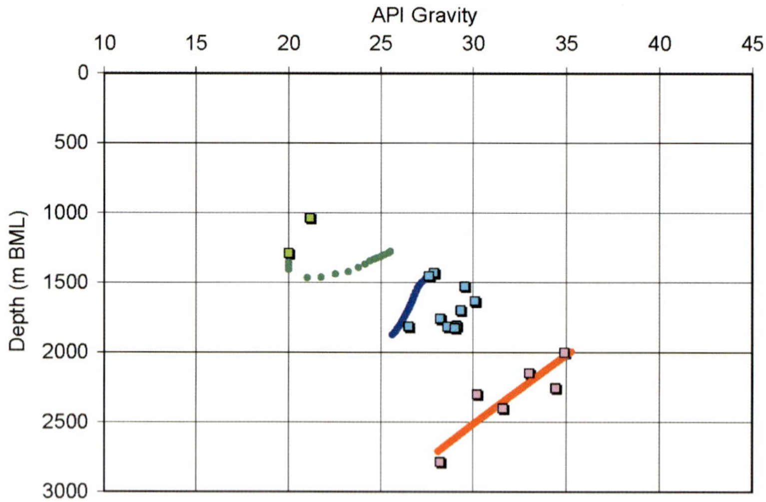


**Fig. 13.** Model predictions of API gravities of oils in a West African Tertiary delta with several stacked reservoirs, colour-coded green, blue and red, compared with data. (The closely spaced and overlapping dots show the modelled trends in the reservoirs, each dot corresponding to increments of oil charge that entered the trap in successive small time steps, whereas the squares are the data.) Because of the simplified reservoir description in the model, it is only possible to compare the general characteristics of the modelled and observed distributions of API gravity.

biodegradation is not a systemic problem, since it takes significant geological time for petroleum to cascade up into the shallowest reservoirs where it is susceptible to biodegradation. In such cases, the effects of biodegradation on API gravity appear to be considerably subordinate to those of source rock maturation.

The second example is of another West African field, but in this case the oil is mainly in one reservoir only. The reservoir filling model, with no mixing, produces a strong and highly curved API depth trend that closely matches the observations (see Fig. 16). As discussed below, this curvature results from subtle interplay between the reservoir geometry and history and the kitchen geometry and history.

## *Depth trends of solution GOR of oil*

The model predictions of solution GOR of oil shown here rely on standard correlations with bubble point pressure, temperature, and the oil and gas gravities (Glaso 1980). We find that such PVT models match field data unexpectedly well when combined with the dynamic trap-filling model with no mixing (e.g. Figs 14 & 15). This implies that more elaborate PVT models for phase separation are superfluous unless they are linked to dynamic models of trap filling.

Our model often predicts oils that are under-saturated with respect to gas, at the present-day reservoir conditions, by 10–20% or more. In some parts of the world very under-saturated oils are common; for example, oils from the deepwater Gulf of Mexico are about 40% under-saturated on average (Dindoruk & Christman 2001). In that area, a major contributing factor to the under-saturation is the gas-leanness of the dominant oil-prone Type II source rocks. However, that is not the only cause of the under-saturation, and in many situations with Type II/III and III source rocks (e.g. the stacked Tertiary reservoir example, Fig. 14), the modelled charge was over-saturated with gas, and the reservoirs had gas caps during much of their filling histories. Our model only saturates the oil at the gas–oil contacts, and these are the only places where the modelled petroleum is in equilibrium with the present-day PT conditions. In such cases, the present-day trend of GOR in the oil column reflects the history of the PT conditions in the trap that each increment of charge experienced during the trap-filling process. (Situations may also arise where a gas cap is

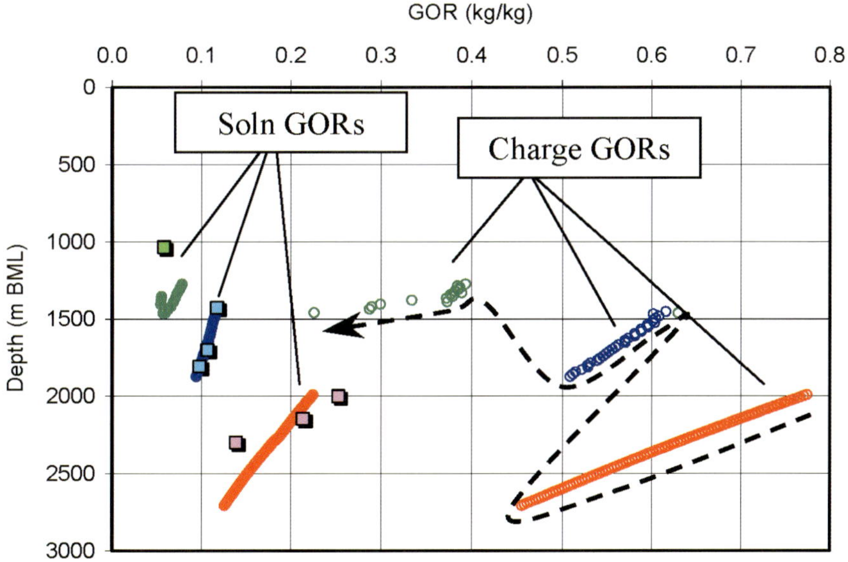


**Fig. 14.** Model predictions of GOR of oils in a West African Tertiary delta with three main stacked reservoirs, compared with data. The modelled GORs (overlapping filled circles) and the observed GORs (squares) are much lower than the corresponding modelled charge GORs of the oil increments shown in their present-day positions (open circles). This suggests that the observed solution GORs, in this case, are controlled mainly by the PVT history of the trap rather than the charge GORs. The dashed line with the arrowhead shows the movement direction of the petroleum during trap filling, with oldest charge now at the deepest part of the green reservoir and the most recent charge at the crest of the red reservoir.

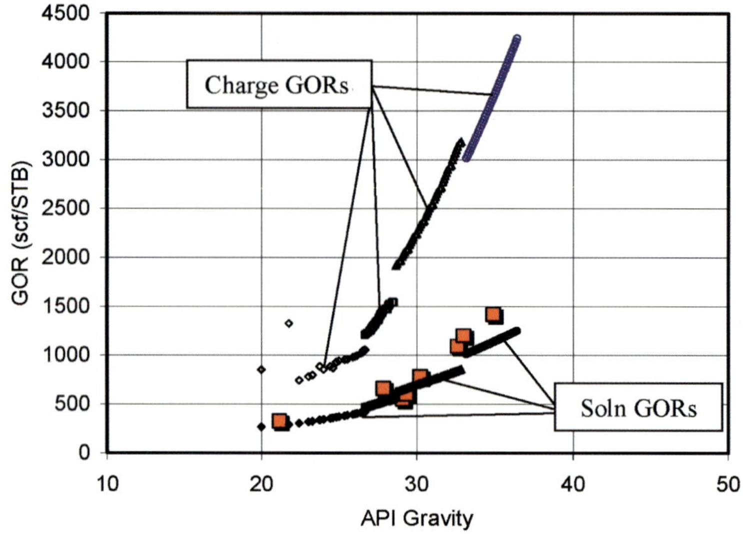


**Fig. 15.** API gravities and GORs of oil in a West African Tertiary delta with several stacked reservoirs. Note the excellent correlation between these for both the model predictions and the data, even though the modelled charge GORs are much greater than the solution GORs. This suggests that the solution GORs are controlled by the PVT evolution of the trap, whereas the APIs are controlled by the maturity evolution of the source rock kitchen.

entirely lost by diffusion and other processes; then further burial would lead to decreased saturation of the oil leg with gas, even in cases where the oil was initially close to saturation.) For reference, Figure 14 also shows the incremental *charge* GORs plotted versus the final present-day depths of the corresponding oil increments rather than the depths of the reservoirs when the charges entered the trap.

In the other example of a predicted compositional gradient (in a West African field charged by an oil-prone Type II source rock, Figs 16 &

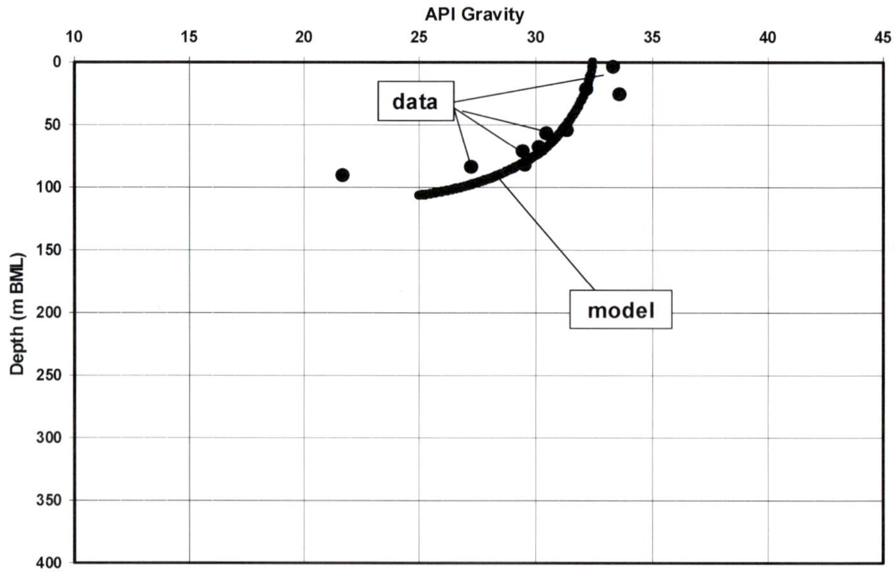


**Fig. 16.** Model predictions of API gravities of oils in a single West African reservoir charged by an oil-prone Type II source rock. The thick line (of closely spaced dots) is the model prediction, whereas the isolated circles are the data, both showing strong gradients.

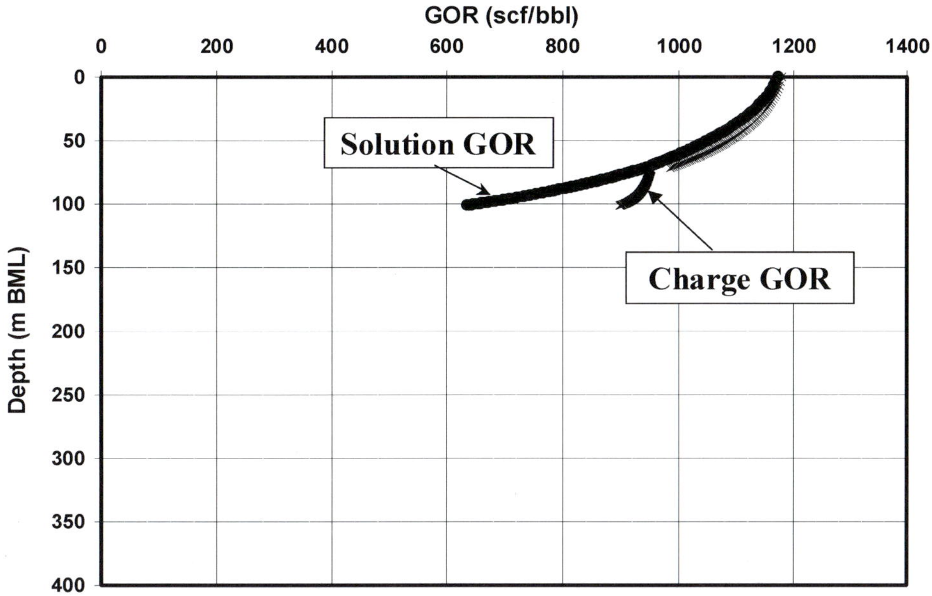


**Fig. 17.** Model predictions of solution GORs of oils (thick line of overlapping dots) in a single West African reservoir charged by an oil-prone Type II source rock, compared with the charge GORs (crosses). In this case, the solution GORs are mainly controlled by the charge GORs.

17), the solution GOR is similar to the charge GOR, showing that in this case the control on the GOR is the charge GOR rather than the PVT history of the trap. Both the API and GOR have curved trends, similar to those shown in Figures 5 and 6. The curvature of these trends results mainly from the interaction between a curved trap volume versus depth curve, like that shown in Figure 18, with a steadily increasing API gravity and solution

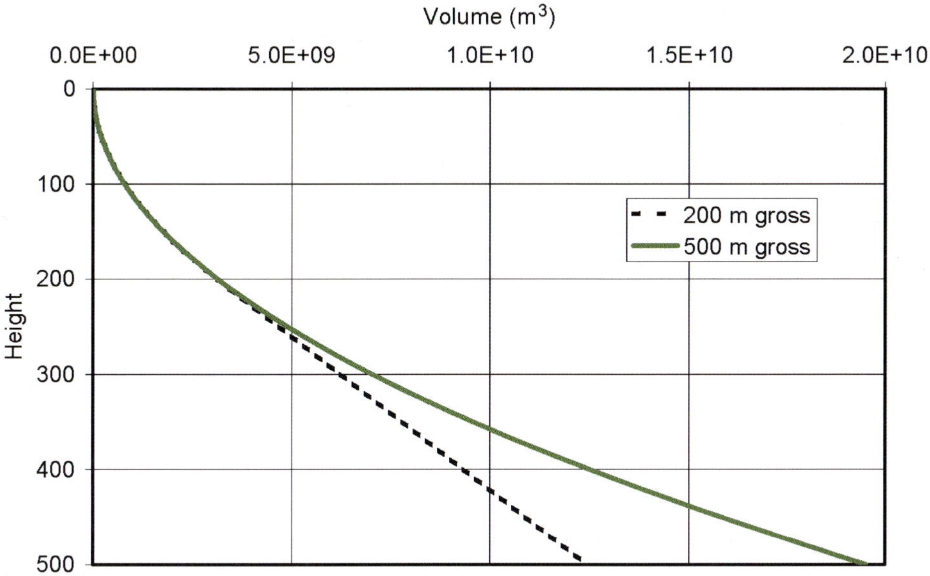


**Fig. 18.** Volume versus depth curve of a spherical-circular trap with a 500 m vertical closure and gross reservoir thicknesses of 200 m (broken curve) and 500 m thick (solid curve).

GOR of the integrated oil charge during the filling of the trap. This is discussed further below.

## Discussion

### *Under-saturated oils*

Oil columns that are truly saturated with gas throughout are rare. Even oil columns (with or without gas caps) that are called 'saturated' often have bubble points around 10% less than the original reservoir pressure. This is because the mixing processes in the trap are not usually fast enough for equilibrium to be achieved throughout a reservoir, even given the geological durations of trap filling, unless the reservoir has large gross thickness or has been filled for a long geological time. A petroleum liquid column usually achieves thermodynamic equilibrium with a vapour column only near the gas–oil contact. The remainder of the column had insufficient time to equilibrate with these conditions, which continually changed during the filling of the trap.

The trap-filling model presented in this paper explains a common paradox of many Tertiary deltas. The oil charge is very gassy, because the source rocks are of mixed Type II/III, even where the source rock kitchens are entirely in the oil window, i.e. no parts have entered the gas window. The outcome of this gassiness is an abundance of reservoirs with gas caps, or ones entirely filled with gas, and abundant evidence of leaked gas in the form of gas chimneys etc. Yet the oils in the traps are often slightly under-saturated with gas regardless of whether they are capped by gas or not.

There has been a tendency to always interpret under-saturated oils as simply reflecting the gas-leanness of an oil-rich charge. Good correlations between GOR and API gravity in individual reservoirs (e.g. Fig. 15), may have been reinforced that view. Such correlations suggest a genetic link between the GOR and API gravity, especially where the oil geochemistry indicates that the main control on the gravity is the maturity of the source rock (rather than an effect of an *in situ* alteration process such a biodegradation).

### *Controls on the API gravity and GOR of oils*

The model presented in this paper suggests, however, that the controls on the API gravity and solution GOR of oil are fundamentally different, except in cases where the oil charge is very gas-lean. The model suggests that:

(a) variations in oil API gravity mainly reflect the history of the maturity of the oil, which is controlled by the maturity evolution of the source rock kitchen during the entire trap-filling process, whereas
(b) variations in the solution GOR of the oil are often controlled by the PT evolution of the reservoir level(s) in the trap, rather than the charge GOR.

A complication is that the present-day GOR often reflects the filling history of *more than one* reservoir, because of spilling and leaking of petroleum from one level to another. A further complication is that gas leaking from deeper reservoirs combines with oil and gas that are spilling from the deeper levels. All these interactions would be virtually impossible to predict without a fully coupled reservoir filling and spilling and leaking model like the one outlined here.

### *Trap and kitchen histories lurking in the compositional trends*

This model explains the wide range of bubble point (and GOR) versus API gravity trends of oils from around the world (e.g. Figs 8 & 9). Every type of combination of trap and kitchen history occurs, from traps whose depths (and temperatures) have varied little during trap filling whilst their kitchens have undergone dramatic changes in temperature (and maturity), and *vice versa*. A correlation that has a small bubble point range and large API range is diagnostic of a trap that has undergone little burial during its filling, whilst the kitchen underwent a large increase in maturity. Such is the case in many structures formed predominantly by 'down-building' (Barton 1933). By contrast, a large bubble point range coupled with a small API range reflects a trap that has undergone a large amount of burial during filling, but whose effective kitchen remained at a similar (integrated) maturity during that time. Kitchens with large planar geometry, whose source rocks straddled the entire oil window during the whole of the trap filling period, create such conditions of near-constant oil maturity.

### *Curvature of API gravity and GOR trends versus depth*

According to the hypothesis presented in this paper, compositional trends in traps result mainly from changes in maturity of the source rock and the PT conditions in the trap during filling, rather than gravity acting on homogenized petroleum.

If this is the case, the shape of an API gravity trend versus depth in a trap is the result mainly of the interaction between two geometries:

(1) the evolving geometry of the source rock kitchen during trap filling, which controls the evolving volume and API gravity of the oil charge integrated over the kitchen area;
(2) the present-day trap geometry, which determines its capacity versus depth (Fig. 18).

These interactions could account for the gross shapes of all the API trends shown in Figure 5.

As we have seen, the controls on the GOR of the oil in the trap may be different from those of the API gravity. If the oil had a gas cap during part of the trap-filling period, the evolving PT conditions in the trap control the GOR, rather than the evolving maturity of the source rock kitchen. Thus, the shape of a GOR trend versus depth in a trap results mainly from the interaction between trap geometry ((2) above), and

(3) the evolving PT conditions during trap filling, which control the instantaneous solution GOR of the oil at the gas-oil contact (GOC).

Interactions (2) and (3) could account for most of the GOR trends shown in Figure 6.

In practice, trap filling is complicated further by spilling (particularly of oil) and leaking (particularly of gas) between reservoir levels, and various *in situ* processes, such as biodegradation of oil. Of course, diffusive mixing also blurs the trapped compositions, to various degrees, as discussed earlier.

### *Biodegradation*

Secondary alteration processes such as biodegradation may also produce compositional gradients in single continuous reservoirs (e.g. Wavrek *et al.* 2001). Larter and co-workers (Horstad & Larter 1997; Larter *et al.* 2000) have proposed that mixing processes may not be fast enough to remove compositional gradients induced by active biodegradation. However, some of the compositional grading that has been ascribed to biodegradation may be simply a maturity variation, with biodegradation acting as an add-on effect that operates in reservoirs that are at shallow depths during filling with oil. In addition, 'biodegradation' probably comprises a suite of processes that operate in shallow reservoirs, including water-washing, gas-washing, selective leaking of light hydrocarbons and biodegradation itself. The explanation for the compositional profiles offered in this paper is simpler than one of biodegradation coupled with diffusion (Larter *et al.* 2000).

Our work on biodegradation supports that of Yu *et al.* (2002). In particular, our models suggest that biodegradation is not necessarily limited to temperatures less than 65 °C; that not all lower temperature reservoirs contain biodegraded oil; and that the dynamic nature of biodegradation results in poor to very poor correlations between API gravity and reservoir depth or temperature.

### *Implications for reservoir geochemical fingerprinting*

Reservoir geochemical fingerprinting is now regarded in the industry as a routine technique for distinguishing continuous reservoirs from those that are compartmentalized, i.e. separated by fluid flow barriers (e.g. Nederlof *et al.* 1994, 1995; Westrich *et al.* 1996, 1999; OilTracers 2002). The most widely used compounds are inter-paraffins that elute between the normal paraffins in gas chromatography. In oils that are in fluid communication, the ratios of the inter-paraffin compounds that are used in the fingerprinting seldom differ between the samples by more than 10%. Less commonly, lighter hydrocarbons are used for the fingerprinting, especially for gas condensates. For example, Halpern (1995) used $C_7$ compounds to assess reservoir continuity in several Saudi Arabian fields.

The main difficulty with the interpretation of reservoir geochemical analyses is deciding on the criteria for determining whether sample compositions are essentially the 'same' (and so interpreted as in the same compartments) or 'different'. Obviously, compositional gradients in continuous oil columns in single reservoir tanks cloud this issue further. However, these gradients are mainly apparent in the gross composition (e.g. API gravity, GOR, bubble point etc.), rather than the specific compounds used for reservoir geochemistry. These compounds have similar molecular weights and other properties to one another, so gravity segregates them little (Kaufman *et al.* 1990). For example, there are strong, gross compositional gradients in the Bullwinkle Field, Gulf of Mexico, yet Westrich *et al.* (1996, 1999) were able to use the fingerprinting technique to show that it is plausible that the reservoir is continuous throughout most of the field.

A logical approach is to estimate the time it would take to homogenize oil in a given reservoir by convection and/or diffusion, and then compare this with the age of trap filling. Thus, Smalley & England (1992, 1994), England *et al.* (1995) and Smalley & Hale (1996) used fingerprint differences to infer barriers within the Forties and Ross Fields, North Sea, given the time since the reservoir filled with oil. Conversely, very young reservoirs, deposited in the past 1 or 2 Ma, may have heterogeneous oil fingerprints within a single reservoir compartment (e.g. Beeunas *et al.* 2000).

## Conclusions

- Oil in reservoirs is often poorly mixed. Two kinds of trend are observed:
  - In stacked oil reservoirs, there is tendency for the average API gravity, GOR and bubble point to increase in successively deeper reservoirs.
  - Within individual reservoirs, the compositional grading usually has the opposite sense, i.e. API gravity, and even GOR and bubble point, tend to decrease with depth. These compositional depth trends are really the vertical components of lateral trends in each reservoir.
- The compositional gradients in individual reservoirs:
  - These are often gravitationally over-stable, i.e. they are stronger than can be explained by equilibrium between diffusion and the gravitational and thermal fields.
  - This suggests that they are not the result of these external fields acting on petroleum that was originally homogenized.
  - These imply that local thermodynamic equilibrium exists only in the vicinity of gas–oil contacts.
- Compositional trends reflect filling history as propounded by England *et al.* (1987), and imply that:
  - Lateral diffusion times are often inadequate for the petroleum column to attain equilibrium.
  - Petroleum in continuous reservoirs rarely undergoes wholesale homogenization followed by gravitational segregation.
  - Petroleum stacks into traps roughly in density sequence.
  - Phase separation takes place not by 'flashing' of the entire petroleum content of the trap, but only in the vicinity of, and in response to, the local PT conditions at the GOC:
- The range of API gravity of oil in a trap reflects the maturity change of the source rock kitchen during the trap-filling, constrained by the capacity of the trap.
- The range of GOR and bubble point of oil in a trap often reflect neither the present-day PT conditions nor the charge GOR, but rather the PT history of the trap.
- The specific shape of an oil API trend versus depth is determined mainly by the interaction between the changing charge composition, which depends on the evolving geometry of the source rock kitchen, and the geometry of the trap (e.g. volume versus depth functions). The shape is not necessarily concave up as in the case of a gravitationally segregated column.
- The specific shape of an oil GOR trend versus depth is determined mainly by the interaction between the geometry of the trap and (a) the evolving PT conditions of the trap, in the case of a gassy charge, or (b) the charge GOR, in the case of a gas-lean charge.
- The usual filling sequence for stacked reservoirs is bottom up (i.e. bottom reservoir first). Petroleum usually migrates into overlying reservoir by (a) spiralling up from deeper reservoirs via faults and other pathways, or (b) by hydraulic leakage from the crest of the underlying reservoir, or (c) by capillary leakage up faults.
- Each reservoir fills downwards:
  - From the top (or GOC if two separate petroleum phases form), if the petroleum charge becomes less dense with time.
  - From the OWC, if the charge becomes more dense with time (which can happen during uplift, etc).
- Biodegradation (and allied processes) only affects petroleum significantly in reservoirs that were at a low temperature ($\lesssim 50\,^{\circ}\mathrm{C}$) during filling.

I would like to thank B. Dindoruk, L. L Garmezy, J. T. Westrich and the two reviewers, William England and Craig Smalley, for their useful criticisms, and Shell International for permission to publish this paper.

## References

Araque, I. E. & Auxiette, G. 2002. Reservoir management of the El Furrial Field, Venezuela, under enhanced oil recovery process. *Society of Petroleum Engineers*, Paper No. **75201**.

Barton, D. C. 1933. Mechanics of formation of salt domes with special reference to Gulf Coast salt domes of Texas and Louisiana. *American Association Petroleum Geologists Bulletin*, **17**(9), 1025–1083.

Bath, P. G. H., Fowler, W. N. & Russell, M. P. M. 1980. The Brent field. A reservoir engineering review. EUR 164. European Offshore Petroleum Conference and Exhibition, London, 21–24 October 1980.

Beeunas, M. A., Hudson, T. A., Valley, J. A., Baskin, D. K. & Clark, W. Y. 2000. Identification of reservoir discontinuities using hydrocarbon compositional analyses, Genesis field (Green Canyon 205), Gulf of Mexico. *American Association of Petroleum Geologists Bulletin*, **84**(13) (Abstract).

Beggs, H. D. 1987. Oil system correlations. *In*: *Petroleum Engineering Handbook*, Ch. 22. Society of Petroleum Engineers.

Carrigan, W. J. *et al.* 1998. Identification of reservoir compartments by geochemical methods: Jauf reservoir, Ghawar. *Saudi Aramco Journal of Technology*, **Summer 1998**, 28–32.

Clinton, G. W. & Ruiz, J. 1968. Performance of Payoa Basal Eocene Block A reservoir. *Journal of Petroleum Technology*, **January 1968**, 31–35.

Creek, J. L. & Schrader, M. L. 1985. East Painter Reservoir: an example of a compositional gradient from a gravitational field. *Society of Petroleum Engineers*, Paper No. **14411**.

Dindoruk, B. & Christman, P. G. 2001. PVT properties and viscosity correlations for Gulf of Mexico oils. *Society of Petroleum Engineers*, Paper No. **71633**.

England, W. A. & Mackenzie, A. S. 1989. Some aspects of the organic geochemistry of petroleum fluids. *Geologische Rundschau*, **78**(1), 291–303.

England, W. A., Mackenzie, A. S., Mann, D. M. & Quigley, T. M. 1987. The movement and entrapment of petroleum fluids in the subsurface. *Journal of the Geological Society of London*, **144**, 327–347.

England, W. A., Muggeridge, A. H., Clifford, P. J. & Tang. Z. 1995. Modelling density-driven mixing rates in petroleum reservoirs on geological timescales, with application to the detection of barriers in the Forties Filed (UKCS). *In*: Cubitt, J. M. & England, W. A. (eds) *The Geochemistry of Reservoirs*. Geological Society, London, Special Publications, **86**, 185–201.

Espach, R. H. & Fry, J. 1951. Variable characteristics of the oil in the Tensleep Sandstone reservoir, Elk Basin Field, Wyoming and Montana. *Transactions of the American Institute of Mechanical Engineers*, **192**, 75–82.

Firoozabadi, A. 1999. *Thermodynamics of Hydrocarbon Reservoirs*. McGraw-Hill, New York.

Firoozabadi, A., Dindoruk, B. & Chang, E. 1996. Areal and vertical composition variation in hydrocarbon reservoirs: formulation and one-D binary results. *Entropie*, **198/199**, 109–118.

Glaso, O. 1980. Generalized pressure–volume–temperature correlations. *Journal of Petroleum Technology*, **32**, 785.

Grant, H. K. 1959. How fluid-property variation in a high-relief oil field affects material-balance calculations. *Oil and Gas Journal*, **57**(35), 93–99.

Halpern, H. I. 1995. Development and applications of light-hydrocarbon-based star diagrams. *American Association Petroleum Geologists Bulletin*, **79**, 801–815.

Havlena, D. 1968. Dynamic reservoir data. 19th Annual Technical Meeting, The Petroleum Society of CIM, Calgary, May 1968.

Hirschberg, A. 1984. The role of asphaltenes in compositional grading of reservoir's fluid column. *Society of Petroleum Engineers*, Paper No. **13171**.

Hoier, L. & Whitson, C. H. 2000. Miscibility variation in compositionally grading reservoirs. *Society of Petroleum Engineers*, Paper No. **63086**.

Hoier, L. & Whitson, C. H. 2001. Compositional grading—theory and practice. *Society of Petroleum Engineers*, Paper No. **74714**.

Holt, T., Lindeberg, E. & Ratkje, S. K. 1983. The effect of gravity and temperature gradients on methane distributions in oil reservoirs. *Society of Petroleum Engineers*, Paper No. **11761**.

Horstad, I. & Larter, S. R. 1997. Petroleum migration, alteration and remigration within Troll Field, Norwegian N. Sea. *American Association Petroleum Geologists Bulletin*, **81**(2), 222–248.

Hunt, J. M. 1979. *Petroleum Geochemistry and Geology*. Freeman, San Francisco.

Jacqmin, D. 1990. Interaction of natural convection and gravity segregation in oil/gas reservoirs. *Society of Petroleum Engineers Reservoir Engineering*, **May 1990**, 233–238.

Kaufman, R. L., Ahmed, A. S. & Elsinger, R. J. 1990. Gas chromatography as a development and production tool for fingerprinting oils from individual reservoirs: applications in the Gulf of Mexico. *In*: Schumaker, D. & Perkins, B. F. (eds) *Proceedings of the 9th Annual Research Conference*. Society of Economic Paleontologists and Mineralogists, October 1990, New Orleans, 263–282.

Larter, S. R., Head, I. & Wilhelms, A. 2000. Implications of slow biodegradation rates in oilfields for crustal biosphere. Goldschmidt Geochemical Conference, September, Oxford, UK.

Mearns, E. W. & McBride, J. J. 1999. Hydrocarbon filling history and reservoir continuity of oil fields evaluated using $^{87}Sr/^{86}Sr$ isotope ratio variations in formation water, with examples from the North Sea. *Petroleum Geoscience*, **5**, 17–27.

Metcalfe, R. S., Vogel, J. L. & Morris, R. W. 1985. Compositional gradient in the Anschutz Ranch East Field. *Society of Petroleum Engineers*, Paper No. **14412**.

Montel, F. & Gouel, P. L. 1985. Prediction of compositional grading in a reservoir fluid column. *Society of Petroleum Engineers*, Paper No. **14410**.

Nederlof, P. J., Gijsen, M. A. & Doyle, M. A. 1994. Application of reservoir geochemistry to field appraisal. *In*: Al-Husseini, M. I. (ed.) *The Middle East Petroleum Geosciences 1994*, vol. 2. Gulf-Petrolink, Bahrain, 709–722.

Nederlof, P. J. R., van der Veen, F. M. & van den Bos, G. A. 1995. Application of reservoir geochemistry in Oman. *In*: Grimalt, J. O. & Dorronsoro, C. (eds) *17th International Meeting on Organic Geochemistry*. San Sebastian, Spain, 329–331.

NUTAKKI, R., BAKER, R. O., GEGUNDE, G., THOMAS, M., PAYNE, D. & GOULD, B. 1996. Three-dimensional simulation of a fractured rich gas-condensate reservoir: a study of Waterton Sheet III. *Society of Petroleum Engineers*, Paper No. **35606**.

OILTRACERS L. L. C. 2002. *Assessing reservoir compartmentalization using oil geochemistry*. www.oiltracers.com

RATULOWSKI, J., FUEX, A. N. & WESTRICH, J. T. 2000. Theoretical and experimental investigation of isothermal compositional grading. *Society of Petroleum Engineers*, Paper No. **63084**.

SAGE, B. H. & LACEY, W. N. 1939. Gravitational concentration gradients in static columns of hydrocarbon fluids. *Transactions of the American Institute of Mechanical Engineers*, **132**, 120–131.

SCHULTE, A. M. 1980. Compositional variations within a hydrocarbon column due to gravity. *Society of Petroleum Engineers*, Paper No. **9235**.

SMALLEY, P. C. & ENGLAND, W. A. 1992. Assessing reservoir compartmentalization during field appraisal: how geochemistry can help. *Society of Petroleum Engineers*, Paper No. **25005.**

SMALLEY, P. C. & ENGLAND, W. A. 1994. Reservoir compartmentalization assessed with fluid compositional data. *Society of Petroleum Engineers Reservoir Engineering*, **August 1994**, 175–180.

SMALLEY, P. C. & HALE, N. A. 1996. Early identification of reservoir compartmentalization by combining a range of conventional and novel data types. *Society of Petroleum Engineers*, Paper No. **30533**.

TISSOT, B. P. & WELTE, D. H. 1979. *Petroleum Formation and Occurrence*. Springer, Berlin.

WADMAN, D. H., LAMPRECHT, D. E. & MROSOVSKY, I. 1979. Reservoir description through joint geologic-engineering analysis. *Society of Petroleum Engineers*, Paper No. **7531**.

WAVREK, D. A., MOSCA, F., CHAOUCHE, A. & JARVIE, D. M. 2001. *Geochemical insights to engineering problems: when does oil heterogeneity really indicate a compartment?* IMOG Convention, September, Nancy, France.

WESTRICH, J. T., KNIGGE, P. O., FUEX, A. N. & HALPERN, H. I. 1996. Evaluating reservoir architecture in the northern Gulf of Mexico using oil and gas chemistry. *Society of Petroleum Engineers*, Paper No. **36541**.

WESTRICH, J. T., FUEX, A. N., O'NEAL, P. M. & HALPERN, H. I. 1999. Evaluating reservoir architecture in the Northern Gulf of Mexico with oil and gas chemistry. *Society of Petroleum Engineers*, Paper No. **59518**.

YU, Z. A., COLE, G., GRUBITZ, G. & PEEL, F. 2002. How to predict biodegradation risk and reservoir fluid quality. *World Oil*, **April 2002**, 63–74.

# An integrated basin modelling study applying asphaltene kinetics from reservoired petroleum in the Snorre Area, northern North Sea

JON ERIK SKEIE[1], ROLANDO DI PRIMIO[2], DAG A. KARLSEN[1] & KNUT BJØRLYKKE[1]

[1]*Department of Geosciences, University of Oslo, P.O. Box 1047 Blindern, 0316 Oslo, Norway*
[2]*GeoForschungZentrum, Potsdam, Telegrafenberg, 14472 Potsdam, Germany*

**Abstract:** The evolution of the petroleum systems in the Tampen Spur area, with main focus on the filling directions of the northern part of Snorre field, was addressed through 2D basin modelling (Petromod V. 4.5 and 7.0). The geochemical classification of the petroleum populations in the area represented the framework for considering the different kitchen areas and migration systems. Results from the basin modelling support, in general terms, the previous geochemical classification and petroleum families in the region. However, a separate well-defined main kitchen area for the Snorre Field was deduced opposed to the multiple kitchen areas having contributed to the filling as proposed in the literature. Our conclusions are based on the quantitative evaluation of the different proposed kitchen areas and the timing and extent of petroleum generation.

Modelling of petroleum generation was performed using asphaltene kinetics determined on petroleum asphaltenes from Snorre oils. This approach was chosen in order to avoid problems associated with the kinetic variability encountered in the Draupne formation. The petroleum asphaltene kinetics was used to delineate the extent of the kitchen area, which reached the time/temperature conditions necessary for the generation of the analysed oil phase. The results thus differ from conventional oil window approximations as we utilize kinetic source rock parameters in the migrated oil for tracing out the generative basin.

Three 2D lines crossing the main kitchen areas were modelled in this study. The models were calibrated to data from eight wells, consisting of measured vitrinite reflectance, corrected well temperatures and pore pressure.

Three main kitchen areas were considered; one to the west and northwest of Snorre field, one directly to the north (Møre basin) and one to the east of the field (34/5 kitchen). Modelling suggests that the kitchen area to the west and northwest of Snorre is largely immature and that the volume of potentially generated petroleum is too small to fill the Snorre structure. In the northern kitchen area, the seismic indicated very thin upper Jurassic deposits, which reaches oil window maturities only at a relatively large distance from the structure. The modelling also demonstrated problems related to the filling of the Snorre structure from the Møre Basin. The combined effect of a thin source rock, which implies a regionally large drainage area to fill the structure, and the large distance to the mature kitchen, lead to the conclusion that the Møre Basin did not contribute significant volumes of petroleum to the Snorre field. In contrast, the kitchen area east of Snorre Field (the 34/5 kitchen) proved in the modelling to be mature and volumetrically large enough to account for the entire filling of the Snorre Field.

Basin modelling attempts to integrate geochemical, geophysical and geological information into a quantitative model of basin evolution (Welte & Yükler 1981; Durand *et al.* 1984; Nakayama & Lerche 1987; Bethke 1989; Hermanrud 1993). This is accomplished by mathematically simulating the processes of heat and fluid flow, compaction, hydrocarbon generation and migration. The reconstruction of the burial history of a sedimentary basin is based predominantly on the physical properties of the sediments and on their evolution with increasing depth of burial. Such models always contain a number of uncertainties and represent a very incomplete picture of the subsurface (Poelchau & Mann 1989; Burley *et al.* 2000). Most of the parameters modelled are only calibrated to today's geological situation and, in addition, geological variations in rock properties such as porosity or permeability are only loosely constrained, thus contributing to uncertainties in the modelling of e.g. fluid flow and migration efficiencies. Despite this, the prime use of these numerical tools focuses on hypothesis testing ('what if' scenarios) and characterization of the influence of the parameters to be evaluated. The performance of so-called sensitivity runs, where model attributes are varied systematically to test their influence on model predictions, aids in recognizing consistent predictions from which confidence in the modelling results can be gained (given that the

*From:* CUBITT, J. M., ENGLAND, W. A. & LARTER, S. (eds) 2004. *Understanding Petroleum Reservoirs: towards an Integrated Reservoir Engineering and Geochemical Approach.* Geological Society, London, Special Publications, **237**, 133–155. 0305-8719/04/$15.00 

sensitivity runs fulfil the constraints of the calibration data).

One main application of basin modelling has been the assessment of timing and extent of petroleum generation in sedimentary basins using chemical kinetics of kerogen degradation determined experimentally on immature source rock samples (Vandenbroucke *et al.* 1999). However, in view of the kinetic variability observed in source rocks (Jarvie 1991; Sundaraman 1995) it is questionable whether such predictions are valid.

The main goal of this study was to constrain the suggested filling directions and kitchen areas responsible for sourcing the Snorre petroleum (Horstad *et al.* 1995), and hence focused on hydrocarbon generation timing and migration. In order to avoid problems related to the selection of kerogen degradation kinetics, especially since the main source rock sequence in the area (Draupne Formation) is characterized by an extreme kinetic variability, we chose to apply the novel concept of asphaltene kinetics (di Primio *et al.* 2000). Such an approach differs from traditional source rock kinetics as it describes the condition of generation and primary migration of the accumulated petroleum. The results should then be more representative as it is derived from actually migrated petroleum in a relatively well-constrained accumulation.

## Geological background

The Snorre field is located in the northern part of the Tampen Spur and is associated with several other large petroleum accumulations (Fig. 1). Major reservoir units are the Triassic Lunde Formation (northern part of the field) and the lower Jurassic Statfjord Formation (southern part of the field). The petroleum is undersaturated with respect to gas and has a generally low gas to oil ratio (GOR) (71–221 $Sm^3/m^3$), typically increasing towards the north. The geological development of the Tampen Spur has been extensively studied and only a brief summary is given here (the reader is referred to Spencer (1984), Faerseth (1996) and Glennie (1998) for a more comprehensive background).

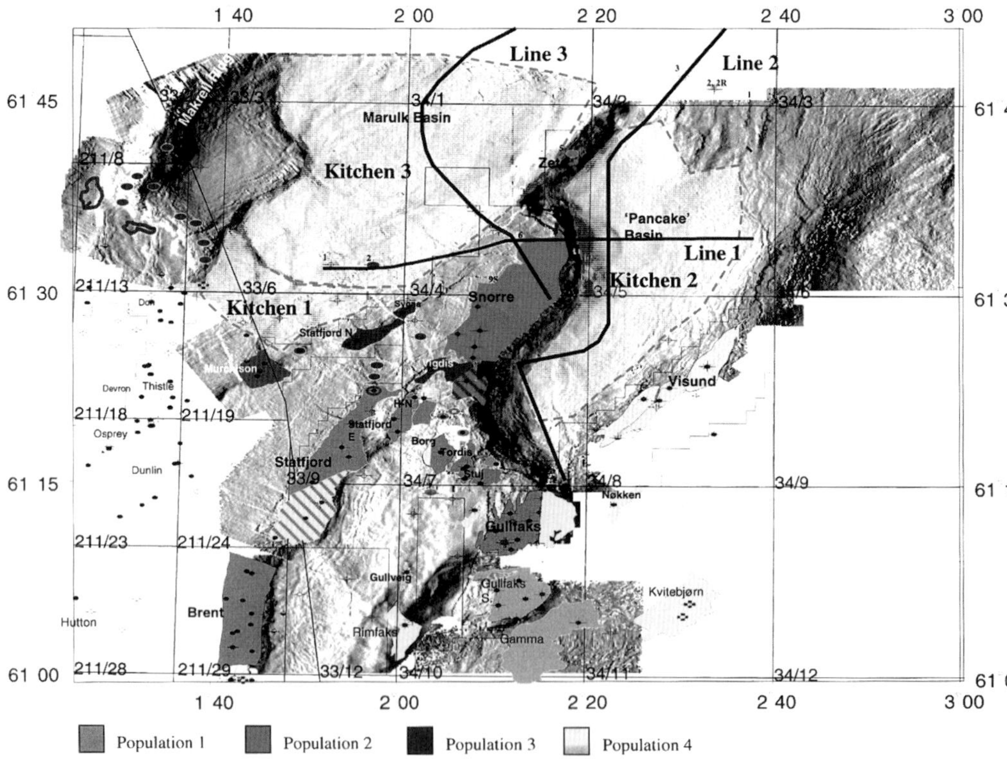


**Fig. 1.** Regional framework in this study. BCU map of the Tampen Spur area, with geochemical classification of petroleum populations (Horstad *et al.* 1995).

**Table 1A.** *The wells used in this study with an overview of available calibration data*

| Well | Temperature (°C) | Vitrinite reflectance (R%) | Pressure (DST) | Year (drilled) | Total depth (m RKB) | Water depth (m) | Geographical location |
|---|---|---|---|---|---|---|---|
| 33/6-1 | | × | | 1979 | 3900 | 377 | Latitude 61°32′49.23″N, longitude 02°16′23.66″E |
| 33/6-2 | × | | × | 1996 | 3950 | 316 | Latitude 61°62′15.97″N, longitude 01°55′33.00″E |
| 34/2-2 | × | | | 1981 | 4074 | 386 | Latitude 61°46′16.83″N, longitude 02°33′11.38″E |
| 34/4-1 | × | | | 1979 | 2961 | 377 | Latitude 61°32′49.23″N, longitude 02°16′23.66″E |
| 34/4-3 | × | | × | 1982 | 4466 | 366 | Latitude 61°36′32.95″N, longitude 02°07′34.42″E |
| 34/4-4 | × | | × | 1983 | 3800 | 345 | Latitude 61°30′20.85″N, longitude 02°14′09.54″E |
| 34/4-6 | × | | × | 1986 | 3282 | 373 | Latitude 61°34′14.09″N, longitude 02°19′19.99″E |
| 34/4-7 | × | | × | 1987 | 2950 | 354 | Latitude 61°31′09.83″N, longitude 02°15′15.45″E |
| 34/4-8 | | × | × | 1994 | 3110 | 363 | Latitude 61°35′04.02″N, longitude 02°10′09.73″E |
| 34/7-1 | × | | × | 1984 | 2905 | 328 | Latitude 61°28′21.80″N, longitude 02°13′25.56″E |
| 34/7-3 | × | | × | 1984 | 3414 | 303 | Latitude 61°25′54.05″N, longitude 02°07′43.95″E |
| 34/7-4 | × | | × | 1985 | 3115 | 319 | Latitude 61°29′04.44″N, longitude 02°08′00.26″E |
| 34/7-6 | × | | × | 1985 | 3685 | 307 | Latitude 61°27′10.85″N, longitude 02°08′17.26″E |
| 34/7-7 | × | | × | 1985 | 3526 | 302 | Latitude 61°26′54.53″N, longitude 02°05′55.69′E |
| 34/7-9 | × | | × | 1986 | 3240 | 330 | Latitude 61°29′11.99″N, longitude 02°11′43.60″E |
| 34/7-10 | × | | × | 1986 | 3000 | 300 | Latitude 61°25′02.64″N, longitude 02°07′36.23″E |
| 34/7-15S | × | × | × | 1990 | 4646 (4333TVD) | 306 | Latitude 61°24′38.59″N, longitude 02°12′53.22″E |

Several pulses of stretching have influenced the Tampen Spur and the Snorre area. The most significant pulses occurred in the Permo-Triassic and Jurassic, culminating in the Jurassic–Cretaceous. It has been proposed that the development of the Viking Graben system was affected by pre-rift geometries inherited from the Caledonian orogenic collapse and sutures, and that the later rifting is partly controlled by the reactivation of these sutures (Faerseth 1996). Furthermore, it has been established that in many cases there is an inverse relationship between Permo-Triassic and upper Jurassic fault throw. This is important in Snorre as this is one of the structures which experienced low throw in the Permo-Triassic stretching, but an estimated throw of 3 km (Faerseth 1996) in the upper Jurassic, causing the Snorre block to be one of the most pronounced structural features in the post-rift basin. This setting is expected to have caused significant input of syn-rift sediments during the deposition of the Heather and Draupne Formations on the flank of the structure during the Late Jurassic (Underhill 1998). Combined with the uplift of the crest of the structurally pronounced Snorre fault block and associated erosion, the model implies the occurrence of thin sand layers in the Draupne and Heather Formation, thus enhancing a lateral direction of secondary migration in this system. The Palaeozoic stretching has also been assigned in the heat flow model used in this study. A high initial basement heat flow (80 $mW/m^2$) is assumed for periods of active stretching and an exponential decrease of heat flow is assumed to present-day heat flow of 55 $mW/m^2$ (McKenzie 1978; Lucazeau & Douaran 1984; Schroeder & Sylta 1993). During Cretaceous the Tampen Spur is assumed to be experiencing passive post-rift subsidence. The deposition of Cromer Knoll Group took place in early Cretaceous, partly draping the Snorre Structure. This was followed by the deposition of the Shetland Group which is the principal cap rock holding the Snorre Petroleum. The leaky characteristics of the field has been proven (Caillet 1993; Leith *et al.* 1993) by pointing to the dynamic character of this accumulation. Mesozoic uplift is recorded in the northern North Sea, followed by significant subsidence and deposition from late Mesozoic to Pliocene (Jordt *et al.* 1995). Available wells in this study and lithological units and properties assigned in the modelling are shown in Table 1 A and B.

The petroleum populations in the Tampen Spur area have been classified geochemically and based on their PVT properties in several studies (Horstad *et al.* 1995; di Primio *et al.* 1998*a*), all indicating an unique petroleum population in the Snorre Field ('Population 3' of Figure 1) with a northward increase in maturity noticed within the field, reflected in biomarker and aromatic maturity parameters and also in terms of PVT properties. Furthermore, the petroleum in the Statfjord, Brent and the Gullfaks oil fields belong to a distinctly different group ('Population 1' of Fig. 1). A third petroleum population is recognized in the Statfjord North and Murchison oil fields ('Population 2' of Fig. 1), while parts of Visund and the lighter

**Table 1B.** *Conceptual framework and principal petroleum system element defined in the basin model*

| Interval | Defined lithology | Age (top) My | Petroleum system element |
|---|---|---|---|
| Quartenary | Silt, sandy | 0 | |
| Pliocene | Silt | 1.8 | |
| Utsira Formation | Silt and sand | 5.3 | |
| Hordalan Graben | Shale and carbonate | 25 | |
| Rogaland Graben | Shale, sandy | 55 | |
| Shetland Graben | Shale (defined specially) | 65 | Cap rock |
| Cromer Chnoll Graben | Lower: marl; upper: shale, silty | 99 | |
| Draupne Formation | Shale (defined specially) | 142 | Source rock |
| Heather Formation | Shale, silty | 155 | |
| Brent Graben | Sandstone | 170 | |
| Dunlin Formation | Shale, silty | 180 | |
| Statfjord Formation | Sandstone | 197 | Reservoir |
| Upper Lunde Formation | Sandstone | 206 | Reservoir |
| Middle Lunde Formation | Sandstone, with carbonate | 207 | Reservoir |
| Lower Lunde Formation | Sandstone, conglomerate | 210 | |
| Lomvi Formation | Sand, silty | 215 | |
| Teist Formation | Sand, silty | 245 | |
| Zechstein Graben | Sand and evaporite | 260 | |
| Permian carbonate layers | Carbonate and evaporates | 270 | |

petroleum in the Gullfaks South define a fourth ('Population 4' of Fig. 1).

## General

Basin modelling was performed using Petromod 2D versions 4.5 and 7.0. Three lines were selected from the 3D seismic dataset available in order to evaluate possible petroleum systems in the area. In this data set three main kitchen areas are conceivable (Kitchens 1, 2 and 3), which accordingly were investigated in relation to maturity and possible petroleum migration (Fig. 1). These lines were assigned horizons, ages and lithologies and they represent the framework for the conceptual model (Fig. 2). The lithological and geochronological framework was calibrated against available wells in the area (Table 1A and B). Line 1 represents the west–east profile of the Snorre fault-block and was used to evaluate petroleum generation and migration from Kitchen 1 and 2. Line 2 is located in the 34/5 kitchen (Kitchen 2) and was used to characterize the evolution of maturation in this area. Line 3 is located north–south from the Møre Basin and into the Snorre structure and defines the Kitchen 3.

## Kinetics

Petroleum generation is commonly simulated using a chemical kinetic approach (Ungerer 1990). Petroleum in traps and carriers are chemically different from extracted petroleum in source rocks (the latter being enriched in asphaltenes and resins) points to the fact that generation and primary migration is of fundamental importance. In basin modelling the calibration to maturity is generally assessed through temperature history and vitrinite reflectance. As most traps are in relation to up faulted or positive structures, the more basinal areas (those generating the petroleum) are often undrilled and therefore uncalibrated. Furthermore the sampled source intervals often exhibit a heterogeneous kinetic and generative signature, impeding the source rock definition in the model.

The different types of source rock exhibit a varied ability to expel petroleum. From published pyrolysis results typical kinetic patterns in different kerogen types is evident. These signatures represent the breakdown of the kerogen structure, but largely ignore the processes involved in the release of petroleum in the source rock. The release of petroleum from different types of kerogen has been discussed (Stainforth & Reinders, 1990; Pepper & Corvi, 1995) where the chemical and physical properties of the kerogen network and its interaction with organic and inorganic surroundings are discussed. From this it is evident that coals are very good generators but poor expellers, and to a large degree gas generative. Type 1 and 2 kerogen are having a less labile kerogen with more narrow activation energies and in rich intervals these kerogen types seem to be excellent generators and expellers.

In order to constrain the uncertainties in relation to Rock Eval maturity presently used in basin modelling an alternative method involving asphaltene kinetics (di Primio *et al.* 2000) is suggested. Asphaltenes can be regarded as a soluble part of kerogen and its kinetic signature will reflect the temperature condition at expulsion, rather than estimated kerogen brake down. Furthermore, these kinetics represent a bulk signature of actually migrated and accumulated petroleum, to a large degree avoiding the problems discussed above.

In the Tampen Spur area the Draupne Formation is considered to be the main source rock (Horstad *et al.* 1995). In all investigated wells in this study the Draupne Formation showed significant vertical heterogeneity (Fig. 3) as indicated from HI and TOC plots in Fig. 4 from well 34/7-15S, sampling both the Draupne and Heather Formations in a basinal structural setting. In order to decrease the uncertainties in the maturation modelling, asphaltene kinetics from the reservoired petroleum was used rather than source rock kerogen, and its kinetic signature used in this modelling study is shown in Table 2. This method uses kinetics determined on petroleum asphaltenes, and represents a straightforward approach to the determination of the time/temperature conditions under which the analysed oil phase was generated (di Primio *et al.* 2000). Asphaltenes, which represent 'soluble particles of source rock (SR) kerogen' (cf. Bandurski 1982; Behar & Pelet 1984; Solli & Leplat 1986), provide useful clues to the maturity of the expelling source rocks. Karlsen & Larter (1991) were the first to utilize a difference in Rock-Eval $T_{max}$ on precipitated asphaltenes from two oils, to argue that the oils differed in maturity.

Basin modelling using asphaltene kinetics was performed to constrain where in the studied basins the required expulsion temperatures as defined by the asphaltene kinetic approach have been met. As discussed in Di Primio *et al.* (2000) a 5% TR was used as the threshold value for determination of the expulsion conditions of the petroleum phase.

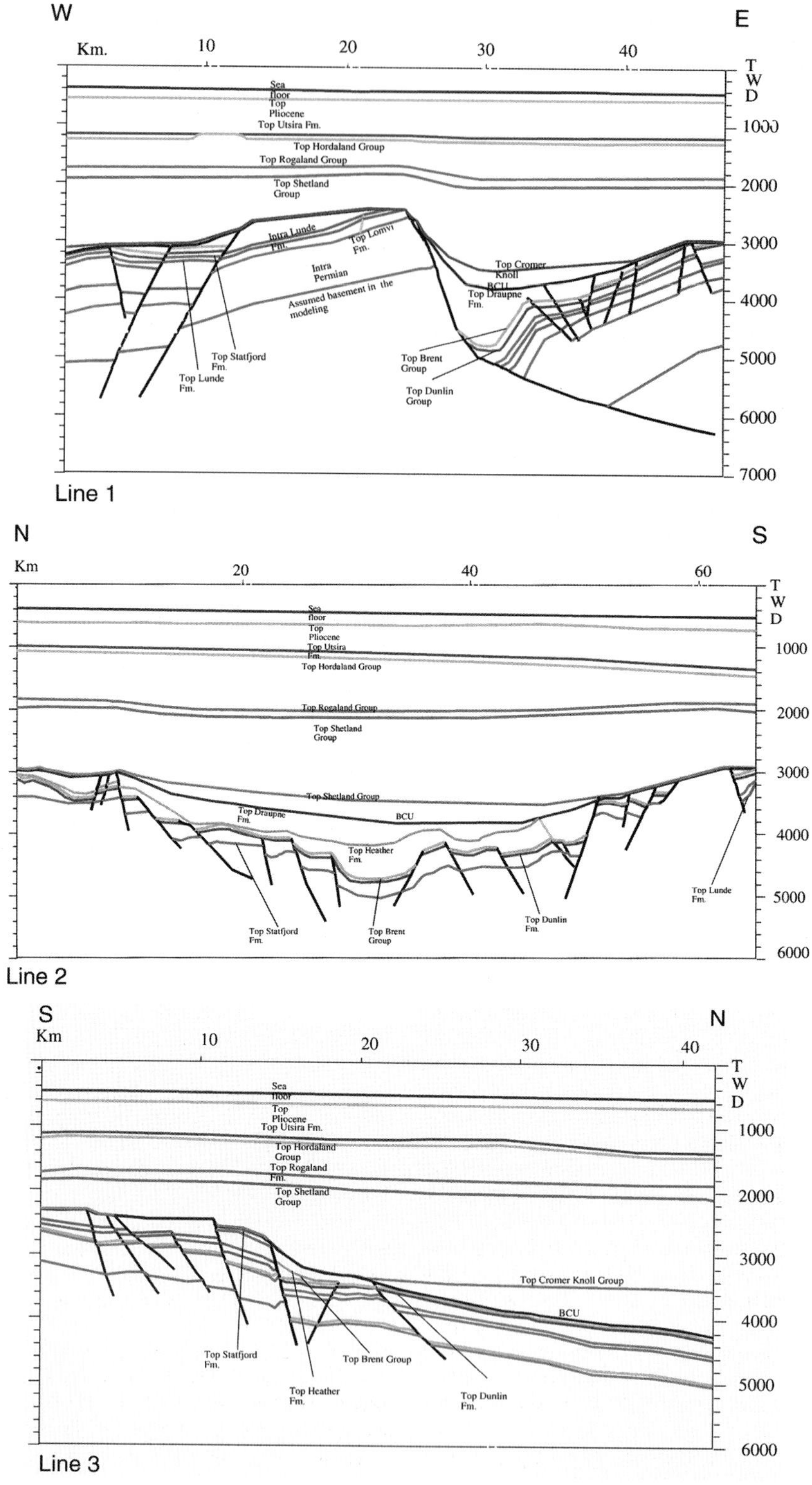


**Fig. 2.** Conceptual models of the three simulated 2D lines. Location are shown on Figure 1.

**Table 2.** *Proposed kinetic models for the Draupne formation, based on kerogen and asphaltenes*

| Kinetic model | Labile kerogen | | Stabile kerogen | | Kinetic from asphaltenes (34/4-7) | | MSSV-pyrolysis (Erdmann, 1999) | |
|---|---|---|---|---|---|---|---|---|
| | $E_a$ | Potential | $E_a$ | Potential | $E_a$ | Potential | $E_a$ | Potential |
| Activation-energy | 52 | 57 | 61 | 12.5 | 52 | 17.7 | 56 | 12.1 |
| (kcal/mol) | 53 | 28 | 62 | 14 | 53 | 33.4 | 59 | 48.1 |
| Partial potential in % | 54 | 13 | 63 | 23 | 54 | 26.8 | 61 | 27 |
| | 55 | 3 | 64 | 2 | 55 | 22.1 | 62 | 9.1 |
| | 59 | 1 | 66 | 4 | | | 63 | 3.7 |
| | | | 69 | 4 | | | | |
| Frequency factor (l/My) | $2.295 \times 10^{28}$ | | $8.596 \times 10^{29}$ | | $1.68 \times 10^{27}$ | | $3.00 \times 10^{29}$ | |

The concept of asphaltene as a tracer of maturity brings us to the concept of 'active kitchen determination'. The asphaltene kinetics delineates the P&T conditions which have been subjected to the source rock and contributed to the accumulation. From numerical modelling these areas are identified, compared to traditional SR kinetics and evaluated from the migration modelling.

## Pressure and migration parameters in basin modelling

The forward modelling of subsidence, erosion and sediment compaction simulates the geometric evolution of the 2D lines. In this respect the process of sediment compaction is a critical factor for two main reasons: Firstly the program uses individual porosity-depth relationships for different lithologies, these can be edited in the program, but they can hardly be assumed to describe the actual compaction of sediments over geological time. Sediment compaction is coupled to the development of palaeo-pressures. The software simulates this process as the difference between geostatic and hydrostatic pressure. This defines the effective stress, a factor controlling the physical compaction of the sediments. Compaction as a major driving force for generation of overpressure was first presented by Dickinson (1953) followed by Dickey *et al.* (1968) and others, who demonstrated that abnormal pressure in the Gulf Coast was controlled by lithofacies distribution of shales, not depth, temperature or stratigraphic age. This model was consistent with the results presented by Hubbert and Rubey (1959) based on the Terzaghi (1923) soil consolidation theory, which is based on the linear relationship between void index and effective stress, leading to the conclusion that overpressure decreases the loss of porosity compared to a hydrostatic situation. The model explained the shape and magnitude of overpressures in the Gulf Coast, assuming a high sedimentation rate and low permeability in shaly sediments (1 nanodarcy).

At present this compaction driven model is still widely used and implemented in modern basin modelling tools. The most important assumption in these models is that the depositional overburden load is the cause of compaction of the deeper layers (IES 1996). The overburden is responsible for the stress field in the basin. The driving force for sediment compaction in these models assuming a single axis stress tensor is the effective stress ($\sigma$) as the difference between overburden ($S_v$) and pore pressure ($P$) (equation (1)). This forms the basis for the compaction law defined in equation (2).

$$\sigma = S_v - P \tag{1}$$

$$\partial_t \phi = -(1 - \phi) C \, \partial_t \sigma \tag{2}$$

where $\phi$ denotes porosity, $C$ is the compressibility of rock matrix and $\sigma$ is the effective stress.

The overburden load causes the compaction process to act strictly in a vertical direction, i.e. only a vertical extension of a volume element will decrease. This is assuming compaction driven flow is slow and can be described by Darcy's law and that a mass balance is maintained between fluid and rock. Furthermore, both the rock material and the pore fluid are ideally incompressible. This means that the deeper elements compact only due to outflow of pore-water; hence a volume of overpressured rock will retain its pore water volume due to lower effective stress maintaining a high porosity during burial. Darcy's law states that the discharge velocity $Q$ of a fluid through a porous material is directly proportional to the gradient of hydrodynamic potential. The pressure difference over a given distance will set up a pressure

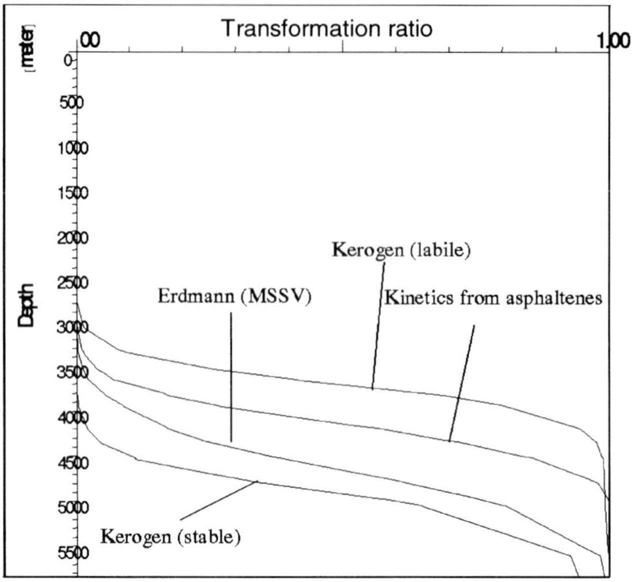


**Fig. 3.** Kinetic signature of the Draupne Source Rock in the study area. The kinetic signature of the asphaltene fraction compared with measured extremes of bulk kerogen kinetic variability (labile and stable) as well as closed system compositional kinetics from Erdmann (1999).

potential ($\nabla P$) where fluid will flow from high potential to lower potential as described in the Darcy equation (equation (3)).

$$Q = -\frac{kA\nabla P}{\mu L} \quad (3)$$

The fundamental assumption in current basin modelling packages is that sediment compaction is controlled by mechanical forces only. Accordingly, the effective stress (overburden – pore pressure) is modelled to drive porosity reduction, with the result that overpressured rocks are modelled to have elevated porosity (undercompaction). As discussed, this can be valid at relatively shallow depth before the onset of significant silica precipitation and most correct during a pulse of sedimentation without an increase in temperature in the sediments. The resulting fluid flow and overpressures are fundamental to the results of the migration modelling, as these are the main driving forces for hydrocarbon migration in the models.

Fluid flow modelling was performed assuming two-phase flow (petroleum and water) based on the Darcy flow equation. This simplification is assumed valid as the petroleum phase encountered in Snorre field is undersaturated.

In this study the permeability definition of the Shetland Group was of critical importance. The formation is considered as the primary cap rock for Snorre, and the permeability of this group to a large degree controls the pressure build up in the reservoir units.

## Results

### *Calibration*

The heat flow model used in this study, as described above, is based on a McKenzie model of exponential decreasing heat flow after the latest tectonic rifting event in late Jurassic–Early Cretaceous.

The simulated vitrinite reflectance (Sweeney & Burnham 1990), is generally in good agreement with the measured values in wells 33/6-1 and 34/4-8, although the calibration against the 34/7-15S well (Fig. 5a) results in higher estimates compared with the measured values. The temperature calibrations show some deviations, but are generally in agreement with the measured values (Fig. 5b). Calibration of formation pressure was performed at the top of the structure in a pseudo-well location near the 34/4-4 well (Fig. 5c). The model showed an excellent fit to the measured values, but the calibration is to a large degree dependent on the permeability/porosity definition of the lithology in the Shetland group, for which we have little control. The calibrated permeability in this study for this sequence was $10^{-6}$ mD at 5% porosity, but an

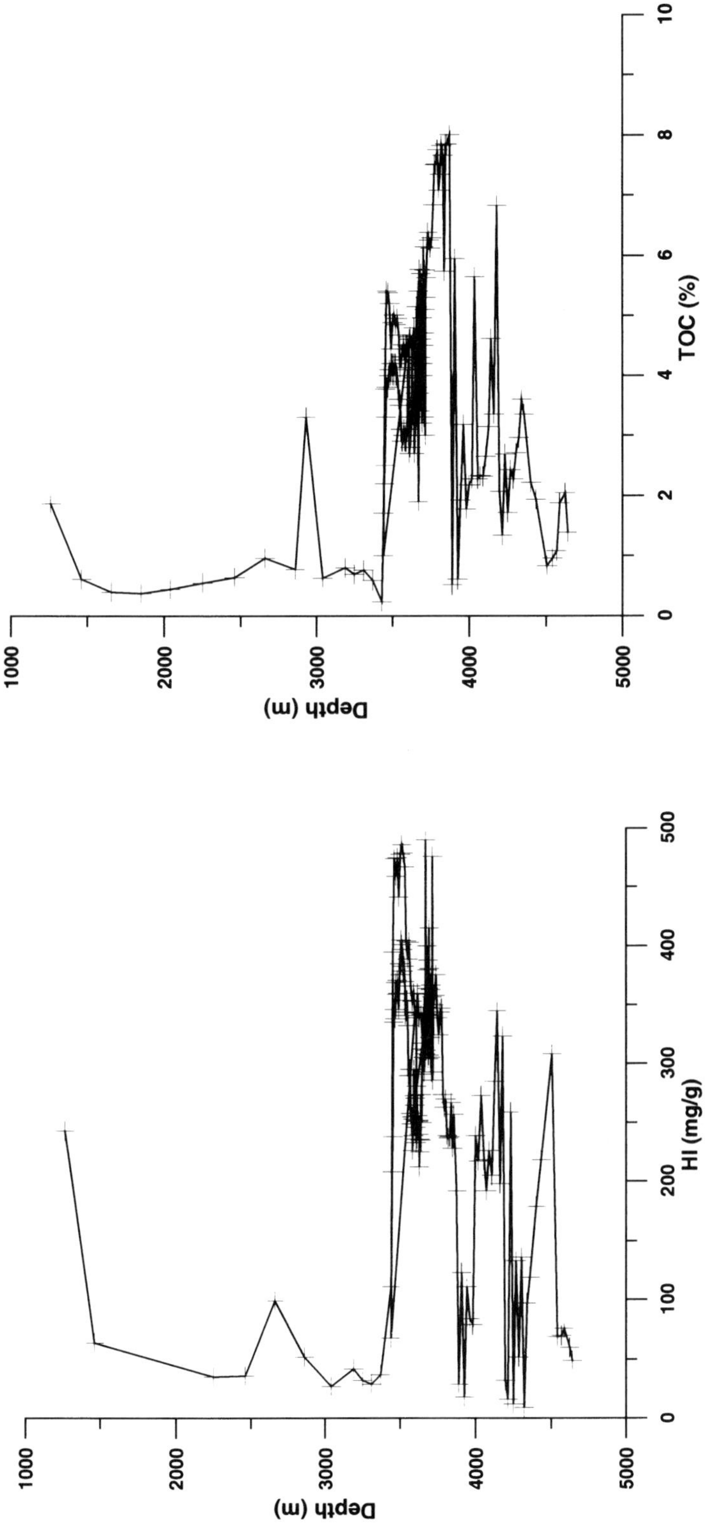


**Fig. 4.** Source rock quality described by TOC and HI in the 34/7-15S well, penetrating the assumed source rock in the area (Draupne Formation).

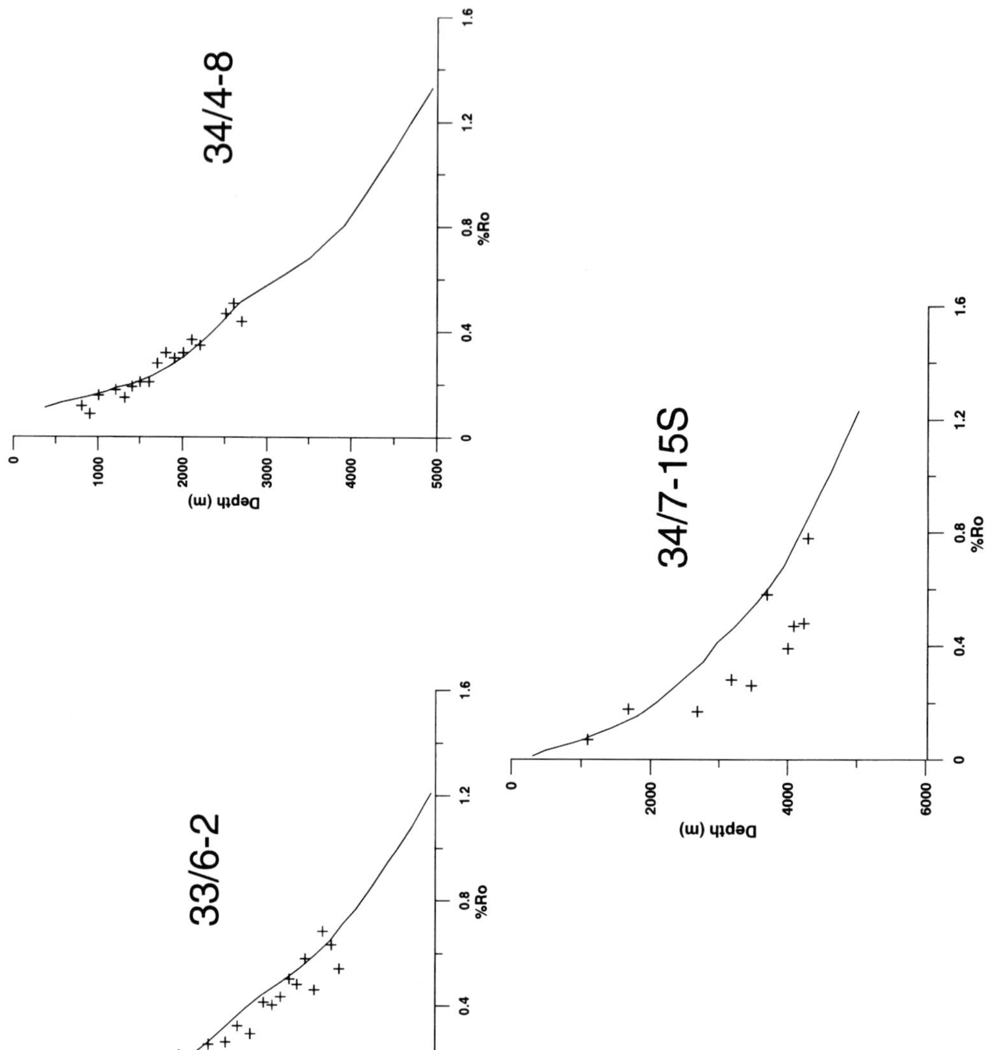


(a)

**Fig. 5.** (a) Simulated vitrinite reflectance against well data (present situation).

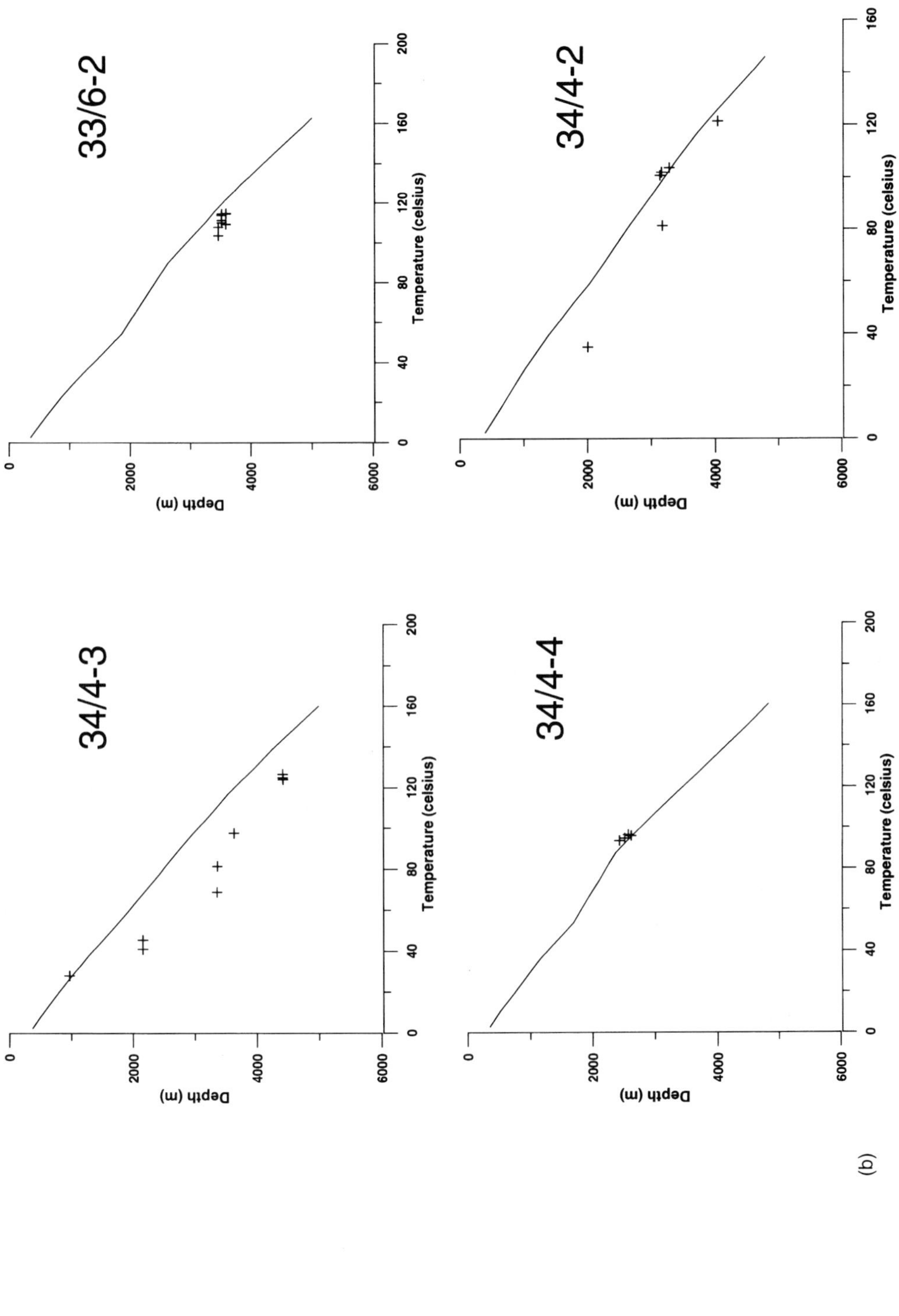


(b)

**Fig. 5.** (b) Simulated temperature against well data (present situation).

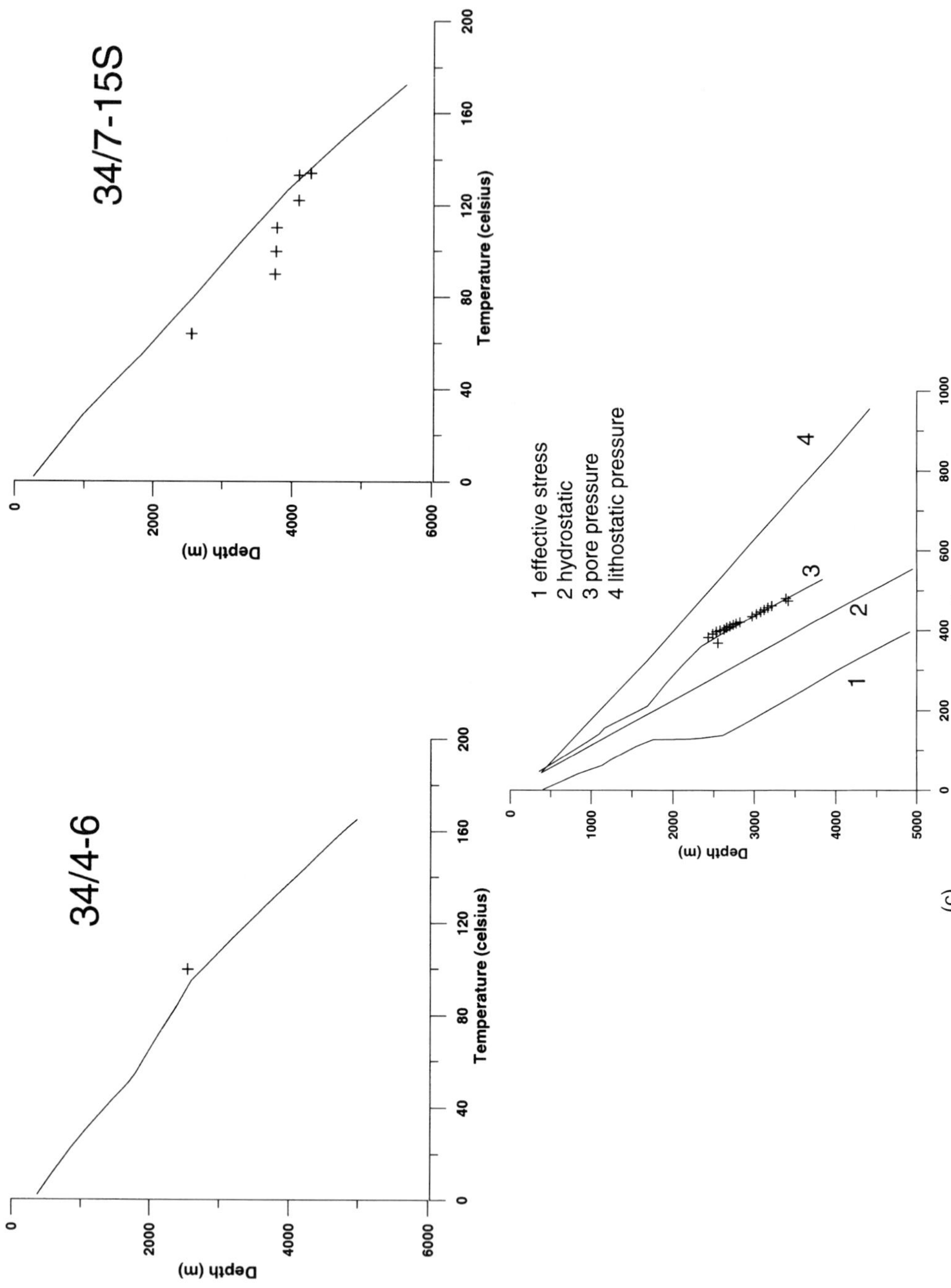


**Fig. 5.** (b) *Continued.* (**c**) Simulated pressure against well data (present situation).

increase in permeability to $10^{-5}$ mD would cause the model to maintain hydrostatic pressure. A decrease to $10^{-7}$ mD would, if no fracture model were applied, give rise to geostatic pressure.

## Active kitchen determination

Active kitchen determination was performed using the calibrated heat flow history discussed above and the asphaltene kinetics of the 34/4-7 DST oil from Snorre field. The calculated transformation ratios allowed an assessment of kitchen maturity. The approach used assumes that the kinetic signature determined on the petroleum asphaltenes represents a prediction of thermal alteration stages not yet achieved by the source rock. Accordingly the onset of asphaltene breakdown predicted by the kinetics give an indication of the time/temperature conditions prevailing during asphaltene phase formation and expulsion (di Primio *et al.* 2000). The threshold value used here is the 5% transformation ratio as calculated using the asphaltene kinetics. A simplified example of the use of this methodology is shown for a pseudo-well situated in the centre of the Pancake Basin (Kitchen 2). Figure 4 shows the evolution of petroleum formation as a function of depth based on the calibrated burial and thermal histories used in this study. The transformation ratios of the most labile and most stable kinetic datasets determined on Draupne formation samples are plotted together with the asphaltene kinetic-based transformation ratio. Based on the asphaltene kinetic transformation ratio the 5% threshold is reached at a burial depth of 3600 m, which represents the depth at which the source rocks burial history has reached the time/temperature conditions for generation and expulsion of the petroleum phase sampled in the reservoir.

These results indicated that the Draupne formation in Kitchen 1 west and northwest of Snorre has not reached the level of transformation required by the asphaltene kinetic data from the Snorre field. Kitchen 1 was modelled to only have reached a transformation ratio of less than 5% (Fig. 6) at present. In addition the limited source rock volume in this area cannot account for the petroleum volumes in the Snorre field. Thus, we would suggest on the basis of our maturity modelling, that the proposed kitchen area west and northwest of the Snorre (Kitchen 1) is immature. Shows in the 33/6-2 well, classified as *in situ* generated, not representative of significant petroleum migration, support this hypothesis. Additionally the fact that the main intra-field maturity trend known in Snorre (Horstad *et al.* 1995) is the increase in biomarker and aromatic maturity parameters from south to north, suggests clearly that Kitchen 1 did not contribute to Snorre as any influx of lower maturity petroleum from

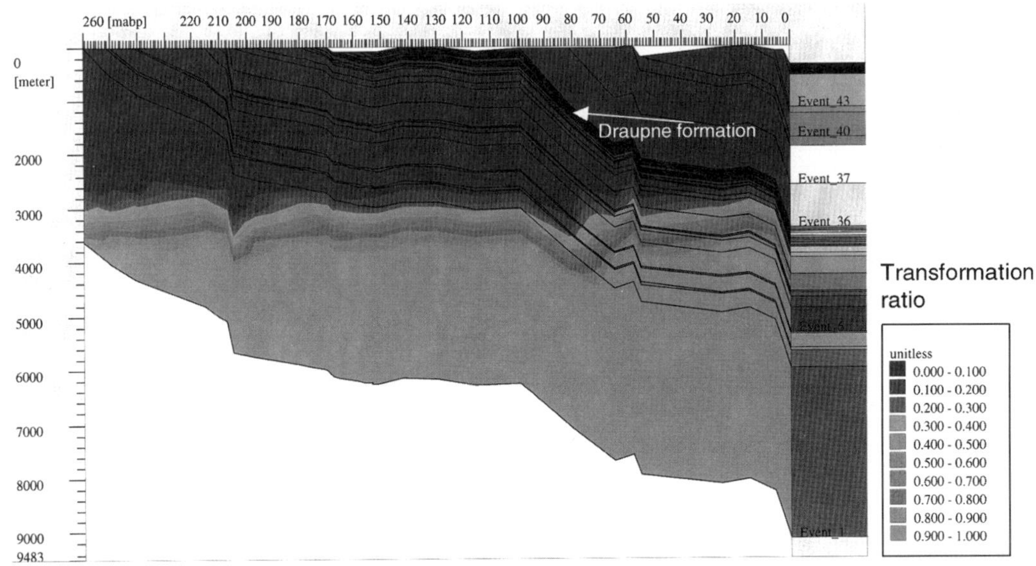


**Fig. 6.** Maturity evolution from asphaltene kinetics in Kitchen 1 defined in the 33/6-1 well, indicating that the area has not reached the required temperature as defined by the asphaltene kinetics.

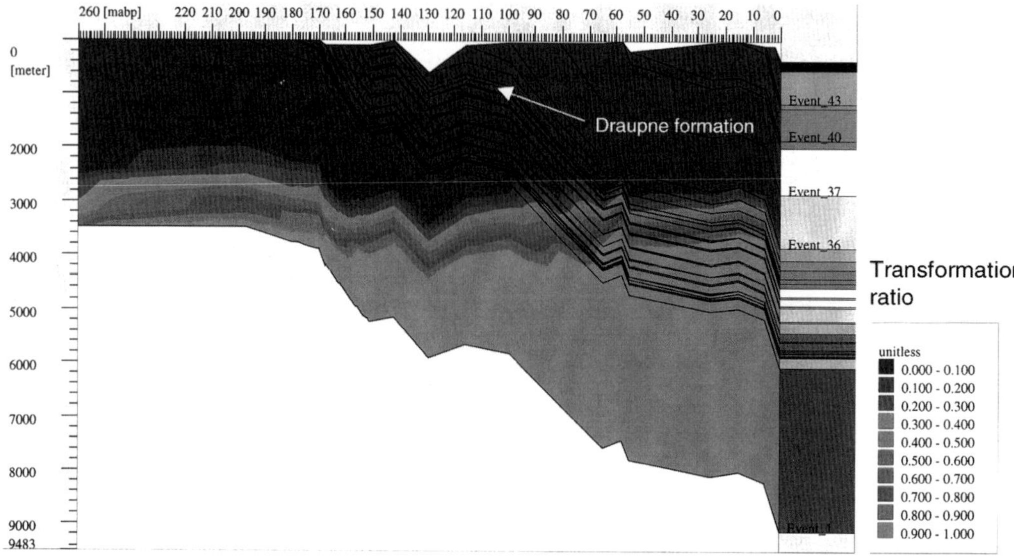


**Fig. 7.** Maturity development in the deepest are in the 34/5 kitchen area, indicating that the source rock is matching (and in the deepest part partly exceeded) the maturity required to match the asphaltene maturity in the reservoired Snorre oil.

west and northwest would have acted to obscure this trend. We therefore conclude that there is no petroleum contribution to the Snorre field from the basin to the west, and that migration from this region can be ruled out.

The basin east of Snorre (Kitchen 2) proved in the modelling to fulfil the asphaltene maturity requirements. At present, in the basin area outlined in Figure 1, the top of the Draupne formation has reached the transformation ratio required (Fig. 6). Figure 7 shows that the active kitchen thus defined is in the immediate vicinity of the Snorre structure, and that hence only short migration distances are required.

The kitchen north of Snorre, the Møre Basin (Kitchen 3) is also modelled to have reached the required maturity for it to contribute to the oil in Snorre, and is from pure maturity consideration, a candidate for filling at least part of the Snorre Field (Fig. 8).

## Migration

However, the active kitchen outlined in Figure 1 lies considerably farther away from the structure than Kitchen 2. This fact combined with a thin source interval indicated from the 3D seismic, the resulting large drainage area required to fill the field and the significant migration distances involved, make this area less likely as a source of the Snorre petroleum. The relatively complex fault lineaments north of Snorre further complicate migration into the field from this area. Modelling backs this up, which show no significant migration of petroleum from this basin south of the 34/4-8 well (Fig. 9). A sensitivity run with a ×10 increased petroleum mobility factor could still not enable lateral migration into the structure from this suggested Kitchen 3. Furthermore, the bulk of this petroleum is expected to exhibit a significantly higher maturity and GOR as compared with the values observed in Snorre. This suggests that the migration across the major fault lineaments north of Snorre (Hitra fault zone) is difficult, and that the generated petroleum is forced in another direction, not represented in this 2D model. It would then seem plausible that some of this petroleum could be trapped in the structurally complex Zeta structure. Still, we are in no position based on our data to rule out the possibility for long-range migration from this area into Snorre. The geometric arrangement of lithologies in this region, the limited SR thickness and the maturity of the Snorre oil collectively make this region less likely as the source kitchen for the Snorre oil.

Kitchen 2, east of Snorre, is favourably positioned geometrically to have contributed to Snorre. This area has, during Tertiary, generated significant amounts of petroleum. In the modelling, the western part of this kitchen is generating petroleum, which clearly could have found its

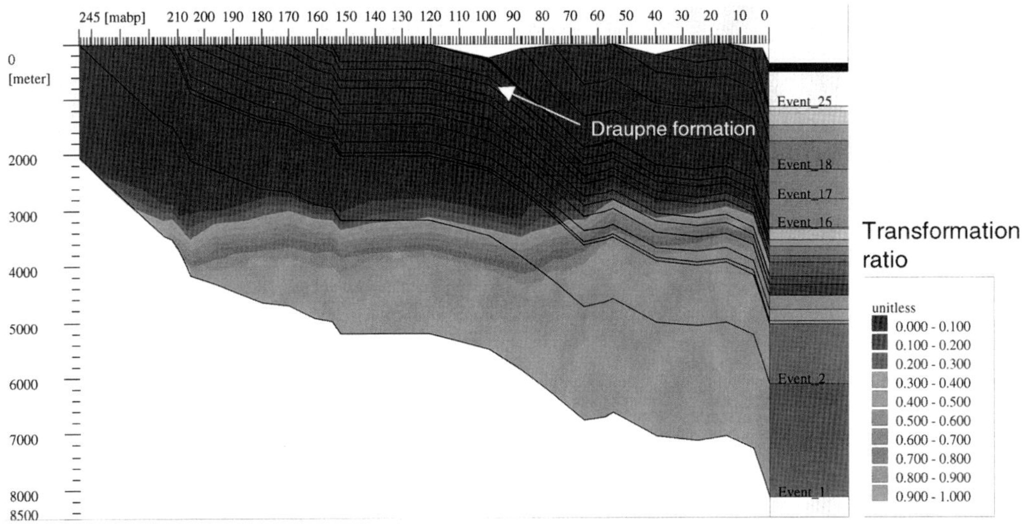


**Fig. 8.** Maturity in the 34/4-3 well from asphaltene kinetics, indicating that the well is matching the required maturity compared with the reservoired Snorre oil.

way into the Snorre structure by short-range up-fault migration (Fig. 10). The eastern part of this basin is found to drain towards the east, but not reaching the top of the northern part of the Visund block. It is only in the deepest part of this kitchen, i.e. ENE of Snorre, that the geometries are favourable for migration directly into Snorre. This basin has at present a maturity in the Draupne Formation fitting very well with the maturity derived from the asphaltenes in Snorre and modelling suggest that that expulsion could have generated petroleum matching the asphaltene kinetic predictions in the Early Tertiary. This could set up the first oil-wet system, thus favouring this migration pattern during progressive burial. This will result in a north, northeastern fill-point of the Snorre Field. Ongoing petroleum influx is thought to cause a backfilling of the reservoir, resulting in an increase in petroleum maturities towards the north in this field. This filling pattern, combined with the relatively leaky cap rock and recent burial (Leith *et al.* 1993; Leith & Fallick 1995), could conserve the low GOR and undersaturated character of the Snorre petroleum, despite the increasing maturity in the deepest parts of the kitchen area (cf. models in Sales 1997) discussed into further detail by di Primio & Skeie (this volume). Simple volume calculations of the western part of the 34/5 basin suggest that the migration efficiency required to fill the Snorre structure from this part of the basin does not exceed 30%. This means that a significant part of the potentially generated petroleum in this part of the kitchen is either not expelled, has not migrated or has followed a different migration path. By comparison, the eastern part of the 34/5 kitchen could prove to have generated more petroleum than the proposed area feeding the Snorre Field. The modelling showed a significant eastward petroleum migration in the Brent Group, but did not predict accumulations in the northern prolongation of the Visund field.

The different petroleum characteristics and the migration modelling indicate filling of the Visund Field from the east, in a similar pattern as described for Snorre. In this case, the deepest part of the kitchen is also to the east of the accumulation, but with a significantly deeper burial (Fig. 11). This will lead to a proposed higher maturity in the Visund Field compared with Snorre, and this is also confirmed by the lighter composition of the petroleum in the eastern part of the accumulation. Furthermore, parts of the complex Visund field do also contain low GOR petroleum, which the authors suggest is a relict from an early filling from the same kitchen area. One can speculate that the structure might have experienced vertical leakage of low GOR petroleum, preferably due to fracturing of the cap rock, as a significant paleo petroleum column exists below the present OWC. Supporting this model is the fact that the structure is far from filled to spill, and a significant amount of shows and petroleum inclusions occur below the present OWC. The influx of high GOR petroleum, which at present represents the bulk accumulation, might be a recent in-fill (Bhullar *et al.* 2000).

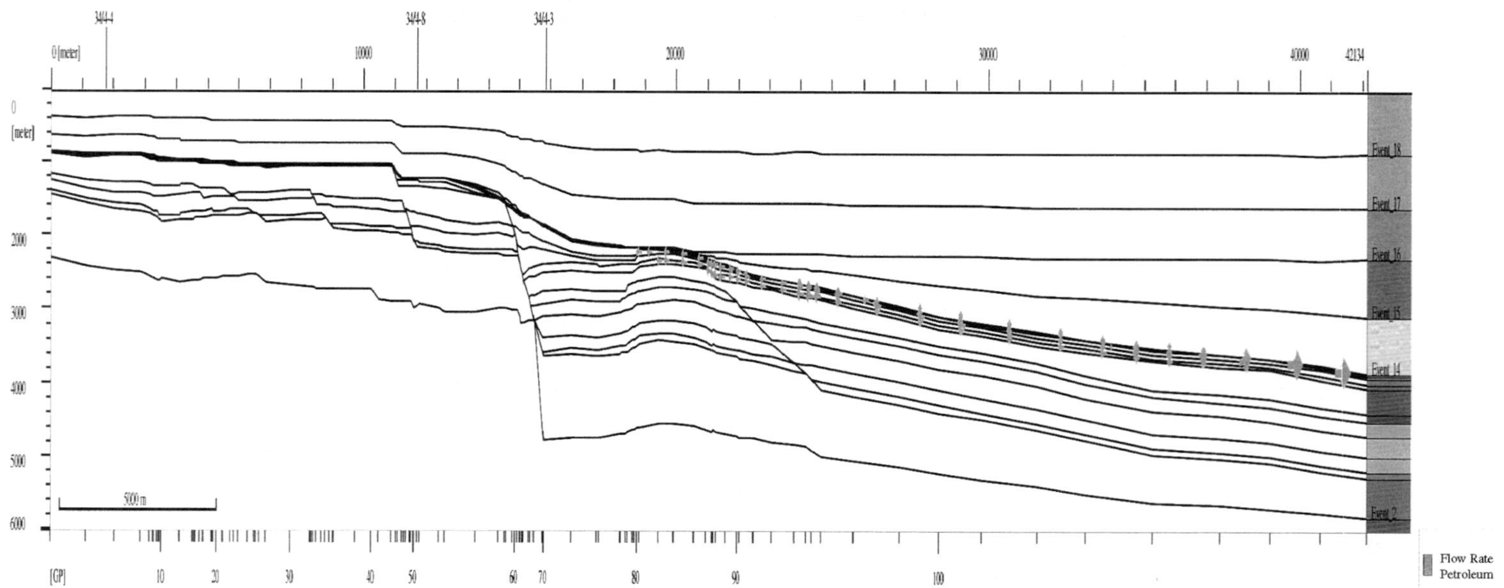


**Fig. 9.** Migration from the north of Snorre shown for 58 Ma and present situation. The migration modelling shows that migration out of the kitchen is limited and not reaching the reservoir.

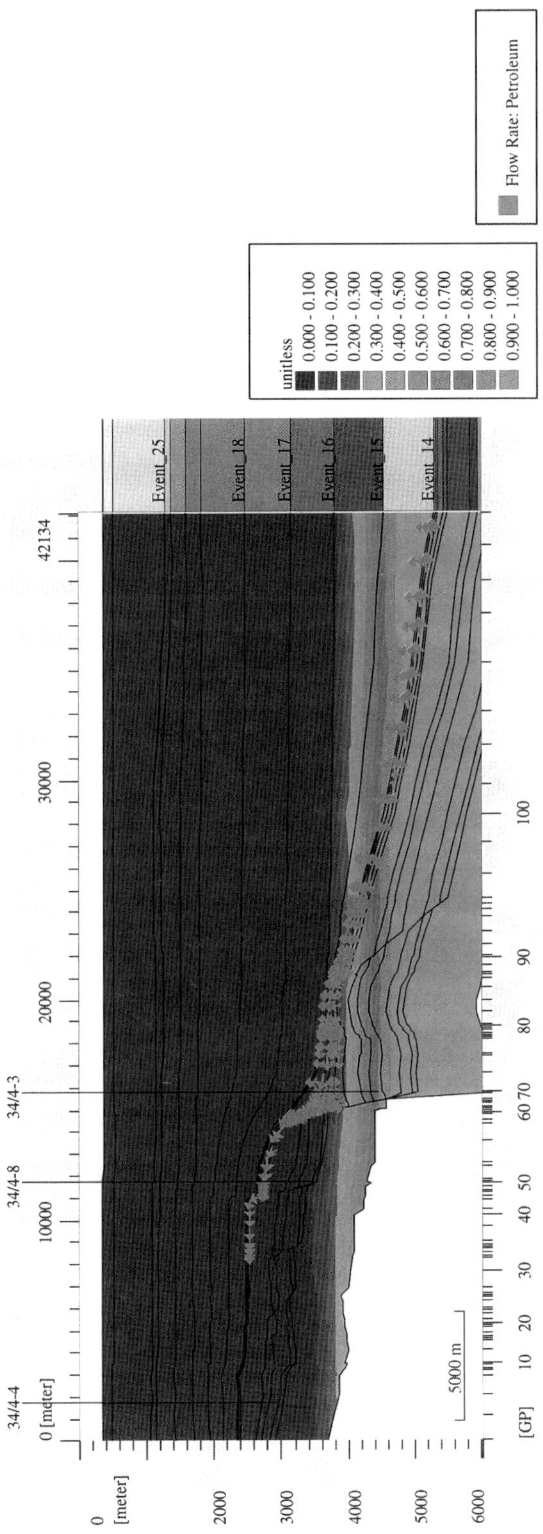


Fig. 9. *Continued*

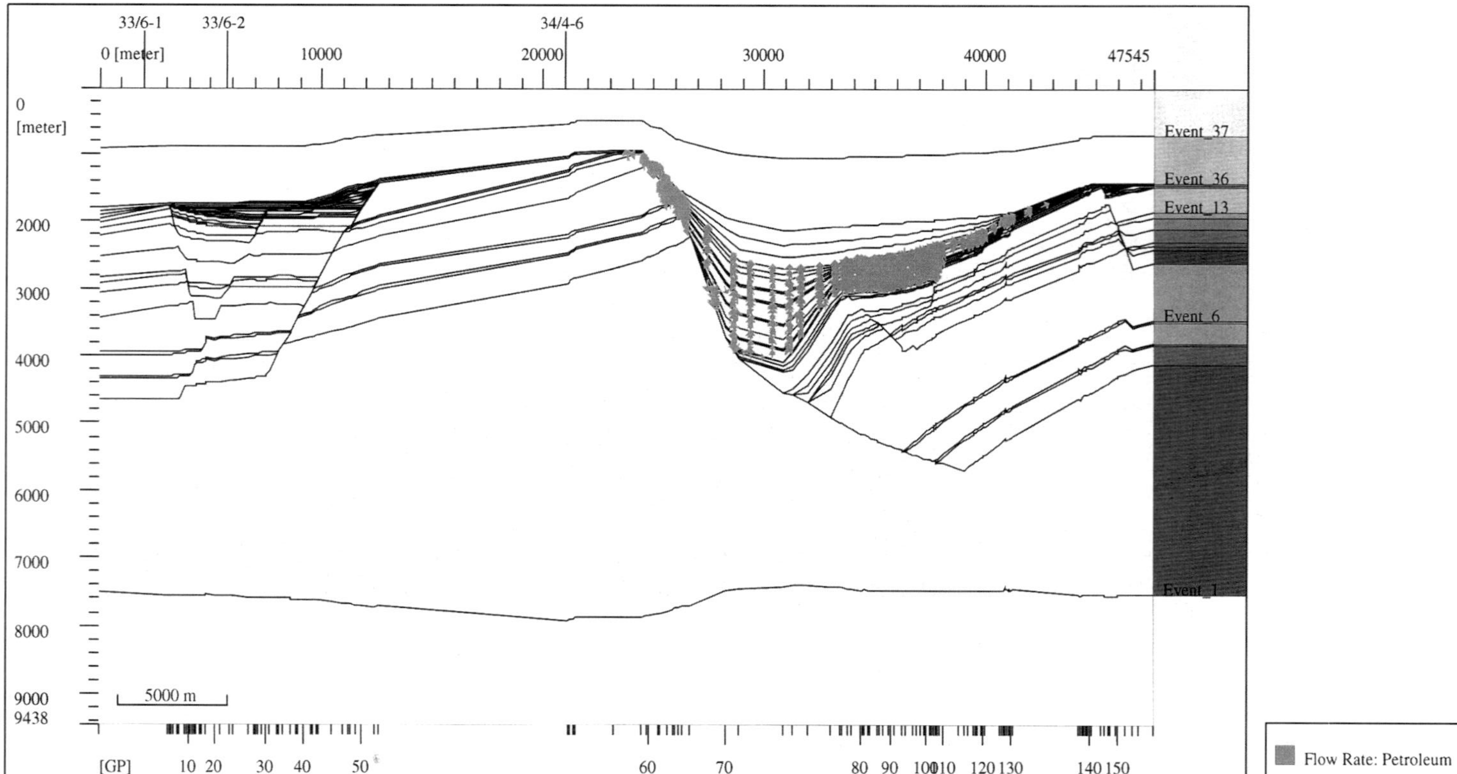


**Fig. 10.** Migration from the east of Snorre for 58 Ma and the present situation show that the Kitchen 2 has generated petroleum reaching the reservoir for more than 50 Ma. It also indicate a possibility of hydrodynamical backfilling of the structure somewhat occluded by the two dimensions in the modelling.

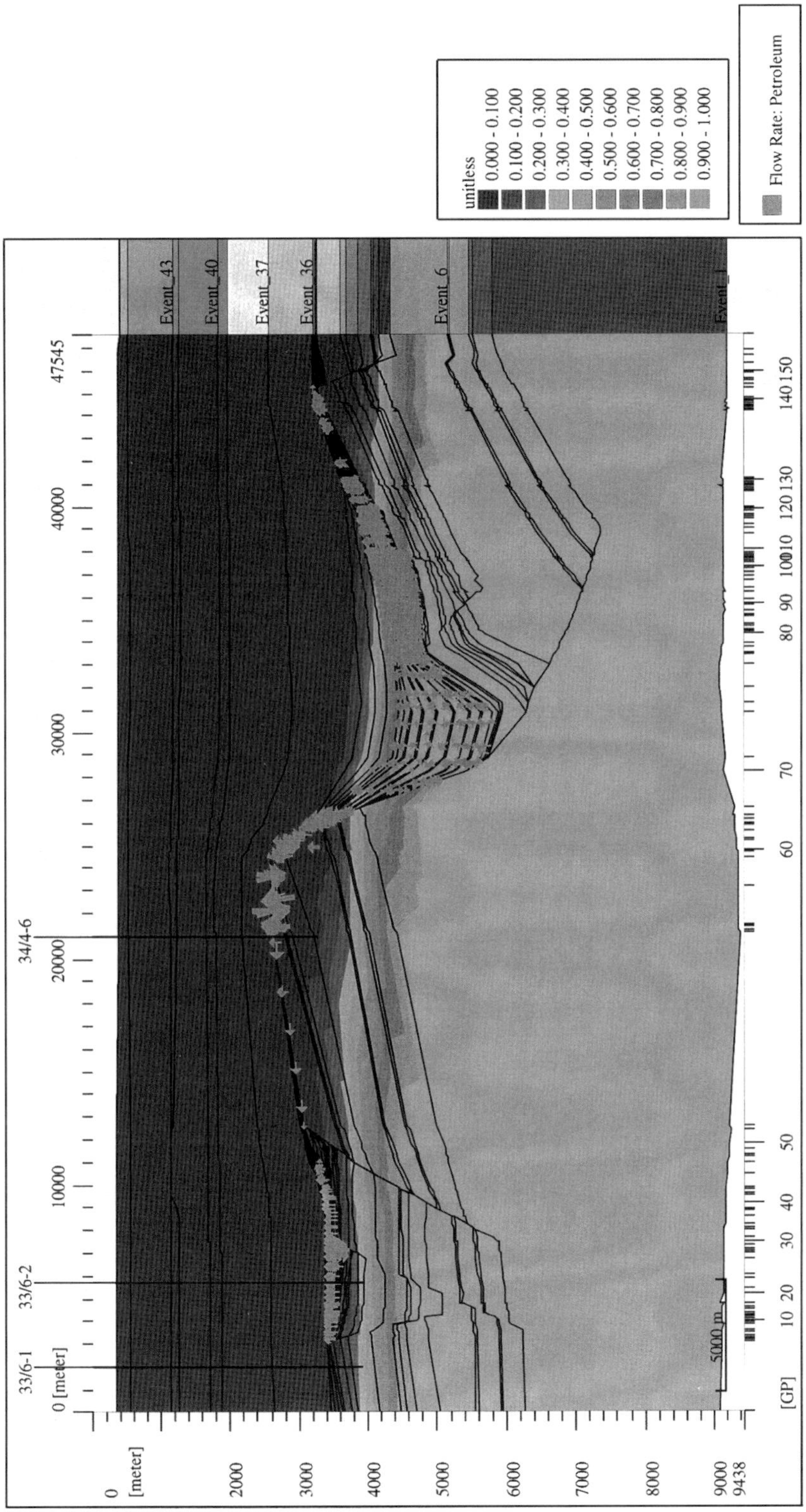


**Fig. 10.** *Continued*

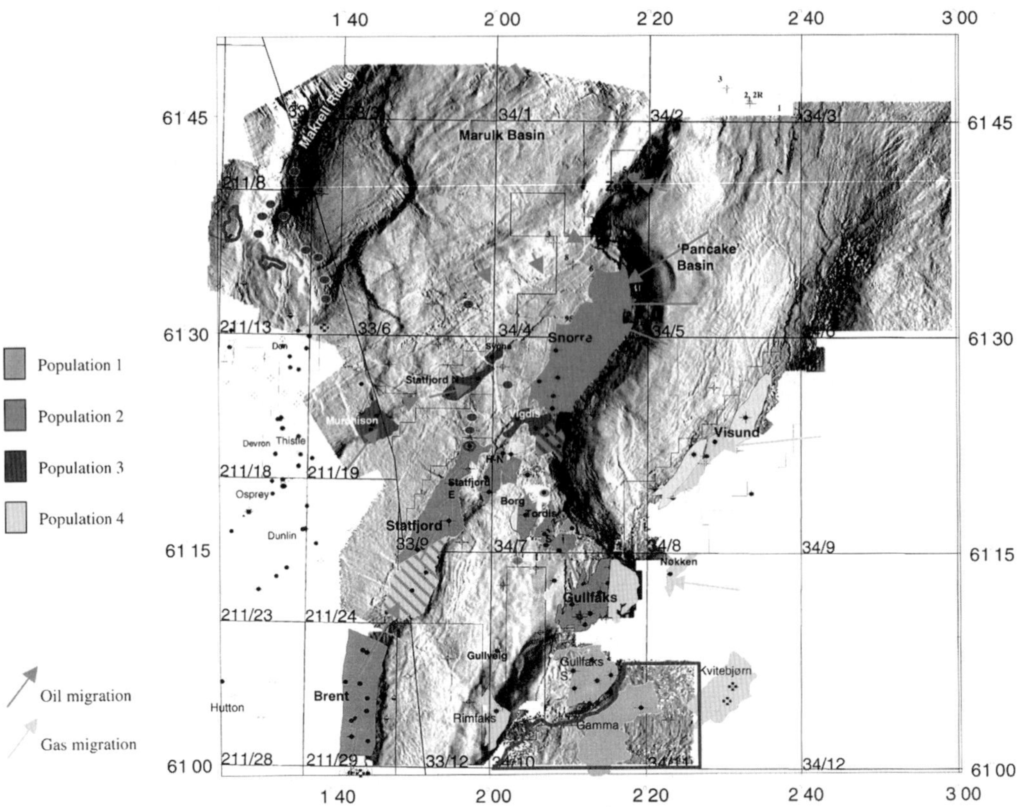


**Fig. 11.** Assumed filling direction from the east into the Snorre structure as concluded in this study. The geochemical classifications by Horstad *et al.* (1995) is supported and the fill spill route from the south into population 1 is maintained.

The immature character and limited source volume in Kitchen 1 makes migration from the west unlikely, thereby suggesting an alternative source for parts of the petroleum population 2, which the authors suggest is dominantly sourced from the East Shetland Basin.

The petroleum population 1 has not been modelled in this study and the fill–spill system from the south, terminating in the Gullfaks field, is maintained. The supposed source area for this population is the North Viking Graben (Thomas *et al.* 1985; Di Primio *et al.* 1998). The concluding filling directions are indicated in Figure 11.

## Further aspects

### *Recent maturity suppression*

One important factor concerning the petroleum system in the area is the low response in maturation to the deeper burial in the Pliocene (Figs 6–8). This is partly explained by the low palaeo-temperature at the sediment–water contact this period, and partly by high sedimentation rates suppressing the conductive heat flux causing a lowering of the geothermal gradient in the basin. This causes a suppression of the maturity development and deepens the oil window by *ca.* 1 km in the area. Sensitivity analyses reveal small changes in this picture even if initial heat flow is varied by ±10% and only a marginal increase in maturity is expected during this period.

### *Modelling of pressure*

The simulated present reservoir pressure (Fig. 5c) indicates significant decrease in effective stress, but below estimated fracture pressure (0.8–0.85 × geostatic pressure) in the Snorre field. An important factor in controlling the estimated porosities in the reservoir sands, due to the

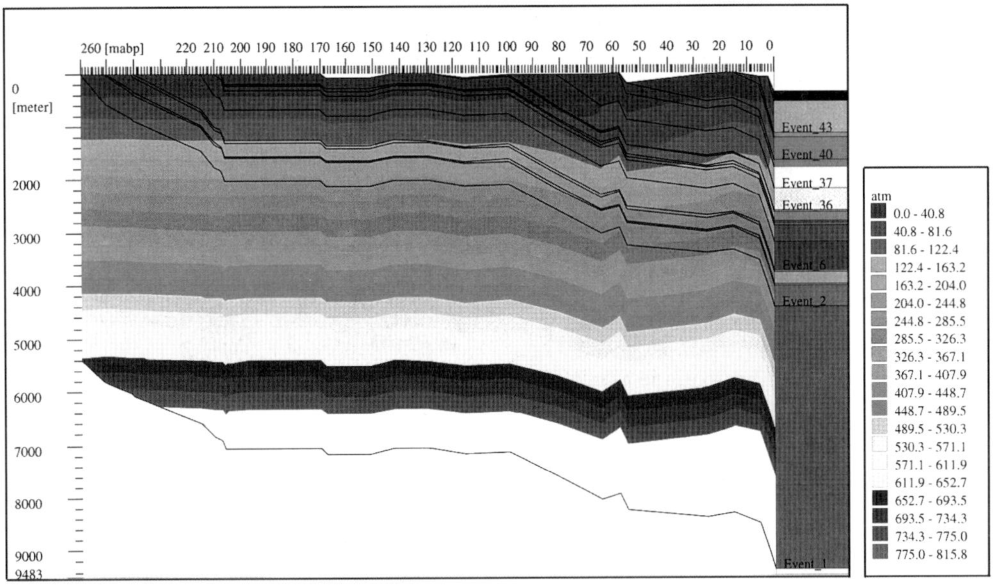


**Fig. 12.** Effective stress evolution in a pseudo-well on the top of the Snorre structure. Notice the rapid drop in effective stress in the reservoir (overpressure build up) during the rapid Pliocene burial.

stress-related compaction model, is the timing of this event. By taking a line through the northern Snorre reservoir, the pressure build up can be assessed by the burial history (Fig. 12). From this type of modelling results it is evident that the generation of overpressure is a recent event, dominantly controlled by the rapid burial during the Pliocene. The overpressure generation causes a drop in the effective stress in the Mesozoic sediments, and the reservoir units do not experience a large increase in mechanical compaction during the last rapid pulse of sedimentation.

### *Undersaturated petroleum imperfect seal—subsidence*

One important consideration is the undersaturated nature of the reservoired petroleum in the Snorre Field, and several other fields on the Tampen Spur. One would expect the source rock to generate petroleum, which by vertical migration to shallower stratigraphic levels eventually should reach the bubble-point pressure and expel excess gas. In a subsiding basin, the progressive maturation of the source rock should cause the petroleum in the Snorre to be saturated with gas, or at least have a significantly higher GOR. The modelling showed an insignificant maturation during the Pliocene, implying that this period is not a major contributor to the present petroleum. Combined with the recent and rapid burial and generation of overpressure, this might explain the presence of undersaturated petroleum in the reservoir.

Another possible explanation of the low GORs in the Field is the imperfect seal over the reservoir. Shows of petroleum resembling the Snorre population have been found several hundreds of metres over the reservoir interval (Caillet 1993; Leith *et al.* 1993; Leith & Fallick 1995), even though the field probably has never reached fracture pressure. It is clear that the poor cap rock quality on the relatively shallow and underfilled Snorre structure makes this a perfect candidate for a 'Type 3' trap as explicitly classified by Sales (1997), i.e. the GOR of the Snorre petroleum charge is determined more by the cap rock properties than by SR quality and its kinetic properties. The oil inferred by Leith *et al.* (1993) may have leaked from the reservoir as the cap rock section has very low pore entry pressures and micro fracturing is also observed in the cap rock section. One can therefore speculate that the leakage may be a prerequisite to maintain black oil in the Snorre Field.

## Conclusions

Results from the modelling shows that the use of asphaltenes as a kinetic parameter is eligible, in particular if the source rock exhibits a heterogeneous character (Fig. 4). It is also suggested that it is a better method for describing the source kinetic as these compounds evidently arise from an actively generative source rock, as opposed to the traditional source rock kinetics most often based upon samples taken from structural heights as the basin areas are rarely drilled. The results of such an approach thus differ from traditional source rock kinetics as they describe the condition of generation and primary migration of the accumulated petroleum.

The 2D migration modelling demonstrated the importance of presumed high permeability layers within the syn-rift sediments of the Draupne and Heather Formations. These thin layers (<5 m) are a prerequisite for any effective migration from the suggested kitchen areas to the Snorre Field. From the lithological distribution in the 34/7-15S well there are many factors pointing towards the importance of these high permeability layers. Their distribution and conductivity have the potential to be a controlling factor for the regional petroleum system, and further studies should be made to investigate their lateral extent. Such layers have also been described in the southern part of the Viking Graben (Isaksen & Ledje 2001).

Combining the results from the fluid flow, kinetic modelling and the accepted geochemical petroleum populations in the area, this study concludes there are evidences pointing towards kitchen 34/5 as uniquely responsible for the Snorre accumulation.

Furthermore, the undersaturated nature of the petroleum is suggested not to be primarily controlled by the maturity of the source rock, but rather by the cap rock properties (Sales 1997) and the burial (pressure) history of the accumulation. The same mechanism is suggested to cause more shallow fields in the area to maintain a 'black oil' preference as lower boiling compounds are rapidly leaking from the relatively poor seals.

Former Saga Petroleum is acknowledged for initializing this project and the support from RWE-DEA Norway and Chevron Norway is greatly appreciated. The paper benefited substantially from review by Johannes Wendebourg.

## References

Bandurski, E. 1982. Structural similarities between oil generating kerogens and petroleum asphaltenes. *Energy*, **6**, 47–67.

Behar, F. & Pelet, R. 1984 Asphaltene characterization by pyrolysis and chromatography. *Journal of Analytical Applied Pyrolysis*, **7**, 121–135.

Bethke, C. M. 1989. Modeling subsurface flow in sedimentary basins. *Geologische Rundschau*, **78**, 129–154.

Bhullar, A. G., Karlsen, D. A., Backer-Owe, K., Le Tran, K., Skalnes, E., Berchelmann, H. H. & Kittelsen, J. E. 2000. Reservoir characterization by a combined micro-extraction/micro-thin-layer chromatography (iatroscan) method: A calibration study with examples from the Norwegian North Sea. *Journal of Petroleum Geology*, **23**, 221–244.

Burley, S. D. *et al.* 2000. New insights on petroleum migration from the application of 4D basin modelling in oil and gas exploration. *Journal of Geochemical Exploration*, **69**, 465–470.

Caillet, G. 1993. The Caprock of the Snorre Field, Norway—a possible leakage by hydraulic fracturing. *Marine and Petroleum* Geology, **10**, 42–50.

di Primio, R., Dieckmann, V. & Mills, N. 1998. PVT and phase behaviour analysis in petroleum exploration. *Organic Geochemistry*, **29**, 207–222.

di Primio, R., Horsfield, B. & Guzman-Vega, M. A. 2000. Determining the temperature of petroleum formation from the kinetic properties of petroleum asphaltenes. *Nature*, **406**, 173–176.

Dickey, P. A., Shriram, C. R. & Paine, W. R. 1968. Abnormal pressures in deep wells of Southwestern Louisiana. *Science*, **160**, 609–615.

Dickinson, G. 1953. Geological aspects of abnormal reservoir pressures in Gulf Coast Louisiana. *AAPG Bulletin*, **37**, 410–432.

Durand, B., Ungerer, P., Chiarelli, A. & Oudin, J. L. 1984. *In*: Ungerer, P., Chiarelli, A. & Oudin, J. L. (eds) *Modélisation de la migration de l'huile. Application à deux exemples de bassins sédimentaires.* World Petroleum Congress, London, 1983, PD1 1, 1–13.

Erdmann, M. 1999. Gas generation from overmature upper Jurassic source rocks, norther Viking graben PhD thesis.

Faerseth, R. B. 1996. Interaction of Permo-Triassic and Jurassic extensional fault-blocks during the development of the northern North Sea. *Journal of the Geological Society, London*, **153**, 931–944.

Glennie, K. W. 1998. *Petroleum Geology of the North Sea: Basic Concepts and Recent Advances*. Blackwell Science, London, 1–636.

Hermanrud, C. 1993. Basin modelling techniques—an overview. *In*: Doré, A. G. *et al.* (eds) Basin modelling: advances and applications. NPF Special Publications, Elsevier, Amsterdam, 1–34.

Horstad, I., Larter, S. R. & Mills, N. 1995. *Migration of Hydrocarbons in the Tampen Spur Area, Norwegian North Sea: A Reservoir Geochemical Evaluation*. Geological Society, London, Special Publications, **86**, 159–183.

Hubbert, K. M. & Rubey, W. W. 1959. Role of fluid pressure in mechanics of overthrust faulting. *Bulletin of the Geological Society of America*, **70**, 115–166.

IES 1996. Petromod Manual. Jülich, Germany.

Isaksen, G. H. & Ledje, K. H. I. 2001. Source rock quality and hydrocarbon migration pathways within the greater Utsira High area, Viking Graben, Norwegian North Sea. *AAPG Bulletin*, **85**, 861–883.

JARVIE, D. M. 1991. Factors affecting Rock-Eval derived kinetic-parameters. *Chemical Geology*, **93**, 79–99.

JORDT, H., FALEIDE, J. I., BJORLYKKE, K. & IBRAHIM, M. T. 1995. Cenozoic sequence stratigraphy of the central and northern North-Sea Basin—tectonic development, sediment distribution and provenance areas. *Marine and Petroleum Geology*, **12**, 845–879.

KARLSEN, D. A. & LARTER, S. R. 1991. Analysis of petroleum fractions by TLC-FID—Applications to petroleum reservoir description. *Organic Geochemistry*, **17**, 603–617.

LEITH, T. L. & FALLICK, A. E. 1995. Organic geochemistry of cap-rock hydrocarbons, Snorre Field, Norwegian North Sea. *American Association of Petroleum Geologists Memoir*, **67**, 115–134.

LEITH, T. L., KAARSTAD, I., CONNAN, J., PIERRON, J. & CAILLET, G. 1993. Recognition of caprock leakage in the Snorre Field, Norwegian North Sea. *Marine and Petroleum Geology*, **10**, 29–41.

LUCAZEAU, S. & LE DOUARAN, S. 1984. Numerical model of sediment thermal history: comparison between the Gulf of Lion and the Viking Graben. *In*: DURAND, B. (ed.) *Thermal Phenomena in Sedimentary Basins*. Editions Technip, Paris, 211–218.

MCKENZIE, D. P. 1978. Some remarks on the development of sedimentary basins. *Earth Planetary Science Letters*, **40**, 25–32.

NAKAYAMA, K. & LERCHE, I. 1987. Basin analysis by model simulation—effects of geologic parameters on one-dimensional and two-dimensional fluid-flow systems, with application to an oil-field. *American Association of Petroleum Geologists Bulletin*, **71**, 1120.

PEPPER, A. S. & CORVI, P. J. 1995. Simple kinetic-models of petroleum formation 3. Modeling an open system. *Marine and Petroleum Geology*, **12**, 417–452.

POELCHAU, H. S. & MANN, U. 1989. Evolution of sedimentary basins—the integrated modeling approach: an introduction. *Geologische Rundschau*, **78**, 1–5.

SALES, J. K. 1997. Seal strength vs. trap closure—a fundamental control on the distribution of oil and gas. *American Association of Petroleum Geologists Memoir*, **67**, 57–85.

SCHROEDER, F. W. & SYLTA, O. 1993. Modelling the hydrocarbon system of the North Viking Graben: a case study. *In*: DORÉ, A. G., AUGUSTSON, J. H., HERMANRUD, C., STEWART, D. J. & SYLTA, O. (eds) *Basin Modelling: Advances and Applications*. NPF Special Publications, Elsevier, Amsterdam, 469–484.

SOLLI, H. & LEPLAT, T. P. 1986. Pyrolysis–gas chromatography of asphaltenes and kerogens from source rocks and coals—a comparative structural study. *Organic Geochemistry*, **10**, 313–329.

SPENCER, A. M. 1984. Petroleum geology of the North European margin. Graham and Trotman, 1–436.

STAINFORTH, J. G. & REINDERS, J. E. A. 1990. Primary migration of hydrocarbons by diffusion through organic-matter networks, and its effect on oil and gas generation. *Organic Geochemistry*, **16**, 61–74.

SUNDARAMAN, P. 1995. Activation-energy distribution—whole-rock versus kerogen. *Abstracts of Papers of the American Chemical Society*, **209**, 18-GEOC.

SWEENEY, J. J. & BURNHAM, A. K. 1990. Evaluation of a simple-model of vitrinite reflectance based on chemical-kinetics. *American Association of Petroleum Geologists Bulletin*, **74**, 1559–1570.

TERZAGHI, K. V. 1923. Die Berechnung der Durchassigkeitsziffer des Tones aus dem Verlauf der hydrodynamishen Spannungerscheinungen, Sitzungbericht Akademie Wissenschaft Wien. *Mathematik Naturwissenschaft*, **132**, 125–138.

THOMAS, B. M., MUELLER-PEDERSENM, P., WHITTAKER, M. F. & SHAW, N. D. 1985. Organic facies and hydrocarbon distributions in the Norwegian North Sea. *In*: *Petroleum geochemistry in exploration of the Norwegian Shelf*. Proceedings of the NPF conference, Stavanger, 3–26.

UNDERHILL, J. R. 1998. Jurassic. *In*: GLENNIE, K. W. (ed.) *Petroleum Geology of the North Sea. Basic Concepts and Recent Advances*. Blackwell Science, London, 245–292.

UNGERER, P. 1990. State-of-the-art of research in kinetic modeling of oil formation and expulsion. *Organic Geochemistry*, **16**, 1–25.

VANDENBROUCKE, M., BEHAR, F. & RUDKIEWICZ, J. L. 1999. Kinetic modelling of petroleum formation and cracking: implications from the high pressure/high temperature Elgin Field (UK, North Sea). *Organic Geochemistry*, **30**, 1105–1125.

WELTE, D. H. & YÜKLER, M. A. 1981. Petroleum origin and accumulation in basin evolution—a quantitative model. *American Association of Petroleum Geologists Bulletin*, **65**, 1387–1396.

# Development of a compositional kinetic model for hydrocarbon generation and phase equilibria modelling: a case study from Snorre Field, Norwegian North Sea

ROLANDO DI PRIMIO[1] & JON ERIK SKEIE[2]

[1]*GeoForschungsZentrum Potsdam, Telegrafenberg, 14473 Potsdam, Germany*
[2]*Institutt for Geologi, Universitetet i Oslo, P.O. Box 1047 Blindern, 0316 Oslo, Norway*

**Abstract:** The possibility to model petroleum composition during hydrocarbon generation as well as the PVT behaviour of the fluids during migration has only recently become available in modern basin modelling software packages. While various compositional kinetic models of petroleum generation have been published in the past few years, none of the studies presented have attempted to match the composition, physical properties and phase state of known petroleum accumulations. Using compositional data from closed-system non-isothermal pyrolysis experiments, we developed a compositional kinetic model of hydrocarbon generation for a marine Type II source rock, which uses 13 components to describe the generated fluid. The data format selected is compatible with the compositional resolution used in reservoir engineering, thus allowing a direct comparison of predicted compositions and phase behaviour with PVT data of natural fluids. Compositional predictions of the model were tuned to a well-documented maturity sequence from the Tampen Spur, Norway, and the calibrated model implemented in a 2D basin modelling study of the Snorre Field, Norway. The results of the modelling led to an excellent correlation between predicted and reported reservoir fluid properties (formation volume factor, GOR and saturation pressure) for the present-day situation. The results indicate that the Snorre reservoir has received a continuous charge since the late Cretaceous–early Tertiary and that it most likely contained a two-phase system prior to the latest Plio-Pleistocene burial and overpressuring event.

It is well known that the compositions of petroleum fluids expelled from source rocks change as a function of increasing maturity (Tissot & Welte 1984 and references therein). Petroleum composition directly controls the physical response of the fluid to changing pressure and temperature conditions (PVT behaviour) during both secondary migration (England *et al.* 1987; Dueppenbecker & Horsfield 1990; di Primio *et al.* 1998; di Primio 2002) and production (Pedersen *et al.* 1989; McCain 1990). The prediction of petroleum composition has up to now focused predominantly on describing the evolution of hydrocarbon compound classes generated and expelled from source rocks. The bulk differentiation and kinetic description of gas versus oil generation as a function of maturity was the first step in predicting hydrocarbon composition (Espitalié *et al.* 1988). Refined analytical methods, involving both closed and open system pyrolysis techniques as well as the combination of both, later allowed the detailed kinetic description of individual compound and compound class generation and destruction (Sweeney *et al.* 1987, 1995; Béhar *et al.* 1991, 1995, 1997; Braun *et al.* 1992; Dieckmann *et al.* 1998; Vandenbroucke *et al.* 1999). The focus of these newer models was to predict the evolution of hydrocarbon composition as defined by the processes of primary generation and secondary cracking (in both the source and reservoir).

The integration of the kinetic predictions with equation of state (EOS) models for phase prediction was, however, not attempted in any of the compositional kinetic studies published to date. Sylta (1991a) demonstrated the general possibility of combining compositional kinetic predictions and migration modelling with equation of state modelling; however, in the synthetic geological study area used in Sylta's study a demonstration of the accuracy of the predictions was not possible.

The correct prediction of the phase behaviour of a migrating hydrocarbon fluid is, however, of paramount importance for the simulation of secondary migration. The physical properties of petroleum fluids change systematically as a function of changing pressure and temperature. Especially upon phase separation, when oil and gas coexist in the carrier or reservoir porosity, the correct assessment of individual phase properties and relative volumetrics are critical for the further migration of the fluid system (di Primio 2002); these again depend on the petroleum composition which is initially controlled by the source rock type and maturity. Depending on relative phase saturation, the physical separation of oil and gas could occur as functions of relative

*From*: CUBITT, J. M., ENGLAND, W. A. & LARTER, S. (eds) 2004. *Understanding Petroleum Reservoirs: towards an Integrated Reservoir Engineering and Geochemical Approach.* Geological Society, London, Special Publications, **237**, 157–174. 0305-8719/04/$15.00 

permeability, differential migration due to density contrasts, capillary failure of the cap rock for a single phase etc. The possibility to model petroleum composition during hydrocarbon generation as well as the PVT behaviour of the fluids during migration has only recently been incorporated into modern basin modelling software packages; however, as stated above, no compositional kinetic model capable of predicting petroleum phase behaviour is yet available.

It was recently demonstrated that compositional information from closed system pyrolysis of source rock samples, when properly calibrated, can be used to reconstruct the regional distribution of hydrocarbons as a function of migration and phase separation using PVT simulation software (di Primio 2002). While this former study was performed using PVT software it indicated the need for the development of phase predictive kinetic models for the correct simulation of secondary migration in basin modelling.

Both the basin modelling (PetroMod, IES GmbH, Germany) and the PVT simulation software used (PVTsim, Calsep a.s., Denmark) applied Soave–Redlich–Kwong type cubic equations of state where fluid systems were modelled at thermodynamic equilibrium.

## Predicting GOR

The prediction of petroleum gas to oil ratio (GOR) was one of the main goals of the first generation of compositional kinetic models. A series of studies performed using different analytical techniques demonstrated that the GOR evolution of fluids generated during increasing levels of maturation in the lab was very similar to those observed for unaltered natural fluids (Dueppenbecker & Horsfield 1990; Muscio *et al.* 1991; Horsfield 1997). GOR predictions from closed system pyrolysis studies performed by Erdmann (1999) on source rock samples from the Upper Jurassic Draupne and Heather Formations correlate very well with the natural fluid GORs observed in the Norwegian North Sea. Figure 1 shows a summary diagram of GOR distribution as a function of reservoir temperature. Reservoir temperature was chosen instead of reservoir pressure or depth in order to include the source rock predictions in this diagram. The shaded portions of Figure 1 show the general GOR distributions observed in the North Sea for both oil and gas for reservoir temperatures up to 150 °C. Reservoirs encountered at temperatures above 150 °C [generally belonging to the high-pressure high-temperature (HPHT) class]

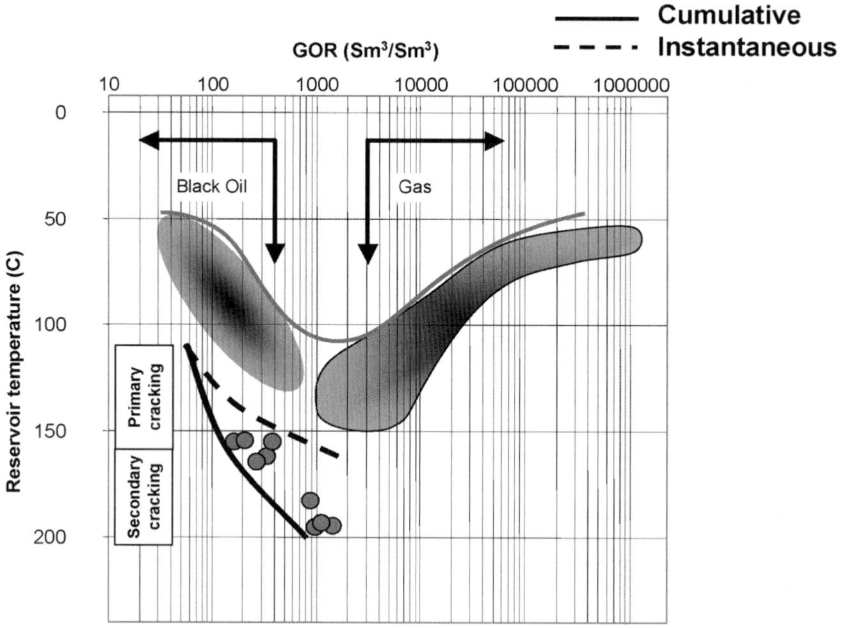


**Fig. 1.** General distribution of GOR as a function of reservoir temperature, North Sea. Shaded areas represent GORs of fluids at reservoir temperatures below 150 °C. Black lines describe GOR evolution predicted by MSSV pyrolysis of source rock samples at geological temperatures. dashed line instantaneous phase, bold line cumulative phase. The circles show fluid GORs of HPHT reservoirs.

are plotted individually. The GOR evolution of pyrolysates generated from a marine Type II Draupne Formation sample (Erdmann 1999) (using non-isothermal MSSV pyrolysis as described by Horsfield *et al.* (1989), at a heating rate of 0.1 K/min and extrapolation to geological temperatures assuming a constant heating rate of 5.3 K/Ma) as a function of increasing temperature is plotted as a line (black). It is interesting to note that when only the primary hydrocarbon generation interval of the closed system pyrolysis results are assessed (geological temperatures roughly up to 160 °C) the match of predicted GORs to actually observed GORs of natural oils found in the North Sea is remarkable (Figure 1). When in-source secondary cracking processes start and, upon increasing temperature, dominate in the microscale sealed vessel (MSSV) tubes (experimental temperatures above 420 °C at 0.1 K/min corresponding to geological temperatures higher than 160 °C at a constant heating rate of 5.3 K/Ma) the predicted GOR evolution closely mimics the spread in GORs encountered in HPHT settings.

The excellent correlation between MSSV-predicted and observed GORs as shown in Figure 1 is only valid for the cumulative compositional prediction, i.e. all products generated and/or cracked in the closed pyrolysis system. When GOR predictions are based on the calculated instantaneous composition (composition generated between two experimental temperatures), they are significantly higher, as exemplified by the dashed line representing instantaneous phase compositions for the realm of primary cracking alone.

The question arises as to why the closed system predictions from MSSV are so much better than those of an open system. One could expect that the instantaneous compositions would better reproduce fluids generated and expelled at increasing levels of maturity, assuming that product generation and expulsion occur more or less simultaneously and that expulsion is very effective. The better fit of the closed system compositional predictions could, however, indicate that a relatively large portion of the generated fluids remain within the source rock during maturation and react further, and that the expelled phase represents aliquots of this cumulative phase at any maturation level, an assumption also proposed by Price & LeFever (1992) and Price (1993). According to this model the best prediction from pyrolysis experiments should lie somewhere between the extremes of a pure cumulative and a pure instantaneous composition, albeit weighted towards the cumulative side. If we look at Figure 1 again with this insight, the cumulative prediction does lie at the lower margin of observed GORs, indicating that a certain degree of open system conditions would be required to make a perfect match. Michels *et al.* (2002) also argued that a semi-open pyrolysis system closely reproduces natural gas compositions. Independent of the reasons why, for the purpose of this paper it is sufficient to document that cumulative GOR predictions from MSSV pyrolysis closely reproduce observed unaltered natural petroleum GORs.

## Compositional predictions from closed system pyrolysis

As mentioned above, the phase behaviour of migrating petroleum fluids is controlled by the fluids composition. The geological conditions upon which a migrating oil separates into oil and gas are strongly controlled by the gas ($C_1$–$C_5$) composition of the fluid (di Primio *et al.* 1998). It should, however, be kept in mind that the composition of the liquid fraction of petroleum fluids also plays an important role in defining its phase behaviour. In this paper, we concentrate on the prediction of black oil phase behaviour (possibly going as far as light oil prediction). In this case gas composition plays the dominant role with respect to saturation pressures and shrinkage behaviour; however, influence of the liquid fraction composition should not be neglected and is discussed in detail below.

High methane contents generally result in phase separation at relatively high pressures, i.e. at great depth. A very wet gas composition results in a much lower saturation pressure ($P_{sat}$). The sensitivity of gas composition on phase behaviour of migrating hydrocarbons has severe implications for the prediction of petroleum phase behaviour during petroleum generation and migration. Mango (1996, 1997) documented the discrepancy between gas compositions generated by pyrolysis of source rocks or oils and natural fluids. As discussed by Mango (1996, 1997) natural fluids display a much stronger predominance of methane in their gas fractions than observed in source rock pyrolysates, but this is not the only difference. The gas compositions from MSSV pyrolysates of a marine Type II source rock (Dieckmann *et al.* 1998) show distinct and characteristic distributions which differ significantly from average gas compositions from natural fluids. Figure 2 shows a comparison of gas compositions from the MSSV experiment (0.1 °C/min heating rate, closed system anhydrous pyrolysis, experimental

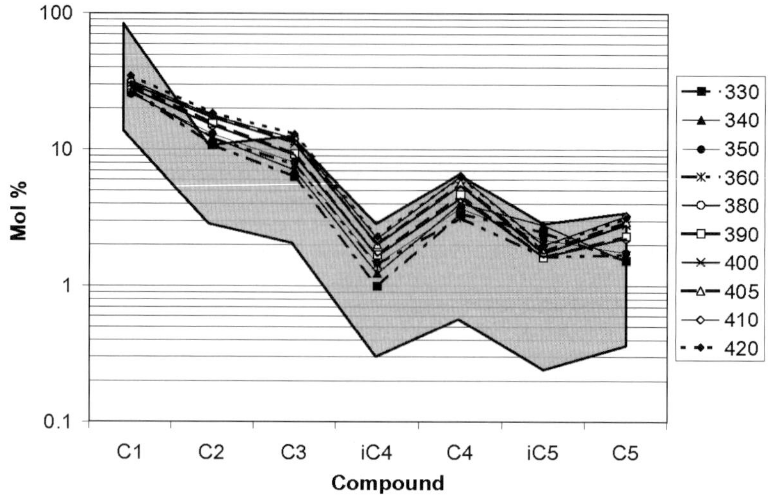


**Fig. 2.** Comparison of closed system gas-composition predictions (bold and dashed lines) and natural fluid compositional ranges (shaded) of undersaturated oils and condensates.

temperatures correspond to the oil window) to those of natural unaltered fluids (PVT data of fluids from the Tampen Spur, as discussed in Javaid (2000), represent close to 60 undersaturated oils and eight gas condensates with a compositional range shown by grey shading). It is interesting to note that while the molar proportions of butanes and pentanes match for both pyrolysates (albeit somewhat high) and natural fluids, propane and especially ethane contents are significantly higher in the pyrolysates. Methane contents of the natural fluids show a wide spread which also encompasses the pyrolysates.

Price & Schoell (1995) account for the compositional differences between gases from pyrolysis experiments and in reservoirs by migration fractionation of an originally wet gas composition. They fail, however, to report how such a mechanism could function in the case of undersaturated black oils, the gas compositions of which, as shown in Figure 2, differ significantly from those of pyrolysates.

Interestingly, the lack of predictive ability of laboratory experiments is common to all experimental approaches—published gas compositions from closed system hydrous pyrolysis (Andresen *et al.* 1995), closed system anhydrous pyrolysis (Béhar *et al.* 1991; Dieckmann 1999; Erdmann 1999; Michels *et al.* 2002) and open system pyrolysis (Berner *et al.* 1995) of Type I, II and III source rocks all show the same systematics.

This lack of compositional predictive capability has important implications for phase behaviour predictions based on pyrolysis results. Matching GORs of pyrolysates and natural fluids have often been used as proof of good compositional reconstructions (Dueppenbecker & Horsfield 1990; Muscio *et al.* 1991; Horsfield 1997). Table 1 shows very similar surface GORs calculated using PVTsim for a natural fluid and a MSSV pyrolysate of a Type II source rock (heating rate of 0.1 °C/min up to a temperature of 400 °C). The samples selected for this comparison were characterized additionally by very similar $C_{7+}$ properties, such that only the detailed gas compositions differed. Based purely on the GOR the pyrolysate could be assumed to be representative of the natural fluid. The calculated

**Table 1.** *Comparison of saturation pressure and volumetrics of phase separation at a pressure of 200 bar for a natural fluid and the product of a closed system pyrolysis experiment of similar GORs*

| | GOR ($Sm^3/Sm^3$) | $P_{sat}$ (bar) | Volume oil (% at 200 bar) | Volume gas (% at 200 bar) |
|---|---|---|---|---|
| Natural fluid | 257 | 298 | 68 | 32 |
| MSSV Type II | 246 | 171 | 100 | |

saturation pressures for these fluids shown in Table 1 demonstrate the influence of gas composition on the physical properties of the fluids. Implications for the simulation of fluid migration are that, depending on which fluid composition is chosen, phase separation could occur in a pressure range from roughly 30 MPa (natural fluid) to 17 MPa (pyrolysate), corresponding roughly to 3 and 1.7 km depth respectively, assuming hydrostatic pressure conditions and a temperature gradient of 30 °C/km.

This vast difference in saturation pressures has severe implications for prospect charge predictions: based on a prediction using the pyrolysate composition a prospect situated at a depth of 2 km (hydrostatic pressure conditions) would contain an undersaturated black oil. If the natural fluid composition is used the same prospect would contain 32 vol% gas and 68 vol% oil, assuming complete trapping of the migrating fluid. If the trap were spilling hydrocarbons, the balance would be shifted towards even higher gas proportions.

Clearly, a good compositional prediction of the generated hydrocarbons is required for an accurate assessment of phase behaviour during migration and entrapment. Currently it appears as though none of the pyrolysis methods available correctly reproduces natural gas compositions. Accordingly, all multi-component kinetic models (i.e. models defining more than two components) based on laboratory experiments should be in error, at least with respect to the $P_{sat}$ behaviour of petroleum liquids. In fact, basin modelling of hydrocarbon generation and migration using established multi-component kinetic models as shown in Table 2 (Ungerer *et al.* 1990; Béhar *et al.* 1997; Vandenbroucke *et al.* 1999) and full PVT calculations leads to a strong predominance of undersaturated black oils in all cases. Even Type III multi-component kerogen kinetics predict the predominance of low $P_{sat}$ black oil. It must, however, be taken into account, that the compositional kinetic models listed above were not developed for phase behaviour modelling.

## Tuning of a compositional kinetic model of hydrocarbon generation

Since laboratory simulation of hydrocarbon generation and destruction seems unable to provide the correct compositional information for the development of multi-component kinetic models a different approach is required. As discussed above, the gross description of oil and gas generation from closed system pyrolysis results, and the surface GORs derived therein, are very similar to the GOR distributions observed in nature. Hence, it appears that the relative gas and oil proportions generated as a function of maturity can be estimated based on laboratory experiments and described using a simple two-component kinetic approach. For phase behaviour modelling, the two components of these kinetic models (gas and oil) would have to be split into several compounds, the relative amounts of which would be determined by calibrating the kinetic models predictions to a natural fluid database. This process is commonly termed 'tuning'.

The minimal compositional resolution required is easily formulated. The PVT data formats commonly used in reservoir engineering include a minimal format: the $C_{7+}$ composition. The compositional resolution of this format is exemplified in Table 3. As shown in Table 3 the data used for PVT evaluations consists of a detailed gas composition and a bulk characterization of a minimum of two compounds belonging to the liquid range ($C_6$ and $C_{7+}$). Non-hydrocarbon gases such as $N_2$, $H_2S$ and $CO_2$ are commonly part of the compositional description of reservoir fluids. Their influence on the phase behaviour of petroleum is important and should not be neglected. For the purposes of phase prediction in basin modelling the properties of these non-hydrocarbon gases can be generalized to affect the phase behaviour of petroleum in a manner akin to methane or ethane. High proportions of $CO_2$, for example, lead to an increase in the saturation pressure, GOR and Bo of a fluid. While $N_2$ and $CO_2$ are

**Table 2.** *Structure of compositional kinetic schemes available to date*

| | Gas components | Oil components |
|---|---|---|
| Espitalié *et al.* (1988) (Type II) | $C_1$, $C_2$–$C_5$ | $C_6$–$C_{14}$, $C_{15+}$ |
| Ungerer *et al.* (1990) (Type II) | $C_1$, $C_2$–$C_5$ | $C_6$–$C_{14}$, $C_{15+}$ |
| Béhar *et al.* (1997) (Type I, II, II-S, III) | $C_1$, $C_2$–$C_5$ | $C_6$–$C_{14}$, $C_{15+}$ |
| Dieckmann *et al.* (1998) (Type II) | $C_1$–$C_5$ | $C_{6+}$ |
| Abu Ali *et al.* (1999) (Type II) | $C_1$, $C_2$, $C_3$–$C_5$ | $C_6$–$C_{14}$, $C_6$–$C_{14}$, $C_{15+}$SAT, $C_{15+}$ARO, NSO |
| Vandenbroucke *et al.* (1999) (Type II, III) | $C_1$, $C_2$, $C_3$–$C_5$ | $C_6$–$C_{13}$SAT, $C_6$–$C_{13}$ARO, $C_{14+}$ISAT, $C_{14+}$NSAT, $C_{14+}$ARO-U, $C_{14+}$NSO |

**Table 3.** *Example of a $C_{7+}$ compositional description commonly used in PVT analysis*

| | Mol % | Mol Wt. | Density |
|---|---|---|---|
| $N_2$ | 0.5 | 28.014 | |
| $CO_2$ | 0.46 | 44.01 | |
| $C_1$ | 43.02 | 16.043 | |
| $C_2$ | 7.53 | 30.07 | |
| $C_3$ | 6.68 | 44.097 | |
| $i$-$C_4$ | 1.09 | 58.124 | |
| $n$-$C_4$ | 3.17 | 58.124 | |
| $i$-$C_5$ | 1.15 | 72.151 | |
| $n$-$C_5$ | 1.96 | 72.151 | |
| $C_6$ | 2.93 | 86.178 | |
| $C_{7+}$ | 31.51 | 225 | 0.863 |

common compounds in petroleums derived from marine Type II source rocks in the North Sea, $H_2S$ is much less common, and hence is disregarded in the further modelling.

The minimal compositional description for phase prediction in basin modelling, therefore, contains nine compounds in the gas range ($N_2$, $CO_2$, $C_1$, $C_2$, $C_3$, $i$-$C_4$, $n$-$C_4$, $i$-$C_5$ and $n$-$C_5$), a pseudo-$C_6$ (corresponding to all compounds eluting between $n$-$C_5$ and $n$-$C_6$ excluding $n$-$C_5$ and including $n$-$C_6$) and a plus fraction, $C_{7+}$, which theoretically contains all other compounds present in the fluid eluting after $n$-$C_6$.

In addition to the $C_{7+}$ amount, the molecular weight and density of this fraction must be defined. In PVTsim the $C_{7+}$ fraction definition is used to calculate a distribution of components representing the total liquid phase. This so-called $C_{7+}$ characterization consists of representing the hydrocarbons with seven and more carbon atoms as a reasonable number (generally 12) of pseudo-components (with specific EOS parameters) whereby a logarithmic relationship between the molar concentration $zN$, of a given fraction and the corresponding carbon number, $CN$, for $CN > 7$ is assumed (Pedersen *et al.* 1989). This characterization leads to the automatic definition of a set of additional compounds of increasing molecular weight (usually up to the molecular weight range corresponding to alkane chain lengths of $C_{80}$–$C_{100}$). The main variables from which the definition of these additional compounds is performed are the amount, molecular weight and density of the $C_{7+}$ fraction. The extrapolated mixture may consist of more than 200 components and pseudo-components. Commercial PVT simulators (in our case PVTsim) can handle a limited number of components such that a reduction of the number of components is required. The component reduction is accomplished through a lumping procedure, where a total of 12 lumped pseudo-components each contain approximately the same weight amount (Pedersen *et al.* 1985).

This type of characterization of the plus fraction is, as yet, not available in the PVT modules of current basin modelling programs. The characterization procedure in the PVT simulation software used leads to a balanced description of the fluid composition, where the gas range is defined based on analytical results and the liquid range is based on a mathematical extrapolation of the $C_{7+}$ properties. The thus characterized fluid contains a distribution of compounds from methane up to $C_{80}$ or $C_{100}$. If, instead of this characterization procedure, a small number of individual compound classes is chosen to represent the liquid fraction of the fluid under study (as assumed in the available multi-component kinetic datasets), the phase behaviour will not be comparable with that of a characterized fluid.

As shown in Table 2 the liquid fraction descriptions in most multi-component kinetic schemes are very limited. In most cases (Béhar *et al.* 1997; Espitalié *et al.* 1988; Ungerer *et al.* 1990) the liquid fraction is described by two compound classes (i.e. two pseudo-compounds for the PVT simulator). The multi-component kinetics of Abu-Ali *et al.* (1999) and Vandenbroucke *et al.* (1999) have a higher resolution of the liquid fraction (five and six pseudo-compounds respectively) but many of these compound groups are very similar with respect to their physical properties, and therefore do not describe the liquid composition adequately for phase behaviour calculations.

For the purpose of our simplified fluid description for phase prediction in basin modelling a detailed $C_{7+}$ characterization such as that performed by PVT simulators may be neglected. A subdivision of the $C_{7+}$ range into 3 pseudo-compounds representing $C_{10}$, $C_{25}$ and $C_{38}$ was determined to be sufficient to model the phase behaviour of the fluid in an adequate fashion. The main drawback in using only three pseudo-compounds is that the calculated API gravities of the fluids tend to be erroneous, both in the PVT simulator of the basin modelling package as well as in PVTsim. This can be avoided by using a larger number of pseudo-compounds including several of higher molecular weight. Expansion of the number of compounds used results, however, in a more difficult tuning procedure and in enhanced processing time for the numerical modelling. Higher resolution fluid models, with an extended compound range describing the liquid composition are currently being tested.

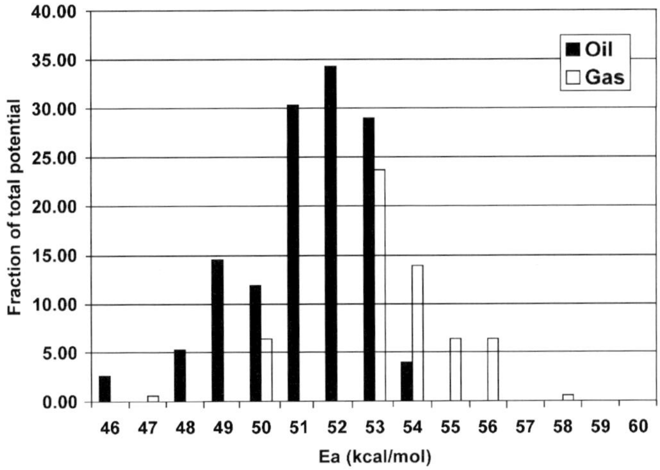


**Fig. 3.** Compositional kinetics of Dieckmann *et al.* (1998).

The minimal fluid description now consists of 13 compounds. The data format selected is compatible with the compositional resolution used in reservoir engineering, thus allowing a direct comparison of predicted compositions and phase behaviour to PVT data of natural fluids.

As a starting point for the tuning of our multi-component kinetic model we used kinetic data from closed-system non-isothermal pyrolysis experiments which describe the generation of oil and gas from a marine Type II source rock (Dieckmann *et al.* 1998). The frequency factors ($A$), activation energy ($E_a$) distributions and hydrocarbon potentials of primary oil and gas generation of Dieckmann *et al.* (1998) were used as the framework for our model (Figure 3). The $E_a$ distributions of oil and gas generation were smoothed for reasons of simplicity, retaining their respective total potentials, onset, maximum and end $E_a$. Kinetics of secondary gas generation by cracking of earlier formed products were defined using the oil to gas cracking kinetics of Dieckmann *et al.* (1998). Only the $C_{10}$, $C_{25}$ and $C_{38}$ compounds were defined as educts for secondary gas generation; the cracking product was methane in all cases.

The kinetics of primary gas formation were assigned equally to all compounds of the gas fraction ($N_2$, $CO_2$, $C_1$–$C_5$) retaining the total gas potential. The same was performed for the primary oil generation kinetics and the compounds of the liquid fraction ($C_6$, $C_{10}$, $C_{25}$ and $C_{38}$). This compositional model was used as the starting point for the tuning procedure.

Compositional predictions ($P_{sat}$, Bo, GOR) of the model were tuned to a well documented maturity sequence from the Tampen Spur, Norway (Horstad *et al.* 1995). The necessary PVT data from the area was taken from Javaid (2000). The reservoirs of the Tampen Spur show a well defined maturity sequence from the Statfjord through Statfjord East and Tordis to Gullfaks fields. The physical properties of fluids from these fields were used as calibration data for the tuning of the kinetic model. Figure 4 shows a cross plot of saturation pressures and formation volume factors (Bo $m^3/Sm^3$) for the Tampen Spur sequence used for calibration purposes. In addition to $P_{sat}$ and Bo also the GOR of the fluids was used for calibration. Predictions of the compositional kinetic model were assumed to be valid if the predicted saturation pressure, Bo and GOR trend fit the trend of the Tampen Spur.

The kinetic model tuning was performed using a fictitious 2D basin model shown in Figure 5. Basin modelling was performed using PetroMod V.7. PVT modelling was performed using the internal PVT simulator of PetroMod as well as PVTsim V.12 for control purposes. The source rock to which the kinetic model was applied was located in the lower left margin of the model. The small structural closures in the model (Fig. 5) trapped only minor amounts of the migrating fluid phase, such that the composition within each reservoir was representative of a composition dominated by the instantaneous phase of the generated fluid. The composition of the fluid in the lowermost reservoir was characteristic for the current source rock maturity at any time step. The entire 2D section was modelled to reflect deposition in a very short time span. Carrier and reservoir properties were constant and relative permeability of hydrocarbon gas and oil phases irrelevant as the

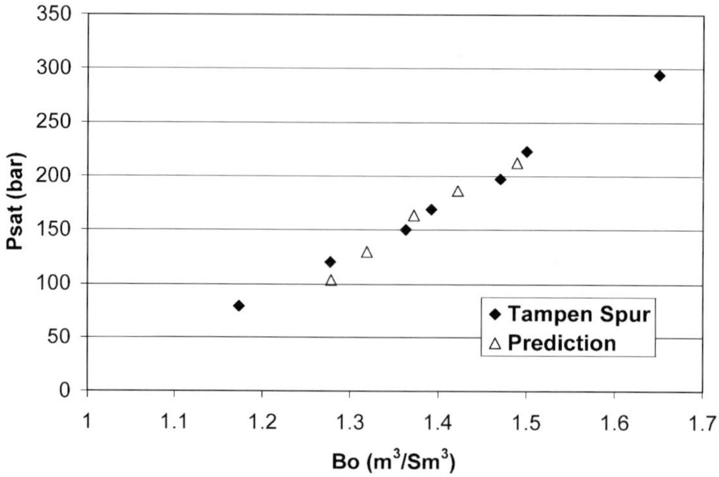


**Fig. 4.** Saturation pressure ($P_{sat}$) versus formation volume factor (Bo) of the Tampen Spur natural fluid calibration sample set as compared with the predicted physical properties of the tuned compositional kinetic model.

pressure conditions of the deepest reservoir cells maintained the fluids in supercritical conditions. Heat flow was increased only when the entire sediment sequence had been deposited and had achieved pressure equilibrium. The heat flow was then increased at a constant rate in order to simulate a heating rate of *ca.* 1 °C/Ma. Maturation progress was monitored every 5 Ma until total kerogen conversion. For details regarding the basin modelling software used, especially the calculation of petroleum migration, the reader is referred to Hantschel *et al.* (2000) and Wygrala *et al.* (2000).

Reservoir fluid compositions at each time step can be obtained for each reservoir in PetroMod. The reported fluid compositions were exported to PVTsim and the physical properties of the fluid calculated. The predicted physical properties were compared with those of the calibration data set and the individual compound kinetics of the multi-component kinetic model iteratively tuned until a good match was achieved (Fig. 4). Figure 6 and Table 4 show the tuned multi-component kinetic model for a marine Type II source rock developed by the method discussed previously.

The iterative tuning of the compositional kinetic model led to significant changes in the oil and gas potentials as compared with the original two-compound kinetics of Dieckmann *et al.* (1998). The final model as shown in Figure 6b demonstrates that the main compositional changes occurring with increasing $E_a$ are a significant increase in methane contents and concordant decrease in $C_{38}$ and $C_{25}$. It should be noted, however, that all compound proportions change systematically. Even small variations in the proportions of individual compounds at any $E_a$ immediately lead to a significant deviation in the physical properties of the generated fluid.

The tuned kinetic model was then implemented in a 2D basin modelling study of the Snorre Field, Norway (Skeie 1999).

## Application to Snorre Field filling history via basin modelling

### *Geological background*

The Snorre Field is located in the northern part of the Tampen Spur (Fig. 7). Major reservoir units are the Triassic Lunde Formation (northern part of the field) and the lower Jurassic Statfjord Formation (southern part of the field). The reservoir contains an undersaturated black oil and has a generally low GOR (62–160 $Sm^3/Sm^3$), typically increasing towards the north. The Snorre Field represents a rotated fault block cut by NNE–SSW and NE–SW faults during Upper Jurassic rifting (Horstad *et al.* 1995). The main source rock in the area is the Upper Jurassic Draupne Formation, a prolific Type II marine source rock. The underlying Heather Formation is also locally developed as a Type II, oil prone source rock. The main kitchen area of the oil encountered in the Snorre Field is situated in the basin directly east and south of the field (Horstad *et al.* 1995; Skeie *et al.* this volume). The Snorre

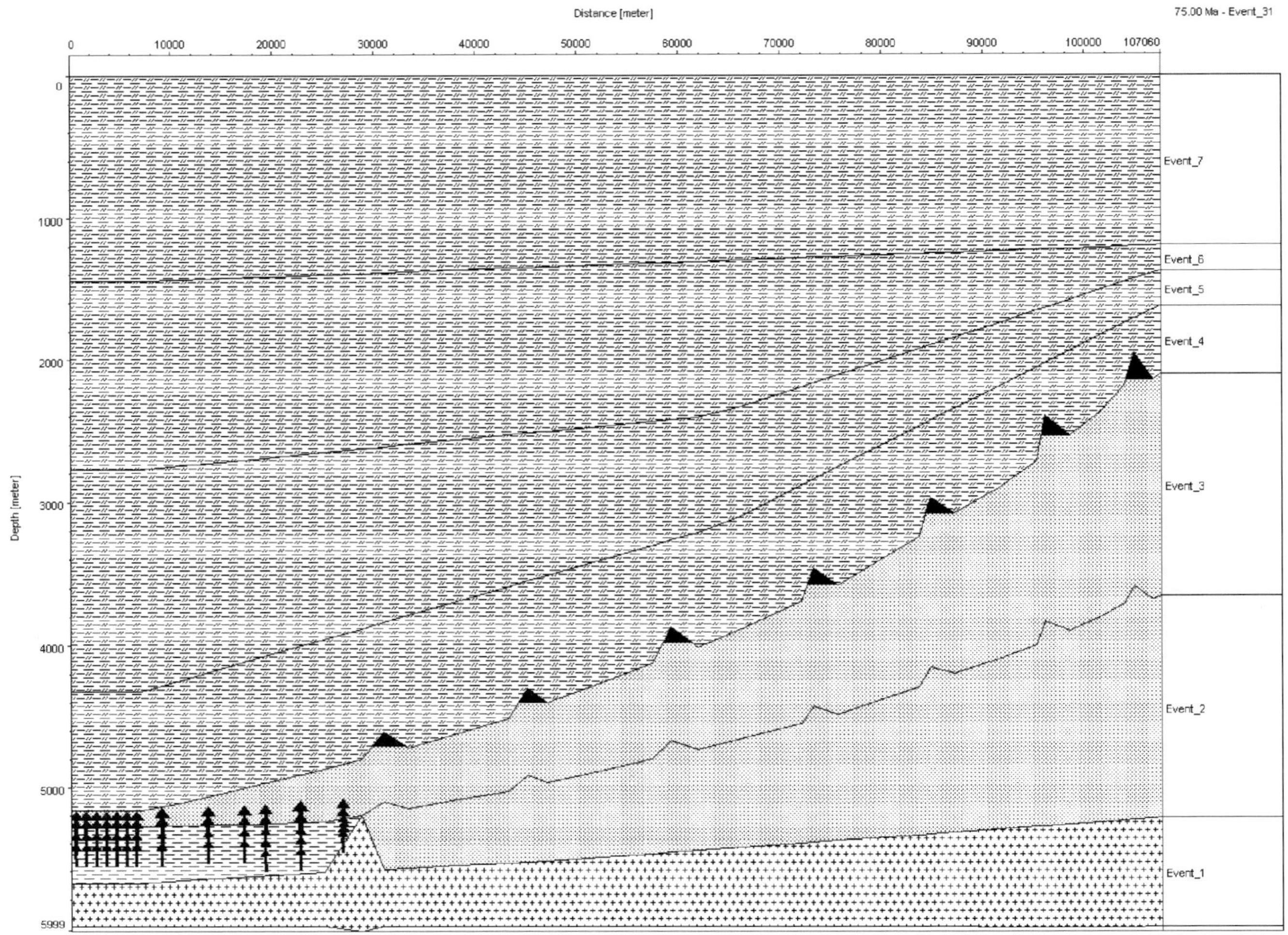


**Fig. 5.** Geological model used for tuning of the compositional kinetics. Black arrows indicate oil migration; oil accumulated in the different structures is also shown in black. The lowermost structure contains a fluid representative of that expelled from the source rock at the current maturity stage.

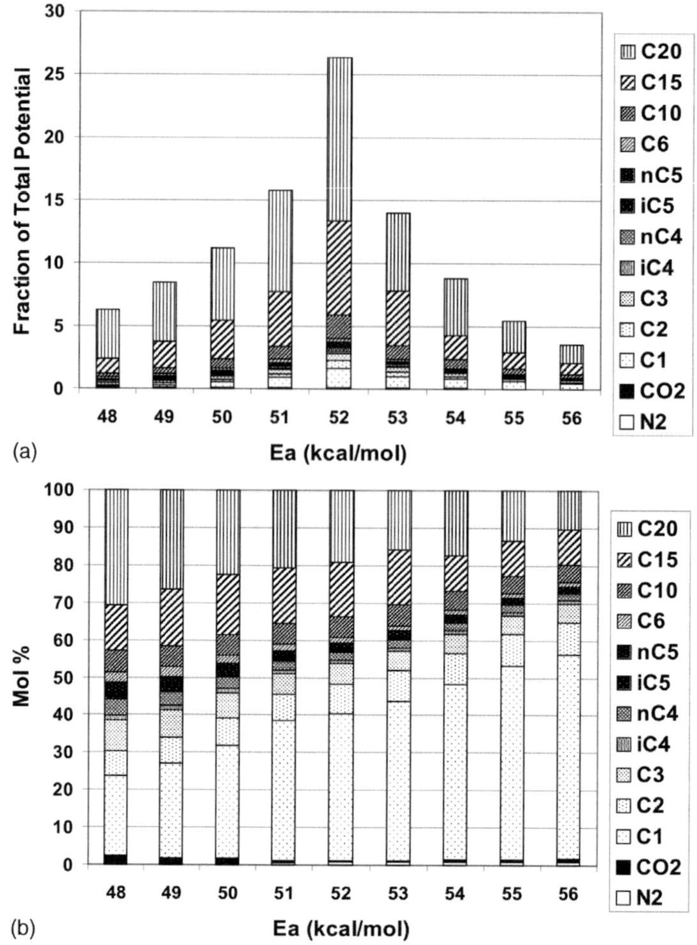


**Fig. 6.** Tuned compositional kinetic model. (**a**) Activation energy distribution as a function of the total potential, (**b**) molar composition of the fluid as defined for each $E_a$.

**Table 4.** *Tuned compositional kinetic model*

| | Relative proportion (%) | | | | | | | | | | | | |
|---|---|---|---|---|---|---|---|---|---|---|---|---|---|
| | 0.2 | 0.3 | 6 | 2.2 | 2.4 | 0.5 | 1.4 | 0.6 | 1.2 | 1.4 | 6.9 | 26.6 | 50.3 |
| $E_a$ (kcal/mol) | $N_2$ (%) | $CO_2$ (%) | $C_1$ (%) | $C_2$ (%) | $C_3$ (%) | $i$-$C_4$ (%) | $n$-$C_4$ (%) | $i$-$C_5$ (%) | $n$-$C_5$ (%) | $C_6$ (%) | $C_{10}$ (%) | $C_{15}$ (%) | $C_{20}$ (%) |
| 48 | 2 | 14.2 | 2.7 | 4.2 | 6.7 | 6.5 | 8.5 | 8.2 | 8.3 | 7.7 | 5 | 4.4 | 7.8 |
| 49 | 3.3 | 13.1 | 4.4 | 6.2 | 8.5 | 8.6 | 9.8 | 10.2 | 10.7 | 10.5 | 7.2 | 7.6 | 9.4 |
| 50 | 5.5 | 17.4 | 7.5 | 9.5 | 11.4 | 11.5 | 12 | 13 | 13.8 | 12.7 | 10.3 | 11.5 | 11.5 |
| 51 | 13.8 | 10.1 | 14.3 | 13.3 | 14.3 | 15.7 | 15.1 | 16.2 | 14.7 | 15.5 | 15.4 | 16.3 | 16.1 |
| 52 | 28.6 | 17.5 | 26 | 26.9 | 24.5 | 25 | 23.3 | 22.7 | 22.5 | 23.4 | 26.7 | 28.2 | 25.8 |
| 53 | 17.6 | 10.1 | 16.3 | 15.7 | 14.1 | 13.9 | 13 | 12.6 | 12.6 | 12.9 | 15.4 | 16.3 | 12.4 |
| 54 | 12.9 | 6.7 | 12.2 | 10.7 | 9.4 | 8.9 | 8.6 | 8.1 | 8.2 | 8.2 | 9.7 | 7.1 | 9.2 |
| 55 | 9.3 | 5.2 | 9.5 | 7.8 | 6.4 | 6 | 5.8 | 5.4 | 5.4 | 5.3 | 6 | 5 | 5 |
| 56 | 7 | 5.7 | 7.1 | 5.7 | 4.7 | 3.9 | 3.9 | 3.6 | 3.8 | 3.8 | 4.3 | 3.6 | 2.8 |

$A = 1.06 \times 10^{27}$/Ma

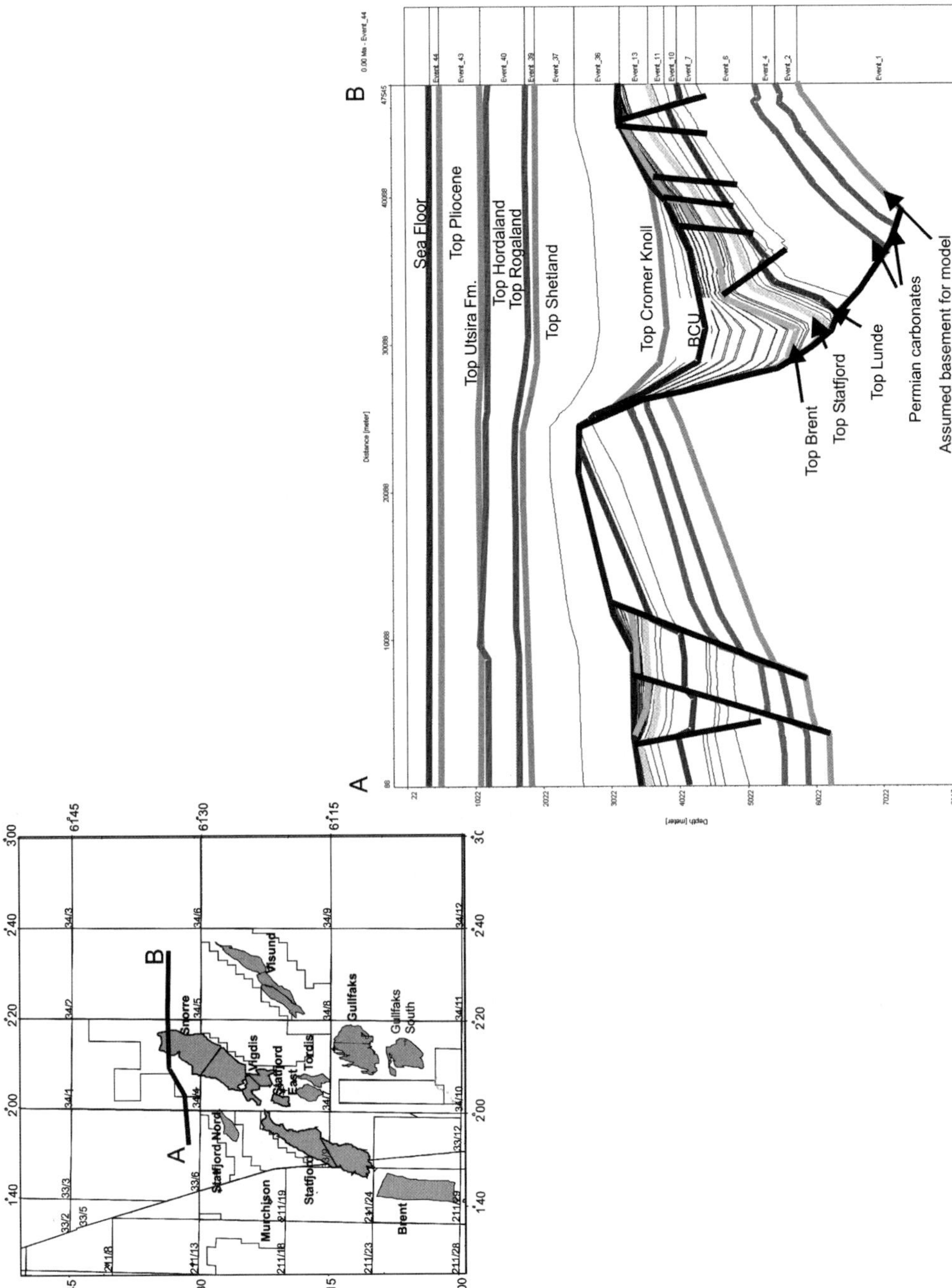


**Fig. 7.** Location map of the Tampen Spur reservoirs and cross section used for basin modelling.

structure experienced an estimated uplift of up to 3 km (Faerseth 1996) in the upper Jurassic, causing the Snorre block to be one of the most pronounced structural features in the post-rift basin. This setting is expected to have caused significant input of syn-rift sediments during the deposition of the Heather and Draupne Formations on the flank of the structure during the Late Jurassic (Underhill *et al.* 1997). Combined with the uplift of the crest of the structurally pronounced Snorre fault block and associated erosion, the model indicates the likelihood of the occurrence of sandy lithologies in the Draupne and Heather Formation, thus enhancing a lateral direction of secondary migration in this system.

A good overview of the geological development of the Tampen Spur is given by Spencer (1987), Faerseth (1996) and Glennie (1998). Skeie *et al.* (this volume) performed a basin modelling study of the Tampen Spur area, focusing also on petroleum migration directions into the Snorre Field. The basin model of the Snorre Field of Skeie *et al.* (this volume) was based on interpretation of a seismic line which crossed the reservoir in a west-to-east direction (Fig. 7). The results of basin modelling were calibrated to temperature, vitrinite reflectance and pressure data, and the calibrated model used for the further evaluations regarding petroleum phase prediction in this study. The reader is referred to Skeie *et al.* (this volume) for details regarding the definition of the basin model.

## *Results*

The tuned kinetic model was applied to the upper Jurassic source rock sequences in the basin and the calibrated model re-run taking hydrocarbon generation and migration into account. The methodology used for migration modelling was a combination of Darcy flow controlled migration in low permeability lithologies and ray tracing in high permeability lithologies (Hantschel *et al.* 2000). In the calculation of Darcy flow in low permeability lithologies the physical formulation of the fluid transport takes all relevant driving forces into account, i.e. buoyancy, pore and capillary pressures, and considers permeability distributions. The interactions of all physical quantities during the time of migration are taken into account. When migrating fluids enter realms of high permeability the model automatically switches to a ray-tracing methodology for flow calculations. In this case fluid flow is modelled as a purely geometric problem, as capillary pressures are generally very low and the entire carrier can be assumed to be flow conductive (Sylta 1991b). The only force to be taken into account is buoyancy and migration is assumed to occur instantaneously relative to geological time. In accumulations fluid leakage occurs either by spilling, whereby ray tracing again tracks the spill pathways, or by capillary failure of the cap rock. When this latter event occurs, flow calculations through the cap rock are again performed assuming Darcy flow.

The degree of kerogen transformation to hydrocarbons of the source rock intervals (transformation ratio) is shown in Figure 8, together with the modelled oil accumulation. Based on the modelling results the onset of petroleum generation and migration occurred in the late Cretaceous. A first significant oil column is modelled to have accumulated at the onset of the Tertiary period, 65 Ma ago. During this time, the Snorre reservoir was at a burial depth of approximately 800 m. The incoming fluid was characterized by a low GOR (around 90 $Sm^3/Sm^3$) and saturation pressure (115 bar) but low reservoir pressure (120 bar) and temperature (64 °C) indicate that the petroleum was most likely saturated, i.e. phase separation of the fluid into an oil leg and a gas cap occurred. These conditions of very similar $P_{sat}$ and reservoir pressure continued until 55 Ma. During the following burial, reservoir pressures increased more rapidly than the petroleum's saturation pressure, indicating that the fluid became undersaturated and that, hence, only a single oil phase was present in the reservoir. During the latest accelerated burial in the past 5 Ma reservoir pressures increased dramatically to reach the present overpressured situation. The oil in the reservoir started to receive fluids with enhanced methane contents due to onset of secondary cracking in the source rock, but saturation pressures remained significantly below reservoir pressure.

The results of migration modelling led to an excellent correlation between predicted and reported reservoir fluid properties (Bo, GOR and saturation pressure) for the present-day situation (Table 5). The modelled composition of the reservoir fluid was tracked through time and the results exported to the PVT simulator with which the fluid properties were determined (Table 5). Figure 9 shows the evolution of saturation pressure of the reservoir fluid as compared with the modelled reservoir pressure evolution during filling.

The thus reconstructed filling history of the Snorre Field matches the data available from the field in a satisfactory manner: the physical properties of the modelled fluid fit within the range of properties monitored in the field and,

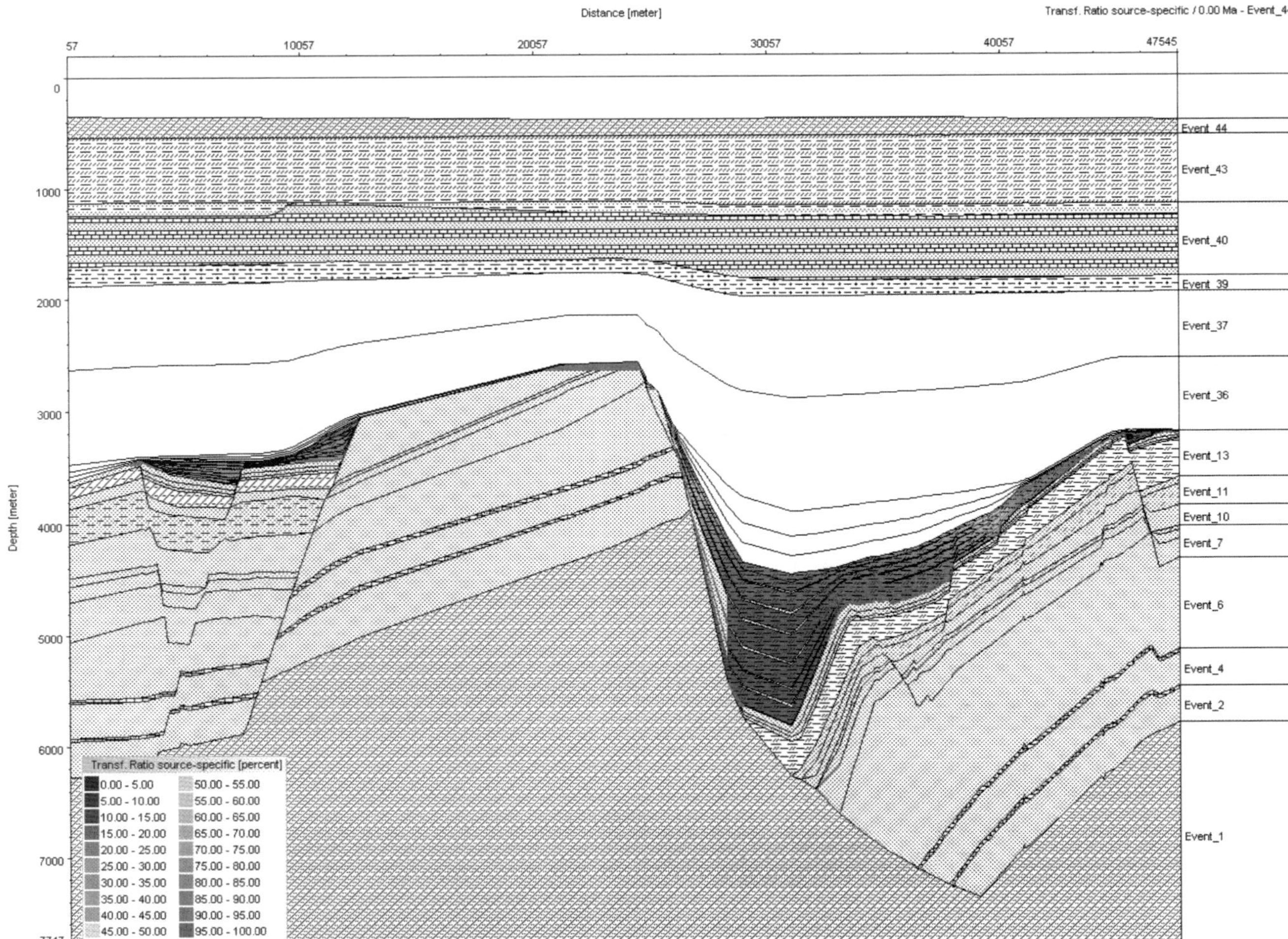


**Fig. 8.** Simulated extent of kerogen transformation in the kitchen area of Snorre Field. The accumulation of undersaturated oil in the reservoir.

**Table 5.** *Comparison of reservoir fluid properties simulated for the Snorre reservoir as a function of time and natural fluid properties from Snorre Field*

| | GOR ($Sm^3/Sm^3$) | Bo ($m^3/Sm^3$) | $P_{sat}$ (bar) |
|---|---|---|---|
| Present | 133.1 | 1.425 | 193.5 |
| 1.8 Ma | 129.4 | 1.416 | 182.58 |
| 5.3 Ma | 124.1 | 1.394 | 173.41 |
| 15 Ma | 120.3 | 1.375 | 164.52 |
| 25 Ma | 116.1 | 1.367 | 160.42 |
| 55 Ma | 97.7 | 1.306 | 129.95 |
| 58 Ma | 92.6 | 1.290 | 121.25 |
| 65 Ma | 88.5 | 1.280 | 115.22 |
| Snorre natural fluid | 59.2 | 1.242 | 89.8 |
| Snorre natural fluid | 81.3 | 1.311 | 112.2 |
| Snorre natural fluid | 121.6 | 1.428 | 154.3 |
| Snorre natural fluid | 157.3 | 1.58 | 199.9 |

hence, we could assume that inferences regarding the filling history are also correct. However, as compared with reality the model contains a significant number of simplifications.

Horstad *et al.* (1995) described a top-down filling as the most likely case for the northern part of the Snorre reservoir. The differences in oil composition monitored for the petroleums hosted in the Statfjord and Lunde Formations respectively indicated already at that point that the Snorre reservoir was compartmentalized. The GOR and maturity variations observed across the field support a highly compartmentalized and sequentially filled reservoir. This is, however, in stark contrast to the assumption of a 'tank of sand' made in the simplified basin model. The predictions concerning the physical properties of the fluid in the model represent the situation of a cumulative filling, in which the composition monitored is an average of all fluid compositions, which reached the reservoir. Figure 10 shows the GOR–$P_{sat}$ trend of the Tampen Spur reservoirs used to tune the compositional kinetic model, as compared with the evolution of fluid properties in the modelled filling history of Snorre. In addition, a few natural fluid properties representing the variability of fluid composition in the Snorre Field are shown together with the fluid properties predicted by the compositional kinetic model in the basin model used for tuning. In general, all the data shown in Figure 10 demonstrates that the kinetic model predictions match the observed fluid properties from both the Tampen Spur reservoirs and the Snorre Field. The evolution of fluid properties based on the basin modelling of the Snorre Field shows the smallest GOR spread, whereas the predictions from the tuning model closely mimic the natural fluid property distributions in the Snorre Field.

The difference in fluid property evolutions between the tuning model and the Snorre Field

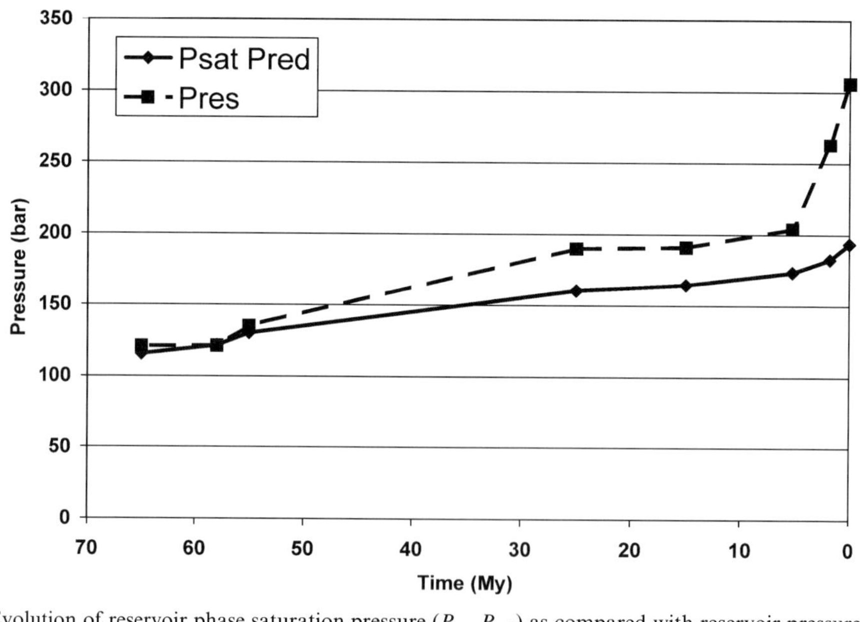


**Fig. 9.** Evolution of reservoir phase saturation pressure ($P_{sat}$ $P_{red}$) as compared with reservoir pressure ($P_{res}$), based on modelling results using the tuned compositional kinetic model.

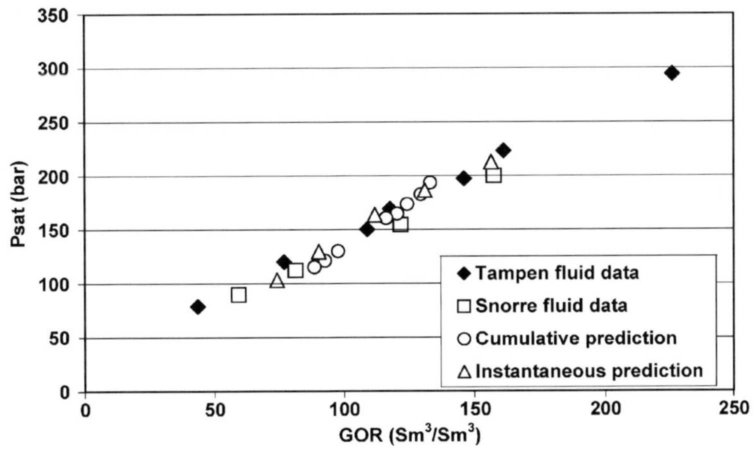


**Fig. 10.** Comparison of petroleum properties from the Tampen Spur calibration set, Snorre Field samples and the predictions of the cumulative and instantaneous tuned models.

model are surely also attributable to differences in the geology and the burial and thermal histories used. However, the much larger spread in fluid properties predicted by the tuning model is most strongly controlled by the use of small traps to catch only minor fluid volumes representative of the phase currently being expelled by the source rock. Accordingly, these fluid predictions can be taken as coming from a relatively open system, where instantaneous phases are monitored. The Snorre Field modelling represents the closed system case, in which the majority of fluids generated accumulate in a single reservoir, which hence contains a cumulative phase. Comparison of the results shown in Figure 10 indicates that the compartmentalized Snorre reservoir contains a fluid distribution, which is better represented by a series of instantaneous phases, which did not mix in the reservoir. Hence, the instantaneous phase properties should better characterize the filling history of the field. Instantaneous phase compositions in the Snorre Field model were defined using average kitchen area maturities per time-step in the Snorre model and retrieving fluid compositions from the tuning model for equivalent maturities. The results with respect to saturation pressure evolution are shown in Figure 11, where the same reservoir pressure evolution as a function of time as shown in

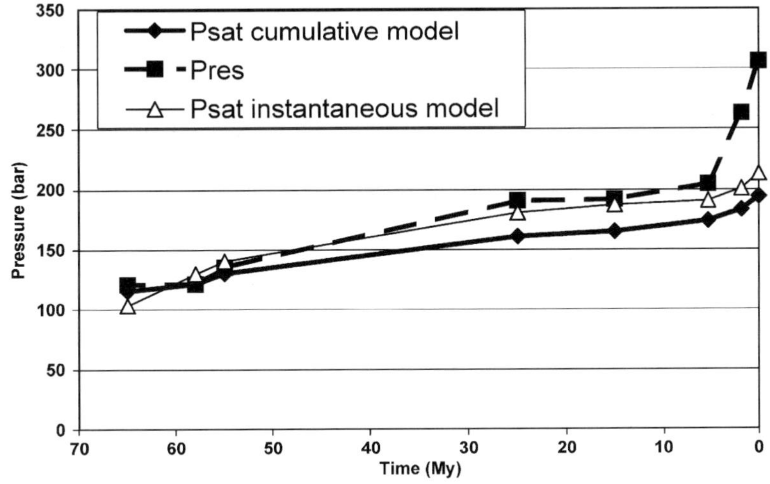


**Fig. 11.** Evolution of reservoir phase saturation pressure of the instantaneous and cumulative phase compositions as compared to reservoir pressure ($P_{res}$).

Figure 9 is plotted together with the instantaneous phase saturation pressures. Figure 11 shows that the evolution of saturation pressures of the instantaneous phase closely follows the reservoir pressure trend until, in the past 5 Ma, reservoir pressures increase sharply. Undersaturation of the highly compartmentalized Snorre reservoir can, in this case, be stated to be a relatively recent event attributable to the latest subsidence and overpressuring event.

The two extremes, using instantaneous and cumulative phase predictions discussed above, provide only the framework for the total variability, which can be expected in the reservoir filling history studied here. Unravelling the evolution of petroleum fluid compositions in the Snorre Field through time would require a model resolution far exceeding what can be handled in reasonable computing time. The approach shown allows, however, a prediction of fluid properties, which is much closer to the natural fluid compositions than previously possible. This compositional kinetic scheme is the first of its kind to allow reasonable petroleum phase behaviour assessment in the simulation of basin evolution and hydrocarbon migration.

## Conclusions

A compositional kinetic model of hydrocarbon generation for a marine Type II source rock was developed based on data from closed-system non-isothermal pyrolysis experiments. The model predictions were tuned to a natural maturity sequence. The compositional data format chosen is compatible with the compositional resolution used in reservoir engineering, and allows a direct comparison of predicted compositions and phase behaviour to PVT data of natural fluids.

2D basin modelling of the Snorre Field (Norway) filling history, using the developed compositional kinetics, resulted in an excellent correlation between predicted and reported reservoir fluid properties for the present-day situation, and indicate that the reservoir contained a two-phase system prior to the latest Plio-Pleistocene burial and overpressuring event.

The compositional kinetic scheme used is the first of its kind to allow meaningful petroleum phase behaviour assessment in the simulation of basin evolution and hydrocarbon migration.

Financial support from RWE-Dea Norway and ChevronTexaco Norway for J. E. Skeie is gratefully acknowledged. This paper benefited greatly from the detailed reviews of Peter Meulbroeck and Nigel Mills.

## References

ABU-ALI, M. A., RUDKIEWICZ, J. L. L., MCGILLIVRAY, J. G. & BÉHAR, F. 1999. Paleozoic petroleum system of central Saudi Arabia. *GeoArabia (Manama)*, **4**(3), 321–336.

ANDRESEN, B., THRONDSEN, T., RAHEIM, A. & BOLSTAD, J. 1995. A comparison of pyrolysis products with models for natural gas generation. *Chemical Geology*, **126**(3/4), 261–280.

BÉHAR, F., UNGERER, P., KRESSMANN, S. & RUDKIEWICZ, J. L. 1991. Thermal evolution of crude oils in sedimentary basins; experimental simulation in a confined system and kinetic modeling. *Revue de l'Institut Français du Pétrole*, **46**(2), 151–181.

BÉHAR, F., VANDENBROUCKE, M., TEERMANN, S. C., HATCHER, P. G., LEBLOND, C. & LERAT, O. 1995. Experimental simulation of gas generation from coals and a marine kerogen. *Chemical Geology*, **126**(3/4), 247–260.

BÉHAR, F., VANDENBROUCKE, M., TANG, Y., MARQUIS, F. & ESPITALIÉ, J. 1997. Thermal cracking of kerogen in open and closed systems: determination of kinetic parameters and stoichiometric coefficients for oil and gas generation. *Organic Geochemistry*, **26**(5/6), 321–339.

BERNER, U., FABER, E., SCHEEDER, G. & PANTEN, D. 1995. Primary cracking of algal and landplant kerogens; kinetic models of isotope variations in methane, ethane and propane. *Chemical Geology*, **126**(3/4), 233–245.

BRAUN, R. L., BURNHAM, A. K. & REYNOLDS, J. G. 1992. Oil and gas evolution kinetics for oil shale and petroleum source rocks determined from pyrolysis; TQMS data at two heating rates. *Energy and Fuels*, **6**(4), 468–474.

DI PRIMIO, R. 2002. Unraveling secondary migration effects through the regional evaluation of PVT data: a case study from Quadrant 25, NOCS. *Organic Geochemistry*, **33**(6), 643–653.

DI PRIMIO, R., DIECKMANN, V. & MILLS, N. 1998. PVT and phase behaviour analysis in petroleum exploration. *Organic Geochemistry*, **29**(1–3) 207–222.

DIECKMANN, V. 1999. Zur Vorhersage der Erdöl- und Erdgaszusammensetzung durch die Integration von Labor- und Fallstudien. Dissertation, RWTH Aachen.

DIECKMANN, V., SCHENK, H. J., HORSFIELD, B. & WELTE, D. H. 1998. Kinetics of petroleum generation and cracking by programmed-temperature closed-system pyrolysis of Toarcian Shales. *Fuel*, **77**(1/2), 23–31.

DUEPPENBECKER, S. & HORSFIELD, B. 1990. Compositional information for kinetic modelling and petroleum type prediction. *In*: DURAND, B. & BÉHAR, F. (eds) *Advances in Organic Geochemistry 1989*; Part I, *Organic Geochemistry in Petroleum Exploration*. Pergamon Press, Oxford, 259–266.

ENGLAND, W. A., MACKENZIE, A. S., MANN, D. A. & QUIGLEY, T. M. 1987. The movement and entrapment of petroleum fluids in the subsurface. *Journal of the Geological Society, London*, **144**, 327–347.

ERDMANN, M. 1999. Gas generation from overmature Upper Jurassic source rocks, Northern Viking Graben. PhD thesis, RWTH Aachen, Germany.

ESPITALIÉ, J., UNGERER, P., IRWIN, I. & MARQUIS, F. 1988. Primary cracking of kerogens; experimenting and modeling $C_1$, $C_2$-$C_5$, $C_6$-$C_{15}$ and $C_{15+}$ classes of hydrocarbons formed. *In*: BAKER E. W. & SCHENCK, P. A. (eds) *Advances in Organic Geochemistry 1987*; Part II, *Analytical Geochemistry*. Proceedings of the 13th International Meeting on Organic Geochemistry, Pergamon Press, Oxford, **13**(4–6), 893–899.

FAERSETH, R. B. 1996. Interaction of Permo-Triassic and Jurassic extensional fault-blocks during the development of the northern North Sea. *Journal of the Geological Society, London*, **153**, 931–944.

GLENNIE, K. W. 1998. *Petroleum Geology of the North Sea: Basic Concepts and Recent Advances*. Blackwell Science, Oxford.

HANTSCHEL, T., KAUERAUF, A. I. & WYGRALA, B. 2000. Finite element analysis and ray tracing modeling of petroleum migration. *Marine and Petroleum Geology*, **17**(7), 815–820.

HORSFIELD, B. 1997. The bulk composition of first-formed petroleum in source rocks. *In*: WELTE, D. H., HORSFIELD, B. & BAKER, D. R. (eds) *Petroleum and Basin Evolution; Insights from Petroleum Geochemistry, Geology and Basin Modeling*. Springer, Berlin.

HORSFIELD, B., DISKO, U. & LEISTNER, F. 1989. The micro-scale simulation of maturation: outline of a new technique and its potential applications. *Geologische Rundschau*, **78**(1), 361–374.

HORSTAD, I., LARTER, S. R. & MILLS, N. 1995. Migration of hydrocarbons in the Tampen Spur area, Norwegian North Sea; a reservoir geochemical evaluation. *In*: CUBITT, J. M. & ENGLAND, W. E. (eds) *The Geochemistry of Reservoirs*. Geological Society, London, Special Publications, **86**, 159–183.

JAVAID, A. A. 2000. PVT modelling of reservoir fluids from the Norwegian North Sea. Cand Scient., Oslo University.

MANGO, F. D. 1996. Transition metal catalysis in the generation of natural gas. *Organic Geochemistry*, **24**(10/11), 977–984.

MANGO, F. D. 1997. The light hydrocarbons in petroleum: a critical review. *Organic Geochemistry*, **26**(7–8), 417–440.

MCCAIN, W. D. J. 1990. *The Properties of Petroleum Fluids*. PennWell Books, Tulsa, Oklahoma

MICHELS, R., ENJELVIN-RAOULT, N., ELIE, M., MANSUY, L., FAURE, P. & OUDIN, J.-L. 2002. Understanding of reservoir gas compositions in a natural case using stepwise semi-open artificial maturation. *Marine and Petroleum Geology* **19**(5), 589–599.

MUSCIO, G. P. A., HORSFIELD, B. & WELTE, D. H. 1991. Compositional changes in the macromolecular organic matter (kerogens, asphaltenes and resins) of a naturally matured source rock sequence from Germany as revealed by pyrolysis methods. *In*: MANNING, D. (ed.) *Organic Geochemistry: Advances and Applications in Energy and the Natural Environment*. Manchester University Press, 662.

PEDERSEN, K. S., THOMASSEN, P. & FREDENSLUND, A. 1985. Thermodynamics of petroleum mixtures containing heavy hydrocarbons. 3. Efficient flash calculation procedures using the SRK equation of state. *Industrial Engineering Chemical Process Design Development*, **24**, 948–954.

PEDERSEN, K. S., FREDENSLUND, A. & THOMASSEN, P. 1989. *Properties of Oils and Natural Gases*. Gulf Publishing Company, Houston.

PRICE, L. C. 1993. Thermal stability of hydrocarbons in nature; limits, evidence, characteristics and possible controls. *Geochimica et Cosmochimica Acta*, **57**(14), 3261–3280.

PRICE, L. C. & LEFEVER, J. A. 1992. Does Bakken horizontal drilling imply a huge oil-resource base in fractured shales? *In*: SCHMOKER, J. W., COALSON, E. B. & BROWN, C. A. (eds) Geological studies relevant to horizontal drilling; examples from western North America. *Rocky Mountain Association of Geologists*, Denver, Colorado.

PRICE, L. C. & SCHOELL, M. 1995. Constraints on the origins of hydrocarbon gas from compositions of gases at their site of origin. *Nature*, **378**, 368–371.

SKEIE, J. E. 1999. Generering, migrsjon og akkumulasjon av petroleum i Snorreomradet, Tampen Spur: et integrert studie ved hjelp av bassengmodellering. Cand Scient., Oslo University.

SPENCER, A. M. 1987. *Habitat of Hydrocarbons on the Norwegian Continental Shelf*. Norwegian Petroleum Society, Graham & Trotman, London.

SWEENEY, J. J., BURNHAM, A. K. & BRAUN, R. L. 1987. A model of hydrocarbon generation from type I kerogen; application to Uinta Basin, Utah. *American Association of Petroleum Geologists Bulletin*, **71**(8), 967–985.

SWEENEY, J. J., BRAUN, R. L., BURNHAM, A. K., TALUKDAR, S. & VALLEJOS, C. 1995. Chemical kinetic model of hydrocarbon generation, expulsion and destruction applied to the Maracaibo Basin, Venezuela. *American Association of Petroleum Geologists Bulletin*, **79**(10), 1515–1532.

SYLTA, O. 1991a. Modelling of secondary migration and entrapment. *In*: ENGLAND, W. A. & FLEET, A. J. (eds) *Petroleum Migration*. Geological Society, London, Special Publications, **59**, 111–122.

SYLTA, O. 1991b. New techniques and their applications in the analysis of secondary migration. *In*: DORÉ, A. G., AUGUSTSON, J. H., HERMANRUD, C., STEWART, D. J. & SYLTA, Ø. (eds) *Basin Modelling: Advances and Applications*, 385–398. Norwegian Petroleum Society Special Publication No. 3, Elsevier, Amsterdam.

TISSOT, B. P. & WELTE, D. H. 1984. *Petroleum Formation and Occurrence*. Springer, Berlin.

UNDERHILL, J. R., SAWYER, M. J., HODGSON, P., SHALLCROSS, M. D. & GAWTHORPE, R. L. 1997. Implications of fault scarp degradation for Brent Group prospectivity, Ninian Field, northern North Sea. *American Association of Petroleum Geologists Bulletin*, **81**(6), 999–1022.

UNGERER, P., BURRUS, J., DOLIGEZ, B., CHÉNET, P. Y. & BESSIS, F. 1990. Basin evaluation by integrated two-dimensional modeling of heat transfer, fluid flow, hydrocarbon generation and migration. *American Association of Petroleum Geologists Bulletin*, **74**(3), 309–335.

VANDENBROUCKE, M., BÉHAR, F. & RUDKIEWICZ, J. L. 1999. Kinetic modelling of petroleum formation and cracking: implications from the high pressure/high temperature Elgin Field (UK, North Sea). *Organic Geochemistry*, **30**, 1105–1125.

WYGRALA, B. P., HANTSCHEL, T. & CARRUTHERS, D. 2000. Hybrid 3D petroleum migration modeling; status and perspectives. *In*: *American Association of Petroleum Geology International Conference and Exhibition*; *abstracts*. American Association of Petroleum Geologists, **84**(9), 1516.

# A petroleum charge model for the Judy and Joanne Fields, Central North Sea: application to exploration and field development

MICHAEL D. LINES & HEATHER A. AULD

*ConocoPhillips (U.K.) Limited, Rubislaw House, North Anderson Drive, Aberdeen, AB15 6FZ, UK*

**Abstract:** There is considerable diversity in petroleum type within the Judy and Joanne Fields of the Central Graben. Superficially, the three major reservoir systems can be considered to contain the following fluid types: gas-condensates in the Pre-Cretaceous, undersaturated black oils in the Chalk and gas-condensates in the Palaeocene. Reality is, however, quite different.

This paper presents the results of several geochemical studies undertaken in the area; oil/condensate analyses to identify differences in maturity and source input, a pressure data review, high resolution GC fingerprinting, strontium isotope analyses to investigate reservoir connectivity/compartmentalization, and 2D basin modelling to determine the timing and extent of petroleum expulsion and migration. These components have been synthesized to produce a composite petroleum charge model which adequately explains the differences observed. This, in turn, enables an asessment to be made of the likely impact, if any, on field development. Furthermore, the petroleum charge model can be applied to predict the charge risk, fluid type and likely petroleum–water contacts in untested parts of the field and in the immediate vicinity.

The results of recent development wells are reviewed in the light of the charge model. Subsequent reservoir geochemistry studies have confirmed the validity of the model and highlight additional applications for reservoir management.

The Judy and Joanne Fields are located in Block 30/7a (J Block) in the central portion of the Central Graben of the UK North Sea (Fig. 1). Figure 2 shows the main tectonic elements of the area. The Judy Horst is the dominant structural feature; it forms the southeasterly extension of the Forties–Montrose High, which separates the eastern and western Central Graben Basins, and continues S–SE into the Josephine Ridge. The horst plunges N–NW and is bounded by major faults. Complex internal faulting results in a mosaic of Pre-Cretaceous fault blocks of locally varying structural dip and sub-crop, as can be seen on the Top Triassic reservoir map (Fig. 3). Zechstein salt remains in small remnant pods, having largely evacuated the crest of the Judy structure and migrated west to form the Joanne salt pillow.

The Judy and Joanne fields comprise three major reservoir intervals: Triassic/Jurassic sands, Cretaceous/Palaeocene chalks and Palaeocene sands. Figure 4 shows the generalized stratigraphy of the area. The deepest of the three reservoirs comprises the fluvial sandstones of the Triassic Skaggerak Formation (Joanne and Judy Sandstone Members), overlain unconformably by Upper Jurassic shallow marine sandstones of the Fulmar Formation. The Joanne Sands are the major reservoir unit in the Judy Field. The productive chalk intervals are the Upper Cretaceous Tor and Palaeocene Ekofisk Formations, located primarily over the Joanne feature. Turbiditic sandstones of the Andrew Member of the Palaeocene Lista Formation form the uppermost reservoir unit and are present over both the Judy and Joanne structures.

The considerable diversity in fluid type within the J Block area can be seen on the petroleum occurrence map (Fig. 5), which is colour-coded both by reservoir age and fluid type. Four gross reservoir intervals are shown: Triassic, Jurassic, Cretaceous (including the Danian Chalk) and Palaeocene.

Within the Triassic reservoirs are the gas-condensate accumulations of Judy, Jocelyn and Jade, and the geochemically distinct volatile oils in Kessog, which are believed to have been generated from a Middle Jurassic source rock. In the Jurassic, the Jacqui fluids range from black oil (30/13-3), to volatile oil (30/13-6), to gas-condensate (30/13-4). Gas-condensates are also present in the Fulmar reservoir in Judy whilst, in Kessog, the Middle Jurassic reservoir contains volatile oils similar to those in the Triassic. The Chalk reservoirs in Block 30/7a contain predominantly black oils, although there are minor gas caps in the Ekofisk and Tor in the Judy structure. In contrast, well 30/6-3 tested high GOR gas-condensates from the Chalk reservoir. In the Palaeocene, gas-condensates are

*From*: Cubitt, J. M., England, W. A. & Larter, S. (eds) 2004. *Understanding Petroleum Reservoirs: towards an Integrated Reservoir Engineering and Geochemical Approach.* Geological Society, London, Special Publications, **237**, 175–206. 0305-8719/04/$15.00 

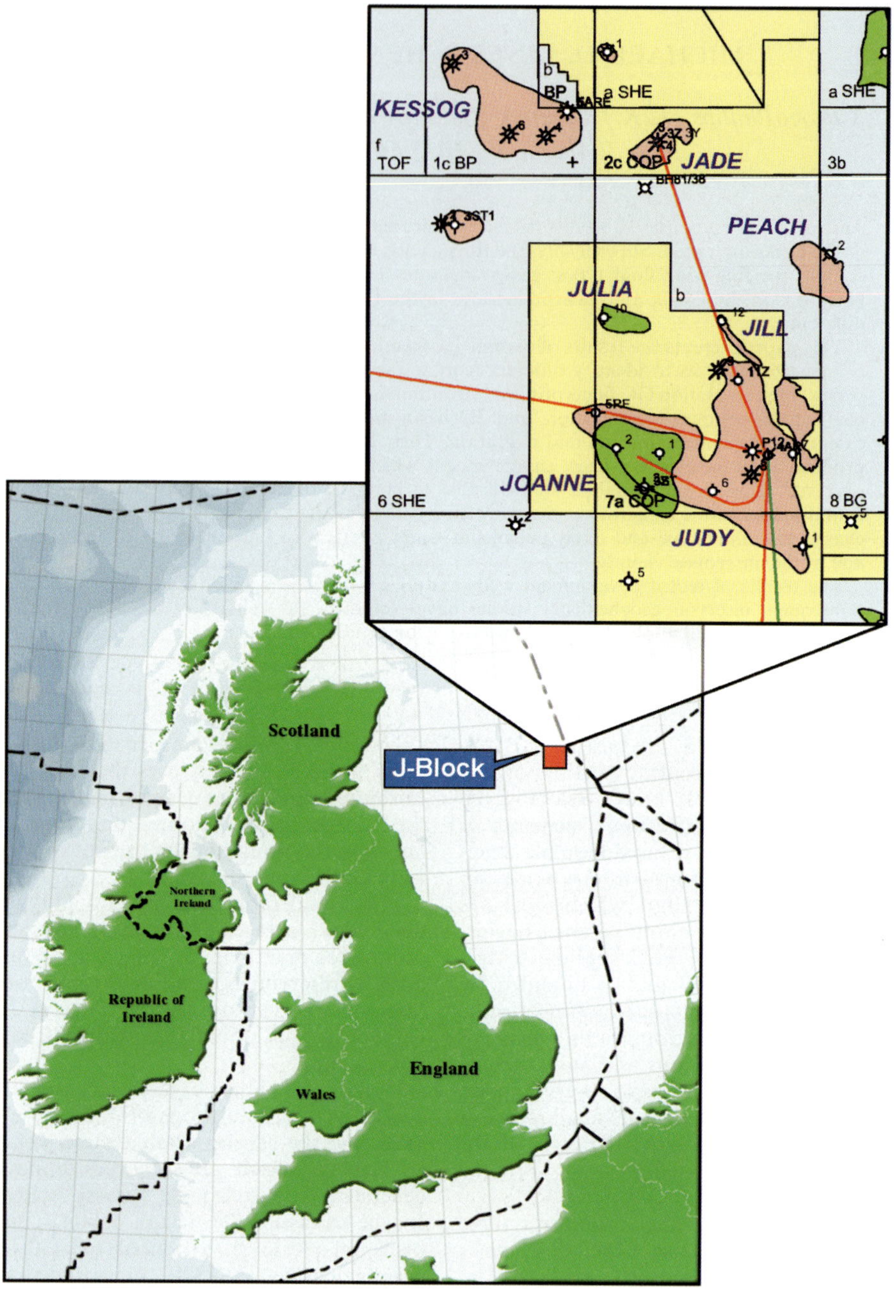


**Fig. 1.** Location map.

present in the Joanne, Peach and 30/6-3 accumulations, whilst black oils are found in the Julia, Josephine and Blane Fields.

Discussion is focused on fluid types within Block 30/7a, primarily within the Pre-Cretaceous reservoirs, where the greatest variability is observed. The major aim of this study is to develop a petroleum charge model that adequately accounts for the observed diversity in fluid type. The charge model can then be used to address the key objectives, which are as follows:

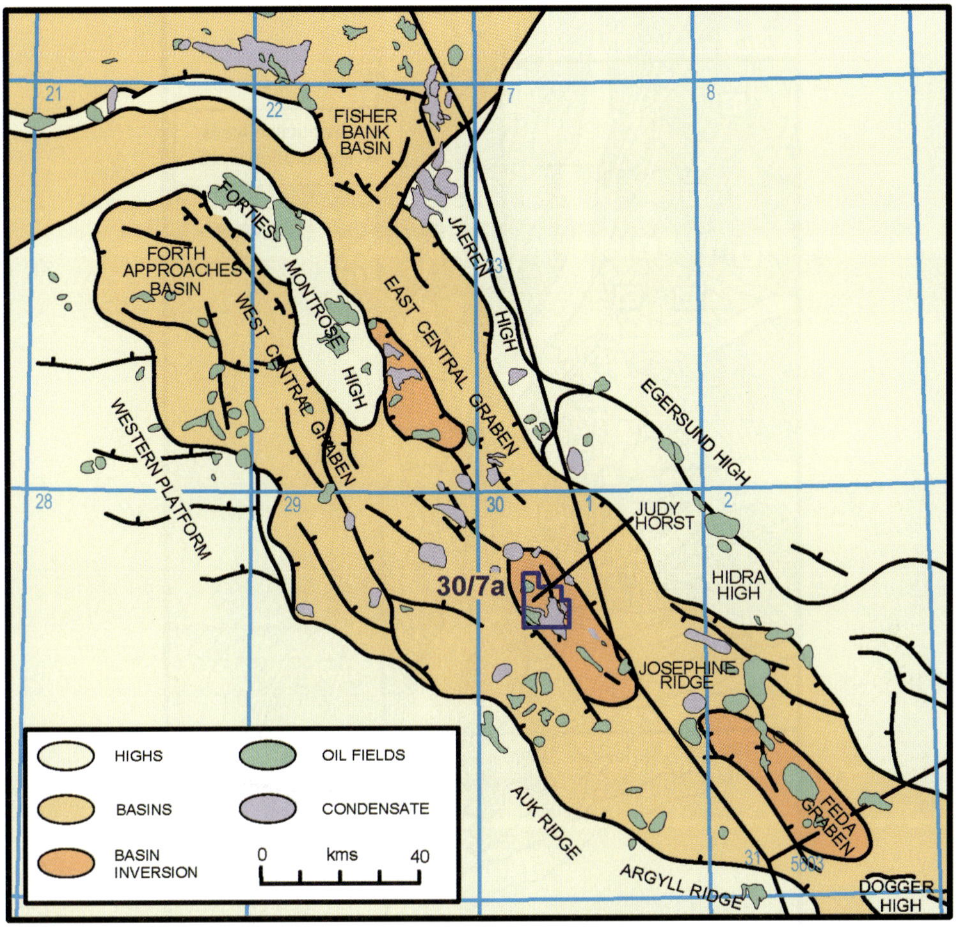


**Fig. 2.** Regional tectonic elements map.

- to provide an understanding of the variability in petroleum type in each of the reservoir intervals;
- to explain the variability of petroleum column heights and hydrocarbon-water contacts in tested fault compartments;
- to assess the impact of the above on field development and prospectivity in untested parts of the field.

## Petroleum characterization within the J Block area

As a group, the J Block fluids show almost as much variation as the entire Central Graben dataset available from in-house semi-regional studies. Oil-source correlation studies have shown that these petroleums are primarily sourced from the Upper Jurassic Kimmeridge Clay and Heather Formations. A crossplot of carbon isotope ratios for the saturate and aromatic fractions is shown in Figure 6. Whilst the majority of the J Block fluids form a relatively close-knit group, Pre-Cretaceous samples from 30/7a-7 and 30/7a-P12 are isotopically much heavier (less negative). In addition, the 30/7a-10 Julia oil sample, despite being more mature than the Palaeocene condensates, has an isotopically light signature (assuming a common source, petroleums generally become isotopically heavier with increasing maturity). Samples that show isotopically light values for the saturate fraction whilst having isotopically heavy aromatic values are from Kessog and are considered to have been sourced from the Middle Jurassic Esk Formation.

Petroleums reservoired within Block 30/7a range from relatively low maturity, low wax condensates in the Palaeocene to high maturity,

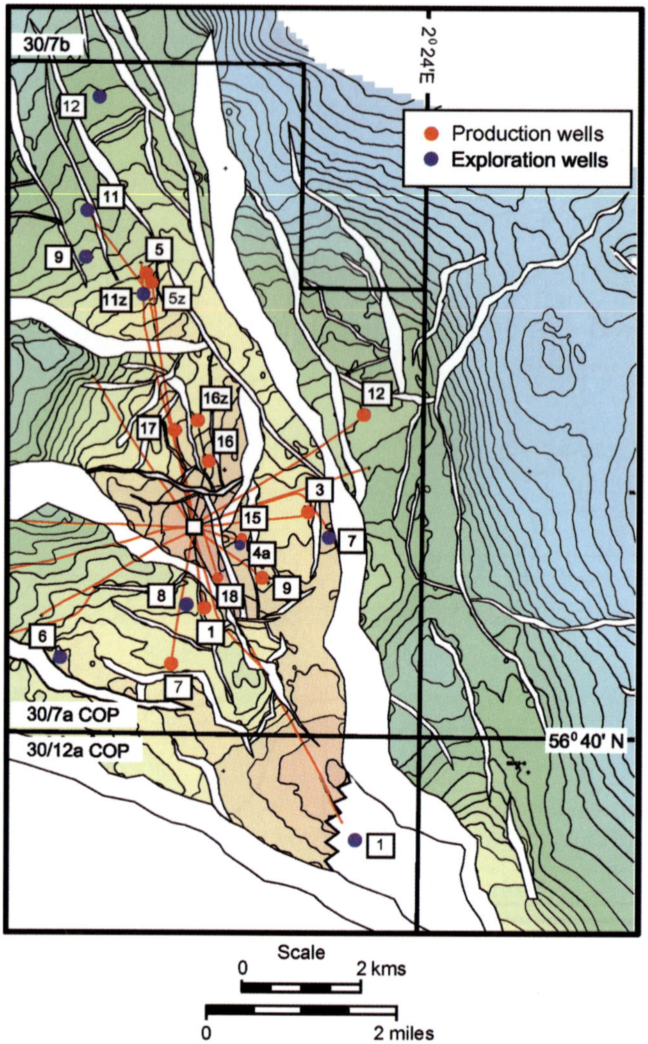


**Fig. 3.** Top Triassic reservoir map.

generally high wax fluids within the Pre-Cretaceous. The Chalk oils are somewhat intermediate between the Palaeocene and the Pre-Cretaceous, having moderate maturities and wax contents. The increase in petroleum maturity from Palaeocene (shown in red) to Chalk (blue) to Pre-Cretaceous (green) can be seen in crossplots of several biomarker maturity parameters, an example of which is shown in Figure 7.

Figure 7 shows a crossplot of an aromatic sterane cracking parameter (A3)[1] and a hopane biomarker maturity parameter (H14)[2]. The A3 parameter increases from a value of 0.3 at *ca.* 0.6% Ro to a value of 1.0 at *ca.* 1.4% Ro equivalent (Peters & Moldowan 1993). The ratio increases as the result of preferential degradation of the long-chain molecules rather than transformation to short-chain homologues (Beach *et al.* 1989). The H14 parameter increases with maturity due to the greater thermal stability of the $C_{30}$ diahopane molecule (Peters & Moldowan 1993). Thus, maturity increases towards the upper right part of Figure 7. The Palaeocene, Chalk and Pre-Cretaceous petroleums are clearly segregated, the only anomaly being the Palaeocene oil from 30/7a-10 (Julia), which has a higher maturity, similar to some of the Pre-Cretaceous samples. The Judy Chalk oils from

[1] A3 = $C_{20}$ triaromatic sterane/($C_{20}$ triaromatic sterane + $C_{28}$ 20R triaromatic sterane).
[2] H14 = $C_{30}$ diahopane as percentage of $C_{30}$ hopane.

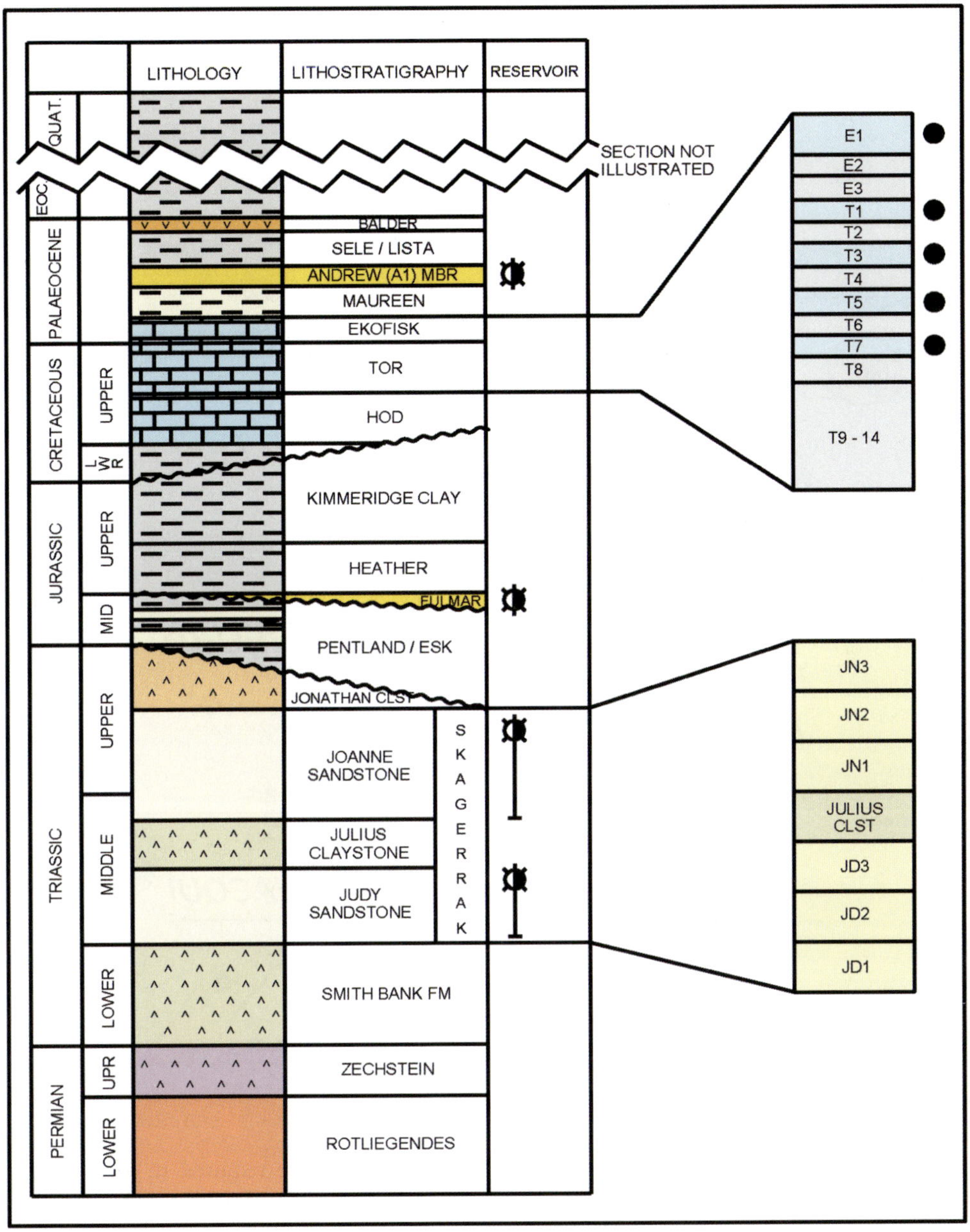


**Fig. 4.** Generalized stratigraphy of the J Block area.

30/7a-4a and 30/7a-6 appear to be more mature than those from the main Joanne accumulation.

Wax contents show considerable variation, ranging from as low as 4% for the Joanne Palaeocene condensate from 30/7a-P13z to almost 20% for the 30/7a-P12 Pre-Cretaceous Jocelyn condensate. An interesting phenomenon is that the wax content of the petroleums increases with maturity (Fig. 8), particularly in the case of the Palaeocene and Chalk fluids. This is rather unusual as wax contents often decrease with increasing maturity as a result of increased oil to gas cracking. We believe that this phenomenon is due to a geochromatographic process, whereby long chain *n*-alkanes are progressively left behind along the migration

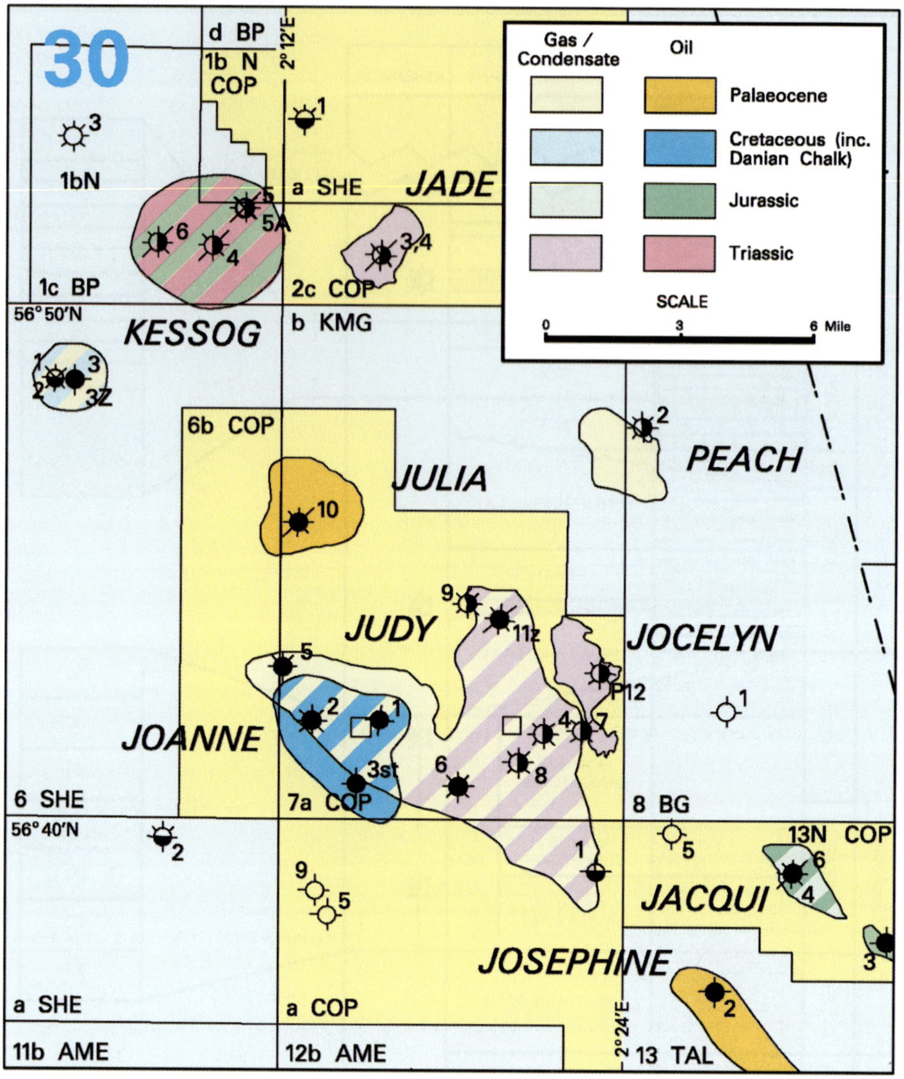


**Fig. 5.** Petroleum occurrence map.

pathway. The Pre-Cretaceous petroleums do not show this trend, possibly because it has been masked by source input variations (see below).

## Pre-Cretaceous charge model

### *Introduction*

The Pre-Cretaceous reservoirs show the greatest variation in fluid type seen in tests from appraisal and development wells. They are also the most structurally complex, being offset by numerous faults of Triassic-Late Jurassic age (Fig. 3). A model is therefore required to account for the variability in fluid type to improve reserves estimation and allow optimization of future development well locations.

### *Petroleum characterization*

*Physical properties.* Gas–oil ratios for the Pre-Cretaceous fluids vary from fairly low values of 1224 and 1691 scf/bbl for wells 30/7a-6 and 30/7a-3 to a maximum of 6287 scf/bbl in 30/7a-11z (Table 1). Most of the petroleums are near-critical point fluids, being either single-phase

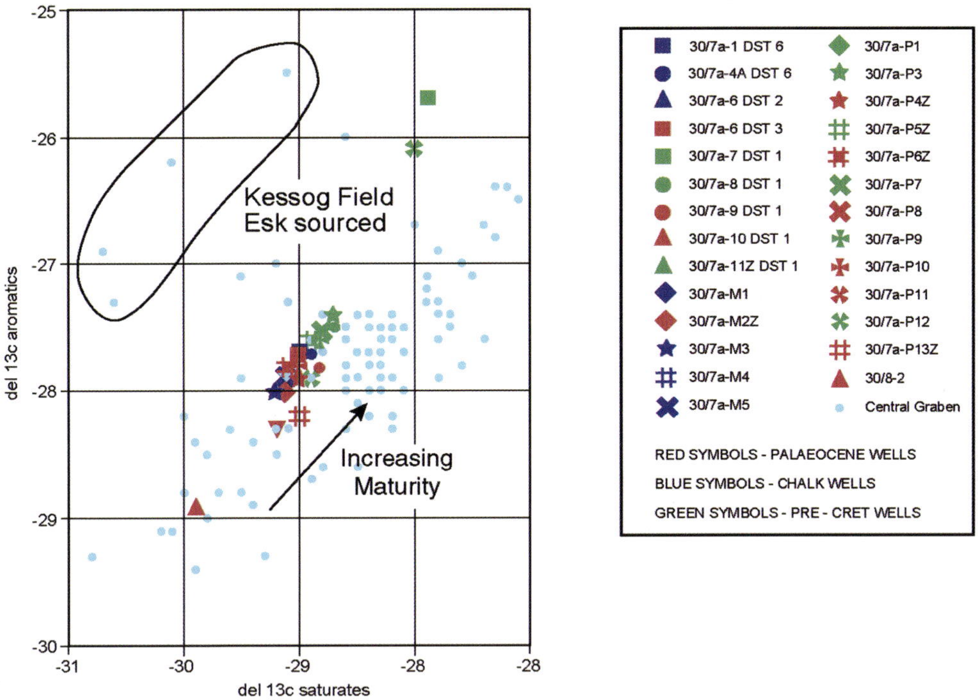


**Fig. 6.** Carbon isotope crossplot for J Block and Central Graben petroleums.

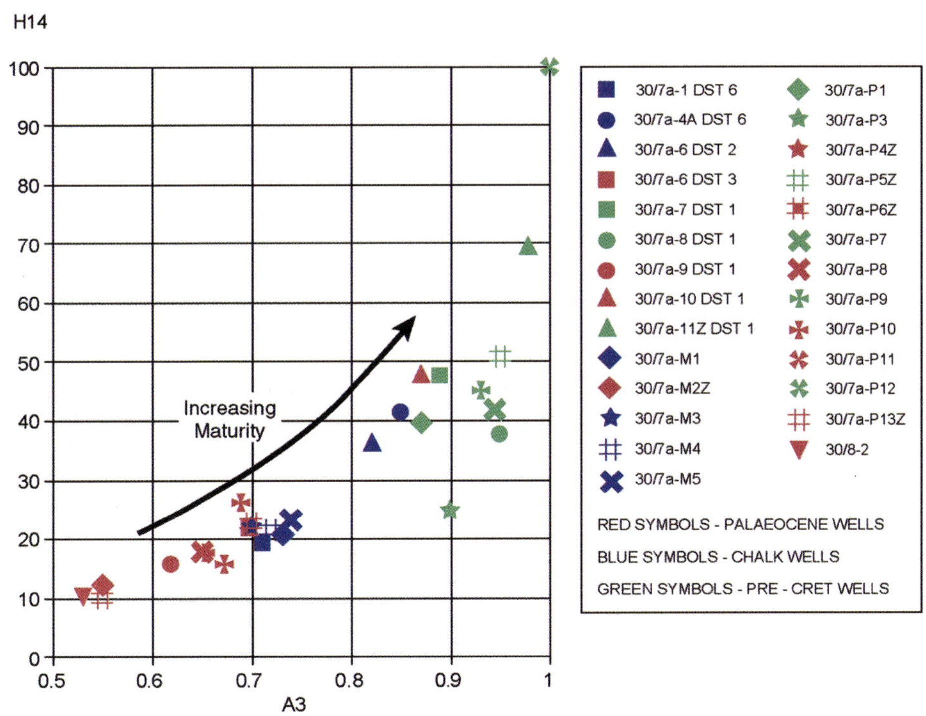


**Fig. 7.** Biomarker maturity crossplot for J Block petroleums.

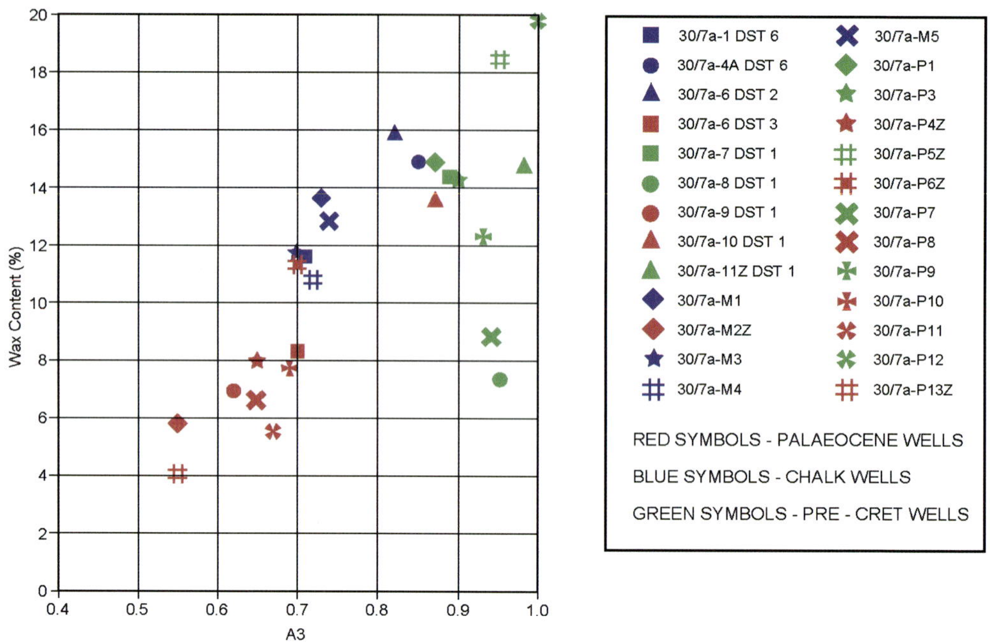


**Fig. 8.** Wax content versus maturity for J Block petroleums.

volatile oils or retrograde gas-condensates. The only exception is in the 30/7a-11z compartment where RFT measurements indicate the presence of a thin (*ca.* 45 ft) oil rim beneath the main gas condensate column tested in the well.

Wax contents are high for the majority of the Pre-Cretaceous petroleums. Exceptions are the samples from wells 30/7a-8 and 30/7a-P7, which have low wax contents (Fig. 8). These two wells lie to the south of a major NW–SE trending fault (Fig. 3) and it is likely that the wells to the southwest of this fault have been sourced from a different kitchen to those on the northeastern side. Note that the high wax fluid from 30/7a-P1 is from an allochthonous section (olisthostrome) of the Joanne reservoir which, from RFT data, appears to be part of the 30/7a-P3, 30/7a-P9 pressure cell (see below). Although not tested, the *in-situ* Joanne reservoir in 30/7a-P1 is predicted to contain a low wax fluid similar to those tested in 30/7a-8 and 30/7a-P7.

*Geochemical characterization.* There are clear geochemical differences between the majority of the Pre-Cretaceous fluids and the samples from 30/7a-8 and 30/7a-P7. These can be seen in a comparison of the saturate GC traces for the 30/7a-P7 and 30/7a-P5z condensates, shown in Figure 9. The low wax sample from 30/7a-P7 is front-end dominated and has an almost concave upwards *n*-alkane profile from *n*-$C_{17}$ onwards. In contrast, the 30/7a-P5z trace is convex upwards and shows abundant long chain *n*-alkanes extending up to *n*-$C_{40}$. The low wax fluids also have higher pristane/*n*-$C_{17}$ and phytane/*n*-$C_{18}$ ratios and lower pristane/phytane ratios than the others. These differences cannot be ascribed solely to maturity variations. It is likely that the petroleums present in the 30/7a-8, 30/7a-P7 and in the *in-situ* Joanne sand in 30/7a-P1, which lie to the southwest of a major NW–SE trending fault, were generated from a different source kitchen to the other Pre-Cretaceous petroleums.

The 30/7a-P12 (Jocelyn) condensate is the most mature of all the Pre-Cretaceous petroleums analysed (Fig. 7). Because of its high maturity, many of the biomarkers have been destroyed and so some parameters cannot be quoted. GC-MS fingerprints of the triterpanes (Fig. 10) indicate that only the most thermally resistant molecules Ts, $C_{29}$Ts and $C_{30}$ diahopane ($\pi$) remain in any abundance. The 30/7a-11z and 30/7a-P5z samples from the northernmost part of the field are also highly mature, being more mature than those from the crestal part of the Judy structure, i.e. 30/7a-7, 30/7a-P3 and 30/7a-P9. The 30/7a-P3 fluid appears less mature than that from 30/7a-P9, despite its location within the same crestal fault block, suggesting the possibility of compartmentalization. The decrease in oil maturity from 30/7a-P12 to 30/7a-P5z to 30/7a-P9 to 30/7a-P3 is shown in the triterpane

**Table 1.** *Pre-Cretaceous test data and fluid properties*

| Well | DST | Depth interval (ft ss) | | Gas rate (MMscfd) | Oil rate (bopd) | GOR | Reservoir P | Depth (Ft ss) | Oil gravity (API) | Gas gravity (SG) | Bubble point P (psia) | Dew point P (psia) |
|---|---|---|---|---|---|---|---|---|---|---|---|---|
| | | From | To | | | (scf/bbl) | (psia) | | | | | |
| 30/7a-3 | 1a | 13742 | 13802 | 0.4 | 256 | 1691 | 10682 | 13772 | 45.0 | 0.79 | | |
| 30/7a-4a | 1 | 11429 | 11474 | 2.6 | 1349 | 1945 | | | 42.7 | 0.74 | | |
| 30/7a-4a | 2 | 11299 | 11404 | 14.4 | 5878 | 2453 | 8798 | 11352 | 40.8 | | 6400 | |
| 30/7a-4a | 3 | 11150 | 11254 | 1.5 | 560 | 2714 | 8826 | 11202 | 46.1 | 0.72 | 6280 | |
| 30/7a-4a | 4a | 11026 | 11088 | 27.0 | 7590 | 3556 | 8729 | 11057 | 39.9 | 0.70 | | 7560 |
| 30/7a-6 | 1 | 12112 | 12135 | 1.8 | 1431 | 1224 | 9185 | 12124 | 40.2 | | 4711 | |
| 30/7a-7 | 1 | 11219 | 11325 | 22.6 | 4168 | 5432 | 8755 | 11272 | 42.5 | | | 7500 |
| 30/7a-8 | 1 | 11608 | 11708 | 13.9 | 3384 | 4108 | 8978 | 11658 | 43.8 | 0.85 | | 6500 |
| 30/7a-8 | 2 | 11550 | 11568 | 19.3 | 4151 | 4646 | 8924 | 11559 | 44.4 | 0.73 | | 6690 |
| 30/7a-8 | 3 | 11309 | 11376 | 11.7 | 1990 | 5880 | 8788 | 11343 | 45.3 | 0.72 | | 6865 |
| 30/7a-11z | 1 | 11570 | 11587 | 23.0 | 4040 | 5693 | 8986 | 11579 | 42.5 | | | 8190 |
| 30/7a-11z | 2 | 11477 | 11494 | 18.0 | 2863 | 6287 | 8973 | 11486 | 42.9 | | | 8165 |
| 30/7a-P1 | 1 | 11095 | 11253 | 13.6 | 5706 | 2385 | | | 42.5 | 0.68 | | |
| 30/7a-P3 | 1 | 11161 | 11300 | 14.3 | 5960 | 2487 | | | | | | |
| 30/7a-P5z | | | | 21.5 | 4326 | 4990 | | | | | | |
| 30/7a-P7 | 1 | 11621 | 11703 | 9.9 | 3838 | 2569 | | | 44.0 | 0.77 | | |
| 30/7a-P9 | 1 | 11184 | 11341 | 18.5 | 7238 | 2555 | 8760 | 11184 | 44.3 | 0.69 | 6135 | |
| 30/7a-P12 | 1 | 13268 | 13407 | 20.3 | 4166 | 4875 | 10199 | 13268 | 48.0 | 0.60 | | 6215 |

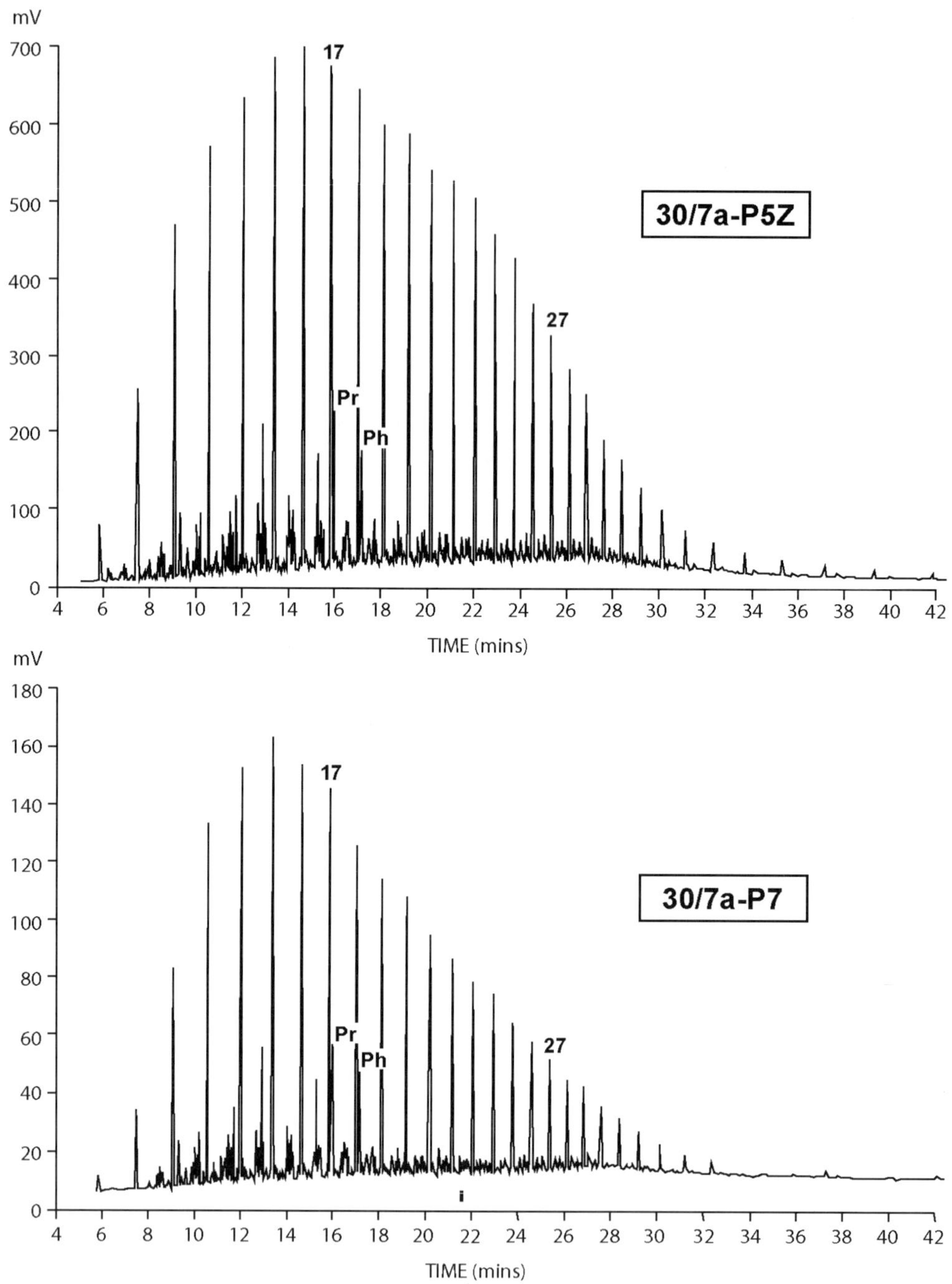


**Fig. 9.** GC traces for Pre-Cretaceous petroleums from 30/7a-P5z and 30/7a-P7.

fingerprints in Figure 10. Key peaks which show an increase in abundance with increasing maturity are $C_{29}$Ts and $C_{30}$ diahopane ($\pi$). The high maturity of the northerly fluids is supported by data from the 30/7a-9 core extracts, which show comparable maturities to the 30/7a-11z and 30/7a-P5z petroleums. A decrease in maturity from north to south suggests that migration of the main Judy petroleums may have been from the north.

A plot of carbon isotope ratios for each of the oil fractions, known as a Galimov plot, is shown

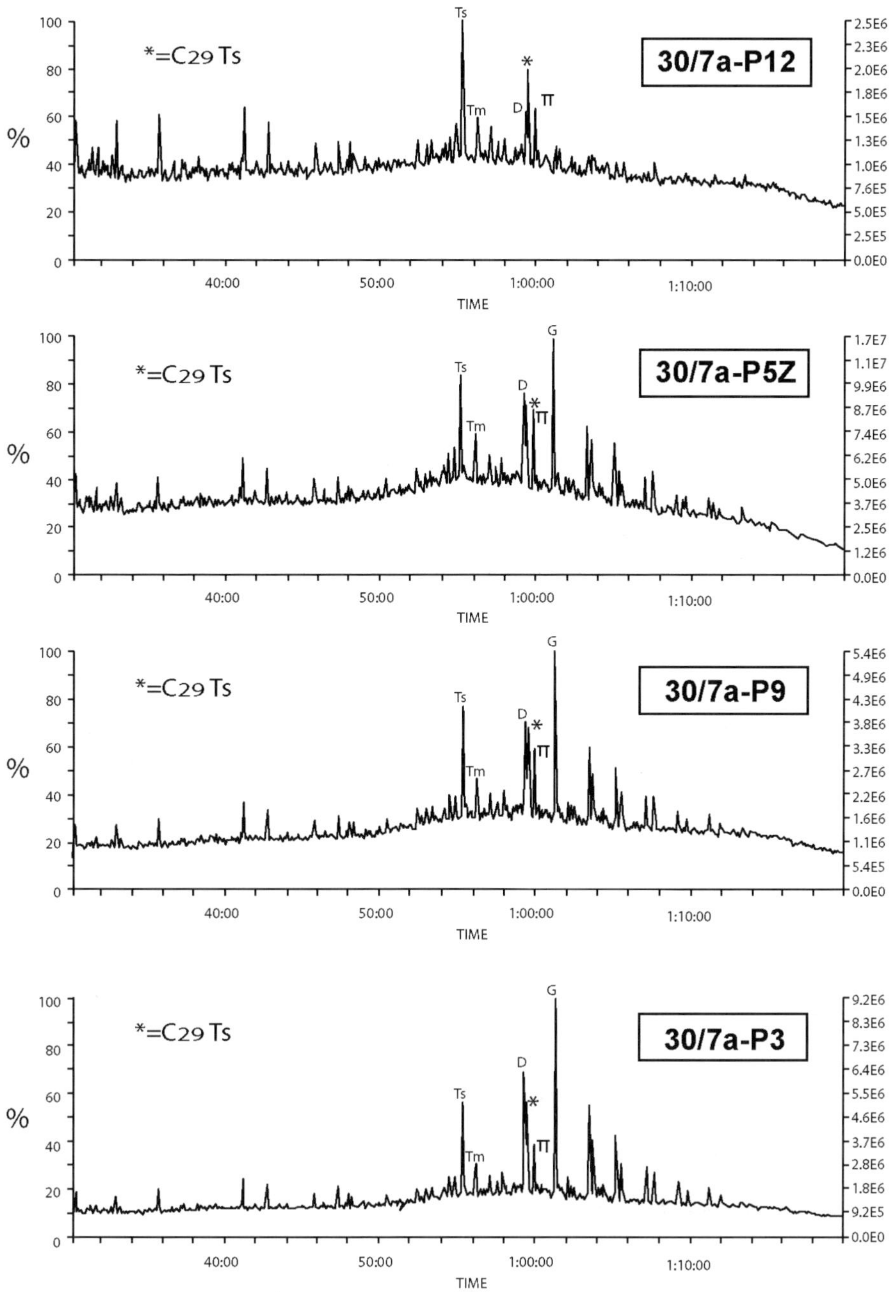


**Fig. 10.** Triterpane fingerprints for Pre-Cretaceous petroleums.

in Figure 11. The Jocelyn (30/7a-P12) and 30/7a-7 condensates are isotopically much heavier (less negative) than the other Pre-Cretaceous fluids. This cannot be explained solely by maturity differences (petroleums generated from a single source tend to become heavier with increasing maturity), because the 30/7a-7 petroleum is less mature than the 30/7a-11z and 30/7a-P5z samples analysed from the northern part of the field. Compound specific isotope analyses (CSIA), which provide the isotopic ratio for individual components, also show clear

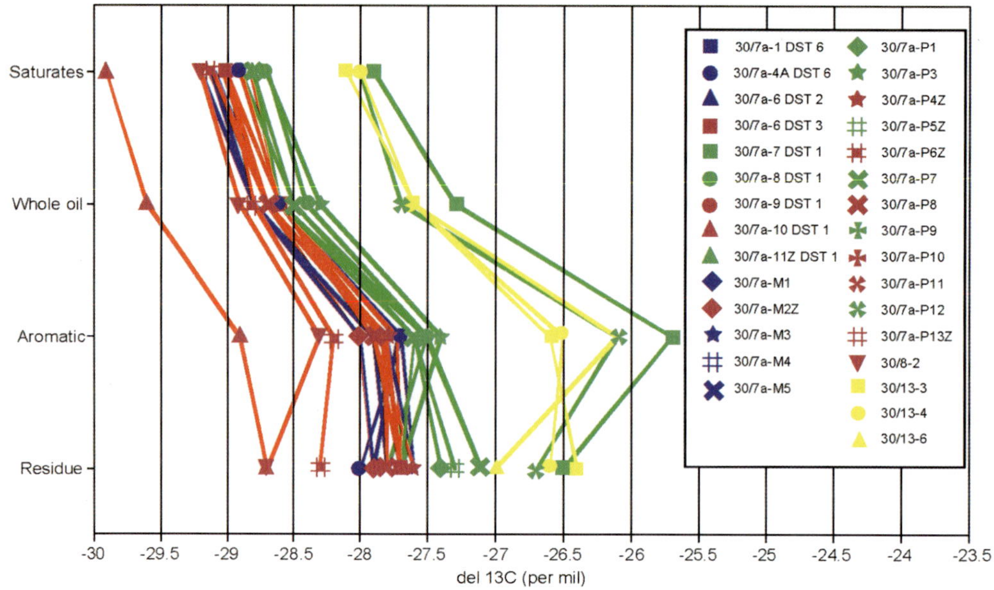


**Fig. 11.** Galimov plot for 30/7a and 30/13 petroleums.

differences between the 30/7a-7 and 30/7a-P12 samples and the other J Block fluids and support the concept of discrete source kitchens. Interestingly, data from Jacqui wells in Block 30/13 to the southeast show very similar isotopic compositions to those in 30/7a-7 and 30/7a-P12 (Fig. 11). A common source facies is probable although they are unlikely to have been expelled from a single source kitchen.

## *Reservoir connectivity*

Compartmentalization within the Pre-Cretaceous reservoirs has been investigated through a review of RFT pressure data. Both intra- and inter-well connectivity issues are addressed.

Available RFT pressure data for the Pre-Cretaceous reservoirs are plotted in Figures 12 and 13. At a gross scale (Fig. 12), the data appear to show a common aquifer for most of the wells, although each has a different petroleum gradient. Only the Jocelyn well (30/7a-P12), which lies on the downthrown side of the major eastern bounding fault to the Judy Field (Fig. 3), appears to be in pressure isolation.

In more detail, however, the picture is somewhat different (Fig. 13). The 30/7a-9 and 30/7a-11z wells, drilled in the northernmost part of the field, appear to have different aquifer gradients. Within individual wells, there is no evidence of pressure discontinuities between the Jurassic Fulmar reservoir and the Triassic Joanne and Judy reservoirs. However, the data show a large pressure discontinuity within the petroleum column between the 30/7a-P1 olisthostrome and the *in-situ* Joanne reservoirs. The *in-situ* Joanne Sands in 30/7a-P1 are part of the southwesterly pressure regime characterized by the 30/7a-8 and 30/7a-P7 wells, whilst the olisthostrome appears to be in pressure communication with the 30/7a-P3 and 30/7a-P9 wells.

Within individual pressure cells, there is a shift in the petroleum gradient across minor faults (see structure map in Fig. 3). The southwesterly cell shows a shallowing of the petroleum–water contact from 30/7a-P7 to 30/7a-8 to 30/7a-P1, i.e. in a northeasterly direction. The data suggest that a pressure differential of *ca.* 10 psi is required for migration to occur across each fault and that the fault acts as a semi-permeable membrane. This pressure difference is referred to as the 'capillary pressure' of the fault. Thus, despite the common aquifer, differences in pressure within the petroleum gradients from well to well indicate that petroleum from individual fault compartments will have to be produced separately.

In the crestal part of the field, the 30/7a-P3 and 30/7a-P9 wells have almost identical trends and appear to be in full pressure communication (Fig. 13). The 30/7a-P1 olisthostrome has slightly higher pressures than the 30/7a-P3 and 30/7a-P9 wells, whilst the 30/7a-7 well has lower pressures.

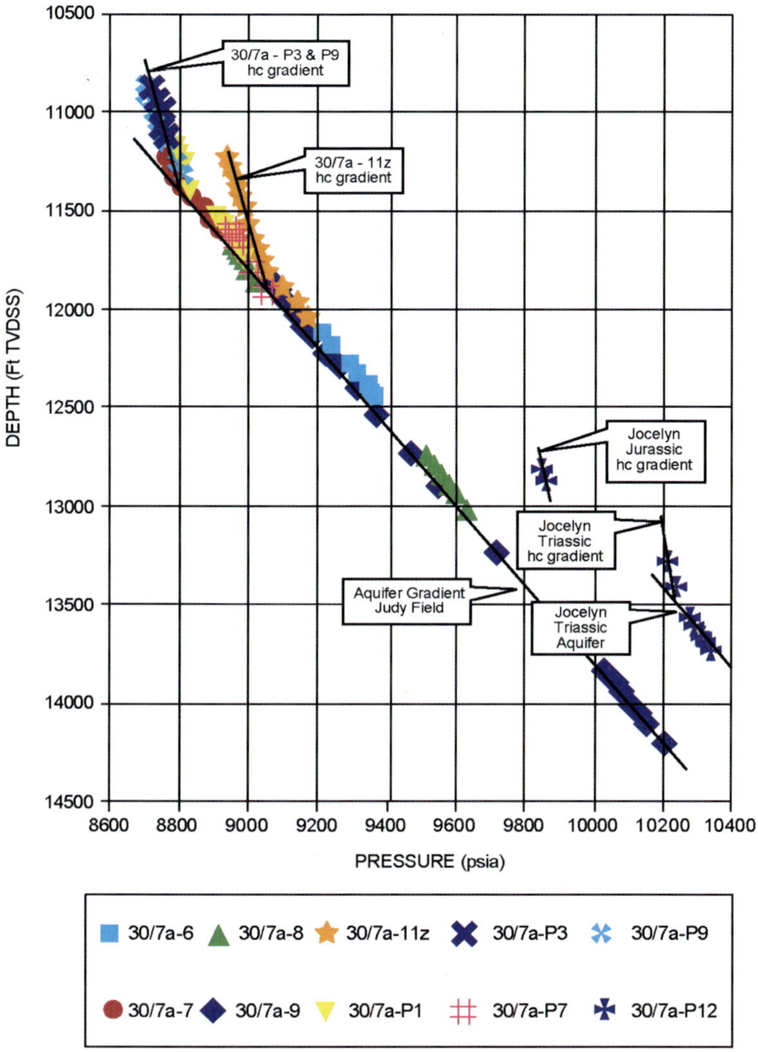


**Fig. 12.** RFT pressure versus depth for Pre-Cretaceous reservoirs—gross scale.

The petroleum gradient in 30/7a-7, however, indicates that a more gas-rich fluid is present (0.195 psi/ft in 30/7a-7 compared with 0.221 psi/ft in 30/7a-P3 and 30/7a-P9). This is consistent with the higher GOR of the 30/7a-7 petroleum (5432 scf/bbl) compared with *ca.* 2500 scf/bbl for the 30/7a-P3 and 30/7a-P9 fluids. It suggests that the 30/7a-7 well tested a gas-condensate accumulation rather than the volatile oil seen in the 30/7a-P3 and 30/7a-P9 wells.

There is further evidence for compartmentalization in that the aquifer for the crestal cell is slightly overpressured (by *ca.* 10 psi) compared with that for wells 30/7a-8, 30/7a-P7 and 30/7a-P1 in the southwest cell (Fig. 13).

## *Petroleum generation, expulsion and migration*

*Introduction.* A study of petroleum generation, expulsion and migration has been undertaken using PetroMod 2D basin modelling software from IES. Porosity, permeability, pressure and temperature data from J Block wells, and temperature and vitrinite reflectance data from 30/8-1 have been used to help calibrate the model. In general, the data show excellent agreement with the calculated values. In particular, the rapid increase in pore pressure within the lower part of the Chalk has been successfully modelled, resulting in a good pressure match

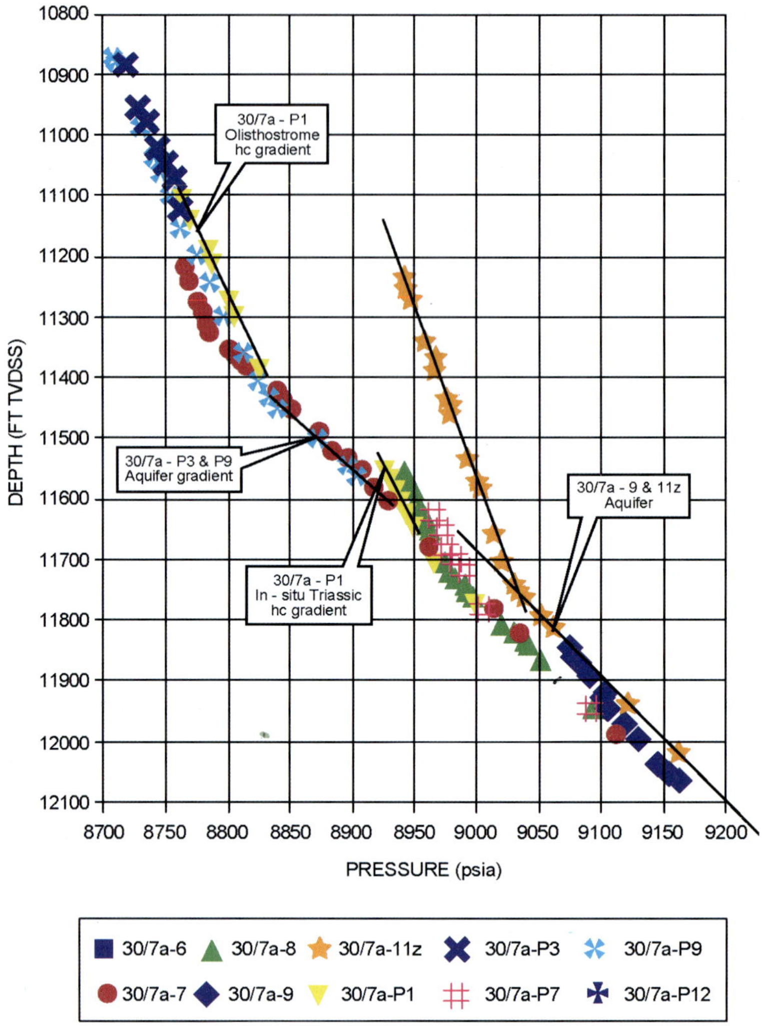


**Fig. 13.** RFT pressure versus depth for Pre-Cretaceous reservoirs – detail.

for the Palaeocene, Chalk and Pre-Cretaceous reservoirs.

Details of the modelling are not included in this paper, but the results are summarized below.

*Generation and expulsion.* In the western source kitchen, petroleum generation from Upper Jurassic source rocks commenced towards the end of the Cretaceous (56.5 Ma bp) and was almost complete by mid Miocene times (10.4 Ma bp) when transformation ratios of 0.9 were reached. In the easterly kitchen, generation from the Kimmeridge Clay Formation (KCF) source rock was only just starting at this time ($TR < 0.3$). At 10.4 Ma bp, no generation has occurred from Upper Jurassic source rocks present over the crest of the Judy/Joanne structure.

At present, the westerly kitchen is totally 'cooked out' ($TR > 0.9$), whilst, in the east, the KCF still retains some potential for petroleum generation, as do source rocks on the crest of the structure.

*Migration.* Petroleum migration into Pre-Cretaceous reservoirs occurs by means of downward expulsion of petroleum, driven by overpressuring of overlying Upper Jurassic source rocks. Once into the Jurassic and Triassic carriers, migration of petroleum is buoyancy driven and controlled by sand-to-sand communication across faults.

The Joanne structure received a petroleum charge during Earliest Miocene times from the western source kitchen. The main carrier from the basin is the Joanne Sand but cross-fault juxtaposition of the Joanne Sand with the Judy Sand in the Joanne structure results in accumulation of oil at the location of the 30/7a-3 well, tested by DST 1a. Continued migration of petroleum from the western kitchen resulted in well 30/7a-6 receiving a petroleum charge by Middle Miocene times. Migration into the Judy structure from the southwest has occurred only recently although, as stated above, migration from the deep kitchen to the east of Judy would have occurred slightly earlier than that for the western kitchen. This is supported by fluid inclusion analyses which suggest that petroleum migration into well 30/7a-11z (charged from the northeast) occurred earlier than in well 30/7a-8 which was sourced from the southwest.

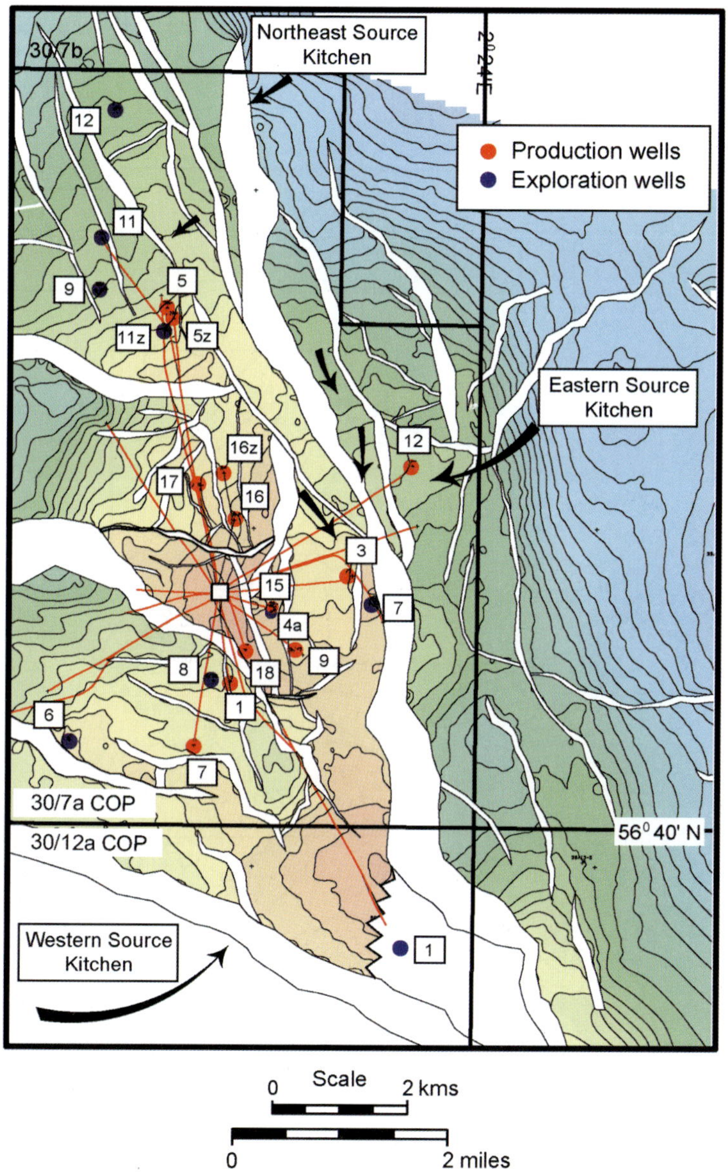


**Fig. 14.** Pre-Cretaceous charge model summary map.

Petroleum would have undoubtedly entered the Judy and Joanne structures as a relatively low GOR oil charge. As the major kitchens to the northeast and west continued to subside, increasing amounts of light oil and gas would have migrated into the structure. This would have resulted in transition from a low GOR oil to a volatile oil or gas-condensate. This is consistent with fluid inclusion data from wells 30/7a-8 and 30/7a-11z, which show the presence of low GOR oil (*ca.* 500 scf/bbl) trapped in inclusions in what are now volatile oil and gas condensate reservoirs.

The charge model is summarized in Figure 14.

## Chalk charge model

### *Introduction*

The Chalk reservoired petroleums of J Block are predominantly undersaturated black oils. The model developed to explain the variation in petroleum type within the Chalk is heavily influenced by the understanding of fluid distribution within the Pre-Cretaceous, as discussed above.

Geochemical data are available for seven Chalk-reservoired oil samples. Five of these are from the Joanne accumulation and two are from wells on the Judy structure. Well locations are shown in Figure 15.

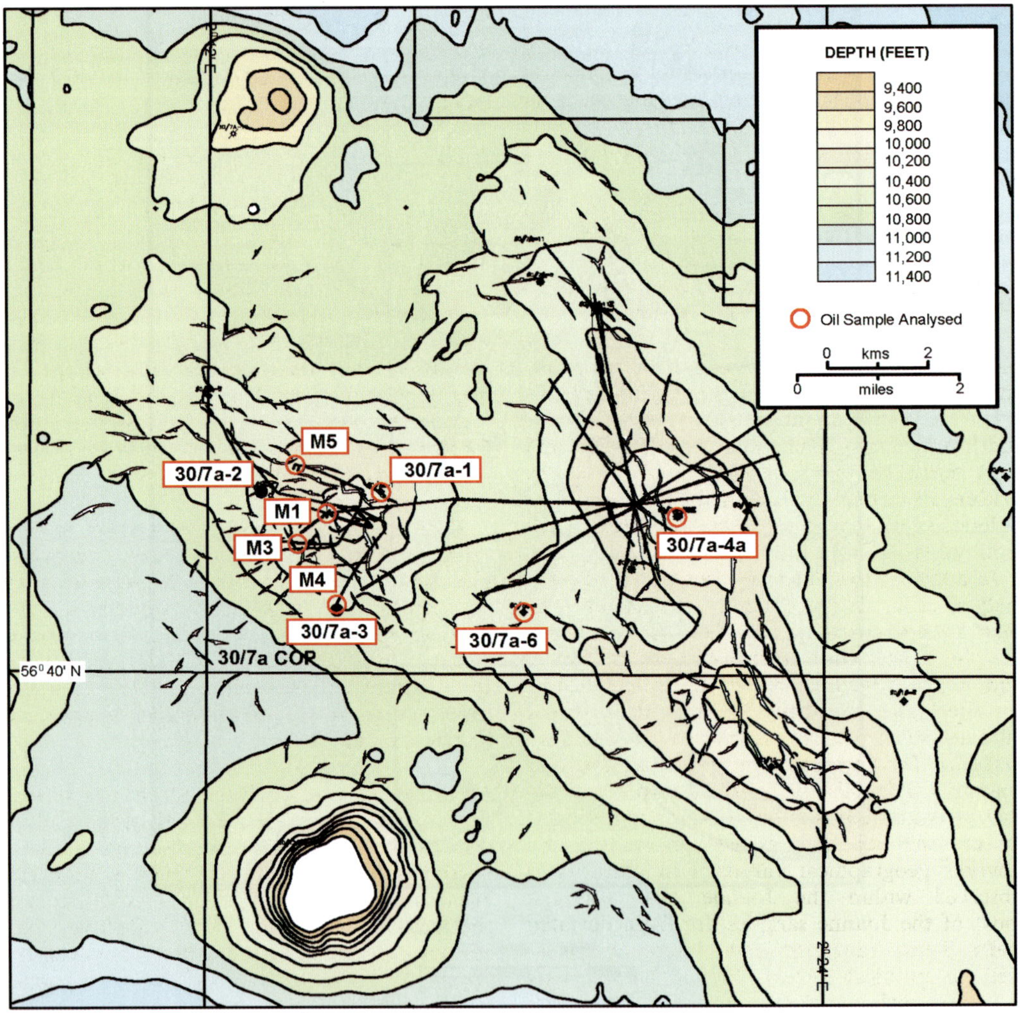


**Fig. 15.** Location map showing Chalk oils analysed.

## *Petroleum characterization*

*Physical properties.* Most of the petroleums tested from the Chalk reservoirs in the Judy and Joanne Fields are undersaturated black oils. Gas–oil ratios are generally in the range 2000 to 2500 scf/bbl and bubble points are approximately 1000 to 1500 psi below present-day reservoir pressures (Table 2). In the Judy structure, well 30/7a-4a tested gas caps in both the Ekofisk and Tor reservoirs. Petroleum tested from the Tor in 30/7a-6 also had a high GOR (4322 scf/bbl) and PVT analysis confirmed the fluid as a gas-condensate (Pd = 7185 psi). However, the dew point pressure is very close to the reservoir pressures obtained from RFT pressure measurements (7150–7200 psi). Furthermore, flow rates were very low and so the test results may not be valid.

Wax contents also show the Joanne and Judy petroleums are somewhat different. The Joanne oils have moderate wax contents ranging from 10.8% to 13.6%, whilst those from 30/7a-4a and 30/7a-6 are slightly higher (14.9% and 15.9% respectively).

Owing to the nature of the Joanne fluid samples analysed—predominantly from development wells, which tested the entire Chalk interval—it is not possible to identify any differences between oils reservoired in the Ekofisk and Tor.

*Geochemical characterization.* The tested petroleums from the Chalk reservoirs are moderately mature and form an intermediate group between the low maturity Palaeocene condensates and high maturity Pre-Cretaceous fluids (Fig. 7). Within the Chalk group itself, there are clear differences between the Joanne oils and the oils from wells 30/7a-4a and 30/7a-6.

In addition to their higher wax contents, the Judy Chalk oils have higher maturities than their Joanne counterparts. This can be seen in the plot of wax content versus maturity (Fig. 8) and in the biomarker maturity crossplot (Fig. 7). The higher maturity oils from 30/7a-4a and 30/7a-6 show similar maturities to the high maturity Pre-Cretaceous petroleums. The five Joanne Chalk oils, in contrast, form a close-knit group and appear to be only slightly more mature than the Palaeocene condensates. No obvious geographical variation in maturity is observed within the Joanne oils, although most of the Joanne samples are from deviated wells. These have significant lateral extent as well as vertical extent, which would tend to mask such variations, e.g. well 30/7a-M5, which tested an interval from 10 325 to 14 490 ft (sub-horizontally), comprising the Ekofisk (E1) and Tor (T1, T3, T5, T7) reservoir zones.

## *Reservoir connectivity*

*Introduction.* The existence or absence of separate compartments within the Ekofisk and Tor reservoirs is a critical issue in determining reserves distribution and optimal production methods. Initially, a fault compartmentalized reservoir model was used to account for the southwesterly deepening oil–water contacts recorded in appraisal wells in the Joanne Field. Within the Ekofisk reservoir, the observed oil–water contact deepens by 280 ft from 9920 ft (tvdss) in 30/7a-1 to 10 200 ft in 30/7a-3; in the Tor, the oil–water contact deepens by almost 400 ft from 10 276 ft in 30/7a-1 to 10 675 ft in 30/7a-3. Following completion of development well drilling, the current model incorporates single tilted oil–water contacts in both the Ekofisk and Tor.

*RFT pressure data.* RFT pressure data for the Chalk wells are shown on Figure 16. Although the data show considerable scatter, several discernible trends can be recognized. The 30/7a-4a and 30/7a-6 wells show the greatest scatter but exhibit different trends from each other and from the Joanne wells. Data for the Ekofisk reservoirs in 30/7a-M1 and 30/7a-M3 appear to be offset from their counterparts within the Tor, highlighting the fact that the Ekofisk and Tor are two separate reservoir systems. In the Ekofisk, well 30/7a-M3 is clearly overpressured with respect to 30/7a-M1 whereas, in the Tor, the reverse is true.

Data for well 30/7a-M1 show a continuous oil gradient with depth suggesting that, over geological time at least, there has been communication between the various Tor porous intervals. Later development wells (e.g. 30/7a-M3) show variable pressure depletion within Tor reservoir zones due to differences in inter-well connectivity; this also demonstrates a lack of vertical communication between zones.

*Strontium isotope analyses.* Strontium isotope data obtained using residual salt analysis (SrRSA) are the $^{87}Sr/^{86}Sr$ isotope ratios of residual salts extracted from reservoir cores (see Mearns & McBride 1999 for a description of the methodology). The $^{87}Sr/^{86}Sr$ isotope ratios of formation waters are modified by diagenesis (water–rock interactions) and/or hydrodynamic processes through time. Once established, differences in the strontium isotope ratio will tend to

**Table 2.** *Chalk test data and fluid properties*

| Well | DST | Depth interval (ft ss) | | Reservoir zone | Gas rate (MMscfd) | Oil rate (bopd) | GOR | Reservoir P | Depth (ft ss) | Oil gravity (API) | Gas gravity (SG) | Bubble point P (psia) | Dew point P (psia) |
|---|---|---|---|---|---|---|---|---|---|---|---|---|---|
| | | From | To | | | | (scf/bbl) | (psia) | | | | | |
| 30/7a-1 | 4 | 10049 | 10113 | T3, T5 | 3.7 | 1560 | 2346 | 7081 | 10051 | 44.0 | 0.71 | 6065 | |
| 30/7a-1 | 5c | 9988 | 10018 | T1 | 8.9 | 4025 | 2201 | 6891 | 9968 | 41.0 | 0.72 | 6075 | |
| 30/7a-1 | 6 | 9988 | 10013 | T1, T3, T5 | 10.6 | 4585 | 2301 | 6829 | 9969 | 41.0 | 0.72 | 6160 | |
| 30/7a-1 | 7 | 9774 | 9818 | E1 | 7.9 | 3050 | 2600 | 6571 | 9754 | 41.7 | 0.73 | 6080 | |
| 30/7a-2 | 2 | 10357 | 10476 | T3, T5, T7 | 5.8 | 2272 | 2531 | 6748 | 10371 | 41.3 | 0.75 | 5455 | |
| 30/7a-2 | 3b | 10289 | 10329 | T1 | 7.4 | 2712 | 2725 | 6959 | 10211 | 41.9 | 0.74 | 5475 | |
| 30/7a-2 | 4 | 10289 | 10476 | T1, T3, T5, T7 | 7.6 | 2178 | 3480 | | 10213 | | | | |
| 30/7a-2 | 5a | 10019 | 10075 | E1 | 5.2 | 2406 | 2174 | 6735 | 10004 | 40.6 | 0.83 | 5647 | |
| 30/7a-3 | 3 | 10458 | 10532 | T3, T5 | 9.0 | 4228 | 2131 | 7086 | 10513 | 40.6 | 0.74 | 5595 | |
| 30/7a-3 | 4 | 10385 | 10425 | T1 | 11.1 | 4973 | 2234 | 7047 | 10410 | 40.4 | 0.74 | 5635 | |
| 30/7a-3 | 5 | 10383 | 10532 | T1, T3, T5 | 10.1 | 4239 | 2371 | | | | | | |
| 30/7a-4a | 6 | 10071 | 10098 | Tor | 2.4 | 1331 | 1788 | 7013 | 9943 | 40.6 | 0.76 | 6300 | |
| 30/7a-4a | 7 | 9894 | 9993 | Tor | 1.0 | 100 | 9710 | 6908 | 9691 | 51.0 | 0.70 | | |
| 30/7a-4a | 7a | 9894 | 9993 | Tor | 0.9 | 148 | 5951 | 6987 | 9719 | 47.2 | 0.71 | | |
| 30/7a-4a | 8 | 9667 | 9726 | Ekofisk | 1.1 | 91 | 12472 | 6951 | 9458 | 51.8 | 0.73 | | |
| 30/7a-5 | 2 | 10363 | 10400 | T1 | 1.9 | 1464 | 1284 | 7088 | 10393 | 42.6 | | | |
| 30/7a-6 | 2 | 10298 | 10423 | Tor | 1.8 | 426 | 4322 | | | 42.0 | | | 7185 |
| 30/7a-M1z | 1 | 10210 | 10217 | T1 | 17.4 | 4324 | 4030 | | | 41.7 | 0.75 | 5875 | |
| 30/7a-M1z | 2 | 9738 | 10217 | T1–T9 | 20.2 | 9485 | 2127 | | | 40.9 | 0.76 | 5875 | |
| 30/7a-M3y | 1 | 9823 | 10450 | T1–T9 | 13.1 | 7630 | 1712 | | | 41.3 | 0.78 | 5335 | |
| 30/7a-M4 | 1 | 9812 | 10347 | T1–T7 | 13.3 | 7036 | 1890 | | | 40.0 | 0.78 | 5575 | |
| 30/7a-M5 | 1 | 9762 | 10114 | T1–T7 | 15.6 | 7390 | 2111 | | | 41.0 | 0.77 | | |

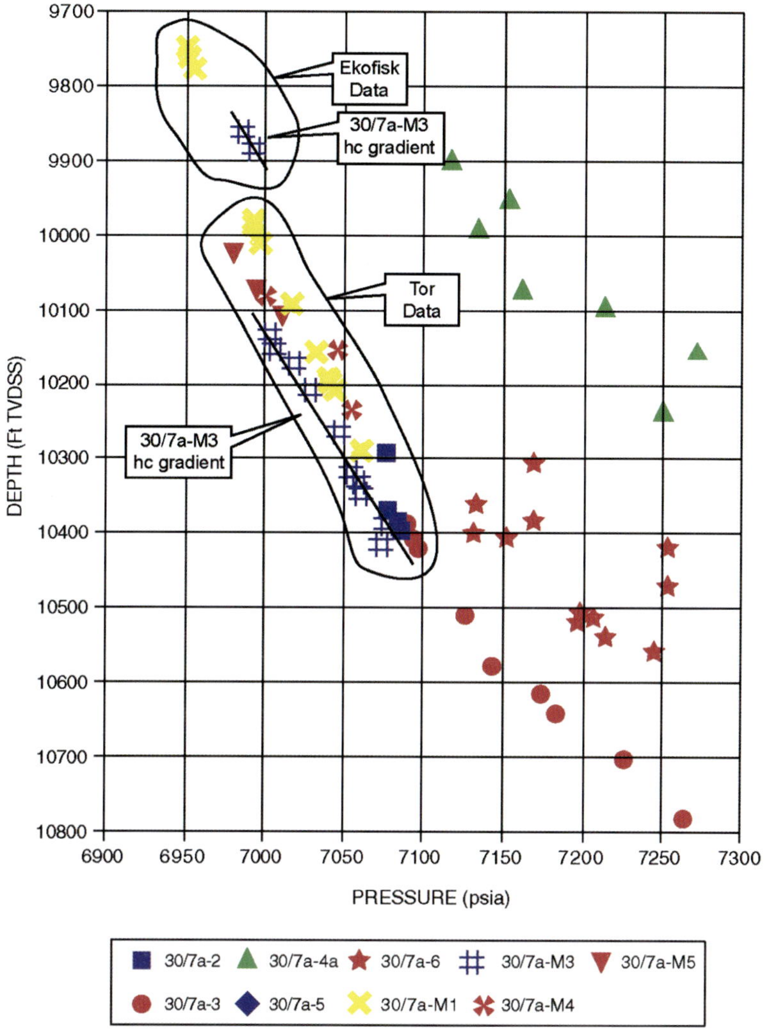


**Fig. 16.** RFT pressure versus depth for Chalk reservoirs.

be minimized over geological time through mixing processes such as diffusion and small-scale fluid flow, unless homogenization is prevented, e.g. by faults or other permeability barriers. Within the petroleum column, the data will record the isotopic composition of the formation water at the time the reservoir was charged. In the aquifer, they will record the present-day strontium isotopic signature. The SrRSA technique therefore provides an excellent means of examining the filling history of a field and identifying whether or not intra-well or inter-well barriers exist which may affect petroleum production.

Core material from six wells has been analysed using the SrRSA technique (30/7a-2, 30/7a-3, 30/7a-4a, 30/7a-5, 30/7a-6 and 30/7a-M1). The $^{87}Sr/^{86}Sr$ data are shown in Figure 17; oil–water contacts are also shown where they occur within the cored intervals.

Data for the 30/7a-4a well, drilled on the crest of the Judy structure, show an increase in isotope ratio with depth, probably reflecting the evolution of the formation water through time as the reservoir filled. All of the other wells show virtually constant $^{87}Sr/^{86}Sr$ ratios with depth, although the actual value varies from well to well (Fig. 17). There is, surprisingly, no change across the oil–water contact in either the Ekofisk or Tor reservoirs. The constant $^{87}Sr/^{86}Sr$ ratios within individual wells are somewhat unusual. Smalley *et al.* (1992) attributed constant

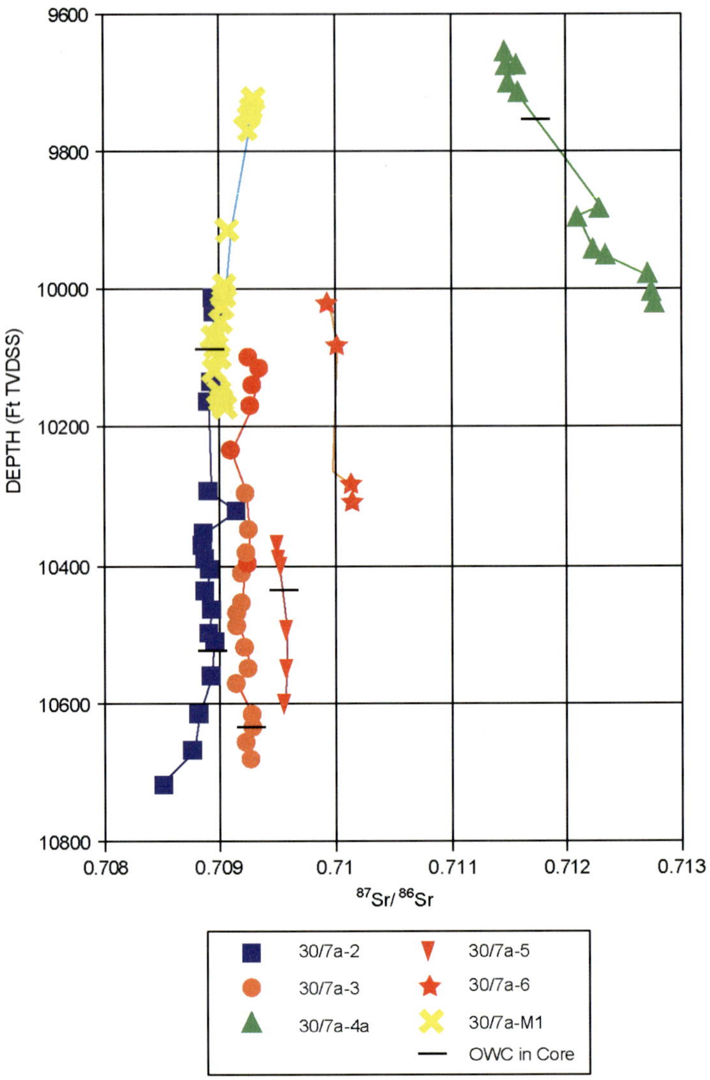


**Fig. 17.** Strontium isotope ratio versus depth for Chalk core extracts.

$^{87}Sr/^{86}Sr$ ratios within the Ekofisk Field as indicating that the strontium in the pore waters was derived from the Chalk itself as they closely matched values from primary calcite and contemporary seawater. The $^{87}Sr/^{86}Sr$ ratios from the Joanne Field vary from *ca.* 0.7088 to 0.7096 and are significantly higher than the value of *ca.* 0.7079 for Late Cretaceous seawater (no primary calcite was analysed for comparison). This suggests the strontium in the Joanne Field pore waters is unlikely to have been derived predominantly from the Chalk.

The constant $^{87}Sr/^{86}Sr$ ratios within individual Joanne wells can be explained in one of two ways—very rapid and very recent filling, or contamination from drilling mud filtrate. With the exception of the 30/7a-2 well, all the wells were drilled with oil-based muds, albeit of different types. The Joanne wells 30/7a-2, 30/7a-3, 30/7a-5 and 30/7a-M1 have $^{87}Sr/^{86}Sr$ ratios very close to that for present-day seawater (0.709185). However, none of these wells (30/7a-M1 excepted) showed any evidence of oil-based mud contamination during Iatroscan and whole oil GC analysis and it is highly unlikely that the water emulsified within the oil-based mud would preferentially invade the formation. The available evidence from the strontium isotope data would therefore tend to suggest recent and rapid filling. The timing of filling of the Joanne structure and the likely effect on $^{87}Sr/^{86}Sr$ ratios is included in the charge model summary below.

### *Petroleum migration*

Migration into the Chalk reservoirs is heavily influenced by the Pre-Cretaceous charge model in terms of explaining the variability in fluid type.

Although petroleum was tested from the Jurassic and Triassic reservoirs in 30/7a-4a, no samples were available for analysis to provide a direct comparison with the Chalk oils. However, the close similarities between the Chalk oil from 30/7a-4a and the Pre-Cretaceous fluids from 30/7a-P3 and 30/7a-P9 (Figs 7 & 8) suggests that leakage of petroleum from the Pre-Cretaceous has occurred over the crest of the Judy structure. Leakage of Pre-Cretaceous petroleums close to their bubble points to lower pressure and temperature regimes in the Chalk would have resulted in exsolution of a separate gas phase and produced the two-phase accumulation seen today. This capillary leakage has probably occurred only from the crestal area of the Judy structure to the north of the major NW–SE trending fault seen on the Top Triassic reservoir structure map (Fig. 3), as the Chalk oils have high wax contents and maturities similar to those in the Pre-Cretaceous oils from 30/7a-P3 and 30/7a-P9 and are quite different from the Pre-Cretaceous oils from 30/7a-8 and 30/7a-P7 (Fig. 8).

It is likely that the Tor-reservoired petroleum in 30/7a-6 migrated within the Tor from the 30/7a-4a area to the northeast. The 30/7a-6 well tested a relatively low GOR oil (1224 scf/bbl) in the Pre-Cretaceous and it is therefore highly improbable that the high GOR oil in the Tor (4322 scf/bbl) would be the result of direct leakage from the underlying reservoir (note that the recorded GOR for the Tor oil is probably an overestimate, possibly as a result of the low flow rates or substantial drawdown during testing).

Migration into the Chalk reservoirs of the Joanne Field could have occurred by one of three conceivable mechanisms:

- direct leakage from the underlying Pre-Cretaceous reservoirs;
- charging from the major source kitchen to the west;
- spillage from the Judy structure to the east.

The applicability of each of these models is addressed below.

Only one of the Joanne wells (30/7a-3) has tested petroleum from the Pre-Cretaceous interval—an oil with a GOR of 1691 scf/bbl. Unfortunately, no analyses were undertaken on this oil due to a lack of sample material. The Chalk oils from wells 30/7a-1, 30/7a-2 and 30/7a-3, in comparison, have uniform GORs ranging from 2131 to 2600 scf/bbl (Table 2). As discussed for the Judy structure, it is highly improbable that leakage of an undersaturated oil from the Pre-Cretaceous would result in a higher GOR fluid in the Chalk.

The oils in wells 30/7a-8 and 30/7a-P7 also provide evidence that direct leakage into the Chalk from the Pre-Cretaceous in Joanne is an unlikely model. The 30/7a-8 and 30/7a-P7 Pre-Cretaceous petroleums have low wax contents and are believed to have been sourced from the kitchen to the west or southwest. It is therefore likely that the Pre-Cretaceous oil in 30/7a-3 would also have a low wax content. Although the wax contents of the Joanne Chalk oils are lower than those from 30/7a-4a and 30/7a-6, they are considerably higher than those in the 30/7a-8 and 30/7a-P7 petroleums (Fig. 8), making it unlikely that they leaked from the underlying Pre-Cretaceous reservoirs.

Petroleum generation from the deep kitchen to the west of Joanne could have resulted in migration vertically upwards into the Chalk reservoirs and hence into the Joanne structure. This kitchen would also have generated the petroleum reservoired in the Ekofisk in 30/6-3. Although no analyses were performed for this study, data available in-house show that the 30/6-3 fluid is highly mature and has a low wax content. GC-MS data confirm the high maturity of the petroleum. In the saturates fraction, the extended triterpanes are virtually absent and Ts, $C_{29}$Ts and $C_{30}$ diahopane are abundant, whilst in the aromatic fraction the triaromatic steranes have been totally cracked. The marked differences between the 30/6-3 petroleum and the Joanne Chalk oils suggest this migration mechanism is not effective.

Migration of petroleum within the Chalk reservoirs from its point of entry on the Judy structure is considered to be the most likely mechanism by which the Joanne Chalk reservoirs are charged. At present, there is no structural spill from Judy to Joanne at either Ekofisk or Tor horizons. However, the Top Ekofisk depth map, flattened on an intra-Oligocene event, shows that the Joanne structure developed considerably later than the Judy structure. It is therefore possible that, during growth of the

Joanne structure, petroleum would have been spilled westwards from the Judy accumulation.

The 'spill' model is consistent with the higher maturities and wax contents seen in the Judy Chalk oils. In general, petroleums become increasingly mature towards the direction from which they were filled, as the region closest to the kitchen receives the most recent and hence the most mature charge (England *et al.* 1987). Leakage of petroleum into the Chalk at Judy followed by migration westwards to Joanne is therefore consistent with both the higher maturities of the Judy Chalk oils compared with the Joanne Chalk oils and the broadly comparable maturities of the Judy Chalk oils and the crestal Pre-Cretaceous petroleums in Judy (minimal migration distance).

It might also be expected during lateral migration within the Chalk that the longer chain *n*-alkanes and other less mobile constituents would be retarded. This 'geochromatographic' effect explains the slightly lower wax contents in the Joanne Chalk oils.

The spill model is also consistent with the strontium isotope data. These suggest that the Joanne Chalk accumulations were filled rapidly. Any spillage may therefore be due to sudden tectonic activity, such as occurred during Oligo-Miocene times, rather than by passive filling and then spilling from the Judy structure. Although the $^{87}Sr/^{86}Sr$ data for the 30/7a-6 well are sparse, rapid filling is also implied by the near vertical trend (Fig. 17). The charging of the 30/7a-6 area also requires spillage from the crestal 30/7a-4a area.

## Palaeocene charge model

### *Introduction*

The Palaeocene petroleums constitute the least mature fluids analysed from Block 30/7a (Fig. 7). In total, 16 petroleum samples were analysed. With the exception of samples from 30/7a-5 and 30/7a-10 (Julia), which are black oils, all of the petroleum samples analysed are condensates. Well locations are shown on Figure 18.

### *Petroleum characterization*

*Physical properties.* With one exception, the petroleums tested from the Andrew Sand in the Joanne Field are gas-condensates. However, even within the condensates, there is considerable variation in the fluid properties (Table 3). Separator condensate-gas ratios (CGRs) vary from 175 bbl/MMscf in 30/7a-4a to 315 bbl/MMscf for the 30/7a-P4 fluid. Well 30/7a-5 tested a black oil with a GOR of 2068 scf/bbl (CGR = 484 bbl/MMscf) and, whilst the flow rates were very low, the test is considered representative of the reservoir fluid (extensive drawdown during testing would only produce a *higher* GOR than that of the actual reservoir fluid).

The other Palaeocene petroleums analysed for the study were a black oil from 30/7a-10 (Julia) and a condensate from 30/8-2 (Peach).

*Geochemical characterization.* The Palaeocene condensates of the Judy/Joanne Field exhibit relatively low maturities and have low wax contents compared with either the Chalk oils or Pre-Cretaceous condensates (Figs 7 & 8). The exception to this is the Julia oil from 30/7a-10, which has a high maturity and moderately high wax content (13.6%). It has a similar maturity and wax content to the Pre-Cretaceous petroleums and is likely to have been sourced by leakage from an underlying Pre-Cretaceous accumulation.

The 30/7a-P13z condensate appears to be one of the least mature samples analysed together with the 30/7a-M2z and the 30/8-2 (Peach) condensates; it also has the lowest wax content of all the J Block petroleums.

In general, samples from the Judy side of the structure (30/7a-6, 30/7a-7, 30/7a-P6z) appear to exhibit higher maturities and have higher wax contents than those from the Joanne side (Figs 7 & 8).

### *Petroleum migration*

The low maturity of the Palaeocene condensates and their relatively low wax contents indicate that they did not evolve in the same way as the Pre-Cretaceous petroleums, i.e. originally an oil charge that became increasingly mature and gas-rich due to a late stage, highly mature gas charge. The Palaeocene condensates are 'phase separated' rather than 'thermal' condensates, i.e. they are the product of physical segregation of a gas phase from a two-phase oil and gas accumulation.

It is highly probable that petroleum has migrated into the Palaeocene reservoir by leakage from the Judy and/or Joanne Chalk reservoirs. Evidence for this is as follows.

- Wells on both the eastern and western flanks of the structure are water-wet at depths above the free water level of 9815 ft subsea as

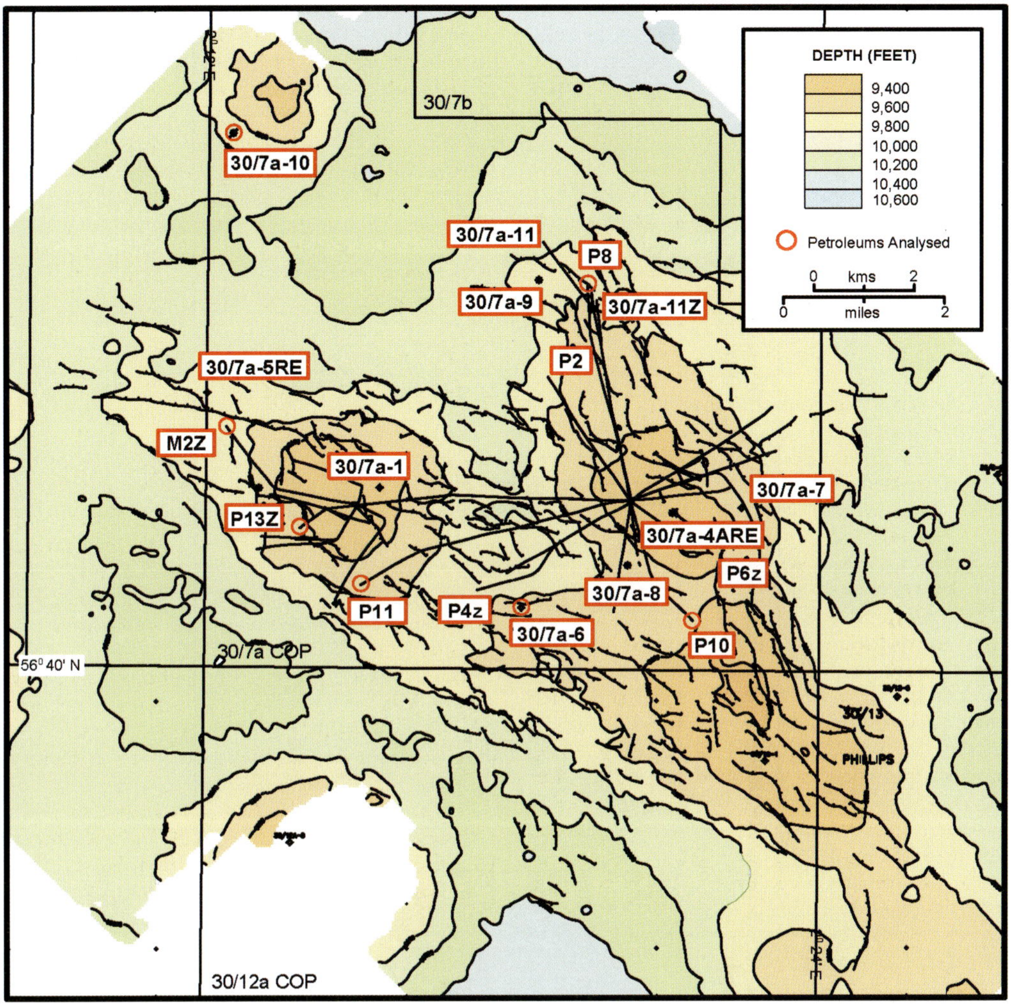


**Fig. 18.** Location map showing Palaeocene petroleums analysed.

interpreted from the 30/7a-11 well. This is thought to be due to a series of faults along the eastern and western margins of the field. The lack of fluids on the eastern and western margins of the field is also interpreted from an AVO study. Lateral migration within the Palaeocene reservoir from outside the confines of the field, either from an easterly or westerly direction, is therefore highly unlikely.

- Migration within the Palaeocene from the northwest is highly unlikely due to the marked difference in petroleum characteristics between the Joanne condensates and the Julia oil in 30/7a-10.
- Migration from the northeast is also unlikely. The Peach (30/8-2) fluid is one of the least mature of all the Palaeocene condensates analysed. Petroleums tend to show an increase in maturity towards the direction from which the field filled and, whilst the Andrew Sand was not penetrated in 30/8-2 (the well reached TD within the Forties Sandstone Member), it is likely that any petroleum within the Andrew reservoir would be similar to that in the Forties.
- The Joanne condensates are low maturity fluids and are the products of phase separation.
- There is a gas cap to the Ekofisk Judy Chalk accumulation in 30/7a-4a, which may represent part of a larger gas cap that has not leaked—alternatively, it might have resulted from additional charging after leakage ceased.

**Table 3.** *Palaeocene test data and fluid properties*

| Well | DST | Depth interval (ft ss) | | Gas rate (MMscfd) | Oil rate (bopd) | Separator GOR | Reservoir P | Reservoir T (F) | Depth (ft ss) | Oil gravity (API) | Gas gravity (SG) | Dew point P (psia) |
|---|---|---|---|---|---|---|---|---|---|---|---|---|
| | | From | To | | | (scf/bbl) | (psia) | | | | | |
| 30/7a-1 | 9 | 9323 | 9496 | 16.0 | 3029 | 5282 | 6650 | | 9410 | 51.0 | 0.72 | 5945 |
| 30/7a-2 | 6 | 9546 | 9578 | 13.9 | 3711 | 3759 | 6723 | | 9562 | 46.9 | 0.74 | 5910 |
| 30/7a-4a | 9 | 9227 | 9300 | 17.4 | 3050 | 5705 | 6640 | | 9264 | 47.6 | 0.73 | 5930 |
| 30/7a-5 | 3 | 9714 | 9728 | 0.6 | 301 | 2068 | | | | | | |
| 30/7a-6 | 3 | 9534 | 9559 | 11.3 | 3224 | 3490 | 6731 | | 9547 | | | 5920 |
| 30/7a-7 | 2 | 9403 | 9442 | 18.6 | 4584 | 4058 | 6639 | | 9423 | | | 6190 |
| 30/7a-8 | 5 | 9494 | 9518 | 12.4 | 2881 | 4317 | 6731 | | 9506 | 46.2 | 0.77 | 6400 |
| 30/7a-9 | 1 | 9720 | 9741 | 14.5 | 3023 | 4782 | 6769 | | 9731 | 47.6 | | 6000 |
| 30/7a-10 | 1 | 9667 | 9697 | 6.5 | 6032 | 1078 | 6751 | | 9682 | 41.0 | 0.77 | |
| 30/7a-M2z | 1 | 9516 | 9653 | 18.4 | 5509 | 3340 | | | | 45.4 | 0.81 | |
| 30/7a-P2z | 1 | 9608 | 9712 | 28.9 | 7432 | 3889 | | | | 41.0 | 0.81 | |
| 30/7a-P4 | 1 | 9656 | 9685 | 12.8 | 4032 | 3175 | | | | 44.9 | 0.77 | |
| 30/7a-P4z | | 9603 | 9675 | 23.0 | 6831 | 3371 | | | | | | |
| 30/7a-P6z | 1 | 9395 | 9456 | 27.0 | 6585 | 4100 | 6745 | 269 | 9450 | 49.5 | 0.70 | 5890 |
| 30/7a-P8 | 1 | 9549 | 9637 | 23.5 | 6585 | 3569 | | | | 48.1 | 0.73 | |
| 30/7a-P10 | 1 | 9460 | 9502 | 37.5 | 8393 | 4468 | 6740 | 270 | 9450 | 47.2 | 0.71 | 5360 |
| 30/7a-P11 | | | | 16.7 | 4688 | 3558 | | | | 48.5 | 0.87 | |
| 30/7a-P13z | | 9271 | 9418 | 25.3 | 5963 | 4241 | | | | 52.2 | 0.76 | |
| 30/8-2 | 1 | 9958 | 10002 | 1.2 | 117 | 10197 | | | | | | |

- Separation of the gas cap would result in a saturated oil remaining within the Chalk reservoir and a saturated gas migrating vertically upwards. Post-accumulation burial would have increased reservoir pressures so that the Chalk oil accumulation is no longer saturated with gas. Migration of the saturated (dew point) gas to shallower depths, and hence lower reservoir temperatures and pressures within the Palaeocene, would result in condensation of liquids along the migration pathway. If these migrated along with the gas, a small oil rim would accumulate beneath the gas accumulation in the Palaeocene (as in 30/7a-5—possibly).
- The Palaeocene condensates are unlikely to be the result of petroleum leakage directly from the Pre-Cretaceous reservoirs. The pre-Cretaceous petroleums are waxy, highly mature thermal condensates, whereas the Palaeocene fluids are the least mature petroleums reservoired in J Block.
- The closer similarities in fluid properties between the low maturity Chalk oils (i.e. those in the Joanne accumulation), and the Palaeocene condensates, suggests that leakage from the Chalk has also occurred over the Joanne structure. Leakage may have occurred during fracturing of the seal as a result of further salt movement or may be due to capillary failure of the Maureen Marl due to the petroleum column within the Chalk.

## Charge model summary

The overall charge model may be summarized as follows. Petroleum is expelled downwards into Upper Jurassic and Triassic reservoirs from source kitchens to both the northeast and southwest of the Judy structure. Petroleum migrates within the reservoir towards the crest of the structure. Leakage into the Chalk occurs at the crest of the Judy structure in the vicinity of the 30/7a-4a well. Lateral migration of petroleums within the Chalk results in spillage into the Joanne structure to the west. Further leakage of petroleum into the Palaeocene then occurs within the confines of the Judy and Joanne structures. Figure 19 shows a schematic cross section of the overall charge model.

Conclusions relating to the various reservoir units are included below.

### *Pre-Cretaceous*

- The Pre-Cretaceous petroleums are the most mature of all those analysed in the study. The Jocelyn condensate from 30/7a-P12 is the most mature, followed by the 30/7a-11z and 30/7a-P5z condensates from the north of the field.
- Two distinct groups of petroleums can be recognized, primarily on the basis of their wax contents. Samples from wells 30/7a-8

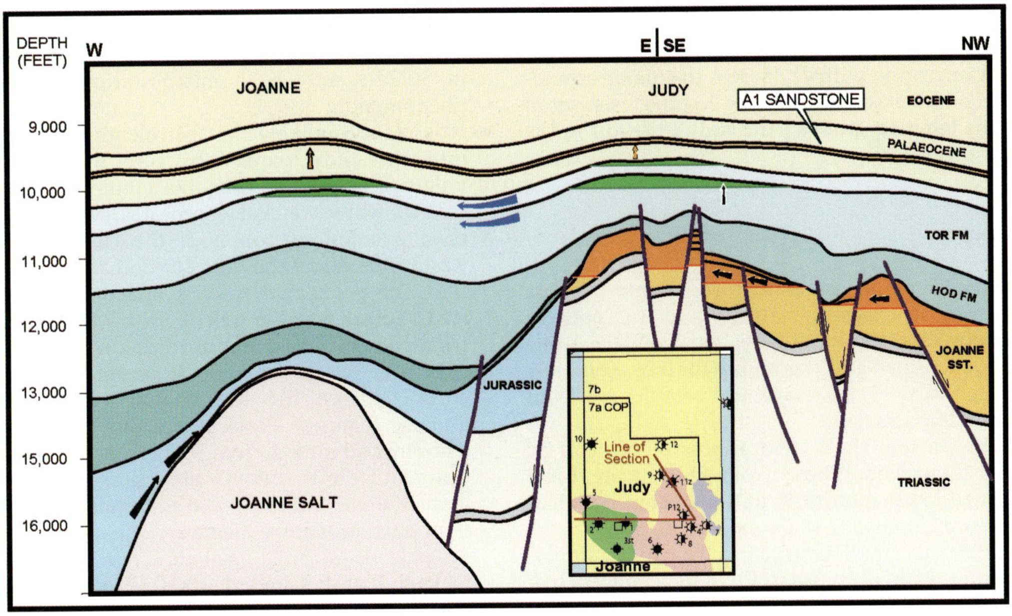


**Fig. 19.** Schematic cross-section showing summarized charge model.

and 30/7a-P7, which lie to the south of a major NW–SE trending fault, have low wax contents, whereas samples from wells to the north of the fault have high wax contents. A third group, comprising the 30/7a-7 and 30/7a-P12 petroleums, can be distinguished on the basis of their heavy carbon isotope signatures.

- The volatile oils encountered in 30/7a-8, 30/7a-P1 (*in-situ* Joanne) and 30/7a-P7 were probably sourced from the deep kitchen to the west and southwest, possibly via leakage around the Jane diapir. RFT data for these wells show a progressive shallowing of the oil–water contact to the northeast across faults, which act as semi-permeable membranes.
- The black oils tested from 30/7a-3 and 30/7a-6 appear to have been isolated from the deep source kitchen to the west by the major bounding fault to the Joanne structure. It is likely that these oils were sourced from locally mature Upper Jurassic source rocks immediately overlying the reservoir.
- The condensates and oils from the northern and crestal parts of the Judy Field were probably sourced from the deep kitchen to the northeast. The petroleums become progressively more mature and have higher GORs towards the north, i.e. the direction from which the field filled. Petroleum–water contacts show a stepwise deepening in this direction.
- The Jocelyn condensate was probably sourced from the deep kitchen to the east. On the basis of the carbon isotope data, the same kitchen probably sourced the condensate in 30/7a-7. A migration pathway across the major bounding fault exists between the Joanne reservoir in the hanging wall and the Judy reservoir in the footwall.

### *Chalk*

- The Chalk oils generally form an intermediate group in terms of maturity and wax content between the low maturity, low wax Palaeocene condensates and the highly mature, generally high wax Pre-Cretaceous volatile oils and condensates.
- Within the Chalk, two groups of oils can be distinguished. Those from the Joanne Field have lower maturities and wax contents than those from wells 30/7a-4a and 30/7a-6.
- The Judy Chalk petroleums have many similarities with the crestal Pre-Cretaceous petroleums from 30/7a-P3 and 30/7a-P9. It is probable that the Judy Chalk reservoirs were charged by direct leakage from the underlying Pre-Cretaceous.
- The Joanne oils are likely to have been spilled from the Judy structure. This has resulted in a decrease in wax content and maturity away from the point of filling. The Joanne oils are unlikely to have been sourced from the southwest as they have higher wax contents than the Pre-Cretaceous oils in 30/7a-8 and 30/7a-P7.
- Strontium isotope data show significant differences between the 30/7a-4a well and the others. The 30/7a-4a data show a steady evolution in $^{87}Sr/^{86}Sr$ ratio with depth — consistent with slow filling of the reservoir. The 30/7a-6 well and Joanne wells have almost vertical profiles, which extend beyond the oil–water contact, implying recent and rapid filling.

### *Palaeocene*

- The Palaeocene petroleums from the Judy/Joanne Field are the least mature and have the lowest wax contents of any fluid samples analysed during this study. They are 'phase separated' condensates resulting from physical segregation of a gas phase from a two-phase oil and gas accumulation.
- It is probable that the Palaeocene reservoir was charged by vertical leakage of petroleum from the underlying Chalk reservoirs. Migration of a saturated (dew point) gas to shallower depths, and hence lower reservoir temperatures and pressures, would have resulted in condensation of liquids. The oil encountered in 30/7a-5 may be a manifestation of this phenomenon.
- It is highly unlikely that petroleum migrated into the Judy/Joanne structure within the Palaeocene from outside the confines of the field as water-wet Palaeocene reservoirs have been encountered on both the eastern and western flanks of the field at depths shallower than the petroleum–water contact of 9815 ft (tvdss) established in well 30/7a-11. Migration from the northwest and northeast can also be ruled out on the basis of geochemical differences between the Judy/Joanne petroleums and the Julia and Peach petroleums.
- The Julia oil sample from 30/7a-10 is markedly different from the Judy/Joanne samples; it has a much higher maturity and wax content and a light carbon isotope signature. The oil in the Palaeocene in Julia probably resulted from leakage from an underlying Triassic accumulation, possibly associated with hydraulic fracturing.

- The Peach petroleum is a very low maturity 'phase separated' condensate and has many similarities with the low maturity Joanne condensates. It was probably charged, either by leakage from an underlying Chalk accumulation or by spillage from another Palaeocene accumulation.

## Application to exploration and development

### *Introduction*

From 1999 to 2001, a re-evaluation of remaining untapped exploration potential within the Judy/Joanne Fields was undertaken. This led to a limited well intervention campaign and the drilling of four additional exploration and appraisal wells (30/7a-12, 30/7a-P16, 30/7a-P17 and 30/7a-P18), targeted at the Pre-Cretaceous reservoirs, and a further Chalk development well (30/7a-M8).

This section documents the impact of the charge model in assessing charge risk and predicting fluid contacts (for improved HCIP estimation and well-path design) for the exploration and appraisal wells. The application of the model to production allocation is also discussed.

### *Petroleum distribution in undrilled areas*

*Northeast area.* The Jill Prospect was drilled in 1999 on the northern margin of the Judy Field. It was predicted to contain a high GOR gas-condensate and have a deeper contact than that seen in the fault block immediately to the south, penetrated by the 30/7a-11z well.

The well successfully tested gas-condensate with a GOR of 11 000 scf/bbl, the highest in the field, and encountered a hydrocarbon–water contact at 12 440 ft (tvdss), some 682 ft below that in the 30/7a-11z well.

*Crestal area.* A number of individual fault blocks, between the Judy platform and the 30/7a-P5z well, remained untested after the initial development drilling. This area is dissected by faults into several compartments, which, on the charge model developed, were likely to be in pressure isolation, and therefore undepleted by production (Fig. 14). The model suggested that these fault blocks would have been charged either by southwards migration of petroleum from the 30/7a-11z area or from the Jill terrace to the northeast. In each case, migration across one or more faults is required. Migration of petroleum is therefore likely to be controlled by the 'seal capacity' of each fault as discussed earlier in the paper, with progressively shallower hydrocarbon–water contacts towards the highest area of the Pre-Cretaceous structure.

Crestal blocks on the structure that have been drilled, i.e. the 30/7a-4a and the 30/7a-P3 and 30/7a-P9 blocks, contain volatile oils at or very close to their bubble points (30/7a-P3, 30/7a-P9) or two phase accumulations as in 30/7a-4a. It was thought that the undrilled blocks to the north might contain two phase accumulations as they are slightly more proximal to the direction of filling and may have received a greater charge of late stage, high maturity gas, or simply have higher GOR volatile oils.

For the 30/7a-P16 fault compartment, an initial sub-vertical well was drilled to define the hydrocarbon–water contact, predicted to occur between 11 444 and 11 758 ft (tvdss). This would then be sidetracked with a sub-horizontal well, about 100 ft above the contact. The pilot hole encountered an undepleted accumulation with a contact close to 11 444 ft (tvdss) and showed the crestal part of the structure had a common contact, but that the bounding faults between the compartments were sealing, at least initially, on a production timescale.

The 30/7a-P17 well was drilled in a separate fault block to the west of P16. It was assumed that the Joanne reservoir would be at virgin pressure. Because a common contact in the crestal area had been demonstrated, a decision was made not to drill a pilot hole. A sub-horizontal well was drilled, with a 100 ft stand-off from the 11 444 ft (tvdss) contact proven in the P16 compartment.

The 30/7a-P18 well was drilled at the most crestal part of the Pre-Cretaceous structure. Questions remained as to whether the magnitude of the faults bounding this compartment precluded an effective charge. The charge risk was significantly reduced by the geochemical evidence, which showed the 30/7a-P1 olisthostrome fluid to be similar to the those from the north of the bounding fault, rather than the 30/7a-8 and 30/7a-P7 fluids to the south (see earlier discussion). The P1 olisthostrome could only have been charged via migration through the P18 fault block. Once again, a crestal contact of 11 444 ft (tvdss) was assumed, and a high angle well was designed to stay 100 ft above the contact. 30/7a-P18 encountered a 'hydrocarbon-down-to' at 11 397 ft (tvdss) in the Judy Sandstone, supporting evidence of a common contact in the crestal part of the field.

*Fluid characterization.* High resolution GC fingerprinting was undertaken to correlate and further characterize the reservoir fluids, both from the original exploration and more recent development wells. High resolution GC fingerprinting (GCF) is a technique widely used in reservoir geochemistry to aid understanding of compartmentalization and reservoir connectivity issues (see Kaufman *et al.* 1990). The approach taken for this study has been to use the $C_8$–$C_{13}$ region on the GC and to use ratios between sequentially eluting peaks only. Reasons for this are as follows. Firstly, the $C_8$–$C_{13}$ region contains a sufficient number of well-resolved peaks, but not too many for co-elution problems to occur. Secondly, by only using sequentially eluting peaks to determine ratios, the impact of phase separation can be minimized. To reduce

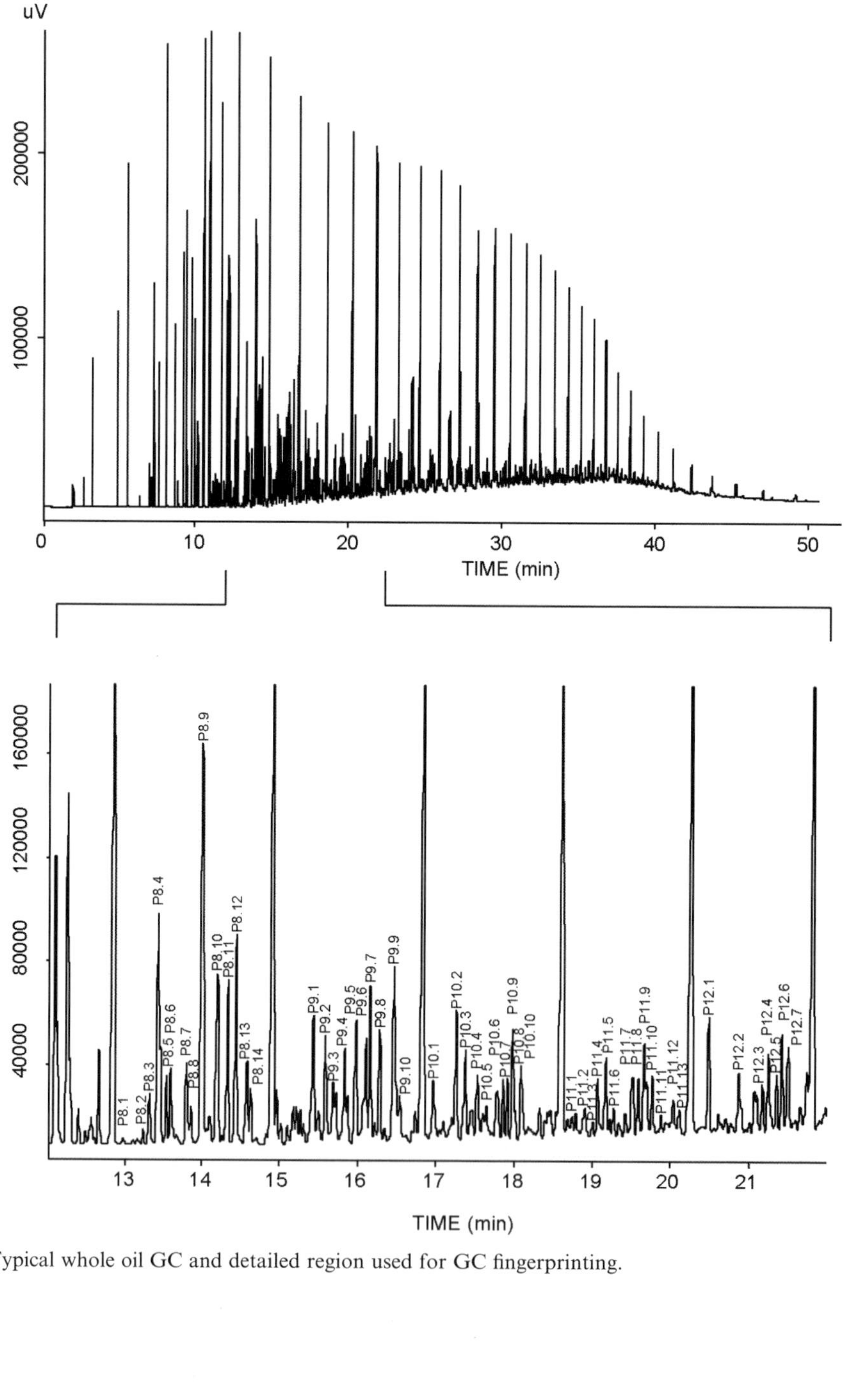


**Fig. 20.** Typical whole oil GC and detailed region used for GC fingerprinting.

any effects of instrument variability, all samples were run in a single batch, together with replicates from 30/7a-11z to define analytical precision. Figure 20 shows a typical whole oil GC, together with the detailed $C_8$–$C_{13}$ fingerprint showing the peaks used in calculating ratios. For example, R8.2 is the ratio of the peaks 8.2 and 8.3 where Peaks 8.2 and 8.3 are the second and third recognizable peaks eluting after $n$-$C_8$ respectively.

Ratios are often shown on 'star' or 'polar' plots, but can be confusing when large numbers of samples are involved. Alternatively, hierarchical cluster analysis (HCA) can be used to group oils into families on the basis of their similarity to one another. The measure of similarity is based on the distance between samples, which is defined using the relationship shown below.

$$d_{ab} = \int_i^m \left[\sum (x_{aj} - x_{bj})^M\right]^{1/M}.$$

For this dataset, Euclidean distances were used, where $M = 2$ in the expression above. The similarity between samples is then defined as follows.

$$\text{similarity}_{ab} = 1 - \left(\frac{d_{ab}}{d_{\max}}\right).$$

Figure 21 shows the resulting dendrogram from the cluster analysis. Replicate samples from well 30/7a-11z allow definition of the limit of subdivision that can be placed on the dataset. Oil families have been defined and colour-coded on the figure; they are also shown on the Judy Field map (Fig. 22). Key features of the plot are as follows. Fluids from wells at the northern end of the field (30/7a-11z and 30/7a-P5z) form a close-knit group; the 30/7a-12 (Jill) fluid can also be considered as part of this group. The low wax fluids from wells to the south of the NW–SE trending crestal fault (30/7a-8 and 30/7a-P7) form a separate group. Fluids from the crestal wells on the northern side of the fault (30/7a-P3, -P9, -P16, -P17, -P18) can all be considered to be part of one large group, although they could be split into two separate groups comprising 30/7a-P1, -P9 and -P18 in one and 30/7a-P3, -P16 and -P17 in the other (Fig. 23).

The sub-group comprising fluids from 30/7a-P1, -P9 and -P18 is very interesting for two reasons. Firstly, the GC fingerprinting confirms the early evidence that the P1 olisthostrome fluid is geochemically similar to the high wax fluids from the north of the NW–SE-trending crestal fault, rather than the 30/7a-8 and -P7 fluids from the area to the south. Secondly, the 30/7a-P1 olisthostrome could only be charged by migration from the 30/7a-P9 compartment through the crestal -P18 fault compartment, which led to a reduced charge risk for the -P18 compartment (see above). The GCF analyses confirm that 30/7a-P1 (olisthostrome), -P9 and

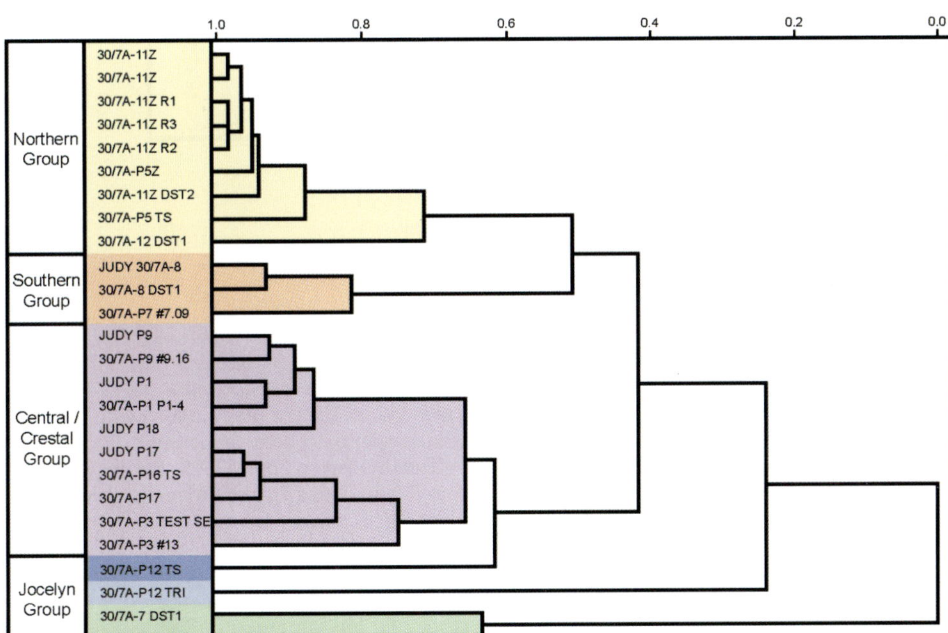


**Fig. 21.** Dendrogram from hierarchical cluster analysis of Judy Field petroleums.

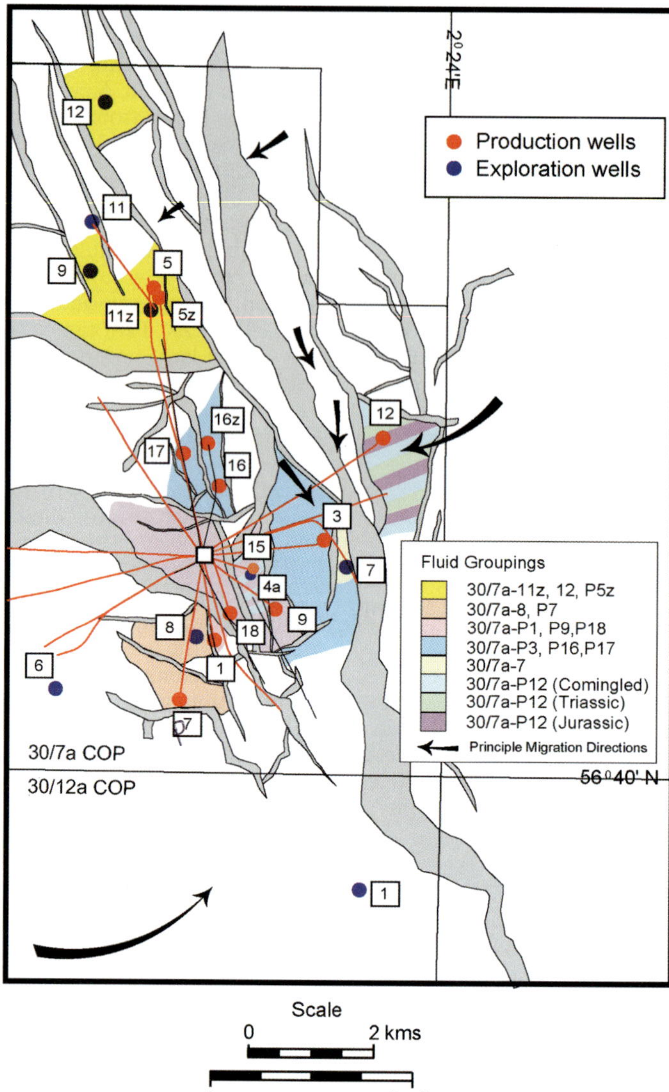


**Fig. 22.** Map showing oil families defined by cluster analysis.

-P18 fluids are all very similar and validates the charge model.

The very close similarity between the 30/7a-P16 and -P17 samples suggests that the fault separating the two compartments may not be sealing, and indeed there is evidence from the drilling operations (lost circulation) that the -P17 block may have been partly depleted due to production from the -P16 well. The 30/7a-P3, -P16 and -P17 fault compartments are separated from the '30/7a-11z oil family' to the north by a NE–SW trending fault. Thus, the geochemistry data, and the available well information (no depletion in 30/7a-P16), suggest that this fault is sealing. The differences seen between the 30/7a-P3 and -P9 fluids suggest there may be a sealing fault within what appears to be a single fault block.

The fluid from 30/7a-7 and the Jocelyn Jurassic sample are defined as belonging to the same group although they are somewhat different. The other two Jocelyn samples form separate groups. Isotopically the Jocelyn and 30/7a-7 fluids were quite distinct from the main Judy Field samples (Fig. 11), having greater affinities with the 30/13 fluids from the Jacqui Field to the southeast. Once again, the GCF data have validated the

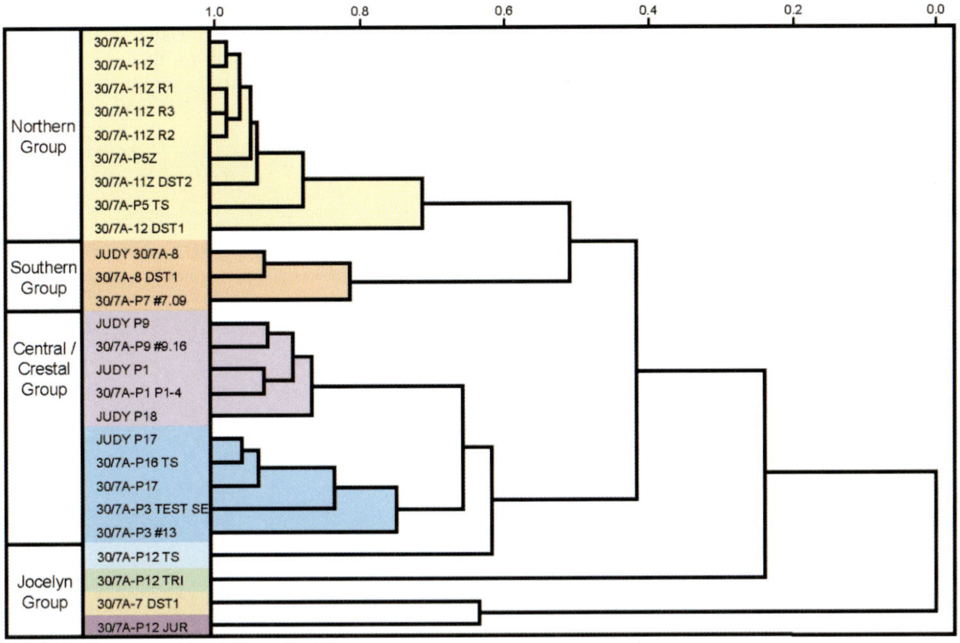


**Fig. 23.** Dendrogram showing alternative clustering of oil families.

early charge model that the Jocelyn and 30/7a-7 fluids may have been derived from a separate source kitchen to the east.

## Conclusions

A petroleum charge model has been developed for the Judy and Joanne Fields that has allowed a much better understanding of the fluid distribution within the three main reservoir units.

The charge model has been used in formulating a strategy for infill drilling to develop additional reserves within the Judy Field. Petroleum–water contacts and fluid types are in line with pre-drilling predictions.

Subsequent reservoir geochemical studies using high resolution GC fingerprinting have confirmed the validity and usefulness of the charge model. More importantly, it has allowed a better understanding of compartmentalization within the field and which faults may or not prove to be barriers to fluid flow on a production timescale.

The authors would like to thank ConocoPhillips (UK) Limited and its partners—Agip (UK) Limited, BG Exploration & Production Limited—for permission to publish this paper. Organic geochemical analyses were provided by LGC Limited and IAS undertook the strontium isotope analyses. Numerous ConocoPhillips colleagues, both past and present are thanked, for many fruitful discussions. Ed Clarke (BG) and Peter Winefield (Shell) are thanked for helpful reviews. The authors would like to point out that the views expressed herein do not necessarily represent those of ConocoPhillips (UK) Limited or its partners.

## References

Beach, F., Peakman, T. M., Abbott, G. D., Sleeman, R. & Maxwell, J. R. 1989. Laboratory thermal alteration of triaromatic steroid hydrocarbons. *Organic Geochemistry*, **14**, 109–111.

England, W. A., Mackenzie, A. S., Mann, D. M. & Quigley, T. M. 1987. The movement and entrapment of petroleum in the subsurface. *Journal of the Geological Society of London*, **144**, 327–347.

Kaufman, R. L., Ahmed, A. S. & Elsinger, R. J. 1990. Gas chromatography as a development and production tool for fingerprinting oils from individual reservoirs: Applications in the Gulf of Mexico. *In*: Schumacher, D. & Perkins, B. F. (eds) *Gulf Coast Oils and Gases*. Proceedings of GCSSEMP Foundation Ninth Annual Research Conference, 263–282.

Mearns, E. W. & McBride, J. J. 1999. Hydrocarbon filling history and reservoir continuity of oil fields evaluated using $^{87}Sr/^{86}Sr$ isotope ratio variations in formation water, with examples from the North Sea. *Petroleum Geoscience*, **5**, 17–27.

PETERS, K. E. & MOLDOWAN, J. M. 1993. *The Biomarker Guide: Interpreting Molecular Fossils in Petroleum and Adjacent Sediments*. Prentice-Hall, Englewood Cliffs, NJ, 363.

SMALLEY, P. C., LØNØY, A. & RÅHEIM, A. 1992. Spatial $^{87}Sr/^{86}Sr$ variations in formation water and calcite from the Ekofisk chalk oil field: implications for reservoir connectivity and fluid composition. *Applied Geochemistry*, **7**, 341–350.

# Compositional grading in the oil column: advances from a mass balance and quantitative molecular analysis

DAVID A. WAVREK[1] & FAUSTO MOSCA[2]

[1]*Petroleum Systems International, Inc., Salt Lake City, UT 84102, USA*
*(e-mail: dwavrek@petroleumsystems.com)*
[2]*Enterprise Oil Italiana S.p.A., Rome 00187, Italy*

**Abstract:** The giant oil fields of the Val D'Agri region (Southern Apennines Internal Thrust Zone, Italy) may rank as the largest onshore accumulations in Europe, but these resources pose special technical challenges due to the secondary alteration process identified as compositional grading. This alteration process is attributed to petroleum system elements and processes that lead to the formation of an unstable oil column. These oil columns are studied within a mass balance perspective that allow identification of key molecular fractionations which can be used to properly diagnose this alteration mechanism in other oil accumulations. The molecular signature is defined in the organic sulphur fraction as well as different hydrocarbon classes, and is completely consistent with the engineering criteria for this reservoir process. This discovery is particularly critical to reservoir compartment studies as the failure to recognize this particular signal will invariably lead to the identification of false compartments.

The central tenet of a reservoir compartment study is that homogenization of an oil column is rapid, particularly in a high-quality well-connected reservoir, unless a barrier is present (e.g. Kaufman *et al.* 1990; Larter & Aplin 1995). The importance of evaluating the validity of this central tenet *before* a reservoir continuity study is attempted cannot be over-emphasized, as our experience indicates that the central tenet is often invalid. That is, the specific reservoir or compartment of interest may actually be in a state of disequilibrium due to a variety of natural and/or anthropogenic causes. For this reason, the initial step in the planning and preparation stage of the reservoir compartment workflow (Fig. 1) is to determine the equilibrium potential of the system. This is normally accomplished with a comprehensive synthesis of existing geological, geochemical and engineering data associated with the hydrocarbon resources.

The equilibrium potential analysis provides critical input to the tool selection process, as each of the available tools (Fig. 1) has particular strengths and weaknesses in application to evaluating reservoir compartments. For example, a methodology that has demonstrated success in the North Sea may not be appropriate for application in the Gulf of Mexico due to different genetic reasons for the reservoir compartment signals, as well as that from competing signals (i.e. noise). Failure to recognize and account for the competing signals invariably leads to the identification of false compartments. The purpose of this paper is to highlight a methodology used to recognize and quantify a signal from an oil column affected by the secondary alteration process of compositional grading. Insights provided in this paper are applicable to the process of deconvoluting the reservoir compartment signal from that associated with compositional grading, but that specific topic is addressed in other communications (Wavrek *et al.* 2001; Mosca *et al.* 2003). The strategy applied here is based on a mass balance approach that is complemented by quantitative molecular analysis of the crude oil fractions. This paper represents the first published case where quantitative molecular chemistry and a mass balance approach is integrated with engineering properties to evaluate hydrocarbon resources affected by compositional grading.

## Compositional grading fundamentals

Compositional grading in the oil column has been known for over 50 years (Sage & Lacey 1938), but the topic received little attention until the 1980s when sufficient analytical technologies became available to probe the phenomenon. These studies are usually biased towards engineering issues, although the topic is equally applicable to the geochemical community. A variety of genetic reasons have been attributed to promoting compositional variation with depth (Schulte 1980; Hirschberg 1988; Hoier and Whitson 2000, 2001a,b), but the most important include:

- an association with thick oil columns in reservoirs with high vertical connectivity,
- a reduction in reservoir pressure (plus a minor temperature effect), and

*From*: CUBITT, J. M., ENGLAND, W. A. & LARTER, S. (eds) 2004. *Understanding Petroleum Reservoirs: towards an Integrated Reservoir Engineering and Geochemical Approach*. Geological Society, London, Special Publications, **237**, 207–220. 0305-8719/04/$15.00 

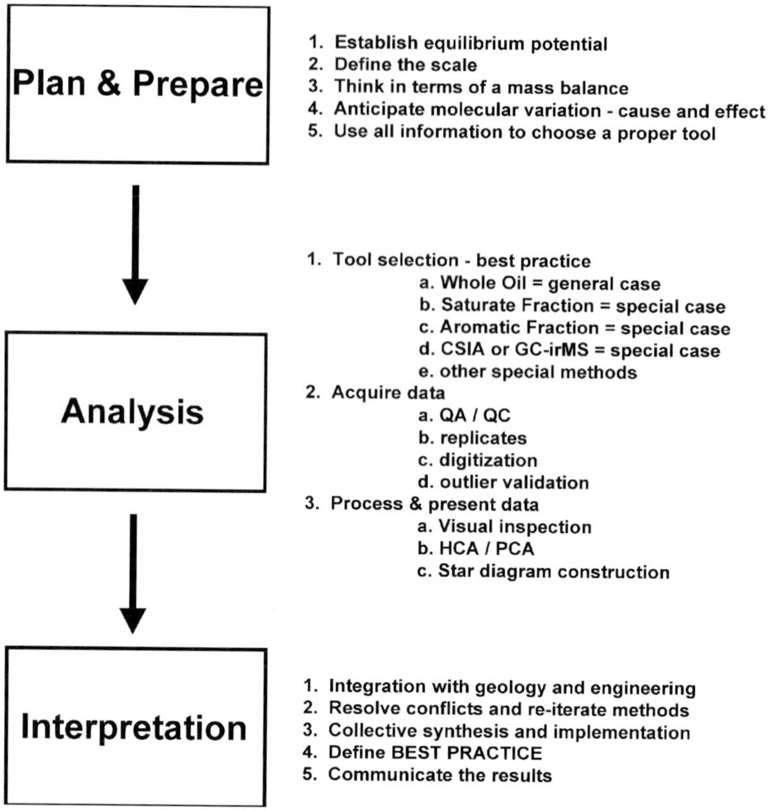


**Fig. 1.** Proposed workflow process for a reservoir compartment study.

- the introduction of volatile-enriched fractions to the liquid phase.

Oil properties that increase with depth include the $C_{7+}$ mole fraction and dewpoint, whereas a decrease with depth is noted for the $C_1$ mole fraction, gas/oil ratio and bubble point. End-member compositional trends include asphaltene precipitation and tar mat formation at the base of the oil column, and concomitant formation of a volatile-rich liquid phase at the top. The evolution of the oil column from a mass balance perspective is accompanied by distinct molecular fractionations that can be used to diagnose the cause for the oil column disequilibrium.

## Petroleum geology synthesis

The geology of Italy is ranked among the most complex and least understood in all of Europe. This is attributed to its location at the margin of several plate boundaries and tectonic regimes. This study focuses on the Southern Apennines Internal Thrust Zone area known as Val D'Agri, near the town of Potenza, and includes the Trend 1 oil fields known as Monte Alpi, Monte Enoc and Cerro Falcone plus the Tempa Rossa field on structural Trend 2 (Fig. 2; Table 1). Despite recorded knowledge of oil seeps in the area since the days of the Roman Empire (e.g. Tramutola), the discovery of these giant fields eluded oil finders until the late 1980s. Persistence, along with advances in seismic acquisition and processing technology, is credited with overcoming difficulties imposed by structure and topography (Armour 1999; Holton 1999a). The total recoverable reserves for the four fields is estimated at just over 1 billion barrels (Holton 1999b) to rank them as the largest onshore oil accumulations in Europe. Commercial production from these fields commenced near the end of 2001 after an assessment and development stage lasting more than a decade.

The Val D'Agri petroleum system has been described by Mosca and Wavrek (2002) to be mainly contained with the buried Apulian

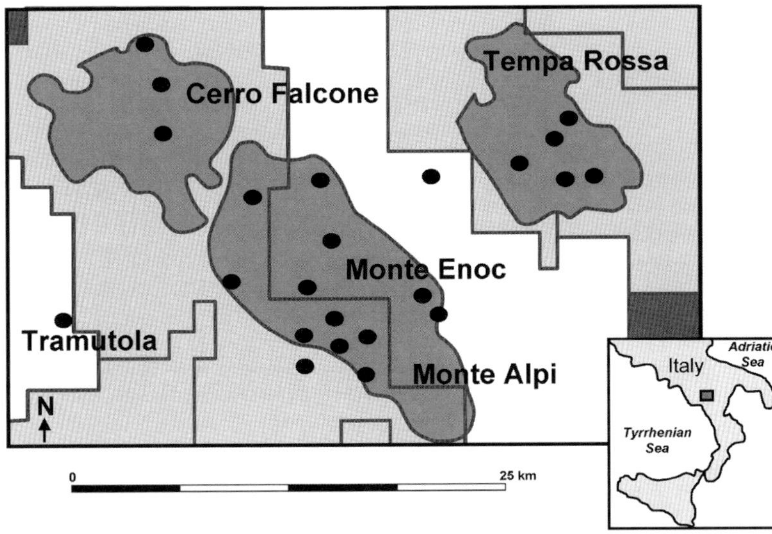


**Fig. 2.** General base map of the Val D'Agri region. The well locations (black dots) represent the surface (spud) location but actual samples are collected various lateral distances away. Angular shapes in the background represent concession boundaries, whereas irregular shapes represent approximate field boundaries. The inset map represents the location in a more regional sense.

Platform (Miocene–Cretaceous), which contains both the source rock and the reservoir interval (Fig. 3). A positive correlation has been established between the produced oils and Upper Cretaceous source rocks deposited in a highly anoxic reducing depositional environment that is dominated by a carbonate lithology. The carbonate reservoirs include karsted vuggy zones, in addition to an extensively developed fracture network. The wells flow at significant rates (3000–12 000 barrels per day) from thick oil columns (600–1000 m). The primary hydrocarbon charge event is modelled to occur in the Pliocene and includes a late influx of light hydrocarbons. Results from tectonic modelling (Sciamanna *et al.* 2000) indicate that the reservoir interval experienced a post-charge physical uplift of 2000 m over a timespan of 2 Ma that is responsible for reservoir pressure reduction on the order of 6000 psi and a concomitant temperature decrease of 50 °C. The entire producing sequence is sealed by the overlying Neogene sequences.

Proposals to explain the disparate oil quality at Val D'Agri have included hydrocarbon charge from different source rocks, as well as charges at different thermal maturities. Neither of these proposed processes, however, can be substantiated with actual data (Figs 4 & 5). Likewise, the search for evidence to support a biodegradation with mixing fresh oil scenario has not been successful, and this topic has even been probed with detailed analysis of the acid fractions (Meredith *et al.* 2000). Instead, a consensus has been reached by the project team scientists and engineers that the accumulations are intensely affected by the compositional grading process that is facilitated by an unstable oil column. The unstable oil column is due to the tectonic uplift with concomitant reservoir pressure and temperature decrease, the reservoir character with extensive fractures and karst intervals, the thickness of the oil column, and the late stage influx of volatile hydrocarbons.

## Results and discussion

### *Bulk oil characterization*

The Val D'Agri dataset includes a total of 30 oil samples (Table 1). Each sample has been collected as part of a PVT study or from surface tubulars during extended production tests. The results of the engineering studies have been incorporated into the geochemical study, and vice versa. The bulk assessment of these oils includes the API gravity, total sulphur content and liquid chromatography. The most peculiar property of these accumulations is illustrated by the dramatic decrease in density with increasing reservoir depth (Fig. 6), which spans a total variation of up to 30 units over a vertical span of 1000 m of oil column. An important

**Table 1.** *Summary of basic knowledge and data for the Val D'Agri sample set*

| Sample code[1] | Average depth (m) TVD SSL[2] | API gravity | Sulphur (%) | $C_{12-}$[3] (%) | Bulk separation[4] | | | | Quantitative GC[5] | | | Type[6] assignment |
|---|---|---|---|---|---|---|---|---|---|---|---|---|
| | | | | | Saturate (%) | Aromatic (%) | Resin (%) | Asphaltene (%) | $nC_5+nC_6+nC_7$ | $nC_{13}+nC_{14}+nC_{15}$ | $nC_{25}+nC_{26}+nC_{27}$ | |
| 1-5-1 | 2583 | 33.1 | 2.6 | 25 | 42.6 | 41.7 | 4.7 | 1.1 | 38141367 | 27136357 | 4600784 | 1L |
| 1-1-1 | 2384 | 34.1 | 2.1 | 35 | 39.3 | 44.7 | 7.3 | 0.3 | 61445213 | 30755191 | 4006472 | 1L |
| 2-2-1 | n/a | 8.2 | 7.3 | 8 | 16.2 | 53.6 | 7.6 | 19.0 | 12238233 | 11637046 | 1967418 | 2H |
| 2-1-1 | n/a | 18.3 | 5.9 | 16 | 20.9 | 47.7 | 8.2 | 21.4 | 20186251 | 15818944 | 2210058 | 2H |
| 2-1-2 | n/a | 14.3 | 6.3 | 14 | 18.3 | 44.6 | 6.2 | 23.0 | 17249599 | 13707435 | 2201531 | 2H |
| 2-1-3 | n/a | 18.3 | 5.5 | 18 | 20.5 | 47.5 | 8.3 | 15.9 | 22464669 | 14715837 | 2305763 | 2H |
| 1-6-1 | 2549 | 31.1 | 2.7 | 27 | 46.4 | 38.0 | 2.7 | 2.4 | 25811762 | 27563422 | 4652924 | 1L |
| 1-1-2 (s) | 2216 | 37.7 | 1.9 | 39 | 35.9 | 42.0 | 3.9 | 1.2 | 57033621 | 34459287 | 4312879 | 1L |
| 1-1-3 (l) | 2734 | 18.7 | 4.6 | 11 | 39.8 | 29.1 | 6.2 | 19.4 | 16722497 | 15696524 | 2493551 | 1H |
| 1-1-4 | 2456 | 36.6 | 2.3 | 36 | 20.8 | 36.1 | 3.1 | 1.0 | 44919867 | 30161831 | 3890439 | 1L |
| 1-1-5 (l) | 2680 | 31.2 | 3.1 | 31 | 36.4 | 43.9 | 3.9 | 2.4 | 49488176 | 24356168 | 3349633 | 1L |
| 1-1-6 (s) | 2286 | 38.4 | 2.1 | 36 | 38.7 | 35.9 | 3.3 | 0.0 | 63016115 | 36592905 | 4661320 | 1L |
| 1-8-1 | 4892 | 16.8 | 5.0 | 17 | 21.0 | 37.7 | 6.3 | 21.0 | 22139387 | 18081711 | 3060187 | 1H |
| 2-1-4 | 4833 | 27.7 | 2.8 | 26 | 39.1 | 35.1 | 6.9 | 0.8 | 24643903 | 31610492 | 4135930 | 2L |
| 2-1-5 | 4919 | 21.6 | 5.5 | 18 | 20.8 | 39.1 | 4.3 | 20.1 | 21110743 | 18933854 | 2544085 | 2H |
| 1-1-7 | 2553 | 33.1 | 2.4 | 34 | 23.8 | 40.1 | 4.6 | 0.8 | 57433452 | 28700025 | 3852536 | 1L |
| 2-1-6 | 4919 | 19.9 | 6.1 | 9 | 14.8 | 34.0 | 4.8 | 22.5 | 13458101 | 13126406 | 2080290 | 2H |
| 0-1-1 | 0 | 8.8 | 7.2 | 0 | 30.0 | 35.7 | 17.9 | 5.3 | n/a | n/a | n/a | 0 |
| 1-1-8 | 2389 | 33.7 | 2.1 | 43 | 32.2 | 25.9 | 13.6 | 0.2 | 63884607 | 25807922 | 3249344 | 1L |
| 1-4-1 | 2894 | 30.4 | 2.2 | 31 | 39.6 | 30.3 | 8.8 | 0.3 | 36393566 | 33071860 | 6109314 | 1L |
| 1-7-1 | 2440 | 31.4 | 2.6 | 39 | 27.6 | 27.4 | 12.5 | 0.4 | 41993213 | 21581255 | 3193314 | 1L |
| 1-5-2 | 2599 | 31.6 | 2.5 | 38 | 32.7 | 27.4 | 11.2 | 0.8 | 38252441 | 29697426 | 4709434 | 1L |
| 2-1-7 | 3790 | 18.1 | 6.1 | 23 | 15.3 | 32.3 | 10.2 | 17.3 | 29549957 | 17723841 | 2518957 | 2H |
| 1-4-2 | 2558 | 32.0 | 2.1 | 39 | 36.1 | 27.0 | 10.2 | 0.0 | 43118827 | 33312176 | 5566186 | 1L |
| 1-1-9 | 2486 | 31.3 | 3.0 | 51 | 45.2 | 47.4 | 5.5 | 0.9 | 43520795 | 38200363 | 4531686 | 1L |
| 1-1-10 | 2503 | 30.1 | 3.2 | 47 | 43.0 | 49.7 | 6.7 | 0.4 | 35582720 | 35733423 | 4383047 | 1L |
| 1-3-1 | 2702 | 33.6 | 2.5 | 53 | 28.6 | 49.2 | 11.4 | 2.2 | 42631870 | 30395667 | 3581741 | 1L |
| 1-3-2 | 2858 | 18.6 | 5.3 | 20 | 19.2 | 32.7 | 8.7 | 25.2 | 7851988 | 21679868 | 3478997 | 1H |
| 1-2-1 | 2996 | 25.8 | 3.1 | 39 | 36.4 | 42.4 | 9.1 | 3.1 | 26292606 | 24768661 | 3816939 | 1L |
| 1-2-2 | 3099 | 16.6 | 4.3 | 25 | 27.4 | 36.3 | 10.1 | 21.3 | 11743923 | 27283062 | 4536116 | 1H |

[1] Sample code: Structural Trend-Reservoir Compartment-Well Number: (s), short string, production from the top of the oil column (light oil); (l), long string, production from the base of the oil column (heavy oil). [2] TVD SSL, total vertical depth sub-sea level. [3] Based on stabilization results prior to SARA separation.
[4] Separation results reported on actual recoveries. [5] Quantification: is direct to 1-hexene for low molecular weight compounds, and squalane for medium and higher ranges; n/a, not available due to biodegradation of surface seep; units are in nanograms/gram.
[6] Assignment summary, i.e. structural trend plus light or heavy classification.

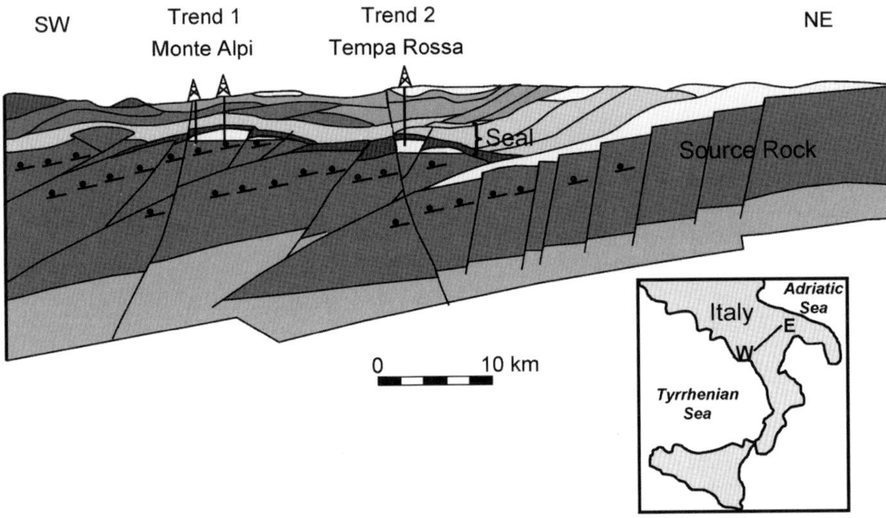


**Fig. 3.** Schematic cross section to illustrate important aspects of the Val D'Agri petroleum system.

implication for the resource assessment is that each reservoir compartment will have its own depth decay function that will establish the total volumes that can be economically produced. Also, the bulk property change is not always linear with increasing depth, but may have a gradual change until a critical point is reached and further decay approximated by a higher-order equation. Although the samples represent a near-continuum in bulk properties such as gas/oil ratio, saturation pressures, or reservoir pressure (Fig. 7), the sulphur content shows a distinct split between the light and heavy oils (Fig. 8). The SARA separation process is preceded by a stabilization procedure that aids in the physical property evaluations (Fig. 9) that

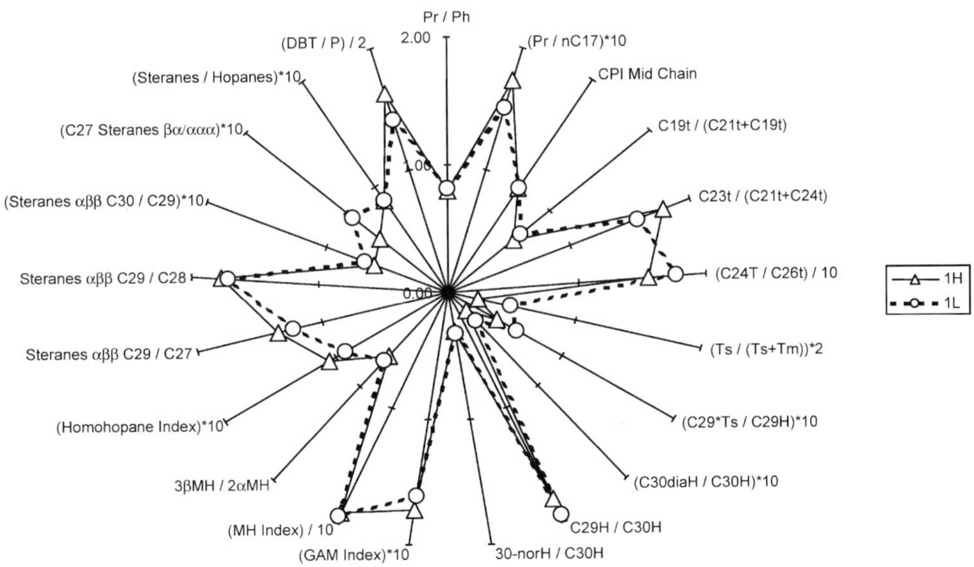


**Fig. 4.** Spider-plot display of geochemical data corresponding to typical light and heavy oils that have correlation significance. The data have been multiplied or divided in this display to provide convenient magnitude for visual comparison. The averaged values with abbreviations are reported in Table 2. These results show that despite the difference in physical properties, the oils were sourced from the same source rock.

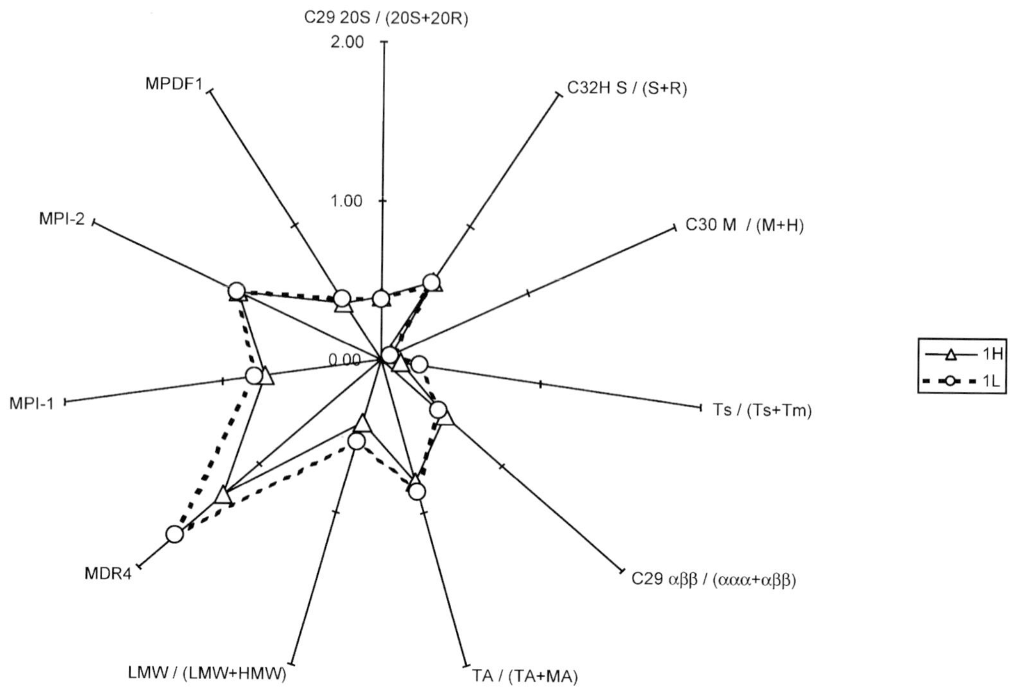


**Fig. 5.** Spider-plot display of geochemical data corresponding to typical light and heavy oils that have thermal maturity significance. The averaged values with abbreviations are reported in Table 3. These results show that despite the difference in physical properties, the oils were generated at similar levels of thermal stress.

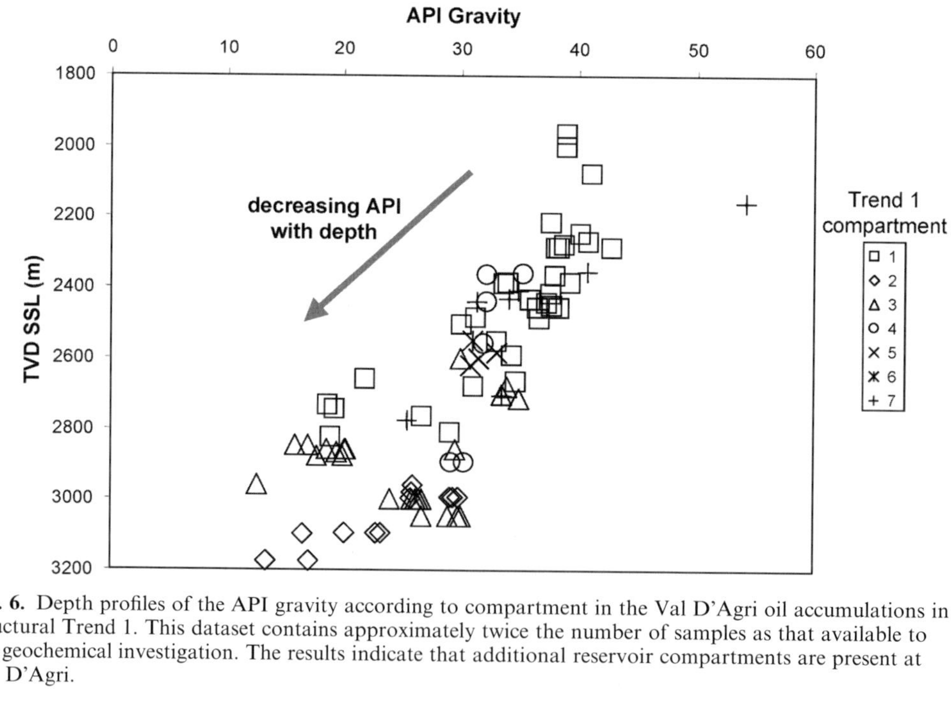


**Fig. 6.** Depth profiles of the API gravity according to compartment in the Val D'Agri oil accumulations in structural Trend 1. This dataset contains approximately twice the number of samples as that available to the geochemical investigation. The results indicate that additional reservoir compartments are present at Val D'Agri.

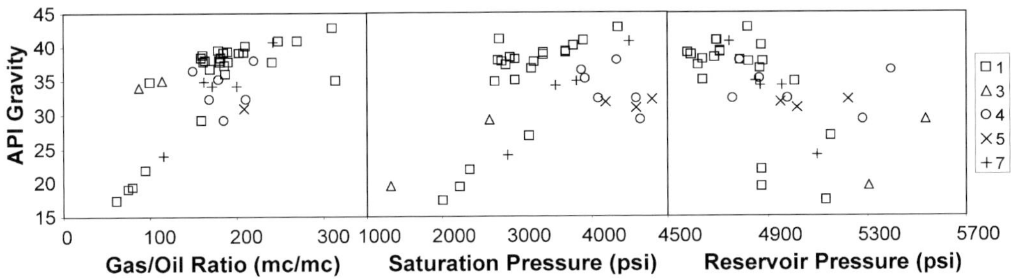


**Fig. 7.** Cross plot of gas/oil ratio, saturation pressure, and reservoir pressure versus depth API gravity that is extracted from the overall engineering dataset. The results show overall consistency for this reservoir system and scatter is probably due to additional complexity (e.g. sub-compartments).

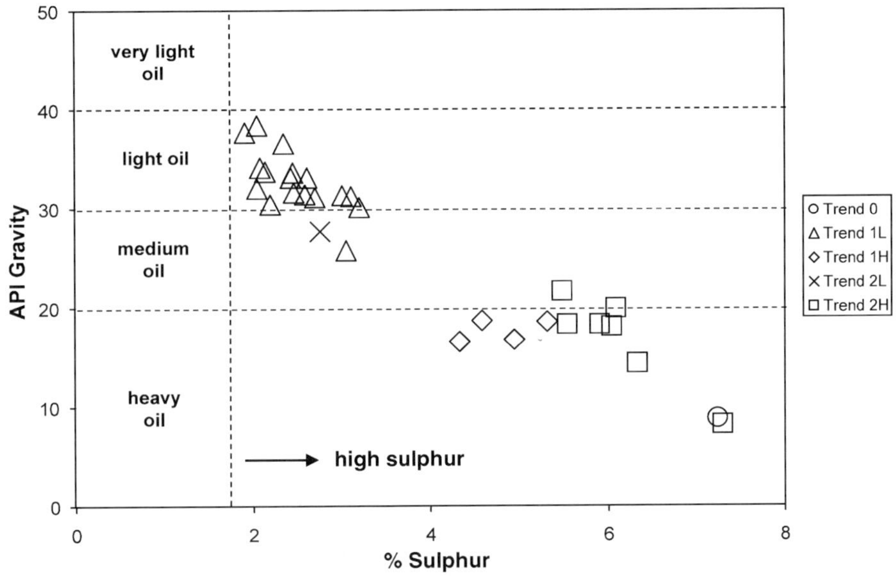


**Fig. 8.** Cross plot of the API gravity and total sulphur content of the Val D'Agri samples. The break between light and heavy oils is a distinctive aspect of the dataset, and provides a stark contrast to the near continuum of many other physical properties.

complement the engineering evaluation (e.g. GOR prediction). Bulk separation and physical property data enable a recombination exercise to be completed to approximate the evolution of the overall mass balance of the light versus heavy oil fractions (Fig. 10). Three features are particularly important in this illustration:

- average volatile fraction varies by a factor of two,
- relative proportion of fractions vary by a factor of two, and
- resin + asphaltene fraction varies by a factor of five.

The implications of these relationships are further investigated, but the emphasis remains that the geochemical observations are in excellent accordance with the engineering criteria of the separation by compositional grading.

## *Molecular characterization: hydrocarbons*

Petroleum consists of a complex mixture of hydrocarbons that range in size from methane to asphaltene, and the relative distribution of these molecules control physical properties. The distribution of these molecules can be used to monitor or predict physical properties of the hydrocarbon phases. The liquid fraction analysis by gas chromatography (GC) is split into three molecular weight ranges, with the quantitative

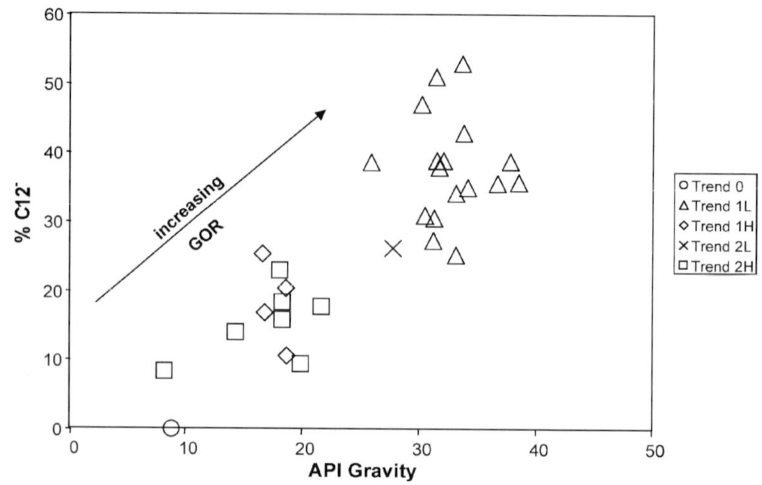


**Fig. 9.** Cross plot of the volatile content ($C_{12-}$) versus API gravity to illustrate the concordance of these physical parameters.

results facilitated by the gravimetric addition of 1-hexene and squalane standards. It is noted, however, that the total number and diversity of compounds used in this study prevented the determination of individual response factor values so each value is reported from the direct response relative to the respective standard. These results (Fig. 11) indicate that the light oils are quantitatively enriched in the volatile components, with sequential less enrichment as carbon number increases. The latter statement is well illustrated by the alkane homologous series (Fig. 12). Between compound classes of a given carbon number, the trends are somewhat less definitive (Fig. 13), but this is interpreted to be a function of subtle variation of surface

**Table 2.** *Averaged molecular ratio data with correlation potential for light and heavy oils*

| Oil type code | 1L | 1H |
|---|---|---|
| Pr/Ph | 0.81 | 0.79 |
| $Pr/nC_{17}$ | 0.15 | 0.17 |
| CPI mid chain | 0.98 | 0.97 |
| $C_{19}t/(C_{21}t+C_{19}t)$ | 0.72 | 0.65 |
| $C_{23}t/(C_{21}t+C_{24}t)$ | 1.56 | 1.77 |
| $C_{24}T/C_{26}t$ | 17.60 | 15.39 |
| Ts/(Ts + Tm) | 0.24 | 0.12 |
| $C_{29^*}Ts/C_{29}H$ | 0.06 | 0.04 |
| $C_{30}diaH/C_{30}H$ | 0.03 | 0.02 |
| $C_{29}H/C_{30}H$ | 1.95 | 1.81 |
| $30\text{-}norH/C_{30}H$ | 0.33 | 0.31 |
| GAM index | 0.16 | 0.17 |
| MH index | 19.46 | 19.21 |
| $3\beta MH/2\alpha MH$ | 0.72 | 0.68 |
| Homohop index | 0.09 | 0.11 |
| $C_{29}/C_{27}\alpha\beta\beta$ | 1.22 | 1.35 |
| $C_{29}/C_{28}\alpha\beta\beta$ | 1.70 | 1.76 |
| $C_{30}/C_{29}\alpha\beta\beta$ | 0.07 | 0.06 |
| $C_{27}$ Steranes $\beta\alpha/\alpha\alpha\alpha$ | 0.09 | 0.07 |
| Sterane/hopanes | 0.09 | 0.09 |
| DBT/P | 2.83 | 3.27 |

Pr, pristine; Ph, phytane; Mid Chain CPI, $(nC_{21}+(2^*nC_2)+(2^*nC_{25})+nC_{27})/(2^*(nC_{22}+nC_{24}+nC_{26}))$; t, tricyclic terpane; T, tetracyclic terpane; Ts, trisnorneohopane (*m/z* 370); Tm, trisnorhopane (*m/z* 370); H, hopane; diaH, diahopane; norH, norhopane; GAM, gammacerane; MH, methylhopane; Homohop, homohopane; Steranes, sum *m/z* 218 $\alpha\beta\beta$ + sum m/z 217$\alpha\alpha\alpha$; Hopanes, sum of $C_{29}$ to $C_{35}$ hopanes; DBT, dibenzothiophene; P, phenanthrene.

**Table 3.** *Averaged molecular ratio data with thermal maturity significance for light and heavy oils*

| Oil type code | 1L | 1H |
|---|---|---|
| $C_{29}$ 20S/(20S + 20R) | 0.38 | 0.39 |
| $C_{32}H$ S/(S + R) | 0.58 | 0.59 |
| $C_{30}$ M/(M + H) | 0.07 | 0.07 |
| Ts/(Ts + Tm) | 0.24 | 0.12 |
| $C_{29}$ $\alpha\beta\beta/(\alpha\alpha\alpha+\alpha\beta\beta)$ | 0.47 | 0.53 |
| TA/(TA + MA) | 0.86 | 0.80 |
| LMW/(LMW + HMW) | 0.54 | 0.41 |
| MDR4 | 1.70 | 1.31 |
| MPI-1 | 0.80 | 0.73 |
| MPI-2 | 1.00 | 0.98 |
| MPDF1 | 0.45 | 0.43 |

M, moretane; H, hopane; Ts, trisnorneohopane (*m/z* 370); Tm, trisnorhopane (*m/z* 370); TA, triaromatic steranes; MA, monoaromatic steranes; LMW, low molecular weight, $C_{20}TA+C_{21}TA$ steranes; HMW, high molecular weight, $C_{26}TA+C_{22}TA+C_{28}TA$ steranes; MDR4, 4MDBT/DBT; MPI-1, 1.5*(2MP + 3MP)/(P + 1MP + 9MP); MPI-2, (3*2MP)/(P + 1MP + 9MP); MPDF1, (2MP + 3MP)/(1MP + 2MP + 3MP + 9MP).

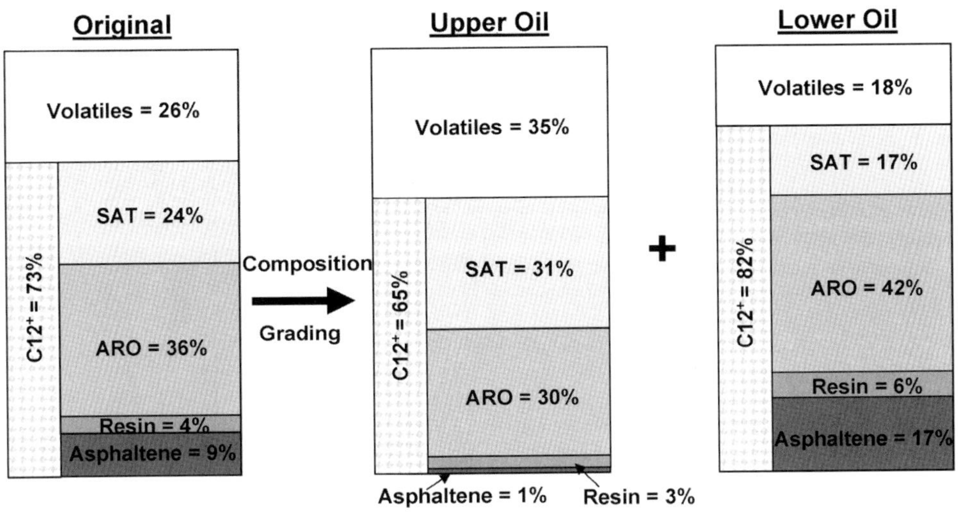


**Fig. 10.** Illustration of the mass balance relationships that occur to individual fractions when the Val D'Agri oil is altered by the compositional grading process. That upper oil is relatively light (>30° API) and contains 2–3% sulphur whereas the lower oil is heavy (<2° API) and enriched in sulphur (4–6%).

charge between the specific molecule and the matrix fluid. The overall conclusion is that quantitative molecular data can be used to monitor and/or predict variation in physical properties of the hydrocarbon fluid (e.g. production monitoring tool).

The current dataset also enables a series of mass balance exercises to be evaluated. Key assumptions include that the free hydrocarbons in the volatile+saturate+aromatic fractions will be amenable to elution from a GC column, and that the resin+asphaltene fractions will contain a negligible amount of GC-amenable compounds. If we assign one unit volume to each oil sample and use the concentration units of ng/g (nanograms per gram), each unit volume will contain one billion nanograms of material. To facilitate the use of these numbers, the abbreviation MM is adopted to represent the term $10^6$. Therefore, there will be 960MM of material to account for in the light oils and 770MM for the heavy oils given the mass balance relationships previously established (Fig. 10). The concentration results (Table 4) are divided into groups according to carbon number, plus that which corresponds to the UCM (unresolved complex mixture). This latter value has the largest potential for error, but special care is taken to adjust the blank runs (i.e. background subtraction) that allow the column bleed to be properly accounted. An item of note is that proper peak shapes (i.e. gaussian) need to be maintained in order to have proper quantification values. It is also noted that the light oils have a larger UCM compared with that of the heavy oils. This, however, is a logical observation within the mass balance process as the light oil has more material to be accounted for within the GC-amenable fraction. The net result (Table 4) indicates that 92% of the light oils can be accounted for in a quantitative molecular mass balance, whereas this number drops to 70% in the heavy oils. This latter value is considered reasonable when evaluated in the context of the input assumptions and the fact that the heavy oil would be expected to have a higher percentage of non-GC-amenable components compared with the light oils.

### *Molecular characterization: non-hydrocarbons*

It is a widely held belief (i.e. albeit not always true) that the bulk of the total sulphur content resides within the asphaltene fractions, and that a reasonable correlation should exist between this fraction and the total sulphur content. This is especially common belief within a single petroleum system where individual samples may be subjected to different degrees of secondary alteration such as differential thermal stress or biodegradation. A simple comparison of the relevant data for Val D'Agri (e.g. Figs 8 & 10) indicates that the total sulphur content usually varies by a factor of two (i.e. heavy versus light), yet the asphaltene fraction differs by more than an order of magnitude, which

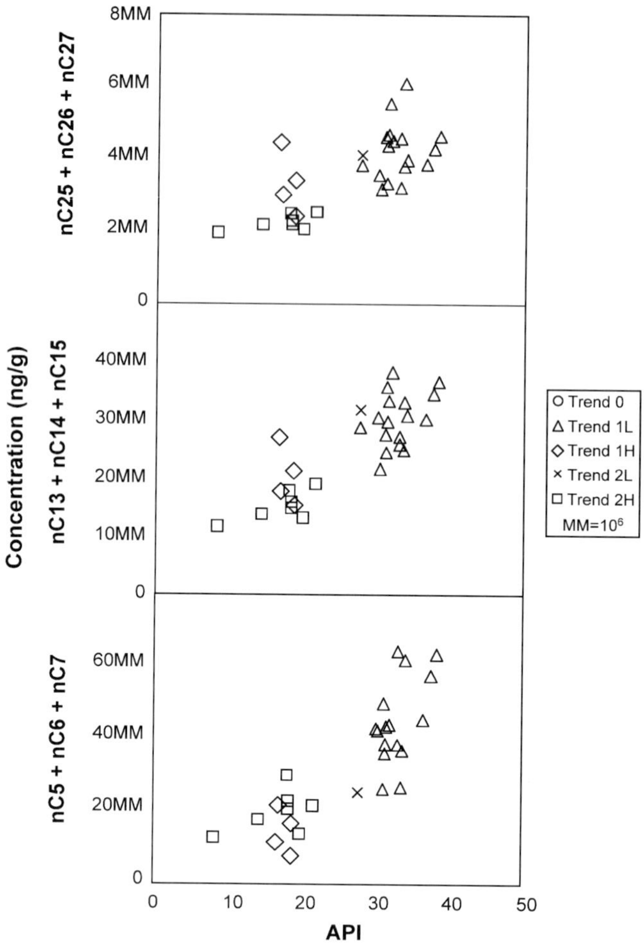


**Fig. 11.** Cross plot of individual quantitative results from three molecular weight ranges in the context of the API gravity. The difference between light and heavy oil pairs decreases with increased carbon number.

highlights an important discrepancy in the sulphur mass balance process. There are two primary hypotheses to explain this discrepancy: that the asphaltene fraction in the heavy oils is contaminated with co-precipitated waxes (i.e. non-sulphur bearing components), or that the light oils had an excessive abundance of free organic sulphur compounds. The former hypothesis was investigated using a special asphaltene isolation procedure (Burger *et al.* 1981; Thanh *et al.* 1999), but found to have a negligible amount of non-asphaltic material present in the precipitated fractions. The GC-FPD (GC-flame photometric detector) fingerprints of the whole oil (Fig. 14), however, show distinct differences between the light and heavy oils in terms of molecular isomers as well as compound class. These data (i.e. BT versus DBT) can be readily explained in the context of the quantitative GC-MS analyses that show the selective enrichment of dibenzothiophene compounds in the light oils (Fig. 15). While this does not *fully* account for the sulphur mass balance discrepancy, it does account for a significant portion. The most important aspect of this non-hydrocarbon investigation remains the documentation that the compositional grading process profoundly affects the organic sulphur compound signature in terms of both isomer distributions and absolute quantitative abundance.

## *Application to reservoir compartmentalization studies*

This paper adds to the database of knowledge regarding the systematic changes on a molecular level as a response to extraneous influences on an

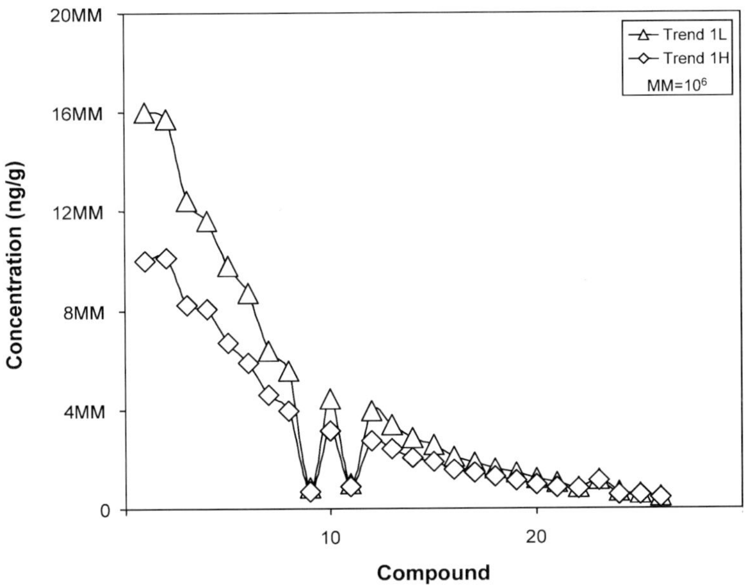


| Compound No. | Name |
|---|---|
| 1 | nC10 |
| 2 | nC11 |
| 3 | nC12 |
| 4 | nC13 |
| 5 | nC14 |
| 6 | nC15 |
| 7 | nC16 |
| 8 | nC17 |
| 9 | Pristane |
| 10 | nC18 |
| 11 | Phytane |
| 12 | nC19 |
| 13 | nC20 |
| 14 | nC21 |
| 15 | nC22 |
| 16 | nC23 |
| 17 | nC24 |
| 18 | nC25 |
| 19 | nC26 |
| 20 | nC27 |
| 21 | nC28 |
| 22 | nC29 |
| 23 | nC30 |
| 24 | nC31 |
| 25 | nC32 |
| 26 | nC33 |

**Fig. 12.** Averaged data display to contrast the light versus heavy oils from Trend 1 reservoirs. The homologous series of *n*-alkanes shows the incremental changes with the addition of each carbon number.

oil system. To integrate this knowledge into the reservoir continuity technologies, this paper advocates the approach introduced by Wavrek *et al.* (2001) to identify the actual molecules (or at least compound class) that are responsible for the observed variation between samples. That is, the peak selection process should contain a degree of subjectivity, as this knowledge can be critical to identify the genetic origin of a particular signal and be subsequently used in the tool

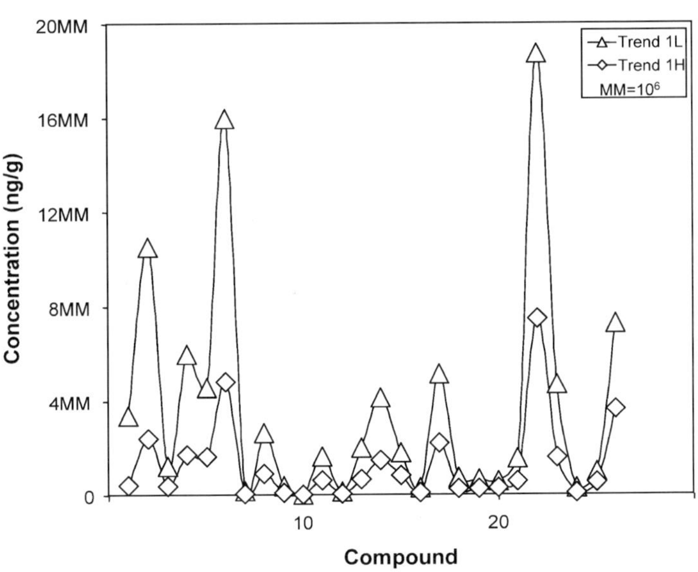


| Compound No. | Name |
|---|---|
| 1 | MB |
| 2 | P |
| 3 | 2,3-diMB |
| 4 | 2-MB |
| 5 | 3-MB |
| 6 | Hx |
| 7 | 2,2-diMB |
| 8 | MCyP |
| 9 | 2,4-diMP |
| 10 | 2,2,3-triMB |
| 11 | benzene |
| 12 | 3,3-diMP |
| 13 | CyHx |
| 14 | 2-MHx |
| 15 | 2,3-diMP |
| 16 | 1,1-diMCyP |
| 17 | 3-MHx |
| 18 | 1c3-diMCyP |
| 19 | 1t3-diMCyP |
| 20 | 3-EP |
| 21 | 1t2-diMCyP |
| 22 | Hp |
| 23 | MCyHp+1c2-diMCyP |
| 24 | 1,1,3-triMP |
| 25 | ECyP |
| 26 | toluene |

**Fig. 13.** Averaged data display for the gasoline range hydrocarbons ($C_5$–$C_7$) for the Trend 1 oil samples (M, methyl; E, ethyl; B, butane; P, pentane; Hx, hexane; Hp, heptane; Cy, cyclo).

**Table 4.** *Results of the mass balance calculations based on quantitative GC analysis*

| Fraction | Oil type | | Reference figures |
|---|---|---|---|
| | Light | Heavy | |
| $C_5$–$C_7$ volatile peaks | 100MM | 50MM | 11, 13 |
| $C_8$–$C_{10}$ volatile peaks | 80MM | 34MM | |
| $C_{10}$–$C_{22}$ mid-range peaks | 300MM | 150MM | 11, 12 |
| $nC_{23}$–$nC_{33}$ peaks | 13MM | 8MM | 11, 12 |
| UCM or hump determination | 390MM | 300MM | |
| Total VSA* | 883MM | 542MM | |
| Theoretical VSA† | 960MM | 770MM | 10 |
| Mass Balance Accounting | 92% | 70% | |

* Sum derived from sample averages from GC-FID analysis.
† Sum derived from sample averages from bulk separation analysis.
UCM, unresolved complex mixture; VSA, volatiles + saturates + aromatics; MM, million.

selection process (e.g. Fig. 1). It is also emphasized that these data should be used in conjunction with a chemometric approach to identify consistent series of peak ratio combinations that are indicative of specific reservoir processes. The combination of these geochemical tools enables the petroleum systems analyst to identify molecular combinations that are attributed to oil column disequilibrium, from those responsible for defining separate reservoir compartments. For example, in this case where an oil reservoir is altered by compositional grading, a consistent increase will be observed in the molecular ratios that correspond to saturate/aromatic pairs, as well as a decrease in the aromatic/saturate pairs, within the evolving light fraction. The inverse relationships will be observed in the evolving heavy fraction. Interestingly, specific non-hydrocarbons will show selective increases or decreases with the compositional grading alteration process, depending on the chemistry of the specific peak versus chemistry of the adjacent peak to which a ratio is applied. These variations are readily identified in whole oil analyses, as well as in isolated saturate or aromatic fractions. The consistency of the numerous ratios provides a converging interpretation within the data analysis scheme to provide a

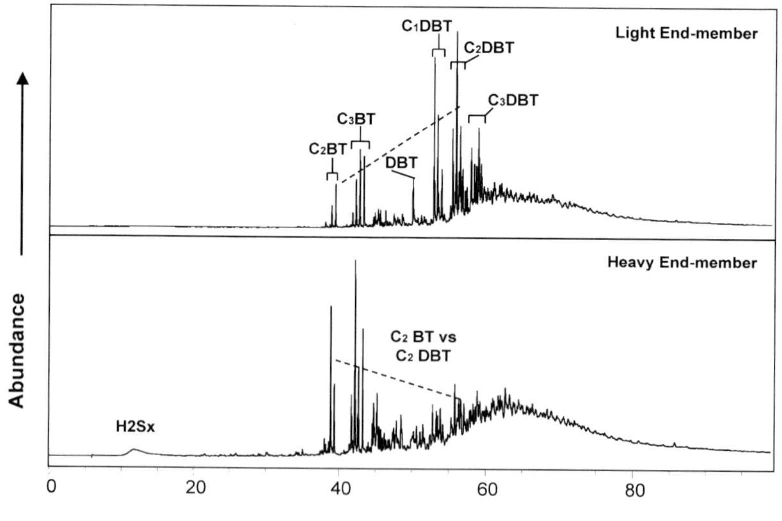


**Fig. 14.** Contrast of organic sulphur compound molecular fingerprints from light and heavy oils to illustrate examples of the systematic fractionation on the molecular and isomer scales (BT, benzothiophene; DBT, dibenzothiophene).

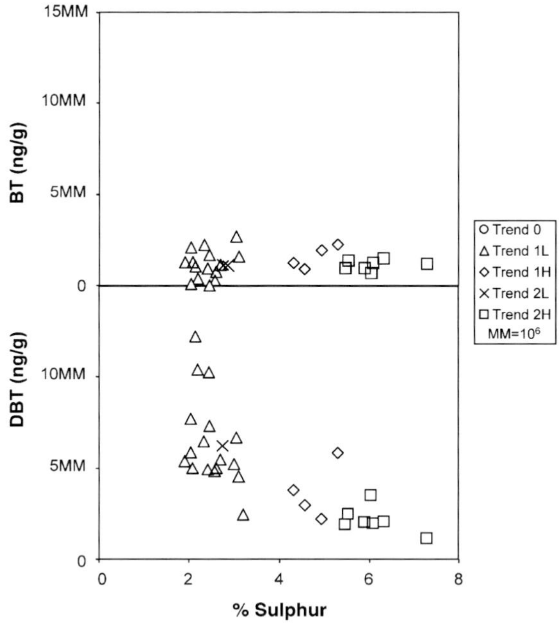


**Fig. 15.** Benzothiophene (BT) and dibenzothiophene (DBT) data from individual samples in Val D'Agri. The results show that the contrasting signatures illustrated in the previous figure is a function of an increased concentration of dibenzothiophenes in the light oils.

high degree of confidence in the overall interpretation and provide new tools for production monitoring. These oil quality ratios can be masked in subsequent analytical steps to search for process-independent molecular ratios that will be diagnostic of individual reservoir compartments. Alternatively, different analytical schemes can be applied to the particular reservoir horizon that will optimize the reservoir continuity signal (i.e. increase signal and/or reduce noise) to enable the reservoir continuity analysis to proceed towards successful completion. This is a function of specific compounds and isomers being part of a systematic fractionation process. This also highlights the importance of this knowledge in early evaluation stages to identify best practice methodology, not only for the reservoir geochemist, but also for the associated engineering team members. The end product of this procedure is to reduce the risk of erroneous interpretations, increase efficiency and reduce overall expenditure by the operator.

## Conclusions

This paper establishes a methodology for linking engineering physical properties and quantitative molecular analysis within a mass balance framework to evaluate hydrocarbon resources affected by the compositional grading phenomena. The results demonstrate that the molecular signature for this secondary alteration process proceeds in a predictable manner. This process is responsible for active disequilibrium in an oil column, which violates the central tenet of establishing the limits of reservoir continuity. Failure to recognize disequilibrium in the oil column will ultimately lead to the identification of false compartments. Petroleum systems elements and processes are identified which are primary contributors to forming an unstable oil column that will be particularly prone to the alteration by compositional grading. Interestingly, these petroleum system circumstances coincide with the physical setting for some of the best candidates for the reservoir continuity process. For this reason, we highlight aspects of this research that are applicable to deconvolution of the respective signals. This methodology is also important to apply at an early stage of field development, where the database is scant and engineering testing budgets are strained by the multitude of unknowns. Successful application of reservoir geochemistry technologies is needed for this discipline to develop to its fullest potential, which will be that of a routine tool used by the reservoir engineer.

We wish to thank Enterprise Oil Italiana S.p.A. for the financial and technical support upon which this work is derived, along with their permission to present and publish this paper. Likewise, Eni S.p.A. is thanked for extending us permission to publish this information as well. A number of individuals and organizations are further acknowledged. D. Jarvie, R. Olson, M. Tobey and M. Dodd from Humble Instruments and Services, Inc.; A. Beccacini, A. Pelliccia, G. Griffi, F. Colacicchi, P. Lynes, G. Lawrence, K. Grassi, R. Gambini (currently OMV), C. Nicolai (currently Shell UK) and C. Nieto from Enterprise Oil Italiana; S. Sciamanna (currently Repsol-YPF), J. Rudkiewicz and W. Sassi from IFP; P. Lemouzy from Beicip Franlab; B. Rohrback and S. Ramos from Infometrix, Inc.; and J. Moore and J.-C. Chao from Petroleum Systems International, Inc. Finally, we thank A. Stephen and C. Whitson for their helpful and constructive reviews of this manuscript.

## References

ARMOUR, A. 1999. Southern European thrust belt exploration—Persistence pays off. *Extended Abstracts, American Association of Petroleum Geologists International Conference*, Birmingham, England.

BURGER, E. D., PERKINS, T. K. & STRIEGLER, J. H. 1981. Studies of wax deposition in the trans Alaska pipeline. *Journal of Petroleum Technology*, **33**, 1075–1086.

HIRSCHBERG, A. 1988. Role of asphaltenes in compositional grading of a reservoir's fluid column. *Journal of Petroleum Technology*, **40**, 89–94.

HOIER, L. & WHITSON, C. H. 2000. Compositional grading—theory and practice. Society of Petroleum Engineers, Paper No. **63085**, 1–16.

HOIER, L. & WHITSON, C. H. 2001a. Compositional grading—theory and practice. *Society of Petroleum Engineers Reservoir Evaluation and Engineering*, **4** (6), 525–535.

HOIER, L. & WHITSON, C. H. 2001b. Miscibility variation in compositionally grading reservoirs. *Society of Petroleum Engineers Reservoir Evaluation and Engineering*, **4**, 36–43.

HOLTON, J. 1999a. Southern Italy; 1, S. Apennines success bodes well for potential off southern Italy. *Oil and Gas Journal*, **97**, 65–68.

HOLTON, J. 1999b. Southern Italy; 2, Four geologic settings dominate oil, gas fields of Italy, Sicily. *Oil and Gas Journal*, **97**, 81–84.

KAUFMAN, R. L., AHMED, A. S. & ELSINGER, R. L. 1990. Gas chromatography as a development and production tool for fingerprinting oils from individual reservoirs. applications in the Gulf of Mexico. *Gulf Coast Section of the Society of Economic Paleontologists and Mineralogists Foundation Annual Research Conference Proceedings*, Austin, 263–282.

LARTER, S. R. & APLIN, A. C. 1995. Reservoir geochemistry: methods, applications and opportunities. *In*: CUBITT, J. M. & ENGLAND, W. A. (eds) *The Geochemistry of Reservoirs*. Geological Society, London, Special Publications, **86**, 5–32.

MEREDITH, W. S., KELLAND, J. & JONES, D. M. 2000. Influence of biodegradation on crude oil acidity and carboxylic acid composition. *Organic Geochemistry*, **31**, 1059–1074.

MOSCA, F. & WAVREK, D. A. 2002. Petroleum system characteristics of the giant oil fields of Val D'Agri region, Southern Apennines, Italy. *American Association of Petroleum Geologists Convention Abstracts*, Houston, TX.

MOSCA, F., WAVREK, D. A. & JARVIE, D. M. 2003. Can the reservoir continuity technique be applied to a compositionally graded oil column? Methodological remarks and case history application. *American Association of Petroleum Geologists Convention Abstracts*, Salt Lake City, UT.

SAGE, B. H. & LACEY, W. N. 1938. Gravitational concentration gradients in static columns of hydrocarbon fluids. *Transactions of the American Institute of Mining and Metallurgical Engineers*, **132**, 120–131.

SCHULTE, A. M. 1980. Compositional variations within a hydrocarbon column due to gravity. Society of Petroleum Engineers, Paper No. **9235**.

SCIAMANNA, S., GAMBINI, R., MOSCA, F., NICOLAI, C., RUDKIEWICZ, J. L. & SASSI, W. 2000. The Southern Apennines thrust belt—forward structural and geochemical modeling since Messinian times. *European Association of Geoscientists and Engineers Conference Proceedings*, Malta.

THANH, N. X., HSIEH, M. & PHILP, R. P. 1999. Waxes and asphaltenes in crude oils. *Organic Geochemistry*, **30**, 119–132.

WAVREK, D. A., MOSCA, F., CHAOUCHE, A. & JARVIE, D. M. 2001. Geochemical insights to engineering problems: when does oil heterogeneity really indicate a compartment? *International Meeting on Organic Geochemistry Proceedings*, Nancy, France.

# Application of geochemistry in the evaluation and development of deep Rotliegend dry gas reservoirs, NW Germany

ERIC MUELLER & TORSTEN SCHOLZ

*RWE Dea AG, Ueberseering 40, 22297 Hamburg, Germany (email: eric.mueller@rwedea.com)*

**Abstract:** The Voelkersen field in northwestern Germany produces thermogenic dry gas from Permian Rotliegend sandstone reservoirs located at depths greater than 4700 m. In addition to this considerable depth Zechstein salt overburden plus the structural complexity and heterogeneity of the reservoirs limit the applicability of conventional methods for definition and characterization of field compartmentalization. The Rotliegend gas is sourced from Westphalian coals of semi-anthracite to anthracite rank with conventional geochemical data suggesting little variation in composition on a field-wide scale. However, detailed analysis of associated condensates and wet-gas hydrocarbons shows considerable differences between compartments and producing horizons, despite the apparent high maturity of the gas. These differences support compartment definition and identification of lateral and horizontal seals. Assuming a quasi-uniform gas composition in communicating reservoirs, variation of carbon and hydrogen isotopic data in particular confirms separation of marginal fault blocks. Additionally, these data also suggest reservoir continuity within a large field compartment and over parts of a major fault system for which the sealing properties so far were previously uncertain.

The RWE Dea AG operated Voelkersen gas field is located at the southern margin of the Southern Permian Basin and produces thermogenic dry gas from deeply buried, highly heterogeneous Permian Rotliegend sandstones. The field is the southwestern extension of a string of gas fields of the Rotenburg–Soltau area where the Rotliegend reservoirs are major targets (Albertsen 1992). Initial gas in place (GIIP) estimate of *ca.* $1 \times 10^{12}$ ft$^3$ (1 tcfg), high productivity of individual wells, surface infrastructure and nearby consumer markets make the fields commercially viable despite the high costs and risks associated with exploration, drilling and development of gas accumulations at these great depths. The considerable depth, overlying Zechstein evaporites and small scale reservoir heterogeneities, limits the potential of subsurface imaging methods for modelling of the reservoirs and their structural/diagenetic details (Gaupp *et al.* 1993). Reservoir models in combination with well pressures and production history, however, provide the basis for field extension and exploitation plans. The validity of the reservoir model and the economic success of individual production wells are verified through production tests after well completion.

The potential of geochemistry as a supporting methodology for solving oil field production problems has been demonstrated in recent publications (e.g. Peters & Fowler 2002, and references therein) and examples of the application to gas and condensate fields have been given by Schoell & Jenden (1993). In general, though, application of geochemistry in gas field production studies has received less attention, possibly due to the limited number of hydrocarbons and molecular classes available for analysis and thus presumed inefficiency of geochemical methods, or simply due to the limited number of appropriate fields to study. This report intends to provide such a case study by demonstrating the application of geochemical analyses to a field that produces apparently fairly homogenous thermogenic and highly mature gas from deep reservoirs, and for which the potential benefits of a geochemical study may not be apparent at first. The basic assumption for the application of geochemistry to field production problems is that the composition of hydrocarbons in continuous reservoirs or communicating reservoir compartments should be similar whilst isolated compartments may show geochemical differences (Larter & Aplin 1995).

## Geological framework and field description

The Voelkersen field is located in the Wuemme Graben bounded in the east by a Carboniferous Horst (Wedehof Platform), to the west by the Schaphusen Platform, and major structural elements in the north and south. Northern and southern gas field boundaries are formed by downdip gas–water contacts (GWC). Separating the field into two blocks is a WNW–ESE striking wrench element (Voelkersen Schwelle) that is assumed to be a northern branch of a regional

*From*: CUBITT, J. M., ENGLAND, W. A. & LARTER, S. (eds) 2004. *Understanding Petroleum Reservoirs: towards an Integrated Reservoir Engineering and Geochemical Approach.* Geological Society, London, Special Publications, **237**, 221–230. 0305-8719/04/$15.00 

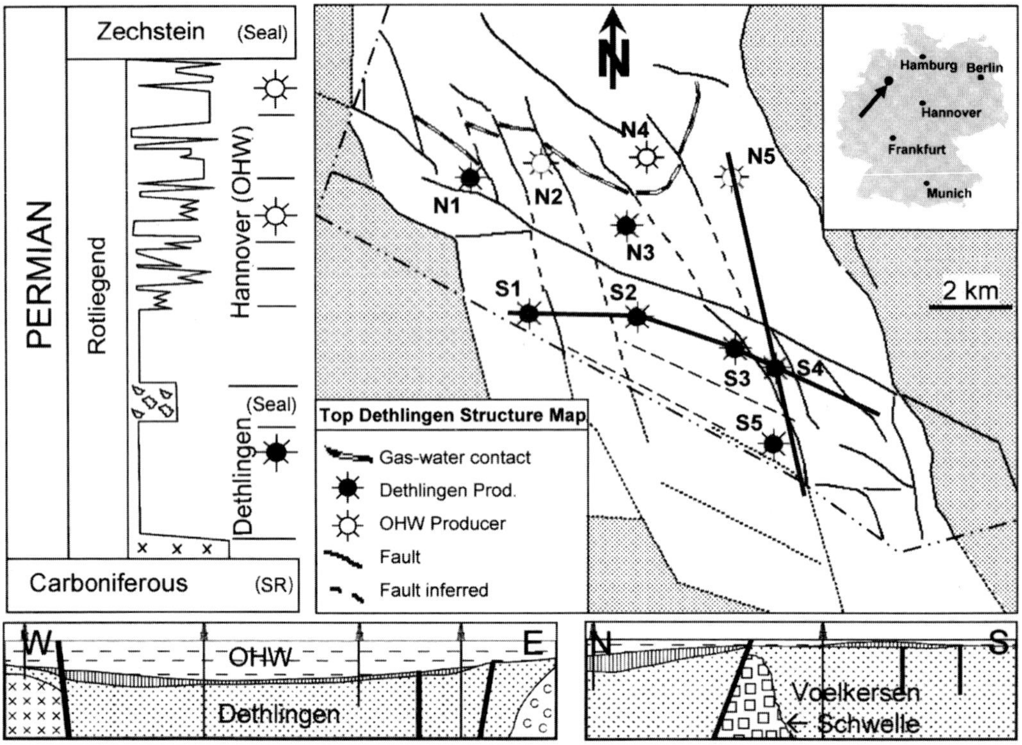


**Fig. 1.** Voelkersen field location (arrow on insert map of Germany, upper right), Upper Paleozoic stratigraphy (SR, source rock), schematic top Dethlingen sandstone structure map, and schematic cross sections (datum is top Rotliegend), Voelkersen field, N Germany (thick dashed line represents concession boundary). Note that well N2 was recently switched to OHW production but gas analysis reported here are for gas in the Dethlingen sandstone.

wrench tectonic zone (Allertal-Lineament) that also causes the NNW–SSE graben strike direction by its sinistral shear (Fig. 1). The Voelkersen Schwelle is an inversion zone which controls local dip directions as well as sediment thicknesses.

In the graben, overlying Carboniferous (source rock) sediments, Lower Permian (Rotliegend) volcanic rocks and sediments are succeeded by alluvial and aeolian sandstones. The aeolian dune/inter-dune sediments have been modified by a variety of diagenetic alterations which cause reservoir properties to range from tight (0.01 mD) to porous and highly permeable (>1000 mD). A major vertical seal for the petroleum system is a proximal alluvial facies (fanglomerate) that caps the lower reservoir sands of the Dethlingen Formation (Fig. 1). The deposits of the overlying Obere Hannover Wechselfolge Formation (OHW) consist of transgressive shale cycles and finer grained aeolian sandstones (Gaupp *et al.* 1993) which are secondary gas reservoirs in the Voelkersen field. Transgressive Upper Permian (Zechstein) evaporites are the top seal of the OHW Formation forming a massive salt dome over the middle and northern part of the field. The top of the main Rotliegend reservoir section (Dethlingen), for which net reservoir thickness can vary between 5 and 150 m, have been drilled to depths of between 4750 and 5200 m (*ca.* 15 600–16 400 ft). Most wells in the Voelkersen field produce from the Dethlingen Formation, except for wells N4 and N5 which are completed in the OHW Formation. Well N2 has recently switched to OHW production; however, analyses in this study are exclusively for gas from the Dethlingen Formation. Only well N3 produces from both Dethlingen and OHW formations.

The Voelkersen field is sub-structured in numerous compartments due to low dip-slip transpressive Riedel shears caused by the Voelkersen Schwelle (Fig. 1). An exception is the uplifted and tilted block location for well S1 which is rather related to the graben margin, hence the Allertal-Lineament. Identified lateral

**Table 1.** *Voelkersen field Rotliegend gas composition (volume %)*

| Well | Reservoir | $C_1$ | $C_2$ | $C_3$ | $i$-$C_4$ | $n$-$C_4$ | $i$-$C_5$ | $n$-$C_5$ | $C_1/C_{2+}$ | $N_2$ | $CO_2$ | He |
|---|---|---|---|---|---|---|---|---|---|---|---|---|
| N1 | D | 87.7 | 0.92 | 0.017 | 0.001 | 0.002 | 0.000 | 0.000 | 92.9 | 11.1 | 0.141 | 0.057 |
| N2 | D | 87.1 | 1.25 | 0.075 | 0.015 | 0.015 | 0.005 | 0.004 | 63.5 | 11.2 | 0.220 | 0.050 |
| N3 | D, OHW | 88.7 | 1.10 | 0.088 | 0.018 | 0.017 | 0.007 | 0.004 | 71.8 | 9.7 | 0.294 | 0.045 |
| N4 | OHW | 86.4 | 0.73 | 0.011 | | 0.001 | 0.000 | 0.000 | 116.2 | 12.5 | 0.210 | 0.059 |
| N5 | OHW | 86.4 | 0.80 | 0.013 | 0.001 | 0.001 | 0.000 | 0.000 | 105.7 | 12.5 | 0.200 | 0.051 |
| S1 | D | 87.0 | 1.11 | 0.037 | 0.003 | 0.004 | 0.001 | 0.001 | 75.4 | 11.6 | 0.143 | 0.055 |
| S2 | D | 87.6 | 1.26 | 0.076 | 0.016 | 0.015 | 0.005 | 0.004 | 63.3 | 10.6 | 0.350 | 0.050 |
| S3 | D | 87.2 | 1.61 | 0.114 | 0.018 | 0.016 | 0.007 | 0.004 | 49.1 | 10.5 | 0.426 | 0.042 |
| S5 | D | 87.9 | 1.64 | 0.118 | 0.026 | 0.022 | 0.011 | 0.005 | 48.0 | 9.6 | 0.589 | 0.048 |

$C_1$, methane; $C_2$, ethane; $C_3$, propane; $i$-$C_4$, *iso*-butane; $n$-$C_4$, $n$-butane; $i$-$C_5$, *iso*-pentane; $n$-$C_5$, $n$-pentane.
D, Dethlingen Sandstone.
OHW, Obere Hannover Wechselfolge

seals are formed by structure (e.g. impermeable faults or cataclastics, Mauthe 2003) or sedimentary facies (e.g. dune type changes). Most compartments drilled revealed a different gas–water contact (GWC). In the case that the GWC could not be drilled, a free-water level (FWL) had been deduced from formation evaluation methods and production tests.

The temperature gradient is constant over the field (~0.03 °C/m) as well as the initial pressure regime within the compartments (~654 bar at 4800 m TVDSS). Compartmentalization has the most significant impact on economical field development and consequently on reserves. Therefore, it is imperative to delineate the extent of compartmentalization fully prior to infill drilling.

Bulk volume composition (%) of the gas produced from nine Voelkersen wells is listed in Table 1. There appears to be little systematic

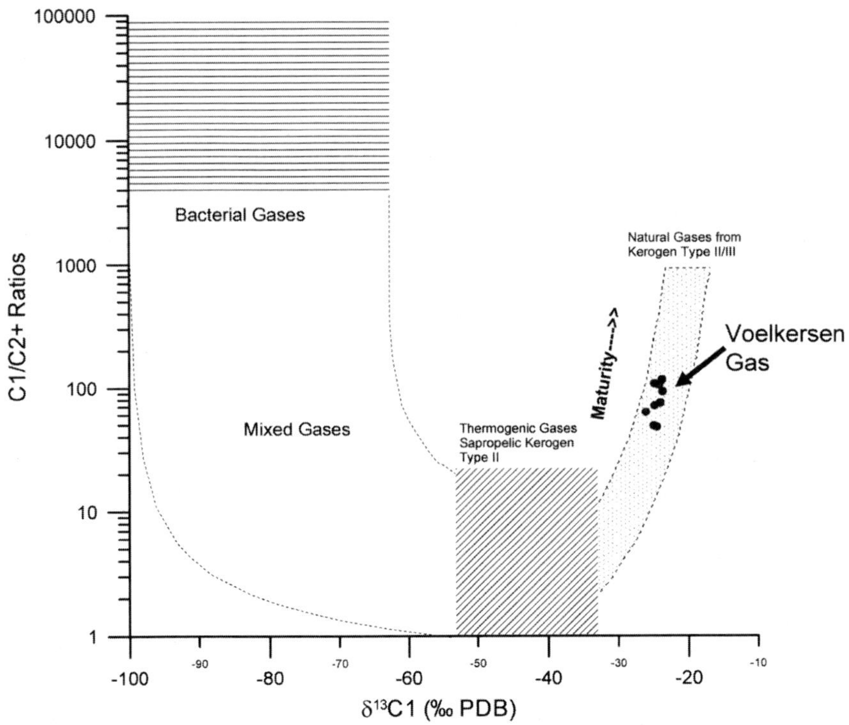


**Fig. 2.** Bernard-type plot for genetical characterization of gas based on methane carbon isotopic composition and gas wetness ($C_1/C_{2+}$).

variation over the field except that the wells producing from OHW sandstones show a comparatively dryer gas enriched in nitrogen ($N_2$). In general, the Voelkersen gas is rich in $N_2$, whilst hydrogen sulphide ($H_2S$) is completely lacking. Minimal volumes of condensates ranging between 85 and 260 barrels/$10^9$ standard $ft^3$ of gas (bbl/bscfg) are only produced from wells completed in the Dethlingen Sandstones. Conventional gas typing using methane stable carbon isotopes and gas wetness do not indicate significant compositional differences between the gases (Fig. 2), but enrichment of $^{13}C$ attests to the high level of maturity of the Westphalian coals from which the gas was generated (Barnard & Cooper 1983). Similarly, relatively high nitrogen content of the gases indicates origin from mature sources (Littke *et al.* 1995).

## Sampling and analytical methods

Application of the Jules–Thompson effect to separate liquids from produced gas requires a number of gas treatment steps after the gas exits the well head. Due to technical, safety and facility constraints, sampling after the cold separator is preferred. In order to investigate potential isotopic fractionation effects one sample was collected before the gas entered the main separation unit (sampling conditions in Table 2). Sampling directly at the well head was not possible due to high temperature and pressure of the gas. Dewatered condensates were collected at the cold separator and stored in glass vials. Gas samples were collected using special gas sample containers which were filled to 1.5 bar.

Condensates were analysed for their *iso*-heptane ($C_7$) composition using a gas chromatograph with the following settings: Chrompack CP 9002, 50 m length, 0.32 mm ID CP-SIL 5 CB column with pre-column 1 m, 0.32 mm ID, carrier gas He; split injection; temperature program 45–300 °C, 4 °C/min; FID detector. Heptane correlation ratios (ratios $C_1$ to $C_5$) were calculated according to the method developed by Halpern (1995).

Carbon and hydrogen isotopic composition of the gas samples was analysed using a Varian 3400 GC–Finnigan MAT 252 gas chromatograph–isotope ratio mass spectrometer (GCIRMS) with the following settings: carbon analysis: Finnigan MAT 252; GC column: Varian Pora-Plot Q; column flow, 1.2 ml/min; purge flow, 30 ml/min; injector, 220 °C; GC program: 50 °C (hold 10 min), 15 °C/min to 220 °C (hold 15 min); combustion oxidation reactor for carbon analysis: 940 °C (three nickel and two platinum wires w/oxygen trickle). $^2H/^1H$ analysis: Delta XL + GC column, Varian Pora-Plot Q; column flow, 1.1 ml/min; purge flow, 10 ml/min; injector, 220 °C; GC program: 50 °C (hold 10 min), 15 °C/min to 220 °C (hold 15 min); combustion Pyrocell for D/H analysis: 1440 °C. Stable carbon $^{13}C/^{12}C$ and $^2H/^1$H ratios were reported in $\delta$-notation ($\delta^{13}$C and $\delta^2$H) per mille relative to the PDB and SMOW standards (see Schoell 1984 for details). Measurements were performed relative to a carbon dioxide ($CO_2$) gaseous standard which was introduced directly into the ion source as a series of pulses such that no interference with sample peaks occurred. Due to the low concentration in the gas samples, carbon isotope ratios of pentanes, as well as hydrogen isotopic values for propane and higher homologues, are not used in the interpretation. Average standard deviations range from 0.04‰ for ethane to 0.47‰ for *n*-pentane in $\delta^{13}$C analysis and 0.93‰ (methane) and 1.13‰ (ethane) in $\delta^2$H analyses.

**Table 2.** *Gas sampling parameters*

| Well | Flow pressure (bar) | Temperature (°C) | Production rate ($m^3$/h) | Separator pressure (bar) | Separator temperature (°C) | Reservoir |
|---|---|---|---|---|---|---|
| N1 | 132 | 61 | 10 000 | 72 | 22 | Dethlingen |
| N2 | 133 | 50 | 7 000 | 69 | 20 | Dethlingen |
| N3 | 314 | 107 | 45 000 | 70 | 13 | Dethlingen |
| N3 sampled before cold separator | 314 | 107 | 45 000 | 70 | 13 | Dethlingen |
| N4 | 105 | 14 | 4 000 | 69 | 18 | OHW |
| N5 | 227 | 45 | 10 000 | 70 | 21 | OHW |
| S2 | 236 | 82 | 30 000 | 70 | 22 | Dethlingen |
| S3 | 335 | 93 | 35 000 | 71 | 8 | Dethlingen |
| S5 | 216 | 98 | 32 000 | 70 | 21 | Dethlingen |

OHW, Obere Hannover Wechselfolge.

## Analytical results

### *Sampling location within the production flow line*

In order to identify possible isotope fractionation effects due to gas sampling location in the production flow line a second sample from well N3 was taken before the gas entered the cold separator. Apparently, there is a maximum difference in the measured isotope ratios of methane of 0.6‰, whilst the isotopic composition of other wet components appears to be almost identical (Fig. 3). $\delta^2$H isotope ratios for methane are similar, whilst ethane values show an almost 7‰ difference. Although these samples by no means represent a statistical evaluation of potential sampling errors, it appears that there are no significant differences such as to overprint or obliterate the differences in isotopic composition reported below. For the time being it is assumed that the gas still carries the isotopic signal of the reservoir contents even after passing through the gas processing unit. Furthermore, gases obtained from well tests show comparable ranges of carbon and hydrogen isotope composition, as do the gases collected from the production line. Efforts to monitor potential temporal changes in composition are currently being made.

### *Condensate analysis results*

Condensates are produced from wells completed exclusively in the Dethlingen sandstones and samples were available from six wells only. Light hydrocarbon based parameters of Thompson (1983) and Mango (1987) confirmed the assumption that the condensate samples originate from a common source. Compound distribution in all samples is dominated by benzene. The heptanes and their $C_1$–$C_5$ correlation ratios plotted on the star diagram of Halpern (1995) show that the well N1 fluid is distinctly different. All ratios except for the $C_4$ value are comparatively low, thus 3,3-dimethylpentane predominates in the heptane distribution (Fig. 4). Additionally, relative concentrations of methylcyclopentane and cyclohexane in N1 were found to be low when compared with other condensate samples. N1 is drilled at the northwestern edge of the Voelkersen field and produces from a compartment that is known to be isolated based on gas–water contact depth and production history.

### *Stable isotope analysis*

Samples from eight producing wells were investigated in terms of the carbon and hydrogen isotopic composition of the methane to pentane components of the gases (Table 3). No samples were available from wells S1 and S6, since the wells did not produce at the time of the sampling campaign. Samples N1, N4 and N5 in Figure 5 show a clearly different carbon isotopic profile and differences in relative $^{13}$C concentration of up to 4.5‰ for ethane. Interestingly, methane carbon isotopes show a range of only 2.5‰ in the samples and apparently do not necessarily

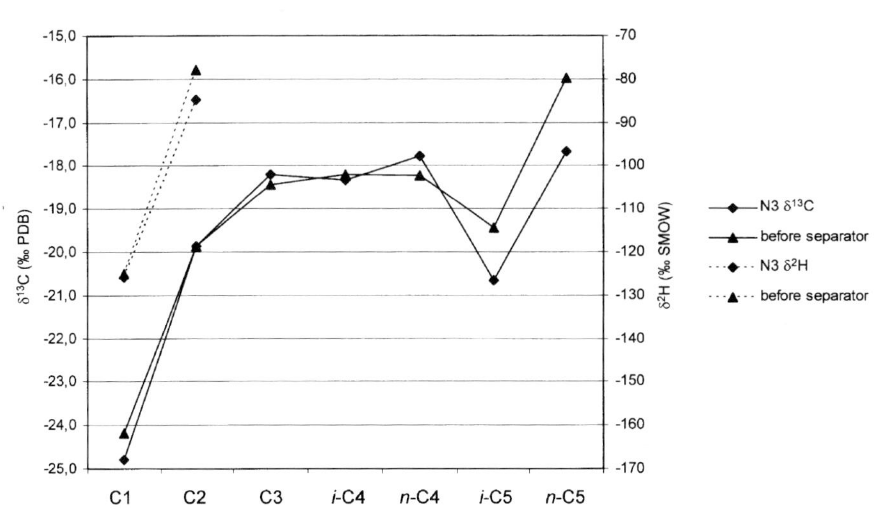


**Fig. 3.** Comparison of carbon and hydrogen stable isotopic composition of gas samples from Voelkersen field well N3 collected before and after the cold separator of the surface gas processing unit.

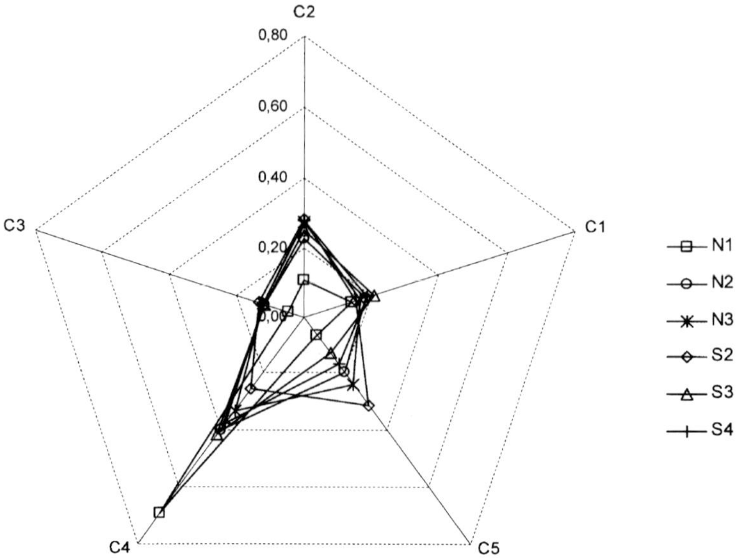


**Fig. 4.** Heptane ratio based oil correlation star diagram (after Halpern 1995) of Voelkersen condensate samples. Note that condensates are recovered only in minimal quantities from Dethlingen producers. Correlation ratios: $C_1$, 2,2-DMP/P3; $C_2$, 2,3-DMP/P3; $C_3$, 2,4-DMP/P3; $C_4$, 3,3-DMP/P3; $C_5$, 3-EP/P3; $P_3$, 2,2-DMP + 3,3-DMP + 2,3-DMP + 2,4-DMP + 3-EP; EP, ethylpentane; DMP, dimethylpentane.

reflect variance in isotopic composition of the wet components of the gas. Propane carbon isotopic composition in wells N1, N4 and N5 is lighter than for ethane, and in addition sample N4 shows again a lighter *n*-butane composition but a relatively heavy *iso*-butane. These irregular isotope ratios cause a distinct pattern in the isotope profiles of Figure 5, easily distinguishable from the regular carbon isotope profiles of the other wells. Because of the decrease in concentration over several orders of magnitude, analysis of the higher carbon number species is difficult and carbon isotopic values for the pentanes is considered unreliable. The hydrogen isotopic composition confirms the grouping of samples and the geochemical separation of wells based on carbon isotope ratios (Fig. 6). Ethane hydrogen isotopes in particular distinguish wells N1, N4 and N5 but also suggest potential heterogeneity of gas produced in the OHW sandstones of N4 and N5. On the other hand, little scatter in the ethane $\delta^2$H ratios of the other wells implies

**Table 3.** *Carbon and hydrogen isotopic composition of Voelkersen Rotliegend gas samples*

| | $\delta^{13}$C ‰ PDB | | | | | | | $\delta^2$H ‰ SMOW | |
|---|---|---|---|---|---|---|---|---|---|
| Well name | $C_1$ | $C_2$ | $C_3$ | $i$-$C_4$ | $n$-$C_4$ | $i$-$C_5$ | $n$-$C_5$ | $C_1$ | $C_2$ |
| N1 | −23.7 | −17.5 | −18.9 | | | | | −125.8 | −53.5 |
| N2 | −24.8 | −20.1 | −17.8 | −17.7 | −18.8 | −19.3 | −14.3 | −126.8 | −77.2 |
| N3 | −24.8 | −19.9 | −18.2 | −18.3 | −17.8 | −20.7 | −17.7 | −125.9 | −84.8 |
| N3 sampled before main separator | −24.2 | −19.9 | −18.4 | −18.2 | −18.2 | −19.4 | −16.0 | −125.2 | −78 |
| N4 | −23.8 | −16.3 | −21.4 | −17.1 | −19.7 | | | −124.5 | −89 |
| N5 | −24.2 | −17.2 | −19.9 | | | | | −123.9 | −68.3 |
| S2 | −26.1 | −20.6 | −18.7 | −18.6 | −18.9 | −20.3 | −19.3 | −123.9 | −87.3 |
| S3 | −24.9 | −20.8 | −18.5 | −18.7 | −18.7 | −20.2 | −18.6 | −126.0 | −86.4 |
| S4 | −24.5 | −20.1 | −18.0 | −18.2 | −18.7 | −19.8 | −18.6 | −125.8 | −78.1 |

$C_1$, methane; $C_2$, ethane; $C_3$, propane; $i$-$C_4$, *iso*-butane; $n$-$C_4$, *n*-butane; $i$-$C_5$, *iso*-pentane; $n$-$C_5$, *n*-pentane.

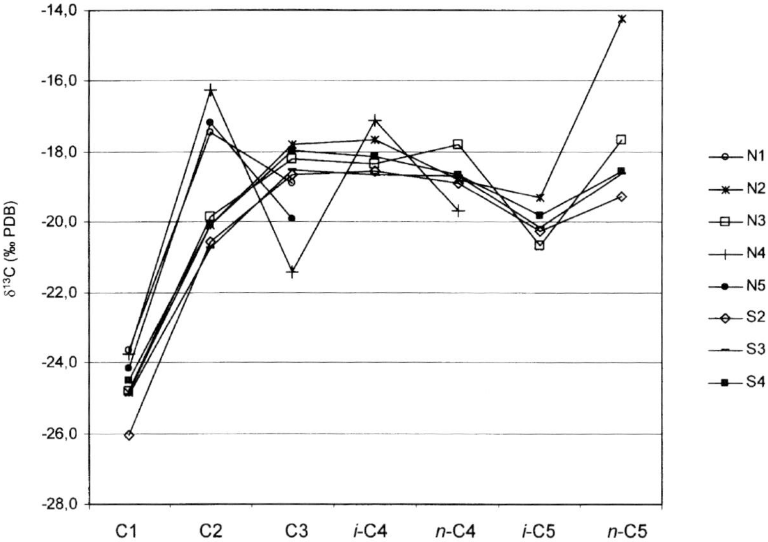


**Fig. 5.** Methane to pentane carbon stable isotope composition of gases produced from wells of the Voelkersen field ($C_1$, methane; $C_2$, ethane; $C_3$, propane; $i$-$C_4$, *iso*-butane; $n$-$C_4$, $n$-butane; $i$-$C_5$, *iso*-pentane; $n$-$C_5$, $n$-pentane).

compositional homogeneity. Again, there is little variance in the range of methane $\delta^2H$ ratios in the samples which range from −124 to −127‰ as compared with −54 to −89‰ for ethane (Table 3). Unfortunately, dynamic range of methane to pentane concentrations and instrument sensitivity limitations did not permit meaningful measurements of the $\delta^2H$ ratios of the higher carbon number compounds but it is conceivable that there may be similar substantial variations considering the larger mass difference between $^1H$ and $^2H$ as opposed to $^{12}C$ and $^{13}C$.

## Implications for reservoir compartmentalization

Geochemistry of the gas produced from well N1 drilled in the western edge of the northern fault block of the Voelkersen field confirms the isolated position of the compartment penetrated by this well. N4 and N5 produce from the OHW sandstones for which obviously no communication with the Dethlingen sandstone producer N3, to the west, exists. This implies that gas mixture ratios can be calculated for wells producing from both intervals if end-member isotopic composition can be established (Bazan 1998), this being a method that will be tested in future well workover projects. However, although well N3 produces from both the Dethlingen and OHW formations (approximately 20% of the produced gas from the OHW according to production logging tool (PLT) runs) there is no mixing apparent from the isotopic composition. Either contribution is still too low to have a measurable impact or the gases are compositionally similar in both formations of this well, implying a lack of a vertical seal. Differences in $\delta^2H$ isotopic composition of ethane on the other hand imply lateral separation of the OHW between N4 and N5. Similarity of gas composition in wells N3 and N2 suggests communication within the Dethlingen reservoir despite structural separation in the central northern block of the Voelkersen field. Similarly, communication over parts of the major WNW–ESE fault system is indicated by comparing the isotopic composition of wells S2, S3 and S4. The proposal of communication over this fault system has a major impact on field reserves estimation, but thus far could not be definitely established based on conventional field characterization methods. Apparently, geochemically the wells producing from the southern part of the Voelkersen field may also represent a continuous reservoir. However, it is known from other wells that even virtually identical isotopic signatures of gas may exist in wells that are clearly draining separate compartments and, as such, the similarity in condensate and gas isotope composition can only provide an indication for possible communication if other parameters are inconclusive or unavailable. A schematic map summarizing the compartment model based on geochemical data is shown in Figure 7. This rather simple approach of using gas isotope

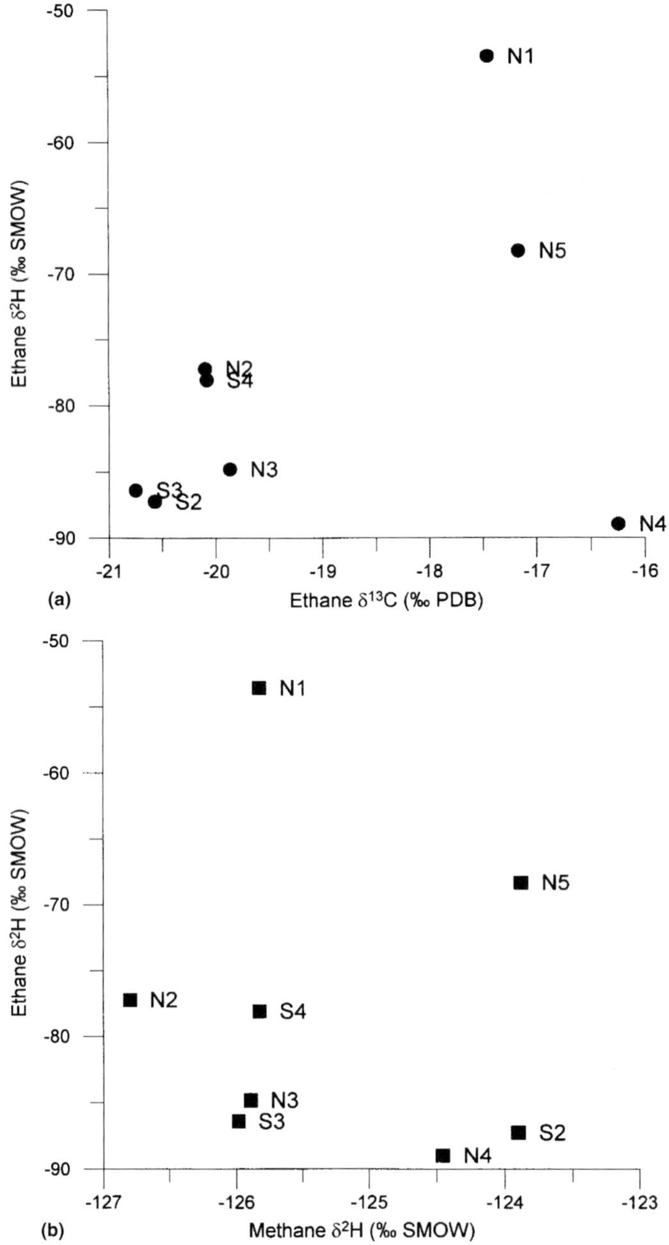


**Fig. 6.** Cross plots of ethane $\delta^2$H isotope composition versus ethane carbon stable isotope ratios (**a**) and $\delta^2$H of methane (**b**) of gas samples from the Voelkersen field ($C_1$, methane; $C_2$, ethane; $C_3$, propane; *i*-$C_4$, *iso*-butane; *n*-$C_4$, *n*-butane; *i*-$C_5$, *iso*-pentane; *n*-$C_5$, *n*-pentane).

composition for compartmentalization interpretation has been used in other Rotliegend fields of NW Germany (Mauthe 2003) where isolated compartments are similarly designated by irregular isotope composition patterns.

## Summary and conclusions

Thermogenic gases generated from anthracitic Westphalian coals and produced from deep heterogeneous sandstone reservoirs in the

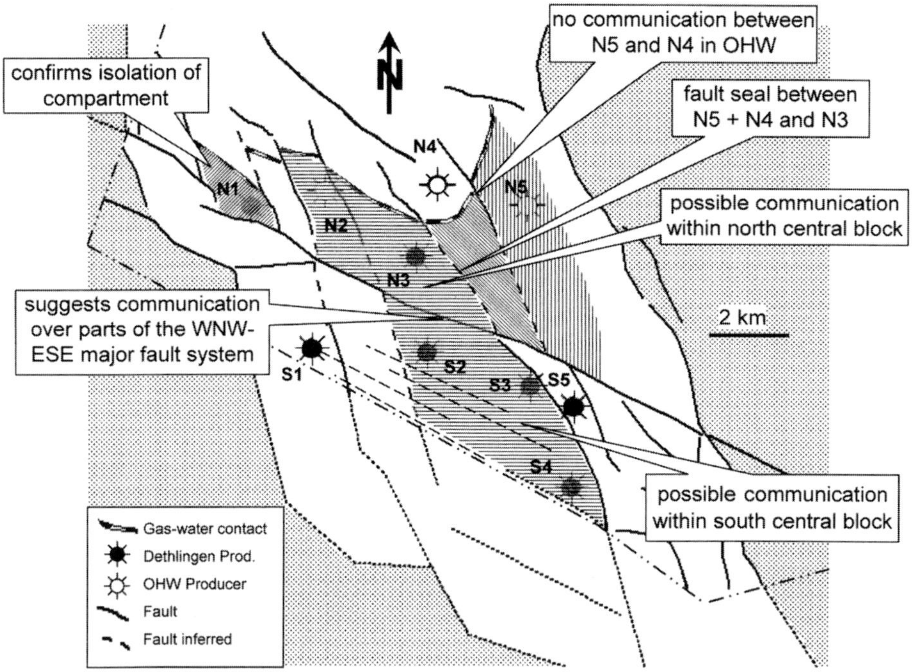


**Fig. 7.** Voelkersen Rotliegend field compartment model based on geochemical data (structure map top Dethlingen sandstone).

Voelkersen Rotliegend gas field show an unexpected variation in condensate and gas isotope composition. These variations correlate well with current field compartmentalization models but also suggest further compartment separation and communication, respectively. In addition to well tests, geochemistry is particularly suitable for study of this type of deep gas field since it can supplement conventional field characterization approaches at comparatively minor additional expense.

So far there is little supporting information to suggest why there are significant differences in the isotopic composition of gases in a field which is thought to be charged by a single source. Geological and structural setting, as well as spatial distribution of compartments showing irregular isotopic signatures, does not indicate different charging directions or sequential filling of the field. Compositional modification due to mixing of produced gases from different horizons can largely be excluded since the wells, with one exception, produce from one reservoir section only. It is considered more likely that geochemical signatures represent specific processes within the compartments, resulting in isotope fractionation effects and modification of the gas and condensate composition. Possibly, isolation of compartments from gas charges continuously migrating into the field area, residence time of gas in an isolated compartment and specific compartment $p/T$ conditions cause modification of the gas composition. Additionally, bitumen present in the high temperature reservoir (150–160 °C) may be responsible for gas composition alteration by *in situ* generation and commingling of catagenetically formed gas although quantification of this process is difficult to gauge at present.

A future application of this study is the analysis of trip gases and gases, adsorbed on or occluded in cuttings during drilling, to provide a first indication about the compartment that the well is entering. Through a combination of isotopic signatures and field production data, gas geochemistry may give first implications about the reservoir and gas content in the compartment that can be expected before the bit reaches projected final depth and before well completion and testing is initiated.

We gratefully acknowledge RWE Dea AG for allocating resources and permission to publish this study. Dr. H.-P. Berner's (RWE Dea AG) suggestions and facilitation of internal review procedures are much appreciated. Dr. R. Eiden and E. Münch of RWE

Dea AG Laboratory Wietze are thanked for sampling and condensate analysis. Timely isotope analysis by Prof. R. P. Philp, University of Oklahoma, is also much appreciated.

## References

ALBERTSEN, M. 1992. Oil and gas in Germany—exploration, production and research. *First Break*, **10**, 225–231.

BARNARD, P. C. & COOPER, B. S. 1983. A review of geochemical data related to the northwest European gas province. *In*: BROOKS, J. (ed.) *Petroleum Geochemistry and Exploration of Europe*. Pergamon Press, Oxford, 19–33.

BAZAN, L. W. 1998. The allocation of gas well production data using isotope analysis. SPE Gas Technology Symposium, Calgary, Canada, 15–18 March 1998, 1–9, Paper SPE **40032**.

GAUPP, R., MATTER, A., PLATT, J., RAMSEYER, K. & WALZEBUCK, J. 1993. Diagenesis and fluid evolution of deeply buried Permian (Rotliegend) gas reservoirs, northwest Germany. *American Association of Petroleum Geologists Bulletin*, **77**, 1111–1128.

HALPERN, H. I. 1995. Development and application of light-hydrocarbon-based star diagrams. *American Association of Petroleum Geologists Bulletin*, **79**, 801–815.

LARTER, S. R. & APLIN, A. C. 1995. Reservoir geochemistry: methods, applications and opportunities. *In*: CUBITT, J. M. & ENGLAND, W. A. (eds) *The Geochemistry of Reservoirs*. Geological Society, London, Special Publications, **86**, 5–32.

LITTKE, R., KROOSS, B., IDIZ, E. & FRIELINGSDORF, J. 1995. Molecular nitrogen in natural gas accumulations: Generation from sedimentary organic matter at high temperatures. *American Association of Petroleum Geologists Bulletin*, **79**, 410–430.

MANGO, F. D. 1987. An invariance in the isoheptanes of petroleum. *Science*, **273**, 514–517.

MAUTHE, G. 2003. Kompartmentbildende Verwerfungen infolge Kataklase (Rotliegend, NW-Deutschland). *Erdöl Erdgas Kohle*, **119**, 12–17.

PETERS, K. E. & FOWLER, M. G. 2002. Applications of petroleum geochemistry to exploration and reservoir management. *Organic Geochemistry*, **33**, 5–36.

SCHOELL, M. 1984. Stable isotopes in petroleum research. *In*: BROOKS, J. & WELTE, D. (eds) *Advances in Petroleum Geochemistry*, Vol. 1. Academic Press, London, 215–245

SCHOELL, M. & JENDEN, P. D. 1993. Isotope analyses of gases in gas field and gas storage operations. SPE Gas Technology Symposium, Calgary, Canada, 28–30 June 1993, 337–344, Paper SPE **26171**.

THOMPSON, K. F. M. 1983. Classification and thermal history of petroleum based on light hydrocarbons. *Geochimica et Cosmochimica Acta*, **47**, 303–316.

# Fluid properties, phase and compartmentalization: Magnolia Field case study, Deepwater Gulf of Mexico, USA

K. S. WEISSENBURGER[1,3] & T. BORBAS[2]

[1]*ConocoPhillips, PR 3024, Box 2197, Houston, TX 77252, USA*
*(e-mail: kate.weissenburger@runbox.com)*
[2]*ConocoPhillips, WL 9082, Box 2197, Houston, TX 77252, USA*
[3]*Present address: 5144 N Academy Blue #639, Colorado Springs, CO 80918, USA*

**Abstract:** The Magnolia Field in the deepwater northern Gulf of Mexico is a Plio-Pleistocene age mixed phase reservoir whose fluids are not in compositional equilibrium. Fluid heterogeneities have arisen principally due to (1) variations in maturity of the source rock from which the hydrocarbons were derived, (2) the extent to which biogenic methane has been incorporated into the fluids and (3) phase fractionation effects. These influences express themselves both in terms of bulk fluid properties such as gas/liquid ratio, API gravity and saturation pressure and minor compositional attributes such as hydrocarbon gas isotopic composition and gasoline range molecular ratios. Significant compositional variations that cannot be ascribed to gravitational fluid segregation occur within reservoirs that are demonstrably in pressure communication. These variations challenge the notion that hydrocarbon fluid mixing is geologically instantaneous and underscore the importance of testing assumptions regarding compositional equilibria in conjunction with reservoir studies. Although the state of disequilibrium impedes compartmentalization assessments at Magnolia, it provides both opportunities for fluid property and phase predictions and potentially a development setting in which geochemical surveillance techniques may be profitably employed.

Since the early 1990s the integration of reservoir fluid geochemical information into reservoir descriptions has become a relatively routine practice. In large part the aim of such studies is to aid in well planning (Hwang *et al.* 1994; Smalley & Hale 1996; Beeunas *et al.* 1999; Edman & Burk 1999; Levaché *et al.* 2000), reconstruction of filling, spilling and alteration histories (Hillebrand & Leythaeuser 1992; Holba *et al.* 1996; Horstad & Larter 1997), and fluid properties characterization and production monitoring and allocation (Kaufman *et al.* 1990, 1997; McCaffrey *et al.* 1996; Nicolle *et al.* 1997; Bazan, 1998). Geochemical data, irrespective of the application, find their greatest utility when interpreted in conjunction with other data types such as those provided by engineering, geological and geophysical studies (Smalley & England 1992; Larter & Aplin 1995; Westrich *et al.* 1999). Enormous advances in computation speeds increasingly allow data-rich high resolution fluid descriptions to be used in reservoir simulation initializations. Additionally, as high resolution secondary migration modelling tools are developed, detailed descriptions of reservoir fluid composition can be used to calibrate forward reservoir filling models (Wilhelms *et al.* 2001), which may find particular utility when used in conjunction with reservoir characterization studies.

The potential utility of fluid geochemical characterization studies during the exploration and field delineation phase in deepwater exploration environments is uncontroversial. Here, well costs are sufficiently high that project decisions are, by economic necessity, often made based on far fewer wells than adequately characterize the fluid character and continuity of most reservoirs. Additionally, project economics in deepwater environments are extremely sensitive to hydrocarbon flow rates which, in the absence of drill stem testing, can only be inferred from conventionally obtained static data such as wireline logs, wireline formation tests and conventional cores. The very high costs in deepwater production settings associated with wellbore intervention also exert a disproportionate influence on project economics. As a consequence, in these environments the value of fluid geochemistry information that may bear on flow rates (e.g. *in situ* viscosity and compartment dimensions) and potential production problems (e.g. wax deposition and asphaltene precipitation) is enormous.

With particular respect to assessments of reservoir continuity, it is critically important to determine the degree to which reservoir fluids have attained compositional equilibrium (Larter & Aplin 1995; Wavrek *et al.* 2001). In order to be able to identify the presence of discontinuities that may influence reservoir performance one must be able to discriminate between ephemeral geochemical variations and those related to the presence of flow barriers.

*From*: CUBITT, J. M., ENGLAND, W. A. & LARTER, S. (eds) 2004. *Understanding Petroleum Reservoirs: towards an Integrated Reservoir Engineering and Geochemical Approach.* Geological Society, London, Special Publications, **237**, 231–255. 0305-8719/04/$15.00 

Although hydrocarbon fluids are likely to establish vertical and lateral compositional equilibrium via density-driven and diffusion processes over relatively short periods (e.g. on the order of $10^4$–$10^6$ years; England *et al.* 1995) equally potentially rapid processes such as those related to hydrocarbon charge (i.e. temporal or spatial variations in charge composition), changing pressure and temperature conditions, biodegradation and leakage may result in hydrocarbon fluid compositional disequilibria in the absence of fluid barriers. Similarly, characterizing the degree of reservoir fluid heterogeneity that may exist due to compositional disequilibrium is a prerequisite for time-series studies. Variations of fluid composition over production time scales have very different implications depending on the extent to which fluids within reservoirs have mixed prior to being produced; changing composition due to production from a heterogeneous hydrocarbon accumulation (e.g. McKinney & Bland 2003) may mimic changing composition due to ongoing hydrocarbon charge into a reservoir over production time scales (e.g. as attributed by Whelan *et al.* 2001).

The discovery of the Magnolia Field in the deepwater Gulf of Mexico in 1998 launched an aggressive geochemical data acquisition programme. The principal objectives of this programme have been to fully characterize the hydrocarbon fluids in order to address issues regarding reservoir continuity and the relationships between hydrocarbon properties and phase behaviour. This paper describes the key geochemical attributes of Magnolia hydrocarbons, the origin of their variability, and their relationship to and control of reservoir pressure/volume/temperature (PVT) properties. Through integration of high resolution mud gas data with formation test information the degree to which field hydrocarbon fluids have established equilibrium can be addressed. Although it will be demonstrated that the nature of Magnolia fluids renders the field particularly challenging, if not unsuitable, for geochemistry-based continuity studies as typically conducted, the heterogeneity evidenced by the hydrocarbon fluids provides a framework for phase behaviour predictions, elucidation of hydrocarbon charge processes and suggests the potential utility of geochemically based surveillance methods during production.

## Magnolia Field geological setting and description

Magnolia was discovered in 1999 with the Garden Banks (GB) 783-1 well that was drilled in 4668 ft of water to a total depth of 16 867 ft. This, and the six appraisal wells drilled during the subsequent two years, are the subject of this paper. Production was sanctioned in 2001 and the first oil is expected in 2004.

ConocoPhillips has a 75% interest in and operates the field; Devon Energy owns the remaining 25%.

The Magnolia Field is located in the Titan mini-basin that lies in Garden Banks Blocks 783 and 784 in the US Federal waters of the Gulf of Mexico approximately 160 miles offshore (Fig. 1). The Titan mini-basin lies in Diegel *et al.*'s (1995) tabular salt and mini-basin province of the deepwater Gulf of Mexico. Titan, which covers an area of approximately 260 square miles, is one of the deepest of the salt-bounded secondary basins in northern Gulf of Mexico containing approximately 38 000 ft (9.0–9.5 sec twt) of Miocene, Pliocene and Pleistocene deepwater siliciclastics at its depocentre in GB 695/696. Patterns of Cenozoic sedimentation in this province are dominated by the very complex interplay of allochthonous salt movements and the deposition of progradational deepwater siliciclastics.

The Titan mini-basin is believed to have formed in a manner analogous to the salt-stock canopy model described by Rowan *et al.* (1994) and Rowan (1995). According to this model, an autochthonous Middle Jurassic (Louann Formation) salt pillow or swell evolved in response to Cretaceous and Palaeogene sedimentation into a deeply rooted 'bulb-shaped' salt stock at approximately the current Titan basin location. Mini-basin formation may have initiated as early as Palaeocene but certainly during the Eocene–Oligocene when the salt source was cut off, thus promoting the accommodation space necessary for initial basin nucleation. Salt withdrawal continued throughout the Miocene, Pliocene and Pleistocene in response to sedimentary loading which resulted in the preservation of the exceptionally thick basin fill. Eventually, continued salt withdrawal resulted in the formation of a deep salt weld at the base of the Titan mini-basin. This weld probably provided access into the mini-basin of hydrocarbons expelled from pre-Tertiary source rocks. Both ongoing salt movement and sedimentation are reflected by the bathymetry in the Titan basin and adjacent areas shown in Figure 2. Additionally, Figure 2 shows an inventory of sea surface natural hydrocarbon slicks, interpreted and mapped from synthetic aperture radar (SAR) imagery, which, among other things, testify as to the present-day activity of the petroleum system in this portion of the northern Gulf of Mexico.

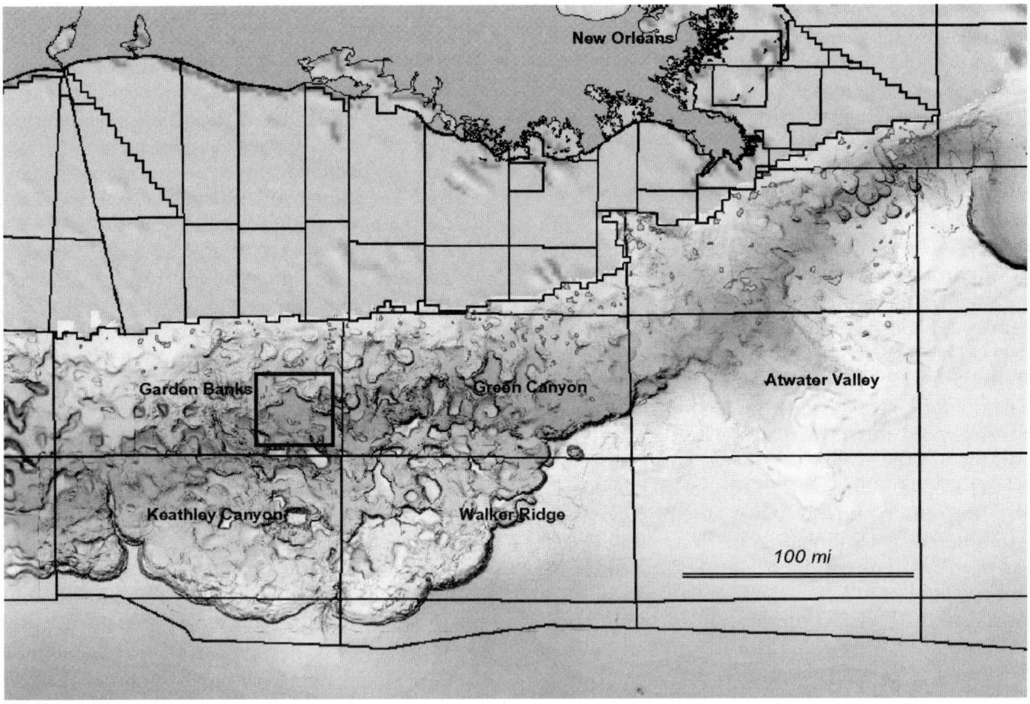


**Fig. 1.** Shaded relief bathymetric image of the north-central Gulf of Mexico. The Titan mini-basin, containing the Magnolia Field in the Garden Banks Federal area, is outlined and shown in Figure 2.

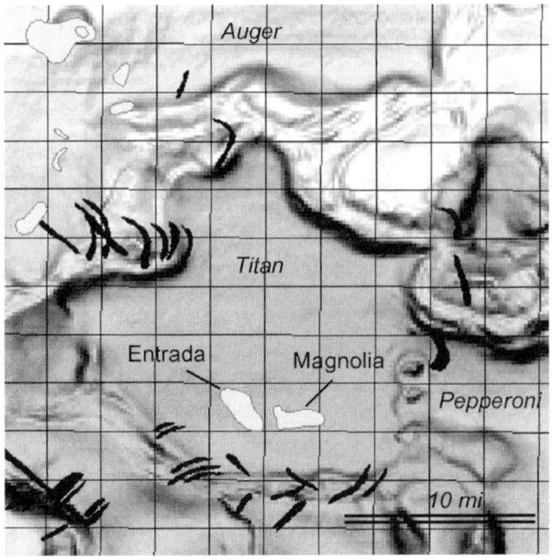


**Fig. 2.** Shaded relief bathymetry in the vicinity of the Titan mini-basin containing the Magnolia and Entrada fields. The adjacent Auger and Pepperoni mini-basins are annotated. Natural hydrocarbon slicks emanating from the basin margins, mapped from synthetic aperture radar (SAR) images, are shown.

Although effective source rock-quality intervals have yet to be penetrated in the immediate vicinity of the Magnolia Field, indirect evidence supports the presence of a prolific Upper Jurassic oil-prone source rock in this portion of the deepwater Gulf of Mexico. Extremely rich oil source rocks of Tithonian age are known to extend from the Campeche Sound to the Tampico–Misantla Basin in Mexico (Santamaria-Orozco *et al.* 1998; Guzmán-Vega & Mello 1999; Romero *et al.* 2000). These were deposited in a restricted carbonate-dominated environment which can be inferred to have extended well offshore by virtue of the character of hydrocarbons recovered by DSDP drilling at the Challenger Knolls (Davis & Bray 1969; Watkins & Buffler 1996) and piston core studies (Holguin *et al.* 2002).

In the northern Gulf of Mexico, the recently drilled Garden Banks 754-1, which lies approximately 100 miles west of the Titan mini-basin, provides direct evidence that a similar source-prone environment existed during the late Jurassic. This well penetrated a thick extremely organic rich interval (Jarvie *et al.* 2002). Biomarker attributes of liquid hydrocarbons recovered in piston cores from the greater Garden Banks 754 area suggest their derivation from a carbonate to marly source rock interpreted as having been derived principally from a Tithonian source rock (unpublished ConocoPhillips data; Cole *et al.* 2001; Hood *et al.* 2002). Based on analysis of reservoired oils and hydrocarbon seeps, an Upper Jurassic (principally Tithonian) hydrocarbon system was delimited by Hood *et al.* (2002) to extend across most of the northern portion of the tabular salt/mini-basin province which includes the greater Titan mini-basin area.

Hydrocarbon-bearing reservoirs at Magnolia are Miocene, Pliocene and Pleistocene in age and were deposited from about 7.0 to 1.4 Ma ago. These siliciclastics were derived principally from the west, northwest and north via long-lived submarine systems and deposited across a pronounced north-protruding salt nose that lies on the southern margin of the Titan mini-basin (Figs 3 & 4). Pliocene deposition was dominated by deepwater ponded facies assemblages (being dominated by confined deposition, *sensu* Prather *et al.* 1998) consisting of sheet sandstones and intervening mudstones. The transition from Pliocene to Pleistocene deposition is marked at Magnolia by a significant unconformity characterized by channel scour morphology which separates Pliocene ponded facies from Pleistocene transitional facies, consisting of a stacked series of amalgamated channels, mudstones and lobe or sheet sandstones. These transitional facies are overlain by a (non-reservoir-bearing) shale-dominated bypass facies assemblage that records a decrease in accommodation space relative to sediment influx during the latest Pleistocene.

Petrographically, the main Magnolia reservoirs are composed of fine grained sand to silt size feldspathic quartzose sandstones. Authigenic components consist of occasional discrete patches of calcite and siderite cements that are unlikely to impact fluid flow characteristics. The sands and silts are otherwise devoid of authigenic components.

The Magnolia trap is formed by a combination of structural and stratigraphic elements. Magnolia hydrocarbon-bearing reservoirs lie within a depth interval extending from 12 600 to 16 800 ft tvd (approximately 8000–12 000 ft below seabed). As would be surmised from the preceding descriptions, sedimentation rates at the field location are high, on average 3500 ft/Ma. Additionally, as is typical of this portion of the northern Gulf

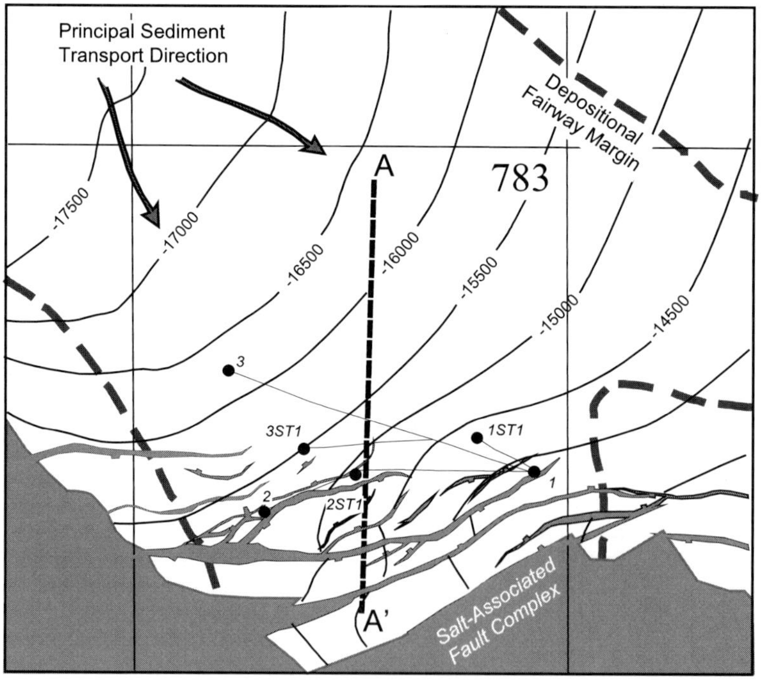


**Fig. 3.** Magnolia Field well locations and generalized fault pattern in relation to bounding allochthonous salt. The location of the seismic line shown in Figure 4 is shown as line A–A′. The depositional fairway margin shown coincides with the stratigraphic extent of the principle Magnolia Field sands. Depth contours of an intra-Pleistocene surface are in feet subsea.

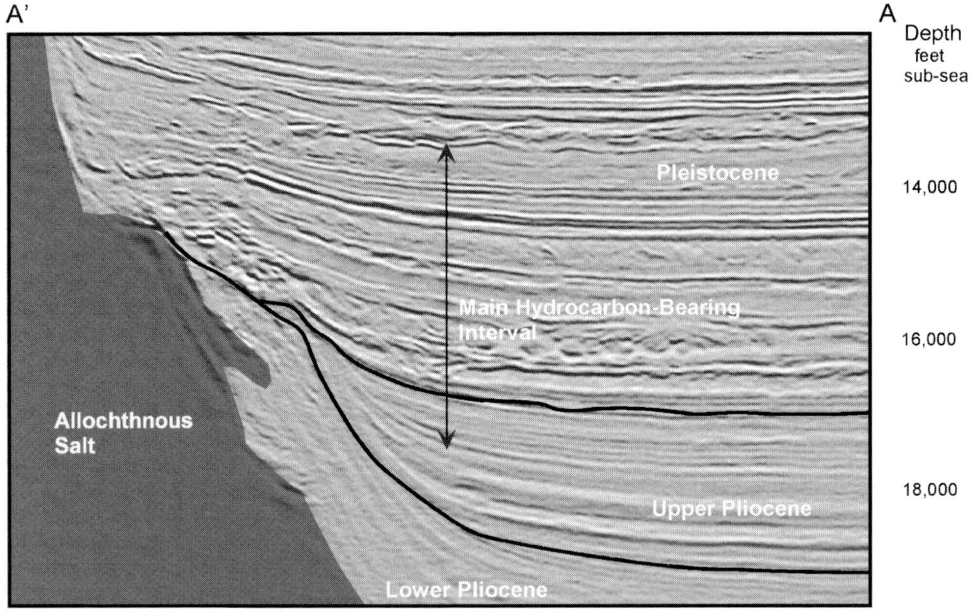


**Fig. 4.** North–south seismic line shown in Figure 3 illustrating the stratigraphic and structural aspects of the Magnolia Field. Miocene-age hydrocarbon-bearing sands are located elsewhere in the field.

of Mexico salt and mini-basin province, present-day geothermal gradients from the seabed to trap are low, on average 1.0 °F/100 ft. Magnolia reservoir temperatures range from 130 to 165 °F. The degree of overpressure is modest and varies between reservoirs, ranging from about 2200 to 5400 psi above hydrostatic. Structural restoration of the field and adjacent strata suggest that trap formation may have initiated as early as the end of the Pliocene.

## Methods

Hydrocarbon fluid samples discussed in this paper were obtained from the seven exploration and appraisal wells drilled between 1999 and 2001. Samples were collected using Schlumberger's MDT wireline formation tester. These were obtained by pumping approximately 15 gallons of filtrate and reservoir fluid, then diverting fluid flow into one gallon and 450 $cm^3$ chambers. Samples were subsequently overpressured downhole prior to bring the tool to the surface. One-gallon samples were tested for contamination on the well site and restored by heating and agitation for 24 h. Following restoration, the one-gallon samples were transferred to smaller, usually 800 $cm^3$ containers for transport. 450 $cm^3$ samples were transported directly to the laboratory where they were restored by heating and agitation for 5 days. 10–15 $cm^3$ subsamples for geochemical analysis from the 800 or 450 $cm^3$ containers were pushed out into an isolated chamber and pressure reduced to atmospheric conditions. A full suite of PVT tests were performed on all samples. The properties discussed in this paper are based on single stage flash analyses of laboratory-recombined samples. These pressurized samples were reduced from reservoir temperature and *ca.* 3000 psi above reservoir pressure to 60 °F and atmospheric pressure.

Cylinders of the flashed gas were analysed for hydrocarbon gas composition on a Carle gas chromatograph (GC) system equipped with both a thermal conductivity detector (TCD) and a flame ionization detector (FID). An off-line preparation system and dual inlet mass spectrometer (MS) were used to analyse the carbon and hydrogen isotopic values of hydrocarbon components. A customized Gow Mac GC was interfaced with a vacuum/combustion system to separate hydrocarbons from other components and combust to $CO_2$ and water that were purified and sealed into Pyrex tubes for isotopic analysis. The $CO_2$ was analysed directly on one of three dual inlet mass spectrometers: Finnigan Delta S, Finnigan Delta + XL or VG SIRA II. The water was reacted with zinc turnings and converted to hydrogen gas, which was analysed on either the Delta S or Delta + XL MS.

Whole oil gas chromatography of the oil and condensate samples was performed using a Hewlett Packard 7673 Autosampler to inject 0.1 μl of whole oil spiked with internal standard into a split/splitless injector of a Hewlett Packard 5890 GC equipped with flame ionization detection. Products were swept into a 60 m J&W Scientific DB-1 capillary gas chromatography column using helium as the carrier gas under constant flow conditions. The oven temperature program started at 30 °C, held for 5 min, ramped to 320 °C at 3 °C/min, and held for 20 min. The injector was held at a constant 275 °C.

Following separation using medium pressure liquid chromatography, saturate and aromatic GC-MS was performed using a Hewlett Packard (HP) 6890 equipped with a HP 7683 autosampler, on-column injector and a HP 5973 mass selective detector (MSD). Saturates were separated using a 60 m DB-5 column using an oven program starting at 100 °C, ending at 320 °C, and having a total run time of 118.5 min. The aromatics were separated using a 60 m DB-1 column using a column oven program starting at 70 °C, ending at 315 °C, and having a total run time of 131.33 min. Both saturate and aromatic GC-MS runs used an on-column injection technique with the injector temperature programmed at 3 °C above that of the column oven. The MSD was operated in the selected ion mode (SIM).

During drilling of one of the Magnolia wells, one-litre CALI-5-Bond™ gas bags were collected from the gas line going from the shale shaker to the mud loggers unit. Mud gas samples from all other wells, excluding the discovery well from which no mud gas samples were taken, were collected using IsoTubes™ also from the gas line. Hydrocarbon composition of the mud gases were measured on an Agilent GC fitted with a customized autosampler and FID and TCD detectors. Isotopic analysis of $C_1$ (and in some cases $C_2$ and $C_3$) were performed when

**Table 1.** *Summary of Magnolia Field fluid properties, corrected for drilling fluid contamination*

| Depth (ft MD) | Depth (ft TVD) | API (°) | Pressure (psi) | $P^*_{sat}$ (psi) | GLR (scf/stb) | Sulphur (%) | TAN (mg KOH/g) |
|---|---|---|---|---|---|---|---|
| *Saturated gas-condensates* | | | | | | | |
| 13209 | 12614 | 43.6 | 7806 | 8650 | 101674 | | |
| 15122 | 14782 | 37.0 | 9660 | 9846 | 4314 | | |
| 16054 | 15540 | 39.1 | 10039 | 9803 | 8288 | | |
| *Saturated oils* | | | | | | | |
| 15341 | 14961 | 26.5 | 9706 | 9512 | 1480 | | |
| 15401 | 15009 | 26.2 | 9721 | 9512 | 1461 | 1.62 | 0.61 |
| 16133 | 15605 | 25.5 | 10054 | 9777 | 1644 | 2.03 | 0.84 |
| *Undersaturated gas-condensates* | | | | | | | |
| 14067 | 14065 | 39.2 | 9158 | 8565 | 7061 | 0.72 | 0.19 |
| 14402 | 14185 | 43.3 | 9200 | 8150 | 17439 | | |
| *Undersaturated oils* | | | | | | | |
| 14965 | 14963 | 36.3 | 10176 | 5439 | 2062 | 0.88 | 0.30 |
| 15197 | 15195 | 37.0 | 11247 | 5096 | 2016 | | |
| 16912 | 16250 | 38.0 | 12516 | 4980 | 2314 | 0.73 | 0.08 |
| 17055 | 16360 | 34.3 | 12553 | 5164 | 1735 | 1.14 | 0.22 |
| 17381 | 16630 | 37.3 | 12825 | 4907 | 2202 | 0.81 | 0.40 |
| 18682 | 15856 | 37.5 | 10880 | 4849 | 2077 | 0.73 | 0.25 |
| 18918 | 16092 | 33.6 | 10950 | 6010 | 1923 | 1.09 | 0.28 |
| 19340 | 16246 | 32.1 | 10860 | 6955 | 1808 | 1.30 | 0.05 |
| 19725 | 16518 | 36.4 | 10972 | 6163 | 2292 | 0.85 | 0.05 |
| 20110 | 16794 | 36.8 | 11048 | 5601 | 2088 | 0.83 | 0.05 |
| 20322 | 16637 | 32.4 | 10545 | 9090 | 2721 | 1.30 | 0.10 |
| 20327 | 16058 | 37.2 | 11060 | 4673 | 1915 | | |
| 20518 | 16766 | 32.6 | 10580 | 8644 | 2352 | 1.35 | |
| 21021 | 16350 | 35.5 | 11150 | 5293 | 1922 | 0.99 | 0.19 |

MD, measured depth; TVD, true vertical depth; $P_{sat}$, saturation pressure; GLR, gas/liquid ratio; TAN, total acid number.

* Saturation pressures calculated from MDT samples are generally somewhat lower than actual saturation pressures due to the combined influence of sample drilling mud contamination and pressure draw-down effects.

gas yields permitted on gas bag and IsoTubes using one of either online systems consisting of an Agilent 6890 GC, a combustion unit and a Finnigan Delta + or Finnigan Delta + XL MS.

In addition to the reservoir fluid samples studied, a sample of the NovaPlus synthetic oil base drilling mud additive used in each well was analysed using the methods described above.

## Results and discussion

### *Fluid property and phase characteristics*

Magnolia Field fluids properties, summarized in Table 1, are strikingly heterogeneous. Figure 5 illustrates this by way of a cross plot of fluid saturation pressure (bubble point and dew point pressures for oil and gas-condensate tests respectively) against measured reservoir pressure for available MDT samples. In very general terms, four fluid types may be defined. Undersaturated oils (solid circles in Fig. 5) are by far the most common. Noteworthy within this family are the two samples that have saturation pressures in excess of 8000 psi (annotated with an asterisk); their laboratory behaviour suggests that they are very close to their critical point, displaying either dew or bubble points depending on sample handling. Saturated oils and their associated gas caps (solid squares proximate to open triangles) occur in some Magnolia reservoirs. A shallow, dry, saturated gas-condensate (whose association with an oil leg is uncertain) was tested in one of the wells (open triangle). Samples taken from a relatively shallow undersaturated gas-condensate reservoir are shown as open circles.

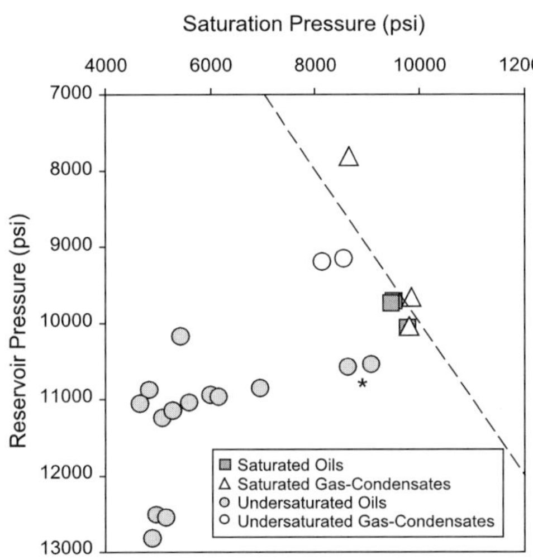


**Fig. 5.** Saturation pressure against reservoir pressure for MDT samples. The diagonal line represents saturation. Fluids are classified in this, and the majority of the figures that follow, based on reservoir phase: undersaturated oils (●), saturated oils (■), saturated gas-condensates (△, which comprise both gas caps of the saturated oils and a shallow dry saturated gas-condensate) and undersaturated gas-condensates (○). Saturation pressures have been corrected for oil-based mud contamination. The two samples marked with an asterisk display behaviour suggesting that they are very close to their critical point.

Depth-dependent generalizations cannot be made across the entire field with respect to trends governing fluid properties (Table 1). The relationship between oil saturation pressure and API gravity shown in Figure 6 is particularly notable in being opposite to that expected in, for example, a simple maturity or biodegradation series. All other things being equal, as the proportion of lighter to heavier molecular weight components increase (e.g. with increasing maturity or decreasing extent of biodegradation) saturation pressures will tend to increase (DiPrimio *et al.* 1998). In Figure 6 the low API gravity saturated oils can be seen to have significantly higher bubble point pressures than those of the highest API oils. Similarly, although the data are sparse, the high API gravity condensates tend to have lower dew point pressures than the low API gravity condensates. Oil and condensate

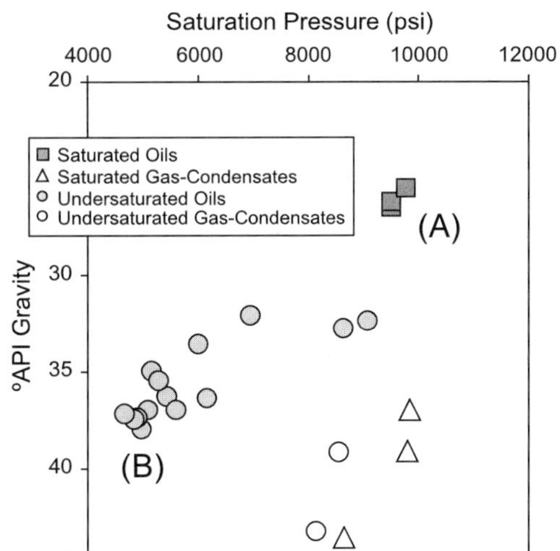


**Fig. 6.** Saturation pressure against oil and condensate API gravity for Magnolia MDT samples. Annotated samples (A) and (B) are shown in Figure 7. Saturation pressures and liquid gravities have been corrected for oil-based mud contamination.

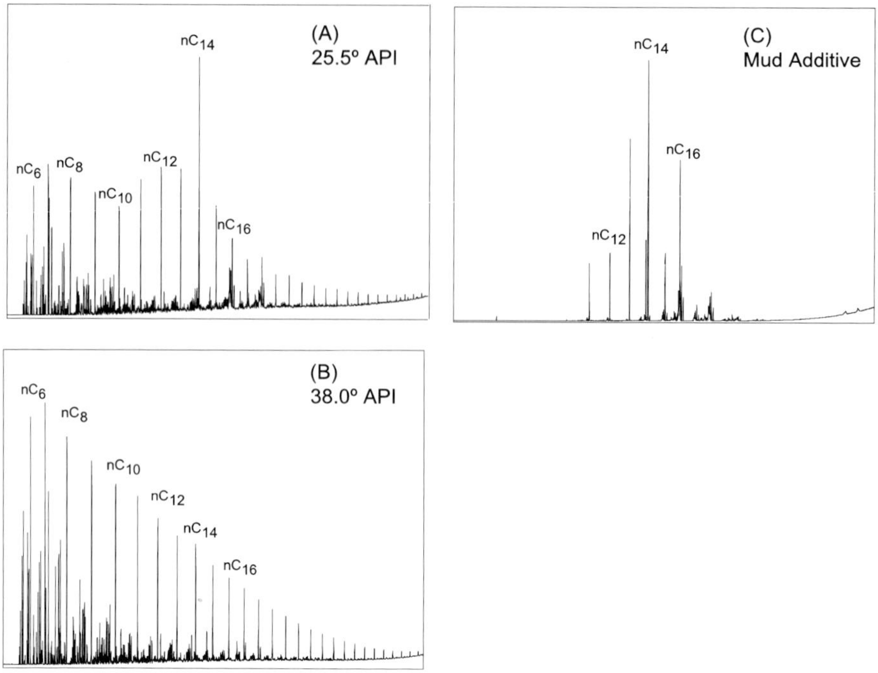


**Fig. 7.** Whole oil gas chromatograms of (**A**) a saturated low gravity oil and (**B**) an undersaturated high gravity oil, each representing oil 'end-members' annotated in Figure 6. API gravity of the saturated sample has been corrected for 5.5% oil-based mud contamination which is also responsible for elevated amounts of $nC_{11}$–$nC_{16}$. The gas chromatogram shown in (**C**) shows the drilling mud additive used in the exploration and appraisal wells.

gravities are positively correlated with sulphur content (Table 1, which ranges from 0.73 to 1.93%, $R^2 = 0.93$). Total acid number (TAN) does not show a strongly linear correlation with API gravity although the saturated oils tend to have a higher TAN than the undersaturated oils.

Illustrative whole oil gas chromatograms of two 'end member' oils (a saturated 25.5° and an undersaturated 38.0° oil) are shown in Figure 7 as well as a typical chromatogram of the synthetic oil-based mud with which the Magnolia wells were drilled. Synthetic oil-based drilling mud contamination of the MDT fluid samples ranged from less than 1% to as much as 30%. Analyses of the base mud used in each of the Magnolia wells shows that this contamination has not materially interfered with biomarker, gas or whole oil data (apart from obscuring the true yields of compounds eluting between *ca.* $nC_{11}$ and phytane). Measured PVT and bulk properties of the heavily contaminated samples were profoundly influenced; all properties shown in Table 1 and discussed in this report have been corrected for oil-based mud contamination.

From a reservoir management standpoint, one of the principal challenges at Magnolia was to understand the origin of the fluid variability in order that reservoir continuity issues and phase prediction could be addressed. The discussion that follows focuses on resolving, to the extent possible using the conventional analysis methods employed, the origin of Magnolia fluid property variations. Towards that end, selected possible sources of variation are treated below in turn: source rock facies, source maturity, water washing, biodegradation, phase fractionation and biogenic gas.

*Source rock facies.* Biomarker data from Magnolia fluids can be used to address the degree to which variations in source rock facies are responsible for or contribute to the heterogeneity of fluid properties discussed above. Sterane and hopane biomarkers of the Magnolia oils compare well with those described by Guzmán-Vega & Mello (1999) from the Sureste basin that were interpreted as having been derived from Tithonian source rocks. Figure 8 illustrates this point showing facies-sensitive hopane ratios from Magnolia oils and condensates against fields defined by

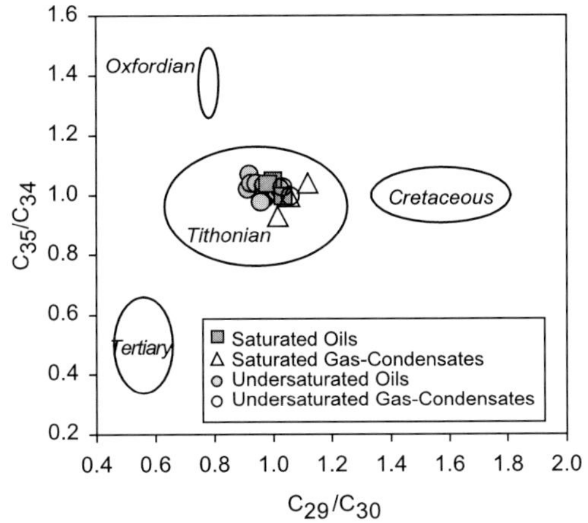


**Fig. 8.** Hopane biomarker ratios $C_{29}/C_{30}$ against $C_{35}/C_{34}$ (Table 2) of Magnolia oils and condensates plotted against the fields defined by Guzmán-Vega & Mello (1999) for oil families and their source rocks in the Sureste basin in Mexico.

these authors (see also Table 2). Not only does the narrow range of Magnolia data suggest minimal variations in source rock, but, by analogy with the Sureste basin Tithonian source rocks, those that charged the Magnolia Field are inferred to have been deposited in a marine carbonate-influenced environment characterized by some degree of anoxia/dysoxia. In the Sureste Basin the oils believed derived from Tithonian source rocks are quite variable in terms of their quality; for example API gravities range from 15 to 47° and sulphur content ranges from 0.1 to 6%. These variations have been interpreted as being principally due to large variations in source maturity and minor variations in the degree of anoxia/dysoxia and siliciclastic input in the source environment (Guzmán-Vega *et al.* 2001).

Oil and condensate gasoline-range facies-sensitive ratios tend to support the interpretation that Magnolia fluids were derived from a source rock of relatively uniform facies. Figure 9 shows a cross plot of P2+N2 versus P3 (see Table 2 for definitions) for Magnolia liquids, parameters believed to be fairly robust, maturity-insensitive, indicators of source rock facies (Mango 1987, 1990; ten Haven 1996). The Magnolia oils and condensates appear, based on their relatively invariant ratio of P2+N2 to P3, to have been derived from a uniform kerogen type. Variations in source rock facies do not appear to be responsible for the compositional differences amongst Magnolia fluids.

*Maturity.* Conventional biomarker maturity indicators suggest that source rock maturity has influenced Magnolia oil properties (e.g. API gravity) but that Magnolia condensates are not a high maturity product. Two maturity-sensitive indices, Ts/Ts+Tm and MPI-1 are shown in Figure 10 (and see Table 2). The condensate sample from the shallow saturated gas zone has been omitted due to very low biomarker yields. Overall, the undersaturated, generally high API gravity oils tend to show modestly higher maturity than the saturated, low API gravity oils. The two saturated oils discussed above (and see Fig. 5) that are near their critical point are anomalous insofar as their Ts/Ts+Tm suggests relatively low maturity. The saturated oils (solid squares) and the condensates of their corresponding gas caps (open triangles) are of similar relatively low maturity. The character of the condensates from the two undersaturated gas-condensates (open circles) suggests their derivation from a source rock of no greater maturity than that which sourced the majority of the undersaturated oils. Vitrinite reflectance equivalent maturity of oil and condensate samples varies within a narrow range, from 0.77 to 0.86%, calculated from MPI-1 after Radke (1988). Other aromatic maturity indicators support the general trends discussed above (Table 2). Sterane isomerization ratios, even those that should operate within the same temperature range as Ts/Ts+Tm (e.g. C29 S/R, Farrimond *et al.* 1998) correlate with neither other biomarker maturity indicators nor Magnolia fluid attributes.

Whole oil data support a maturity influence on Magnolia oil properties. As compared with the high API gravity undersaturated oils, the low API gravity saturated oils tend to be enriched in isoprenoid relative to normal alkanes (e.g. $iC_9/nC_9$ and $iC_{10}/nC_{10}$ shown in Table 2 which are uninfluenced by the oil-based drilling mud contamination); this is consistent with source rock maturity influences (Connan & Cassou 1980; Zumberge 1993). Condensate isoprenoid : normal attributes do not suggest their derivation from a particularly mature source rock. Similarly, there is a strong correlation between the $C_7$ maturity-sensitive parameter 2,4-DMP/2,3-DMP (Mango 1994, 1997) and oil API gravity (the 2,4-DMP/2,3-DMP ratio of the undersaturated relatively high API gravity oils average 0.40; that of the saturated relatively low API gravity oils average 0.22, Table 2) whereas there is little correlation between this parameter and condensate gravity. Generation temperatures calculated in accordance with the scheme proposed by Mango (1997) fall within a narrow range for the oils: from 116 to 129 °C; temperatures calculated for Magnolia condensates all lie within this range.

**Table 2.** *Selected Magnolia fluid geochemical attributes discussed in text*

| Depth (ft TVD) | $C_{35}/C_{34}$ | $C_{29}/C_{30}$ | Ts/(Ts + Tm) | MPI-1 | DBT/ 4 + 1-MDBT | Tol/$nC_7$ | $nC_9$/MCH | $iC_9/nC_9$ | $iC_{10}/nC_{10}$ | 2,4-DMP/ 2,3-DMP | Bz/MCH | N2 + P2 | P3 | SF |
|---|---|---|---|---|---|---|---|---|---|---|---|---|---|---|
| *Saturated gas-condensates* | | | | | | | | | | | | | | |
| 12614 | 1.04 | 1.12 | 0.375 | 0.648 | 0.149 | 0.32 | 2.42 | 0.21 | 0.64 | 0.23 | 0.01 | 2.26 | 0.76 | 1.268 |
| 14782 | 0.93 | 1.02 | 0.422 | 0.724 | 0.587 | 0.44 | 2.03 | 0.15 | 0.30 | 0.28 | 0.09 | 3.06 | 0.81 | 1.125 |
| 15540 | 1.00 | 1.06 | 0.423 | 0.683 | 0.632 | 0.14 | 2.61 | 0.20 | 0.45 | 0.20 | 0.02 | 2.90 | 1.00 | 1.149 |
| *Saturated oils* | | | | | | | | | | | | | | |
| 14961 | 1.05 | 1.00 | 0.422 | 0.711 | 0.588 | 0.40 | 1.69 | 0.19 | 0.31 | 0.22 | 0.09 | 1.95 | 0.70 | n.a. |
| 15009 | 1.04 | 0.98 | 0.428 | 0.706 | 0.588 | 0.19 | 1.73 | 0.19 | 0.33 | 0.23 | 0.08 | 2.17 | 0.77 | 1.074 |
| 15605 | 1.00 | 1.03 | 0.411 | 0.637 | 0.616 | 0.18 | 2.28 | 0.21 | 0.39 | 0.20 | 0.04 | 2.67 | 0.95 | 1.073 |
| *Undersaturated gas-condensates* | | | | | | | | | | | | | | |
| 14065 | 1.02 | 1.04 | 0.447 | 0.714 | 0.219 | 0.28 | 1.34 | 0.54 | 0.68 | 0.38 | 0.08 | 4.28 | 1.51 | 1.162 |
| 14185 | 0.99 | 1.06 | 0.443 | 0.786 | 0.410 | 0.18 | 1.93 | 0.18 | 0.73 | 0.32 | 0.07 | 3.91 | 1.32 | 1.203 |
| *Undersaturated oils* | | | | | | | | | | | | | | |
| 14963 | 1.03 | 0.95 | 0.461 | 0.751 | 0.350 | 0.33 | 1.61 | 0.14 | 0.25 | 0.46 | 0.21 | 3.34 | 1.09 | 1.136 |
| 15195 | 1.02 | 0.92 | 0.464 | 0.761 | 0.406 | 0.34 | 1.61 | 0.13 | 0.24 | 0.50 | 0.23 | 2.37 | 0.78 | n.a. |
| 16250 | 1.07 | 0.92 | 0.473 | 0.794 | 0.444 | 0.37 | 1.51 | 0.12 | 0.23 | 0.45 | 0.19 | 3.42 | 1.07 | 1.138 |
| 16360 | 1.04 | 0.93 | 0.433 | 0.748 | 0.424 | 0.32 | 1.63 | 0.14 | 0.25 | 0.37 | 0.15 | 3.02 | 0.97 | 1.123 |
| 16630 | 1.04 | 0.95 | 0.457 | 0.775 | 0.436 | 0.37 | 1.53 | 0.12 | 0.23 | 0.44 | 0.17 | 3.26 | 1.03 | 1.134 |
| 15856 | 0.98 | 0.97 | 0.457 | 0.740 | 0.348 | 0.46 | 1.63 | 0.12 | 0.24 | 0.45 | 0.21 | 3.18 | 0.98 | 1.138 |
| 16092 | 1.01 | 0.97 | 0.437 | 0.711 | 0.313 | 0.33 | 1.71 | 0.14 | 0.26 | 0.37 | 0.16 | 3.21 | 0.99 | 1.120 |
| 16246 | 1.03 | 0.99 | 0.434 | 0.732 | 0.313 | 0.35 | 1.79 | 0.14 | 0.26 | 0.34 | 0.13 | 2.99 | 0.95 | 1.111 |
| 16518 | 1.03 | 0.97 | 0.460 | 0.732 | 0.344 | 0.33 | 1.69 | 0.12 | 0.23 | 0.39 | 0.16 | 2.97 | 0.93 | 1.133 |
| 16794 | 1.00 | 0.97 | 0.459 | 0.721 | 0.349 | 0.31 | 1.68 | 0.12 | 0.23 | 0.41 | 0.17 | 3.09 | 0.95 | 1.129 |
| 16637 | 1.01 | 1.02 | 0.422 | 0.710 | 0.318 | 0.23 | 1.89 | 0.15 | 0.29 | 0.29 | 0.09 | 2.70 | 0.91 | 1.114 |
| 16058 | 0.97 | 0.96 | 0.467 | 0.715 | 0.336 | 0.43 | 1.62 | 0.13 | 0.24 | 0.43 | 0.19 | 3.31 | 1.06 | 1.131 |
| 16766 | 1.02 | 1.03 | 0.408 | 0.685 | 0.398 | 0.20 | 2.00 | 0.15 | 0.29 | 0.29 | 0.12 | 2.76 | 0.88 | 1.109 |
| 16350 | 1.03 | 0.97 | 0.438 | 0.717 | 0.328 | 0.31 | 1.71 | 0.13 | 0.24 | 0.46 | 0.21 | 3.26 | 1.04 | 1.131 |

$C_{35}$,17α(H),21β(H),22(S)-pentakishomohopane + 17α(H),21β(H),22(R)-pentakishomohopane; $C_{34}$,17α(H),21β(H),22(S)-tetrakishomohopane + 17α(H),21β(H),22(R)-tetrakishomohopane; $C_{29}$,17α(H),21β(H)-30-norhopane; $C_{30}$,17α(H),21β(H)-hopane; Ts, 18a(H)-22,29,30-trisnorneohopane; Tm, 17a(H)-22-29-30-trisnorhopane; MPI-1, [methylphenanthrene index] 1.5(2-methylphenanthrene + 3 methylphenanthrene)/(phenanthrene + 1-methylphenanthrene + 9-methylphenanthrene); DBT, dibenzothiophene; 4 + 1-MDBT, 4-methyldibenzothiophene + 1-methyldibenzothiophene; Tol, toluene; MCH, methylcyclohexane; 2,4-DMP, 2,4-dimethylpentane; 2,3-DMP, 2,3-dimethylpentane; Bz, benzene; N2, 1,1-dimethylcyclopentane + *cis*-1,3-dimethylcyclopentane + *trans*-1,3-dimethylcyclopentane; P2, 2-methylhexane + 3-methylhexane; P3, 2,2-dimethylpentane + 2,4-dimethylpentane + 3,3-dimethylpentane + 2,3-dimethylpentane + 3-ethylpentane; SF, slope factor (see Fig. 13 caption for calculation).

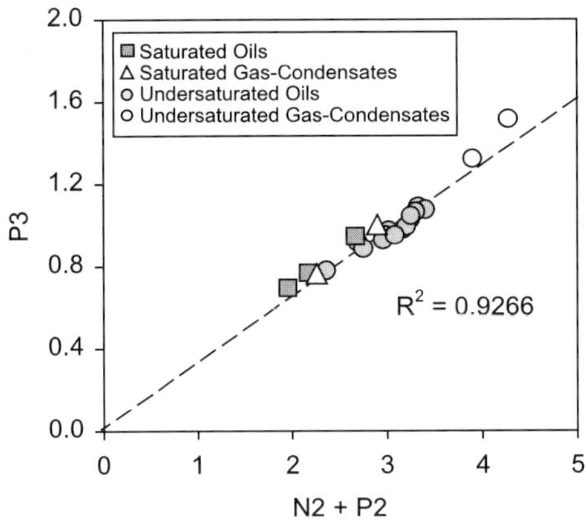


**Fig. 9.** Magnolia oil and condensate facies-associated $C_7$ molecular indices. The regression line shown was calculated using all samples. For ratio definitions see Table 2.

Carbon isotopic composition of all associated and free gases are shown in Table 3. Isotopic profiles of four representative gas samples are shown in Figure 11, which also shows modelled isotopic composition using the approach of Clayton (1991). This approach allows isotopic composition to be estimated using a Raleigh fractionation scheme as a function of the fraction of gas potential realized ('gas generation index', or GGI) and the $\delta^{13}C$ of the parent kerogen. For the Magnolia samples (for which we have no direct evidence regarding the isotopic character of the source rock) the parent kerogen was assumed to have a $\delta^{13}C$ of −24.5‰ in order to match, using a range of GGI from 0.1 to 1.0, the measured range of butane $\delta^{13}C$ (−28.7 to −25.5‰). The carbon isotopic compositions shown in Figure 11 of the $C_{2+}$ components are strikingly consistent with the maturity-based modelled isotopic trends; from $nC_2$ to $nC_4$ gases become progressively lighter at a rate consistent with the maturity evolution of the source rock from which the gases were derived. These data are entirely consistent with gasoline-range maturity parameters such as 2,2-DMP/2,3-DMP (discussed above) as illustrated in Figure 12. The methane carbon isotopic compositions of Magnolia free and associated gases are invariably significantly lighter than that predicted from a simple maturity sequence. This is discussed below in the context of biogenic gas.

Finally, with respect to the influence of source rock maturity, it is instructive to examine the distribution of $C_{7+}$ pseudo-components in Magnolia oils. Representative component profiles of cleaned samples from the two end-member Magnolia oils (see Fig. 7) are shown in Figure 13. Component distributions are in effect perfectly exponential and, presuming the saturated and undersaturated oils to be derived from a uniform source facies, suggest their derivation from source rocks of different maturity (Thompson, 2002). Slope factors (SF, defined in Fig. 13 caption) of the remaining oil samples vary between the end member values of 1.07 and 1.14 and positively correlate with other maturity indicators. In contrast, the relatively high slope factors evidenced by the gas-condensate samples (1.12–1.27) do not correlate with any of the maturity-influenced indicators discussed above, suggesting that the gas-condensates are not a high maturity product.

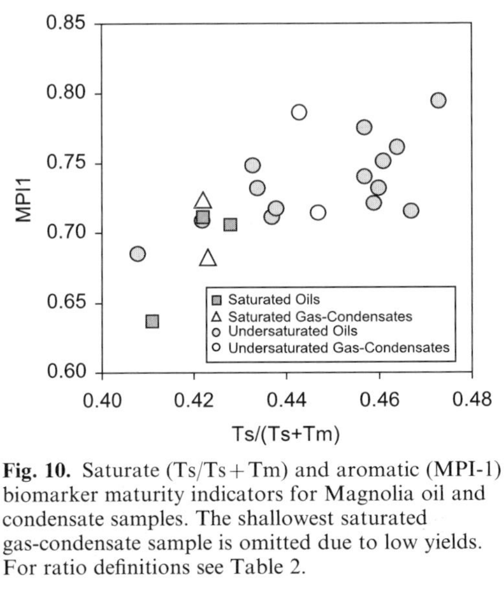


**Fig. 10.** Saturate (Ts/Ts + Tm) and aromatic (MPI-1) biomarker maturity indicators for Magnolia oil and condensate samples. The shallowest saturated gas-condensate sample is omitted due to low yields. For ratio definitions see Table 2.

*Water washing.* The degree to which Magnolia fluids have been influenced by water washing is addressed in Figure 14 which shows a cross plot of benzene/methylcyclohexane (Bz/MCH) against dibenzothiophene/(4-methyldibenzothiophene + 1-methyldibenzothiophene), DBT/(4 + 1-MDBT). Being extremely water soluble, depletion of benzene and dibenzothiophene has been proposed as good indicators of water washing (Palmer 1993; Kuo 1994). In Figure 14 the two Magnolia fluid samples that are known to have been taken adjacent to water legs are annotated. Only two samples, both from gas-condensates, show consistent evidence of water washing (i.e. relative depletion of both benzene and dibenzothiophene). Inspection of Tables 2 and 3 shows that these ratios in the remaining gas-condensate and oil samples generally mimic the trends seen in the maturity-sensitive

**Table 3.** *Magnolia Field free and associated gas properties*

| Depth (ft TVD) | $\delta D_{C_1}$ (‰) | $\delta^{13}C_{C_1}$ (‰) | $\delta^{13}C_{C_2}$ (‰) | $\delta^{13}C_{C_3}$ (‰) | $\delta^{13}C_{iC_4}$ (‰) | $\delta^{13}C_{nC_4}$ (‰) | $C_1$ (mol%) | $C_2$ (mol%) | $C_3$ (mol%) | $iC_4$ (mol%) | $nC_4$ (mol%) | $iC_5$ (mol%) | $nC_5$ (mol%) |
|---|---|---|---|---|---|---|---|---|---|---|---|---|---|
| *Saturated gas-condensates* | | | | | | | | | | | | | |
| 12614 | −187 | −67.9 | −38.2 | −30.0 | −31.2 | −28.3 | 98.1 | 0.4 | 0.4 | 0.1 | 0.2 | 0.1 | 0.1 |
| 14782 | −203 | −64.2 | −34.8 | −29.7 | | −27.7 | 89.7 | 3.3 | 2.6 | 0.6 | 1.1 | 0.5 | 0.5 |
| 15540 | −198 | −64.6 | −38.7 | −31.0 | | −28.7 | 90.4 | 2.0 | 2.3 | 0.5 | 0.9 | 0.4 | 0.5 |
| *Saturated oils* | | | | | | | | | | | | | |
| 14961 | −198 | −65.5 | −36.8 | −30.5 | | −28.3 | 89.6 | 2.4 | 2.8 | 0.6 | 1.3 | 0.5 | 0.6 |
| 15009 | −196 | −65.5 | −37.1 | −30.6 | | −28.3 | 91.0 | 2.1 | 2.4 | 0.5 | 1.1 | 0.5 | 0.6 |
| 15605 | −200 | −64.1 | −38.5 | −31.0 | | −28.7 | 89.0 | 2.6 | 3.0 | 0.6 | 1.2 | 0.5 | 0.6 |
| *Undersaturated gas-condensates* | | | | | | | | | | | | | |
| 14065 | −196 | −55.9 | −33.3 | −28.8 | −29.5 | −27.0 | 89.7 | 4.1 | 2.8 | 0.6 | 1.1 | 0.4 | 0.4 |
| 14185 | −198 | −56.3 | −33.8 | −29.3 | −30.1 | −27.6 | 90.3 | 3.6 | 2.2 | 0.5 | 1.0 | 0.4 | 0.5 |
| *Undersaturated oils* | | | | | | | | | | | | | |
| 14963 | −200 | −57.1 | −30.7 | −27.4 | −28.4 | −25.9 | 82.6 | 8.1 | 4.8 | 1.0 | 1.6 | 0.6 | 0.6 |
| 15195 | −202 | −52.4 | −29.9 | −26.9 | −28.0 | −25.5 | 81.1 | 10.9 | 4.6 | 0.8 | 1.2 | 0.3 | 0.3 |
| 16250 | −204 | −51.5 | −29.9 | −26.9 | −28.2 | −25.7 | 76.0 | 10.5 | 6.2 | 1.4 | 2.5 | 0.9 | 0.8 |
| 16360 | −207 | −53.8 | −31.3 | −28.1 | −29.1 | −26.7 | 76.6 | 9.2 | 6.2 | 1.4 | 2.8 | 1.0 | 1.0 |
| 16630 | −206 | −52.4 | −30.4 | −27.2 | −28.5 | −26.0 | 75.0 | 10.4 | 6.4 | 1.5 | 2.8 | 1.0 | 1.0 |
| 15856 | −208 | −52.7 | −30.8 | −27.3 | −28.4 | −26.0 | 74.3 | 11.0 | 6.6 | 1.5 | 2.8 | 1.0 | 1.0 |
| 16092 | −209 | −58.5 | −32.6 | −28.6 | −29.4 | −26.9 | 79.0 | 7.8 | 5.4 | 1.2 | 2.5 | 1.0 | 1.1 |
| 16246 | −204 | −60.6 | −33.4 | −29.1 | −29.7 | −27.3 | 84.0 | 5.5 | 4.3 | 0.9 | 1.9 | 0.7 | 0.8 |
| 16518 | −209 | −57.4 | −31.7 | −28.0 | −28.9 | −26.3 | 81.0 | 7.4 | 4.7 | 1.0 | 2.1 | 0.8 | 0.9 |
| 16794 | −211 | −55.8 | −31.6 | −27.9 | −29.0 | −26.5 | 78.9 | 8.5 | 5.4 | 1.2 | 2.4 | 0.9 | 1.0 |
| 16637 | −206 | −62.2 | −34.9 | −29.8 | −30.2 | −27.7 | 87.2 | 4.2 | 3.4 | 0.7 | 1.5 | 0.6 | 0.7 |
| 16058 | −214 | −53.7 | −30.6 | −27.7 | −28.6 | −26.2 | 36.3* | 17.2 | 12.4 | 4.7 | 10.4 | 3.5 | 3.5 |
| 16766 | −203 | −61.4 | −34.7 | −29.7 | −30.1 | −27.7 | 85.6 | 4.8 | 3.8 | 0.8 | 1.7 | 0.7 | 0.8 |
| 16350 | −213 | −55.7 | −31.7 | −27.8 | −28.6 | −26.2 | 85.0 | 9.5 | 3.3 | 0.4 | 0.7 | 0.2 | 0.2 |

* Gas yields reflect cylinder leakage.

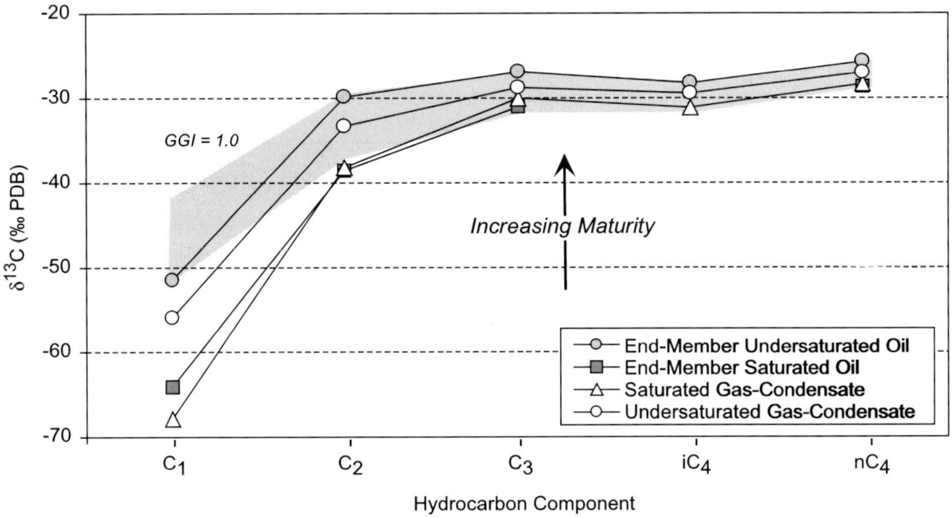


**Fig. 11.** Methane carbon isotopic profiles of associated gases from end-member undersaturated and saturated oils and representative gases from gas-condensate reservoirs. The grey region represents range of carbon isotopic compositions calculated after Clayton (1991) as a function of maturity extending from GGI = 0.1 (lower limit) to GGI = 1.0 (annotated), assuming the parent kerogen $\delta^{13}C$ to be −24.5‰ PDB.

parameters discussed above, exemplified by the strong correlation ($R^2 = 0.92$) between Bz/MCH and 2,4-DMP/2,3-DMP. These data indicate that although some of the Magnolia fluids may have experienced a degree of water washing, its influence was not a first-order control on fluid properties, such as API gravity, gas/liquid ratio and saturation pressure.

*Biodegradation.* Biodegradation is well known to influence some northern Gulf of Mexico hydrocarbon accumulations (e.g. Whelan *et al.* 1993;

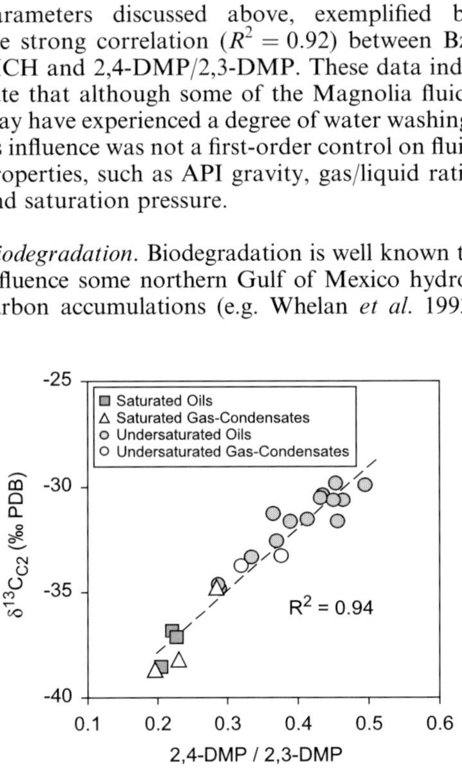


**Fig. 12.** Condensate and oil maturity-sensitive ratio of 2,4-dimethylpentane to 2,3-dimethylpentane versus ethane carbon isotopic composition of corresponding free and associated gases.

Holba *et al.* 1996; Edman & Burk 1999). Within the same field, lightly to severely biodegraded oils often co-exist with pristine unaltered oils, resulting in large within-field hydrocarbon property variations. Available data, discussed below, do not suggest that biodegradation has influenced Magnolia hydrocarbon fluids.

Sulphur content and total acid number (TAN) of Magnolia oils and condensates inversely correlate with liquid API gravity (Table 1). Although sulphur and TAN are known to increase as a consequence of biodegradation (Behar & Albrecht 1984; Meredith *et al.* 2000; Watson *et al.* 2002), both may be influenced by source rock maturity (Dias *et al.* 2002). In view of the discussion above, low maturity is viewed as being more likely responsible for the modestly elevated sulphur and TAN content in some of the Magnolia fluids.

In terms of gross aspect, oil and condensate whole oil gas chromatograms (e.g. those from the oil end-members shown in Fig. 7) do not suggest microbial alteration. Susceptible low molecular weight normal alkanes are abundant and there is no evidence of the characteristic 'hump' of unresolved compounds. Table 2 contains $iC_9/nC_9$ and $iC_{10}/nC_{10}$ ratios (which are unaffected by oil-based mud contamination) and shows that the low API gravity saturated oils are modestly enriched in iso- relative to normal alkanes. This could be either due to biodegradation (Connan 1984; Palmer 1993) or the influence of source rock maturity (Zumberge 1993).

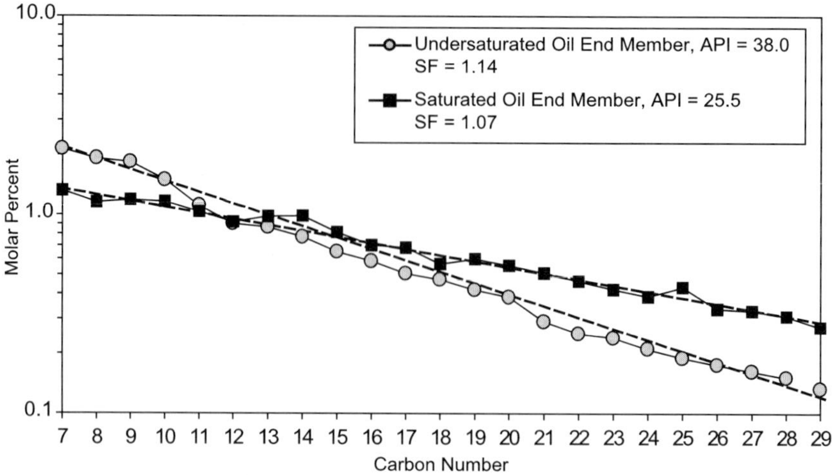


**Fig. 13.** Representative $C_{7+}$ pseudo-components profiles from end-member Magnolia oils (Fig. 7). Samples have been mathematically cleaned in order to remove drilling mud-associated artifacts. The slope factors are calculated in accordance with Thompson (2002): $SF = e^{-a}$, where $a$ represents the slope of the exponential best fit line.

The isotopic character of associated and free gases from saturated and undersaturated oil and gas-condensate reservoirs are presented in Table 3 and shown in Figure 11 (discussed above in the context of maturity). Gases show no evidence of biodegradation. Evidence of early-stage effects are the preferential removal of $C_{3+}$ normal alkanes, leading to both an increase in the residual $\delta^{13}C_{C_3}$ and $\delta^{13}C_{nC_4}$ by as much as 10–20‰ and an increase in iso-alkanes to normal alkanes (James & Burns 1984; Katz *et al.* 2002).

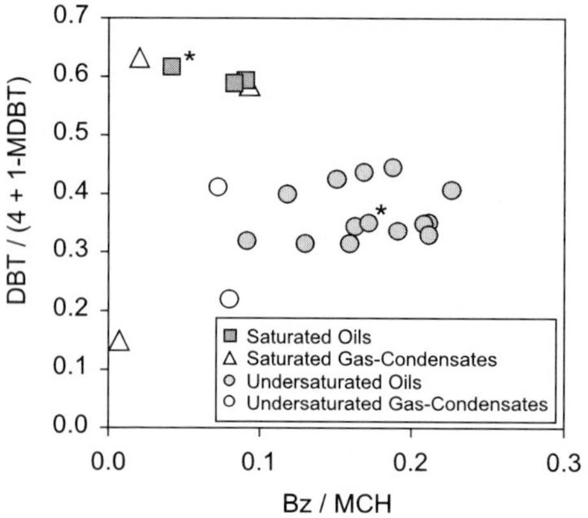


**Fig. 14.** Benzene/methylcyclohexane (Bz/MCH) against dibenzothiophene/1 + 4-methyldibenzothiophene (DBT/4 + 1-MDBT) of oil and condensate samples. Asterisks denote samples that are known to be proximate to water legs.

The distribution of $C_{7+}$ pseudo-components in Magnolia fluids similarly suggests that biodegradation is not responsible for the oil quality variations. The representative low API gravity oil end-member component profile shown in Figure 13 is in effect perfectly exponential, showing no depletion of lower molecular weight compounds, unlike profiles produced by biodegradation (Meulbroek *et al.* 1998; Losh *et al.* 2002).

*Phase fractionation.* Hydrocarbon accumulations whose characteristics are attributed to phase fractionation have been documented in a number of northern, principally shallow water, Gulf of Mexico fields (Thompson 1987, 1988; Curiale & Bromley 1996; Meulbroek *et al.* 1998; Losh *et al.* 2002). This process has been credited with producing a spectrum of alteration products, most notably 'residual' aromatic-rich oils that are depleted in low molecular components and 'fractionated' paraffinic gas-condensates, both of which are common in the northern Gulf of Mexico.

Figure 15 shows a plot of toluene/$nC_7$ against $nC_7$/methylcyclohexane for Magnolia oil and condensate samples. This has been annotated using the interpretation scheme of Thompson (1987) that reflects the various potential influences of phase fractionation, biodegradation and maturation on these ratios. Liquids from the saturated and unsaturated oil reservoirs are relatively uniform and similar to oil samples from the Jolliet field (unpublished in-house data), which lies

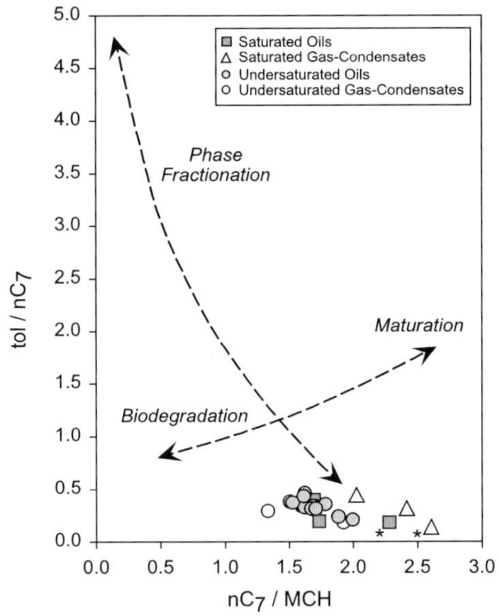


**Fig. 15.** $nC_7$/methylcyclohexane ($nC_7$/MCH) versus toluene/$nC_7$ (tol/$nC_7$) of Magnolia oils and condensates showing the interpretation scheme of Thompson (1987). Asterisks denote a saturated oil–gas cap pair.

approximately 50 miles to the northeast. Jolliet field oils were interpreted by Losh *et al.* (2002) as being effectively unfractionated, in contrast with other (shallower water) northern Gulf of Mexico samples comprising their dataset that showed evidence of gas washing and attendant *n*-alkane depletion. Samples from Magnolia saturated and undersaturated oil reservoirs thus appear to represent pristine unfractionated oils. The saturated gas-condensates appear to be somewhat paraffinic relative to the other fluids. This is consistent with their derivation as a fractionated product. The character of the saturated gas-condensate–oil pair that is annotated in Figure 15 (*) could be reflecting a very modest degree of *in situ* phase fractionation associated with the formation of this gas cap. Large compositional differences such as those observed in systems characterized by extreme evaporative fractionation processes (e.g. Thompson 1988; Dzou & Hughes 1993; Curiale & Bromley 1996) are not evident in Magnolia molecular ratios.

Component profiles of the saturated and undersaturated oils (e.g. Fig. 13) do not show evidence of extensive evaporative fractionation. Residual oils formed by this process may be depleted in lower molecular weight components and deviate from the expected exponential distribution (Meulbroek *et al.* 1998; Losh *et al.* 2002; Thompson 2002). Magnolia Field gas-condensate molar slope factors (SF, Table 2) are low, ranging from 1.12 to 1.26, and correlate very strongly with reservoir pressure ($R^2 = 0.92$). A similar relationship between SF and pressure was described by Thompson (2002) and attributed to 'migration depletion' such that molar slope factors are a simple consequence of the affinity of hydrocarbons to the vapour versus liquid phase as a function of pressure. Although Magnolia fluids appear to be largely unaltered by extreme evaporative fractionation processes that typify much of the shallow water gas-prone portion of the northern Gulf of Mexico, phase fractionation associated with pressure reduction (and/or changing composition, discussed below) appears to have been responsible for formation of the saturated and undersaturated gas-condensates in the field.

*Biogenic gas.* As shown in Table 3, Magnolia free and associated gases are quite variable in terms of their composition and character, discussed above in the context of source rock maturity (and see Fig. 11). In addition, Figure 16 shows gas dryness plotted against methane carbon isotopic composition ($\delta^{13}C_{C_1}$) along with generalized interpretation fields taken from Clayton (1991). These data point towards a mixed origin of free and associated Magnolia gases: dry, isotopically light biogenic methane has mixed with thermogenic methane. The departure of $\delta^{13}C_{C_1}$ from its expected maturity-based composition (shown in Fig. 11) is due to admixture of this biogenic component. The presence of very small amounts of extremely isotopically light biogenic ethane, similar to that described by Edman *et al.* (2001) from the northern Gulf of Mexico Mississippi Canyon 348 field, may contribute to the very slightly $^{13}C$ depleted ethane in the samples that contain the most biogenic methane. Methane hydrogen isotopic character is relatively invariant and heavy (Table 3), suggesting that the biogenic component formed by $CO_2$ reduction rather than acetate fermentation (Whiticar *et al.* 1986). Fractionation associated with diffusion does not appear to have influenced Magnolia gas compositions; a strong linear relationship exists between the gas ratio $C_2/C_1$ and $\delta^{13}C_{C_1}$ suggesting that mixing processes are dominant (Prinzhofer & Pernaton 1997; Prinzhofer *et al.* 2000).

Many hydrocarbon reservoirs in the deepwater Gulf of Mexico are known to contain a biogenic methane component (Rice & Claypool 1981; Bissada *et al.* 1990, 2002; Kornacki *et al.* 1996; Westrich *et al.* 1999; Kornacki & Nelson 2000; Beeunas *et al.* 2001; Edman *et al.* 2001)

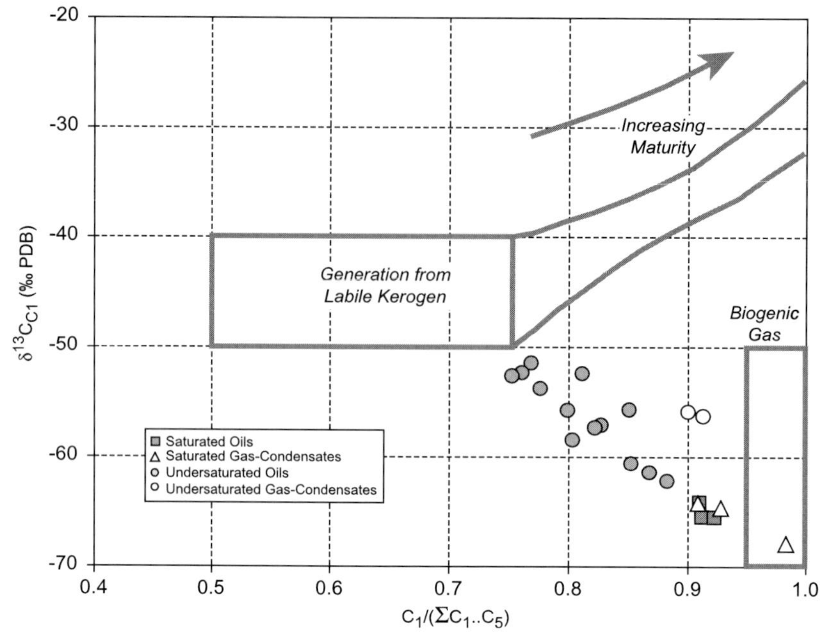


**Fig. 16.** Dryness ($C_1/\sum C_1 \ldots C_5$) versus methane carbon isotopic composition ($\delta^{13}C_{C_1}$, ‰ PDB) for Magnolia associated and free gases. Interpretation fields are after Clayton (1991).

whose presence is usually attributed to production in and migration from adjacent immature organically lean sediments. Although biogenic methane is known to form as a by-product of microbial degradation of reservoired thermogenic hydrocarbons (Larter *et al.* 1999; Pallasser 2000), as none of the tested Magnolia liquids show evidence of being biodegraded, the most probable origin of the Magnolia biogenic gas component is primary production in adjacent argillites whose present-day temperature conditions (<180 °F) are still hospitable to methanogens (Rice & Claypool 1981; Zhang & Chen 1985). Transport and accumulation may have been promoted by the geometry and size of the drainage area (Fig. 3) and by the field's (relatively) long-lived trapping geometry.

## *Origin of geochemical and phase property variations*

Magnolia fluid property variations are principally controlled by differences in the maturity of emplaced thermogenic hydrocarbons, variable biogenic methane content and phase fractionation processes.

Oil API gravity, sulphur content and TAN are most notably influenced by the maturity of the source rock from which they were derived. Maturity variations evidenced by Magnolia fluids are probably related to the fact that (1) individual reservoirs have access to different portions of a source system of variable maturity and/or that (2) fluid compositions reflect the time at which reservoirs were charged, such that reservoirs containing relatively low maturity fluids were charged earlier than those containing higher maturity fluids. Limitations in seismic resolution preclude an analysis of source rock distribution and maturity and potential hydrocarbon migration pathways and filling directions at Magnolia. Empirically, it can be observed that reservoir depth alone is an extremely poor predictor of fluid maturity and corresponding properties. When depth is considered together with reservoir age and overpressure, a reasonably good correlation can be found between a linear combination of these parameters and fluid maturity ($R^2 = 0.71$) as proxied by $\delta^{13}C_{C_3}$. Overpressure alone, though an imperfect predictor of fluid maturity, is by far the most influential of the three parameters; the more highly overpressured reservoirs tend to contain the highest maturity fluids. This suggests that reservoir continuity has played a role in hydrocarbon charge patterns at Magnolia but does not go terribly far toward resolving the relative control of spatial versus temporal influences.

The incorporation of biogenic methane into Magnolia reservoirs is the single most important control of fluid saturation pressures. Due to its high vapour pressure, methane, more so than other hydrocarbon gases, exerts a disproportionate influence on saturation pressure. Excess methane (e.g. 75 mol% of the reservoir fluid in one of the oil zones tested) has resulted in saturation pressures that are at odds with oil characteristics (e.g. Fig. 6). As regards the distribution of biogenic gas within the field, neither variations in sandstone texture nor composition appear to be implicated. Similarly, the correlation between reservoir depth and methane content (and saturation pressure) is poor. As in the case of fluid maturity, however, when depth is considered together with reservoir age and overpressure, a fairly good correlation is found between a linear combination of these parameters and biogenic gas ($R^2 = 0.76$), as proxied by $\delta^{13}C_{C_1}$. Overpressure again is the most influential of the three parameters. The relationship between the accumulation of biogenic gas and overpressure is relatively easy to rationalize. The degree to which Magnolia unconsolidated sands develop and maintain overpressure is influenced, if not controlled, by their lateral continuity. It follows that the least overpressured sands have had the greatest access to biogenic gas formed within adjacent compacting shales in the associated biogenic gas 'fetch' areas.

In the dataset comprising this study there is a strong correlation between biogenic gas content and fluid maturity; reservoirs containing the lowest maturity fluids tend to contain the greatest proportion of biogenic methane. Fluid composition data suggest that this correlation is due in part to the character of the initial thermogenic charge. Molar profiles of Magnolia fluids (from recombined PVT samples) reveal that the lower maturity fluids contain relatively low absolute amounts of $C_{2+}$ gases. This correlation is illustrated in Figure 17 that shows the carbon isotopic composition of ethane plotted against the mol% of ethane in Magnolia fluids. Low maturity samples contain less $C_{2+}$ gases than high maturity samples. This is unsurprising in view of the large body of empirical and experimental data that shows as source rock maturity progresses the gas/liquid ratio of generated products increases. Similarly, methane yield is influenced by source rock maturity such that at low maturities source rocks generate little methane in proportion to other hydrocarbon products (e.g. Behar *et al.* 1997). It follows that in the case of Magnolia fluids the presence of biogenic isotopically light methane has exerted a disproportionate influence on the lower maturity fluids (containing relatively small amounts of thermogenic methane) than on the higher maturity fluids.

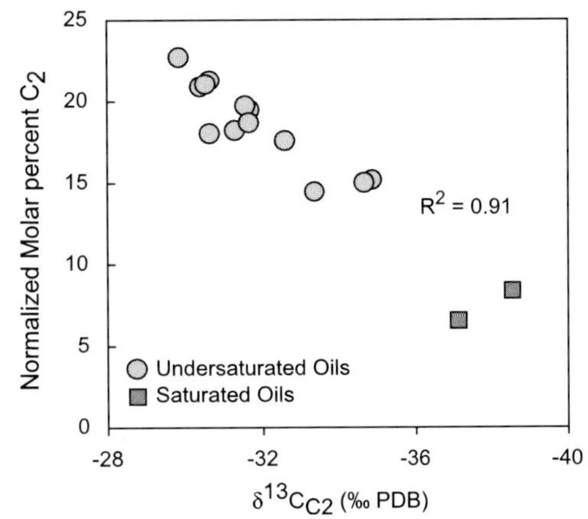


**Fig. 17.** Ethane $\delta^{13}C$ versus ethane content of associated gases from Magnolia undersaturated and saturated oils. Ethane percent has been calculated relative to $C_{2+}$ components and pseudo-components (i.e. methane is not included). Samples have been mathematically cleaned in order to remove the effects of oil-based drilling mud contamination.

These simple mixing considerations do not, however, account for the fact that the Magnolia fluids that contain the isotopically lightest methane contain a very high proportion of methane relative to $C_{2+}$ components, ranging from 54 mol% in the undersaturated high maturity oils to as much as 75 mol% in the saturated oils. This is not consistent with the notion that the field fluids contain a relatively consistent biogenic methane content whose isotopic expression is simply amplified in the reservoirs containing the lowest maturity fluids. The correspondence between biogenic gas content and fluid maturity is thus real and not a mixing artefact; the ultimate connection between the two, however, which probably relates to migration pathway and charge access, remains elusive.

Phase fractionation is responsible for development of the saturated gas-condensates at Magnolia and has resulted from the mixture of biogenic methane and thermogenic fluids. The relatively low pressure reservoirs that contain a large biogenic gas component are saturated. Fractionation in these cases probably occurred *in situ*, as undersaturated thermogenic fluids mixed with previously-emplaced biogenic gas (of unknown but probably low saturations). Two phase systems formed when, as a result, saturation pressures exceeded reservoir pressures.

The undersaturated gas-condensates are somewhat enigmatic. These fluids occur in a relatively shallow, low pressure low temperature reservoir. Despite their very different gas-liquid ratios

(Table 1) both were derived from a source rock at a maturity level similar to that which charged the undersaturated oil reservoirs in the field (Figs 10–12). These undersaturated gas-condensates contain little biogenic gas but a relatively large proportion of methane; their dryness versus $\delta^{13}C_{C_1}$ attributes deviate from the simple mixing trend evidenced by the other Magnolia samples (Fig. 16). They are not particularly paraffinitic relative to the other samples (Fig. 15), and are somewhat, but not consistently, depleted in water soluble compounds (Fig. 14). Isoprenoid : normal ratios, though again inconsistent, suggest that they tend towards *n*-alkane depletion (Table 2). We speculate that these gas-condensates developed due to emplacement of multiple thermogenic charges of different maturity (and/or gas/liquid ratio) into a low saturation biogenic gas accumulation. In addition, the array of characteristics demonstrated by these fluids suggests that some degree of secondary leakage and fractionation has occurred within the reservoir during or subsequent to charge episodes.

## *Fluid continuity and reservoir characterization*

The fluid samples and companion PVT analyses provide ample data with which to assess the origin of the fluid variability at Magnolia. They do not alone, however, address questions regarding the degree to which Magnolia fluids have established compositional equilibrium. These issues are pivotal to assessments of reservoir continuity and fluid connectivity which bear, in turn, on potential field development strategies. In this section the integration of high resolution mud gas and pressure data with fluid analyses (1) illustrates how these data may be used to evaluate the extent to which Magnolia fluids are in compositional disequilibrium and (2) provides a case example addressing notions of mixing rates. In addition, the compositional heterogeneity of the Magnolia fluids allows fluid property predictions to be made using empirical relationships; this approach allows reservoir fluid properties to be characterized when tested fluid samples are unavailable or their properties equivocal.

Mud gas samples were obtained using gas bags or IsoTubes during drilling of all but the Garden Banks 783-1 well. Though (as is typical of mud gas samples) they were extremely contaminated by air, sufficient sample was usually obtained that the carbon isotopic composition of methane (and less often ethane and propane) could be measured without enrichment. The well profile shown in Figure 18 illustrates the downhole correspondence between IsoTube gas composition and isotopic character

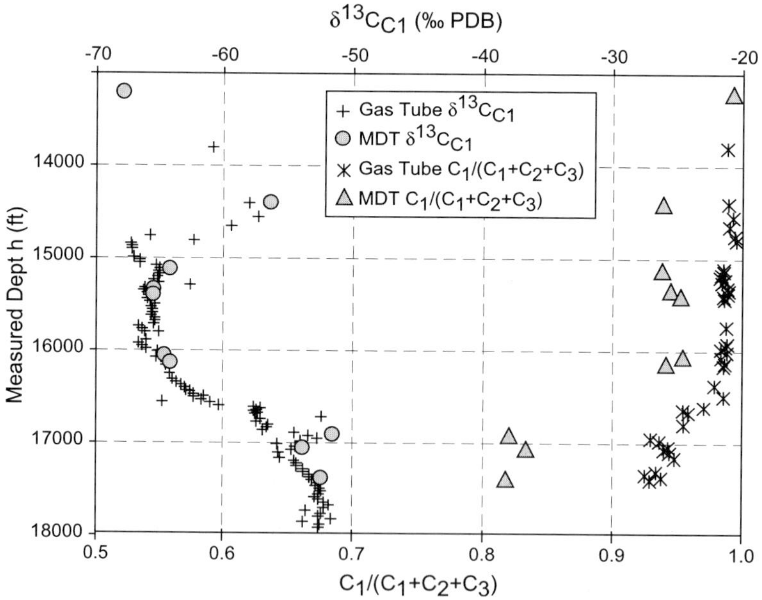


**Fig. 18.** Depth plot for a Magnolia Field appraisal well showing gas composition and $\delta^{13}C_{C_1}$ obtained from mud gas (IsoTubes) samples against those of associated gases of ten MDT samples obtained.

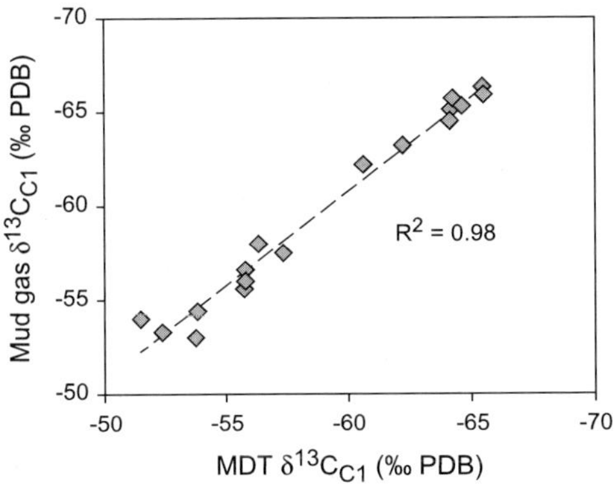


**Fig. 19.** $\delta^{13}C_{C_1}$ of Magnolia MDT samples plotted against the $\delta^{13}C_{C_1}$ of mud gas samples (gas bag and IsoTube) collected from the same interval.

($C_1/\sum C_1 \ldots C_3$ and $\delta^{13}C_{C_1}$) and that of MDT samples. Mud gas compositions at Magnolia and elsewhere in the deepwater Gulf of Mexico are often much drier than the composition of corresponding MDT samples. $\delta^{13}C_{C_1}$, on the other hand, correlates extremely well with the isotopic composition of MDT samples (Fig. 19). A similar comparison of mud gas $\delta^{13}C_{C_2}$ against that of corresponding MDT samples is also quite strong ($R^2 = 0.97$). The limited number of mud gas propane measurements comprising this dataset show larger scatter than that of methane and ethane when compared with corresponding MDT samples ($R^2 = 0.74$).

One of the interesting aspects of the Magnolia Field that is that hydrocarbon fluids are not well mixed. This presents an enormous obstacle to the identification of flow barriers through application of geochemical fingerprinting methods. A particularly data-rich example provided by one of the Magnolia wells and its sidetrack is discussed below.

Figure 20 shows a log display of one of the Magnolia Field pay zones. In addition to conventional wireline log and lithology data, the $\delta^{13}C_{C_1}$ of associated gases from two MDT samples are shown. The dissimilarities between the two samples, taken only 240 ft apart, are striking. Methane carbon isotopic composition, for instance, differs by as much 5.8‰. Sample API gravity and saturation pressure, annotated in Figure 20, are also significantly different. Calculated fluid densities at reservoir pressure however differ by only 0.005 g/cm$^3$ between the two samples (being 0.639 and 0.644 g/cm$^3$ for the shallow and deep sample respectively). Three gas bags were obtained across this zone whose methane carbon isotopic composition at the approximate MDT sampling points compare well with that of the fluid samples (Fig. 20). Although a barrier to vertical fluid migration would be suspected between the two MDT sample depths based on the degree to which the fluids differ, this is not supported by analysis of measured formation pressures (Fig. 21) obtained across the pay interval. Analysis of these data suggests fluid continuity within the zone.

A sidetrack to the borehole shown in Figure 20 was drilled about 100 ft from the original well in order to obtain whole cores. Figure 22 shows a log plot from the sidetrack. MDT samples were not obtained from the pay interval, but the opportunity was taken to obtain very high resolution mud gas information from this borehole. Mud gas (IsoTube) samples were collected during coring across the pay zone and the adjacent interval at 25 ft intervals; their methane carbon

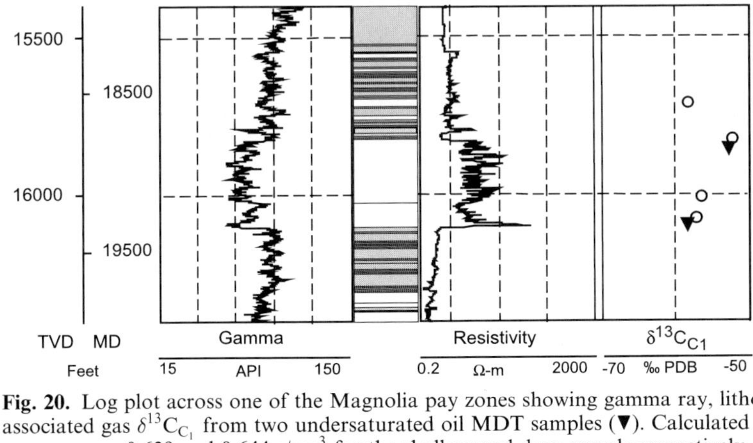


| TVD | API | PSat | $\delta^{13}C_{C1}$ |
|---|---|---|---|
| ft | ° | psi | ‰ |
| 15,856 | 37.5 | 4849 | -52.7 |
| 16,092 | 33.6 | 6010 | -58.5 |

**Fig. 20.** Log plot across one of the Magnolia pay zones showing gamma ray, lithology, resistivity, and associated gas $\delta^{13}C_{C_1}$ from two undersaturated oil MDT samples (▼). Calculated fluid densities at reservoir pressures are 0.639 and 0.644 g/cm$^3$ for the shallow and deep sample respectively. The $\delta^{13}C_{C_1}$ of the few gas bag samples taken over this interval are also shown (O).

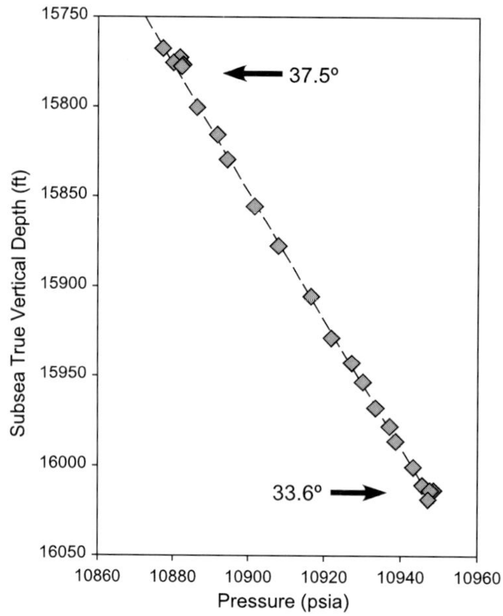


**Fig. 21.** Depth plot of pressures across the sand zone shown in Figure 20. The API gravities of the MDT fluid samples are annotated. $R^2$ for the linear correlation is 0.998.

isotopic composition is plotted in Figure 22. The values annotated (−52.5 and −58.6‰) represent the mud gas $\delta^{13}C_{C_1}$ at the depths equivalent to MDT samples in the original well (−52.7 and −58.5‰) with which they are isotopically effectively identical. The most revealing aspect is, however, the gradual vertical decrease in methane carbon isotopic composition from top to base of the sand. The isotopic gradient across this interval is quite high: 2.5‰ per 100 ft. The absence of isotopic discontinuities with depth provides fairly compelling evidence that the pressure data have, in fact, correctly reflected fluid continuity across this vertical interval. The character of sedimentary structures observed in whole cores obtained from this interval provides additional support that laterally extensive barriers are not present, though discontinuous baffles may exist. Pressures in this instance have equilibrated; hydrocarbon fluids of variable maturity and biogenic gas content have not been emplaced sufficiently long to mix.

The interval shown in Figures 20–22 also provides an example in which the vertical hydrocarbon fluid maturity pattern is clearly not in accordance with a simple top-down least-mature-to-most-mature filling scenario. Here the least mature thermogenic hydrocarbons lie at the base of the interval; maturity increases gradually upwards. Although the reason for this vertical pattern remains obscure, these data provide a useful reminder that charge timing, migration directions and consequent mixing patterns are likely to be strongly influenced by sedimentary fabrics and other, potentially sub-seismic, discontinuities, particularly in stratigraphically and/or structurally complex depositional systems such as those that typify many deepwater settings. In these (and arguably most) settings, reservoir filling studies clearly must incorporate petrophysical descriptions that capture vertical and lateral variations in flow unit properties.

Although hydrocarbon composition variations within compartments challenge our ability

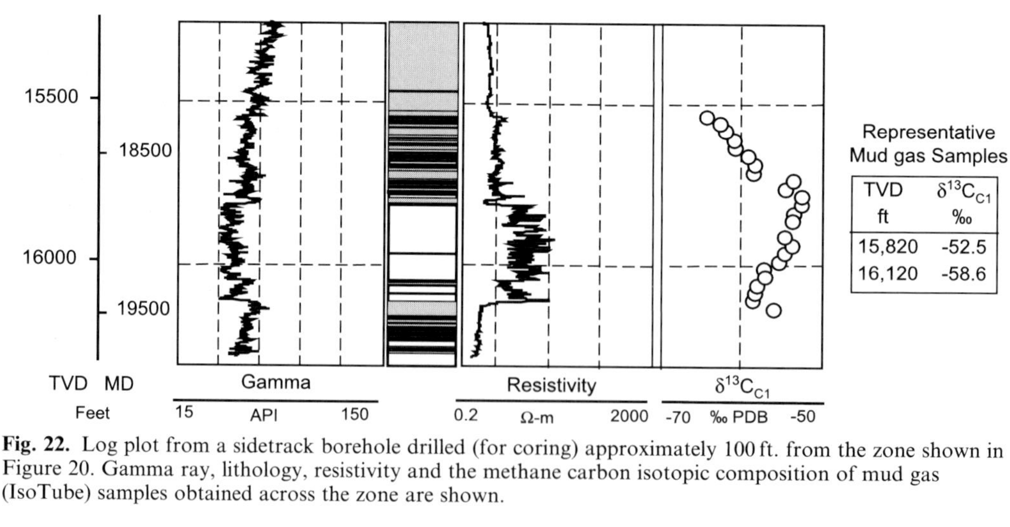


**Fig. 22.** Log plot from a sidetrack borehole drilled (for coring) approximately 100 ft. from the zone shown in Figure 20. Gamma ray, lithology, resistivity and the methane carbon isotopic composition of mud gas (IsoTube) samples obtained across the zone are shown.

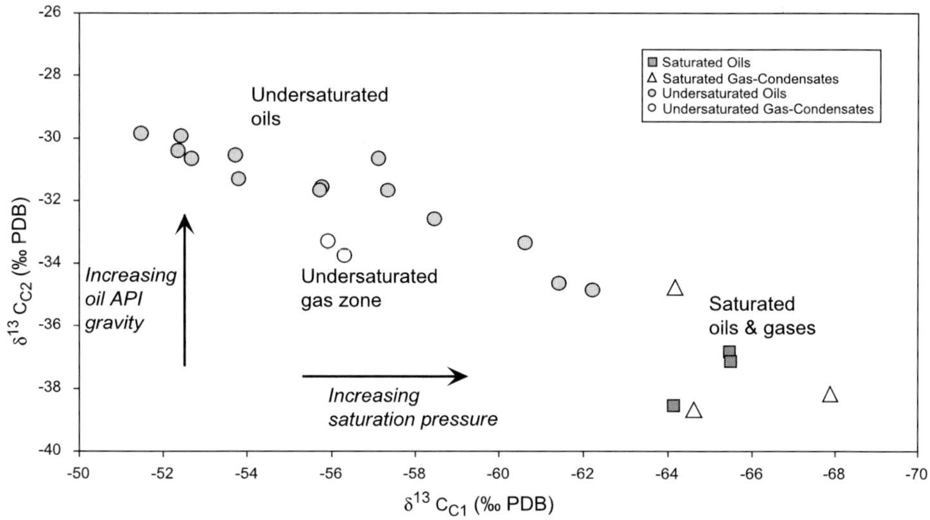


**Figure 23.** $\delta^{13}C_{C_1}$ versus $\delta^{13}C_{C_2}$ for Magnolia MDT sample associated gases. Fields are annotated in accordance with general fluid characteristics. Discrimination of the undersaturated gas zone is best accomplished by also integrating gas dryness, e.g. $C_1/\sum C_1 \ldots C_5$, as shown in Figure 16.

to infer the presence of flow barriers using fluid attributes at Magnolia, the relationships that exist between fluid properties and gas isotopic character provide an unusual opportunity for fluid predictions. Figure 23 shows a cross plot of $\delta^{13}C_{C_1}$ versus $\delta^{13}C_{C_2}$ of associated and free gases from the oil and gas-condensate reservoirs. These gas data have been annotated in order to illustrate a few of the empirical relationships observed to exist between hydrocarbon gas character and fluid properties in the field. These general trends, as well as those shown in Figure 16, can also be observed in mud gas data, although, as demonstrated in Figure 18, mud gas dryness must be adjusted to account for the excess methane typically recovered from these samples. Once mud gas data have been mapped in accordance with empirical schemes developed using MDT data (such as that shown in Fig. 23), predictions of phase and fluid properties can be made for untested zones from which mud gas samples have been taken. The inclusion of additional information when available, e.g. wireline log data, show information and particularly pressure data, enormously improves the accuracy of the fluid property predictions made using this empirical approach.

The compositional heterogeneities observed to exist at Magnolia offer, in addition, the potential that geochemical attributes may be useful to discern fluid production patterns and problems. In addition to the data acquired from the six exploration and appraisal wells discussed in this paper, currently drilling development wells are being extensively sampled both by conventional MDT methods and very high resolution mud gas samples. Mapping of the vertical and lateral geochemical characteristics within fluid compartments may provide a framework for the interpretation of changing fluid characteristics over the production life of the field. These characteristics can be monitored using relatively low cost gas and whole oil analyses and may provide insights into fluid production efficiencies and patterns as well as mechanical problems such as tubing leaks that may arise.

A number of factors have contributed to the heterogeneous and unmixed state of the Magnolia fluids. Thermogenic hydrocarbons of varying composition were emplaced relatively recently. Based on the time of trap formation, emplacement certainly occurred since the end of the Pliocene; it is possible, in view of the presence of extremely active hydrocarbon seepage along the mini-basin margins, that hydrocarbon charge is ongoing. Reservoir temperatures are relatively cool and *in situ* hydrocarbon fluid viscosities are correspondingly high, which slows mixing rates. Cool temperatures and ample indigenous immature organic material in Plio-Pleistocene shales sequences adjacent to the Magnolia Field sands (and their basinal equivalents) promoted the production of biogenic methane that occurs in variable amounts in Magnolia fluids. Finally, high sedimentation rates have resulted in rapidly changing pressure and temperature conditions in the field that have undoubtedly promoted fluid disequilibria.

These factors are hardly unique to the Magnolia Field or even to the deepwater Gulf of Mexico. Deepwater exploration settings globally are often characterized by young reservoirs, cool temperatures and high sedimentation rates that together will tend to promote compositional disequilibria in hydrocarbon reservoirs. Reservoir studies conducted in these environments must explicitly address the degree to which fluids have mixed in order that the implications of fluid variations can be correctly interpreted. It is not defensible in dynamic hydrocarbon systems (or arguably in any system) to dismiss the possibility that hydrocarbon fluids are unmixed in continuous reservoirs based on the well-worn argument that mixing processes are 'geologically instantaneous'.

## Conclusions

Magnolia hydrocarbon fluid compositional variations are controlled by a number of processes. Hydrocarbon fluids were derived from a Tithonian source rock at various levels of maturity. Low maturity products are typified by oils which have low API gravity, contain a relatively large fraction of sulphur (though not in excess of 2%) and organic acids and contain both $C_{2+}$ gases and liquids whose molecular characteristics are suggestive of low source rock maturity at the time of expulsion. High maturity products have, by comparison, higher (though variable) API gravity, are sulphur and TAN poor and contain $C_{2+}$ gases and liquids whose molecular characteristics suggest a higher source rock maturity at the time of expulsion. Although all fluids contain a biogenic methane component, the lowest overpressure reservoirs that are presumably well connected and open to a large adjacent biogenic source kitchen contain the greatest amounts. Saturation pressures are very strongly influenced by excess biogenic methane to the extent that its presence obscures expected relationships between charge maturity and phase behaviour. Some of the reservoirs have experienced a minor degree of water washing which has not profoundly influenced phase properties. *In situ* phase fractionation in some of the reservoirs resulted from mixing of biogenic methane with thermogenic liquids (and associated gases). By increasing fluid saturation pressures the admixture of methane resulted in the transformation of single-phase (undersaturated) to two-phase (saturated) oil and gas-condensate accumulations. Finally, one reservoir, containing an undersaturated gas-condensate, may record a complex history of multiple episodes of thermogenic charge emplacement and leakage.

Hydrocarbon fluids in some of the Magnolia reservoirs are not well mixed. This is most evident in hydrocarbon gas character whose vertical variations are revealed by high resolution mud gas data. From MDT and mud gas data across one of the reservoirs, a vertical methane carbon isotope ($\delta^{13}C_{C_1}$) gradient of as much as 2.5‰ per 100 ft can be demonstrated to exist in the absence of a pressure discontinuity. Other fluid properties such as density and saturation pressure are correspondingly variable in this instance in the absence of compartmentalization. Similar, though usually not as abundantly sampled, examples of hydrocarbon compositional heterogeneities occur throughout the field.

Empirical relationships between gas character and fluid properties allow the prediction of the phase and fluid properties of untested (and sometimes unlogged) hydrocarbon-bearing zones. The $\delta^{13}C$ and dryness (e.g. $C_1/\sum C_1 \ldots C_3$) of associated and free gases correlate with a number of fluid properties, including phase, API gravity and saturation pressure. These same trends can be recognized in the mud gas hydrocarbon $\delta^{13}C$ and dryness data and, once calibrated with the MDT information, predictions can be made for untested zones. When integrated with ancillary information (when available) such as reservoir pressure, show information and wireline log data, these empirically based phase and fluid predictions are extremely accurate.

The factors that contribute to the heterogeneous and unmixed state of Magnolia fluids include recent and possibly ongoing charging, thermogenic fluids of variable maturity, cool reservoir temperatures, admixture of biogenic methane and rapidly changing pressure and temperatures within the field. Similar conditions are typical of many deepwater petroleum exploration and production settings where it is consequently of paramount importance that reservoir characterization studies explicitly address the degree to which hydrocarbon fluids have attained compositional equilibrium.

We wish to thank ConocoPhillips and our partner Devon Energy for permission to publish this work. Reviewers Steve Thompson and Sheng-Tai Lee are thanked for their constructive comments. Particular gratitude is extended to our colleagues for discussions and support of this work, especially Albert Holba, Mike Lines, Lung-Chuan Kuo, Eric Michael, Bruce Reitz, Susan Young, Fraser Ross, Patty Travis and Lloyd Brown.

## Conversions to SI units

ft × 0.3048 = metre
mile × 1.609 = kilometre
sq mile × 2.59 = $km^2$
°F − 32 × 9/5 = °C
°F/100 ft × 1.82 = °C/100 m
psi × 0.0689 = bar
scf/stb × 0.178 = $m^3/m^3$
141.5/(API + 131.5) = $g/cm^3$

## References


BAZAN, L. W. 1998. The allocation of gas well production data using isotope analysis. *Society of Petroleum Engineers*, Paper No. **40032**.

BEEUNAS, M. A., HUDSON, T. A., VALLEY, J. A., BASKIN, D. K. & WESLEY, Y. C. 1999. Reservoir continuity and architecture of the Genesis Field, Gulf of Mexico (Green Canyon 205): An integration of fluid geochemistry within the geologic and engineering framework. *Transactions of the Gulf Coast Association of Geological Societies*, **49**, 90–95.

BEEUNAS, M. A., SCHOELL, M. & ZUMBERGE, J. 2001. Habitat of natural gas in the Gulf of Mexico shelf. *American Association of Petroleum Geologists Bulletin*, **85**, supplement.

BEHAR, F. & ALBRECHT, P. 1984. Correlations between carboxylic acids and hydrocarbons in several crude oils. Alteration by biodegradation. *Organic Geochemistry*, **6**, 597–604.

BEHAR, F., VANDENBROUCKE, M., TANG, Y., MARQUIS, F. & ESPITALIÉ, J. 1997 Thermal cracking of kerogen in open and closed systems: determination of kinetic parameters and stoichiometric coefficients for oil and gas generation. *Organic Geochemistry*, **26**, 321–339.

BISSADA, K. K., KATZ, B. J., BARNICLE, S. C. & SCHUNK, D. J. 1990. On the origin of hydrocarbons in the Gulf of Mexico Basin—a reappraisal. *In*: SCHUMACHER, D. & PERKINS, B. F. (eds) *Gulf Coast Oils and Gases: Their Characteristics, Origin, Distribution, and Exploration and Production Significance*. Society of Economic Paleontologists and Mineralogists Research Conference, 163–171.

BISSADA, K. K., CALLEJON, A., QUINONES, N. H., GARCIA, E. C. & GALLARDO, H. R. 2002. Biogenic methane in the Gulf of Mexico Basin—significant resource or secondary contaminant? American Association of Petroleum Geologists 2002 Annual Convention, Abstract.

CLAYTON, C. 1991. Carbon isotope fractionation during natural gas generation from kerogen. *Marine and Petroleum Geology*, **8**, 232–240.

COLE, G. A. ET AL. 2001. The deepwater GOM petroleum system: insights from piston coring seepage versus anomalies versus background. Gulf Coast Society of Economic Paleontologists and Mineralogists Research Conference, 315–342.

CONNAN, J. 1984. Biodegradation of crude oils in reservoirs. *In*: BROOKS, J. & WELTE, D. H. (eds) *Advances in Petroleum Geology*, Vol. 1. Academic Press, London, 299–335.

CONNAN, J. & CASSOU, A. M. 1980. Properties of gases and petroleum liquids derived from terrestrial kerogen at various maturation levels. *Geochimica et Cosmochimica Acta*, **44**, 1–23.

CURIALE, J. A. & BROMLEY, B. W. 1996. Migration of petroleum into Vermillion 14 field, Gulf Coast, U.S.A.—molecular evidence. *Organic Geochemistry*, **24**, 563–579.

DAVIS, J. B. & BRAY, E. E. 1969. Analysis of oil and cap rock from the Challenger (Sigsbee) Knoll. *Initial Reports of the Deep Sea Drilling Project, Leg* **10**, 415–500.

DIAS, R. F., FREEMAN, K. H., LEWAN, M. D. & FRANKS, S. G. 2002. $\delta^{13}C$ of low-molecular-weight organic acids generated by the hydrous pyrolysis of oil-prone source rocks. *Geochimica et Cosmochimica Acta*, **66**, 2755–2769.

DIEGEL, F. A., KARLO, J. F., SCHUSTER, D. C., SHOUP, R. C. & TAUVERS, P. R. 1995. Cenozoic structural evolution and tectono-stratigraphic framework of the northern Gulf Coast continental margin. *In*: JACKSON, M. P. A., ROBERTS, D. G. & SNELSON, S. (eds) *Salt Tectonics: A Global Perspective*. American Association of Petroleum Geologists Memoir, **65**, 109–151.

DI PRIMIO, R., DIECKMANN, V. & MILLS, N. 1998. PVT and phase behavior analysis in petroleum exploration. *Organic Geochemistry*, **29**, 207–222.

DZOU, L. I. P. & HUGHES, W. B. 1993. Geochemistry of oils and condensates, K field, offshore Taiwan: a case study in migration fractionation. *Organic Geochemistry*, **20**, 437–462.

EDMAN, J. D. & BURK, M. K. 1999. Geochemistry in an integrated study of reservoir compartmentalization at Ewing Bank 873, offshore Gulf of Mexico. *Society of Petroleum Engineers*, Paper No. **57470**.

EDMAN, J. D., JAMESON, E. W., PERKINS, T. M. & GUTHRIE, G. E. 2001. Mississippi Canyon 348: some reservoir engineering considerations regarding development of a deepwater biogenic gas accumulation. *Society of Petroleum Engineers*, Paper No. **71611**.

ENGLAND, W. A., MUGGERIDGE, A. H., CLIFFORD, P. J. & TANG, Z. 1995. Modelling density-driven mixing rates in petroleum reservoirs on geological timescales, with application to the detection of barriers in the Forties Field (UKCS). *In*: CUBITT, J. M. & ENGLAND, W. A. (eds) *The Geochemistry of Reservoirs*. Geological Society, London, Special Publications, **86**, 185–201.

FARRIMOND, P., TAYLOR, A. & TELNAES, N. 1998. Biomarker maturity parameters: the role of generation and thermal degradation. *Organic Geochemistry*, **29**, 1181–1197.

GUZMÁN-VEGA, M. A. & MELLO, M. R. 1999. Origin of oil in the Sureste Basin, Mexico. *American Association of Petroleum Geologists Bulletin*, **83**, 1068–1095.

GUZMÁN-VEGA, M. A. *ET AL.* 2001. Classification and origin of petroleum in the Mexican Gulf Coast Basin: an overview. *In*: BARTOLLINI, C., BUFFLER, R. T. & CANTU-CHAPA, A. (eds) *The Western Gulf of Mexico Basin: Tectonics, Sedimentary Basins and Petroleum Systems*. American Association of Petroleum Geologists Memoir, **75**, 127–142.

TEN HAVEN, H. L. 1996. Applications and limitations of Mango's light hydrocarbon parameters in petroleum correlation studies. *Organic Geochemistry*, **24**, 957–976.

HOLBA, A. G., DZOU, L. I. P., HICKEY, J. J., FRANKS, S. G., MAY, S. J. & LENNEY, T. 1996. Reservoir geochemistry of South Pass 61 Field, Gulf of Mexico: compositional heterogeneities reflecting filling history and biodegradation. *Organic Geochemistry*, **24**, 1179–1198.

HILLEBRAND, T. & LEYTHAEUSER, D. 1992. Reservoir geochemistry of Stockstadt oilfield: compositional heterogeneities reflecting accumulation history and multiple source input. *Organic Geochemistry*, **19**, 119–131.

HOLBA, A. G., DZOU, L. I. P., HICKEY, J. J., FRANKS, S. G., MAY, S. J. & LENNEY, T. 1996. Reservoir geochemistry of South Pass 61 Field, Gulf of Mexico: compositional heterogeneities reflecting filling history and biodegradation. *Organic Geochemistry*, **24**, 1179–1198.

HOLGUIN, N., LARA, J., MEDRANO, L., BERNARD, B., BROOKS, J. M. & ZUMBERGE, J. 2002. PEMEX's surface geochemical exploration studies in the southern Gulf of Mexico. American Association of Petroleum Geologists 2002 Annual Convention, Abstract.

HOOD, K. C., WENGER, L. M., GROSS, O. P. & HARRISON, S. C. 2002. Hydrocarbon systems analysis of the northern Gulf of Mexico: delineation of hydrocarbon migration pathways using seeps and seismic imaging. *In*: SCHUMACHER, D. & LESHACK, L. A. (eds) *Surface Exploration Case Histories: Applications of Geochemistry, Magnetics and Remote Sensing*. American Association of Petroleum Geologists Studies in Geology, **48**, 25–40.

HORSTAD, I. & LARTER, S. R. 1997. Petroleum migration, alteration, and remigration within Troll field, Norwegian North Sea. *American Association of Geologists Bulletin*, **81**, 222–248.

HWANG, R. J., AHMED, A. S. & MOLDOWAN, J. M. 1994. Oil composition variation and reservoir continuity: Unity field, Sudan. *Organic Geochemistry*, **21**, 171–188.

JARVIE, D. J., MORELOS, A., SASSEN, R., CHENET, P.-Y. & BRAME, J. W. 2002. Hydrocarbon charge assessment, Gulf of Mexico: rates of oil/gas generation from source rocks and oil asphaltenes. American Association of Petroleum Geologists 2002 Annual Convention, Abstract.

KAUFMAN, R. L., AHMED, A. S. & ELSINGER, R. J. 1990. Gas chromatography as a development and production tool for fingerprinting oils from individual reservoirs: applications in the Gulf of Mexico. Gulf Coast Society of Economic Paleontologists and Mineralogists Research Conference, 315–342.

KAUFMAN, R. L. *ET AL.* 1997. Characterizing the greater Burgan field: use of geochemistry and oil fingerprinting. *Society of Petroleum Engineers*, Paper No. **37803**.

KORNACKI, A. S. & NELSON, K. C. 2000. Assessing the mix and the quality of crude oil and natural gas in the deepwater Gulf of Mexico. *In*: Models for Understanding Risk. American Association of Petroleum Geologists Hedberg Research Conference, 1998, Galveston, Texas.

KORNACKI, A. S., ALLIE, A. D. & HOLMAN, W. E. 1996. Wet bacterial gas in the northern Gulf of Mexico Basin. *Transactions of the Gulf Coast Association of Geological Societies*, **46**, 476.

KUO, L.-C. 1994. An experimental study of crude oil alteration in reservoir rocks by water washing. *Organic Geochemistry*, **21**, 465–479.

LARTER, S. R. & APLIN, A. C. 1995. Reservoir geochemistry: methods, applications and opportunities. *In*: CUBITT, J. M. & ENGLAND, W. A. (eds) *The Geochemistry of Reservoirs*. Geological Society, London, Special Publications, **86**, 5–32.

LARTER, S. *ET AL.* 1999. When biodegradation preserves petroleum: North Sea oil rimmed gas accumulations. *In*: Natural Gas Formation and Occurrence. American Association of Petroleum Geologists Hedberg Research Conference, 1999, Durango, Colorado.

LEVACHÉ, D., MONTEL, F. & WALGENWITZ, F. 2000. Deep offshore fluid evaluation/connectivity study. *Society of Petroleum Engineers*, Paper No. **62904**.

LOSH, S., CATHLES, L. & MEULBROEK, P. 2002. Gas washing of oil along a regional transect, offshore Louisiana. *Organic Geochemistry*, **33**, 655–663.

MANGO, F. D. 1987. An invariance in the isoheptanes of petroleum. *Science*, **237**, 514–517.

MANGO, F. D. 1990. The origin of light hydrocarbon in petroleum: a kinetic test of the steady state catalytic hypothesis. *Geochimica et Cosmochimica Acta*, **58**, 895–901.

MANGO, F. D. 1994. The origin of light hydrocarbons in petroleum: ring preference in the closure of carbocyclic rings. *Geochimica et Cosmochimica Acta*, **58**, 895–901.

MANGO, F. D. 1997. The light hydrocarbons in petroleum: a critical review. *Organic Geochemistry*, **26**, 417–440.

MCCAFFREY, M. A., LEGARRE, H. & JOHNSON, S. J. 1996. Using biomarkers to improve heavy oil reservoir oil management: an example from the Cymric field, Kern County, California. *American Association of Petroleum Geologists Bulletin*, **80**, 898–913.

MCKINNEY, D. & BLAND, K. 2003. Time lapse geochemistry at the Ursa field: a surveillance tool for understanding complex, heterogeneous reservoir fluids. 21st International Meeting on Organic Geochemistry, 8–12 September 2003, Krakow, Poland, p. I/096.

MEREDITH, W., KELLAND, S.-J. & JONES, D. M. 2000. Influence of biodegradation on crude oil acidity and carboxylic acid composition. *Organic Geochemistry*, **31**, 1059–1073.

MEULBROEK, P., CATHLES, L. & WHELAN, J. 1998. Phase fractionation at South Eugene Island Block 330. *Organic Geochemistry*, **29**, 223–239.

NICOLLE, G., BOIBIEN, C., TEN HAVEN, H. L., TEGELAAR, E. & CHAVAGNAC, P. 1997. Geochemistry: a powerful tool for reservoir monitoring. *Society of Petroleum Engineers*, Paper No. **37804**.

PALLASSER, R. J. 2000. Recognising biodegradation in gas/oil accumulations through the $\delta^{13}C$ compositions of gas components. *Organic Geochemistry*, **31**, 1363–1373.

PALMER, S. E. 1993. Effect of biodegradation and water washing on crude oil composition. *In*: ENGEL, M. J. & MACKO, S. A. (eds) *Organic Geochemistry Principles and Applications*. Plenum Press, New York, 511–533.

PRATHER, B. E., BOOTH, J. R., STEFFENS, G. S. & CRAIG, P. A. 1998. Classification, lithologic calibration, and stratigraphic succession of seismic facies of intraslope basins, deep-water Gulf of Mexico. *American Association of Petroleum Geologists Bulletin*, **82**, 701–728.

PRINZHOFER, A. & PERNATON, E. 1997. Isotopically light methane in natural gas: bacterial imprint or segregative migration? *Chemical Geology*, **142**, 193–200.

PRINZHOFER, A., MELLO, M. R. & TAKAKI, T. 2000. Geochemical characterization of natural gas: a physical multivariable approach and its applications in maturity and migration estimates. *American Association of Petroleum Geologists Bulletin*, **84**, 1152–1172.

RADKE, M. 1988. Application of aromatic compounds as maturity indicators in source rocks and crude oils. *Marine and Petroleum Geology*, **5**, 224–236.

RICE, D. D. & CLAYPOOL, G. E. 1981. Generation, accumulation, and resource potential of biogenic gas. *American Association of Petroleum Geologists Bulletin*, **65**, 5–25.

ROMERO, M. A., RUIZ, J. M., MEDRANO, L. M., DURAN, J. G., ROJAS, R., GUTIERREZ, I. & BALTAZAR, O. 2000. Revitalized Tithonian–Cretaceous petroleum system (!), offshore Campeche, Mexico. *Transactions of the Gulf Coast Association of Geological Societies*, **50**, 663–668.

ROWAN, M. G. 1995. Structural styles and evolution of allochthonous salt, central Louisiana outer shelf and upper slope. *In*: JACKSON, M. P. A., ROBERTS, D. G. & SNELSON, S. (eds) *Salt Tectonics: A Global Perspective*. American Association of Petroleum Geologists Memoir, **65**, 199–228.

ROWAN, M. G., WEIMER, P. & FLEMINGS, P. B. 1994. Three dimensional geometry and evolution of a composite multi-level salt system, western Eugene Island, offshore Louisiana. *Transactions of the Gulf Coast Association of Geological Societies*, **44**, 641–648.

SANTAMARIA-OROZCO, D., HORSFIELD, B., DI PRIMIO, R. & WELTE, D. H. 1998. Influence of maturity on distribution of benzo- and dibenzothiophenes in Tithonian source rocks and crude oils, Sonda de Campeche, Mexico. *Organic Geochemistry*, **28**, 423–139.

SMALLEY, P. C. & ENGLAND, W. A. 1992. Assessing reservoir compartmentalization during field appraisal: how geochemistry can help. *Society of Petroleum Engineers*, Paper No. **25005**.

SMALLEY, P. C. & HALE, N. A. 1996. Early identification of reservoir compartmentalization by combining a range of conventional and novel data types. *Society of Petroleum Engineers*, Paper No. **30533**.

THOMPSON, K. F. M. 1987. Fractionated aromatic petroleums and the generation of gas-condensates. *Organic Geochemistry*, **11**, 573–590.

THOMPSON, K. F. M. 1988. Gas-condensate migration and oil fractionation in deltaic systems. *Marine and Petroleum Geology*, **5**, 237–246.

THOMPSON, K. F. M. 2002. Compositional regularities common to petroleum reservoir fluids and pyrolysates of asphaltenes and kerogens. *Organic Geochemistry*, **33**, 829–841.

WATKINS, J. S. & BUFFLER, R. T. 1996. Gulf of Mexico deepwater frontier exploration potential. *Transactions of the Gulf Coast Association of Geological Societies*, **46**, 79–92.

WATSON, J. S., JONES, D. M., SWANNEL, R. P. J. & VAN DUIN, A. C. T. 2002. Formation of carboxylic acids during aerobic biodegradation of crude oil and evidence of microbial oxidation of hopanes. *Organic Geochemistry*, **33**, 1153–1169.

WAVREK, D. A., MOSCA, F., CHAOUCHE, A. & JARVIE, D. M. 2001. Geochemical insights to engineering problems: when does oil heterogeneity really indicate a compartment? *In*: 20th International Meeting on Organic Geochemistry. Nancy, France, September 2001, 123–124.

WESTRICH, J. T., FUEX, A. M., O'NEAL, P. M. & HALPERN, H. I. 1999. Evaluating reservoir architecture in the northern Gulf of Mexico with oil and gas chemistry. *Society of Petroleum Engineers*, Paper No. **36541**.

WHELAN, J. K., KENNICUTT, M. C., BROOKS, J. M., SCHUMACHER, D. & ELGINTON, L. B. 1993. Organic geochemical indicators of dynamic fluid flow processes in petroleum basins. *Organic Geochemistry*, **22**, 587–615.

WHELAN, J. K., EGLINTON, L., KENNICUTT, M. C. & QIAN, Y. 2001. Short-time-scale (year) variations of petroleum fluids from the U.S. Gulf Coast. *Geochimica et Cosmochimica Acta*, **65**, 3529–3555.

WHITICAR, M. J., FABER, E. & SCHOELL, M. 1986. Biogenic methane formation in marine and freshwater environments: $CO_2$ reduction vs. acetate fermentation—isotope evidence. *Geochimica et Cosmochimica Acta*, **50**, 693–709.

WILHELMS, A., REIN, E., ZWACH, C. & STEEN, A. S. 2001. Application and implication of horizontal well geochemistry. *Petroleum Geoscience*, **7**, 75–79.

ZHANG, Y. G. & CHEN, H. J. 1985. Concepts on the generation and accumulation of biogenic gas. *Journal of Petroleum Geology*, **8**, 405–422.

ZUMBERGE, J. E. 1993. Organic geochemistry of Estancia Vieja Oils, Rio Negro Norte Block: correlation with other Nequen Basin, Argentina, oils. *In*: ENGEL, M. J. & MACKO, S. A. (eds) *Organic Geochemistry Principles and Applications*. Plenum Press, New York, 481–471.

# High temperature quartz cementation and the timing of hydrocarbon accumulation in the Jurassic Norphlet sandstone, offshore Gulf of Mexico, USA

THOMAS TAYLOR,[1] RICHARD STANCLIFFE,[2] CALUM MACAULAY[1] & LORI HATHON[1]

[1] *Shell International Exploration and Production, Bellaire Technology Centre, P.O. Box 481, Houston, Texas 77001, USA (e-mail: thomas.taylor@shell.com)*

[2] *Shell Exploration and Production Company, New Orleans, Louisiana 70161, USA*

**Abstract:** Jurassic Norphlet Formation sandstone reservoirs in Mobile Bay (offshore Alabama, USA) produce gas from great depths (>6.4 km) and elevated temperatures (>200 °C). Quartz cement is concentrated at the top of these aeolian reservoirs forming a low porosity 'tight-zone' of widely variable thickness (3–58 m) above a more porous reservoir section. The extent of the tight-zone is independent of depositional facies and its thickness strongly influences well performance. Intergranular porosity in the Norphlet has been preserved by inhibition of quartz cementation due to the occurrence of robust grain-coating chlorite. Quantitative petrographic data reveal that chlorite grain-coat coverage is less in the tight-zone sands (mean = 92%) than in the reservoir sands (mean = 99%). Burial history and quartz precipitation kinetics modelling indicate that this seemingly minor difference in the completeness of grain coatings is sufficient to produce the observed differences in cementation and porosity. Quartz cementation to form the tight-zone took place under conditions of deep burial and high temperature. It followed in time the onset of pressure solution, emplacement of liquid hydrocarbons, and the precipitation of a solid hydrocarbon film (pyrobitumen) on the walls of the intergranular pores. Fluid inclusion microthermometry data indicate that volumetrically significant quartz cement precipitated at temperatures of 150 °C or greater from highly saline aqueous fluids. Hydrocarbon-bearing inclusions are notably absent in quartz cement of the tight-zone, implying that the pore fluids were predominantly brine during precipitation. Oil and gas associated with pyrobitumen evidently escaped from Norphlet traps prior to tight-zone cementation. Gas presently found in Norphlet reservoirs of Mobile Bay represents a relatively recent accumulation and is not the product of *in situ* thermal cracking of oil.

Reservoir sandstones of the Jurassic Norphlet Formation in Mobile Bay (Gulf of Mexico, offshore Alabama, USA) produce natural gas from depths in excess of 6.4 km (21 000 ft). Reservoir temperature and effective stress generally exceed 200 °C and 70 MPa, respectively. These conditions approach the upper boundaries of the arbitrary division between diagenesis and low-grade metamorphism. The detection more than 20 years ago of sandstones with porosities approaching 20% under such extreme diagenetic conditions led many in the oil industry to reconsider the depths and temperatures to which commercial sandstone reservoirs are likely to occur. Numerous ideas and hypotheses for preservation of depositional porosity and creation of secondary porosity have been proposed (Dixon *et al.* 1989; Lock & Broussard 1989; Kugler & McHugh 1990; Thomas *et al.* 1993) but none fully explain the distribution of porosity within the Norphlet section.

This study examines the Norphlet reservoir at Fairway Field and surrounding areas of Mobile Bay (Fig. 1). Analyses of seismic, well logs, fluid samples and core material provide data and information that are essential for understanding both the history and present-day properties of these deeply buried reservoirs. The diagenetic evolution of the Norphlet with respect to time, temperature and effective stress, along with the timing and retention of hydrocarbon charge, have important implications for exploration, field development and production decisions.

## Geological setting

Sedimentation during the Triassic and Jurassic in the Gulf of Mexico Basin was controlled primarily by rift margin tectonics (Mancini *et al.* 1985, 1990; Salvador 1987). The regional tectonic setting was

*From*: CUBITT, J. M., ENGLAND, W. A. & LARTER, S. (eds) 2004. *Understanding Petroleum Reservoirs: towards an Integrated Reservoir Engineering and Geochemical Approach*. Geological Society, London, Special Publications, **237**, 257–278. 0305-8719/04/$15.00 

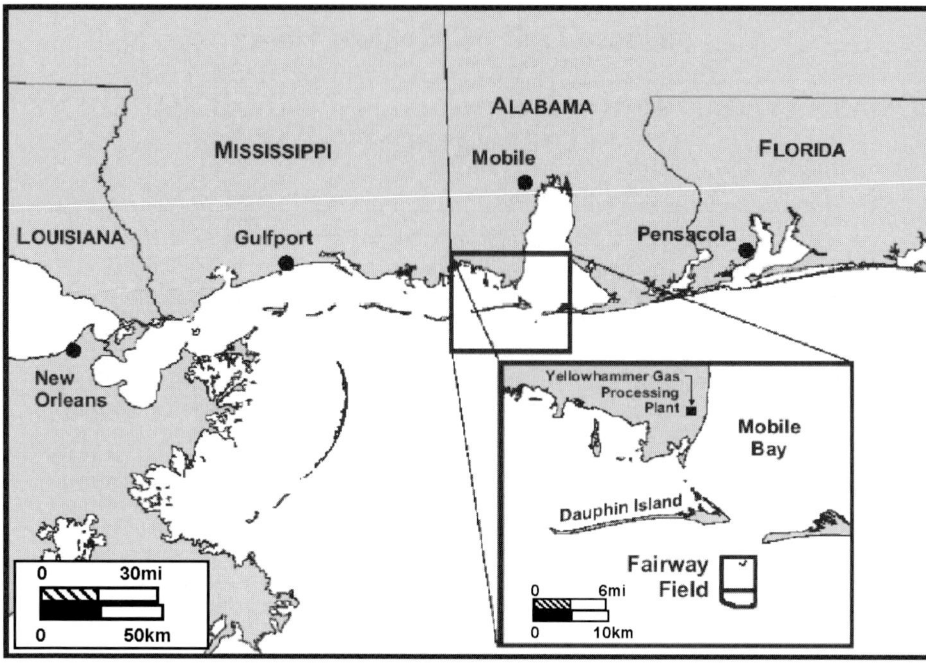


**Fig. 1.** Location of Fairway Field in the northern Gulf of Mexico, offshore Alabama, USA.

dominated by rifting and the Norphlet Formation in Mobile Bay was deposited within a large, east–west trending, pull-apart graben system.

The Norphlet sands were derived primarily from erosion of the southern Appalachian Mountains located to the north (Mancini *et al.* 1985, 1990). In the Mobile Bay area, the Norphlet is interpreted as an aeolian deposit composed of large dune complexes and associated interdune deposits (Mancini *et al.* 1985; Marzano *et al.* 1988; Dixon *et al.* 1989). The Jurassic-age Pine Hill Anhydrite and the Louann Salt lie directly beneath the Norphlet and the Oxfordian marine carbonates of the Smackover Formation occur directly above (Fig. 2).

The overall geometry of the Norphlet stratigraphic interval in Mobile Bay is characterized by a series of sub-parallel, NW–SE trending, thick sand lenses separated by intervening areas where the section is thinner than seismic resolution (Fig. 3). The lenticular sand bodies are generally 120–180 m (400–600 ft) thick at the centre whereas the inter-lens areas are estimated to be less than ~30 m thick. The thick sand masses are notably depressed into the underlying salt (Fig. 4), suggesting some deformation of the Louann substrate during Norphlet deposition (Story 1998).

The overlying Smackover carbonates are thin over the crest of Norphlet sand lenses and thick in the inter-lens areas (Fig. 4). This indicates that considerable relief existed on the top of the Norphlet during Smackover deposition. The basal Smackover carbonates were presumably

| SERIES | STAGE | ROCK UNIT |
|---|---|---|
| UPPER JURASSIC | TITHONIAN | Cotton Valley Group |
| | KIMMERIDGIAN | Haynesville Formation |
| | OXFORDIAN | Smackover Formation |
| | | **Norphlet Formation** |
| | | Pine Hill Anhydrite |
| MIDDLE JURASSIC | CALLOVIAN | Louann Salt |
| | | Werner Formation |

**Fig. 2.** Generalized stratigraphic section for the Jurassic in southwestern Alabama and Mobile Bay.

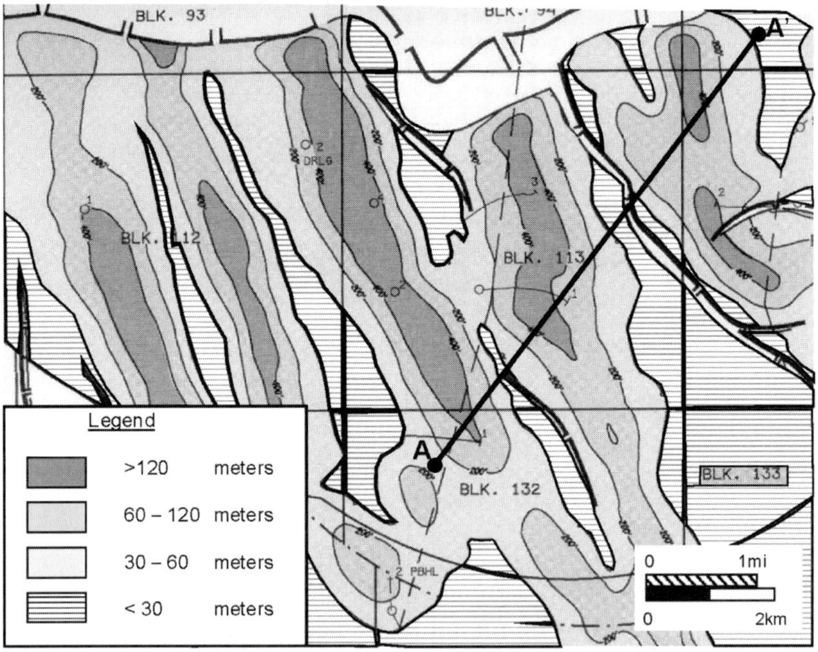


**Fig. 3.** Isopach map of the Norphlet Formation in Fairway Field and surrounding areas, Gulf of Mexico, offshore Alabama, USA. Position of cross section A–A′ in Figure 4 is shown.

deposited under very low energy conditions and consequently only minor erosion or reworking of the upper Norphlet occurred. The contact between the top of the Norphlet and the overlying Smackover Formation is typically sharp and easily picked from the resistivity log due to the concentration of pyrite within the basal Smackover (Fig. 5). The lower Smackover is organic rich and considered the most probable source of oil and gas in Norphlet and other Jurassic reservoirs of the region (Sassen *et al.* 1987; Claypool & Mancini 1989).

An east–west cross-section through Fairway Field shows the current structure and pre-production distribution of gas and brine (Fig. 4). Much of the present day relief (~40–75 m) is

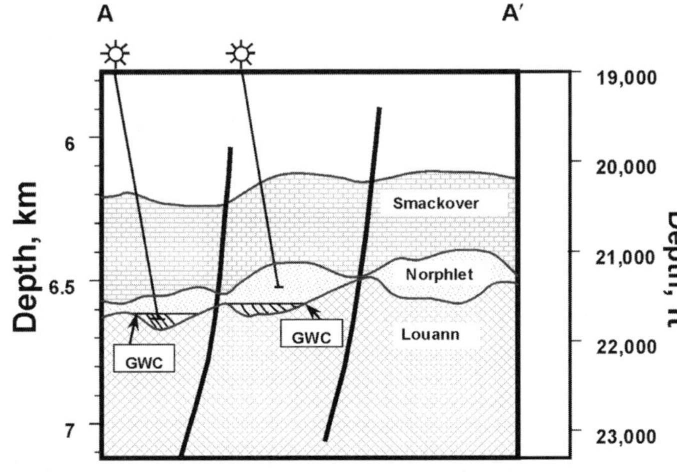


**Fig. 4.** Generalized cross section through Fairway Field along A–A′ in Figure 3. The two gas accumulations designated by well penetrations have independent gas/water contacts (GWC).

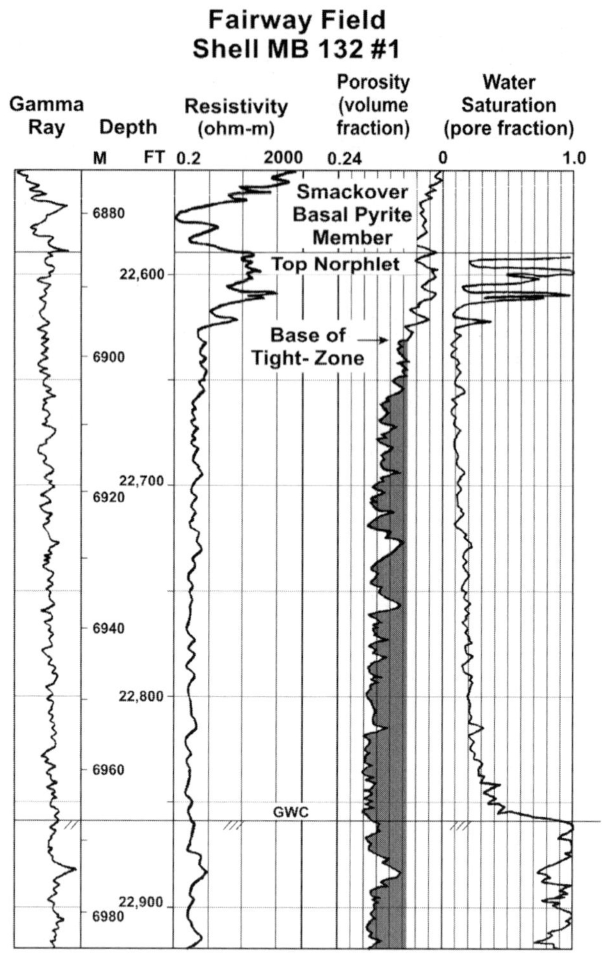


**Fig. 5.** Wireline logs showing a typical well penetration in Fairway Field. A low resistivity pyrite-rich zone characterizes the lower Smackover. The tight-zone extends from the top of the Norphlet to the first significant occurrence of porosity >8%. The Norphlet reservoir zone extends to the bottom of the well beneath the current gas/water contact (GWC). Several low porosity streaks, 1–3 m in thickness, occur within the reservoir interval.

attributed to palaeotopography although some structural tilting is evident. The two bulbous sand lenses shown in Figure 4 have separate gas–water contacts as indicated by pressure data, gas composition, and later by development drilling. The gas-water contact in the eastern accumulation is ~40 m higher than in the western accumulation, which is gas-filled to the spill point. A separate sand lens to the northeast with a different water level comprises the third discrete gas accumulation in Fairway Field.

Present-day burial depth for the Norphlet in Fairway Field ranges from ~6.6 to 7.0 km (21 600–23 000 ft). Corrected bottom hole temperatures are roughly 200–220 °C with a geothermal gradient of approximately 33°/km. The two Fairway Field reservoirs shown in Figure 4 are moderately geopressured with vertical effective stress, the difference between vertical overburden stress and fluid pressure, equal to approximately 58 MPa. The third sand lens to the northeast is essentially hydropressured with vertical effective stress of approximately 79 MPa.

## Porosity distribution from wireline logs and cores

In Fairway Field and throughout most of Mobile Bay, the typical vertical well penetration through a Norphlet sand lens encounters a low porosity

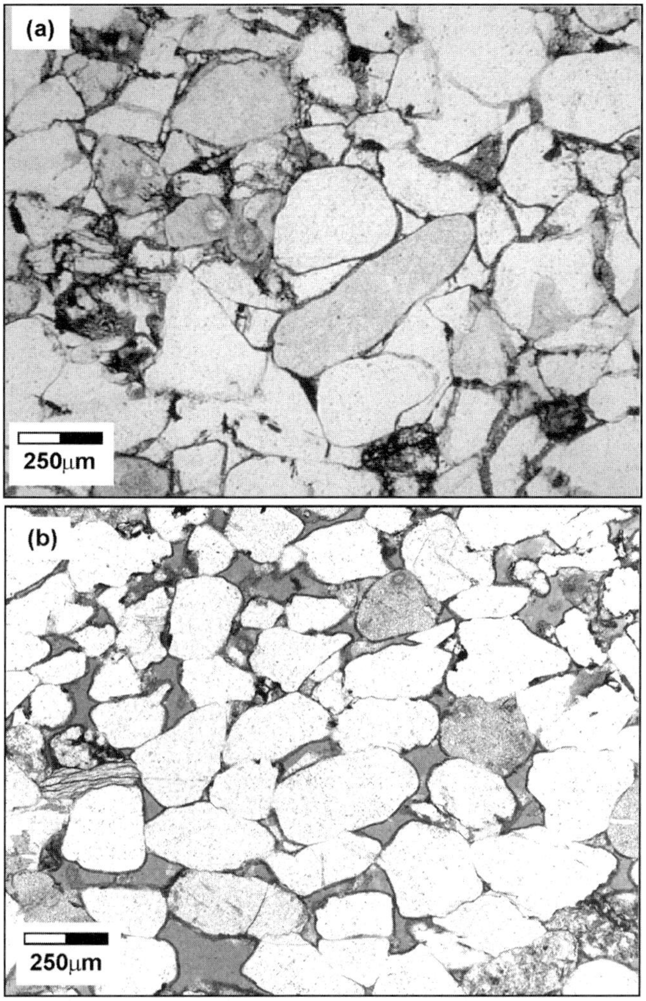


**Fig. 6.** Thin section photomicrographs of Norphlet Formation sandstones from Mobile Bay. (**a**) The tight-zone occurs at the top of the Norphlet. Almost all intergranular pore space is filled with quartz cement which encases grain-coating chlorite and pyrobitumen (black). (**b**) Porous reservoir sandstones occur beneath the tight-zone and are characterized by open intergranular pores that are lined with chlorite and pyrobitumen (black).

($\phi = 2$–$8\%$), quartz cemented 'tight-zone' of variable thickness at the top of the section (Marzano *et al.* 1988; Dixon *et al.* 1989; Thomas *et al.* 1993). Beneath the tight-zone, porous reservoir sands ($\phi = 10$–$20\%$) are prevalent generally to the base of the section (Fig. 6). The thickness of the tight-zone limits net pay and strongly influences the rate of natural gas production from an individual well.

The tight-zone is defined as the interval from the top of the Norphlet to the first occurrence of wireline log porosity greater than 8% (Fig. 5). In Fairway Field wells, the tight-zone ranges in thickness from 12 to 28 m (39–91 ft) and averages approximately 5% porosity. Examination of wireline logs from 22 wells in Mobile Bay reveal a range in tight-zone thickness from 3 to 58 m (10–190 ft). The Norphlet reservoir sands underneath the tight-zone in Fairway Field have an average porosity of 14% with a maximum of 20%. Low porosity sandstones are occasionally encountered within the reservoir interval, but rarely exceed several metres in thickness.

In Fairway Field and in the large majority of wells examined in this study, the transition from

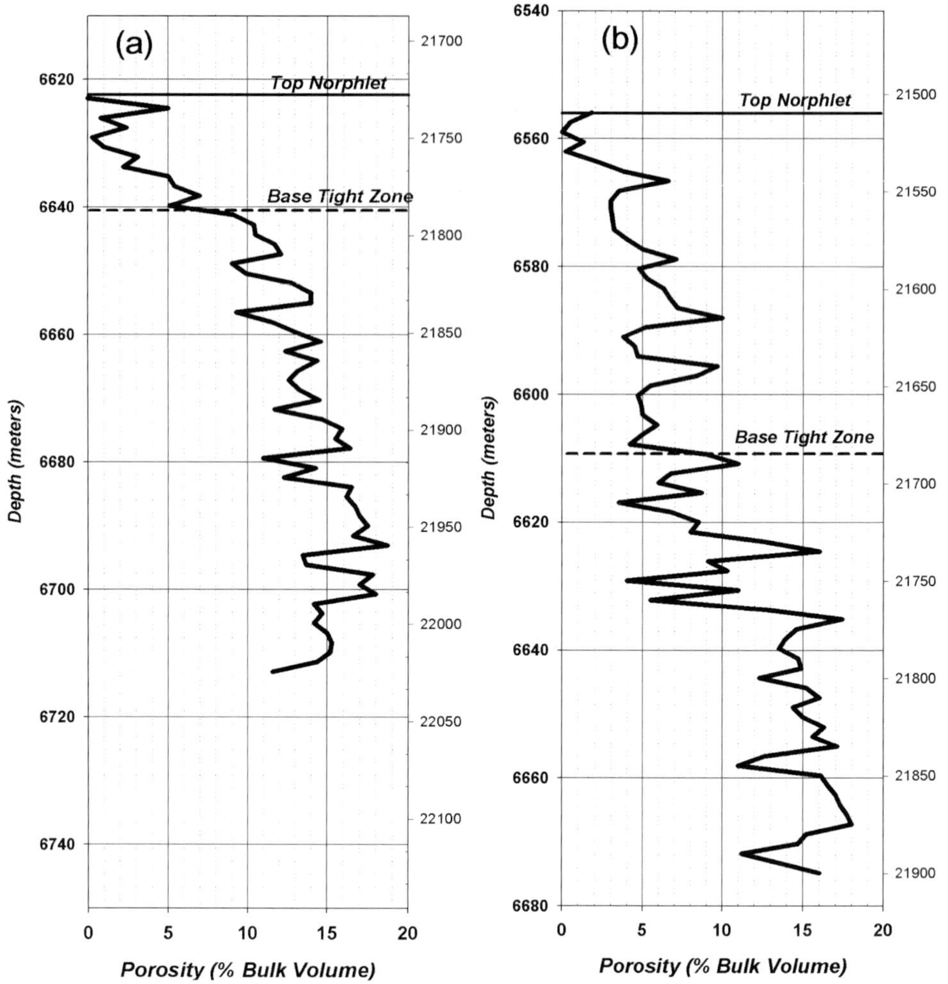


**Fig. 7.** Log porosity versus depth for the Norphlet in two wells from Mobile Bay. (**a**) This well encountered an 18 m tight-zone. The transition from to porous reservoir rock is rapid occurring over the distance of a few metres. (**b**) The transition from tight-zone to reservoir in this well occurs over a long interval making the determination of the tight-zone base problematic. The interpreted base yields a tight-zone thickness of approximately 64 m for this well.

tight-zone to reservoir sand is sharp, occurring over an interval of less than 10 m (Fig. 7a). In a few cases the transition is gradual, with porosity repeatedly increasing above and dropping below the 8% cut-off over an interval of 40 m or more (Fig. 7b).

## Lithology, reservoir quality and fluids

### *Stratification and environmental setting*

Information about Norphlet facies and stratification from cores is important for understanding the potential effects of reservoir heterogeneity. The Norphlet in Mobile Bay is interpreted to represent a preserved series of thick complex linear dunes and associated thin interdune areas. Each linear dune represents a composite accumulation resulting from the vertical aggradation of successive dune bedforms having a longitudinal or oblique character. Net sand transport in such a system is subparallel with the longitudinal axis of each linear dune complex. Such an interpretation accounts for the relative segregation of aeolian facies (dune versus interdune) and allows one to relate preserved sand thickness to depositional trends.

The sedimentary features observed in Norphlet cores are identical to those observed in modern aeolian desert dune fields (Glennie 1970; Kocurek & Dott 1981; Ahlbrandt & Fryberger 1982). In cores from the thick portions of sand lenses, the Norphlet is almost exclusively composed of aeolian dune deposits. Individual dune bedsets range from less than 0.5 m to ~10 m in thickness and consist of laminated and graded medium- to fine-grained sand. The dune bedsets exhibit both low angle wind-ripple stratification and higher angle avalanche cross-stratification in vertical succession (Kocurek & Dott 1981). In cores from the edges of sand lenses where the Norphlet section is much thinner, interdune deposits are interbedded with aeolian dune deposits. Interdune sands are characterized by wavy and contorted laminations produced by periodic wetting and evaporation near the depositional surface and are interpreted as having been deposited in shallow water ponded adjacent to aeolian dunes. The interdune deposits consist of fine-grained sand and range from approximately 0.1 to 2 m in thickness.

Excluding the highly cemented tight-zone, reservoir quality can generally be related to aeolian facies and stratification type (Marzano *et al.* 1988; Dixon *et al.* 1989; Thomson & Stancliffe 1990). All examples of the interdune facies have extremely low porosity and permeability and do not contribute to hydrocarbon production (Fig. 8). Sands exhibiting dune avalanche cross-stratification have, as a population, higher porosity and permeability than dune wind-ripple laminated sandstones (Fig. 8). In practice, heterogeneity related to facies variations (dune versus interdune) has the potential to compartmentalize the reservoir. However, the distribution of the interdune deposits appears mostly limited to the Norphlet sand lens margins, thus reducing their potential to form reservoir wide baffles.

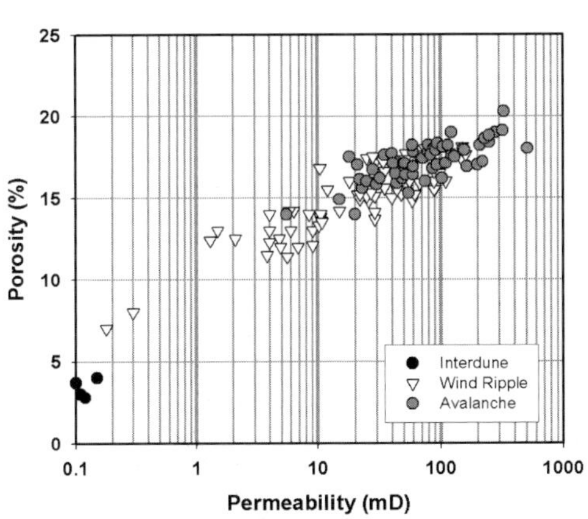


**Fig. 8.** Porosity versus air permeability for Norphlet sandstones from Mobile Bay. The data plotted are from the reservoir interval and are divided according to stratification type (determined by examination of core). Tight-zone samples have porosity <8% and air permeability <1 mD regardless of stratification type and are excluded from this plot.

## *Formation water and natural gas composition*

Norphlet formation waters produced from three Fairway Field wells were analysed for major components and for selected minor cations as part of normal production operations (Table 1). The samples were taken from platform separators during production. Multiple samples from individual wells taken over a period of several years show only minor variations in composition. The waters are extremely saline brines with total dissolved solids between 300 and 373 g/l. Chlorine is the dominant anion in these brines, and sodium (71.6–89.8 g/l), calcium (33.0–41.6 g/l) and potassium (12.9–17.6 g/l) are the major cations. Norphlet formation waters at Fairway are also characterized by relatively high concentrations of Mg, Ba, Sr, Fe, Mn, Li and Pb. Brines of similar composition are produced from deep Norphlet reservoirs and other Mesozoic strata of onshore Mississippi (Carpenter *et al.* 1974; Kharaka *et al.* 1987).

The Norphlet produces dry gas at Fairway Field. The gas consists mostly of methane with smaller amounts of ethane, propane and $CO_2$. The concentration of $H_2S$ varies among the separate accumulations within the field. For example, the western accumulation contains 0.02% $H_2S$ whereas the eastern accumulation contains 1.4%. The Fairway gases also contain unusually high concentrations of liquid diamondoids, carbon compounds that have a spherical structure. High concentrations of both liquid and solid diamondoids are reportedly common in Norphlet gases from other fields in Mobile Bay but are scarce in the vast majority of Gulf of Mexico gases.

# Petrographic analysis

## *Detrital composition*

From more than 300 thin sections examined in this study, 45 were chosen for detailed modal analysis by thin section point counting. The framework grain mineralogy is primarily composed of quartz, K-feldspar and plagioclase.

**Table 1.** *Chemical composition (mg/l) of Norphlet Formation waters, Fairway Field, offshore Alabama*

| Well | S/L 531 JC | S/L 531 JC | S/L 531 JD | S/L 531 JD | S/L 531 JD |
|---|---|---|---|---|---|
| TDS | 324012 | 301820 | 340302 | 353875 | 373564 |
| Cl | 190209 | 178196 | 202222 | 210231 | 220242 |
| B | 245 | 270 | 442 | 382 | 409 |
| Na | 80716 | 71658 | 79721 | 83902 | 89779 |
| K | 15637 | 13105 | 12928 | 13989 | 17643 |
| Li | 91 | 77 | 140 | 149 | 142 |
| Ca | 33000 | 36000 | 41600 | 39440 | 38880 |
| Mg | 2187 | 1215 | 1774 | 3353 | 4010 |
| Sr | 820 | 134 | 112 | 1052 | 1068 |
| Si | 28 | 35 | 71 | 39 | 40 |
| Ba | 635 | 710 | 852 | 883 | 904 |
| Fe | 65 | 30 | 3 | 6 | 5 |
| Mn | 256 | 265 | 275 | 271 | 271 |
| Al | 112 | 113 | 151 | 167 | 159 |
| Pb | 11 | 12 | 11 | 11 | 12 |

Lithic fragments are dominantly of metamorphic and plutonic origin. The majority of samples are classified as arkoses (Folk *et al.* 1970) with an average QFR ratio of 68:22:10. The detrital feldspar grains consist of approximately equal amounts of K-feldspar and plagioclase. The sandstones are extremely clean and lack any recognizable detrital clay matrix.

### *Solid carbonaceous residue*

Throughout much of the Norphlet section, the pores are lined with a non-extractable, solid carbonaceous residue (pyrobitumen) that gives the sandstones their dark grey colour in core (Dixon *et al.* 1989). The occurrence of this material implies the former presence of liquid hydrocarbon pore fluids. In several of the cores examined in this study, pyrobitumen is present in the uppermost portion of the section and disappears with depth. The transition may represent a previous oil/water contact within the Norphlet. It does not correspond with or bear any consistent relationship to present-day water levels or the base of the tight-zone.

### *Authigenic components*

Although a large number of diagenetic phases are evident, only chlorite, quartz and K-feldspar are on average volumetrically important. Other authigenic minerals present include albite, anhydrite, calcite, dolomite and pyrite. The order of formation of authigenic minerals (paragenesis) relative to grain-coating chlorite and pore-lining pyrobitumen can be determined with reasonable certainty by detailed petrographic examination (Fig. 9).

Small patches of anhydrite cement mark the earliest phase of diagenetic mineral precipitation observed. Anhydrite formed prior to significant compaction, as demonstrated by the open packing of detrital grains enclosed in poikilotopic anhydrite cement. Anhydrite probably precipitated as gypsum nodules during early diagenesis in an arid climate. A minor amount of fine, rhombic dolomite cement is observed in some fine-grained sands of the wet interdune facies.

The entire Norphlet section, including the tight-zone, contains grain-coating chlorite (Fig. 10). Chlorite occurs as a network of radial hexagonal plates that forms a rim around detrital grains. The Norphlet chlorites are enriched in magnesium (Fe:Mg$\approx$2:3) and most likely formed from some type of iron-oxide/clay precursor (Kugler & McHugh 1990). Brines associated with Louann evaporites presumably supplied the magnesium present in Norphlet chlorites. The precipitation of K-feldspar overgrowths on detrital K-feldspar grains overlaps in time with the formation of chlorite coatings.

The emplacement of pyrobitumen follows all the authigenic phases mentioned above. Where present, the residue lines intergranular pores and is observed directly on top of both grain-coating chlorite (Fig. 11a,b) and K-feldspar overgrowths. Rounded, semi-spherical masses of pyrobitumen, ~30–60 μm in diameter occur on the surfaces of detrital grains in some samples within the tight-zone (Fig. 11c). These appear to have once been viscous liquid oil droplets that were later encased in quartz cement. Albite overgrowths are often separated from detrital

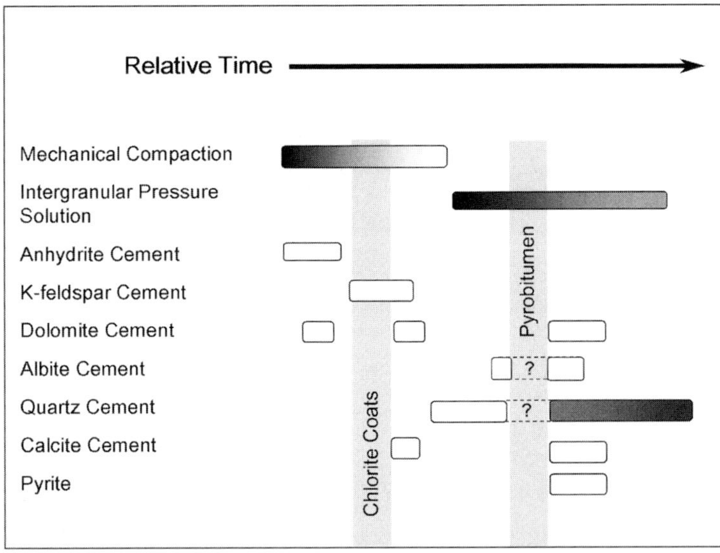


**Fig. 9.** General paragenesis for the Norphlet Formation at Fairway Field and surrounding areas. The timing of compaction, intergranular pressure solution and authigenic cements are plotted relative to the occurrence of grain-coating chlorite and pyrobitumen.

plagioclase grains by a thin coating of pyrobitumen. In some cases the coating is discontinuous beneath the albite overgrowth indicating that albite precipitation overlapped in time with the arrival of liquid hydrocarbons. Ferroan and non-ferroan calcite cements also occur following chlorite. Examples were found of separate calcite phases that formed before and after pyrobitumen but with the exception of a few intervals all carbonate cements are of minor abundance. Pyrite forms as a late diagenetic phase, and in some cases, large portions of the rock are entirely impregnated with pyrite cement. Pyrite cement may represent replacement of early-formed nodular anhydrite cement.

The most abundant authigenic mineral phase is late diagenetic quartz cement. Quartz cement averages 7.0% within the tight-zone and 0.6% throughout the rest of the reservoir. The bulk of quartz cement in the tight-zone precipitated on top of both chlorite and solid pyrobitumen coatings (Figs 10 & 11). Authigenic quartz can usually be shown to be in optical continuity with one of the surrounding grains despite the presence of chlorite and pyrobitumen. In some samples, minute prismatic crystals of quartz, 10–30 μm in length, are coated with pyrobitumen indicating that some quartz cementation, albeit very minor, preceded the initial arrival of hydrocarbons (Fig. 11d). However, subsequent quartz

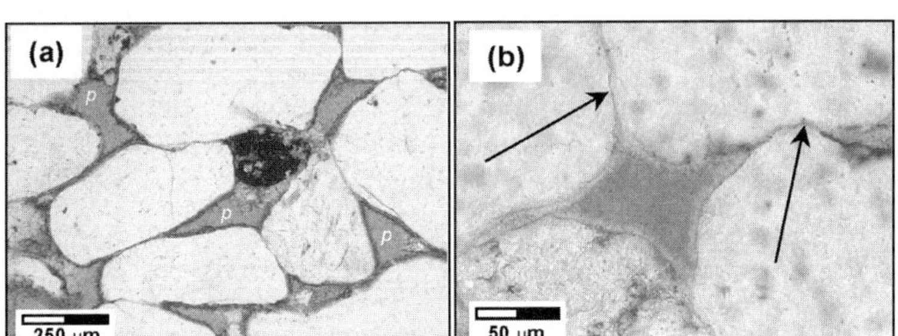


**Fig. 10.** Thin section photomicrographs of grain-coating chlorite in Norphlet sandstones. (**a**) Intergranular pores (*p*) are lined with a dense array of chlorite crystals forming a uniform coating on detrital grains. (**b**) Chlorite occurs along the contacts of detrital grains (arrows) indicating that chlorite formed before a high degree of mechanical compaction and the onset of intergranular pressure solution.

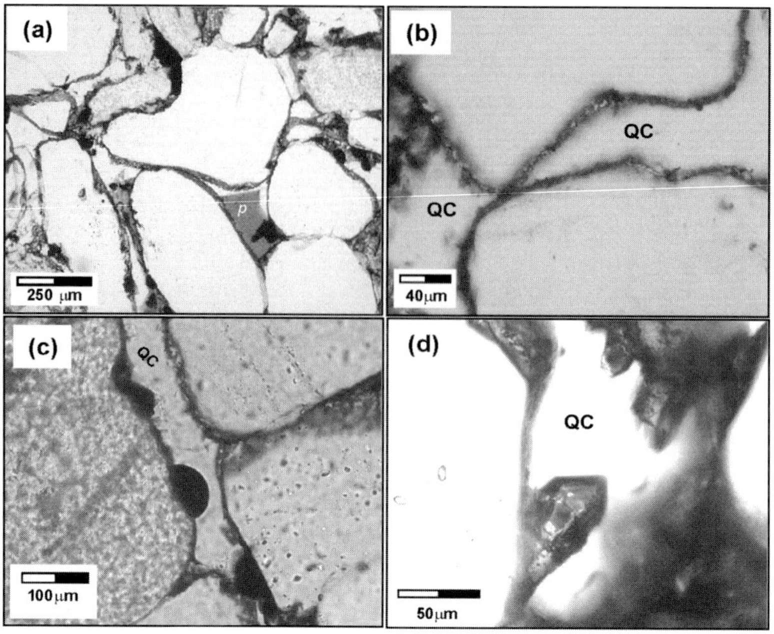


**Fig. 11.** Photomicrographs depicting pore-lining carbonaceous residue (pyrobitumen) in the Norphlet of Mobile Bay. (**a**) Quartz overgrowths encompass chlorite rims and pyrobitumen (black). (**b**) Pyrobitumen (black) forms a thin film over pore-lining chlorite. The pore space has subsequently been filled with authigenic quartz cement (QC). (**c**) Semi-spherical bodies of pyrobitumen lining the edge of what was intergranular pore space prior to formation of pore-filling quartz cement (QC). The bead-like shape suggests that some pyrobitumen was trapped within quartz in a semi-liquid state. (**d**) Minute prismatic crystals of quartz cement grew on some detrital quartz grains prior to emplacement of pyrobitumen. Extensive post-pyrobitumen quartz cementation (QC) followed.

cementation completely fills the adjacent pore space.

Nodular white patches observed in core represent a somewhat unusual mode of quartz cement occurrence. These millimetre-scale, globular bodies occur both in the tight-zone and within the porous reservoir sands. Thin section examination of these uncommon patches reveals that intergranular porosity is completely filled with quartz cement and that the detrital grains are loosely packed compared with adjacent areas of porous sand (Fig. 12a). Chlorite coatings and

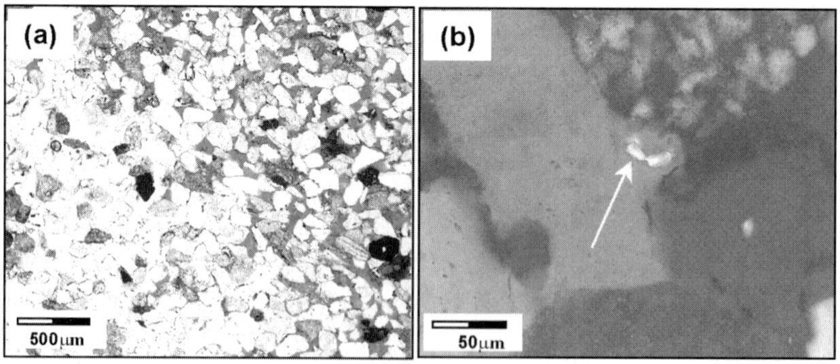


**Fig. 12.** (**a**) Nodular patches of quartz cement, 1–3 mm in diameter, occur within the Norphlet tight-zone and reservoir. The contact between the cemented sand and porous sand runs diagonally through the image from upper left to lower right. Detrital grains within the cemented sand are significantly less compacted that the adjacent porous sand. (**b**) Inclusions of anhydrite (arrow) are commonly observed in quartz cement within these globular parches when viewed under high magnification and crossed polars.

pyrobitumen are absent from the surfaces of the detrital grains, unlike the neighbouring grains outside of the nodules. Detailed petrographic examination reveals the occurrence of numerous small inclusions of remnant anhydrite within quartz overgrowths (Fig. 12b). These are interpreted as evidence that nodular quartz cement formed as a replacement of early formed, minute anhydrite nodules, which preceded the development of chlorite coats and emplacement of pyrobitumen. The timing of replacement cannot be determined on the basis of petrographic evidence alone.

## *Compaction*

Norphlet sandstones examined in this study display clear evidence of mechanical compaction and intergranular pressure solution (Fig. 13). A high proportion of long grain-to-grain contacts are observed along with sutured and interpenetrating quartz–quartz and quartz–feldspar contacts (Fig. 13a,b). Sutured contacts between adjacent quartz grains are common within the tight-zone and the underlying reservoir, demonstrating that intergranular pressure solution had progressed appreciably by the time significant quartz cementation had begun.

The intergranular volume (IGV), equal to the sum of intergranular porosity, cements and matrix, is used to evaluate sandstone compaction (Houseknecht 1987; Paxton *et al.* 2002). Mechanical compaction of rigid grain-rich sandstones such as the Norphlet can reduce IGV from depositional values (~40–45%) to a hypothetical minimum of ~26%, the value associated with closest packing of spherical grains of equal diameter. Any further reduction would theoretically require grain fracturing or intergranular pressure solution. In nature, grain fracturing and pressure solution may begin at slightly higher or lower IGV values depending on variations in grain sorting, shape and composition. A plot of IGV versus total cement volume (Fig. 14) illustrates several points with respect to compaction. The

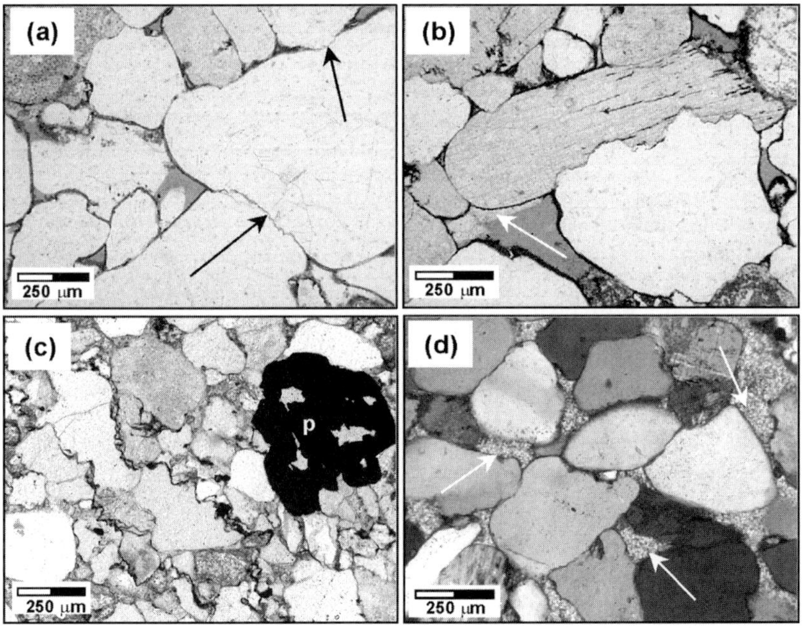


**Fig. 13.** (**a**) Intergranular pressure solution (IPS) is frequently observed in the Norphlet sandstones from Mobile Bay both within the tight-zone and the reservoir intervals. Sutured, interpenetrating contacts between quartz grains (arrows) contain a thin sheet of chlorite. Note that a film of pyrobitumen (black) lines the pore but is absent from contacts involved in pressure solution. (**b**) Dissolution of plagioclase and quartz as a result of IPS is evident in this photomicrograph and is commonly observed in the Norphlet. Note that pyrobitumen is not present along the sutured contact between quartz and plagioclase, but lines the adjacent intergranular pores. Albite cement (arrow) has precipitated following pyrobitumen. (**c**) Stylolites are most often associated with finely bedded or laminated strata. Highly deformed and sutured grains of quartz, K-feldspar, and plagioclase suggest that significant material has been dissolved along stylolites. Nodular pyrite (*p*) is also seen in this image. (**d**) Dense, authigenic muscovite (arrows) fills pores adjacent to zones of intense stylolitization.

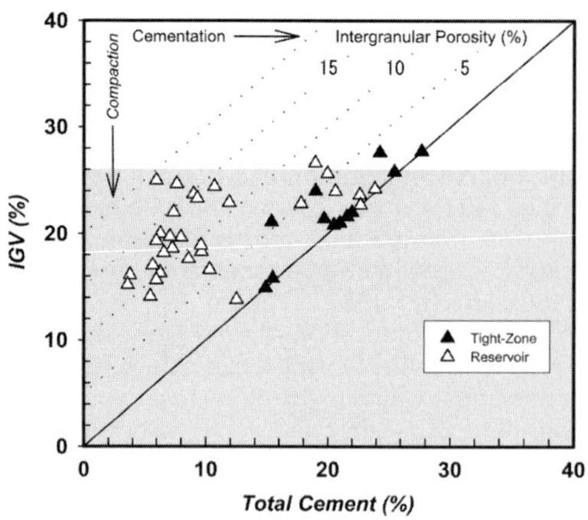


**Fig. 14.** Intergranular volume (IGV) versus total cement for Norphlet sandstones from Mobile Bay. The diagonal lines represent intergranular porosity. The majority of samples plot in the shaded area representing IGV < 26%, indicating that intergranular pressure solution has played a significant role in reducing IGV.

obvious majority of Norphlet sandstones have IGV values less than the theoretical minimum of 26%, many under 20%. Evidence of minor amounts of grain fracturing is apparent in thin sections examined by SEM-CL. Although it is difficult to quantitatively estimate volume loss due to intergranular pressure solution, the IGV values of these Norphlet sandstones clearly indicate that it is significant. With the exception of the quartz cemented nodules, no apparent difference in the extent of compaction is seen when comparing low porosity with higher porosity samples (Fig. 14). Evidently the level of compaction is no different in the tight-zone than in the underlying porous reservoir sands.

Stylolites are relatively common (Fig. 13c), further indication of the importance of pressure solution in the Norphlet. They are most abundant in finer-grained intervals and laminae, occur both within the tight-zone and reservoir sections, and their distribution influences reservoir quality (Thomas *et al.* 1993). A volumetrically minor but potentially important feature is the occurrence of coarse, authigenic muscovite crystals and muscovite-cemented horizons adjacent to stylolites (Fig. 13d). The detrital grains within the muscovite-cemented horizons show evidence of intergranular pressure solution and stylolitization. These horizons are approximately 0.2–0.8 m thick and have exceptionally low porosity (0.8–2.6%) and permeability (0.04–0.15 millidarcies). Depending on their lateral extent, these highly compacted intervals could influence well performance. Thomas *et al.* (1993) used $^{40}Ar/^{39}Ar$ dating of authigenic muscovite to determine that active stylolitization occurred at burial depths of ~4–5.5 km in parts of the Norphlet.

### *Porosity*

Intergranular porosity is the dominant pore type observed in thin sections from the cores examined in this study. The amount of porosity that can be attributed to dissolution of framework grains in the Norphlet is volumetrically insignificant (Table 2). Although partial dissolution of early anhydrite is noted, no evidence for dissolution of significant volumes of pore filling cements was found in this study.

## Fluid inclusion data

Samples containing quartz cement from both the Norphlet tight-zone and the reservoir interval in four separate wells were scrutinized for the presence of fluid inclusions under both plane-polarized and ultraviolet (UV) light. Although not abundant, two-phase aqueous inclusions were found within quartz cement in numerous samples. No examples of hydrocarbon bearing inclusions were revealed under UV light. Microthermometric measurements were made using a modified US Geological Survey, gas-flow stage designed for heating and freezing of inclusions. Homogenization temperatures ($T_h$) were measured for 98 aqueous inclusions and ice final melting temperatures ($T_m$) were obtained for 26 of these.

Fluid inclusions in quartz cement are situated both in close proximity to the boundary between detrital quartz grains and cement as well as within the overgrowths themselves. The inclusions chosen for microthermometry are interpreted as primary inclusions that were trapped during growth of quartz cement. Inclusions located in healed fractures were not selected for analysis.

For primary inclusions in quartz cement the $T_h$ values obtained range from 85 to 225 °C (Fig. 15). The conspicuous majority of $T_h$ values fall above 150 °C with the greatest number clustering between 190 and 225 °C. There is no distinction between the range of $T_h$ values measured in quartz cements of the tight-zone and the reservoir interval. In addition, no difference was observed in $T_h$ between nodular quartz cement attributed to anhydrite replacement and the more common post-pyrobitumen

**Table 2.** *Summary of petrographic point-count data for Norphlet samples, Mobile Bay*

| | Tight-zone ($n$ = 15) | | | | Reservoir ($n$ = 30) | | | |
|---|---|---|---|---|---|---|---|---|
| | Average | Max. | Min. | Standard deviation | Average | Max. | Min. | Standard deviation |
| Mean grain size (μm) | 177.6 | 222.9 | 147 | 21.9 | 183.2 | 260.9 | 141.6 | 27.3 |
| Grain coating (%) | 92.2 | 98.2 | 85.2 | 4.4 | 99.6 | 100.0 | 95.8 | 0.9 |
| Quartz (%) | 52.5 | 58 | 48 | 3.3 | 51.6 | 59.7 | 45.7 | 3.7 |
| K-feldspar (%) | 9.3 | 12.6 | 7 | 1.4 | 10.0 | 14.0 | 7.7 | 1.8 |
| Plagioclase (%) | 9.1 | 12.6 | 7 | 1.5 | 9.6 | 13.3 | 5.7 | 2.0 |
| Igneous lithics (%) | 1.5 | 5 | 0.3 | 1.3 | 2.8 | 7.7 | 0.3 | 1.5 |
| Sedimentary lithics (%) | 0.3 | 1.3 | 0 | 0.4 | 0.2 | 1.0 | 0.0 | 0.3 |
| Metamorphic lithics (%) | 3.6 | 5.3 | 1 | 1.4 | 4.6 | 8.3 | 2.0 | 1.4 |
| Detrital clay (%) | 0.1 | 1 | 0 | 0.2 | 0.3 | 1.3 | 0.0 | 0.4 |
| Chlorite (%) | 3.4 | 9.3 | 0.3 | 2.5 | 4.5 | 11.7 | 0.6 | 2.7 |
| Quartz cement (%) | 7.0 | 12 | 3.3 | 3.1 | 0.6 | 4.0 | 0.0 | 0.9 |
| K-feldspar cement (%) | 2.2 | 3.7 | 0 | 1.2 | 2.3 | 5.7 | 0.6 | 1.1 |
| Albite cement (%) | 1.3 | 3.3 | 0 | 1.2 | 0.6 | 4.3 | 0.0 | 0.9 |
| Calcite cement (%) | 0.1 | 0.7 | 0 | 0.2 | 0.2 | 2.0 | 0.0 | 0.4 |
| Dolomite cement (%) | 0.6 | 1.7 | 0 | 0.5 | 0.2 | 1.3 | 0.0 | 0.4 |
| Anhydrite cement (%) | 0.0 | 0 | 0 | 0.0 | 0.1 | 0.7 | 0.0 | 0.2 |
| Pyrite (%) | 0.2 | 1.7 | 0 | 0.4 | 0.1 | 0.7 | 0.0 | 0.2 |
| Pyrobitumen (%) | 6.5 | 17.1 | 1.7 | 3.9 | 1.1 | 6.7 | 0.0 | 1.8 |
| Intergranular porosity (%) | 1.6 | 5.7 | 0 | 2.1 | 10.4 | 19.0 | 3.7 | 3.6 |
| Intragranular porosity (%) | 0.1 | 0.7 | 0 | 0.2 | 0.3 | 2.0 | 0.0 | 0.4 |
| Lab measured porosity (%) | 4.5 | 8.3 | 1.6 | 2.2 | 14.3 | 19.8 | 9.4 | 3.0 |
| Air permeability (mD) | 0.3 | 0.92 | 0.064 | 0.3 | 69.7 | 500.0 | 1.6 | 90.8 |
| IGV (%) | 22.7 | 27.7 | 14.9 | 3.4 | 20.3 | 26.6 | 14.1 | 3.4 |
| Total cement (%) | 21.1 | 27.7 | 14.9 | 3.7 | 9.9 | 20.0 | 3.6 | 4.6 |

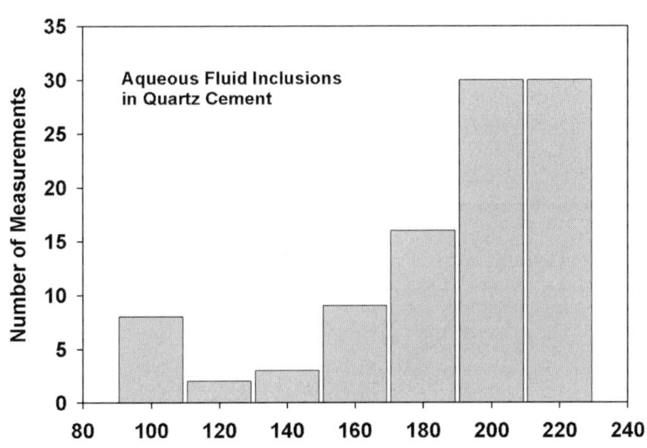


**Fig. 15.** Homogenization temperatures ($T_h$) for aqueous fluid inclusions in quartz cement–Norphlet sandstones, Mobile Bay.

quartz cements. The $T_m$ values measured in this study range from $-12$ to $-32^\circ$ C and as in the case of $T_h$ data show no correlation with mode of quartz cement occurrence. Salinities of ~160–300 g/l NaCl equivalent are calculated from the $T_m$ data. Eutectic temperatures of $\sim -57\,^\circ$C were measured in several inclusions indicating the presence of significant concentrations of divalent cations (e.g. $Ca^{2+}$, $Mg^{2+}$). These inferred salinities and compositions are consistent with the analyses of produced formation waters discussed above (Table 1).

## Discussion

### *Burial history*

Taking into account the heat flow history from the opening of the Gulf of Mexico to the present, the burial history of the Norphlet at Fairway Field was modelled using Shell proprietary basin modelling software. The generalized temperature history for the middle of the Norphlet within the western sand lens is shown in Figure 16 and is fundamentally similar to burial histories presented in other studies of Mobile Bay (Dixon *et al.* 1989; Thomas *et al.* 1993). At the Fairway Field location the Norphlet reached about 80 °C at roughly 110 Ma and continued on a path of increasing temperature to the present-day value of approximately 220 °C. This is noteworthy in that the onset of significant quartz cementation in sandstones is most commonly reported at temperatures of 80–90 °C (Walderhaug 1994; Giles *et al.* 2000). The Norphlet was

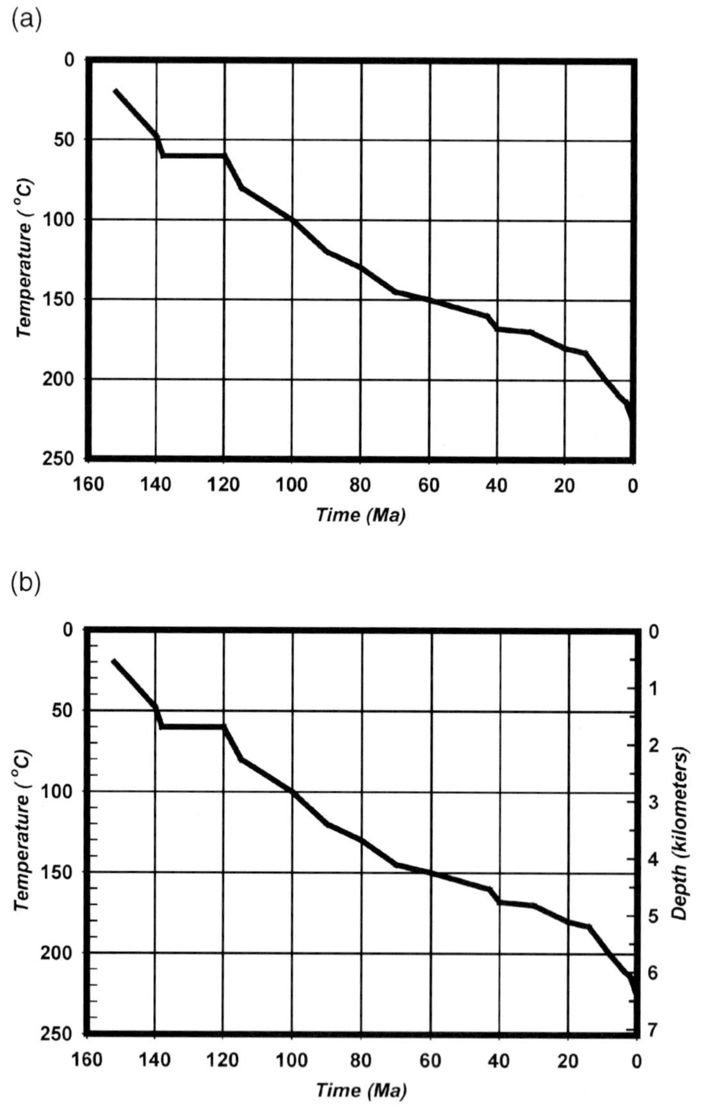


**Fig. 16.** Burial history model for the Norphlet Formation at Fairway Field, offshore Alabama. (**a**) Temperature versus time. (**b**) Depth versus time.

therefore in the window for potential quartz cementation for more than 100 million years.

## *Origin of pyrobitumen*

The process by which pore-lining carbonaceous residue in the Norphlet is derived from liquid oil is a matter of some debate. Isotopic analysis, biomarkers or C/H ratios are unlikely to yield useful information due to excessive heating resulting from deep burial. Several investigators have proposed that the carbonaceous residue represents an asphaltene precipitate resulting from *in situ* cracking of liquid oil to gas (Dixon *et al.* 1989; Mudford *et al.* 1995). Alternatively, the residue may be a product of thermal sulphate reduction (TSR) involving the reaction of liquid oil with anhydrite to produce $H_2S$ gas (Dixon *et al.* 1989; Heydari 1997). Based on the distribution of $H_2S$ in onshore Louisiana fields, Heydari (1997) concluded that TSR in the Smackover Formation occurred at temperatures exceeding 140–150 °C.

Several lines of evidence suggest the involvement of TSR. Partially dissolved remnants of early diagenetic anhydrite cement are found embedded in quartz cement (Fig. 12). Fluid inclusion data indicate that nodular quartz cement with anhydrite inclusions (Fig. 15) formed at elevated temperatures, suggesting that anhydrite dissolution may have coincided with high temperature TSR. Post-pyrobitumen carbonate and pyrite cements and nodules are also present in the Norphlet and may represent late diagenetic precipitates associated with TSR. Pyrite cement from the Norphlet has sulphur isotope ratios that are identical to those of the Pine Hill Anhydrite (Table 3) and within the range reported for Jurassic seawater (Hoefs 1997). Dixon *et al.* (1989) report similar $\delta^{34}S$ for $H_2S$ from the Norphlet reservoir in Mary Ann Field. Norphlet and Smackover reservoirs from onshore Mississippi also contain pyrite, anhydrite and $H_2S$ with $\delta^{34}S$ compositions that are similar to Jurassic marine values, indicating that TSR has occurred in these locations as well (Shew 1992).

## *Origin of the tight-zone*

Perhaps the most perplexing issue regarding diagenesis of the Norphlet in Mobile Bay has been finding a plausible explanation for the occurrence of the tight-zone (Dixon *et al.* 1989; Kugler & McHugh 1990; Thomas *et al.* 1993). Although a variety of mineral cements are present, it is chiefly the difference in the volume of quartz cement that differentiates reservoir from non-reservoir. Any model for the origin of the tight-zone must account for its confinement to the top of the Norphlet section, the apparent late diagenetic timing of quartz cementation, and the relationships among chlorite coatings, pyrobitumen and authigenic quartz precipitation.

Several hypotheses can be eliminated based on core and petrographic evidence. Suggestions that the quartz cement in the tight-zone is related to destruction of syndepositional clay/iron oxide precursors of chlorite as a result of sediment reworking during the Smackover marine transgression are not supported by evidence from cores (Dixon *et al.* 1989; this study). The extent of the tight-zone is independent of depositional facies or stratification type. No significant difference in detrital framework composition (i.e. amount of quartz or abundance of polycrystalline quartz) that might influence the precipitation of quartz cement is observed in the petrographic data. In addition as noted above, chlorite lines the perimeter of detrital grains within the tight-zone as well as in the more porous reservoir sands (Figs 10 & 11). Evidence for dissolution of pre-existing carbonate or halite cements within the Norphlet reservoir is lacking and therefore is not considered to have significantly influenced the distribution of porosity. Partial dissolution of anhydrite cement is noted but is not judged to have been volumetrically abundant.

**Table 3.** *Sulphur isotope data*

| | $\delta^{34}S$ (‰ CDT) | Source |
|---|---|---|
| Pine Hill anhydrite | 15.49 | This study |
| Pyrite (Norphlet) | 15.28 | This study |
| Pyrite (Norphlet) | 15.66 | This study |
| Pyrite (Norphlet) | 16.66 | This study |
| $H_2S$ gas (Mary Ann Field) | 14.50 | Dixon *et al.* (1989) |
| Upper Jurassic Marine | ~16–17 | Hoefs (1997) |

The formation of chlorite rims on detrital quartz grains has been demonstrated to be an important mechanism for inhibiting quartz cementation and preserving porosity in many sandstone reservoirs (Heald & Larese 1974; Thomson 1979; Pittman *et al.* 1992; Ehrenberg 1993). The evidence from the Norphlet sands of Mobile Bay would seem to be contradictory. Based on the observational evidence presented above (Figs 10 & 11), the pervasively cemented tight-zone appears to possess robust chlorite coatings that are in all ways similar to those in the porous reservoir sands. However, detailed petrographic inspection under high magnification reveals that minor breaks in the continuity of chlorite coatings occur more frequently in the quartz cemented tight-zone (Fig. 17). Discontinuities in grain coatings are best observed in 60–90 μm thick, polished thin sections prepared for fluid inclusion analysis. Focusing beneath the surface of the thin section allows for a greater proportion of grain–cement contact area to be examined and for breaches in the coatings that occur beneath the upper plane of the sample to be observed. The width of the uncoated surface exposed at these grain-coat discontinuities ranges from ~5 to 50 μm. Pore filling quartz overgrowths are present in all cases where a gap in grain coating chlorite was observed, whether in the tight-zone or within the underlying reservoir sands. The location of the observed breaches in chlorite coats do not correspond to healed grain fractures such as those described by Chuhan *et al.* (2002).

In order to quantify the inferred differences in the degree of grain coating between the tight-zone and reservoir, the 2D proportion of coated and uncoated surface was measured for 75–100 quartz grains per thin section in 41 samples. Although time consuming to collect, these data clearly demonstrate the relationship between chlorite coat continuity and quartz overgrowth cement volume (Fig. 18a). Tight-zone sands (<8% total porosity) have grain coatings that average 92% (±4%) complete whereas the most porous sands generally have chlorite coatings that average 99% (±1%) complete (Fig. 18b). Given the high temperatures and extensive amount of geological time over which quartz cementation has had the opportunity to form, even small discontinuities in grain coats have apparently provided adequate surface area for significant quartz cement nucleation. As noted above, chlorite coatings are absent in quartz-cemented nodules formed following dissolution of early-formed anhydrite cement,

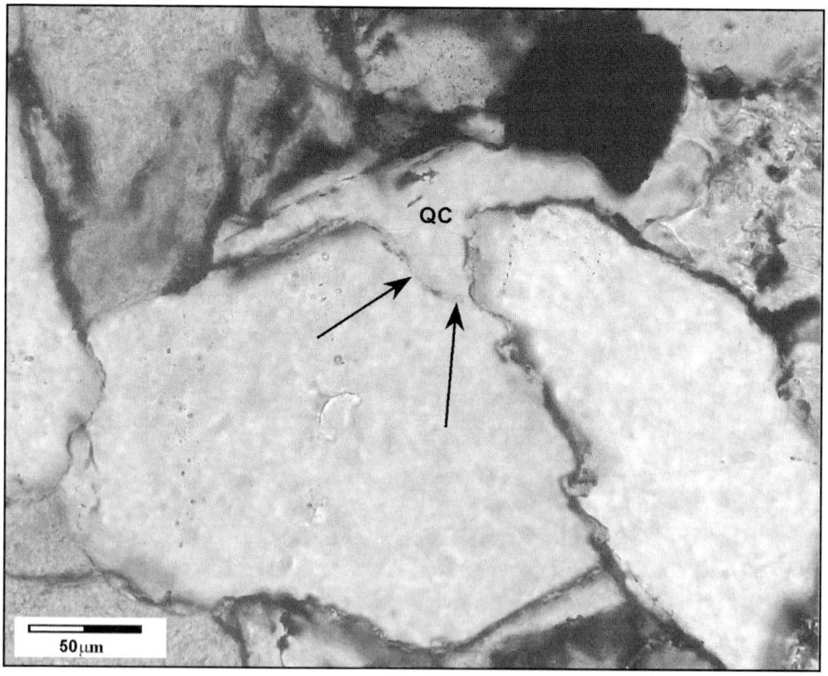


**Fig. 17.** Small discontinuities in chlorite grain coats are common in the tight-zone but very rare in the porous reservoir sandstones. Minor breaks in grain coatings (arrows) are almost without exception linked to quartz overgrowths (QC) in the adjacent intergranular space.

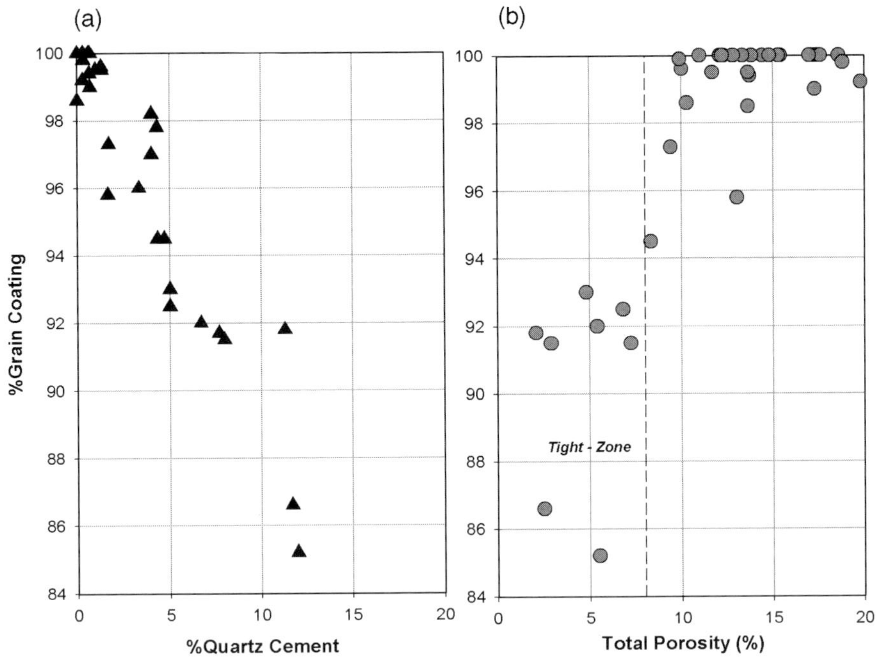


**Fig. 18.** (**a**) Quartz cement volume versus the percentage of quartz grain surface that is coated with chlorite. (**b**) Total porosity versus the percentage of quartz grain surface that is coated with chlorite.

presumably as a result of TSR. This constitutes further evidence that quartz overgrowths formed wherever grain coats were lacking.

To test whether the observed differences in chlorite coat coverage can explain the differences in quartz cement volumes in the Norphlet, Touchstone version 3.1 (developed by Geocosm LLC) was used to simulate sandstone compaction and cementation as a function of geological time. Detailed petrographic data and burial history models are required as input for these simulations. The Touchstone compaction model simulates decline in IGV through time as a function of the vertical effective stress history and bulk mechanical strength of the sandstone. IGV declines exponentially with effective stress in a similar fashion to that described by Lander & Walderhaug (1999). Touchstone's bulk mechanical strength model considers the influence of framework grain types and mechanical properties as well as the types, mechanical properties and timing of development of pore-filling matrix and cements (R. Lander, pers. comm. 2003). The Touchstone quartz cementation model assumes that the precipitation rate per unit nucleation surface area is described by an Arrhenius kinetic expression (Walderhaug 2000) and that the nucleation surface area for quartz overgrowths is a function of grain size and sorting, quartz grain abundance, grain coating coverage, compaction state and extent of pore occlusion by matrix and cements (Lander & Walderhaug 1999). The simulations also account for precipitation of other cement types and secondary porosity development according to paragenetic constraints related to the burial history. Parameter values describing the mechanical properties of framework grains, matrix and cements and the activation energy for quartz precipitation were obtained using a Genetic Algorithm that provided an optimal fit to point count data from 38 calibration samples.

The results of the Touchstone simulations for the Norphlet using Fairway Field burial history models (Fig. 19) strongly suggest that minor differences in grain coat coverage can indeed explain the observed volumes of quartz cement. The compaction model predicts measured IGV within $\pm 4\%$ for more than 90% of the samples (Fig. 19a). The volume of quartz cement is predicted within $\pm 3\%$ for all but three samples (Fig. 19b) using optimized kinetic parameters constrained by the results of proprietary studies of sandstones of a wide range of ages and burial histories from throughout the world. No accelerated or suppressed rates of quartz cementation due to thermal or hydrocarbon pore fluid

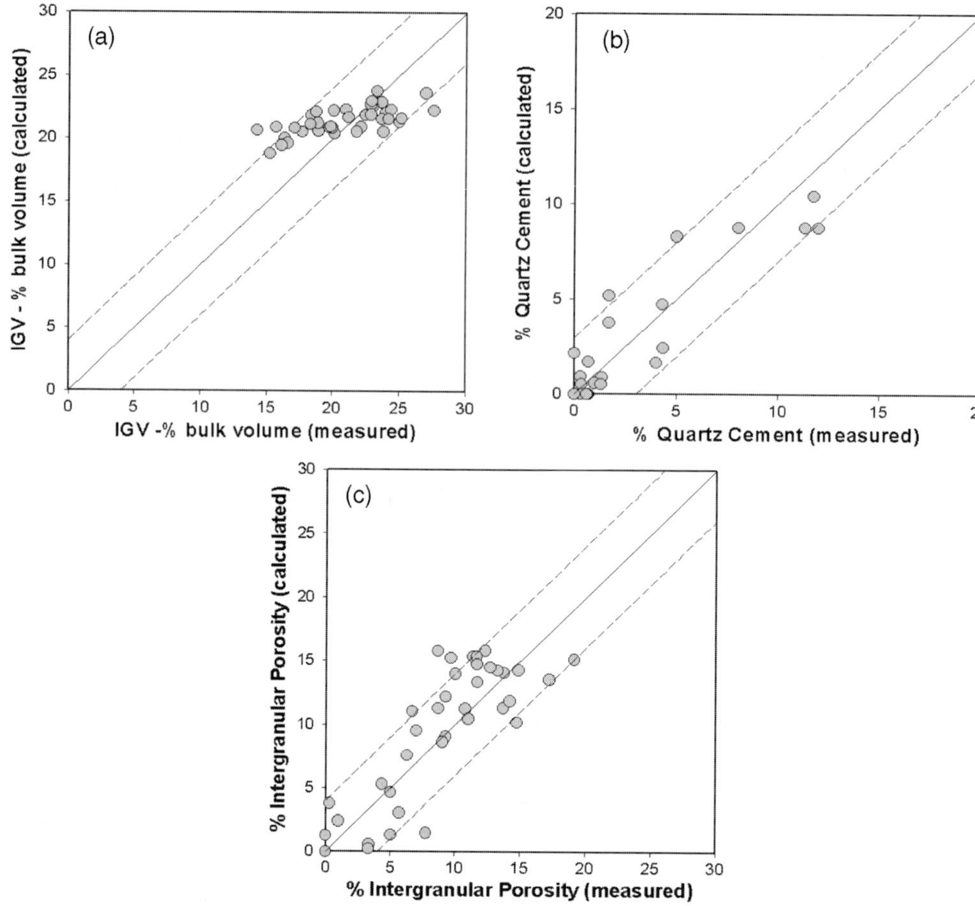


**Fig. 19.** Results of compaction and quartz cementation models calculated using quantitative petrographic data, Touchstone and the thermal history shown in Figure 16. (**a**) Calculated versus observed IGV ± 4% tolerance. (**b**) Calculated versus observed quartz cement volume ± 3% tolerance. (**c**) Calculated versus observed intergranular porosity ± 4% tolerance.

effects are implied by the modelling results. Model predictions of intergranular porosity match the measured values within ±4% for 87% of the samples used in the simulation (Fig. 19c).

The model results indicate that the amount of quartz cement in the tight-zone is limited by the amount of intergranular pore space and essentially ceases when the pores are filled. In the underlying reservoir interval, the amount of quartz cement is limited by the exceedingly small amount of detrital quartz surface area available for cement nucleation and growth.

An explanation as to why the upper part of the Norphlet sand lenses has slightly less continuous chlorite coatings on detrital grains is not readily apparent from the data and observations presented in this study. Seismic and well data suggest that the base of the tight-zone is roughly flat in relation to the top of the Norphlet. This may indicate that the transition from the tight-zone to porous reservoir is related to a previous fluid contact. In the case of quartz-cemented nodules described above, early-formed anhydrite cement apparently prevented the formation of chlorite coatings by isolating grain surfaces from diagenetic pore fluids. Subsequent dissolution of anhydrite exposed these uncoated surfaces for nucleation of quartz overgrowths. It is proposed here that anhydrite (originally gypsum) nodules and cements formed very early in diagenesis due to evaporation of seawater during periodic wetting and drying cycles. Although not abundant, these cements were presumably concentrated in the upper Norphlet above the water table. Much later, under conditions of deep burial and high temperature, anhydrite cements were dissolved during thermal sulphate reduction. A greater percentage of uncoated quartz surfaces were thereby exposed in the uppermost Norphlet leading to a significantly greater degree of quartz cementation.

The distribution of pyrobitumen within present-day Norphlet reservoirs ostensibly mimics an earlier liquid hydrocarbon column. The location of the contact between bitumen-stained and unstained sandstone varies within the cores examined in this study, falling at different levels within the tight-zone or the reservoir but above the present-day gas/water contact. In one relatively thin, 45 m Norphlet section (Exxon State Lease 624 #1), the Norphlet contains pore-lining pyrobitumen from top to bottom. Thomas *et al.* (1993) report pyrobitumen staining extending 12 m below the gas/water contact in a thick Norphlet section penetrated in a well to the south-east of Fairway Field.

It has been generally assumed that gas present today in the Norphlet was, to a great extent, a product of *in situ* cracking of liquid oil to natural gas plus pyrobitumen (Dixon *et al.* 1989; Thomas *et al.* 1993; Mudford *et al.* 1995). Geochemical evidence linking present-day gas to pyrobitumen is presumably unattainable, as noted above. Calculations by Dixon *et al.* (1989) suggest that cracking of oil within the Norphlet could not account for the entire volume of gas present in the reservoir, indicating that additional gas charge must have occurred (Lock & Broussard 1989).

The fact that quartz cement in the tight-zone clearly postdates pyrobitumen (Figs 10 & 11) raises several issues with regard to cementation in the presence of hydrocarbon pore fluids and the timing of gas accumulation in Norphlet reservoirs in Mobile Bay. Petrographic evidence shows that the precipitation of solid hydrocarbon followed the onset of intergranular pressure solution. In addition, the likely involvement of this organic material in high temperature TSR clearly points to deep burial for the timing of formation. The homogenization temperatures of aqueous fluid inclusions verify a high temperature origin for quartz cement in the tight-zone.

If the widely held scenario of *in situ* thermal cracking of oil occurred within the Norphlet and the gas generated was retained within the present day traps, quartz cementation in the tight-zone would have occurred in the uppermost portion of a large hydrocarbon column. Based on the principles of capillarity and capillary experiments on Norphlet core samples, it is estimated that the pore system would have attained gas saturation of ~90% or more at the top of the Norphlet sand bodies, the exact location of tight-zone cementation. Present-day gas saturation in the Norphlet at Fairway Field and other Mobile Bay fields commonly reaches 85–92% in the porous sands directly beneath the tight-zone. Under the assumption of closed-system generation and retention of gas, quartz cementation in the tight-zone would be required to form in a relatively thin water phase constituting only 10% of the pore volume. Numerous studies have documented the precipitation of authigenic quartz from formation waters in the presence of hydrocarbons as evidenced by the existence of oil and/or gas bearing fluid inclusions in quartz cement (e.g. Saigal *et al.* 1992; Ramm 1992; Walderhaug 1996; Rossi *et al.* 2002). Implicit to cementation within a hydrocarbon accumulation is a local source of silica (i.e. pressure solution of quartz, feldspar dissolution etc.). Silica transport to the site of precipitation is assumed to occur by means of diffusion within the continuous water phase of the hydrocarbon column. Marchand *et al.* (2000, 2001) propose suppression of the rate of quartz cementation due to hydrocarbon pore fluids to explain differences in quartz cement volumes between oil and water legs in reservoirs from the North Sea. Worden *et al.* (1998) present theoretical arguments suggesting that, under conditions of very high oil or gas saturation, silica transport via diffusion within the low saturation water phase is sufficiently slowed as to essentially halt quartz cementation.

Evidence presented here is insufficient to prove whether or not hydrocarbon pore fluids may have temporarily inhibited quartz cementation in the Norphlet. As noted above, no hydrocarbon-bearing fluid inclusions were found in this study of quartz cement in the tight-zone. The absence of oil or gas fluid inclusions does not conclusively prove that the pore fluids were hydrocarbon-free during precipitation of quartz cement. However, it seems unlikely that quartz cement would be entirely void of liquid oil or gas inclusions and contain only aqueous inclusions if the pore system contained mostly hydrocarbons during cementation. The fluid inclusion data at the very least imply that the pore fluids were dominantly saline brine during quartz cementation of the tight-zone. This casts serious doubt on the common assumption that gas generated from thermal cracking of a large oil column was retained to the present day within the Norphlet reservoir/trap system.

A more plausible explanation is that mobile oil and gas escaped from the Norphlet leaving behind an immobile bitumen precipitate, which has subsequently been thermally degraded to its present form. Periodic microfracturing of the overlying Smackover seal during various stages of source rock maturation may have provided a mechanism for hydrocarbons to leak from the Norphlet traps (Thomas *et al.* 1993). Quartz cement precipitated from highly saline brines, the dominant pore fluid during formation of the tight-zone. The integrity of the top seal was re-established as fractures healed under conditions of much higher effective stress. The gas presently trapped within Norphlet reservoirs in Mobile Bay in all probability represents a secondary, and relatively recent accumulation.

## Conclusions

Variation in the thickness of the low porosity tight-zone (3–58 m) at the top of the Norphlet Formation in Mobile Bay limits net pay and strongly influences the rate of natural gas production in individual wells. The proportion of tight-zone to porous reservoir sandstone in the Norphlet is unrelated to depositional facies, sediment reworking, geopressures, development of secondary porosity, or inhibition of diagenesis due to hydrocarbon pore fluids.

The occurrence of porous sandstone in the Norphlet at depths, temperatures and effective stresses that border low-grade metamorphism is in large part related to the formation of robust grain-coating chlorite. Inhibition of quartz cementation by the formation of clay coats is a well-established mechanism for porosity preservation but appeared to have been ineffective in the low porosity tight-zone. However, petrographic observations and data demonstrate that small discontinuities in chlorite coats are common in the quartz-cemented sands of the tight-zone but extremely rare in the porous reservoir sands. The results of compaction and quartz cementation modelling indicate that ~98% or greater grain coverage is required to prevent extensive quartz cementation given the burial history of the Norphlet.

Quartz cementation in the tight-zone took place under conditions of deep burial and high temperature. It followed in time the retention of liquid hydrocarbons and the precipitation of a solid hydrocarbon film (pyrobitumen) on the walls of the intergranular pores. Data from aqueous fluid inclusions in quartz cement indicate that most quartz cement precipitated at temperatures of 150 °C or greater from extremely saline fluids. Hydrocarbon-bearing inclusions are notably absent from quartz cement in the tight-zone, implying that the pore fluids were dominantly brine during precipitation. Gas presently found in Norphlet reservoirs of Mobile Bay was trapped following tight-zone cementation and most likely represents a relatively recent accumulation.

The authors gratefully acknowledge Shell International Exploration & Production and Shell Exploration & Production Company for permission to publish this work. The data and ideas presented in this paper benefited greatly from the participation and advice of the following colleagues: G. Wilson, R. McNeil, G. Macleod, R. Garner, D. Hiser, R. Shew, R. Lander, L. Bonnell, I. Espejo, J Scheibal, T. Boronow and I. Yanushevska.

## References

Ahlbrandt, T. S. & Fryberger, S. G. 1982. Introduction to eolian deposits in sandstone depositional environments. *In*: Scholle, P. A. & Spearing, D. (eds) *Sandstone Depositional Environments*. American Association of Petroleum Geologists, Memoir, **31**, 11–47.

Carpenter, A. B., Trout, M. L. & Pickett, E. E. 1974. Preliminary report on the origin and chemical evolution of lead- and zinc-rich oil field brines in central Mississippi. *Economic Geology*, **69**, 1191–1206.

Chuhan, F. A., Kjeldstad, A., Bjørlykke, K. & Høeg, K. 2002. Porosity loss in sand by grain crushing. Experimental evidence and relevance to reservoir quality. *Marine and Petroleum Geology*, 19, 39–53.

Claypool, G. E. & Mancini, E. A. 1989. Geochemical relationships of petroleum in Mesozoic reservoirs to carbonate source rocks of Jurassic Smackover Formation, southwest Alabama. *American Association of Petroleum Geologists Bulletin*, **73**, 904–924.

Dixon, S. A., Summers, D. M. & Surdam, R. C. 1989. Diagenesis and preservation of porosity in Norphlet Formation (Upper Jurassic), southern Alabama. *American Association of Petroleum Geologists Bulletin*, **73**, 707–728.

Ehrenberg, S. N. 1993 Preservation of anomalously high porosity in deeply buried sandstones by grain-coating chlorite: examples from the Norwegian continental shelf. *American Association of Petroleum Geologists Bulletin*, **77**, 1260–1286

Folk, R. L., Andrews, P. B. & Lewis, D. W. 1970. Detrital sedimentary rock classification and nomenclature for use in New Zealand. *New Zealand Journal of Geology and Geophysics*, **13**, 937–968.

Giles, M. R., Indrelid, S. L., Beynon, G. V. & Amthor, J. 2000. The origin of large-scale quartz cementation: evidence from large data sets and coupled heat-fluid mass transport modelling. *In*: Worden, R. H. & Morad. S. (eds) *Quartz Cementation in Sandstones*. International Association of Sedimentologists, Special Publication, **29**, 21–38.

GLENNIE, K. W. 1970. Desert Sedimentary Environments. *Developments in Sedimentology*, **14**, Elsevier, Amsterdam.

HEALD, M. T. & LARESE, R. E. 1974. Influence of coatings on quartz cementation. *Journal of Sedimentary Petrology*, **44**, 1269–1274.

HEYDARI, E. 1997. The role of burial diagenesis in hydrocarbon destruction and $H_2S$ accumulation, Upper Jurassic Smackover Formation, Black Creek Field, Mississippi. *American Association of Petroleum Geologists Bulletin*, **81**, 26–45.

HOEFS, J. 1997. *Stable Isotope Geochemistry*. Springer, Berlin, Heidelberg.

HOUSEKNECHT, D. W. 1987. Assessing the relative importance of compaction processes and cementation to reduction of porosity in sandstones. *American Association of Petroleum Geologists Bulletin*, **71**, 633–642.

KHARAKA, Y. K., MAEST, A. S., CAROTHERS, W. W., LAW, L. M., LAMOTHE, P. J. & FRIES, T. L. 1987. Geochemistry of metal-rich brines from central Mississippi Salt Dome basin, U.S.A. *Applied Geochemistry*, **2**, 543–561.

KOCUREK, G. & DOTT, R. H. JR. 1981. Distinctions and uses of stratification types in the interpretation of eolian sand. *Journal of Sedimentary Petrology*, **51**, 579–595.

KUGLER, R. L. & MCHUGH, A. 1990. Regional diagenetic variation in Norphlet Sandstone: implications for reservoir quality and the origin of porosity. *Gulf Coast Association of Geological Societies Transactions*, **40**, 411–423.

LANDER, R. H. & WALDERHAUG, O. 1999. Predicting porosity through simulating sandstone compaction and quartz cementation. *American Association of Petroleum Geologists Bulletin*, **83**, 433–449.

LOCK, B. E. & BROUSSARD, S. W. 1989. The Norphlet reservoir in Mobile Bay: origins of deep porosity. *Gulf Coast Association of Geological Societies Transactions*, **39**, 187–194.

MANCINI, E. A., MINK, R. M., BEARDEN, B. L. & WILKERSON, R. P. 1985. Norphlet Formation (Upper Jurassic) of southwestern and offshore Alabama: environments of deposition and petroleum geology. *American Association of Petroleum Geologists Bulletin*, **69**, 881–898.

MANCINI, E. A., MINK, R. M., BEARDEN, B. L., MANN, S. D. & BOLIN, D. E. 1990. Desert environments and petroleum geology of the Norphlet Formation, Hatters Pond Field, Alabama. *In*: BARWIS, J. H., MCPHERSON, J. G. & STUDLICK, J. R. J. (eds) *Sandstone Petroleum Reservoirs*. Springer, New York, 153–180.

MARCHAND, A. M. E., HASZELDINE, R. S., MACAULAY, C. I., SWENNEN, R. & FALLICK, A. E. 2000. Quartz cementation inhibited by crestal oil charge: Miller deep water sandstone, UK North Sea. *Clay Minerals*, **35**, 210–210.

MARCHAND, A. M. E., HASZELDINE, R. S., SMALLEY, P. C., MACAULAY, C. I. & FALLICK, A. E. 2001. Evidence for reduced quartz cementation rates in oil-filled sandstones. *Geology*, **29**, 915–918.

MARZANO, M. S., PENSE, G. M. & ANDRONACO, P. 1988. A comparison of the Jurassic Norphlet Formation in Mary Ann Field, Mobile Bay, Alabama to onshore regional Norphlet trends. *Gulf Coast Association of Geological Societies Transactions*, **38**, 85–100.

MUDFORD, B., LUNDEGARD, P. & LERCHE, I. 1995. Timing of hydrocarbon generation and accumulation in fault-bounded compartments in the Norphlet Formation, offshore Alabama. *Marine and Petroleum Geology*, **12**, 549–558.

PAXTON, S. T., SZABO, J. O., AJDUKIEWICZ, J. M. & KLIMENTIDIS, R. E. 2002. Construction of an intergranular volume compaction curve for evaluating and predicting compaction and porosity loss in rigid-grain sandstone reservoirs. *American Association of Petroleum Geologists Bulletin*, **86**, 2047–2067.

PITTMAN, E. D., LARESE, R. E. & HEALD, M. T. 1992. Clay coats: occurrence and relevance to preservation of porosity in sandstones. *In:* HOUSEKNECHT, D. W. & PITTMAN, E. D. (eds) *Origin, Diagenesis, and Petrophysics of Clay Minerals in Sandstones*, Society of Exploration Paleontologists and Mineralogists, Special Publication, **47**, 241–264.

RAMM, M. 1992. Porosity-depth trends in reservoir sandstones: theoretical models related to Jurassic sandstones, offshore Norway. *Marine and Petroleum Geology*, **9**, 553–567.

ROSSI, C., GOLDSTEIN, R. H., CERIANI, A. & MARFIL, R. 2002. Fluid inclusions record thermal and fluid evolution in reservoir sandstones, Khatatba Formation, Western Desert, Egypt: a case for fluid injection. *American Association of Petroleum Geologists Bulletin*, **86**, 1773–1799.

SAIGAL, G. C., BJORLYKKE, K. & LARTER, S. 1992. The effects of oil emplacement in diagenetic processes—examples from the Fulmar reservoir sandstones, Central North Sea. *American Association of Petroleum Geologists Bulletin*, **76**, 1024–1033.

SALVADOR, A. 1987. Late Triassic–Jurassic paleogeography and origin of the Gulf of Mexico Basin. *American Association of Petroleum Geologists Bulletin*, **71**, 419–451.

SASSEN, R., MOORE, C. H. & MEENDSEN, F. C. 1987. Distribution of hydrocarbon source potential in the Jurassic Smackover formation. *Organic Geochemistry*, **11**, 379–383.

SHEW, R. D. 1992. Origin and variability of $H_2S$ concentrations in siliciclastic and carbonate reservoirs, Smackover and Norphlet Formations of central and eastern Mississippi. *American Association of Petroleum Geologists Bulletin*, **76**, 118.

STORY, C. 1998. Norphlet geology and 3-D geophysics: Fairway Field, Mobile Bay, Alabama. *The Leading Edge*, **17**, 243–248.

THOMAS, A. R., DAHL, W. M., HALL, C. M. & YORK, D. 1993. $^{40}Ar/^{39}Ar$ analyses of authigenic muscovite, timing of stylolitization, and implications for pressure solution mechanisms: Jurassic Norphlet Formation, Offshore Alabama. *Clays and Clay Minerals*, **41**, 269–279.

THOMSON, A. 1979. Preservation of porosity in the deep Woodbine/Tuscaloosa trend, Louisiana. *Gulf Coast Association of Geological Societies Transactions*, **30**, 396–403.

THOMSON, A. & STANCLIFFE, R. J. 1990. Diagenetic controls on reservoir quality, eolian Norphlet formation, South State Line Field, Mississippi. *In*: BARWIS, J. H., MCPHERSON, J. G. & STUDLICK, J. R. J. (eds) *Sandstone Petroleum Reservoirs*. Springer, New York, 205–224.

WALDERHAUG, O. 1994. Temperatures of quartz cementation in Jurassic sandstones from the Norwegian continental shelf–evidence from fluid inclusions. *Journal of Sedimentary Research*, **64**, 311–323.

WALDERHAUG, O. 1996. Kinetic modelling of quartz cementation and porosity loss in deeply buried sandstone reservoirs. *American Association of Petroleum Geologists Bulletin*, **80**, 732–745.

WALDERHAUG, O. 2000. Modelling quartz cementation and porosity in middle Jurassic Brent Group Sandstones of the Kvitebjorn Field, northern North Sea. *American Association of Petroleum Geologists Bulletin*, **84**, 1325–1339.

WORDEN, R. H., OXTOBY, N. H. & SMALLEY, P. C. 1998. Can oil emplacement prevent quartz cementation in sandstones? *Petroleum Geoscience*, **4**, 129–137.

# Evidence of reservoir compartmentalization by calcite cement layers in deepwater sandstones, Bell Canyon Formation, Delaware Basin, Texas

SHIRLEY P. DUTTON[1] & WILLIAM A. FLANDERS[2]

[1] *Bureau of Economic Geology, The University of Texas at Austin, Austin, Texas 78713, USA (e-mail: shirley.dutton@beg.utexas.edu)*

[2] *Transpetco Engineering, 110 N. Marienfeld Place, Suite 525, Midland, Texas 79701, USA*

**Abstract:** Sandstones of the Upper Permian Bell Canyon Formation were deposited by turbidity currents in a basin-floor setting. The sandstones were deposited in a channel-levee system that terminated in broad lobes; overbank splays filled topographically low inter-channel areas. Diagenesis and reservoir quality of the sandstones were examined in cores from East Ford field, which is undergoing a $CO_2$ flood. Porosity and permeability are controlled by calcite cement, mainly concentrated in layers ranging from 5 to 40 cm in thickness. In a new infill well, initial production was of a high gas volume that contained a high concentration of $CO_2$ from the interval beneath several low-permeability, calcite-cemented layers. The $CO_2$ was most likely from an injector well and was trapped below the calcite layers. Geophysical log correlations support the interpretation that some calcite layers are laterally continuous over a distance of at least 300 m, causing vertical compartmentalization in the reservoir.

Upper Permian (Guadalupian) Delaware Mountain Group strata compose a 1400 m succession of slope and basin deposits in the Delaware Basin in west Texas and southeast New Mexico (Fig. 1). The Delaware Mountain Group is divided, from oldest to youngest, into the Brushy Canyon, Cherry Canyon and Bell Canyon Formations. Sandstones of the Delaware Mountain Group in west Texas and southeast New Mexico contained an estimated 1.8 Bbbl of original oil in place (Dutton *et al.* 2003). Many of the fields producing from Bell Canyon sandstones have low primary recovery and are candidates for enhanced oil recovery. East Ford field (Fig. 1), which recovered only 16% of the original oil in place during primary production, is now undergoing a $CO_2$ flood.

East Ford field produces from the upper Bell Canyon Ramsey sandstone. The reservoir is divided into an upper (Ramsey 2) and lower (Ramsey 1) interval by a laterally continuous siltstone (SH1) (Fig. 2). A depositional model of the East Ford unit was developed using data from Bell Canyon outcrops and subsurface data (Dutton *et al.* 2003). Stratigraphic relations in outcrop indicate that Bell Canyon sandstones were deposited in a basin-floor setting by a system of leveed channels having attached lobes and overbank splays that filled topographically low interchannel areas (Barton & Dutton 1999). Individual channel-levee and lobe complexes stack in a compensatory fashion and are separated by laterally continuous, laminated siltstones that were deposited by the settling of marine organic matter and airborne silt during periods when coarser particles were prevented from entering the basin.

## Evidence of diagenetic reservoir compartmentalization

The $CO_2$ flood of East Ford field began in 1995. A new well, the East Ford Unit (EFU) 41R, was drilled and cored in 2000 to replace well 41 (Fig. 1). Petrographic analysis of the reservoir sandstones was conducted by point counts of 30 thin sections made from end trims of core-analysis plugs. The reservoir is a well-sorted, very fine grained arkose. Calcite is the most abundant authigenic mineral, having an average volume of 7% and ranging from 0 to 30%. Calcite cement is concentrated mainly in layers ranging from 5 to 40 cm in thickness. Core-plug porosity ranges from 4.5 to 24.6%, and permeability to air from 0.01 to 78 mD. Because the sandstones have a narrow grain-size range and contain no detrital clay, volume of calcite cement is the dominant control on porosity and permeability.

Cemented zones are most common near the top and base of sandstones. Areas having high percentages of calcite-cemented sandstone (>20%) occur along the margins of the sandstones, in levee, overbank and lobe deposits, where the sandstone pinches out into siltstone. The areas having the lowest percentage of calcite-cemented sandstone (<10%) occur where the sandstone is thickest, in the channel facies. The calcite layers may be associated with pulses of turbidite deposition. The source of

*From*: CUBITT, J. M., ENGLAND, W. A. & LARTER, S. (eds) 2004. *Understanding Petroleum Reservoirs: towards an Integrated Reservoir Engineering and Geochemical Approach.* Geological Society, London, Special Publications, **237**, 279–282. 0305-8719/04/$15.00 

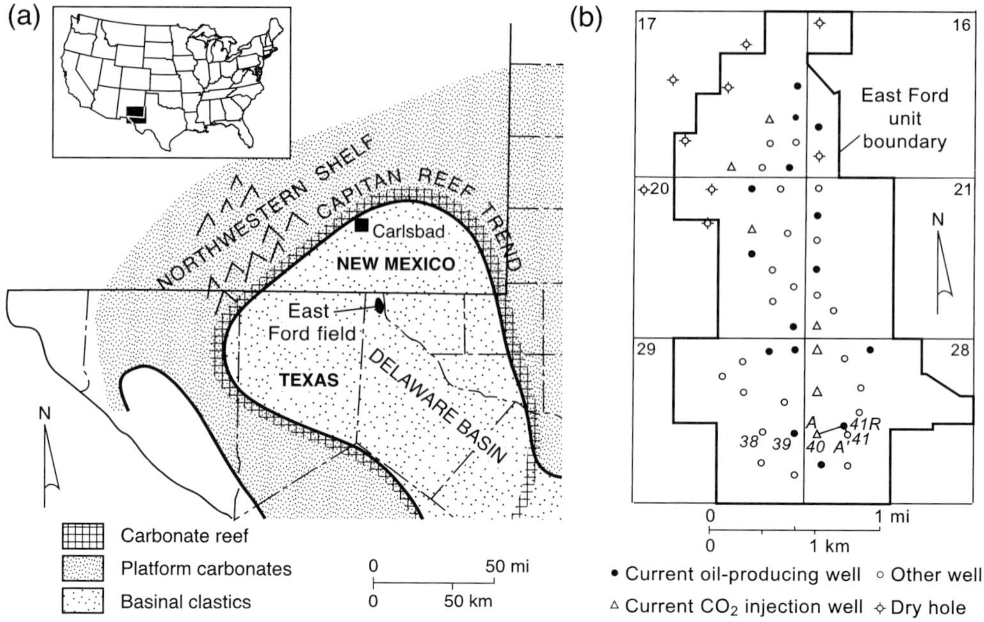


**Fig. 1.** (**a**) Location of East Ford field in the Delaware Basin and palaeogeographic setting during the Late Permian (modified from Silver & Todd 1969). (**b**) Map of East Ford field (modified from Dutton *et al.* 2003). Cross section A–A′ shown in Figure 3.

calcium carbonate is probably dissolution and reprecipitation of detrital carbonate rock fragments and fossils that occur in both the sandstones and surrounding siltstones.

Sonic and neutron logs from the EFU 41R well showed a gas effect in the lower 3 m of the Ramsey 1 sandstone, the same interval in which the calcite-cemented layers occur (Fig. 2). No

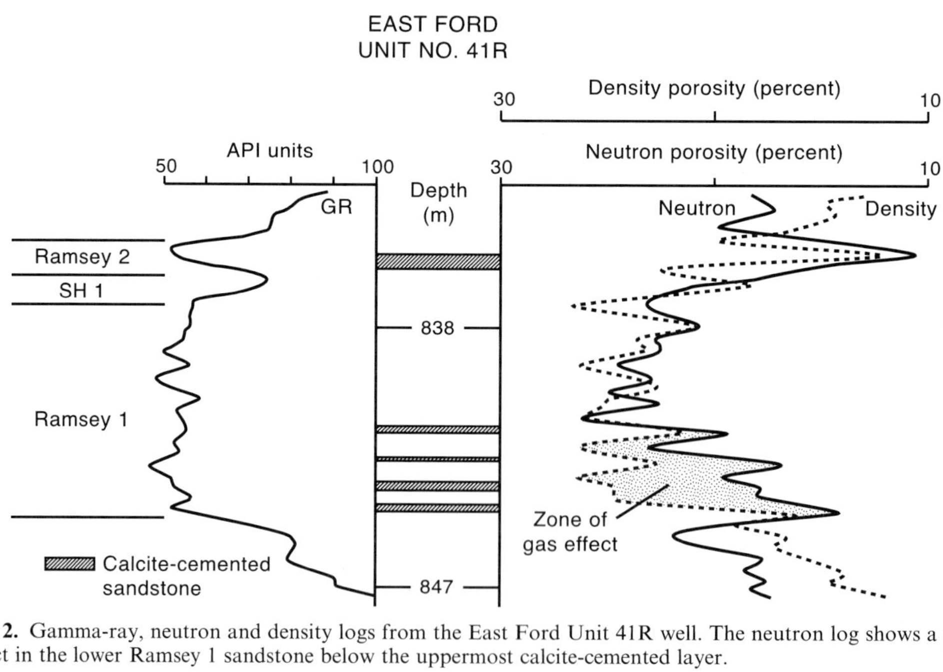


**Fig. 2.** Gamma-ray, neutron and density logs from the East Ford Unit 41R well. The neutron log shows a gas effect in the lower Ramsey 1 sandstone below the uppermost calcite-cemented layer.

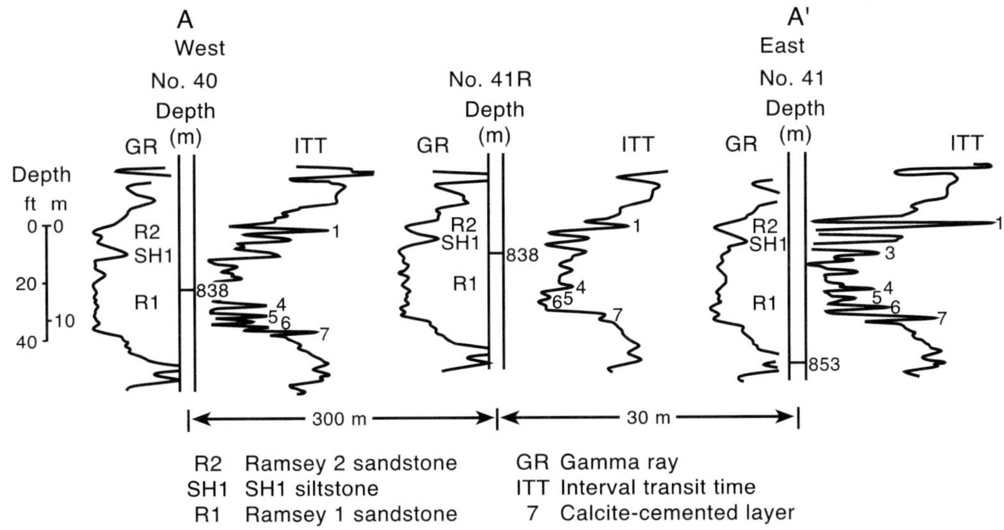


**Fig. 3.** Cross section A–A′ of south end of East Ford field. Four calcite-cemented layers (labelled 4–7) in the lower Ramsey 1 sandstone can be correlated in the EFU 40, 41 and 41R wells. The layer at the base of the Ramsey 1 sandstone (7) and another near the top of the Ramsey 2 sandstone (1) appear to be continuous across the south part of East Ford field.

gas effect was seen above the uppermost calcite layer. When the well was first completed, it produced a high volume of gas containing more than 90% $CO_2$. Production and temperature logs confirmed the gas effect by indicating that inflow to the well bore was essentially all occurring in the bottom 3 m of the Ramsey 1 sandstone. $CO_2$ from nearby injector well EFU 40 was apparently trapped in the bottom part of the Ramsey 1 sandstone, below the low-permeability, calcite-cemented layers. The trapping suggests that one or more of the calcite layers are laterally continuous between wells 40 and 41R, causing vertical compartmentalization in these Ramsey 1 turbidite lobe sandstones.

$CO_2$ was injected in the EFU 40 well above and below the calcite-cemented layers, but the previous producing wells, EFU 39 and 41, had no perforations in the Ramsey 1 sandstone below the calcite layers. The $CO_2$ that was produced in the EFU 41R well probably represents banked-up energy that gave a first flush of $CO_2$ when the well was completed. $CO_2$ injected below the calcite-cemented layers in well EFU 40 became trapped, resulting in significantly higher reservoir pressure (energy). The $CO_2$ could not dissipate to other portions of the reservoir, and there was no production below the cemented layers.

Calcite layers can be identified on interval transit time (ITT) logs as prominent deflections. Spikes on the EFU 41R ITT log in the lower part of the Ramsey 1 sandstone appear to correlate to those on the EFU 40 and EFU 41 ITT logs (Fig. 3), further evidence of lateral continuity of the cement layers. Because the distance between well EFU 40 and wells EFU 41 and 41R is about 300 m, the four calcite layers observed in the EFU 41R core are interpreted as having a lateral extent of at least that distance. Most of these layers did not extend to wells EFU 39 and 38 (Fig. 1), which contain fewer calcite-cemented layers in the Ramsey sandstone. However, the cemented zones at the base of the Ramsey 1 sandstone and near the top of the Ramsey 2 sandstone (Fig. 3) both appear to be continuous across the south part of East Ford field. Similar laterally extensive calcite-cemented layers in turbidite lobe sandstones have also been reported in turbidite reservoirs in Brazil (Moraes & Surdam 1993).

## Conclusions

It is difficult to determine the lateral extent of calcite-cemented layers observed in geophysical logs or core. Thin zones of calcite-cemented sandstone are commonly assumed to be concretions that extend no more than a few metres laterally, thus having little effect on production. Engineering production data from East Ford field provide evidence that some calcite layers extend hundreds of metres laterally and form vertical compartments in these deepwater turbidite lobe sandstones.

This research was funded by the US Department of Energy under contract no. DE-FC22-95BC14936 and by the State of Texas under State Match Pool Project 4201 and as part of the State of Texas Advanced Resource Recovery project. We thank J. Cubitt and the anonymous reviewers, whose comments improved the manuscript. Published by permission of the Director, Bureau of Economic Geology, The University of Texas at Austin.

## References

BARTON, M. D. & DUTTON, S. P. 1999. Outcrop analysis of a sand-rich, basin-floor turbidite system, Permian Bell Canyon Formation, West Texas. *In*: HENTZ, T. F. (ed.) *19th Annual Bob F. Perkins Research Conference*. Transactions, Gulf Coast Section SEPM 5–8 December, Houston, 53–64.

DUTTON, S. P., FLANDERS, W. A. & BARTON, M. D. 2003. Reservoir characterization of a Permian deepwater sandstone, East Ford field, Delaware Basin, Texas. *American Association of Petroleum Geologists Bulletin*, **87**, 609–627.

MORAES, M. A. S. & SURDAM, R. C. 1993. Diagenetic heterogeneity and reservoir quality: fluvial, deltaic, and turbiditic sandstone reservoirs, Potiguar and Reconcavo rift basins, Brazil. *American Association of Petroleum Geologists Bulletin*, **77**, 1142–1158.

SILVER, B. A. & TODD, R. G. 1969. Permian cyclic strata, northern Midland and Delaware Basins, west Texas and southeastern New Mexico. *American Association of Petroleum Geologists Bulletin*, **52**, 2223–2251.

# Spatial variations in the composition of formation waters from the central North Sea: implications for fluid flow in the deep high-pressure high-temperature hydrocarbon play

R. A. McCARTNEY,[1] P. WINEFIELD,[2] P. WEBB[2] & O. KUHN[3]

[1] *GeoScience Limited, Falmouth Business Park, Bickland Water Road, Falmouth TR11 4SZ, UK*
*(e-mail: mccartney@geoscience.co.uk)*

[2] *Shell U.K. Exploration and Production Ltd, 1 Altens Farm Road, Nigg, Aberdeen AB12 3FY, UK*

[3] *Shell China Exploration and Production Co. Ltd, China World Trade Centre, No. 1 Jian Guo Men Wai Daije, Beijing 100004, P.R. China*

**Abstract:** Formation water composition data and formation pressure data from oil, gas and gas condensate fields covering an area of approximately 15000 $km^2$ in the Central Graben Area of the UK North Sea have been analysed. The purpose was to determine whether large-scale barriers to flow influence the compositional distribution of formation waters and how this aids understanding of overpressure distribution and potential hydrocarbon migration pathways. Analyses (Na, K, Mg, Ca, Sr, Ba and Cl) of water and Elemental Residual Salt Analysis (ERSA) samples from both the water and hydrocarbon legs of wells were analysed statistically to reveal eight different water types in the study area. These were located in distinct geographical areas. Aquifer overpressure from formation pressure data for numerous wells in the area were plotted on large-scale regional maps to reveal the distribution of pressure 'cells' or compartments. An integrated interpretation of the location of pressure cells, salinity variations and water types provided information on the current and past fluid flow across the pressure cell boundaries, the operational longevity of these boundaries and the nature of the Mesozoic aquifer. The implications of the work are important for understanding: (a) the usefulness of the new ERSA technique for estimating formation water compositions; (b) the effectiveness of integrating fluid composition and pressure data to understand compartmentalization; (c) fluid flow dynamics in the HPHT sector of the Central North Sea; and (d) likely hydrocarbon migration routes and the character and distribution of overpressured basin compartments.

The Central Graben of the North Sea (Fig. 1) is one of the world's most prolific hydrocarbon producing provinces, with major accumulations of oil, gas and condensate present throughout the stratigraphic pile, from the Devonian to the Tertiary (Glennie 1990). Hydrodynamically, the graben can be broadly divided into an upper hydropressured regime, generally occurring at depths below sea bed less than 3 km and including the main Tertiary sandstone reservoirs, and a deeper (4–5 km) overpressured regime, much of which is high temperature, that includes Upper Jurassic, Mid-Jurassic and Triassic reservoirs (Cayley 1987; Darby *et al.* 1996; Holm 1998) (Fig. 2). Hydrocarbon accumulations within these reservoirs exhibit a range of pore pressures, with the more extreme approaching the lithostatic pressure gradient at depth.

Significant lateral variations in pore pressure are recognized within the heartland of the high-pressure, high-temperature (HPHT) play and are attributed to large-scale barriers (or baffles) to fluid flow. Regionally, the highest pore pressures are seen in the deepest parts of the Central Graben with pressures becoming less towards the shallower margins (Gaarenstroom *et al.* 1993). Flow barriers or baffles are believed to exist as otherwise pressures would have dissipated in the deeper graben. In essence, these barriers must effectively compartmentalize the deep section into a number of different compartments (or cells) that are characterized by significantly different aquifer pressures (Cayley 1987; Holm 1998). Therefore, understanding the distribution and nature of large-scale barriers, and of the pressure compartments or cells they surround, is of critical importance in attempting to predict pressure distribution and hydrocarbon migration pathways.

Existing techniques for recognizing and characterizing these basin compartments commonly use the regional seismic mapping of reservoirs

*From*: CUBITT, J. M., ENGLAND, W. A. & LARTER, S. (eds) 2004. *Understanding Petroleum Reservoirs: towards an Integrated Reservoir Engineering and Geochemical Approach*. Geological Society, London, Special Publications, **237**, 283–303. 0305-8719/04/$15.00 

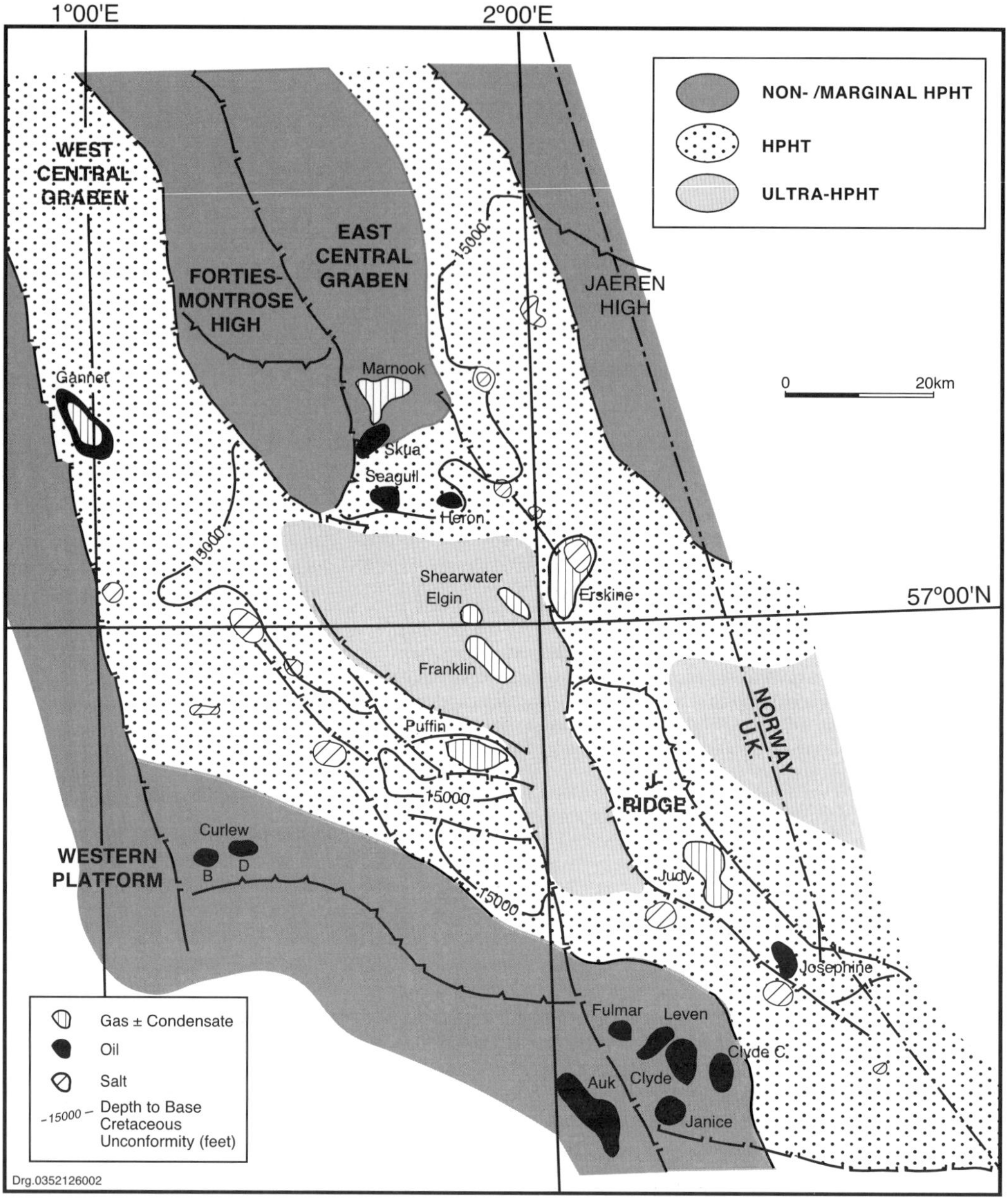


**Fig. 1.** Location and structural elements of the Central Graben, North Sea. Also shown are fields from which data were obtained, and key structural and geological elements of the Eastern Central Graben and surrounding areas.

(aquifers) and seals (aquitards) in combination with formation pressure data from wells in the area. This approach works well in characterizing the present-day pressure regime in areas with sufficient well penetrations. However, a key uncertainty in understanding trap evolution and in mapping hydrocarbon migration pathways is the location and temporal character of potential barriers to flow and of the basin compartments themselves.

The objective of this study was to compile an extensive formation water database for the HPHT play area and to determine whether large-scale barriers to flow may have influenced the compositional distribution of formation waters. It was also hoped that the results would

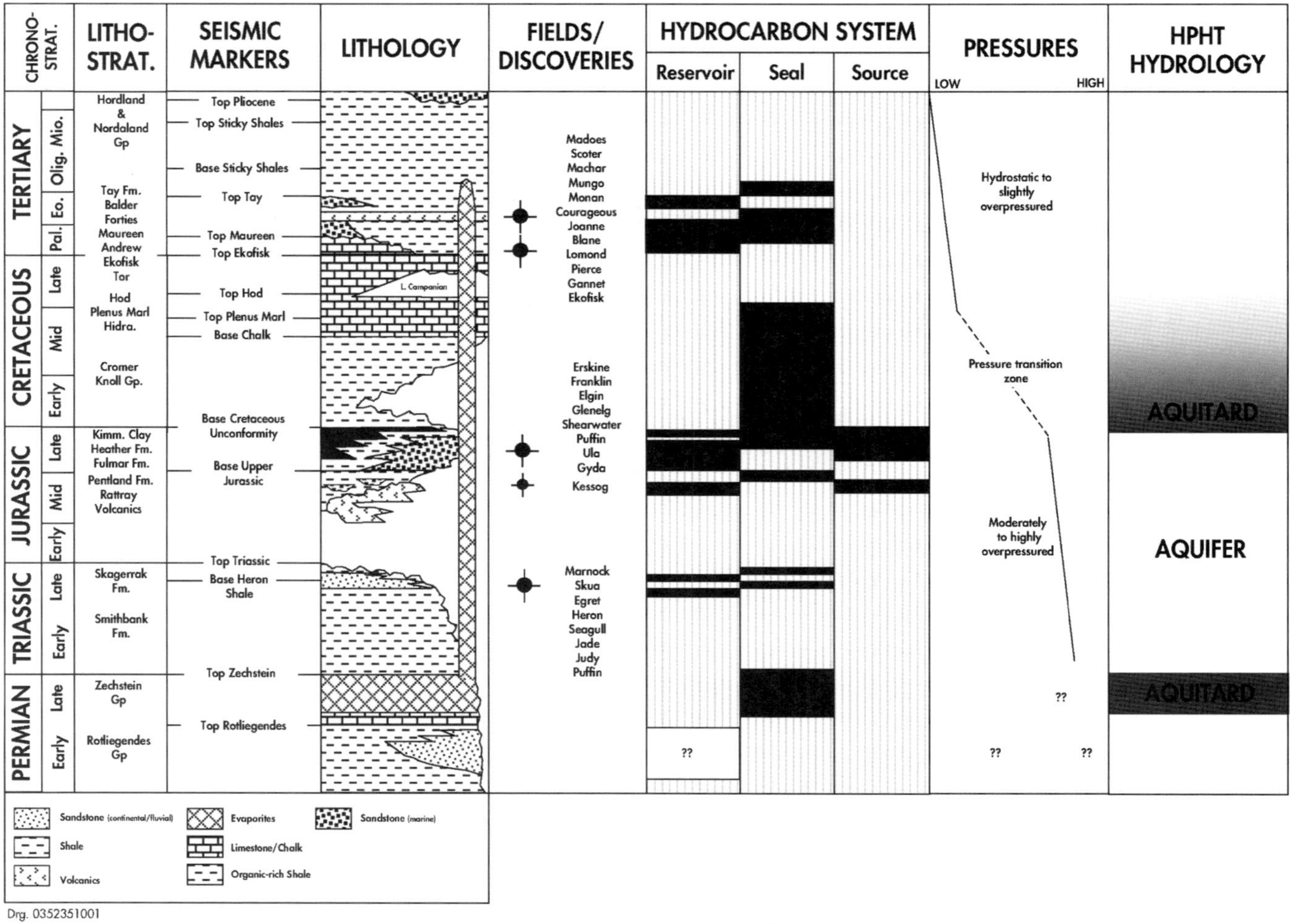


**Fig. 2.** Regional stratigraphic column for the Central North Sea highlighting potential aquifers and aquitards.

highlight if there were additional aquitards within the system, e.g. stratigraphically separating the main Jurassic and Triassic reservoirs. The latter would have significant impact in supporting the presence of possible stacked hydrocarbon accumulations in deeper, untested parts of the graben.

## Regional setting

For a general review of North Sea geology, the reader is referred to Glennie (1990) and also Holm (1998) for an overview of the geology of Central Graben with relation to the distribution and origin of overpressure. The Central Graben of the North Sea is a failed rift trending NW–SE, separating the Caledonides of Scotland to the west from the basement areas of Norway to the east. The Central Graben is further subdivided into a number of structural and geological elements (Fig. 1). The timing of rifting in the Central North Sea is uncertain, but Cornford (1994) argues that the principal phase of rifting occurred in the Late Jurassic following a period of thermal upwelling during the Early to Mid-Jurassic. Following rift development, a phase of subsidence associated with thermal collapse began in the Cretaceous and continued through Tertiary to the present day.

Critical elements in understanding the fluid 'plumbing' system of the deep HPHT section of the Graben are the main aquifers (reservoirs) and aquitards (seals). The main reservoirs or aquifers within the deep section are sandstones of the Skagerrak, Pentland and Fulmar Formations (Fig. 2). The main aquitards for the overpressured system are most likely the underlying Zechstein Group, which forms the base or seat seal, and the overlying Cromer Knoll and Chalk Groups that act as a highly effective top seal. Extreme overpressures in the deep graben are believed to be related to rapid basin subsidence in the late Tertiary combined with the restriction of vertical fluid flow through the Cromer Knoll and Chalk Groups and of lateral flow due to large-scale structural lineaments or barriers (Cayley 1987). As an additional factor the generation and input of hydrocarbons as well as diagenetic effects (cementation) might have further increased the amount of pressure (Holm 1998). In the deepest basin compartments where pore pressure is close to or at the lithostatic gradient (Fig. 1), the pressure is thought to have periodically leaked off by hydraulic fracturing of the overlying formations (Gaarenstroom *et al.* 1993).

## Approach

As we wanted to determine whether large-scale barriers to flow might have influenced the compositional distribution of formation waters, we evaluated two sets of data: formation water analyses and formation pressure data. Regional differences in formation pressure from one locality to another can be maintained by the presence of barriers to flow (e.g. Cayley 1987). Also, previous studies have shown that interpretation of formation water composition data can be used to identify such barriers. For example, $^{87}Sr/^{86}Sr$ RSA can be interpreted for this purpose on a well/field scale (e.g. Smalley *et al.* 1992; Smalley & England 1994; Mearns & McBride 1999). The interpretation of salinity data can provide the same information on a regional scale (e.g. Hanor & Sassen 1990; Bjørlykke & Gran 1994; Funayama & Hanor 1995; Nikiel & Hanor 1999). Chemical and isotopic data can provide information on mixing between different waters (e.g. Egeberg & Aagaard 1989; Cai *et al.* 2001; Ziegler *et al.* 2001) and, where it can be demonstrated that mixing has not occurred in a particular area, the possible presence of a barrier to flow can be inferred.

Formation pressure data were routinely obtained from wells drilled in the Eastern Central Graben Area (ECGA) and adjacent areas. Also, a large number of formation water analyses were available for this area having been originally obtained for other purposes (e.g. reservoir compartmentalization, scaling assessments, $R_w$ characterization etc). Because this was a regional study, we were limited to the use of Na, K, Mg, Ca, Sr, Ba and Cl analyses from Jurassic and Triassic reservoirs (Fulmar, Pentland and Skagerrak) of the ECGA and adjacent areas. These were the only data available for a large number of wells and they were of reasonably good quality (see below). We analysed the formation pressure data to determine the location of current barriers to flow and we analysed the formation water analyses to determine the distribution of different water types and location of zones of significant changes in salinity. These results have then been compared to determine the relationship between the barriers and formation water compositions in the ECGA and adjacent areas.

## Formation water analysis and results

### *Data*

Formation water composition data derived from water and elemental residual salt analysis

**Table 1.** *Source of data used in the study*

| Field | Hydrocarbon type | Well No. | Formation | Sample location | Sample type | Depth (TVDSS, ft) | |
|---|---|---|---|---|---|---|---|
| | | | | | | Top | Bottom |
| **Wells in the ECGA** | | | | | | | |
| Commander | Water | 22/30a-16 | Pentland | Water leg | ERSA | 18 009.73 | 18 022.30 |
| Elgin | Gas-condensate | 22/30c-13 | Fulmar | Gas leg | ERSA | 17 506.35 | 18 088.69 |
| Elgin | Gas-condensate | 22/30c-13 | Fulmar | Water leg | ERSA | 18 114.74 | 18 411.39 |
| Elgin | Gas-condensate | 22/30c-8 | Fulmar | Gas leg | ERSA | 17 842.85 | 18 001.96 |
| Elgin | Gas-condensate | Elgin well‡ | Fulmar | ? | Water | No depth | |
| Erskine† | Gas-condensate | Erskine well‡ | Pentland | GWC | Water | 15 600.00 | |
| Franklin | Gas-condensate | 29/5b-6,6Z | Fulmar | Gas leg | ERSA | 17 080.34 | 17 734.82 |
| Franklin | Gas-condensate | 29/5b-8 | Fulmar | Gas leg | ERSA | 18 141.87 | 18 374.19 |
| Franklin | Gas-condensate | 29/5b-8 | Fulmar | Water leg | ERSA | 18 390.29 | 19 054.66 |
| Franklin | Gas-condensate | Franklin well‡ | Fulmar | GWC | Water | 18 390.00 | |
| Heron | Oil | 22/29-5RE | Skagerrak | Oil leg | CERSA | 14 100.00 | 15 110.60 |
| Heron | Oil | 22/30a-2 | Skagerrak | ? | Water | No depth | |
| Heron | Oil | 22/30a-2 | Skagerrak | Water leg | CERSA | 15 529.75 | |
| Heron | Oil | 22/30a-6 | Skagerrak | Water leg | CERSA | 15 579.07 | |
| Heron | Oil | 22/30a-6 RE | Skagerrak | Water leg | Water | 15 699.00 | |
| Juno | Water | 22/29-6S2 | Fulmar | Water leg | ERSA | 15 490.04 | 15 532.77 |
| Juno | Water | 22/29-6S2 | Fulmar | Water leg | Water | 15 458.29 | |
| Kessog | Gas | 30/1c-3 | Pentland | Gas leg | Water | 14 522.22 | |
| Marnock | Gas-condensate | 22/24a-3Z | Skagerrak | Gas leg | Water | 10 383.53 | |
| Marnock | Gas-condensate | 22/24b-5 ST1 | Skagerrak | Gas leg | Water | 11 732.27 | |
| Marnock | Gas-condensate | 22/24b-5 ST1 | Skagerrak | Gas leg | Water | 11 732.27 | |
| Marnock | Gas-condensate | 22/24b-5 ST1 | Skagerrak | Water leg | Water | 11 872.06 | |
| Marnock | Gas-condensate | 22/24b-5 ST1 | Skagerrak | Water leg | Water | 11 872.06 | |
| Martha | Water | 22/30a-1 | Fulmar | Water leg | ERSA | 15 163.60 | 15 222.80 |
| Puffin | Gas-condensate | 29/10-3 | Skagerrak | Gas leg | ERSA | 14 519.15 | 14 653.53 |
| Puffin | Gas-condensate | 29/4a-1A | Fulmar | ? | ERSA | 14 975.00 | |
| Puffin | Gas-condensate | 29/5a-3 | Skagerrak | ? | Water | No depth | |
| Puffin | Gas-condensate | 29/5a-3 | Skagerrak | Gas leg | ERSA | 14 099.61 | 14 396.77 |
| Puffin | Gas-condensate | 29/5a-3 | Skagerrak | Gas leg | Water | 14 446.54 | |
| Puffin | Gas-condensate | 29/5a-3 | Skagerrak | Water leg | ERSA | 14 457.06 | 14 523.78 |
| Puffin | Gas-condensate | 29/5a-3 | Skagerrak | Water leg | Water | 14 505.53 | |
| Puffin | Gas-condensate | 29/5a-5 | Pentland | Water leg | Water | 14 500.00 | |
| Puffin | Gas-condensate | 29/5a-7 | Skagerrak | Gas leg | ERSA | 14 576.08 | |
| Puffin South East | Water | 29/9a-5 | Fulmar | Water leg | Water | 14 820.61 | |
| Seagull | Oil | 22/29-2 | Skagerrak | Oil leg | Water | 13 782.54 | |
| Seagull | Oil | 22/29-3 | Skagerrak | Oil leg | Water | 13 715.04 | |
| Shearwater | Gas-condensate | 22/30b-11 | Fulmar | Gas leg | ERSA | 16 391.87 | 16 548.39 |
| Shearwater | Gas-condensate | 22/30b-11 | Fulmar | GWC | Water | 16 900.00 | |
| Shearwater | Gas-condensate | 22/30b-15 | Pentland | Gas leg | Water | 16 670.00 | |
| Shearwater | Gas-condensate | 22/30b-15S1 | Fulmar | Gas leg | ERSA | 16 392.45 | 16 889.84 |
| Shearwater | Gas-condensate | 22/30b-15S1 | Pentland | Water leg | ERSA | 17 199.45 | 17 212.29 |
| Shearwater | Gas-condensate | 22/30b-A2 | Fulmar | Gas leg | ERSA | 16 199.49 | 16 270.36 |
| Shearwater | Gas-condensate | 22/30b-A6 | Pentland | Gas leg | ERSA | 15 826.16 | 16 016.64 |
| Shearwater | Gas-condensate | 22/30b-A6 | Pentland | GWC | Water | 16 670.00 | |
| Skua | Oil | 22/24b-7 | Skagerrak | Oil leg | ERSA | 11 665.16 | 12 032.53 |
| Skua | Oil | 22/24b-9 | Skagerrak | Oil leg | ERSA | 11 782.18 | 11 970.00 |
| Skua | Oil | 22/24b-9 | Skagerrak | Water leg | ERSA | 12 117.96 | 12 133.87 |
| Skua | Oil | 22/24b-9 | Skagerrak | Water leg | Water | 12 305.00 | |
| Skua | Oil | 22/24b-9 | Skagerrak | Water leg | Water | 12 413.15 | |
| Turnstone | Oil | 22/23b-6 | Fulmar | Oil leg | Water | 14 949.84 | |
| Turnstone | Oil | 22/23b-6 | Fulmar | Water leg | Water | 15 142.85 | |
| **Wells outside the ECGA** | | | | | | | |
| Bittern | Oil | 29/1b-5 | Skagerrak | Water leg | Water | 8 958.00 | |
| Bittern | Oil | 29/1b-5 | Skagerrak | Water leg | Water | 8 961.00 | |
| Clyde* | Oil | 30/17b-A01 | Fulmar | Water leg | Water | 12 900.00 | |

**Table 1.** *continued*

| Field | Hydrocarbon type | Well No. | Formation | Sample location | Sample type | Depth (TVDSS, ft) | |
|---|---|---|---|---|---|---|---|
| | | | | | | Top | Bottom |
| Clyde* | Oil | 30/17b-A09 | Fulmar | Water leg | Water | 12 813.77 | |
| Clyde* | Oil | 30/17b-A10 | Fulmar | Water leg | Water | 12 900.00 | |
| Curlew B | Gas-condensate | 29/7-4 | Fulmar | ? | Water | No depth | |
| Durward | Oil | 21/16 | Fulmar | OWC | Water | 7 300.00 | |
| Fulmar* | Oil | Fulmar well‡ | Fulmar | OWC | Water | 10 840.00 | |
| Gadwall | Oil | 21/19-6 | Fulmar | Oil leg | ERSA | 11 847.12 | 11 947.90 |
| Gadwall | Oil | 21/19-6 | Skagerrak | Oil leg | ERSA | 11 900.98 | 12 015.35 |
| Guillemot A | Oil | GUAP2 | Fulmar | Water leg | Water | 8 460.00 | |
| Guillemot A | Oil | GUAP4 | Fulmar | Water leg | Water | 8 460.00 | |
| Guillemot A* | Oil | Guillemot A well‡ | Fulmar | Water leg | Water | 8 460.00 | |
| Gyda* | Oil | Gyda well | Fulmar | OWC | Water | 13 684.00 | |
| Gyda* | Oil | N1/03-3A | Fulmar | OWC | Water | 13 684.00 | |
| Gyda* | Oil | N2/01-6A | Fulmar | OWC | Water | 13 684.00 | |
| Gyda* | Oil | N2/1-6 | Fulmar | OWC | Water | 13 684.00 | |
| Kittiwake | Oil | KA01S1 | Fulmar | Oil leg | Water | 10 228.07 | 10 228.07 |
| Kittiwake | Oil | KA05S1 | Fulmar | Oil leg | Water | 10 124.58 | 10 124.58 |
| Kittiwake | Oil | KA08 | Fulmar | Oil leg | Water | 9 949.43 | |
| Kittiwake* | Oil | Kittiwake well‡ | Fulmar | Water leg | Water | 10 440.00 | |
| Mallard | Oil | 21/19-4 | Fulmar | Water leg | Water | 12 870.00 | |
| Molly (S of Puffin) | Water | 29/10-2 | Fulmar | Water leg | Water | 13 811.81 | |
| Structure (N of Acorn) | Water | 29/8a-4 | Skagerrak | Water leg | Water | 13 809.58 | 13 951.42 |
| Teal | Oil | 21/25-8 | Fulmar | Water leg | Water | 10 874.00 | |
| Ula* | Oil | Ula well‡ | Fulmar | OWC | Water | 12 431.00 | |

* Data from Warren and Smalley (1994).
† Data from Jordan *et al.* (2000).
‡ Well number not known.
OWC, oil water contact.
GWC, gas water contact.

(ERSA) samples of 59 wells in 29 oil, gas and gas-condensate fields were compiled from internal, proprietary Shell databases and Warren & Smalley (1994). All data used were quality checked and contaminated samples or samples of dubious origin were excluded. Table 1 summarizes the fields, wells, sample location and formations from which the different types of sample were obtained. Field locations are shown in Fig. 1. Data were obtained from sandstone and sandstone/shale sequences at depths between 7300 and 19 000 ft TVDSS. At these depths, reservoir temperatures and pressures are between 107 and 191 °C and between 323 and 1136 bars, respectively.

Water samples were obtained through down-hole sampling (RFT, MDT, bottom hole samplers), well testing (DST, production tests) and actual production. The sampling density for most water samples was one per well with most of these samples obtained from the water-leg.

ERSA samples are generated by completely extracting the precipitated salt residues of evaporated formation water from core material with deionized water. All water and ERSA leachate samples were analysed by standard methods for major and minor cations and anions and $^{87}Sr/^{86}Sr$, although only Na, K, Ca, Mg, Ba, Sr and Cl data were used in this study. Ideally, the leachate analyses will represent formation water diluted with pure water. Therefore, ratios of constituent concentrations calculated from the leachate analyses (e.g. Na/Cl) should be equivalent to the same ratios in the original formation waters. This conclusion is only valid if (a) the formation water is not contaminated with drilling mud, (b) the composition of the formation water does not change during retrieval of the core from the reservoir and during storage (e.g. due to reduction of pressure and temperature and associated gas loss), (c) all evaporated salts are removed during the leaching

**Table 2.** *Summary of data reproducibility and consistency of reproducibility*

| Constituent | Water samples (17 No) | | All ERSA/CERSA samples (20 No) | | Elgin ERSA samples (4 No) | |
|---|---|---|---|---|---|---|
| | Mean RSD (%) | St Dev RSD (%) | Mean RSD (%) | St Dev RSD (%) | Mean RSD (%) | St Dev RSD (%) |
| Na | 4.0 | 3.2 | 20.6 | 19.2 | 15.7 | 16.9 |
| K | 7.0 | 11.2 | 13.6 | 12.7 | 8.3 | 11.1 |
| Mg | 5.6 | 7.5 | 18.9 | 16.7 | 18.9 | 23.1 |
| Ca | 4.7 | 4.5 | 15.4 | 12.6 | 13.3 | 9.2 |
| Sr | 5.1 | 4.8 | 22.0 | 16.5 | 36.6 | 12.1 |
| Ba | 12.1 | 14.6 | 28.2 | 34.2 | No data | No data |
| Cl | 3.0 | 3.1 | 16.8 | 15.1 | 15.3 | 14.8 |
| Na/Cl | 2.0 | 1.9 | 6.8 | 7.9 | 1.9 | 2.1 |
| Ca/Cl | 4.7 | 5.2 | 8.5 | 5.3 | 9.2 | 4.7 |
| K/Cl | 6.1 | 11.4 | 8.2 | 6.9 | 7.5 | 10.7 |
| Mg/Cl | 5.3 | 8.3 | 15.1 | 14.9 | 33.5 | 23.3 |
| Ba/Cl | 10.9 | 12.2 | 20.5 | 34.0 | No data | No data |
| Sr/Cl | 3.6 | 3.4 | 12.6 | 10.9 | 24.5 | 17.2 |
| Na/Ca | 5.2 | 5.4 | 14.4 | 13.1 | 9.6 | 6.9 |

process, (d) the leachate does not react with the core minerals, (e) systematic contamination does not occur during the leaching process, and (f) the leachate analyses are accurate. Issue (a) was minimized by sampling wells where oil-based drilling mud was used to obtain the core samples. The laboratory procedures adopted (see Appendix 1) minimized (e) and (f). Therefore, it is believed that uncertainties only exist about assumptions (b), (c) and (d). ERSA is a technique that is still under development (E. Mearns, pers. comm. 2002) and the use of such data in this study should still be considered exploratory.

For the Heron Field, original formation water concentrations (corrected elemental residual salt analyses: CERSA) were back-calculated using ERSA leachate analyses, the volume of the core sample, and values for core sample porosity and $S_w$ from nearby core samples. Further corrections were made for $H_2O$ condensation from the oil-phase to formation water using a Shell in-house PVT package. The total dataset used in this study included analyses of 478 samples (189 water samples, 274 ERSA samples, 15 CERSA samples).

## *Data quality*

Based on available records, and for the constituents of interest to this study, the water samples and associated analyses are of good quality, i.e. free from significant modification due to contamination from other fluids, materials and pre-sampling scale deposition, with the majority of sample analyses having ion balances of less than $\pm 6\%$. Duplicate sample analyses were used to obtain an estimate of the reproducibility and consistency of reproducibility of the water sample data (see Table 2). Generally, these were good with the variation reflecting uncertainties in sampling as well as analytical conditions. For example, all duplicate water samples were collected independently usually with a period of time between the collection of each 'duplicate'. In the case of Ba, greater uncertainty in the results reflects the low levels of Ba in some of the duplicate samples. Analytical uncertainties were probably more significant for these samples and they were probably more susceptible to minor sample contamination.

The quality of the ERSA and CERSA data is less clear although, for this application, analysis of the data appears to provide results consistent with the good quality water sample data (see Appendix 2). Most of the ERSA data have ion balances less than $\pm 6\%$ and are consistent with those observed in the Central North Sea (e.g. see Fig. 3). For these samples, evaluation of the $^{87}Sr/^{86}Sr$ RSA and core permeability data suggests that mud contamination is probably low or negligible. Many of the Elgin and Franklin analyses have ion balances greater than $\pm 6\%$ and variable ion ratios that are not always consistent with Central North Sea water sample data (e.g. see Fig. 3). The reproducibility, and consistency of reproducibility, of ERSA/CERSA data is poorer than for water samples (Table 2). Where ERSA, CERSA and water sample data were closely located (i.e. $\pm 200$ ft), depending on the samples and analyses

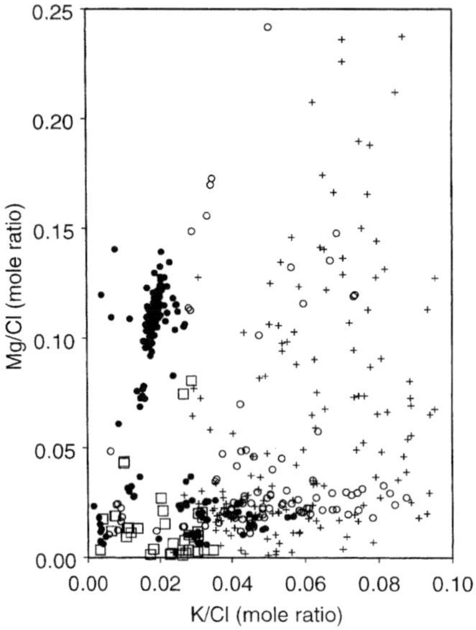


**Fig. 3.** Comparison of water (●), Central North Sea (Warren & Smalley 1994; □), Elgin/Franklin ERSA (×), and remaining ERSA/CERSA (○) sample K/Cl and Mg/Cl mole ratios.

compared, a bias may or may not exist between them (e.g. Fig. 4). Although some validation studies have been undertaken on the ERSA and CERSA methods, further studies are required to assess whether processes occurring during coring, core retrieval, core storage and core leaching may have affected these data.

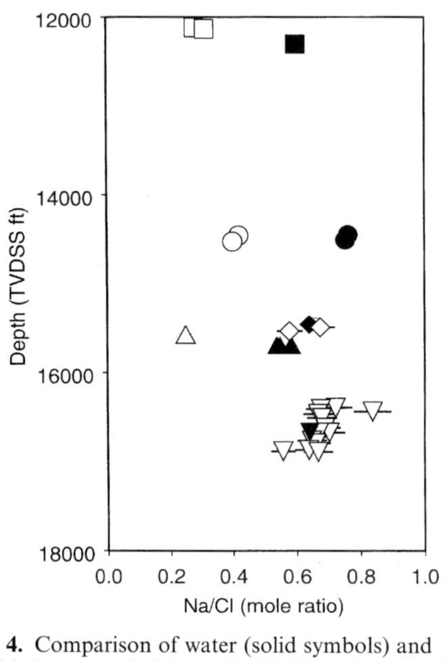


**Fig. 4.** Comparison of water (solid symbols) and ERSA (open symbols) sample Na/Cl mole ratios for wells in the ECGA.

## *Data analysis*

Based on the expectation that the presence of barriers to flow would result in significant changes in formation water composition from one location to another, the objectives of data analysis were (a) to group wells with formation waters of similar composition and (b) to identify zones where significant changes in salinity occurred over relatively short lateral distances.

To meet the first objective, the water sample data were subjected to principal components analysis (PCA) and cluster analysis (CA) (see Appendix 2). These calculations were used to group formation waters into 'water types'. A water type is defined in this paper as a formation water with a distinctive composition in terms of mole ratios of dissolved ions (rather than absolute concentrations of dissolved ions). Eight water types were identified and Table 3 shows the water type of each sampled well in the study area.

For illustrative purposes, Table 4 and Figure 5 summarize the average relative compositions of each water type based on water sample data. Although the average relative compositions of water types 6 and 7 appear similar based on these data, they are easily separated via the PCA/CA results.

To meet the second data analysis objective, water sample Cl data were contoured (Fig. 6). The waters in the study area are Na-Cl brines with salinities varying from 90 to 395 g/l TDS (54–240 g/l Cl). The largest increases over shortest lateral distances are located between Franklin and Puffin, Franklin and Elgin, Shearwater and Erskine, Shearwater and Elgin–Commander–Martha, Erskine and Commander–Martha, and between Heron–Skua and Turnstone–Seagull. Note that although contouring has helped identify these zones, it is not known whether the changes in these areas are gradational or step changes. This is discussed further below.

Figure 7 shows the geographical extent of each of the water types which occur in limited and well-defined areas. Water type 1 covers much of the western and northwestern area adjacent to the Central Graben. Water type 2 is present in the south in the 30/6 graben. In contrast, water type 4 covers much of the Central Graben (other than in the north where no data were available). Water types 3 and 5–8 occur within

**Table 3.** *Grouping of wells based on PCA/CA results for water sample data*

| Water type | Cl (g/l) | Wells* |
|---|---|---|
| 1 | 101–102 | Bittern 29/1b-5 (S) |
| | 119–134 | Guillemot A GUAP 4 (F), Guillemot A well (F) |
| | 132 | Guillemot A GUAP 2 (F) |
| | 132–141 | Kittiwake well (F) |
| | 132–163 | KA01S1* (F), Kittiwake KA05S1* (F), Kittiwake KA08* (F) |
| | 150 | Durward 21/16 (F) |
| | 172–181 | Mallard 21/19-4 (F) |
| 2 | 98 | Fulmar well |
| | 111–122 | Clyde 30/17b-A01 (F), Clyde 30/17b-A09 (F), Clyde 30/17b-A10 (F) |
| 3 | 203–206 | Puffin 29/5a-3 (S), Puffin 29/5a-5 (P) |
| | | Puffin 29/5a-3*§ (S), Puffin 29/5a-5*§ (S), Puffin 29/10-3*§ (S) |
| 4 | 152–163 | Curlew B 29/7-4 (F) |
| | 153–171 | Ula well (F), Gyda N1/03-3A (F) |
| | 156 | Teal 21/25-8‡ (F) |
| | 161–171 | Seagull 22/29-2 (S), Molly 29/10-2 (F) |
| | 165–184 | Turnstone 22/23b-6† (F), Gyda well (F), Gyda N2/01-6A (F), Gyda N2/1-6 (F) |
| | 171–193 | Seagull 22/29-3* (S), NOA structure 29/8a-4 (S) |
| | 189 | Puffin south-east 29/9a-5 (F) |
| | 199–223 | Marnock 22/24a-3z* (S*), Marnock 22/24b-5ST1† (S†) |
| | 211 | Heron 22/30a-2 (S) |
| | 211–218 | Heron 22/30a-6RE (S) |
| | 213–240 | Skua 22/24b-9 (S) |
| | | Heron 22/30a-2§ (S), Heron 22/30a-6§ (S), Skua 22/24b-9†§ (S) |
| 5 | 54 | Erskine well low salinity water‡ (P) |
| | 91 | Erskine well high salinity water (P) |
| | 99 | Kessog 30/1c-3* (P) |
| | 140–142 | Shearwater 22/30b-A6 (P), Shearwater 22/30b-15* (F) |
| | 169 | Juno 22/29-6S2‡ (F) |
| | | Martha 22/30a-1§ (F), Heron 22/29-5RE*§ (S), Commander 22/30b-11§ (P) |
| | | Shearwater 22/30b-A2*§ (F), Shearwater 22/30b-15S1†§ (F, P) |
| | | Shearwater 22/30b-A6*§ (P), Juno 22/29-6S2§ (F) |
| 6 | 174 | Elgin well (F) |
| | | Elgin 22/30c-8*§ (F), Elgin 22/30c-13†§ (F) |
| 7 | 142–145 | Shearwater 22/30b-11 (F) |
| 8 | 54 | Franklin well‡ (F) |
| | | Franklin 29/5b-8†§ (F), Franklin 29/5b-6,6Z*§ (F) |

S, Skagerrak; P, Pentland; F, Fulmar Formations.
All water leg samples except * (hydrocarbon leg sample only).
Wells grouped on the basis of water sample data and shown on the same row have identical mole ratio compositions (within data uncertainties).
† Water leg and hydrocarbon leg samples.
‡ Uncertainty in water type membership (see Appendix 2).
All based on water sample data except § (based on CERSA/ERSA sample data).

the area bounding water type 4. Water type 7 lies within the Shearwater Field and given the regional nature of this study it will not be discussed further. Despite grouping the formation waters of wells on their mole ratio compositions, water types 1, 4 and 5 contain formation waters with a significant range of Cl concentrations (and hence concentrations of the other constituents; see Table 3 and Figs 6 & 7). There is no apparent correlation between sampled formation and water type or salinity (see Fig. 7). For example, the Fulmar, Skagerrak and Pentland formations produce water with 54–180, 100–240 and 90–165 g/l Cl respectively. Also, water type 1 can be found in the Fulmar and Skagerrak formations, and water types 4 and 5 can be found in the Fulmar, Skagerrak and Pentland formations.

**Table 4.** *Average compositions of water types*

| Water type | %Na | %K × 10 | %Mg × 10 | %Ca | %Sr × 100 | %Ba × 10 | %Cl | No. samples |
|---|---|---|---|---|---|---|---|---|
| 1 | 24.0 ± 2.2 | 13.3 ± 2.7 | 45.3 ± 11.1 | 6.3 ± 0.7 | 15.8 ± 3.6 | 0.1 ± 0.03 | 63.6 ± 1.1 | 116 |
| 2 | 31.3 ± 0.9 | 20.0 ± 1.0 | 4.4 ± 0.5 | 4.6 ± 0.3 | 30.2 ± 9.3 | 0.1 ± 0.06 | 61.4 ± 0.5 | 4 |
| 3 | 29.8 ± 0.2 | 33.9 ± 3.7 | 7.6 ± 0.7 | 5.1 ± 0.4 | 25.6 ± 0.7 | 0.2 ± 0.02 | 60.7 ± 0.1 | 6 |
| 4 | 23.4 ± 1.8 | 20.2 ± 7.4 | 9.8 ± 3.9 | 11.9 ± 1.6 | 30.4 ± 6.0 | 1.4 ± 1.2 | 61.3 ± 0.7 | 37 |
| 5 | 28.4 ± 2.9 | 3.9 ± 4.0 | 4.7 ± 2 | 8.7 ± 2.5 | 46.3 ± 20.0 | 10.2 ± 2.0 | 60.6 ± 0.9 | 8 |
| 6 | 27.7 ± 0 | 27.5 ± 0 | 8.6 ± 0 | 6.9 ± 0 | 68.9 ± 0 | 12.7 ± 0 | 59.8 ± 0.0 | 1 |
| 7 | 28.0 ± 0.9 | 20.8 ± 0.9 | 2.7 ± 0.1 | 7.8 ± 0.3 | 70.3 ± 2.7 | 10.4 ± 0.6 | 60.1 ± 0.5 | 4 |
| 8 | 34.6 ± 0 | 3.3 ± 0 | 3.9 ± 0 | 3.3 ± 0 | 77.7 ± 0 | 11.1 ± 0 | 59.4 ± 0.0 | 1 |

Compositions given as concentration (mg/kg) of cation or anion as a percentage of TDS (mg/kg).

## Regional pressure data

Formation pressure data were obtained from repeat formation tester (RFT) or modular dynamic tester (MDT) measurements of numerous deep wells in the Central North Sea. These data were used as the primary pressure dataset as they are the most accurate pressure measurements possible down-hole. The MDT/RFT wireline tool takes a pressure reading within a permeable formation, by setting a seal at a precise depth determined by using an accompanying gamma ray tool for depth correlation. Drill stem test (DST), mudweight data and 'kick' (influxes of pore fluids into the wellbore) information was also used where RFT or MDT data were not available or of very poor quality. A summary of the various approaches used to derive formation pressures is provided by Holm (1998).

Data from over 100 wells that penetrate the pre-Cretaceous were examined as part of a larger review of overpressures within the deeper section of the Central Graben. For this purpose, only those data measurements that could be quantified as of good quality were included. When comparing pressure data between wells spatially, the relative overpressure (difference between the hydrostatic and aquifer pressure gradients) should always be quoted. The hydrostatic gradient was taken as 0.45 psi/ft from surface as standard and, where necessary, the pressure data were also corrected to remove the effects of hydrocarbon buoyancy.

Once overpressures are calculated for each reservoir interval within a well, they can then be plotted onto a map and inferences made about the spatial variation of the pressures. Significant differences between wells over relatively short distances are commonly observed indicating reservoir or aquifer compartmentalization and the approximate location of potential cell boundaries. The power of integrating the well data with regional seismic mapping is that it allows quick comparison of the structural framework of the graben with pressure cells, i.e. areas with similar overpressures (Fig. 8).

## Discussion

### *Location and nature of barriers*

From maps such as Figure 8, inferences about the position of the main barriers or baffles to fluid flow can be made and these can then be characterized using more detailed seismic and geological analysis. The main pressure cell boundaries, i.e. those separating pressures by more than 500 psi, are characterized as being either major structural lineaments or the coincidence of high salt, faults and 'pod/interpod' margins. The latter serve to reduce the thickness of the Jurassic/Triassic aquifer in these areas (Fig. 9). These boundaries have presumably acted as baffles to fluid flow for a considerable period of geological time as the main build-up of pressure is thought to be due to late, rapid Tertiary burial of the sequence (Swarbrick *et al.* 2000). Faulting or stratigraphic sealing within pressure cells may also cause lesser pressure differences (<500 psi), but in general Jurassic and Triassic reservoirs are broadly in pressure communication.

### *Generation and dissipation of variations in formation water composition and the role of barriers to flow*

Before discussing the relationship between the location of pressure cell boundaries and variations in formation water composition, it is important to consider how compositionally different formation waters in adjacent areas can be generated and how they can be dissipated.

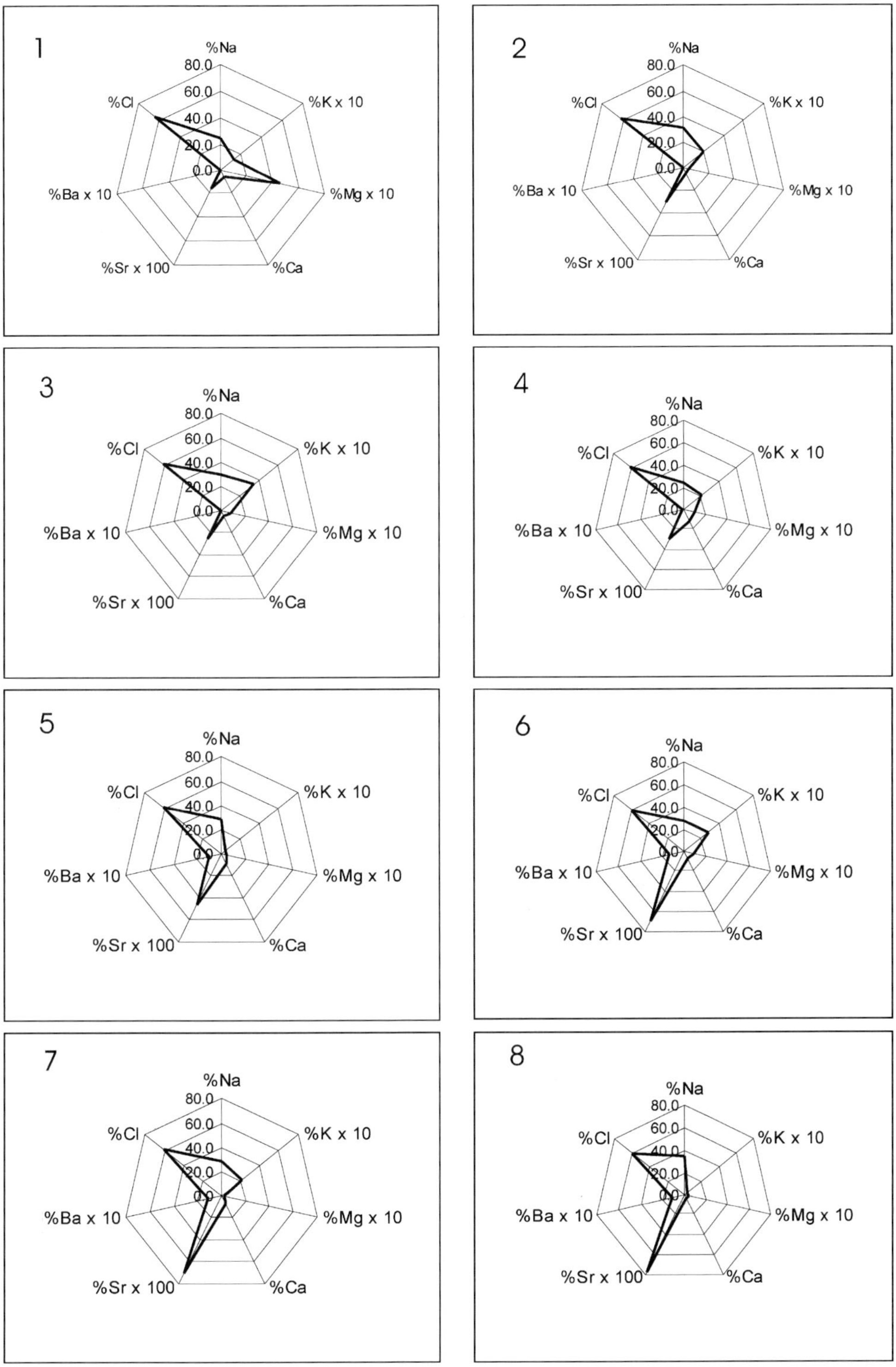


**Fig. 5.** Star plots showing the average relative compositions (from Table 4) of each water type.

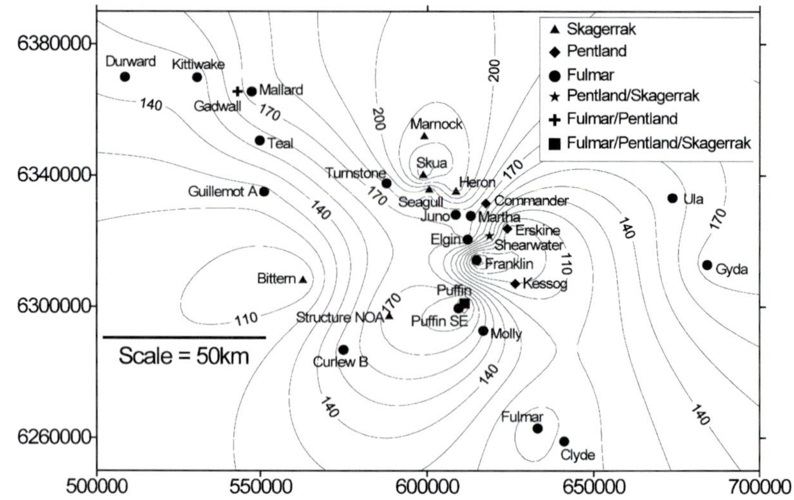


**Fig. 6.** Contoured distribution of formation water Cl (g/l) based on water sample data. Symbol shapes identify producing formation.

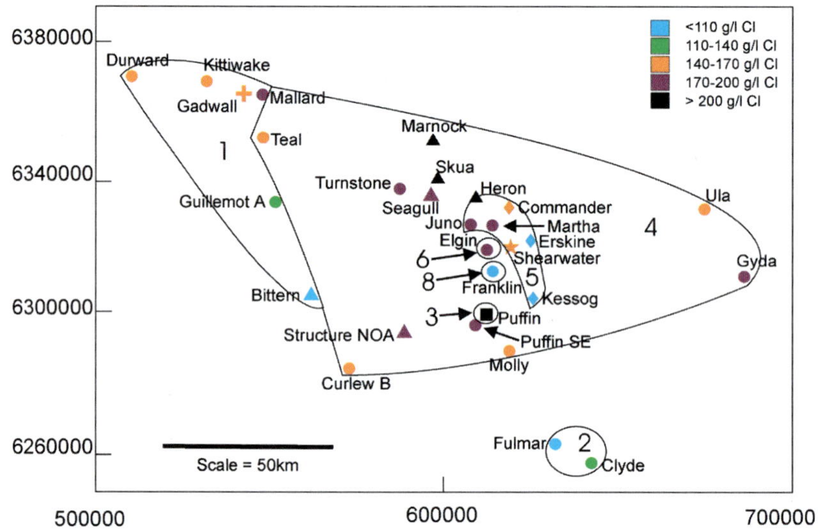


**Fig. 7.** Geographical distribution of the water types (numbered). Symbol shapes identify producing formation (see legend to Fig. 6). Symbol colours indicate formation water Cl (g/l).

Of particular interest to this study is the role of barriers to flow in these processes.

Compositionally different formation waters can be generated in adjacent areas by two general mechanisms. Firstly, the water–rock interactions occurring in each area may be different. This may be because the areas have different mineralogy, grain size/texture, porosity, $CO_2$ partial pressure, temperature, pressure, effective stress etc. Secondly, the waters in each area may have different origins/ages. Evidently, a combination of the two might also explain the differences in water composition in adjacent areas.

Barriers to flow may have a role in the generation processes. For example, if the barrier is a fault it might define the transition from one set of water–rock interactions to another as the rock type changes across the fault. Also, where waters of different origin/age are entering an area from different directions, the barrier might

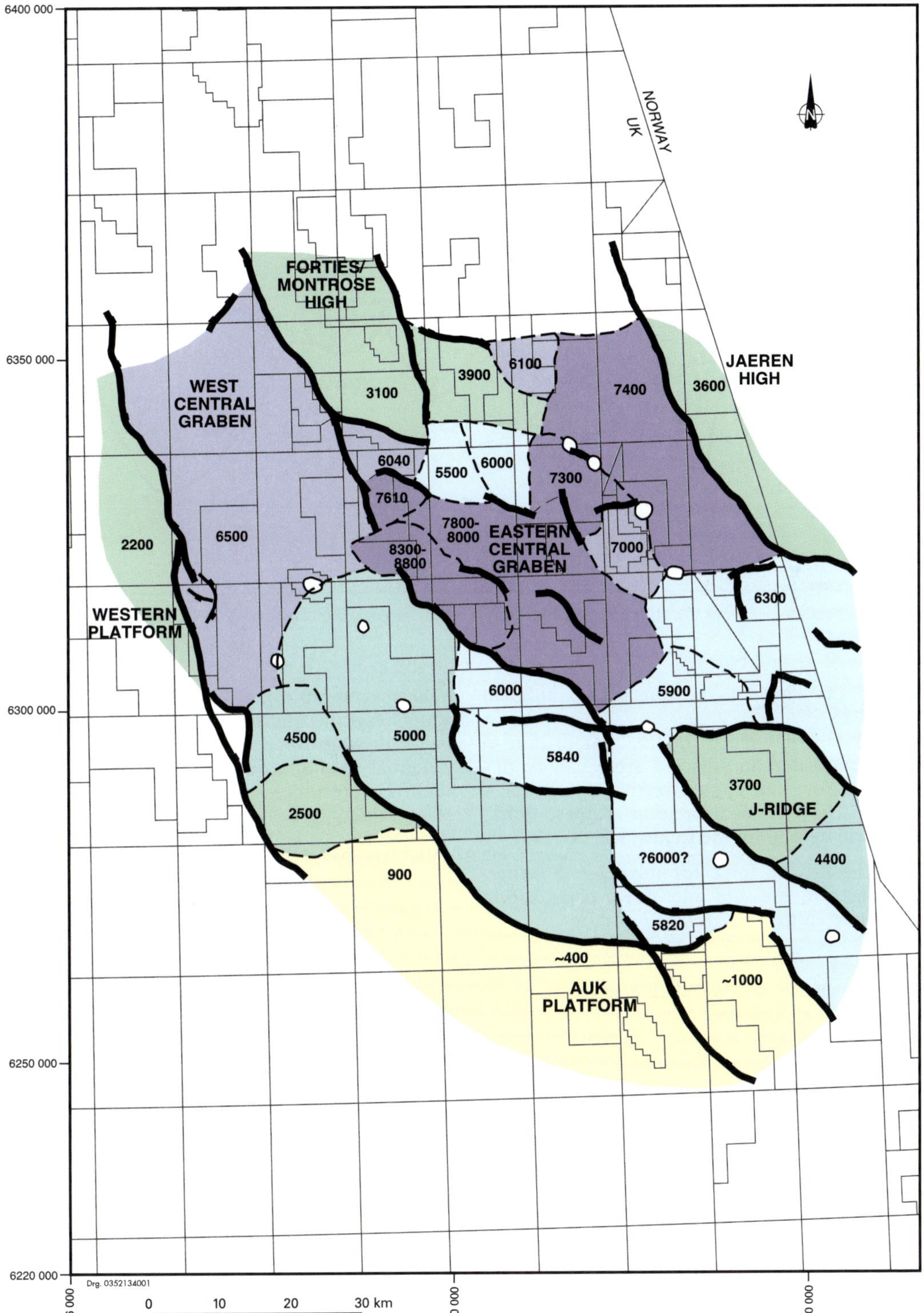


Fig. 8. Overpressure distribution map for the study area.

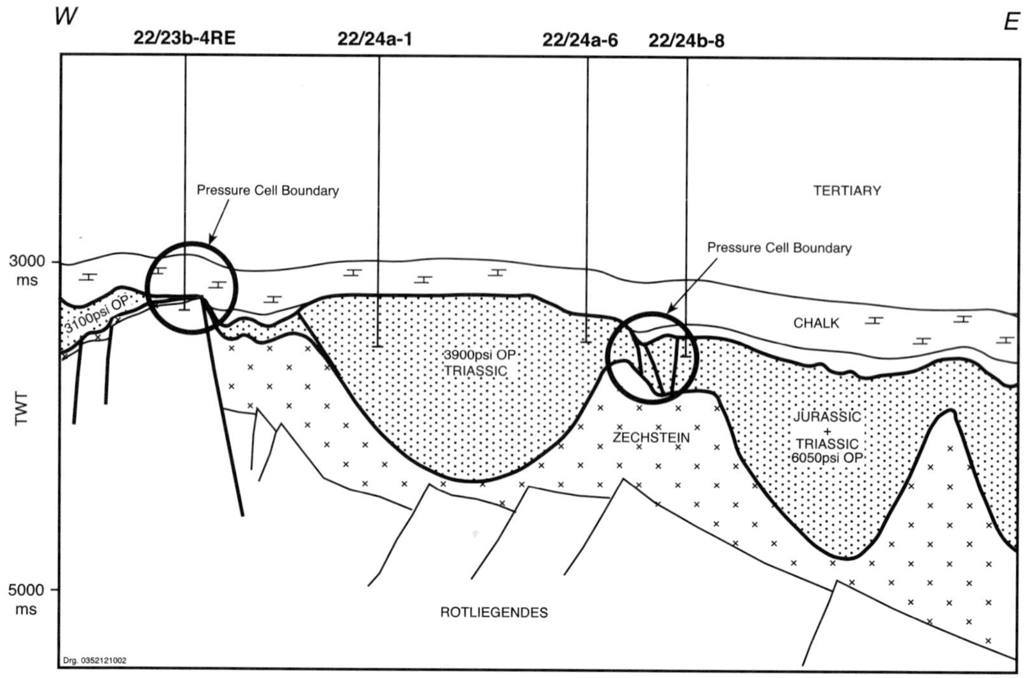


**Fig. 9.** Schematic detailing the characteristics of fluid barriers or pressure cell boundaries within the study area.

cause these waters to be directed to different localities.

Significant differences in the composition of formation waters in adjacent areas can be dissipated over time by advective or diffusive mixing. The rate of dissipation of these differences will depend on the following:

(a) The driving force for mixing (e.g. differences in pressure and water composition between the two areas).
(b) The permeability of the rock between the areas.
(c) The amount of fluid that needs to be transported (a function of porosity, formation thickness, and distance between the two areas).
(d) The influence of water–rock interactions. For example, where different water–rock interactions dominate the compositions of waters in each area, if the rates of these reactions are fast relative to the rate of mixing, the effects of mixing may be masked.

Significant differences in composition can be dissipated more quickly where the driving force for mixing is high, permeability between the areas is high, fluid volumes transported are low and water–rock interaction effects are negligible. Where compositionally different formation waters exist in adjacent areas, there will always be a driving force for mixing (i.e. there is potential for diffusion if not advection). In these circumstances, the compositional differences can continue to exist where:

(i) there is a barrier to flow (low permeability zone) between the areas that prevents mixing;
(ii) mixing is occurring but different water–rock interactions in each area mask the effects of mixing;
(iii) mixing is occurring but insufficient time has elapsed since the compositional differences were generated for the differences to be removed.

On a regional scale, pressure cell boundaries will prevent significant mixing because they have sufficiently low permeability to maintain pressure differences (Smalley *et al.* this volume). Therefore, where the location of pressure cell boundaries and changes in water composition coincide, this will be an argument in favour of (i) above. Where changes in water composition occur in locations where pressure cell boundaries are absent, explanations (ii) or (iii) would be favoured. These conclusions refer to current conditions and past conditions at these locations should also be considered. Evidence for the past behaviour of current barriers might be derived from formation water composition data across the barrier. Where mixing has not occurred in the past, a step-change in composition would be expected whereas evidence of gradational changes would be expected if mixing had occurred (i.e. the barrier leaked in the past).

### *Relationships between pressure cell boundaries, water type and salinity*

Comparison of the location of pressure cell boundaries, changes in formation water type and significant changes in salinity revealed five different relationships summarized in Table 5. Examples of where these relationships can be observed are shown in Figure 10.

There are several locations where pressure cell boundaries and changes in formation water type coincide. For example, separating Mallard, Guillemot A and Bittern from Teal and Curlew B, separating Puffin from Puffin SE, Molly, Franklin, Kessog and Structure North of Acorn, and separating Juno, Martha and Commander from Seagull and Heron (see Fig. 10, locations marked 'A'). As discussed above, mixing of the water types at these locations will not be significant at present. There is also some evidence to suggest that these barriers to flow have been active and have maintained the differences in water type for a significant time. Water in the hydrocarbon leg is likely to represent formation water trapped at the time of hydrocarbon emplacement (e.g. Rosenbaum & Coleman 1998; Worden *et al.* 1998; Mearns & McBride 1999; Heasley *et al.* 2000; Marchand *et al.* 2000; Ziegler *et al.* 2001). In areas where water types 4, 5, 6 and 8 are present, in all but one case the same water type was observed in both the water and hydrocarbon legs of wells. The implication is that, in these areas at least, the chemical characteristics of the formation waters have not changed significantly between the start of hydrocarbon emplacement and the present day. If these barriers to flow maintain the separation of different water types, a further implication is that these barriers could have been active since at least the onset of hydrocarbon emplacement and probably earlier. The exceptional case identified above is Puffin 29/5a-3 ERSA data. In this case, water and hydrocarbon leg data are not of the same water type. This may be related to the use of ERSA data or natural causes might explain this anomaly (e.g. influx of a different water type after hydrocarbon emplacement).

Current models of overpressure for the Central Graben include a static model where pressure cell boundaries are maintained without leakage (Hunt 1990). A dynamic model has also been proposed (e.g. Darby *et al.* 1996; Holm 1998) whereby fluids intermittently flow (leak) across pressure cell boundaries over time. With no leakage, we would expect to see a step-change in water type across the barrier. If leakage is not currently occurring but has in the past, we would probably see a step-change in composition at the barrier with gradual changes in composition and water type either side of the barrier due to past mixing events. Based on the locations of fields on either side of the barriers in the examples above ('A' in Fig. 10), the change in water type occurs over distances of less than 4–7.5 km (i.e. approximately 2–3.75 km either side of the barrier). Evidently, with no data within this 4–7.5 km wide zone, either model may be valid in these locations. Smalley *et al.* (this volume) report initial attempts to quantify homogenization times of chemical differences by diffusion over different distances. Similar calculations might be undertaken in future for conditions more appropriate for these barriers to determine the maximum cumulative time that they have leaked in the past. Refinement of these leakage times and use of formation water composition data to distinguish between the models of overpressure will require acquisition of more closely spaced data across the barriers.

**Table 5.** *Relationships between locations of pressure cell boundary, changes in water type and significant changes in salinity*

| Relationship | Presence of pressure cell boundary? | Change in water type? | Significant change in salinity present? |
|---|---|---|---|
| A | Yes | Yes | – |
| B | Yes | No | Yes |
| C | Yes | No | No |
| D | No | No | Yes |
| E | No | Yes | – |

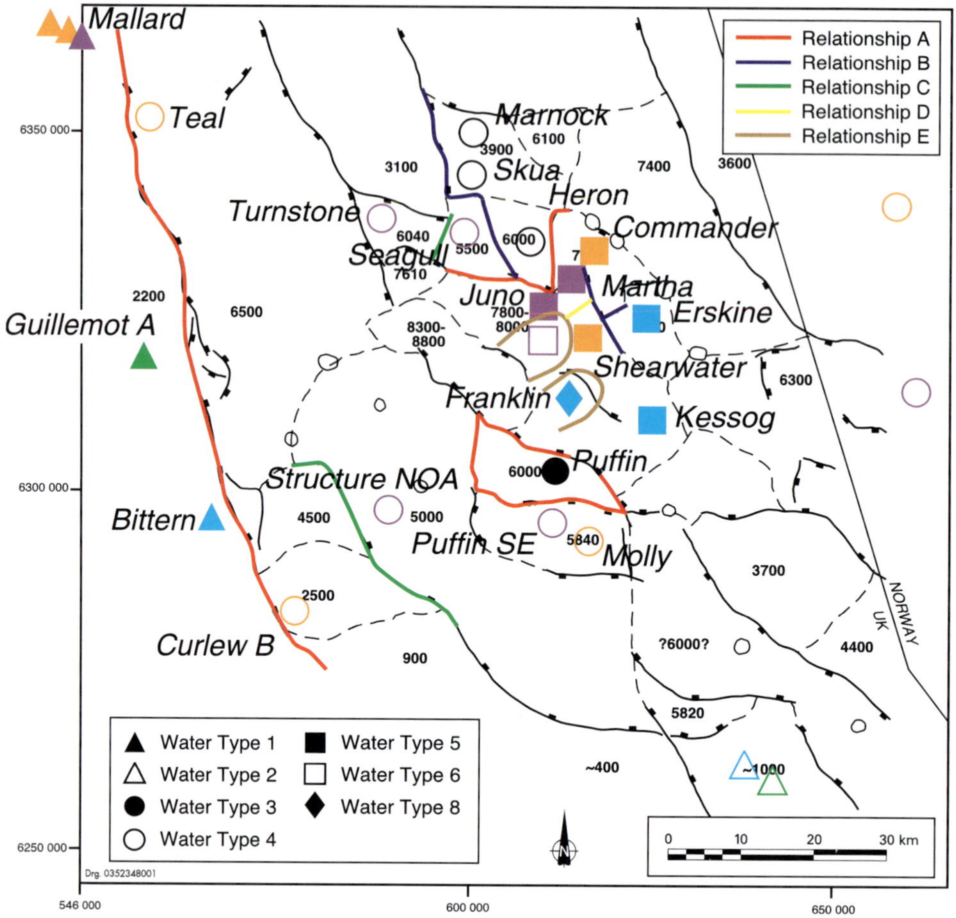


**Fig. 10.** Examples of relationships between pressure cell boundaries, changes in water type and significant changes in salinity (see Table 5). Symbol colours depict formation water Cl concentrations (see Fig. 7 for legend).

There are also locations where pressure cell boundaries coincide with significant changes in formation water salinity but not with a change in water type. For example, between Heron–Skua and Seagull–Turnstone, between Erskine and Shearwater–Commander–Martha, and between Shearwater and Commander (see Fig. 10, locations marked 'B'). As discussed above, these barriers will currently be preventing mixing but based on the salinity data there are uncertainties about their sealing ability in the past. This is because the significant changes in salinity occur over distances of up to 3.0–13 km and we do not know whether these salinity changes are step-changes or gradational changes in this zone.

Examples of locations where pressure cell boundaries are present but are not associated with either a change in water type or significant change in salinity include (locations identified 'C' in Fig. 10) those cell boundaries between Turnstone and Seagull and between Curlew B and Structure North of Acorn. Although these barriers are currently tight, there are several possible reasons why the boundary may not have apparently affected the compositions of the formation waters as follows:

(1) The barrier has been very leaky in the past (these fields are 10–20 km apart).
(2) Water–rock interactions are dominant in these areas resulting in similar formation water compositions on either side of the boundary. In this case, the barrier may or may not have been active for significant periods in the past.
(3) If water type 4 flowed into these areas at some time in the past, the barrier was either

not active or, if it was, it failed to prevent the water flowing either side of the barrier.

In some locations, there are significant changes in salinity from one area to another but no change in water type occurs and no pressure cell boundary is present (location identified 'D', Fig. 10, between Shearwater and Martha). This suggests that mixing is probably occurring at present across this 9 km zone. Whatever the processes that originally generated the salinity contrast, its continued presence might be explained by there being insufficient time for mixing to dissipate the salinity contrast since the contrast was generated. For example, the rate of mixing may be low due to the low pressure differences across this zone or its low permeability. Equally, a barrier may have been active in this zone in the more recent past. Finally, the dominant flow direction either side of the zone might be vertical, allowing only lateral diffusional mixing. It is less likely that the salinity contrast is being maintained by different water–rock interactions either side of the zone (for example, dissolution of halite on the Martha side of the zone but not on the Shearwater side) because the water type does not change. On the other hand, each of these explanations can account for the relationship observed at those locations (identified by 'E' in Figure 10) where there is a change in water type but no pressure cell boundary is present. Examples include the areas between Elgin and Juno–Martha–Shearwater and between Franklin and Shearwater–Kessog.

### *Salinity variations within areas occupied by individual water types*

Related to the above discussion is the occurrence of salinity variations observed in those areas occupied by individual water types. For example, the western and northwestern area adjacent to the Central Graben occupied by water type 1, and much of the Central Graben occupied by water types 4 and 5. The existence of salinity variations in these areas may reflect the influence of barriers to flow and time constraints on mixing between lower and higher salinity waters. Barriers to flow appear to have had less influence and mixing appears to have been more complete in those areas occupied by water types 2 (30/6 graben) and 3 (Puffin field) where little variation in salinity is observed. As most of the salinity data is obtained from water leg locations, it is not possible to constrain when mixing in these areas has occurred. It may have occurred at any time in the past. In the case of Heron 22/29-5RE, however, Cl gradually increases with depth in the hydrocarbon leg (31–235 g/l Cl). This suggests that, in this area, mixing appears to have been occurring during or prior to hydrocarbon emplacement.

### *Nature of the Mesozoic aquifer*

In areas where water types 3, 4 and 5 are present (covering much of the Central Graben), the same water types were observed in combinations of formations penetrated in various wells (e.g. Skagerrak and Pentland, Fulmar and Skagerrak, and Pentland and Fulmar respectively, see Table 3). Given the wide variety of conditions that exist in this area, it is unlikely that the same water type could be generated in adjacent formations by similar water–rock interactions occurring in each formation. More likely is that, in these areas at least, the Mesozoic formations have acted as a single aquifer and there is no evidence that stratigraphic or intra-formational boundaries in these areas have acted as regional barriers to flow. This is in agreement with the formation pressure data.

## Conclusions

This exploratory study has shown the potential for formation water composition data, obtained for a variety of other purposes, to be integrated with formation pressures and regional geological models to better characterize deep basin compartments and their associated boundaries. A unique benefit of using formation water composition data from water and hydrocarbon legs is that these can also provide information on past fluid flow across boundaries and the period of effectiveness of a barrier to flow (i.e. during and/or since hydrocarbon emplacement). The timing of formation of these barriers records the compartmentalization of the deeper section of the Central Graben and has a significant impact on models of hydrocarbon migration in the area and in predicting the likely hydrocarbon composition of undrilled prospects or field extensions. By characterizing the aquifer chemistry of various basin compartments, it has been possible to show that the Jurassic and Triassic reservoirs are essentially part of a single Mesozoic aquifer consistent with observations from formation pressure data. This has implications with considering the possibility of deeper hydrocarbon accumulations under existing fields and untested prospects, and also in well design. It would be useful to apply the numerical techniques outlined by Smalley *et al.* (this volume) to better quantify past fluid flow across the pressure cell boundaries and to apply the techniques used in this study to other similar datasets to (a) better calibrate this integrated approach and (b) increase the understanding of the palaeohydrodynamics of hydrocarbons provinces.

The authors would like to thank Shell UK Exploration and Production Limited and ExxonMobil International for permission to publish this work, Euan Mearns and John McBride (ISOtopic Limited) for providing the majority of the ERSA data used in this study, and Total Fina Elf and partners in the Elgin/Franklin development for permission to use their data from Elgin and Franklin. We are grateful for the careful and insightful reviews provided by Craig Smalley and Harald Johansen which resulted in significant improvements to this paper.

## Appendix 1. ERSA sampling technique

ERSA samples were obtained from the core of 21 wells drilled using oil-based mud. Pieces from the centre of the core were dried and weighed, before being disaggregated in an agate mortar. At this point, any pre-existing formation water in the core was assumed to have evaporated and deposited evaporative salts. For Elgin and Franklin well samples 200 ml pure water leachate was then added and mixed thoroughly with the core pieces at room temperature for 1 min to dissolve these salts and then removed for analysis. For samples from the remaining wells, 30 ml pure water leachate was used with a leaching time of between 10 and 20 min and the leachate was filtered through 0.2 μm pore size cellulose nitrate membranes before analysis. ERSA core samples were taken from both the water and hydrocarbon legs with a sampling density of one sample per few feet or tens of feet of core. Blanks were run with each batch of samples. The leachate was analysed using standard methods.

## Appendix 2. Principal component analysis and cluster analysis

To assess the robustness of any conclusions drawn from the Principal Components Analysis (PCA) results, four sets of calculations were undertaken using different types of data and different calculation methods:

(a) water and CERSA sample concentration data (Na, K, Mg, Ca, Sr, Ba, Cl) and the varimax normalized rotated method;

(b) water and CERSA sample concentration data (Na, K, Mg, Ca, Sr, Ba, Cl) and the rotated method;

(c) water, CERSA and ERSA sample mole ratio data (Na/Cl, K/Cl, Ca/Cl, Mg/Cl, Sr/Cl, Ba/Cl, Na/Ca) and the varimax normalized rotated method;

(d) water, CERSA and ERSA sample mole ratio data (Na/Cl, K/Cl, Ca/Cl, Mg/Cl, Sr/Cl, Ba/Cl, Na/Ca) and the rotated method.

For the water sample data only, factor scores for the first six principal components (PC 1-6, accounting for >96% of the data variance) obtained from PCA calculations were standardized and then subjected to cluster analysis (CA; single linkage, Euclidean distances). The CA results were reviewed and linkage distances selected such that wells with similar factor scores were grouped together. Wells with similar factor scores have formation waters with the same 'water type'. That is, these wells have formation waters that have similar ratios of dissolved ions although concentrations of these ions may vary. Table 3 shows the wells and water types (1–8) identified using water sample data. Slightly different results were obtained

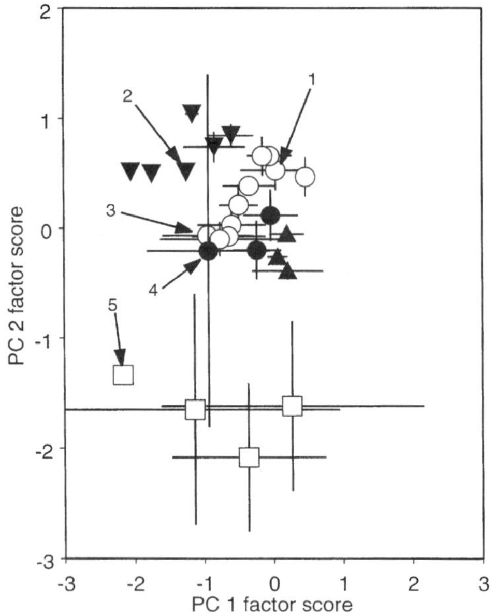


**Fig. 11.** Mean PC 1 and PC 2 factor scores for ERSA/CERSA data and the resulting water types (▲, A; ◆, B; ○, C; ●, D; □, E). Error bars represent 1 standard deviation about the mean. Inconsistent wells marked with arrows (1, Gadwall 21/19-6; 2, Puffin 29/4a-1; 3, Shearwater 22/30b-11; 4, Elgin 22/30c-8; 5, Puffin 29/5a-3; water-leg).

from the use of the different water sample data and PCA methods, so that in some cases there was some uncertainty in the water type of wells. Franklin well could belong to water type 5. Erskine well low salinity water may represent a unique water type in the area. Teal 21/25-8 and Juno 22/29-6s2 may or may not belong to water types 4 and 5 respectively. If they do not, the results did not indicate that they should be placed in any other water type.

Due to the variability of the ERSA and CERSA data, CA was not applied to the PCA factor scores for these data. Instead, means and standard deviations for the factor scores of PC 1 and PC 2 for each well were calculated. PC 1 and PC 2 accounted for 70–75% of the data variance. The means and standard deviations about the means were plotted (see Fig. 11) and the wells were visually grouped into water types A to E allowing for the uncertainties about the mean values (see Table 6).

Comparing Table 6 and the results of the water sample data (Table 3), it can be seen that, where water and ERSA/CERSA data are available for individual wells, there is a reasonable correlation between wells allocated to water types based on (a) water sample data and (b) CERSA and ERSA sample data. For example, those wells with formation water of water type A, B and C based on ERSA and CERSA sample data also have formation waters of water type 3, 4 and 5 respectively based on water sample data. That Elgin and Franklin wells have been allocated to their own water types based on ERSA and CERSA sample data is consistent with the water sample data results implying that water types D and E may correlate with water types 6 and 8. There is, however, some uncertainty over the inclusion of Elgin 22/30c-8 in water type D due to the larger variability of these data. The inclusion of Martha 22/30a-1, Heron 22/29-5RE, Commander 22/30a-16 and Shearwater 22/30b-A2 in water type C is consistent with the geographical distribution of water type 5 wells (see main text). There are some discrepancies to the water type correlation:

(a) the inclusion of Puffin 29/5a-3 (water leg) in water type E;
(b) the inclusion of Shearwater 22/30b-11 in water type C;
(c) the geographical location of Gadwall 21/19-6 and Puffin 29/4a-1 relative to other wells in with water type B and C respectively.

**Table 6.** *Water types derived from visual grouping of ERSA/CERSA PCA PC 1 and PC 2 factor scores*

| Water type | Wells | Water type from water sample data |
|---|---|---|
| A | Puffin 29/5a-3 (G) | 3 |
| | Puffin 29/5a-5 (G) | 3 |
| | Puffin 29/10-3 (G) | |
| B | Heron 22/30a-2 (W) | 4 |
| | Heron 22/30a-6 (W) | 4 |
| | Skua 22/24b-9 (O, W) | 4 |
| | Skua 22/24b-7 (O) | |
| | Puffin 29/4a-1 (?) | |
| C | Shearwater 22/30b-15S1 (G, W, S, P) | 5 |
| | Shearwater 22/30b-A6 (G) | 5 |
| | Juno 22/29-6S2 (W) | 5 |
| | Shearwater 22/30b-11 (W) | 7 |
| | Commander 22/30a-16 (W) | |
| | Shearwater 22/30b-A2 (G) | |
| | Martha 22/30a-1 (W) | |
| | Gadwall 21/19-6 (O, F, S) | |
| | Heron 22/29-5RE (O) | |
| D | Elgin 22/30c-8 (G) | |
| | Elgin 22/30c-13 (G, W) | |
| E | Franklin 29/5b-8 (G, W) | |
| | Franklin 29/5b-6,6Z (G) | |
| | Puffin 29/5a-3 (W) | 3 |

O, oil leg; G, gas leg; W, water leg; ?, sample location not known; F, Fulmar; S, Skagerrak; P, Pentland.

Given these results, the overall impression is that the groupings based on water samples and ERSA/CERSA samples are similar. The causes of the discrepancies noted above are not known. They may be the result of only using PC 1 and PC 2 factor scores to group the data as these account for only approximately 70–75% of the data variance. Alternatively they may reflect our visual interpretation of the membership of the associations (e.g. Shearwater 22/30b-11) or they may represent individual cases of large bias between the different types of sample data. Equally, they may have physical explanations. For example, the formation water in Puffin 29/4a-1 may have a common origin to that observed in Skua Field and parts of Heron Field.

Based on the above results, for clarity only the water types identified from the water sample data (i.e. 1–8) are discussed in the main text. Other than for the inconsistencies noted above, it is assumed that wells with formation water of water type A, B, C, D and E actually have formation water of water type 3, 4, 5, 6 and 8 respectively. It is also worth noting that due to the reasonable correlation between water types based on water sample and ERSA and CERSA sample data, the uncertainties associated with the latter data (see main text) appear not to be significant for most wells in terms of this application.

## References

BJØRLYKKE, K. & GRAN, K. 1994. Salinity variations in North Sea formation waters: implications for large-scale fluid movements. *Marine and Petroleum Geology*, **11**, 5–9.

CAI, C., FRANKS, S. G. & AAGAARD, P. 2001. Origin and migration of brines from Palaeozoic strata in Central Tarim, China: constraints from $^{87}Sr/^{86}Sr$, $\delta^2H$, $\delta^{18}O$ and water chemistry. *Applied Geochemistry*, **16**, 1269–1284.

CAYLEY, G. T. 1987. Hydrocarbon migration in the Central North Sea. *In*: BROOKS, K. & GLENNIE, J. (eds) *Petroleum Geology of Northwest Europe*. **1**, Geological Society, London, 549–555.

CORNFORD, C. 1994. Mandal–Ekofisk(!) petroleum system in the Central Graben of the North Sea. *In*: MAGOON, L. B. & DOW, W. G. (eds) *The Petroleum System—from Source to Trap*. American Association of Petroleum Geologists Memoir, **60**, 537–571.

DARBY, D., HASZELDINE, R. S. & COUPLES, G. D. 1996. Pressure cells and pressure seals in the UK Central Graben. *Marine and Petroleum Geology*, **13**, 865–878.

EGEBERG, P. K. & AAGAARD, P. 1989. Origin and evolution of formation waters from oil fields on the Norwegian Shelf. *Applied Geochemistry*, **4**, 131–142.

FUNAYAMA, M. & HANOR, J. S. 1995. Pore water salinity as a tool for evaluating reservoir continuity and fluid migration pathways in the Wilcox Group of Central Louisiana. *Transactions of the Gulf Coast Association of Geological Societies*, **45** 195–201.

GAARENSTROOM, L., TROMP, R. A. J., DE JONG, M. C. & BRANDENBURG, A. M. 1993. Overpressures in the North Sea: implications for trap integrity. *In*: PARKER, J. R. (ed.) *Petroleum Geology of Northwest Europe*. Proceedings of the 4th Conference, Geological Society of London, 1305–1314.

GLENNIE, K. W. 1990. *Introduction to Petroleum Geology of the North Sea*. Blackwell Scientific Publications.

HANOR, J. S. & SASSEN, R. 1990. Evidence for large-scale vertical and lateral migration of formation waters, dissolved salt, and crude oil in the Louisiana Gulf Coast. *In*: SCHUMACHER, D. & PERKINS, B. F. (eds) *Gulf Coast Oil and Gases: Their Characteristics, Origin, Distribution, and Exploration and Production Significance*. Proceedings of the 9th Annual Research Conference of the Gulf Coast Section of the SEPM Foundation, 283–296.

HEASLEY, E. C., WORDEN, R. H. & HENDRY, J. P. 2000. Cement distribution in a carbonate reservoir: recognition of a palaeo oil–water contact and its relationship to reservoir quality in the Humbly Grove field, onshore, UK. *Marine and Petroleum Geology*, **17**, 639–654.

HOLM, G. M. 1998. Distribution and origin of overpressure in the Central Graben of the North Sea. *In*: LAW, B. E., ULMISHREK, G. F. & SLAVIN, V. I. (eds) *Abnormal Pressures in Hydrocarbon Environments*. American Association of Petroleum Geologists Memoir, **70**, 123–144.

HUNT, J. M. 1990. Generation and migration from abnormally pressured fluid compartments. *American Association of Petroleum Geologists Bulletin*, **74**, 1–12.

JORDAN, M. M., SJURSAETHER, K., EDGERTON, M. C. & BRUCE, R. 2000. Inhibition of lead and zinc sulphide scale deposits formed during production from high temperature oil and condensate reservoirs. Paper SPE **64427**, presented at *SPE Asia Pacific Oil and Gas Conf.*, Brisbane, Australia, 16–18 October.

MARCHAND, A. M. E., HAZELDINE, R. S., MACAULAY, C. I., SWENNEN, R. & FALLICK, A. E. 2000. Quartz cementation inhibited by crestal oil charge: Miller deep water sandstone, UK North Sea. *Clay Minerals*, **35**, 201–210.

MEARNS, E. W. & MCBRIDE, J. J. 1999. Hydrocarbon filling history and reservoir continuity of oil fields evaluated using $^{87}Sr/^{86}Sr$ isotope ratio variations in formation water, with examples from the North Sea. *Petroleum Geoscience*, **5**, 17–27.

NIKIEL, A. M. & HANOR, J. S. 1999. Spatial variations in formation water salinities, South Pelto and South Timbalier areas, eastern Louisiana continental shelf. *Transactions of the Gulf Coast Association of Geological Societies*, **49**, 396–403.

ROSENBAUM, J. M. & COLEMAN, M. L. 1998. Complexity in the North Sea Basin: Trace waters in oilfields in the Central North Sea. *Mineralogical Magazine*, **62A**, 1291–1292.

SMALLEY, P. C. & ENGLAND, W. A. 1994. Reservoir compartmentalization assessed with fluid compositional data. *SPE Reservoir Engineering*, **August 1994**, 175–180.

SMALLEY, P. C., LONOY, A. & RHEIM, A. 1992. Spatial $^{87}Sr/^{86}Sr$ variations in formation water and calcite from the Ekofisk chalk oil field; implications for reservoir connectivity and fluid composition. *Applied Geochemistry*, **7**, 341–350.

SMALLEY, P. C., ENGLAND, W. A., MUGGERIDGE, A. H., ABACIOGLU, Y. & CAWLEY, S. J. (this volume). Rates of reservoir fluid mixing: Implications for interpretation of fluid data.

SWARBRICK, R. E. *ET AL.* 2000. Integrated study of the Judy Field (Block 30/7a)—an overpressured Central North Sea oil/gas field. *Marine and Petroleum Geology*, **17**, 993–1010.

WARREN, E. A. & SMALLEY, P. C. 1994. *North Sea Formation Water Atlas*. Geological Society, London, Memoirs, **15**.

WORDEN, R. H., OXTOBY, N. H. & SMALLEY, P. C. 1998. Can oil emplacement prevent quartz cementation in sandstones? *Petroleum Geoscience*, **4**, 129–137.

ZIEGLER, K., COLEMAN, M. L. & HOWARTH, R. J. 2001. Palaeohydrodynamics of fluids in the Brent Group (Oseberg Field, Norwegian North Sea) from chemical and isotopic compositions of formation waters. *Applied Geochemistry*, **16**, 609–632.

# Petroleum migration, faults and overpressure. Part II. Case history: The Haltenbanken Petroleum Province, offshore Norway

DAG A. KARLSEN,[1] JON ERIK SKEIE,[1] KRISTIAN BACKER-OWE,[1] KNUT BJØRLYKKE,[1] RICHARD OLSTAD,[2] KARI BERGE,[3] MARCELLO CECCHI,[3] EIRIK VIK[4] & RAINER G. SCHAEFER[5]

[1]*Petroleum Geochemistry Program, Department of Geology P.O. Box 1047, Blindern, N-0316 Oslo 3, Norway*

[2]*ChevronTexaco Norway AS, Karenslyst Alle 2, P.O. Box 97, N-0212 Oslo, Norway*

[3]*Enterprise Oil Norge Ltd., P.O. Box 399, N-4001 Stavanger, Norway, now at A/S Norske Shell, P.O. Box 40, N-4098 Tananger, Norway*

[4]*Statoil Research Centre, Trondheim, N-7005 Trondheim, Norway*

[5]*Institute of Petroleum and Organic Geochemistry, Forschungszentrum Jülich D-52425 Jülich, Germany*

**Abstract:** Petroleum inclusion and geochemical data from core extracts were applied to deduce a model for oil migration, overpressure development and palaeo-leakage of oil from currently dry structures in the Haltenbanken Vest area. The existence of fluorescent oil type inclusions in quartz in the Smørbukk (Åsgard-2) field suggest that oil migrated into this structure 70–50 million years before present (Ma bp). This is also the case for the dry structures 6506/12-4, 6506/11-3 and 6506/11-1, west of the main Smørbukk Fault Zone. Black oil inclusions with medium gas/oil ratio (GOR) occur in these fields together with condensate-type petroleum inclusions. This suggests that the dry structures transformed from containing oil to condensate before leakage. Petroleum extracted from inclusions in these structures and in nearby fields have identical marine type II kerogen signatures. Source rocks at the Spekk Formation level in the current drainage area of Smørbukk and these dry structures, were immature 70–50 Ma bp and the Smørbukk Sør (Åsgard-3) field did not fill at this early time. Thus, oil must initially have entered into Smørbukk from areas to the W–SW, through the currently pressure sealing Smørbukk Fault Zone which today marks the westward limit of the Smørbukk field. Diagenesis in this fault zone caused the much later overpressure development and petroleum was lost from the 6506/12-4, 6506/11-3 and 6506/11-1 structures as overpressure built up regionally. Petroleum loss from these structures with their often thick seals must have occurred via self-propagating open-fracture-induced mechanisms. Lack of petroleum in the Cretaceous strata above these structures suggests that leakage occurred to even shallower strata. This could imply that the Cretaceous strata in Halten Vest were overpressured at the time of leakage. In contrast, the oil in the Cretaceous Lysing and Lange Formation (above the Jurassic reservoirs in Smørbukk and Smørbukk Sør) most likely originated (based on geochemistry and GORs) from the Jurassic reservoirs below and not from Cretaceous strata. This migration event would have been facilitated if it occurred before these sands became overpressured as they are today. Modelling suggests that the Spekk Formation became mature in the Smørbukk Sør region <10 Ma bp and microthermometry of oil inclusions from Smørbukk Sør supports filling during the past 10 Ma. This implies that caprock failure in the Halten Vest structures 6506/12-4, 6506/11-1 and 6506/11-3 most likely occurred after filling of the Smørbukk Sør and 6406/3-1 structures. Rapid regional burial during the past 10 Ma caused local migration of oil into Smørbukk Sør, Smørbukk and 6406/3-1 structures, and generation of high GOR oils in the deeper Halten Vest region. High GOR petroleum inclusions in the Halten Vest structures signify this event and suggests that caprock fracturing occurred after a gas-condensate had replaced oil in these traps. Rapid burial during the past 3 Ma is likely to have caused the current overpressure and associated leakage in Halten Vest. The fact that these traps did not later refill in this progressively subsiding and maturing basin must be related to trap pressures remaining too close to the actual fracture pressures.

## Aims and structure of the paper

The prime message contained in this paper concerns the use of organic geochemistry from traps in the form of hydrocarbon (HC) gas, bitumen and oils to deduce palaeo- and more recent movement of petroleum into traps, and by using petroleum inclusions, to 'time' these events of

*From*: CUBITT, J. M., ENGLAND, W. A. & LARTER, S. (eds) 2004. *Understanding Petroleum Reservoirs: towards an Integrated Reservoir Engineering and Geochemical Approach.* Geological Society, London, Special Publications, **237**, 305–372. 0305-8719/04/$15.00 

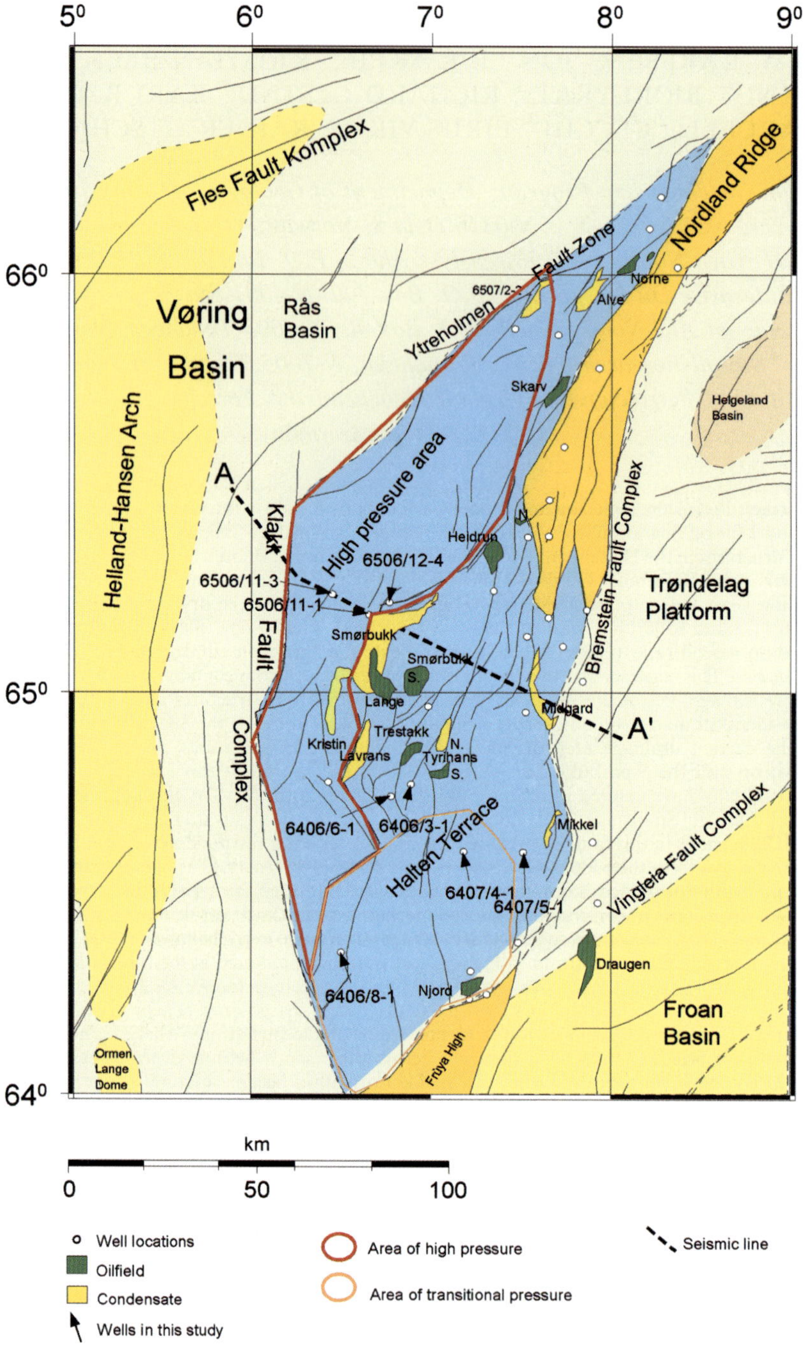


**Fig. 1.** (**a**) The overpressure in the Haltenbanken Area is generally confined to the more deeply buried and extensively faulted regions to the west. Most oil fields occur along a SSW–NNE trend where the Jurassic type II marine Spekk Formation (Fig. 3) is in the oil window. Only the Draugen and the Midgard field have lateral migration distances of more than 15–20 km. Line A–A′ is the cross section in Figure 1b. The 'Halten Vest High Pressure Dry Hole Region' follows in a general sense the eastern red-line perimeter of the outlined 'Area of High Pressure'. Also cf. Figures 2, 5 and 27 for more details on wells and overpressure respectively. Figure 1a is based on a structural map from Blystad *et al.* (1995).

migration. We believe this method to be helpful in basin modelling efforts by constraining models for source rock (SR) maturation relative to expulsion and identification of the SR actually responsible for filling given traps. By doing so we have over the years build models for basin scale migration patters off Mid-Norway.

In this paper we are particularly focusing on the role of faults and how these may shift in time from generally open for petroleum migration to eventually becoming restraining features for fluid flow and their role in overpressure build up. The main fault, now separating the 'hydrostatic' Smørbukk trap from the overpressured 'Halten Vest' region, is discussed in detail as organic geochemical evidences show clearly that this fault was open to petroleum migration 70–50 Ma bp. We focus on when dysmigration took place from the so called 'Halten Vest Dry Hole Region' (cf. Fig. 1a) and present evidence to suggest that this region was until very recently a prolific petroleum province, as is the rest of the Haltenbanken Region.

'Paper I' in this series, published in *Journal of Petroleum Geology*, focuses more on the general principles deduced from the setting discussed in this 'Paper II' and consolidating work over *ca.* 13 years in this region. Thus, this current paper is in this sense a case history; however, we hope with in-depth discussions of general principles or relevance for exploration in this and similar settings.

The structure of this paper follows a set-up with a regional introduction to the study area and relevant published works, followed by a discussion of known characteristics of source rocks of the region, general principles of productivity, maturity and expulsion from these known source rocks in the region, a review of the published existing geochemistry on petroleum in the regional traps before we turn to the current data set, methods and results/discussion.

Emphasis is placed on understanding the dynamic nature of traps in subsiding multi-SR basins, and how we can use structural, organic geochemical and diagenetic mineral data points in an integrated fashion to better understand basin-scale petroleum migration patterns as these changed in response to progressive burial. The petroleum content of traps represents in our perspective vital information to the understanding, not only of petroleum families and SR relationships, but also to the understanding of fault zones as migrating petroleum represents the most mobile element in a generally static lithological/geological framework.

## Introduction

### *Study area, structural elements and geological evolution*

The Haltenbanken region offshore Mid-Norway is a highly faulted oblique-slip tectonically induced basin with a series of SSW–NNE trending main fault zones, separating platforms, terraces and basins (cf. Bøen *et al.* 1984; Bugge *et al.* 1984; Gabrielsen *et al.* 1984; Gowers & Lunde 1984; Hollander 1984; Bukovics & Ziegler 1985; Blystad *et al.* 1995; Gabrielsen & Doré 1995). From a generally compressional stress regime culminating with the Caledonian Oregony, extension prevailed from late Devonian to Eocene, after which a weak compression regime took over (cf. Blystad *et al.* 1995).

The central part of Haltenbanken, the Halten Terrace (Figs 1a,b & 2), is separated from the Trøndelag Platform to the east by the Kristiansund–Bodø Fault Zone. To the west, the West Haltenbanken High separates the Halten Terrace from the deeper Møre-Vøring Basin (cf. Blystad *et al.* 1995). Recent explorational drilling in the deep-water regions to the west found dry biodegraded gas in the Nyk High, whilst the Vema Dome and Gjallar structures were dry. These structures occur tentatively 150 km (Nyk) to less than 30 km (Gjallar) away from the massive volcanic province, the Vøring Margin High in the west which is genetically related to the break-up of the North Atlantic and associated sea-floor spreading (cf. Blystad *et al.* 1995). It is potentially the case that sills associated with the magmatic activity have intruded and matured organic rich rocks at Mid-/Upper-Cretaceous and Tertiary levels in the vicinity of the Vøring Margin High. Such intrusive activity is, however, likely to yield only gas (Dahlgren *et al.* 1998).

The Norwegian Sea has become a highly prolific hydrocarbon province (Rønnevik 2000) following the initial discovery of condensate in well 6506/11-1 in 1981 (cf. Elvsborg *et al.* 1985; Mo *et al.* 1989), and total remaining and reported future resources in the whole Norwegian Sea may fall in the range of $6 \times 10^9$ Sm$^3$ oil as compared with $9 \times 10^9$ Sm$^3$ oil for the more heavily explored and mature North Sea (Norwegian Petroleum Directorate statistics).

The siliciclastic lithologies of the Jurassic Båt, Fangst and Viking groups make up the main reservoirs of the Halten Terrace, with some additional reserves discovered above Smørbukk and Smørbukk Sør in sands of the otherwise shaley Cretaceous Lysing and Lange formations (Fig. 3). A recent review of exploration in the Norwegian Sea by Rønnevik (2000) reports

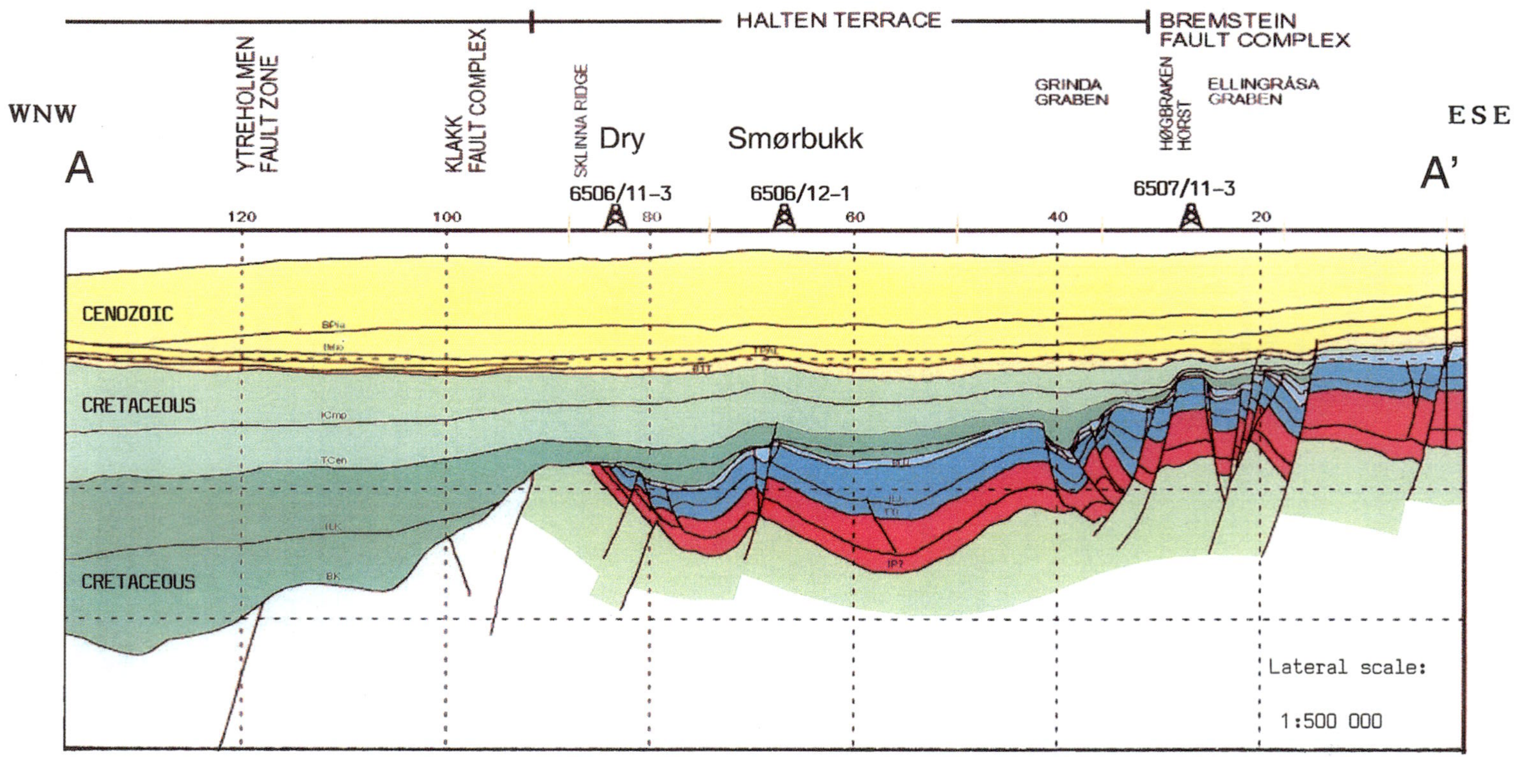


**Fig. 1.** (**b**) The large scale westward dipping and faulted configuration of the greater Haltenbanken Area controls hydrocarbon generation and flow properties in this basin. Lack of continuous strata in an updip eastward direction illustrates why overpressure may develop in e.g. dry structure 6506/11-3 in contrast to in the Smørbukk field (6506/12-1) which has pressure communicating via eastern updip strata. Jurassic strata are marked in blue and contain the main reservoirs in the region; Triassic strata are in red. Note that today Smørbukk is called Åsgard II as it has been developed together with Midgard (6507/11-3, the horst), now called Åsgard I, and with Smørbukk Sør, now called Åsgard III. Figure adapted from Blystad *et al.* (1995).

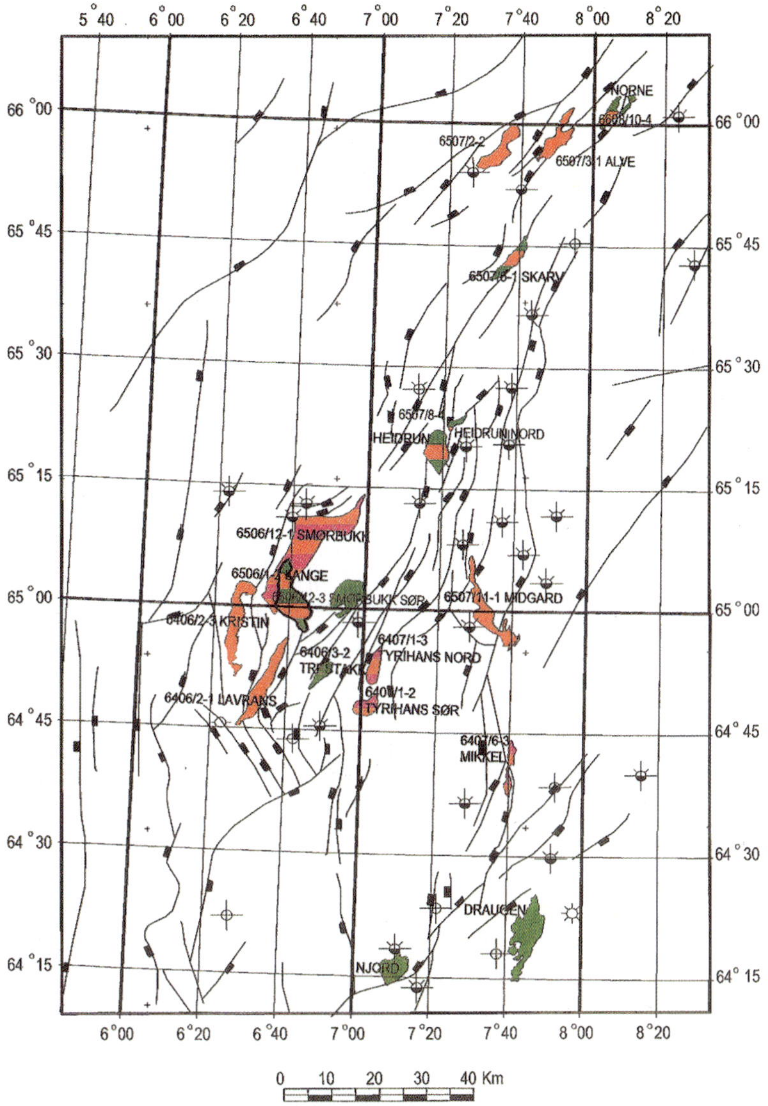


**Fig. 2.** Illustration of the general trend of the faults. Please note that some faults are only tentatively mapped. Courtesy of the Norwegian Petroleum Directorate (NPD).

that the main scientific achievement of exploration in this region was the invalidation of two paradigms—that commercial porosity did not occur below 4500 m, and that structures in the overpressured areas were incapable of holding hydrocarbons.

The overpressured Cretaceous and Lower Tertiary sequence constitutes the caprock over the main part of the Haltenbanken area (Vik *et al.* 1991). This rock sequence is believed to have developed overpressure by rapid burial during the extremely rapid Pliocene to Recent burial (Vik *et al.* 1991; Lilleng & Gundesø 1997). The Jurassic Smørbukk Sør (3160 m) and Smørbukk (3380 m) main reservoirs are both normally pressured, whilst there is an overpressure of *ca.* 140 bar at the Lysing level (Fig. 3). Above Smørbukk in the Lange Formation is there an overpressure of *ca.* 186 bar.

Undersaturated oil occurs in the Cretaceous strata both above Smørbukk and Smørbukk Sør, e.g. the saturation pressure in 6506/11-2 (Smørbukk) is 434 bar with a reservoir pressure of 630 bar.

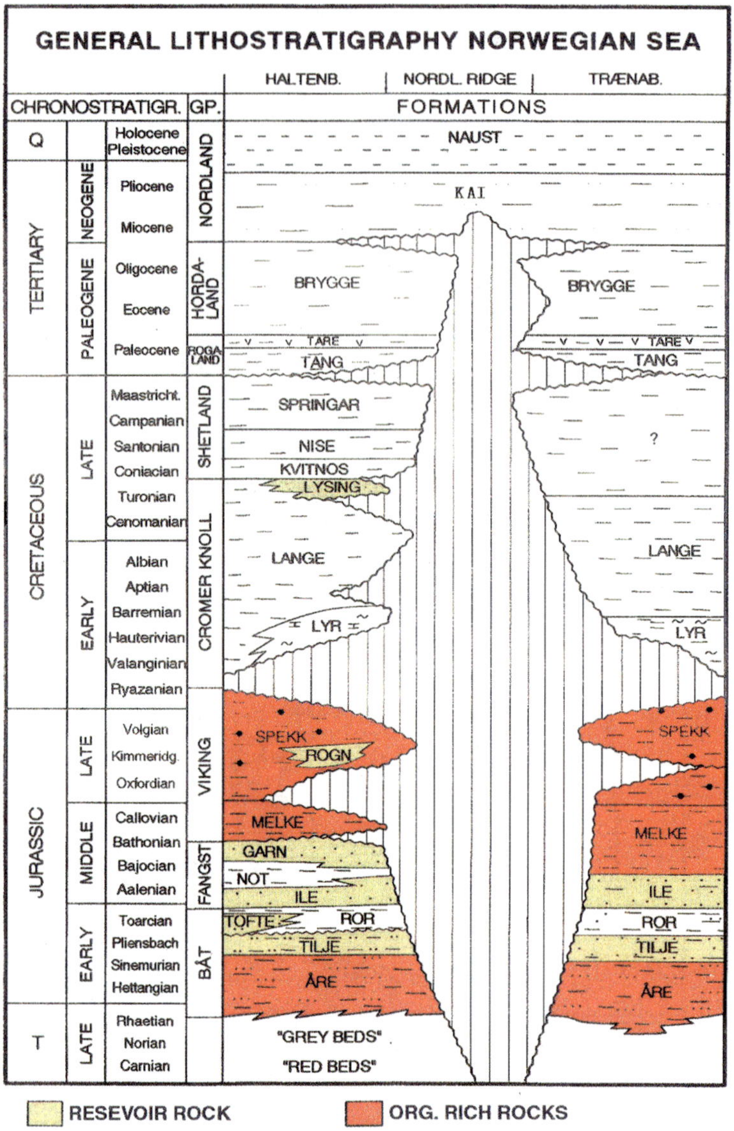


**Fig. 3.** Organic rich rocks in Haltenbanken include the coals and shales in the Åre Formation (Brent equivalent), the Melke (Heather equivalent) and the Spekk (Kimmeridge Clay equivalent) Formations. Petroleum in Haltenbanken is sourced mainly from the Spekk/(Melke) Formation. Petroleum from the Åre coals and shales may have increased hydrocarbon saturation in carrier rocks long before the onset of oil expulsion from the Spekk Formation.

The Haltenbanken region can tentatively be divided into three pressure regions (Fig. 1) with high pressures generally confined to the deeper western regions. Pressures here may approach and exceed estimated fracture pressures (Fig. 4). Most structures in the western high-pressure regions (Fig. 1a,b) are today dry (e.g. 6506/11-3, 6506/11-1, 6506/12-4, all three west of Smørbukk, and 6406/3-1 and 6406/6-1 farther south), but most of these, as discussed below, earlier held oil. The high pressure Kristin field with its condensate is an exception to the general Halten Vest 'dry hole scenario'. The nearby Lavrans field, occurring to the east across a major fault, is 'normally pressured' (Fig. 5).

The border zone between the overpressured regions in the west and the normally pressured regions in the east can be diffuse as in the

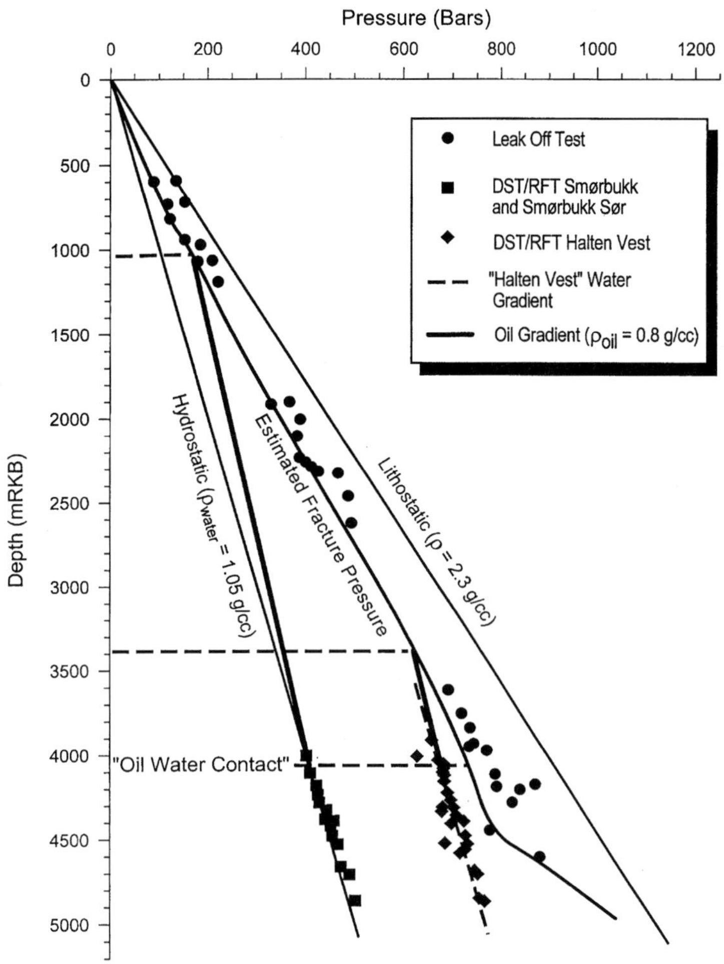


**Fig. 4.** Whilst the Smørbukk and Smørbukk Sør fields (Fig. 1a) today have pressures on the oil gradient just above the hydrostatic gradient, pressures in structures in Halten Vest are close to the estimated fracture pressures (from leak-off tests). Most wells in the overpressured region are dry, unless traps are sufficiently deep to avoid the fracture gradient, e.g. Kristin (Figs 1a & 5), and this region coincides roughly with the area where the Spekk Formation is in the very last part of the oil window, or already over-mature.

southern part of Haltenbanken (with arguably also less well control), or better defined as along the fault zone delimiting the Smørbukk field to the west (Figs 1a,b & 5). It is suggested from Figures 1a,b, 2 & 5 that the regional pressure cells follow the generally trending N–S faults systems, and we infer there to be a genetic linkage between the faults and overpressure development. Still, to further complicate matters, Jurassic reservoir strata in Smørbukk Sør are hydrostatic, whilst the shallower Cretaceous strata are overpressured, both holding black oil. We suggest that the structural configuration with generally updip draining areas in the east may help to establish pressure communication towards the east (Figs 1a,b, 2 & 5). In contrast, the presently dry and overpressured structures in the west, e.g. 6506/11-3 (Figs 1a, 1b & 5), apparently lack updip continuous pressure communicating strata towards the east.

According to the model illustrated in Figure 6, petroleum in deep overpressured reservoirs may be preserved in the traps provided that pressure communication exists in the updip direction (cf. Olstad *et al.* 1997). This ‘pressure-vent’ acts to keep the reservoir pressure of the deeper structures below the fracture pressure if leakage occur only from the shallowest structure in the pressure system.

Figure 7 depicts another situation which may allow oil to remain in the trap, provided that a thin, non-perfect, caprock ‘semi-continuously’

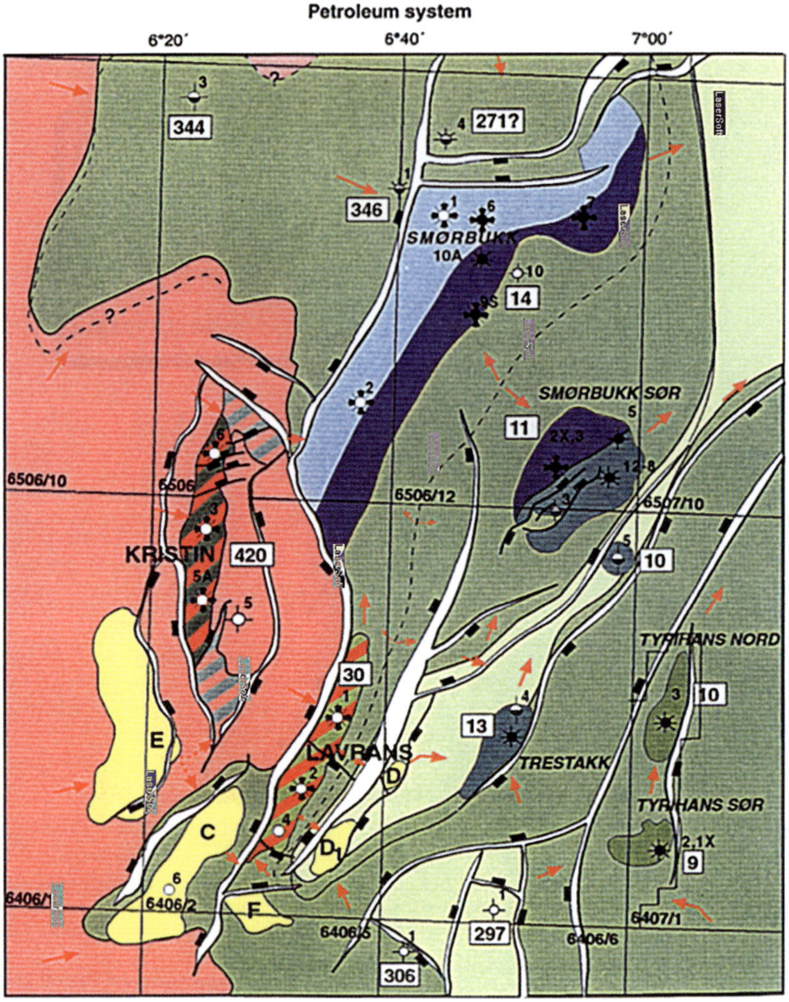


**Fig. 5.** Structural details and pressure domains of the Smørbukk region, reproduced from Rønnevik (2000) with courtesy of Elisabeth Holter of the Norwegian Petroleum Society (NPF). Arrows indicate filling directions and the green and red colours represent low/medium- and high-pressure domains respectively. Numbers in boxes represents overpressure in bar. Note the often sharp and dramatic changes in pressure over short lateral distances.

leaks preferentially water-soluble methane, ethane and propane through the caprock, possibly through the films of interconnected and continuous water (adsorbed water) on water wetting minerals constituting the caprock.

In this model the water film may actually transfer the pressure through the caprock whilst oil cannot move due to capillary restricting effects in the water wetting mineralogy.

We also assume, in a similar model, that pore-size related molecular sieving of methane and ethane with their identical and minute kinetic diameters may allow these petroleum constituents to move by diffusion through the caprock (Karlsen 1987; Karlsen *et al.* 1988).

However, larger alkanes with more than two carbon atoms which have kinetic diameters of more than 4.3 Å, cyclohexane with a kinetic diameter of 6.0 Å, and even larger and more viscous oil constituents like polyaromatic compounds, steranes, hopanes, resins and asphaltenes, are unable to move through the caprock, partly clogging the available effective permeability, and hence remain in the trap (cf. Karlsen 1987;

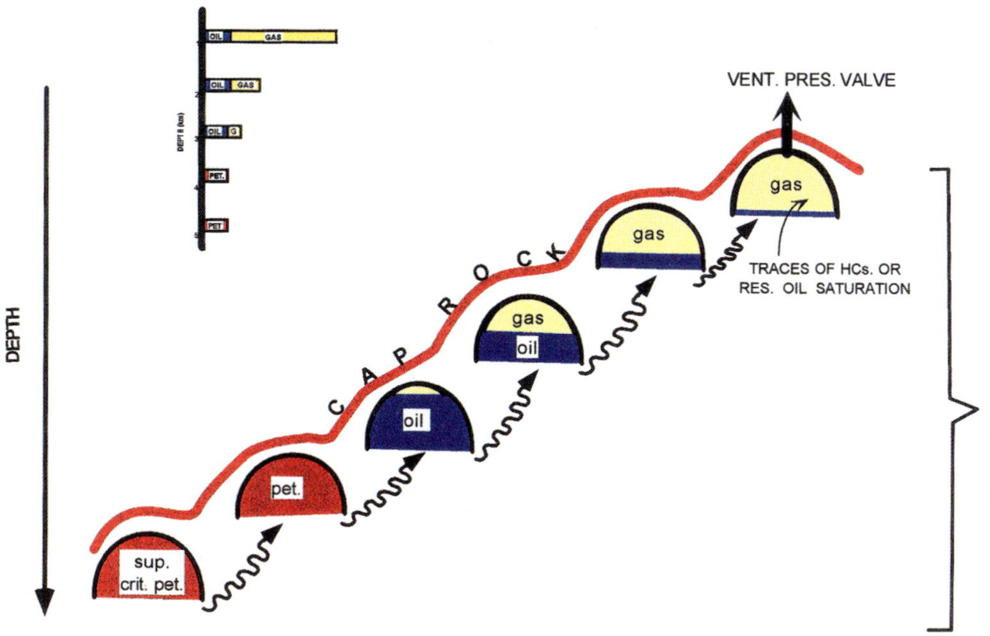


**Fig. 6.** Lack of pressure-leakage results in overpressure development in the terminal trap in this diagram. Leakage from the terminal structure acts to preserve petroleum in the deeper structures. Partly leaking thin caprocks, permeable for gas (dynamic trap), or superthick tight and strong caprocks in deep prospects (static trap) where advanced diagenesis has occurred, represent two other configurations able to retain petroleum for some time.

Karlsen *et al.* 1988; Karlsen and Larter 1989; Hodgson *et al.* 1992).

It is obvious from Figure 1b, that both the degree of faulting and the throw on the faults are similar in the western and eastern regions of Haltenbanken. Thus, it cannot be the lack of juxtapositioning of permeable strata across faults (Figs 1b, 8) or more extensive clay smear/ fault gauge formation *sensu stricto,* that results in reduced updip pressure communication in the strata in Halten Vest. However, burial is deeper in Halten Vest, suggesting that more extensive diagenesis in fault planes (see Fig. 8) may have resulted in pressure cells in Halten Vest by reducing the flux of penelateral compaction driven fluid flow.

In this model the reduced permeability at faults is caused by advanced quartz diagenesis which is exponential in temperature (cf. Fig. 9), leading to constrained fluid flow and thereby overpressure development. Traps which earlier contained oil will, in the normal case of continued communication with a subsiding basin, eventually receive high GOR petroleum in the form of a condensate, i.e. a gas phase, from the subsiding deep source rocks. The replacement of the oil with the low density gas phase, in shallower sub-critical traps, may exert an additional pressure on the caprock, and the trap could fail (cf. Hodgson *et al.* 1992). This is particularly critical if the traps occur at depths close to the regional fracture gradient (Fig. 4). Much deeper traps may avoid the fracture gradient due to their depth, and the petroleum density in these traps would in any case approach that of the initial oil even if petroleum with relatively more methane migrated into the trap due to increased maturity of the source rock basin.

The additional overpressure development in shallow and intermediately deeply buried traps, caused by the influx of a high API gravity gas phase, may have resulted in catastrophic leakage from traps with thick tight caprocks which lacks a pressure valve (cf. Figs 6 & 7) like those in Halten Vest.

Traps with such thick caprocks may not be able to 'bleed-off' the excessive pressure by capillary leakage, as may more readily occur from thinner, partly eroded and more silty and sandy caprocks as often found off Mid-Norway along the proximal basin perimeter, i.e. along the Nordland Ridge or the Frøya High (Fig. 1a). Examples of such traps include Heidrun, Skarv, Norne, Svale and Falk, but also traps in

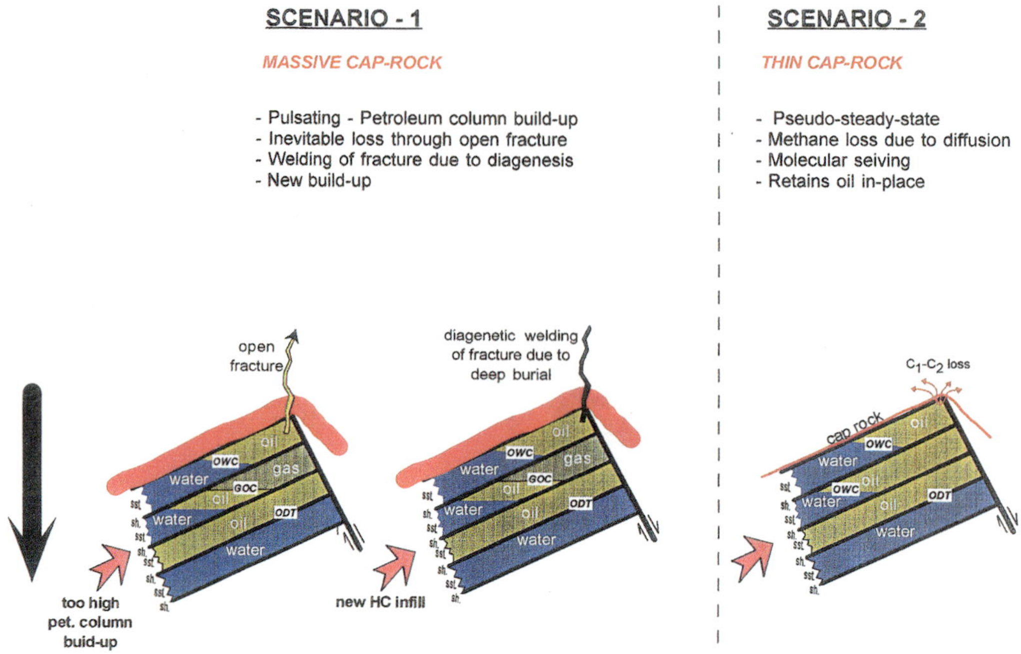


**Fig. 7.** Is a bad caprock in fact a good caprock if oil is the target? Pseudo-steady-state loss of gas by diffusion through a silty/thin/partly eroded caprock may allow room for oil in a structure (an example of a type III trap, Sales 1993, 1997), whilst a tight caprock will retain gas and allow little room for oil (Sales type I to partial type II trap). The 'tight caprock' scenario may result in petroleum (or gas) columns becoming eventually too high for the caprock strength, resulting in catastrophic open fracture dysmigration. We believe 'Scenario-2' to be a good exploration model for oil plays, whilst 'Scenario-1' will be gas plays if shallow or very deep (i.e. below the regional fracture gradient) or generally 'dusters' at intermittent depths.

other regions like Njord, Draugen, Snorre, Gullfaks, Statfjord, Oseberg, Visund and Ekofisk, the latter containing in addition oil wetting mineralogy which facilitates a 'semi-steady-state depletion mechanism. (See Paper I, figure 7.)

Catastrophic leakage could be expected to be recurring (Schowalter 1979; Chen *et al.* 1990; Ortoleva 1994; Wensaas *et al.* 1994) as loss of fluids and gas from the structure would only continue until the elevation of the HC–water contact is sufficient for the pressure to reduce and the fracture to close (cf. Roberts & Nunn 1995). Cycles with expulsion events lasting 1/100 to 1/5000 of the time period leading up to the expulsion event were suggested by Roberts & Nunn (1995) and Ortoleva (1994) indicated that the cycles could repeat on the order of every 100 000 years. Still, it is clear from principles discussed in Paper I (cf. figures 6–9, 12 & 13 in Paper I) that the cycling rate of the 'fill–drain–recharge–drain' processes will depend on, and be rate controlled by the filling rate. This filling rate is determined by the parameters discussed in Paper I, and include, besides HC generation rate and expulsion mode *sensu stricto*, 'the degree of focus' of secondary migration and its overall efficiency.

Such 'dynamic overpressure systems' (*sic.*) with traps experiencing cyclic fill–drain periods was proposed by Leonard (1993) and invokes intermittent breaching of the seal and subsequent annealing due to petroleum column loss, progressive burial and diagenetic cementation.

However, some overpressured traps, such as Kristin (Figs 1 & 5), with a substantial column of condensate exist in Halten Vest and preservation of petroleum in such high pressure 'isolated' fault-bounded reservoirs is fundamentally no different from a hydrostatic system, i.e. the occurrence of the accumulation simply reflects that the fracture pressure at the given depth is higher than the actual reservoir pressure (cf. general principles discussed in Holm 1996). Thus, Kristin is presently 'sufficiently deep' to avoid the fracture gradient and has also a very thick caprock (cf. Figs 4 & 7). No obviously visible gas chimney exists above Kristin, whilst such is indicated over the Lavrans trap.

The presence of gas chimneys as evidence of leaking traps or traps which did leak in the past

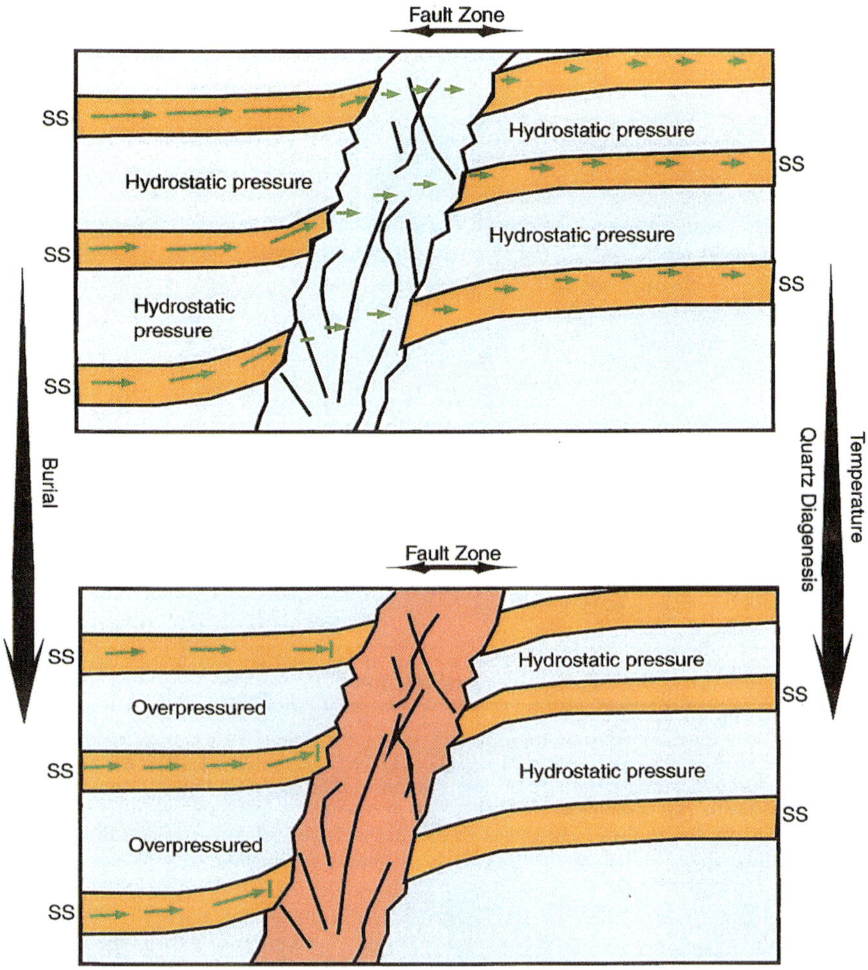


**Fig. 8.** Movement on a fault plane may result in juxtaposition of high and low permeability sandstones, or reconnection of two high poroperm sandstones, previously disconnected. Conversely microfabric rearrangements of clay minerals, grain breakage and fault gouge formation may reduce the relative permeability for oil versus water across and in the fault zone. Burial-induced diagenesis in the fault zone, as in the deeper parts of Halten Vest, will eventually reduce/constrain/disrupt fluid flow (flux) communication and induce regional fault controlled pressure cells (cf. Figs 1a & 5).

(cf. Leonard 1993) is difficult to use, in particular in overpressured regions. This is not only due to problems in rationalizing the effect the gas has in scattering seismic energy, depending on the gas occurring in fine grained porosities or in fractures (cf. Hodgson *et al.* 1992; Holm 1996), but more importantly no free gas will form until the bubble-point is reached during the ascent of the petroleum towards the surface.

Thus, undersaturated petroleum could migrate over larger vertical distances, without developing a free gas phase. In contrast, a saturated petroleum entering hydrostatic strata would 'immediately' exsolve a free gas phase and a 'seismically-recognizable' gas chimney would appear even at only 5–10% gas saturation.

If, on the other hand, tighter and well cemented lithologies are present in the overburden, well defined open fractures could form and exsolving gas and oil would move, for considerable vertical distances, in these open fractures to shallower depth without producing a broad 'gas chimney', until reaching shallower depths. The well known regionally significant 'gas brightening' in the Lysing Formation off Mid-Norway may be a result of this particular effect.

Modelling oil migration and pressure evolution in a basin requires real calibration points

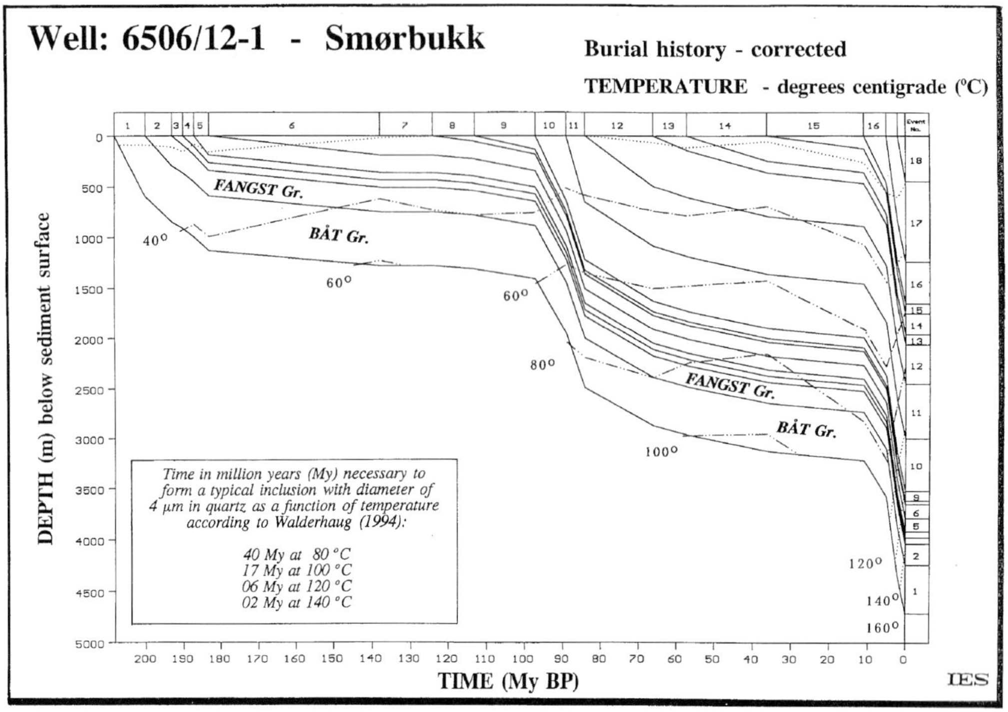


**Fig. 9.** The temperature for fluid inclusion formation in quartz is translated into time using the burial history of Smørbukk crestal well 6506/12-1 (Figs 2 & 27). Oil occurred in Smørbukk at least as early as 70–50 Ma bp. The biomarkers in the inclusions are similar to those found in Halten Vest (see Fig. 24). Inclusion formation is exponential in temperature (see insert) and the GOR is set at the time of inclusion closure. Biomarkers will in contrast represent an average, though biased towards the earliest low maturity oil as such oil contains far higher absolute concentrations of biomarkers than do more mature and high GOR petroleums like 'condensates'.

in time. It is obvious that the hydrocarbon and water communication across faults, e.g. like west of Smørbukk and Lavrans (Figs 1a,b & 5), have changed during time in response to movement on the fault and level of diagenesis (cf. Fig. 9). Fluid movement may become slowed down and impeded in fault zones. This may occur by the fault movement reducing or snapping off fluid flow communication across the fault (cf. Fig. 8), due to production of clay smear or fault gouge, reducing effective permeabilities and fluxes of oil, whilst water may still move. Eventually, movement of oil and water may cease in the fault zone due to the progress of ongoing temperature dependent diagenetic permeability-destructive mineral transformations (Bjørlykke *et al.* 1986; Ehrenberg 1990; Walderhaug 1990, 1994, 2000; Ehrenberg *et al.* 1992; Bjørkum 1996; Muhuri *et al.* 2003). The rate of quartz cementation is exponential in burial temperatures (cf. insert in Fig. 9). We believe this process to be the only one which both qualitatively and quantitatively can, at about 3–4 km, suddenly transform faults from being open for lateral flow, to become flow/flux restrictors, and eventually define isolated pressure cells.

Disconnection of strata across a fault and production of fault gauge are *linear* processes, determined by the actual relative fault movement, whilst mineral reactions like quartz diagenesis are *exponential* with temperature, i.e. reflecting absolute burial depth and temperature.

The generally deeper burial in Halten Vest, as compared with eastern regions, has therefore not so much resulted in more faulting and fault gauge formation, as higher temperatures and hence exponentially increased porosity reducing mineral transformation and quartz diagenesis in the fault zones. Diagenesis of the fault zones is thus the most likely cause of 'isolated pressure cells' in the western part of the Halten Terrace and in Halten Vest.

Lateral HC migration at Haltenbanken is short, e.g. mostly *ca.* 15-20 km except for Midgard (Åsgard-1) and Draugen (Karlsen *et al.* 1995). Thus, fields in this region generally fill by short distance migration from separate sub-basins. This conclusion is deduced from the organic geochemical molecular parameters in the reservoired petroleum.

Fluid communication across faults and within faults may thus be genetically linked to both petroleum migration and overpressure development (cf. Figs 8 & 10). In particular, 'palaeomigration' of petroleum from the Åre source

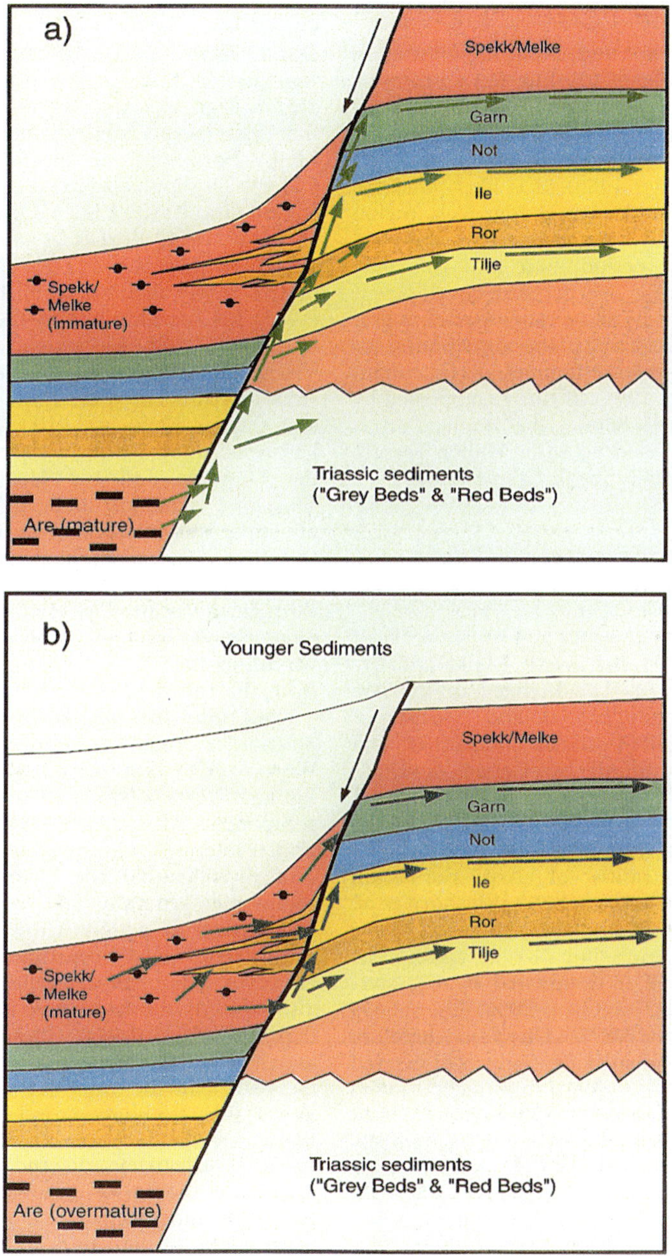


**Fig. 10.** Early expulsion from a deeper source rock, e.g. the Åre Formation (**a**), may have established a network of hydrocarbon and asphaltene wetted conduits that may have resulted in much higher migration efficiencies than assumed for a later generated and migrating oil charge from shallower source rocks such as the Spekk Formation (**b**). This effect may be overlooked in basin modelling.

rock may have increased the 'oil-wettability' in lateral carriers so that increased migration efficiencies are obtained when the Spekk Formation later expels petroleum (Fig. 10). Thus, it is the intricate relationship between fault properties and episodes of petroleum migration and trap filling, traced in traps, that we investigate in this paper.

The Halten Terrace underwent rapid burial in 'Mid'-Cretaceous times (cf. Fig. 9, i.e. from *ca.* 110 to 80 Ma bp), whilst the entire Trænabanken–Haltenbanken region underwent rapid subsidence in Late Pliocene–Pleistocene times (Jackson & Hastings 1986). It is likely that the fault planes in the deeper Halten Vest region were exponentially diagenetically modified during this latter period, and that this may have caused both the overpressure development and also the relatively short range lateral migration patterns revealed by the distribution of Spekk derived petroleum families at Haltenbanken (Karlsen *et al.* 1995). If this is supported by the data presented below, the implication is that overpressure build-up at the Halten Terrace occurred, 'geologically speaking', yesterday.

## *Known regional source rocks and their geochemical signature*

The main organic-rich rocks and source rocks of this region belong to the Spekk, Melke and Åre Formations (Fig. 3). The kerogen of the Åre Formation is heterogeneous and consists of rooted coals, drifted coals, mudstones and shales, and was found by Cohen & Dunn (1987) to have $\delta^{13}C$ values of about $-26.1 \pm 1.1$ ($n = 11$), and are comparable to data published by Elvsborg *et al.* (1985) ($-25.8 \pm 0.7$, $n = 4$). Pristane/phytane ratios of Åre Formation extracts fall in the range 1.4–9.5 (Elvsborg *et al.* 1985; Cohen & Dunn 1987) and we have observed values exceeding 10 in coals. The sterane distribution is dominated by $C_{29}$ steranes, e.g. $C_{27} = 15\%$, $C_{28} = 20\%$, $C_{29} = 65\%$ (Cohen & Dunn 1987). There is therefore no problem in identifying Åre coal-derived hydrocarbons ($C_{15+}$). Vanadium/nickel ratios [$V(V + Ni)$] for the coals range from 0.5 to 0.95 even within a metre of coals, most likely reflecting the extremely varied Eh and pH conditions during deposition (Karlsen *et al.* 1995).

The $V/(V + Ni)$ of Spekk extracts from 6407/5-1 are much more uniform (*ca.* 0.9) than in the Åre Formation (0.5–0.9; see Karlsen *et al.* 1995).The Spekk Formation kerogen is isotopically lighter than found for the Åre Formation with $\delta^{13}C$ values of $-28.5 \pm 1.6$ ($n = 10$, Cohen & Dunn 1987), thus closely matching the values ($-28.2 \pm 2.7$, $n = 3$) found by Elvsborg *et al.* (1985). The pristane/phytane ratio of extracts from this formation fall in the range of typical type II marine shales, e.g. 0.6–1.60 (Elvsborg *et al.* 1985; Cohen & Dunn 1987) and with a uniform sterane distribution centring around $C_{27} = 34\%$, $C_{28} = 33\%$, $C_{29} = 33\%$ (Cohen & Dunn 1987).

Available data tentatively suggest that there are, in a gross sense, substantial similarities in the organic geochemistry between the Spekk/Melke Formations in the Haltenbanken area, and the Draupne/Heather Formations from the Viking Graben (cf. Cohen & Dunn 1987; Espitalié *et al.* 1991; Forbes *et al.* 1991). The $\delta^{13}C$ values of kerogen in the Spekk Formation match closely the data published by Huc *et al.* (1985) for the Draupne in the Viking Graben. Huc *et al.* (1985) reported in a study from the Viking Graben that both the $\delta^{13}C$ and the $\delta^{2}H$ values became heavier, from −30‰ to −25‰ and from −125‰ to −100‰, respectively, in kerogens isolated from a 100 m section through the Draupne and top Heather Formations. This, and reduction in bisnorhopane and clay minerals, was interpreted by Huc *et al.* (1985) to reflect a more proximal terrestrially derived organic matter input with increasing depth through this section (also cf. Scotchman 1991). Similarly, Bailey *et al.* (1990) described vertical variations in $\delta^{13}C$ of kerogens and pyrolysates from the top of the Hot Shale (mostly −30 to −29‰) into the Heather, Brent Dunlin and Statfjord Formations (mostly −27 to −24‰). We postulate that the Spekk/Melke transition similarly reflects an increase in the terrestrial component with depth and we assume that similar lateral variations occur in Haltenbanken. Thus, it is possible that the Spekk/Melke source rocks may generally take on a more dysoxic and terrestrial signature towards the east, i.e. towards the palaeo-coastline (Walther's Law). However, a number of deeper local basins exist also towards the east where it may be assumed that the Spekk/Melke formations are thick, anoxic and productive and the geochemistry of the petroleum in traps like Skarv, Norne and Svale (Fig. 1a) suggests indeed that such subbasins are highly generative in this region (cf. Winterstad 2003; Xu 2003). In particular Skarv is evidently sourced from a superior source rock facies more commonly associated with the highly prolific source rocks in the Trestakk, Tyrihans, Smørbukk Sør region (Sletten 2003) whilst the Svale and Falk oils belong to the general Norne sub-family, albeit progressively biodegraded in an updip direction, i.e. closer to aquifers which were active during the low sea level of the last glaciations (Xu 2003).

Some of the shallow wells in the deep water region which encountered reservoirs with mainly gas, and in some cases thin oil rims, have been suggested to contain petroleum derived from Cretaceous or Tertiary strata. A problem with some of these rather shallow traps and their content of the age-specific biomarkers among the nordiacholestanes, which may give Tertiary or Cretaceous ages, is the potential for these compounds to have been generated *in situ,* and examination of some of these petroleum phases and core extracts have convinced us that these compounds are in cases having the same *in situ* origin as co-occurring non-isomerized hopanes and steranes, i.e. these compounds are not allochthonous. Still, the gaseous compounds are certainly allochthonous. It may therefore, in this setting, be generally unsound to conclude about a Tertiary or Cretaceous oil play based solely on age specific compounds. Furthermore, the nordiacholestanes may tend to give 'young' ages in immature sediments (Holba *et al.* 1998a,b), and little is known about how these age-specific biomarkers respond to biodegradation, which often is a factor in the shallow and cool traps in the deep-water province. If not supporting age-specific compounds of Cretaceous/Tertiary age, such as oleanane, is found in these petroleum systems, it may be difficult to conclude that these traps contain allochthonous petroleum from Tertiary or Cretaceous strata. Despite this we are aware of three traps off Mid-Norway which clearly contain non-Spekk/Melke derived petroleum. One of these is the Idun trap in which the $C_{20-}$ condensate in the separate reservoir units has strong terrigenous characteristics (including heavy $\delta^{13}C_{total\,oil}$ values, mostly heavier than $-27‰$, and high Pr/Ph values, mostly from 3 to 5) suggesting Idun to have been sourced at least partly from the Åre Formation (Xu 2003). The nearby Skarv trap, on the other hand, contains all the characteristics of Spekk-derived cogenetic gas and oil (Winterstad 2003; Xu 2003).

Concerning unusual extracts and source rocks from the general region, Karlsen *et al.* (1995) described a very *n*-alkane rich, isoprenoids-poor, isotopically very light (-32‰) bitumen staining with only 10% total aromatic compounds from a clean sandstone at *ca.* 2.5 km in well 6609/11-1. The vanadium/nickel ratio suggests a non-marine origin and a potential Triassic/Palaeozoic origin was suggested. Bugge *et al.* (2002) describe their interpretation of this sample as well as their own very extensive database from shallow drilling, and provide irrefutable evidence for the Upper Permian to Lower Triassic SR play of the Ravnefjeld Formation type (known from Greenland) off Mid-Norway. They also provide compelling evidences for this facies to represent Permo-Triassic generative basin play scenarios in proximal regions off Mid-Norway where the Jurassic section is too shallow to be generative for oil. A tentative regional model for the existence of this facies all along the Norwegian Continental Shelf from the Central Graben and into the Barents Sea was first published by Surlyk *et al.* (1984) and these data put together suggest strongly that palaeo-migration of Palaeozoic derived oil might have occurred over large portions of the NOCS.

### *Source rock productivity, HC generation and expulsion*

The kerogen of the most prolific 'Hot Shale' Kimmeridge equivalent part of the Spekk Formation is a highly favourable source for oil with a gas/oil generation index (GOGI $= C_1$–$C_5/C_6$–$C_{36+}$) of about $0.19 \pm 0.05$ ($n = 8$) (Cohen & Dunn 1987). The Åre Formation contains coals and shales that are rated at having generated about 135 kg of petroleum/t rock (gas $\approx 30\%$, condensate $\approx 50\%$, oil $\approx 20\%$) during burial to about 4% $R_c$ (Heum *et al.* 1986). Shales and coals in the Lower Jurassic Åre Formation have GOGI in the range of about $0.33 \pm 0.06$ ($n = 5$), i.e. giving rise to a mainly 'mixed oil and gas' (Cohen & Dunn 1987). Espitalié *et al.* (1991) rated the petroleum potential ($10^6$ t of HCs/km$^2$) of the coals in the Åre Formation to be four times higher (HCs/km$^2$) than that of the Viking Group (Spekk + Melke), due to the high productivity of the Åre Formation and also due to its greater cumulative thickness.

Espitalié *et al.* (1991) suggest that the Åre Formation expelled about 2.5 times more hydrocarbons than the Spekk and Melke Formations. The suggested breakdown between coals and shales in Åre in terms of expelled hydrocarbons is proposed to be about 17:1, while the ratio between expulsion for the Spekk and Melke Formations is about 2.4:1 (Espitalié *et al.* 1991). Forbes *et al.* (1991) ascribed most of the oil generation in the drainage area of the Smørbukk Sør accumulation to the Åre Formation, claiming that the Åre Formation is about six times as important, volumetrically with respect to expelled $C_{15+}$ hydrocarbons, compared with the Spekk plus the Melke Formations.

The expulsion of oil from humic coals like these may, however, be seriously questioned. The data referred to above are based on extensive pyrolysis experiments. It is well known that pyrolytic conditions, i.e. the use of kerogen versus source rock particles, open versus closed pyrolysis, hydrous versus anhydrous conditions, as well as temperature profile/heating rate and the size of the pyrolytic cell, are all parameters affecting chemical reactions and products. Furthermore, high temperature generation profiles for isolated kerogens or source rock (SR) samples are not by any standard a relevant analogue for SR expulsion. The latter involves generation and mass transfer, whilst HC generation *sensu stricto* is merely a temperature sensitive expression of the second law of thermodynamics.

Coals are generally good generators but poor expellers (Isaksen *et al.* 1998) and recent pyrolysis work on coarse grained coal particles suggests that humic and brown coals like the Åre coals basically will not expel liquid range HCs. This was also inferred for torbanites, lamosites, marinites and liptinite-rich humic coals (Inan *et al.* 1998).

Likewise Isaksen *et al.* (1998) suggested that there is too low a concentration of $C_{8+}$ precursors in the Middle Jurassic Hugin coals in the North Sea to cause expulsion of non-volatile black oils, and they also suggested that neither the hydrogen index (HI) nor the resinite content correctly reflect the ability of a given coal to expel black oil.

Absorption and adsorption of petroleum in the solid organic matrix ('equal dissolves equal') in SRs means that, unless the generative potential of the kerogen is high, expulsion is delayed and the probability for early expulsion from these SRs is reduced (cf. Pepper & Corvi 1995). A minimum HI of *ca.* 200 mg/HC/g TOC was rated as a minimum for the petroleum products to overcome the sorptive capability of the organic matrix, and Pepper & Corvi (1995) concluded that the initial oil expulsion threshold could increase from 110 °C (best possible SR) to 135 °C in the worst case for oil producing source rocks in a basin subsiding at a rate of 2 °C/Ma.

Shifting the initial part of the oil window 'down' by 20 °C can have *severe* consequences in risking prospects, not only in terms of trap filling as such, but also for GOR predictions, as GOR increases with delayed expulsion (increase in temperature). It follows that detailed knowledge about the kerogen quality, quantity and geometric organization in the SR is needed to determine if marginally mature basins are prospective for oil.

It is for these reasons that the distribution of oil and gas discoveries provide important calibration for understanding the actual expulsion and migration status of a basin.

The Melke Formation has insignificant oil potential compared with the Spekk Formation and the deeper Åre Formation (cf. Heum *et al.* 1986; Cohen & Dunn 1987). Heum *et al.* (1986) rated the Melke Formation as potentially releasing no more than 1.5 kg of petroleum/t rock (mainly gas $C_{5-}$ and some condensate $C_{6-}$–$C_{14}$), whereas approximately 30–35 kg of petroleum/t rock (gas ≈ 15%, condensate ≈ 45%, oil, i.e. $C_{15+}$ ≈ 40%) was estimated to have been released from the Spekk Formation during progressive burial to a maturity of about 4% $R_c$. It is still likely that the oil potential of the Melke Formation may vary significantly laterally as well as vertically in off Mid-Norway, and more systematic investigations on facies variations are needed especially from deeper parts of the basins where the formation may have developed as 'anoxic ponds'.

Peak oil generation from the Spekk Formation was estimated by Leadholm *et al.* (1985) to have occurred at Haltenbanken at about 3400 m of burial. Initial oil expulsion was estimated by Cohen & Dunn (1987) to have occurred at about 120 °C, equivalent in their dataset to about 3100 m of burial, while Heum *et al.* (1986) have 0.7% vitrinite reflectance equivalence (VRE) as the threshold for primary migration from this formation. In their dataset 0.7% VRE corresponds to about 3900 m of burial, whilst this depth is considered by Cohen & Dunn (1987) to be associated with initial oil to gas cracking ($T \approx 150$ °C). Initial refractory gas generation may occur at 4600 m (Cohen & Dunn 1987) and Heum *et al.* (1986) similarly put the end of the oil window at 1.0% VRE, or 4700 m.

Our database on biomarkers from the Spekk Formation contains good quality mature core samples in only one extensive core—6407/5-1 in southern Haltenbanken. The biomarker distribution of these extracts at 3900 m reveals a maturity equivalent to that found in the most mature petroleums from Haltenbanken, except in 6407/4-1, Kristin and Lavrans. These limited data may indicate that significant oil expulsion has occurred from the Spekk Formation shallower than 3900 m. There is also likely to be a difference in temperature or depth for expulsion from leaner SRs like the Melke as discussed above. Thus, more extensive generation and higher maturities are necessary for primary migration to occur in lean source rocks lacking a 2D or 3D continuous kerogen network. This difference in temperature for expulsion is important with respect to depth and GOR of the expelled products (cf. Durand *et al.* 1987; Durand 1988; England and Mackenzie 1989; Stainforth and Reinders 1990; Pepper and Corvi 1995). We suggest that the Spekk Formation at Haltenbanken is generally more prolific and thicker in the non-drilled basinal depocentres than is commonly assumed. We also propose that expulsion occurs shallower than some of the contemporary models seem to indicate.

Over much of the Haltenbanken area the Åre Formation is of 0.2–0.5% $R_c$ higher in maturity than the overlying Spekk Formation (cf. Leadholm *et al.* 1985; Heum *et al.* 1986; Whitley 1992). Still, great variations exist, due to differential burial depth of the units, e.g. the difference between the tops of these two formations varies from about 500 m in the Midgard Horst to approximately 1000–1500 m 25 km farther south (Mo *et al.* 1985). Petroleum generation is thus assumed to have been initiated significantly earlier in the deeper Åre Formation than in the shallower Spekk Formation, i.e. during Mid- to Late Tertiary times (Champbell & Ormaasen 1987). Heum *et al.* (1986) suggested that expulsion in the Spekk Formation is initiated at a maturity of about 0.7% $R_c$, and at 0.5–0.7% $R_c$ and at 0.85% $R_c$ in the deeper Åre Formation coals and shales, respectively. Forbes *et al.* (1991) proposed that expulsion from the Spekk and Melke Formation in the Smørbukk Sør region became significant as late as about 5 Ma bp, whilst in the Åre Formation expulsion probably occurred as early as 40–50 Ma bp.

We assume that petroleum products from the Åre Formation could conceivably have helped in saturating laterally extensive migration avenues in the basins, so that secondary migration from the Spekk Formation, when 'turned on' millions of years later, would proceed which much increased efficiency (cf. Fig. 10, and figure 8 in Paper I).

## *Maturity and geochemical signatures of petroleum in traps in the study region*

Karlsen *et al.* (1995) concluded, on the basis of a database of 33 petroleums from the Haltenbanken region, that the $C_{10+}$ fractions of the oils and condensates of this region were all sourced from the Spekk Formation. These petroleums represented the majority of petroleum accumulations at that time, apart from Smørbukk. This interpretation was based on biomarkers, medium range alkanes and aromatic hydrocarbons as well as on individual molecular isotope patterns. An east–west systematic (bulk $C_{10+}$) carbon isotope trend (Fig. 11) was found for the petroleums. This was interpreted to reflect possibly a more terrestrial input to the source rock kerogen in the proximal eastern regions, whereas in the distal western regions the inferred kerogen was assumed to be more amorphous and deposited under generally more reducing conditions. Extension of this work (Angard 1996; Steinhoff 1996; Storvoll 1999; Storvoll *et al.* 2002; Sletten 2003; Winterstad 2003; Xu 2003) has allowed expansion of the dataset to include newer discoveries, all of which fall along the trend shown in Figure 11. The maturities of the petroleums generally follow a systematic east–west trend, being more mature and less terrestrially influenced in the western portion of the basin.

This trend basically follows a maturity map based upon the depth to base Cretaceous and it is hence inferred that lateral migration distances for the presently reservoired hydrocarbons is short, i.e. shorter than 15–20 km, with notable exceptions concerning migration of the hydrocarbons found in Midgard and Draugen.

The following maturity sequence is proposed:

$$
\begin{aligned}
\text{Kristin} &= \text{Lavrans} > 6407/4\text{-}1 \\
&\quad > 6507/2\text{-}2\ \text{DST}\ 2(C_{20-}\ \text{fraction}) \\
&= \text{Idun}(C_{20-}\ \text{fraction}) \\
&\quad > \text{Smørbukk(condensates)} \\
&= \text{Alve} \gg \text{Smørbukk Sør}\ 6506/12\text{-}3 \\
&\quad > \text{Midgard} \\
&\quad > \text{Smørbukk Sør}\ 6506/12\text{-}5 \\
&\quad > \text{Trestakk} \gg \text{Njord} \\
&= \text{Norne} \\
&= \text{Svale} \gg \text{Heidrun} \\
&= \text{Skarv} > \text{Draugen} > \text{Tyrihans Nord} \\
&\quad > \text{Tyrihans Sør} \gg \text{Mikkel}.
\end{aligned}
$$

Dahle (1997), who studied light hydrocarbon distributions ($C_2$–$C_8$) of 33 petroleums from Haltenbanken and additional petroleums from the Central and Viking Graben, concluded that no evidence could be found to suggest that the Haltenbanken petroleums had an origin from other than an anoxic type II source rock like the Spekk/Draupne equivalent or, alternatively, an anoxic Melke/Heather analogue. Dale (1997) concluded, furthermore, that the light hydrocarbon maturity trend of the petroleums parallels to a large extent the maturity trend based upon biomarker and medium range hydrocarbon

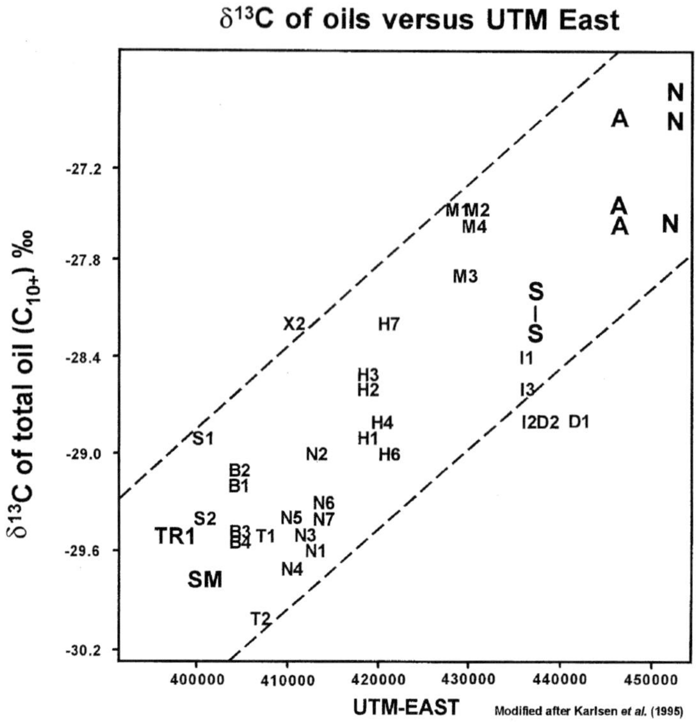


**Fig. 11.** The $\delta^{13}C$ values of the $C_{10+}$ fraction from fields in Haltenbanken. Light carbon isotope values like −30 to −29‰ (PDB) dominate in the western fields like Smørbukk (SM), Smørbukk Sør (B & S), Trestakk (TR1), Tyrihans Nord (T1) and Sør (T2). In the more shallowly buried regions in the east are the $\delta^{13}C$ values towards −29 to −27.4‰), e.g. Mikkel (I1, I3), Midgard (M1–M4), Norne (N), Alve (A), Skarv (S). Intermediate values occur in the middle of the basin, e.g. Njord (N1, N3), Draugen (D1, D2), with long range migration from west and in Heidrun (H1–H4). The trend reflects generation from more terrestrially influenced and dysoxic source rock intervals in the proximal eastern regions closer to the palaeo-coastline, whilst more algal/bacterial/amorphous kerogens are inferred to occur in the western regions.

parameters suggested by Karlsen *et al.* (1995). This suggests that the reservoired petroleums are all cogenetic in terms of light hydrocarbons, medium range hydrocarbons and biomarkers except from the Idun and the 6507/2-2 discoveries.

The more recent work of Odden *et al.* (1998, 2000), using molecular distributions and individual carbon isotopes of light ($C_4$–$C_{13}$) hydrocarbons mobilized by thermal evaporation and pyrolysis from Spekk, Melke and Åre source rocks, strongly suggests that the bulk of these compounds on the Halten & Dønna Terrace have been sourced from the Spekk Formation. However, these authors suggested that the Norne oil could have been derived from a SR with more terrestrial components than the bulk of the Mid-Norwegian oil. This overall view is supported by Patience (2003) who expresses doubt as to the existence and nature of any reservoired gas or liquid contributions from the Åre Formation and who suggests three tentative explanations, none of which are supported by our observations.

Since coals are so rich in organic material, they have very high ability to dissolve liquid and gaseous hydrocarbons and associated organic compounds, i.e. cf. the standard chemical rule of thumb: 'equal dissolves equal'.

The microporosity of coals involves highly complex pores of various 'openness' and sizes and generally with a very high retention capacity of organic molecules (van Krevelen 1961; Hanbaba, *et al.* 1968; Peters & Jüntgen 1968; Given *et al.* 1986; Durand *et al.* 1987; Pepper 1991). Different molecular signature were extracted from open versus closed pores, or from more or less strongly adsorbed organic matter (Beletskaya & Syrova 1972; Beletskaya 1978; Sagjó *et al.* 1983; Price & Clayton 1992), suggesting a lack of free exchange within the coals. Wilkins & George (2003) suggest in their excellent review paper that interconnected pores do not exist in vitrinite macerals in coals, and concluded oil expulsion from coals in general to be an extraordinarily rare phenomenon, a view supported by our own work during the past 15 years on the NOCS.

We believe that the abundant aromatic carbon content in coals with associated $\pi$-electrons available in the coaly matrix is also probably effective in promoting retention and absorption of polar compounds and aromatic hydrocarbons, but also nonpolar material like alkanes are adsorbed in large proportions (cf. Sandvik *et al.* 1991; Isaksen *et al.* 1998).

This retention of humic coals will delay expulsion from the coals until higher maturities, i.e. emphasizing to a dramatic degree the general difference between *generation* and initial *expulsion* or following the nomenclature of Pepper & Corvi (1995) between generation, expulsion and primary migration. This could suggest that the bulk of the generated hydrocarbons are retained in the coals until higher temperatures cause cracking to gas (cf. Sandvik *et al.* 1991). This type of gas generation may have promoted trap seal failure and could be partly responsible for the numerous gas chimneys and gas escape structures observed at Haltenbanken and even the gas brightening seen regionally in the Lysing Formation. This latter gas might, however, have an *in situ* Cretaceous origin due to the exponential increase in gas generation from OM before the 'peak oil window' (cf. Schaefer & Leythaeuser 1984; Karlsen 1987).

We, however, suggest that the time gap between generation and expulsion from the deep Åre Formation relative to the shallower Spekk Formation, and time dependent HC loss from traps, i.e. the 'pseudo-steady-state' model of loss and replenishment discussed in this paper and in Paper I, is the reason why we today generally do not observe type III/coal derived gas in the traps off Mid-Norway. Thus, the time lag between, even delayed expulsion from the Åre Formation, and expulsion from the shallower Spekk Formation is simply too significant in light of the 'dynamic trap model' (cf. Sales 1993, 1997—trap type II and III), for these traps to hold on to petroleum generated e.g. 50 Ma bp.

Furthermore, we feel convinced that the relative enrichment of toluene, xylenes and methyl-cyclohexane in Norne, used by Odden *et al.* (1989, 2000) as indicative of a coaly source rock, originates from palaeo-biodegradation of Norne petroleum. These aromatic and cyclic molecules are less susceptible to biodegradation than normal alkanes and their relative enrichment in Norne may simply reflect palaeo-biodegradation which occurred before Norne became buried deeper than *ca.* 2.2 km and biodegradation ended due to the increased temperature and/or cut-off of oxygenated water as faults became more sealing. These compounds occur in 'normal' concentrations in the nearby Alve discovery which is buried *ca.* 1 km deeper than Norne and which thus was sufficiently deep when trap filling was initiated to avoided biodegradation.

It is clear that 'meteoric-pumping' would have occurred during the 30–40 or so periods of glaciations–deglaciations which were initiated *ca.* 3 Ma bp and that this mechanism would provide an effective means of mobilizing not only oxygenated water (during glaciations) but also, during the deglaciations, influx of marine sulphate containing water for anaerobic oxidation. Thus, the issue of anaerobic versus aerobic oxidation of commercial quantities of oil, such as found incipiently in Draugen, in variable but generally more severe development in Heidrun (where also influx of non-degraded oil occurred), and in more extreme degree in Svale and Falk (even with demethylated homohopanes) respectively (Xu 2003), must be understood in a more dynamic basinal/climatic sense and in relation to the water fluxes set up by the glaciations/deglaciations, than, e.g. 'burial resulting in palaeo-pasteurization' as suggested by Wilhelms *et al.* (2000). These elements are discussed more fully in 'Paper I'.

The Norne trap with its thinner, partly eroded and imperfect caprock, currently buried *ca.* 1 km shallower than Alve with its much thicker caprock, is in our opinion a good example of a 'Sales type 3' trap, and the maturity of its oil, which clearly is of the same organic facies as found in Alve, suggests that it originated from a 'kitchen area' buried *ca.* 1 km shallower than the 'kitchen' that generated the condensate in Alve. With its thicker caprock, Alve, on the other hand, is a good example of a 'Sales type 1' trap (cf. Sales 1993, 1997).

Concerning 'partially leaking caprocks' and biodegradation, it is a fact in the study area that this phenomenon is restricted to proximal traps with bad caprock properties, buried currently at depths less than *ca.* 2.2 km (70–80 °C).

Thus, it seems that some kind of 'openness' of the reservoir itself to water movement, including possibly the caprock with interconnected water films on the grains (cf. Nordgård Bolås *et al.* 2003), and not only 'large scale regional openness', i.e. lateral connectivity of communicating migration avenues or aquifers, as suggested by Ahsan (1993) and Ahsan *et al.* (1997), are important elements for understanding in-reservoir biodegradation. This 'openness' in of interest

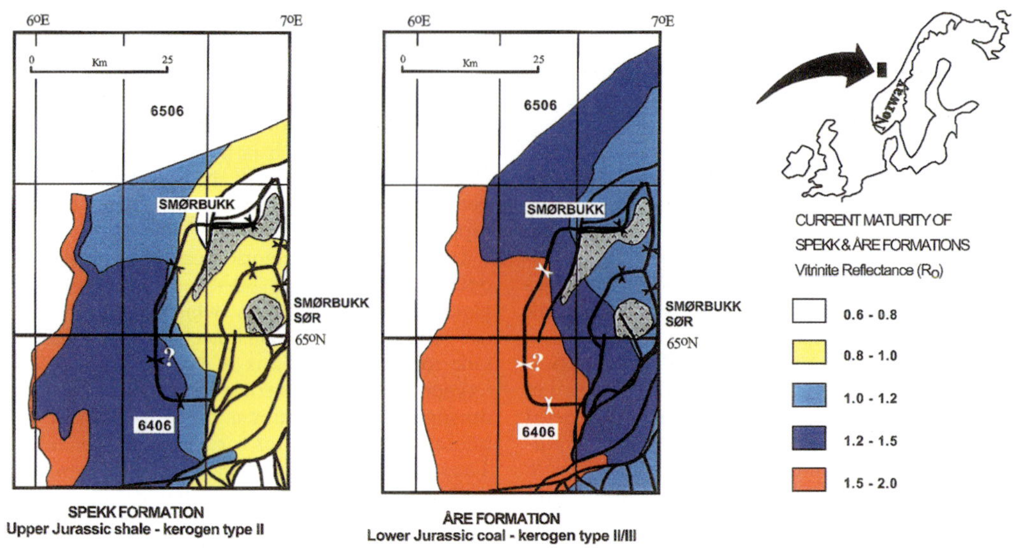


**Fig. 12.** Current Spekk Formation maturities in the Smørbukk region illustrate that oil which entered Smørbukk 70–50 Ma bp must have originated from western regions. See also Figure 9.

concerning the repeated development of hydraulic heads of meteoric oxygenated waters during glaciations (aerobic oxidation) and influx of marine sulphate containing waters during deglaciations (anaerobic oxidation).

There is no published systematic (non-commercial) database on gas isotopes from oils and condensate associated gases from the Norwegian Sea. Our own data on methane to propane from Smørbukk, Draugen, Norne, Alve and Skarv show isotope values in the following ranges: $\delta^{13}C_1 = -38$ to $-46$‰, $\delta^{13}C_2 = -28$ to $-35$‰ and $\delta^{13}C_3 = -25$ to $-30$‰, with the heavier values representing the eastern accumulations Norne and Alve. These fields also contain isotopically heavier oil/condensate than was generally found in the western accumulations. Relatively heavy values for ethane and propane ($\delta^{13}C_1 = -41.6$‰, $\delta^{13}C_2 = -27$‰ and $\delta^{13}C_3 = -26$‰) were reported for the Midgard gas by Elvsborg *et al.* (1985). The $\delta^{13}C$ difference between methane and ethane in Midgard is greater than in Smørbukk, Norne and Alve, where this difference is *ca.* 10‰, i.e. indicating 100% cogenetic gas isotope signatures.

The gas isotope values at Mid-Norway (unpublished data) are not systematically different from North Sea gases in which the $\delta^{13}C_1$ range mostly from −39 to −44. There is general agreement that the gases as well as the oils in the Central & Viking Graben are derived from the Upper Jurassic marine shales, in particular the Draupne/Kimmeridge SR sequence.

Thus, despite the undisputed high generation capacity of the Åre Formation, there is hardly any evidence for coal-derived hydrocarbons in fields and discoveries at the Halten & Dønna Terrace. The methyl-anthracenes found in the Midgard condensate and in some of the Heidrun and Draugen oils discussed in Karlsen *et al.* (1995) might constitute weak circumstantial evidence to suggest palaeo-HC contributions from the Åre Formation. These compounds are known to occur in coals (M. Radke, pers. comm.) and it is possible that such relatively polar compounds could have remained in the migration system or in the traps from an earlier Åre-derived condensate.

This, and the deeper burial of the Åre Formation relative to the Spekk Formation (cf. Figs 1b & 12), could indicate that the Haltenbanken traps are indeed dynamic traps which lost all of their Åre-derived gases and condensates before the Spekk Formation matured. As the Spekk Formation became mature more recently, its petroleum 'wiped out' most of the potential Åre-derived molecular signatures from the traps.

The Spekk Formation with its type II kerogen and its overall lower maturity, compared with the Åre Formation, would be expelling petroleum with much higher relative concentrations of biomarkers. This effect will also act to reduce possible trace signals from palaeo-Åre-derived petroleum (gas-condensate) in the traps.

### *Samples*

87 core samples from the following wells were available for this study. Dry wells: 6406/3-1, 6406/6-1, 6506/11-1, 6506/11-3 and 6506/12-4. Petroleum-filled structures: Smørbukk 6506/12-1 and Smørbukk Sør 6506/12-5. These wells are discussed in detail below. Fluid inclusions were studied in these samples. 70 additional core samples from these wells and 66 drill stem test (DST) fluids from the general Haltenbanken region were analysed and discussed by Angard (1996), Steinhoff (1996), Dale (1997) and Olstad (1997), and results from these studies are integrated in this paper. 16 core samples analysed for fluid inclusions and core extracts were also available from the dry wells 6506/11-3 (4100 m), 6406/8-1 (4380 m) and 6507/5-1 (4220 m), the condensate containing 6407/4-1 structure (4000 m) plus the oil fields Trestakk 6406/3-2 (3950 m), Tyrihans Sør 6407/1-2 (3670 m) and Tyrihans Nord 6407/1-3 (3630 m). The data on these wells are utilized here more as auxiliary regional control points for the interpretations. Inorganic and organic coatings in Kristin and Lavrans were discussed by Storvoll (1999) along with core extract compositions, mineral diagenesis and maturity signatures. 16 DST samples and 31 core samples from Heidrun were examined geochemically (Angard 1996; Steinhoff 1996). Furthermore, results of geochemical investigations of core extracts, oils, gases and petroleum inclusions by Sletten (2003), Winterstad (2003) and Xu (2003) have been implemented in this study. Thus, whilst a limited dataset is discussed in depth here, more data were available for the interpretations offered. Details of the methods and analytical conditions are found in the appendix.

## Results and discussion

### *Core extracts of dry wells compared with fluids and core extracts in Smørbukk*

Quantitative extract data are shown as vertical well profiles in Figure 13. The four wells 6506/12-1 (Smørbukk), 6406/3-1 (dry), 6506/11-1 (dry) and 6506/12-4 (dry) hold typical migrated hydrocarbons, i.e. core extract compositions typified by 65–80% saturated hydrocarbons, 10–20% aromatic hydrocarbons and polar compounds typically in the range of only 10–22%. The extract amounts are mostly in the range 0.2–1.3 mg/g rock with most values in the range 0.3–0.7 mg extract/g rock. The values are comparable with those found in the Smørbukk field (6506/12-1). Extract yields and compositions like these are empirically found for oil and condensate producing Jurassic sandstones at similar depths offshore Norway (cf. Horstad *et al.* 1990; Karlsen & Larter 1989, 1991; Karlsen *et al.* 1993; Skålnes 1993; Horstad 1995; Angard 1996; Steinhoff 1996). However, higher yields are also common and in Trestakk, Tyrihans Sør and Nord values range from 3 to 5.6 mg/g rock.

Well 6506/12-1, the Smørbukk discovery well, today holds condensate at four levels in Garn and Ile with GORs between *ca.* 975 and 1550 $m^3/m^3$ (Fig. 14), but fluid inclusions, discussed below, suggest that oil was present at earlier times in the strata that at present contain gas-condensate.

Core extracts from the dry well 6406/6-1 deviate from those of the other wells in being dominated by polar compounds and very low yields. Such signatures are not representative of allochthonous bitumen or oil. In the dry well 6506/12-4 the extracts are more enriched in polar compounds, i.e. resins and asphaltenes, compared with the two other wells including the Smørbukk well 6506/12-1. The Lysing/Lange extract from 3142.75 m in well 6506/12-4 is even more enriched in polar compounds, reaching about 45% (Fig. 13) and the extract yield is also very low, i.e. *ca.* 0.12 mg/g rock. This value is in the range of two of the extracts from the Tilje Formation in well 6506/11-1 at 4264.5 and 4504.75 m. These latter core extracts have compositions which correlate with the core extracts from well 6406/6-1. Extract yields are poor in this well, i.e. <0.09 mg extract/g rock, and the extracts are dominantly polar, i.e. typical signatures of *in-situ* generated bitumen. Even in relatively homogeneous high-energy-deposited sandstones like the Garn Formation at Haltenbanken, some trace amounts of organic matter occur in the dust rim on sandstone grains and also in more silty lithologies. Such organic matter, being of a largely oxidized type with low hydrogen to carbon ratios, will still at depths of 3–4 km inevitably produce trace amounts of extractable organic matter. Generally, this material, as any *in-situ*-derived organic matter, is easily recognized by its very polar composition (cf. Tissot & Welte 1984).

In three out of four dry wells there is clear evidence for the presence of bitumen in proportions indicating that the reservoir rocks contained up to the same hydrocarbon saturations as the reference well 6506/12-1 from the Smørbukk field (Fig. 27). Note that the depth trend through the Garn and the Ile Formation in the Smørbukk 6506/12-1 well is very similar to that of the dry

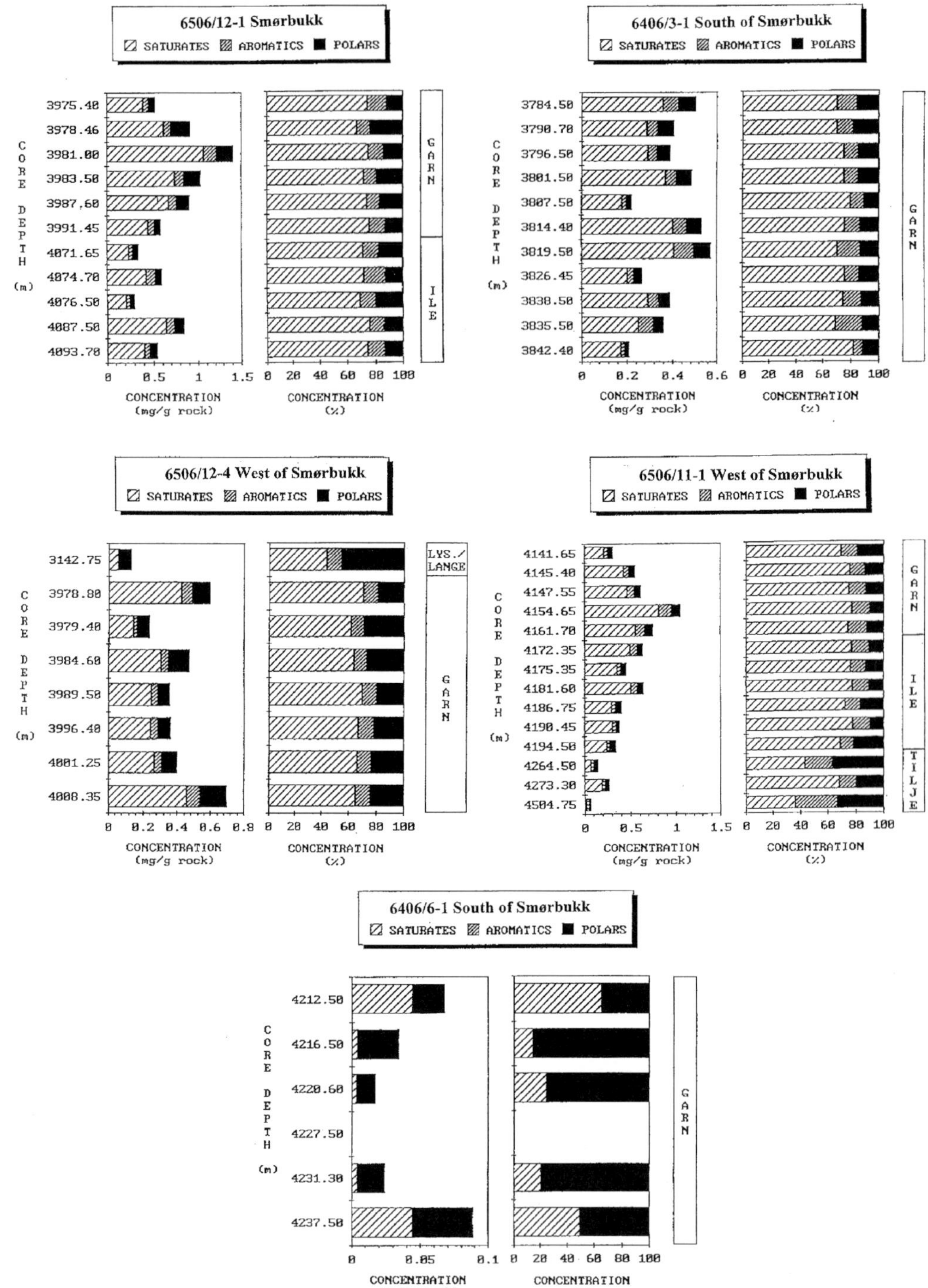


**Fig. 13.** The dry structures 6506/12-4 and 6506/11-1 west of Smørbukk contain core extracts of the same amount and gross composition as found in the Smørbukk well 6506/12-1. Well 6406/3-1, south of Smørbukk (Fig. 1a), is today dry whilst the core extracts indicate clearly that this well earlier held oil, in contrast to well 6406/6-1 which only contains small amounts of polar and saturated compounds (cf. Figs 1a & 27).

| Smørbukk DST Results | | | | | |
|---|---|---|---|---|---|
| **Zone** | **6506/12-1** | **6506/12-6** | **6506/12-7** | **6506/12-9S** | **6506/11-2** |
| **Lysing** | | | | | Oil $\delta^{13}C$=-29.3 API=43.7 |
| **Lange** | | | | | Gas/Cond. $\delta^{13}C$=-29.5 API=42.2 |
| **Garn 3** | | Water | | | |
| **Garn 2** | | Water | | | |
| **Garn 1** | Gas/Cond. GOR 1280 $\delta^{13}C$=-28.4 API=47.6 | Water | Tight | | |
| **Ile 3** | | Gas/Cond. GOR 1850 $\delta^{13}C$=-29.7 API=47.7 | Gas/Cond. GOR 1570 $\delta^{13}C$=-28.8 API=46.4 | | |
| **Ile 2** | | Gas/Cond. | Gas/Cond. | Gas/Cond. GOR 1380 | Gas/Cond. GOR 1440 $\delta^{13}C$=-29.0 API=47.8 |
| **Ile 1** | Tight | | | Gas/Cond. | Gas/Cond. |
| **Tofte 2** | | Tight | | | Tight |
| **Tofte 1** | Gas/Cond. GOR 1550 $\delta^{13}C$=-30.2 API=47.6 | Tight | | | Tight $\delta^{13}C$=-29.0 API=31.5 |
| **Tilje 6** | | Water | Gas/Cond. GOR 860 $\delta^{13}C$=-28.7 API=42.4 | | Gas/Cond. GOR 915 $\delta^{13}C$=-28.1 API=41.00 |
| **Tilje 5** | Gas/Cond. GOR 975 $\delta^{13}C$=-29.2 API=45.5 | | | Tight | Gas/Cond. |
| **Tilje 4** | | Oil GOR 410 $\delta^{13}C$=-29.3 API=39.6 | Oil GOR 490 $\delta^{13}C$=-28.7 API=41.2 | Tight | |
| **Tilje 3** | Tight | Oil | | Oil GOR 290 | |
| **Tilje 2** | | | | | |
| **Tilje 1** | Gas/Cond. GOR 1340 $\delta^{13}C$=-28.9 API=47.6 | | | Oil GOR 390 | Gas/Cond. 1390 $\delta^{13}C$=-28.2 API=41.0 |
| **Åre** | Tight | | | Gas/Cond. GOR 620 | |

GOR in $Sm^3/Sm^3$; $\delta^{13}C$ values in ‰ (PDB); API values in degrees

**Fig. 14.** The highly segmented nature of the 'multi-stacked play zones' in the Smørbukk field is illustrated in this tentative grid arranged by tentative reservoir horizons. Note the highly varied GORs and the more uniform carbon isotope values. In well 12-1 there is producible condensate today, whilst residual oil saturation in Garn (in the other wells) indicates palaeo-leakage. The dramatically different GORs illustrate the complex and long filling history of this trap, which was sourced from basically a rather uniform and progressively maturing source rock facies, as suggested by the more uniform carbon isotope values.

well 6506/11-1. In contrast, there is no depth trend in 6406/3-1 and 6506/12-4. In the dry well 6506/11-3 west of Smørbukk and in the dry well 6406/8-1 south of Lavrans and west of Njord (Fig. 1a), the core extracts reach 0.8–1.7 mg/g rock and 0.4 mg/g rock, respectively. In well 6407/5-1 west of Mikkel (Fig. 1a), the amount of core extracts is even lower, i.e. *ca.* 0.3 mg/g rock. The amount of saturated hydrocarbons in both 6407/5-1 and 6506/11-3 is about 70% with *ca.* 17% polar compounds, whilst in well 6406/8-1 the extract composition is very different, being dominated by 47% saturated compounds and 47% polar compounds,

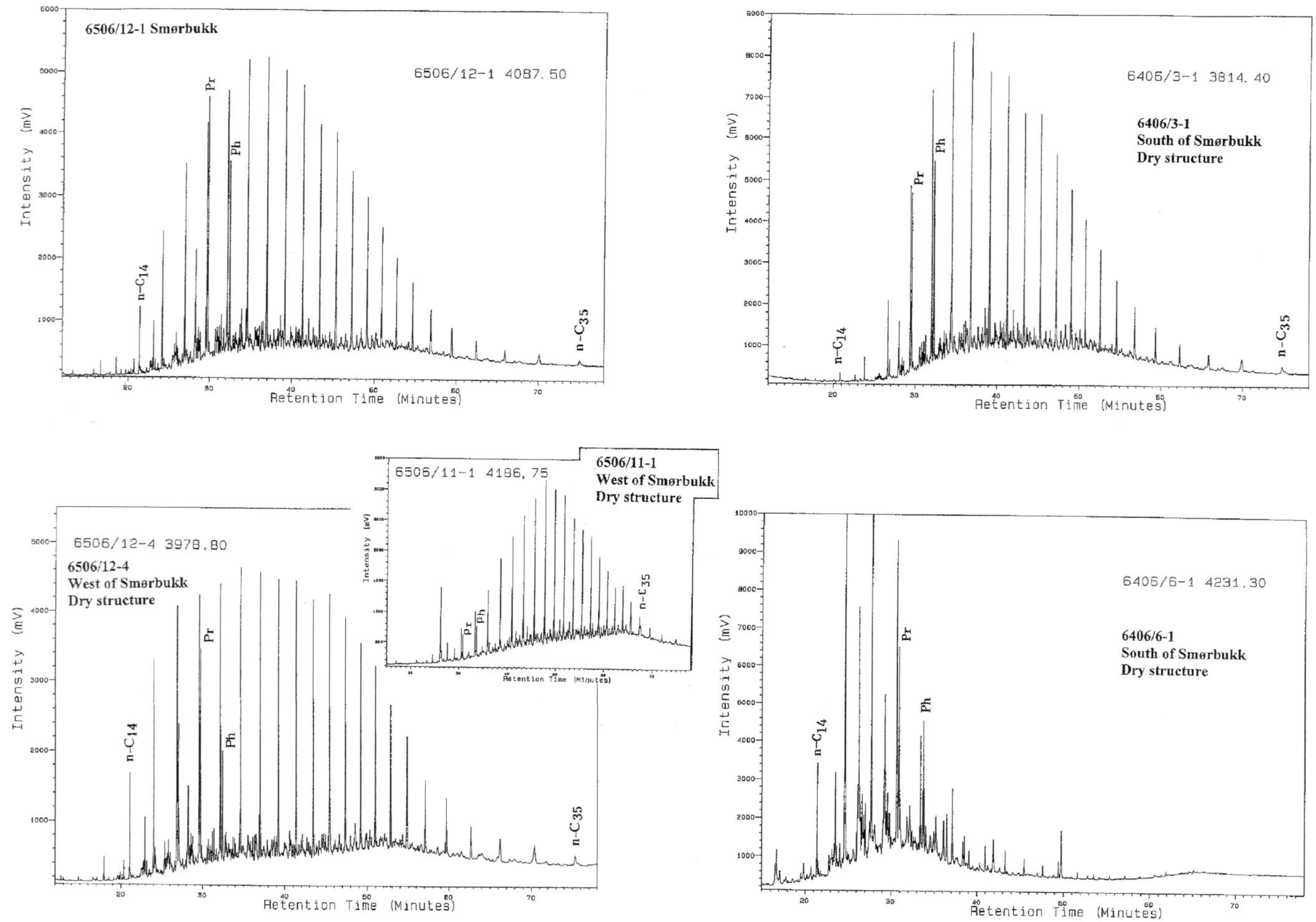


**Fig. 15.** The diesel GC-FID signature of the core extracts from well 6406/6-1 contrasts sharply with the other petroleum type core extracts. All core extracts have, however, suffered severe evaporative losses of hydrocarbons in the $C_{20-}$ range during core retrieval and core storage.

i.e. similar to the extract composition from well 6406/6-1 discussed above.

These additional regional data points on extract amounts and composition support that well 6406/6-1 and 6406/8-1 never contained oil, whilst petroleum existed in well 6407/5-1 and 6506/11-3.

Some representative GC-FID chromatograms from these five wells are shown in Figure 15 and it is indicated that the presently dry well 6406/6-1 did not contain allochthonous thermogenic hydrocarbons of the same type as found in the other wells. The extracts from this well, which all had very polar signatures on Iatroscan TLC-FID, have alkane compositions similar to that observed in diesels and recycled diesel-based drilling fluids. The other chromatograms show that core extracts have suffered severe losses of light hydrocarbons, e.g. 6506/11-1, but retain the characteristic petroleum signatures. The GC-FID chromatograms of the core extracts from the dry wells all substantiate the observations made from the Iatroscan TLC-FID measurements. Undoubtedly, there is in 11-1 pristine/phytane a large loss of $C_{24-}$ components caused by core bleeding during sample retrieval from the high pressure and deep reservoir conditions. Due to the loss of light ends, the pristine/phytane (Pr/Ph) ratios are also severely modified and therefore unsuitable for direct correlative purposes. This applies also for the 12-1 sample in Figure 15. Extrapolation of the *n*-alkane ($C_{30}$–$C_{20}$) and isoprenoid profile to that of the 6506/12-1 DST 7 (whole oil GC-FID, Fig. 16, Tables 1 & 2) sample from the Garn Formation may still allow for a more realistic tentative estimate of the Pr/Ph ratios also in the other core extracts. In this condensate sample (DST 7) the Pr:Ph 1.56 and the Pr:$nC_{17}$ and the Ph:$nC_{18}$ ratios are 0.81 and 0.62, respectively. The extrapolation is possible as the relative loss of isoprenoids relative to *n*-alkanes from the Smørbukk core sample will not differ from that in the dry wells. The absolute and relative losses of these compounds reflect the relative vapour pressures of these compounds and the PVT conditions in the trap, in addition to core porosity/preservation conditions. We suggest tentatively the original alkane/isoprenoid ratio to be less than unity, and the Pr/Ph ratio to be between 1 and 2 for the evaporatively influenced core extracts in Figure 15.

## *Fluorescent petroleum inclusions*

The time for filling of Smørbukk is best evaluated in the context of petroleum inclusions and the burial history of Smørbukk (Fig. 9) using the methods of Karlsen *et al.* (1993). Figure 17 shows ample evidence for fluorescent petroleum inclusions in Smørbukk as well as in the dry wells 6506/12-4, 6506/11-1 and 6406/3-1. Figures 18 & 19 illustrate depth profiles of the amount of fluorescent inclusions in these wells. The petroleum inclusions occur mostly in the cement and along the dust rim of the detrital grains, albeit some secondary inclusions in fractures are also recorded. The size of the gas bubble in the petroleum inclusions is an indicator of the GOR of the entrapped petroleum fluid (Fig. 20). In the Halten Vest samples from well 6506/12-4 and 6506/11-1, two types of inclusion are often found. One type has rather small gas bubbles, i.e. 10–20%, and a second type has gas bubbles constituting *ca.* 70–80% of the inclusion area. This could indicate that the sampled reservoir strata have experienced an increase in GOR during inclusion growth. The inclusions with the larger gas bubbles have blue fluorescent colours whilst the petroleum inclusions with smaller gas bubbles have greenish-yellowish fluorescing colours (cf. Fig. 17). The apparent contradiction of higher maturity oils exhibiting higher fluorescence intensity at shorter wavelengths compared with lower maturity oils which contain higher absolute concentrations of chromophores, i.e. molecules with double bonds, was explained by Bertrand *et al.* (1985). They demonstrated the higher aliphatic to aromatic content in the more mature oils results in a 'dilution' effect which increases the fluorescence intensity and shortens the wavelength, i.e. dilution to 10% increases the intensity by one order of magnitude and shifts the maximum about 80 nm towards the lower wavelength (from *ca.* 560 to 480 nm). Similarly, removal of asphaltenes, which acts as a fluorescence quencher, increases the intensity by *ca.* an order of magnitude and acts to lower the wavelength of the fluorescence maximum by *ca.* 40 nm (Bertrand *et al.* 1985). Using the correlation between fluorescent colours and API by Bodnar (1990), Sellwood *et al.* (1993) and Emery & Robinson (1993), the high GOR bluish fluorescing inclusions can be tentatively assigned an API of 40–47°, whilst the greenish-yellowish lower GOR inclusions may have APIs in the range 30–37°. We do not here present microthermometric evidence to prove that the inclusions with small gas bubbles formed before those with larger gas bubbles. Still, we observe in quartz that petroleum inclusions with large gas bubbles occur at shallower depths in the overgrowths, i.e. towards the porosity, whilst the inclusions with the smaller gas bubbles occur

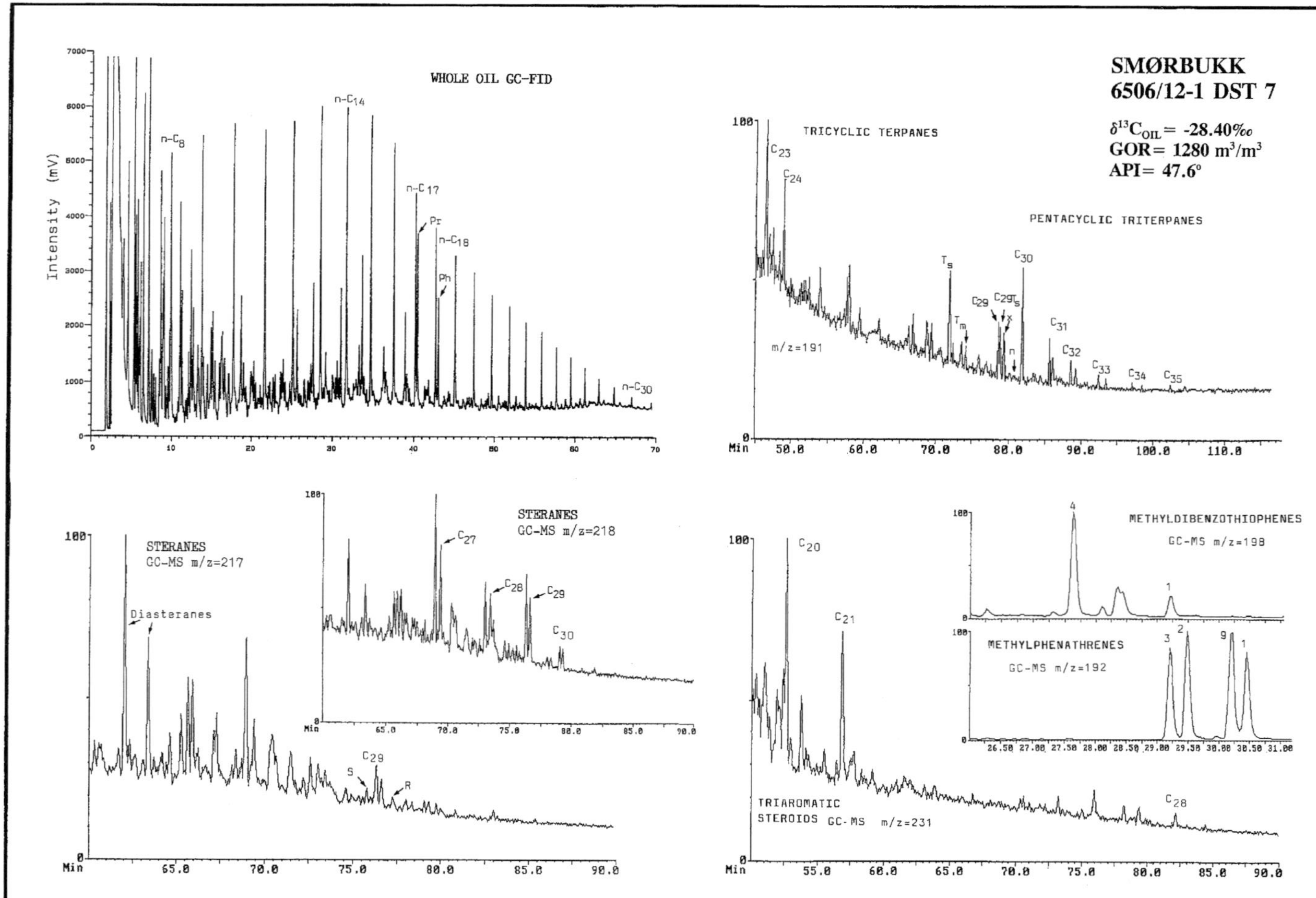


**Fig. 16.** 6506/12-1 DST 7 (Garn Formation) exemplifies molecular distributions in the Garn condensate. The condensate represents geochemically a mixture between an early arriving less mature oil, and a later arriving more mature condensate (with higher GOR). See text for discussion.

**Table 1.** *Variation in gross composition, API and GOR in Smørbukk. The evaporation loss is determined (as the $C_{10-}$ loss) during Iatroscan TLC-FID analysis of the stock tank oil and correlates with GOR and API (cf. Karlson et al.* 1995)

| Well | Sample | SAT (%) | ARO (%) | POL (%) | SAT/ARO | Evaporation loss (%) | API (°) | GOR ($m^3/m^3$) |
|---|---|---|---|---|---|---|---|---|
| 6506/11-2 | DST 6 | 81.48 | 15.99 | 2.53 | 5.10 | 47.88 | 43.7 | 170 |
| 6506/11-2 | DST 5 | 78.15 | 19.71 | 2.14 | 3.96 | 41.38 | 42.2 | 513 |
| 6506/11-2 | DST 4 | 74.14 | 19.91 | 4.49 | 3.80 | 68.22 | 47.8 | 1440 |
| 6506/11-2 | DST 2 | 77.08 | 20.85 | 2.07 | 3.70 | 43.17 | 41.1 | 915 |
| 6506/11-2 | DST 1A | 75.97 | 20.97 | 3.06 | 3.62 | 52.10 | 40.9 | 1392 |
| 6506/12-1 | DST 7 | 77.22 | 20.11 | 2.67 | 3.84 | 50.03 | 47.6 | 1280 |
| 6506/12-1 | DST 5D | 76.47 | 21.89 | 1.64 | 3.49 | 66.23 | | |
| 6506/12-1 | DST 5B | 76.94 | 21.79 | 1.28 | 3.53 | 49.50 | 47.6 | 1551 |
| 6506/12-1 | DST 4B | 75.67 | 22.20 | 2.14 | 3.41 | 50.79 | | |
| 6506/12-1 | DST 4 | 74.39 | 22.33 | 3.28 | 3.33 | 50.23 | 45.5 | 973 |
| 6506/12-1 | DST 3 | 71.33 | 27.23 | 1.44 | 2.62 | 35.23 | | |
| 6506/12-1 | DST 2C | 77.07 | 19.89 | 3.04 | 3.88 | 61.26 | | |
| 6506/12-1 | DST 2B | 69.86 | 28.93 | 1.21 | 2.42 | 26.50 | | |
| 6506/12-1 | DST 2 | 77.63 | 20.40 | 1.97 | 3.64 | 59.90 | 47.6 | 1338 |
| 6506/12-6 | DST 3 | 83.64 | 13.67 | 2.69 | 6.12 | 35.46 | 47.7 | 1850 |
| 6506/12-6 | DST 1 | 75.74 | 21.30 | 3.82 | 3.52 | 41.24 | 39.6 | 410 |
| 6506/12-7 | DST 3 | 76.09 | 20.91 | 3.00 | 3.64 | 60.54 | 46.4 | 1570 |
| 6506/12-7 | DSP 2 | 74.26 | 22.36 | 3.38 | 3.32 | 56.19 | 42.4 | 860 |
| 6506/12-7 | DST 1 | 74.90 | 21.58 | 3.53 | 3.47 | 45.68 | 41.2 | 489 |
| 6506/12-9S | DST 5 | 79.10 | 17.82 | 3.08 | 4.44 | 64.30 | 46.4 | 1379 |
| 6506/12-9S | DST 3 | 74.90 | 21.03 | 4.06 | 3.56 | 43.09 | 39.7 | 292 |
| 6506/12-9S | DST 2 | 71.64 | 24.83 | 3.53 | 2.89 | 23.55 | 34.9 | 390 |
| 6506/32-9S | DST I | 77.17 | 20.39 | 2.43 | 3.78 | 52.21 | 42.3 | 622 |

**Table 2.** *Variation in pristane to phytane (Pr/Ph), pristane to* $n$-$C_{17}$ and $n$-$C_{18}$ *and the carbon preference index (CPI) in Smørbukk. Maturity and phase fractionation are the main factors responsible for the variation*

| Well | Sample | Pt/Ph | Pr/$n$-$C_{17}$ | Ph/$n$-$C_{18}$ | CPI |
|---|---|---|---|---|---|
| 6506/11-2 | DST 6 | 1.58 | 0.64 | 0.47 | 0.99 |
| 6506/11-2 | DST 5 | 1.66 | 0.73 | 0.54 | 0.95 |
| 6506/11-2 | DST 4 | 1.63 | 0.69 | 0.56 | 0.95 |
| 6506/11-2 | DST 2 | 1.61 | 0.75 | 0.50 | 0.99 |
| 6506/11-2 | DST 1A | 1.56 | 0.64 | 0.45 | 0.99 |
| 6506/12-1 | DST 7 | 1.56 | 0.81 | 0.62 | 0.99 |
| 6506/12-1 | DST 5D | 1.59 | 0.70 | 0.52 | 1.02 |
| 6506/12-1 | DST 5B | 1.66 | 0.71 | 0.52 | 0.97 |
| 6506/12-1 | DST 4B | 1.56 | 0.71 | 0.51 | 0.98 |
| 6506/12-1 | DST 4 | 1.65 | 0.71 | 0.50 | 0.98 |
| 6506/12-1 | DST 3 | 1.44 | 0.50 | 0.43 | 1.12 |
| 6506/12-1 | DST 2C | 1.58 | 0.68 | 0.50 | 1.00 |
| 6506/12-1 | DST 2B | 1.56 | 0.48 | 0.39 | 1.06 |
| 6506/12-1 | DST 2 | 1.67 | 0.64 | 0.45 | 1.00 |
| 6506/12-6 | DST 3 | 1.66 | 0.77 | 0.57 | 0.99 |
| 6506/12-6 | DST 1 | 1.57 | 0.71 | 0.51 | 1.02 |
| 6506/12-7 | DST 3 | 1.59 | 0.76 | 0.61 | 0.99 |
| 6506/12-7 | DST 2 | 1.57 | 0.76 | 0.53 | 0.98 |
| 6506/12-7 | DST 1 | 1.63 | 0.68 | 0.44 | 0.98 |
| 6506/12-9S | DST 5 | 1.61 | 0.75 | 0.56 | 0.98 |
| 6506/12-9S | DST 3 | 1.58 | 0.66 | 0.45 | 0.98 |
| 6506/12-9S | DST 2 | 1.49 | 0.61 | 0.44 | 0.98 |
| 6506/12-9S | DST 1 | 1.67 | 0.74 | 0.51 | 1.00 |

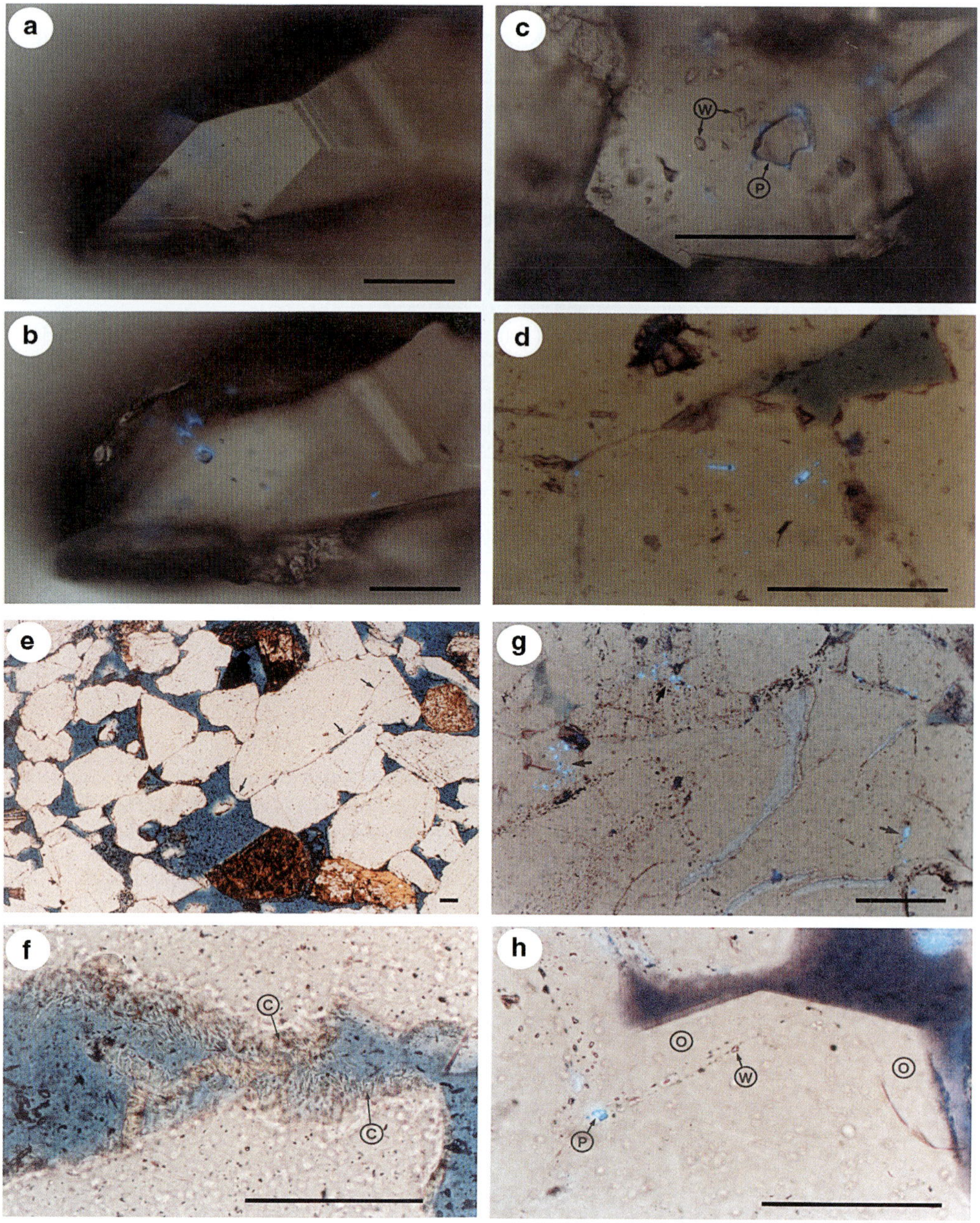


**Fig. 17.** UV-photomicrographs of thin sections and cleansed sandstones. All depths are core depths. Scale bar = 100 μm. (**a**) Euhedral overgrowths on quartz grain from cleansed sand from 4186.75 m in the dry well 6506/11-1. (**b**) Same as (a), but with 'deeper' focus, revealing a fluorescent inclusion with rather big (*ca.* 50 vol%) gas bubble. Due to refraction in the quartz crystal, the inclusion appears as three individual inclusions. The large gas bubble signifies that the trap contained a high GOR fluid before caprock failure. (**c**) The dry 6506/12-4 well at 3989.50 m. Quartz grain from cleansed sand, with water inclusions (w) and a petroleum inclusion (p) with fluorescent petroleum lining the inside of the mainly gas filled inclusion. The inclusion represents a high GOR (40–47° API) petroleum which existed in the trap at the time the inclusion closed and before leakage occurred. (**d**) Two large and several smaller fluorescent inclusions with rather small gas bubbles (5–10 vol%) appearing in a quartz grain in thin section from 4145.25 m in well 6506/11-1. The petroleum inclusions represent medium GOR black oil. (**e**) Thin section from 3805.00 m in well 6406/3-1,

closer to the detrital surface. This textural relationship suggests that the petroleum inclusions with the larger gas bubbles formed after the other population (Fig. 20).

A progressively maturing source rock would give rise to petroleums of increasing GOR due to maturity (England & Mackenzie 1989). Such higher GOR petroleums will, when entering a reservoir, either displace or mix with earlier reservoired petroleum with lower GOR, due to the differences in density between the two fluids. It is irrelevant whether mixing or pure displacement actually occurs, as the important element here concerns the fact that arrival of the higher GOR petroleum will in either case result in increased GOR for the trap and the formation of inclusions with larger gas bubbles (Fig. 20). Similar observations were recorded in the Hild field in the Viking Graben (cf. Skålnes 1993). In that case, three types of inclusion were found: greenish-yellowish with small 5% gas bubbles, bluish inclusions with gas bubbles ranging around 5–10%, and light blue fluorescent petroleum inclusions with large gas bubbles of 40–80%. It was also concluded, on the basis of microthermometry and textural relationships, that this could indicate entrapment of petroleum from earlier-formed low GOR petroleums at *ca.* 60 Ma bp, towards higher GOR petroleum of significantly higher APIs and also higher maturities. Furthermore, Johansen (1997), who studied fluorescent inclusions in the presently dry but bitumen-rich uplifted structure 7219/9-1 from the Barents Sea, found mostly low–medium GOR inclusions with *ca.* 5–20% gas bubbles, indicating that oil had been present in this structure. However, the presence of very few petroleum inclusions with large gas bubbles of about 85% was noted. This could indicate that this structure experienced a high GOR petroleum influx before onset of uplift and caprock failure.

Data on the onset of quartz diagenesis in Haltenbanken (Bjørlykke *et al.* 1986, 1989; Ehrenberg 1990; Walderhaug 1990; Ehrenberg *et al.* 1992; Storvoll *et al.* 2002), and generally offshore Norway (Walderhaug 1994, 1995, 2000; Walderhaug *et al.* 2000) indicates that quartz from pressure solution is effectively precipitating and forming petroleum inclusions from about 80–90 °C. The lower level of homogenization temperatures for inclusions in quartz at Haltenbanken and specifically Smørbukk fall mostly in the 90–100 °C range, for both water and petroleum inclusions with maximum, often in the range 110–120 °C. Quartz brought into solution from sites of pressure solution is moving along water films on interconnected grains to sites of precipitation (Nedkvitne *et al.* 1993). Karlsen *et al.* (1993) concluded, in the case of overlapping homogenization temperature histograms for water and oil inclusions, that both systems are saturated with respect to gas. This means that pressure corrections are not necessary for estimating the trapping temperatures. Microthermometric distributions for petroleum inclusions in deep Haltenbanken fields and in Halten Vest generally show a tail of inclusions formed at higher temperatures, i.e. towards the maximum reservoir temperatures of about 150–160 °C.

The relatively smaller proportion of later formed inclusions may reflect the presence of fewer lattice defects in quartz at the higher temperatures when diagenesis has proceeded for substantial time. Thus, diagenesis has already 'covered up' irregularities on the detrital grains with their mixed wettabilty characteristics where inclusions tend to form (Karlsen *et al.* 1993; Nedkvitne *et al.* 1993). The net result of this process would be the reduced possibility of petroleum inclusion entrapment.

Oil inclusions formed in Smørbukk at temperatures of about 90–100 °C confirm that oil was present in the reservoir strata at least as early as corresponding to this temperature. The burial history of Smørbukk (Fig. 9) suggests that oil was present in the Smørbukk field at least as early as 70–50 Ma bp. The Garn Formation is part of the Fangst Group and experienced temperatures exceeding 80 °C 70–60 Ma bp, whilst the deeper Tilje Formation in the Båt Group experienced such temperatures by 90–80 Ma bp.

No fluorescent inclusions were recorded in well 6406/6-1, as can be seen in Figure 18, which display the relative number of fluorescent petroleum inclusions in the samples from the five wells. The relative number of inclusions is a subjective estimate of the number of fluorescent inclusions

demonstrating reduced quartz cementation due to chlorite coatings (cf. Ehrenberg 1991). The euhedral quartz surfaces and dust rims (arrows) prove that quartz overgrowths are present. Thus, petroleum inclusions could have formed had oil ever existed in this trap. **(f)** Chlorite coatings (c) on quartz grains in thin section from 3805.00 m in well 6406/3-1 (same sample as in photo (e)). **(g)** Clusters of small, fluorescent inclusions (arrows) in quartz in thin section from 4145.25 m in well 6506/11-1. APIs of the inclusions may fall in the range 30–37° (cf. Bodnar 1990). Note that colours of the reproductions deviate from those originally observed. **(h)** Petroleum inclusion (p) and water inclusion (w) occurring in the dust rim between the detrital quartz grain and the euhedral overgrowth (o). From 3996.00 m in well 6506/12-1.

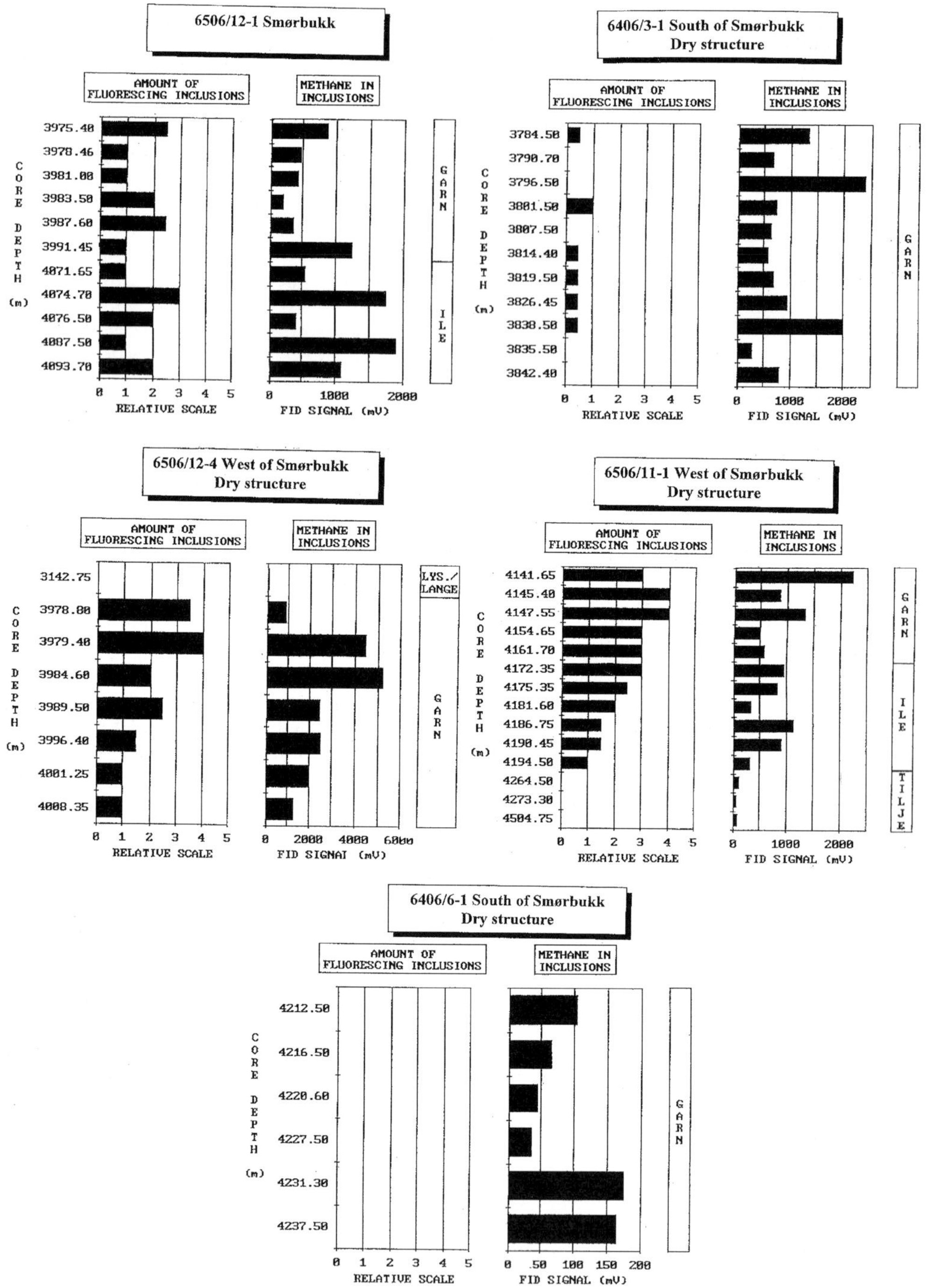


**Fig. 18.** The amount of methane in bulk crushed cleansed sand grains and the relative amount of fluorescent petroleum inclusions on an arbitrary scale from 0 to 5 illustrate that all the dry structures, apart from 6406/6-1, previously contained oil. See text for discussion.

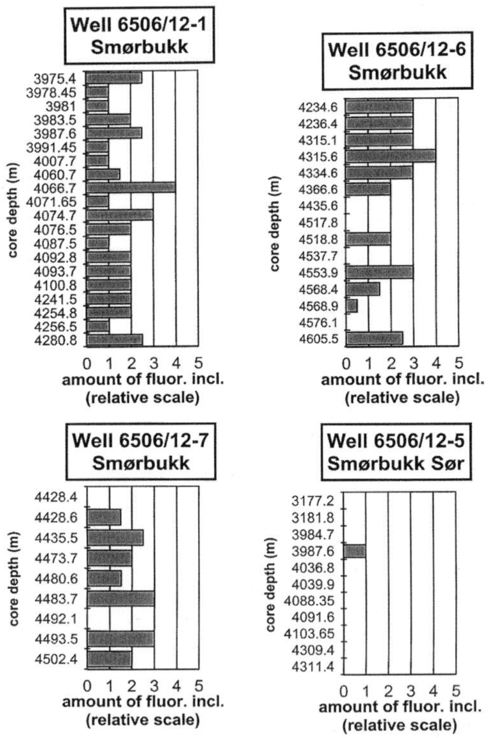


**Fig. 19.** The amount of fluorescent petroleum inclusions in several Smørbukk wells and one well from Smørbukk Sør. Note that fewer inclusions occur in Smørbukk Sør, a number on a par with that in 6406/3-1 (Fig. 16). This indicates tentatively that oil arrived at more or less the same recent time in these two structures.

observed during UV-microscopy. In all the other wells fluorescent petroleum inclusions are observed. The amount of such fluorescent inclusions in the dry wells 6506/12-4 and 6506/11-1 is as high as in the Smørbukk well (6506/12-1). These observations substantiate the observations made from Figures 13 and 15, i.e. that no petroleum occurred in the presently overpressured structure 6406/6-1, whilst the other dry wells did contain petroleum earlier.

According to Walderhaug (1994), it takes tentatively on the order of 17 Ma to form a 4 µm inclusion in quartz overgrowth at 100 °C, whilst at 120 and 140 °C the inferred formation time is on the order of 6 and 2 Ma, respectively. One may speculate if the decreasing amount of inclusions with increasing depth in well 6506/12-4 and 6506/11-1 (Fig. 18) could reflect a progressive lowering of the oil/water contact of a palaeo-petroleum. The stacked and multi-layer structure of these traps, schematically illustrated in Figure 21, complicates this simple interpretation. However, in 6506/12-4 this trend occurs within the uniform Garn Formation and it is therefore likely that the high frequencies of inclusions in the upper part of this formation could reflect that the top of the structure held oil for a longer time than the deeper sands as the pressure exerted by the oil columns will be able to force oil into smaller openings near the crest (top) of any such structure (cf. Nedkvitne *et al.* 1993), resulting in increased petroleum inclusion formation.

Similarly, in well 6506/11-1, the inclusion abundance may reflect that oil entered the deeper Ile Formation later than the time for the petroleum invasion in the Garn Formation, which constitutes the apex formation. The petroleum inclusion abundance in the best strata in the 12-4 and 11-1 structures is comparable with that in the Smørbukk 12-1 well. We find approximately the same frequencies in the other Smørbukk wells 6506/12-6 and 6506/12-7 (Figs 18 & 19). This indicates tentatively that petroleum has occurred in the sands penetrated by these wells for approximately the same length of time.

This would imply that caprock failure occurred recently for the dry 12-4 and 11-1 structures. Along this line of reasoning, we suggest that the dry 6406/3-1 contained oil for a much shorter time than the 12-4 and the 11-1 structures. The number of fluorescent oil inclusions in the Smørbukk Sør field (Fig. 19) is similar to that found in the 6406/3-1 dry trap (Fig. 18). This tentatively indicates that the Smørbukk Sør field filled relatively recently and that it has a totally different filling history from that of the nearby Smørbukk field (Fig. 1a).

We could not detect fluorescent inclusions in the other dry wells, 6406/8-1, 6506/11-3, nor in well 6507/5-1, all of which contain quartz overgrowths, suggesting that fluorescent petroleum inclusions could have been entrapped had petroleum occurred in these structures. However, a lack of observable fluorescent inclusions alone cannot be taken as evidence for lack of petroleum in the investigated sandstones. The conclusions drawn are based upon comparison, in all wells, of inclusions in the same or related formations, at virtually the same depths. The samples selected for the analysis are uniform in grain size, degree of polygrains and overall sedimentological characteristics.

## *Gas from inclusions*

Oil type inclusions, i.e. fluorescent inclusions constituting at room temperatures a liquid

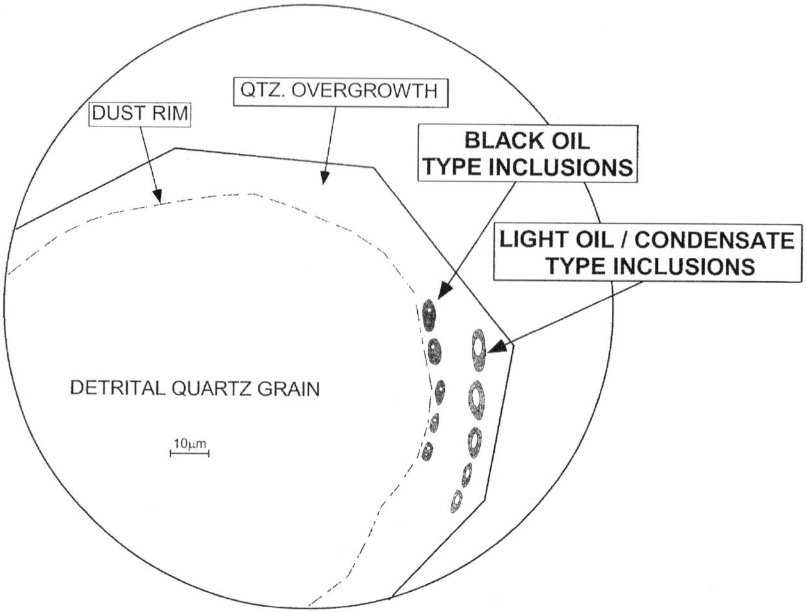


**Fig. 20.** Fluid inclusions in diagenetic minerals like quartz may represent 'hermetically sealed' testers of the fluid existing in fields and dry structures at the time of inclusion formation. Chemical analysis of hydrocarbon gas and oil in such inclusions may reveal the maturity and source of the hydrocarbons as the trap has been 'seeing down' into the generative basins during progressive burial. API and GORs of the petroleum in the inclusions are tentatively estimated from the fluorescence colour and the relative size of the gas bubble.

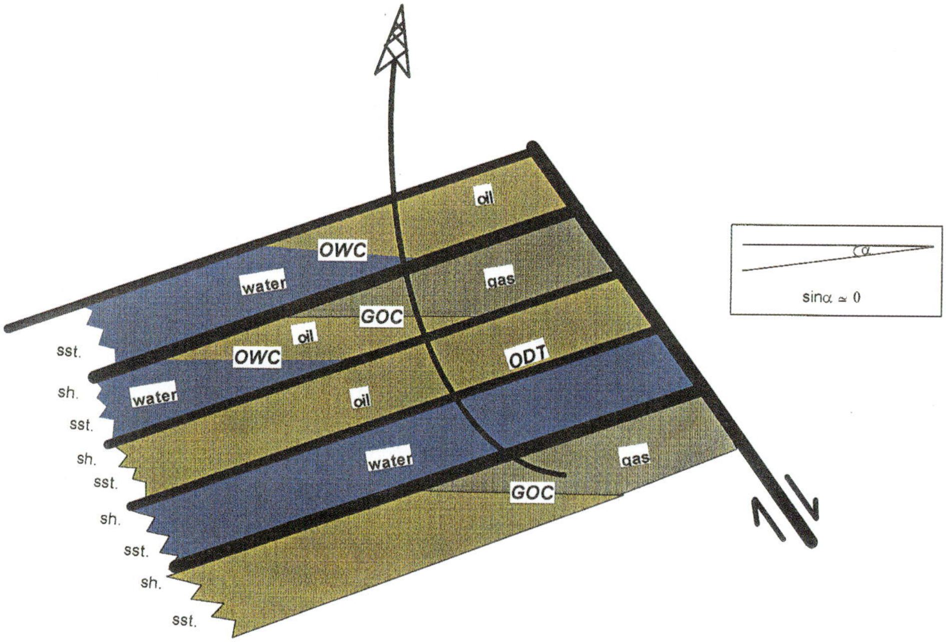


**Fig. 21.** Tilted rotated fault blocks with repetitive clastic lithologies of sandstones, shales and coals causes a multitude of down-to contacts, e.g. oil down-to contacts (ODC) and true oil–water contacts are seldom intersected by the well path. A slight tilting of the block results in significant mobilization of OWCs, gas–oil contacts and gas–water contacts, i.e. hydrocarbons get mixed within the individual compartments.

phase with a minor gas phase, are easily detected during microscopy. Fluorescent inclusions homogenizing to liquid are termed oil type inclusions. Inclusions which at ambient temperature contain gas as the major phase may be more difficult to detect, especially if the gas phase constitutes more than 95% of the inclusion volume. Such gas-rich inclusions may represent entrapment of condensate, wet gas or dry gas and are progressively more difficult to detect and differentiate from water/gas inclusions during microscopy. Detection of any type of petroleum containing inclusions, not only detectable fluorescent inclusions, is important for understanding petroleum filling of reservoirs. For this reason, cleansed sand grains (Karlsen *et al.* 1993) should always be crushed so that liberated hydrocarbon gas can be analysed by GC-FID methods. Figure 18 shows the amount of methane liberated from such crushed inclusions and these depth trends can be easily compared with the amount of fluorescent inclusions. In well 6406/6-1 methane is recorded in quantities higher than found in magmatic inclusions (cf. Karlsen *et al.* 1993). Some ethane and propane is also present in the 6406/6-1 samples (Fig. 22), in addition to trace amounts of butanes. The amount of these gas constituents is so low, however, that it cannot be excluded that these, along with the methane, are occurring in true water solution in the inclusions. Magmatic inclusions may hold higher boiling hydrocarbons than methane (cf. Galimov 1973; Karlsen *et al.* 1993). Still, it is more likely that the bulk of the methane, ethane, propane and butane in the 6-1 samples originated from a source rock, as the 6-1 trap occurs in a petroliferous basin. The ratio of iso- to n-butane isomers is approximately one in the samples from this well. A value in the range 0.45–0.55 is more typical for gas generated in the oil window (cf. Reznikov 1969). Higher values may indicate immature conditions, a less likely interpretation in this case. If we assume the water solubility of the butane isomers to be comparable, then the unit ratio between them potentially could suggest that the entrapped gaseous species occur in water inclusions. We have no other data to support this hypothesis, but we have on other occasions made similar observations of approximately unit ratios between the butane isomers in low yield inclusion samples from dry structures that lack evidence of ever having contained oil or producible gas.

The samples from the other wells contained over an order of magnitude more butane than 6406/6-1 (Fig. 22). The gas in all these other inclusion samples was wetter and had iso- to normal-butane ratios typically found for mature gas (cf. Reznikov 1969; Durand & Espitalié 1972; Astafev *et al.* 1973; Alexander *et al.* 1981). The gas composition of the inclusions in the Smørbukk 6506/12-1 well (Fig. 22) shows an undisputed oil associated thermogenic origin, with $C_{2+}$ (wetness) in the range 15–37% (cf. Schoell 1983). This same general wet gas signature is observed in the inclusions from wells 6506/12-4, 6506/11-3 and 6506/11-1 which have $C_{2+}$ wetnesses in the range 25–27%. We may thus conclude that the bulk of the inclusions in the samples from the Smørbukk (6506/12-1), 6506/11-1, 6506/12-4 and 6506/11-3 structures, all contain the same type of oil-associated gas. This supports the hypothesis that these structures received oil from similar types of source rocks at similar maturities. Drier gas is observed in the inclusions from wells 6406/3-1, where the $C_{2+}$ content falls in the 10–12% range, i.e. representing gas associated with a condensate (cf. Schoell 1983), whilst wells 6406/6-1 and 6407/5-1 represent even drier systems with $C_{2+}$ content in the range 7–9%.

Well 6407/5-1 occurs west of the condensate-containing Mikkel field. The core extracts from the 5-1 structure represent lean allochthonous bitumen, and the gas inclusions thus support a model in which the investigated strata represent either a migration avenue or a leaked structure. The most likely hydrocarbon phase to have existed in these sands is a gas condensate. The core samples from the 6406/6-1 structure contained drilling fluid (diesel) and the gas composition and the trace amount yields from the inclusions suggest that the gaseous hydrocarbons occur only in water solution, i.e. this was always a dry structure. The 6406/3-1 structure contains allochthonous bitumen and has much higher gas inclusion yields. This gas has a normal mature iso- to normal-butane ratio (Fig. 22). The wetness is only 5% $C_{2+}$, similar to that found in inclusions from Tyrihans Nord and Sør. These traps do not contain abundant fluorescent oil-type inclusions and the dry gas composition suggests that most of the inclusions contain only water with dissolved hydrocarbons. This indicates a very recent oil arrival to these traps, i.e. the few petroleum inclusions are outnumbered by water inclusions—hence the dry gas signatures. The 6406/8-1 structure only contained trace amounts of *in-situ* generated bitumen and the gas from inclusions constitutes trace amounts and is very dry with only 4% $C_{2+}$. Thus, this trap was always dry. Another characteristic element of the lean dry inclusion gas distributions in these two dry traps is the fact that the iso- to normal-butane ratios deviate

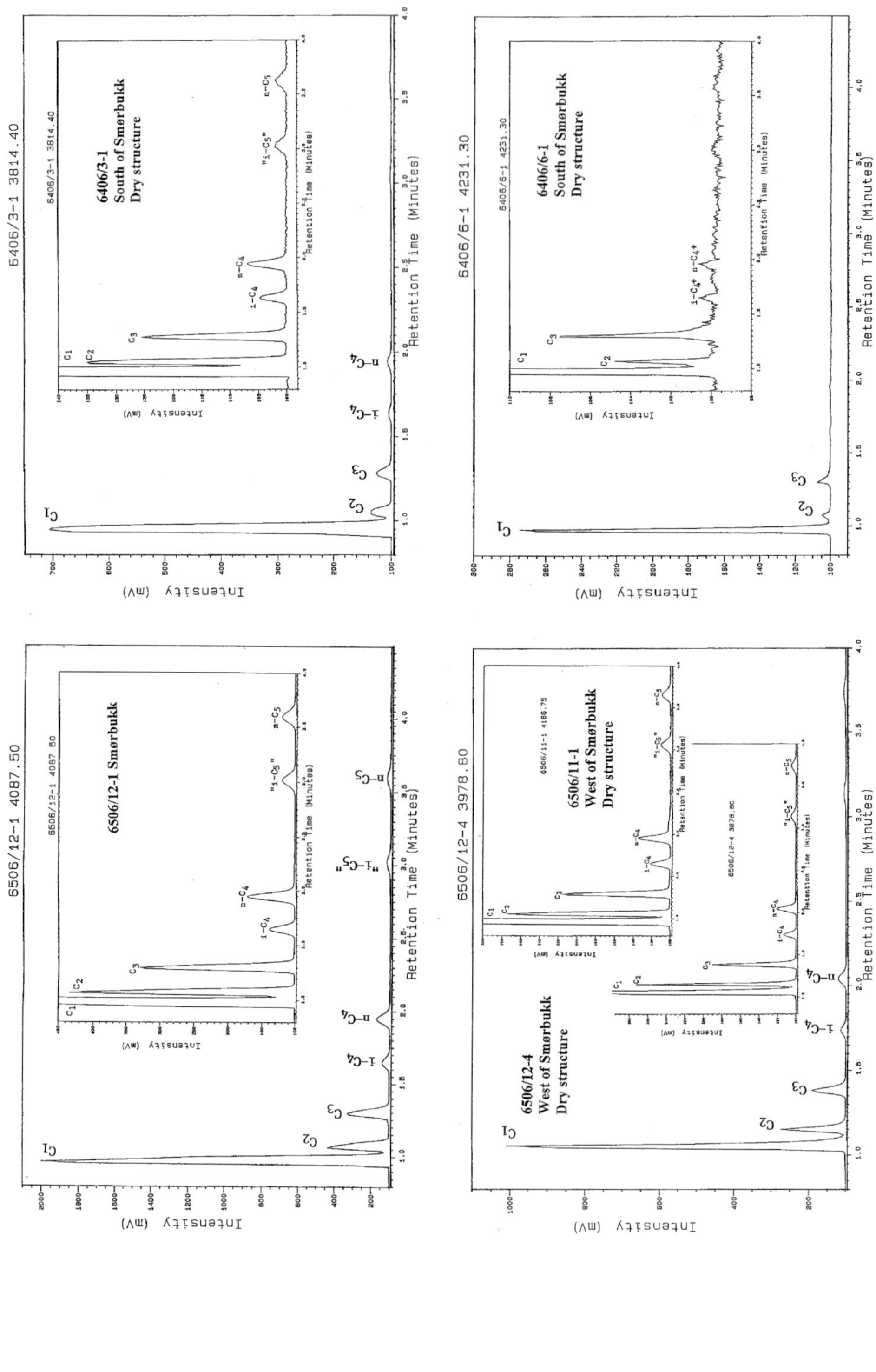


**Fig. 22.** The gas liberated from inclusions in cleansed sand grains from the presently dry structures in Halten Vest has the same wetness as gas liberated from inclusions in Smørbukk (e.g. 6507/12-1). Trace amounts of dry gas signatures are observed inclusions in well 6406/6-1. This well does not contain observable fluorescent inclusions (Fig. 18), nor core extracts representing migrated oil (Figs 15 & 18). See Figures 1a & 27 for well locations.

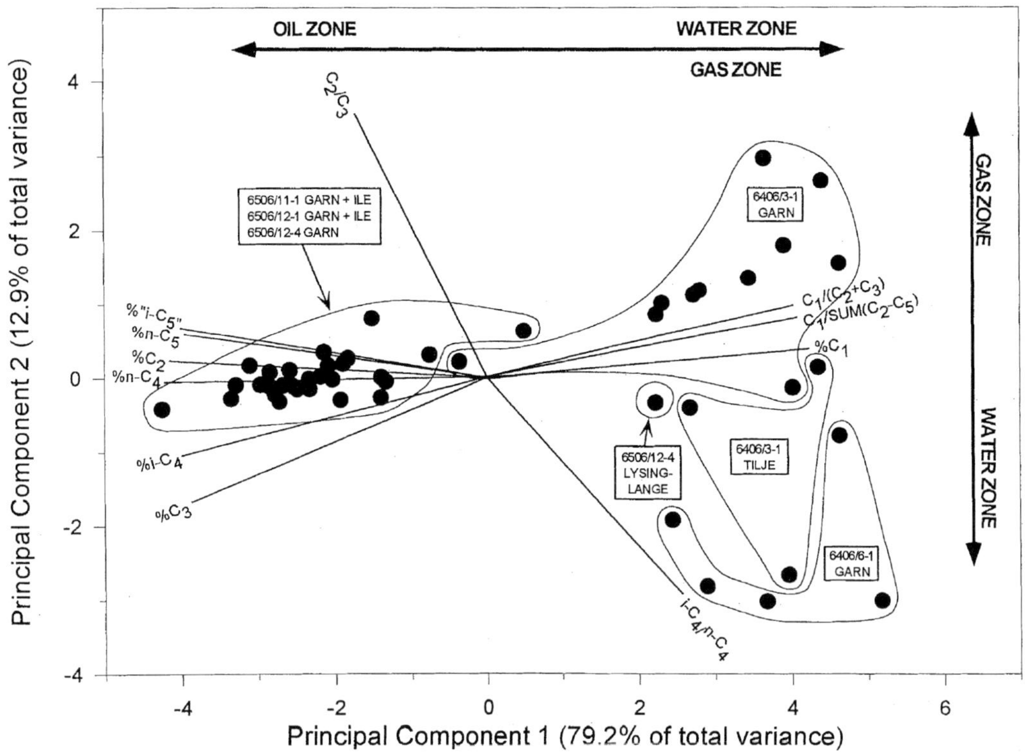


**Fig. 23.** Principal component plot of the composition of the gas liberated from inclusions. Note that the main variation occurs along the horizontal axis representing gas wetness. The inclusions in sandstones from the present oil zone in Smørbukk (12-1) group together with samples from the dry structures 11-1 and 12-4 west off Smørbukk, signifying that inclusions in these dry structures contain the same gas type. Hence these different traps may earlier have been part of the same petroleum province and hydrocarbon machinery. A drier gas signature is observed in the structure with few inclusions (3-1) and in the 'duster' 6-1.

from the typical values around 0.5. This may be due to this ratio being controlled, in these traps, by the actual solubility in water of these compounds, i.e. free gas did not exist in these traps.

Figure 23 is a principal component plot of the gas content of the inclusions from Smørbukk and Halten Vest. Most of the variation in the diagram is due to variation in wetness. Note that all the Garn and Ile Smørbukk 12-1 samples plot in the same region as the Garn and Ile samples from the dry wells 11-1 and 12-4. This indicated that these structures contained the same type of gas during inclusion growth.

The dry signatures of the Garn inclusion gas samples from 3-1 and 6-1 is also illustrated and the separation between the two dry wells in terms of gas composition is mainly due to the iso- to normal-butane ratio. The Tilje samples from 3-1 have gas signatures more similar to the gas signatures in 6-1, i.e. Tilje in 3-1 most likely never contained gas, nor oil zones. The gas composition of the Cretaceous Lysing/Lange sand in 6506/12-4 is very different from the Garn samples in the same well and reflects a lack of allochthonous petroleum in these strata.

## *Biomarkers from inclusions*

Crushing of cleansed sands under an organic solvent allows extraction of hydrocarbons from inclusions in quartz and feldspars (cf. Karlsen *et al.* 1993). Figure 24 illustrates the biomarker parameters (Table 3) recorded from DSTs and fluid inclusions from Smørbukk, Smørbukk Sør and Halten Vest.

The main variation in the data set occurs along the inferred maturity axis. Thus, in general, the DST samples show the same facies signatures as the extracts from the petroleum inclusions, but at a higher maturity. This could correspond

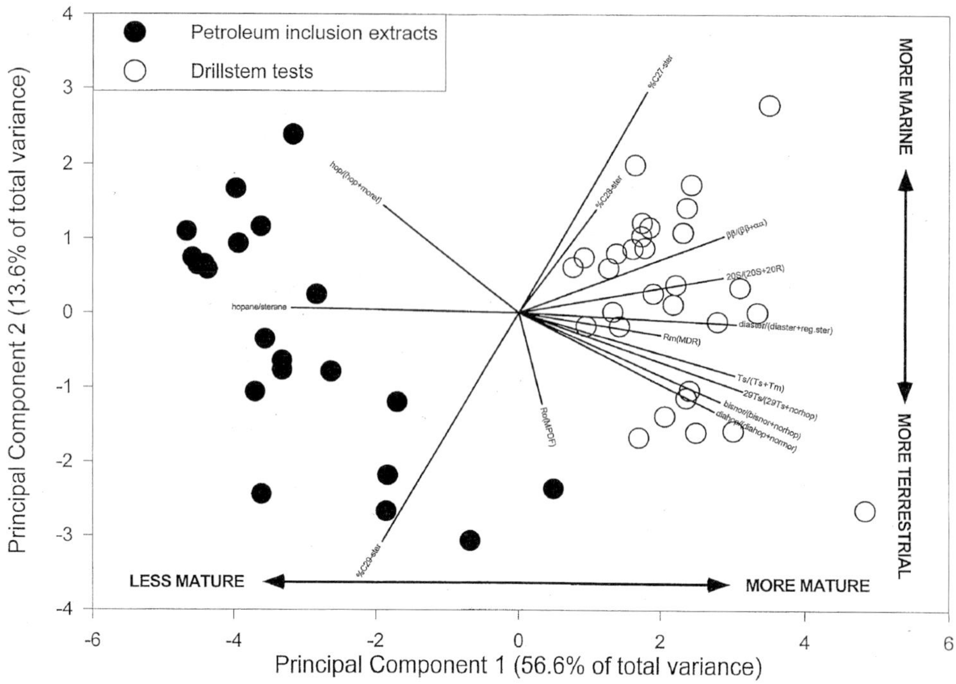


**Fig. 24.** Geochemical evidence suggest that the oil found today in petroleum inclusions in Smørbukk as well as in Halten Vest, which started to form 70–50 Ma bp, originated from the Spekk Formation This petroleum is of the same facies as found in Smørbukk and Smørbukk Sør, but less mature.

**Table 3.** *Biomarker parameters*

1. $T_s/(T_s + T_m)$ (Seifert & Moldowan 1978)
2. Diahopane/(diahopane + normoretan) (Cornford *et al.* 1986)
3. 22S/(22S + 22R) of $C_{31}$ 17$\alpha$ (H),21$\beta$ (H)-hopanes (Mackenzie *et al.* 1980)
4. $C_{30}$-hopane/($C_{30}$-hopane + $C_{30}$-morethane) (Mackenzie *et al.* 1985)
5. $29T_s/(29T_s$ + norhopane)
6. Bisnorhopane/(bisnorhopane + norhopane)
7. $C_{23}$-$C_{29}$ tricyclic terpanes/$C_{30}$ $\alpha\beta$-hopane (modified from Mello *et al.* 1988)
8. $C_{24}$ tetracyclic terpane/$C_{30}$ $\alpha\beta$-hopane (Mello *et al.* 1988)
9. Hopane/sterane (Mackenzie *et al.* 1984)
10. $\beta\beta/(\beta\beta + \alpha\alpha)$ of $C_{29}$ (20R + 20S) sterane isomers (Mackenzie *et al.* 1980)
11. 20S/(20S + 20R) of $C_{29}$ 5$\alpha$ (H),14$\alpha$ (H),17$\alpha$ (H) sterane isomers (Mackenzie 1984)
12. Diasteranes/(diasteranes + regular steranes)
13. % $C_{27}$ of $C_{27}$ + $C_{28}$ + $C_{29}$ $\beta\beta$-steranes (Mackenzie *et al.* 1985)
14. % $C_{28}$ of $C_{27}$ + $C_{28}$ + $C_{29}$ $\beta\beta$-steranes (Mackenzie *et al.* 1985)
15. % $C_{29}$ of $C_{27}$ + $C_{28}$ + $C_{29}$ $\beta\beta$-steranes (Mackenzie *et al.* 1985)
16. $C_{20}/(C_{20} + C_{28})$ triaromatic steroides (TA) (Mackenzie *et al.* 1985)
17. $C_{28}$ TA/($C_{28}$ TA + $C_{29}$ MA) (Mackenzie *et al.* 1985)
18. Methyl phenanthrene ratio (MPR) (Radke *et al.* 1982b)
19. Methyl phenanthrene index 1 (MPI1) (Radke *et al.* 1982a)
20. Methyl phenanthrene distribution fraction 1 (MPDF (F1)) (Kvalheim *et al.* 1987)
21. Methyl dibenzothiophene ratio (MDR) (Radke 1988)
22. Calculated vitrinite reflection, $R_m = 1.1 \times \log_{10}$ MPR + 0.95 (Radke 1988)
23. Calculated vitrinite reflection, $\%R_c = 0.60 \times$ MPI1 + 0.40 (Radke 1988)
24. Calculated vitrinite reflection, $\%R_o = 2.242 \times$ F1 − 0.166 (Kvalheim *et al.* 1987)
25. Calculated vitrinite reflection, $R_m = 0.073 \times$ MDR + 0.51 (Radke 1988)
26. 3-methyl phenanthrene/4-methyl dibenzothiophene (Hughes *et al.* 1995)

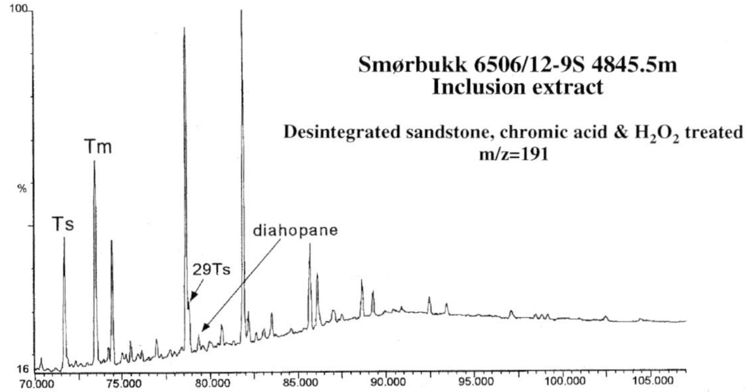


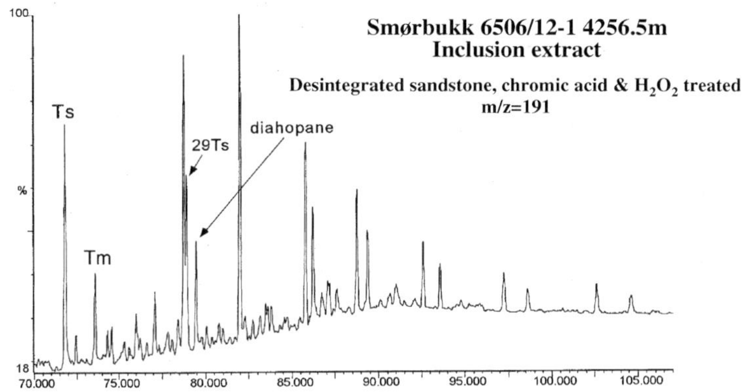


**Fig. 25.** Biomarkers extracted from inclusion in different wells in Smørbukk show varying maturities. This reflects the very long filling history of this complex field. See Figure 27 for well location.

with the inclusions having trapped a palaeopetroleum generated when the source rocks were shallower and were expelling petroleums of lower maturity.

Figure 25 illustrates the type of maturity variation occurring between individual fluid inclusion extracts in Smørbukk. It is clear from this variation that inclusions in quartz does not as a rule contain only very low maturity signatures. However, the maturity signature will depend on the actual burial and filling history of the field.

The sterane distribution of DST and inclusions samples from Halten Vest, Smørbukk and Smørbukk Sør (Fig. 26) suggests that both inclusions and DSTs have been sourced from type II marine source rocks, i.e. the Spekk Formation. There is no statistically relevant difference between the biomarker parameters extracted from inclusions in 6506/12-1, 6506/11-1, 6506/12-4 and Smørbukk Sør. This conclusively shows that inclusions formed in these structures all trapped petroleum from the Spekk Formation.

Above, we have presented evidence to suggest that Smørbukk contained oil as early as 70–50 Ma bp. It was suggested tentatively by Olstad *et al.* (1997) that Smørbukk could have filled across the present sealing fault zone. The geochemical data on the inclusions presented here and the minor amounts of fluorescent inclusions in Smørbukk Sør and 6406/3-1 plus the lack of petroleum inclusions in 6406/6-1 support this conclusion. At this time, the drainage areas in the vicinity of Smørbukk were immature, implying that early filling occurred from distant and deeper western regions (see Fig. 12). This means that migration took place across the presently sealing fault zone west of Smørbukk (Fig. 27). In this context, the amount of fluorescent inclusions and their gas content in the 6506/12-4 and the 6506/11-1 structures is also suggestive of early filling. This evidence implies that overpressure development and sealing of

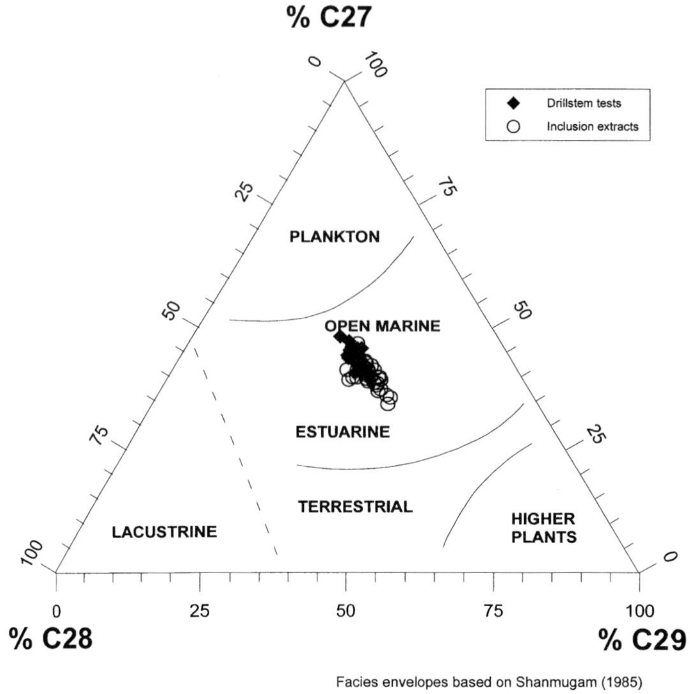


**Fig. 26.** The sterane distribution for fluid inclusions from Halten Vest is similar to the variation observed for the inclusions and present DSTs from the Smørbukk and Smørbukk Sør petroleum systems. This suggests strongly that their SR facies origin is the same, i.e. the same type II marine Spekk Formation The trend along the $C_{27}$–$C_{29}$ axis is representing phase fractionation/evaporative fractionation plus maturity effects (cf. Karlsen *et al.* 1995).

the main fault zone west of Smørbukk significantly post-dates initial filling and that Smørbukk and Halten Vest at that time constituted one unified migration regime.

## Smørbukk—a highly compartmentalized structure with a long filling history

The current Spekk maturity of 0.8–1.0% $R_c$ in the vicinity of Smørbukk (Fig. 12) suggests that the Smørbukk proximal regions were non-expelling 40–70 Ma bp when these Spekk shales were approximately 2 km shallower (cf. Figs 1b & 3). The Spekk Formation east of Smørbukk only acquired its 'oil-window' maturities during the past 10 Ma, with an accelerated maturation during the past 5 Ma (cf. Fig. 9). Thus, if Smørbukk received oil from Spekk type source rocks *ca.* 70–50 Ma bp, as indicated by the data discussed above, then this oil had to have come from distant western locations. Moreover, the petroleum had to cross the presently pressure sealing main Smørbukk fault (Fig. 27).

Such early initial filling of Smørbukk would imply that the trap has received petroleum of progressively increasing maturity and GOR (cf. England & Mackenzie 1989) during a very long time interval. This would be expected to show up in the core extracts and DSTs. The span in maturities, representing 70–50 Ma of progressively increasing maturities of the SR, may be the reason for the variation observed in Ph/$n$-$C_{18}$ versus Pr/$n$-$C_{17}$ ratios in Smørbukk (Table 2). Figure 28 suggests that most of this variation is simply due to maturity and not to facies, and supporting evidence for this is found in the sterane distributions (Fig. 26).

Figure 14 illustrates the spatial variation in GORs in Smørbukk and the location of water zones, tight non-producible zones and petroleum-producing reservoir sections. Large variations in GORs are often found in surprising proximity, both vertically and laterally, and this is irrefutable evidence of the highly segmented nature of this accumulation. It is clear from these highly variable GORs that all these

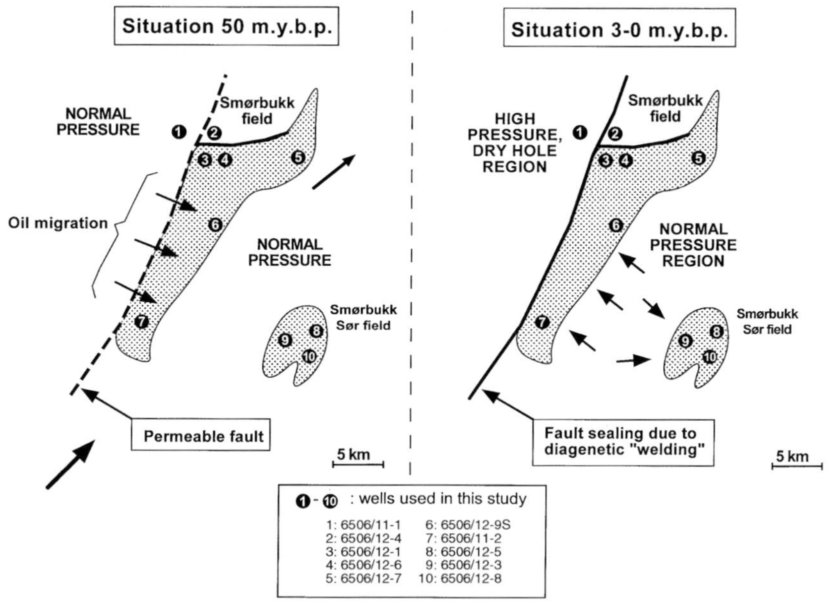


**Fig. 27.** Our conclusive model for filling of Smørbukk. Migration from the western regions, where the Spekk Formation is currently over-mature, must have occurred through the now pressure sealing Smørbukk main fault zone. The low abundance of petroleum inclusions in Smørbukk Sør testify to its late filling from the local basin between Smørbukk and Smørbukk Sør.

gas-condensates and oils cannot at the present time be in communication with each other, nor with an expelling SR basin. A central question concerns how this giant field filled.

Rotated fault blocks like Smørbukk and the Halten Vest high-pressure structures have a general morphology as illustrated in Figure 21 and the shales separating the reservoir sands

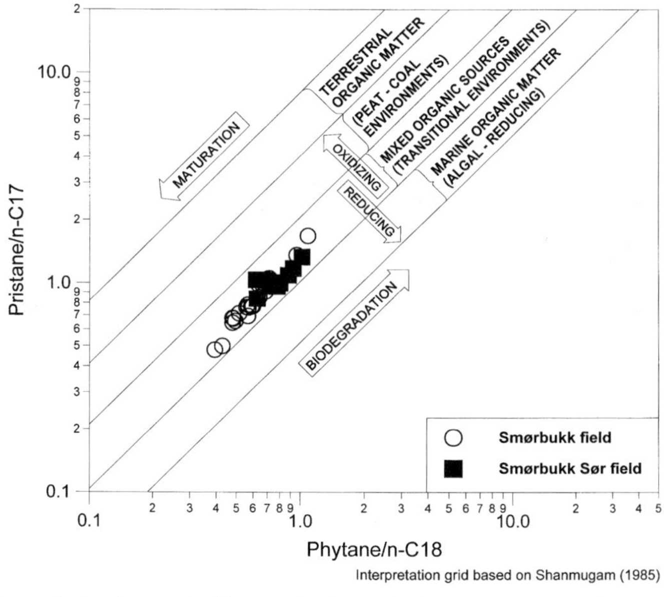


**Fig. 28.** The maturity variation in Smørbukk DST fluid samples is much more significant than facies variations (cf. Connan & Cassou, 1980; Shanmugam 1985). Also observe Figure 25 concerning the maturity span versus variation in organic facies.

Garn, Ile, Tile and Åre horizontally segment the structures. Extraction of reservoir sandstones, reservoir siltstones, mudstones and intra-reservoir shales reveal that the invasion signatures of reservoir oil can only be followed into the siltstones. Thus, mudstones and shales have only *in-situ* generated hydrocarbon signatures, irrespective of their thickness (cf. Olstad 1997). Hence, the different reservoir units can accommodate petroleums of different GOR, perfectly sealed off vertically from shallower and deeper reservoir units by intra-reservoir shales and coals. In this respect, Smørbukk represents a series of stacked accumulations laterally sealed off due either to lithological heterogeneities, intra-reservoir faults, or diagenesis (cf. Fig. 8). If we assume early initial filling, as advocated above, we may assume that certain, early cemented, reservoir units became gradually inaccessible to later-arriving petroleum charges due to progressive diagenesis.

The petroleums in Smørbukk vary in maturity, $\delta^{13}C$ values and GOR. Spatial variation in oil characteristics even within uniform reservoirs is well known since Espach & Fry (1951), and such differences may be interpreted in the context of a model for reservoir filling. It is not possible to discuss in detail here the geochemical variation in Smørbukk. Some of this variation was outlined above, e.g. cf. Tables 1 & 2 plus Figure 28.

It is clear, however, that most of the fluids in Smørbukk bear evidence of being mixed. A quick examination of the chromatograms in Figure 16 exemplifies this. It can be observed in Figure 16 that the ratio of pristane to $n$-$C_{17}$ and phytane to $n$-$C_{18}$ indicates only a relatively low maturity whilst the steranes in $m/z = 217$ indicate a higher maturity, as does the ratio of hopane X (diahopane) to normoretane and the ratio of $T_s$ to $T_m$ in $m/z = 191$. The triaromatic steroid ion chromatogram $m/z = 231$ reveals a ratio of the $C_{20}$ to the $C_{28}$ extended triaromatic steroid that again indicates a high maturity. It can be argued further, on the basis of the $n$-alkane profile in the whole oil GC-FID chromatogram, that the decrease in the peak heights in the range from $n$-$C_8$ to $n$-$C_{14}$ could indicate either phase fractionative loss of compounds or mixing between a light 'condensatic' fluid and a heavier oil, possibly already reservoired at an earlier stage. The reduced peak height in the $n$-$C_8$–$n$-$C_{14}$ range could also have been interpreted as incipient biodegradation, as an analogue to the Draugen field (cf. Karlsen *et al.* 1995), implying initial reservoir filling when the reservoir was substantially shallower. This DST 7 from Smørbukk contains no identifiable monoaromatic steroids, suggesting complete conversion to triaromatic steroids, again an indication of high maturity. This DST thus contains considerable evidence of mixed maturity, and it is likely that this fluid sample represents a mixture between an earlier low maturity oil and a later higher maturity fluid with higher GOR. Today, this fluid sample is a condensate at reservoir conditions.

We could have similarly gone through the carbon isotope values of the $C_{10+}$ fractions in more detail. The carbon isotope value of the DST 7 sample ($\delta^{13}C_{total\,oil}$) is −28.40‰. Other Smørbukk petroleums (Fig. 14) have $\delta^{13}C_{total\,oil}$ values ranging from −28.1‰ to −30.2‰ with a mean value of −29.0‰. This mean value is in line with western petroleums like Trestakk, Smørbukk Sør, Tyrihans Nord and Sør (−29.0‰ to −30.0‰, Fig. 11). The pristane to phytane (Pr/Ph) values of this latter group of oils is generally lower than that found in Smørbukk (Table 2), i.e. mostly in the range 1.2–1.5 (Karlsen *et al.* 1995). Lower relative concentrations of bisnorhopane may, in combination with the higher Pr/Ph ratios (1.5–1.6) and the carbon isotope values, suggest that the Smørbukk petroleum system is of higher maturity and possibly also of a slightly less anoxic facies compared with the petroleum in Trestakk, Smørbukk Sør, Tyrihans Nord and Sør.

Concerning GOR, the existing data suggest that the Smørbukk gas-condensates and oils have experienced significant phase fractionation and also that the low GOR oils found in Tilje in 12-6, 12-7 and 12-9S represent oil legs of condensates. Phase fractionation or evaporative fractionation (cf. Silverman 1965; England *et al.* 1987; England & Thompson 1987, 1988; Mackenzie 1989; Thompson *et al.* 1990; Larter & Mills 1992), was suggested by Karlsen *et al.* (1995) as the main causal origin of the condensates at Haltenbanken, apart from the 6407/4-1 condensate with its very high maturity. As a gas becomes less soluble in oil and vice versa, the constituents of the original petroleum will have to redistribute between the exsolving gas phase and the residual oil (Fig. 29), in accordance with the new temperature and pressure regimes encountered as the petroleum migrates to shallower depths. Note in Figure 30 that most petroleums off Mid-Norway seem to follow rather well the 'phase diagram' from Karlsen *et al.* (1995) and that these oils and condensates/gases originated primarily not due to low versus high SR maturities, but due to migration-induced phase fractionation. Collectively, the effect of this process on individual hydrocarbons and related compounds in the migrating phases are referred to as evaporative

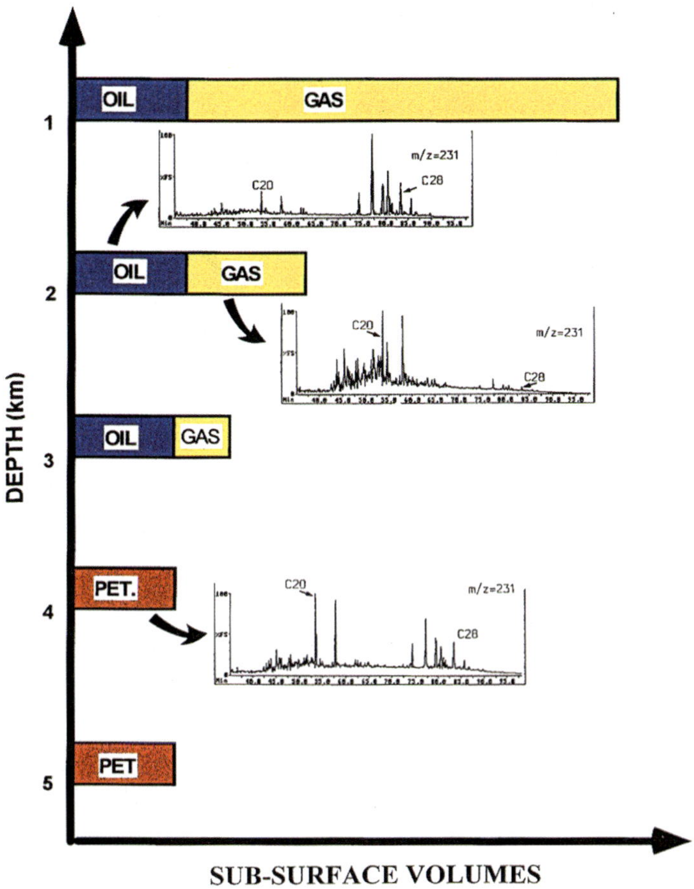


**Fig. 29.** Haltenbanken fluids are phase separated reflecting the vertical migration, and molecular patterns of triaromatic steroids, for example, are influenced by phase fractionation, hence the $C_{20}$ triaromatic steroid will be enriched relatively to the $C_{28}$ compound in the gas phase, whilst the $C_{28}$ compounds will be partitioned with relative enrichment into the oil phase (oil leg). These effects often puzzle interpreters, but are readily rationalized in relation to vapour pressure of the compounds (cf. Karlsen *et al.* 1995). Figure modified after England & Mackenzie (1989).

fractionation/retrograde condensation or partitioning. This process will, to varying degrees, modify many of the chemical parameters that are measured on petroleums to deduce facies and maturity. Interpretations of the origin and maturity, i.e. genetic relationships between petroleums, is thereby complicated but can be accounted for (cf. Karlsen *et al.* 1995).

Phase fractionated condensates and oils from Haltenbanken have an effect on the gross composition, GOR and API of the petroleums (Karlsen *et al.* 1995). The ratio of hopanes to steranes, triaromatic steroid maturity parameters, and, in extreme cases, also the Pr/Ph, Pr/$n$-$C_{17}$ and Ph/$n$-$C_{18}$ ratios and the carbon isotope values are altered by this effect (Karlsen *et al.* 1995). The effect on the latter is due to the intramolecular carbon isotope effect (Galimov 1973, 1985) which makes isoprenoids isotopically heavier than normal alkanes, isoprenoids also having more spherical molecular dimensions and hence higher vapour pressure than normal alkanes of the same carbon number.

It is still difficult to account for the degree of mixing in the Smørbukk fluids without taking into account an element of mechanical overturning, the most likely being tilting. Tilting of a gently dipping fault block like Smørbukk (Fig. 21) will result in significant re-mobilization of the contacts, e.g. the gas–oil contacts and oil–water contacts. Thus, gas-condensates may redissolve biomarkers from the oil legs as the field is buried deeper. This complicates interpretations of biomarkers and other facies and maturity

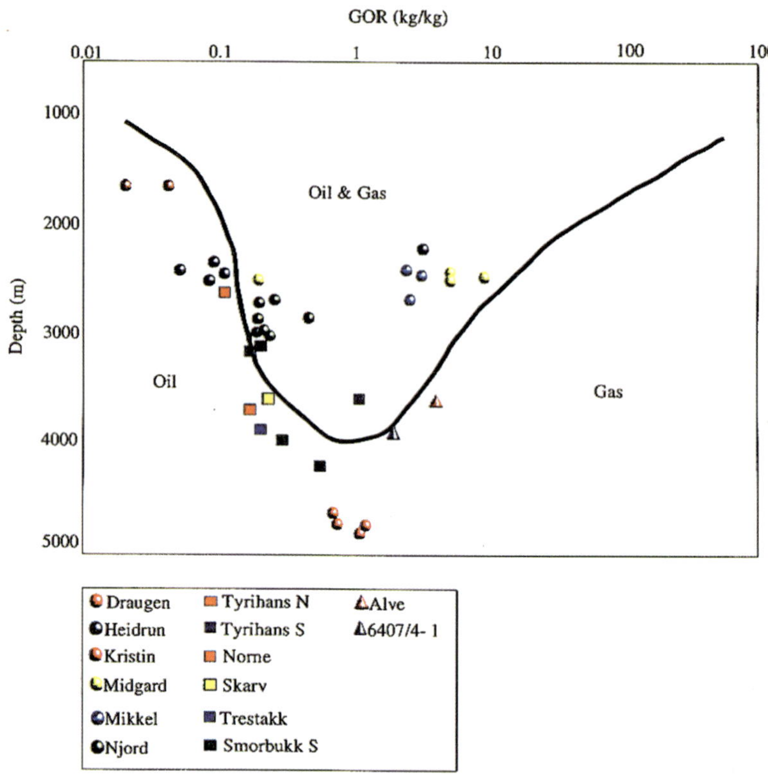


**Fig. 30.** GOR in petroleums off Mid-Norway. The 'phase envelope' represents an empirical envelope from the Viking Graben (see Karlsen *et al.* 1995; England & Mackenzie 1989) and suggests that condensates/gases of the region originated by migration induced fractionation rather than high or super-high maturity.

parameters (cf. Fig. 16). This complex relationship between mixed maturity signatures of light hydrocarbons, medium range hydrocarbons and heavy biomarker range maturities in Smørbukk oils and core extracts suggests that significant physical mixing must have taken place during filling and before pseudo-lateral segmentation of the structure (cf. Fig. 14). Evidence from burial histories suggests that Smørbukk at an earlier stage during Cretaceous times had its highest point not in the north as today, but rather in the south, in the area of 6506/11-2 (G. Butenco, Norsk Hydro, pers. comm.). Filling of the trap with gas and oil 70–50 Ma bp, i.e. before a later tilting episode, would clearly have brought about significant internal mechanical overturning of the structure and could be a factor in explaining the highly varied maturity signatures observed today in Smørbukk even in singular petroleums.

The large variation in maturity and GOR and the minor variation in facies and carbon isotope values within Smørbukk are best explained as reflecting trapping and partial mixing over a long time interval (cf. Angard 1996; Steinhoff 1996).

## *Mechanism for filling of Smørbukk*

The reduced abundance of petroleum inclusions in Smørbukk Sør compared with Halten Vest and Smørbukk suggests that Smørbukk filled initially from the west or southwest. The occurrence of biomarkers in petroleum inclusions from the Halten Vest region, and their geochemistry, which are of the same maturity and facies as in Smørbukk, suggests that Smørbukk initially filled from the west (cf. Fig. 27). The geometry of Smørbukk suggests that initial filling must have occurred through the fault zone (Fig. 31a,b) rather than from the area between Smørbukk and Smørbukk Sør (cf. Fig. 27) as the latter regions were immature at the time inclusions with petroleum formed both in Halten Vest and in Smørbukk.

However, more recent influx (<10 Ma) of hydrocarbons from the area between Smørbukk

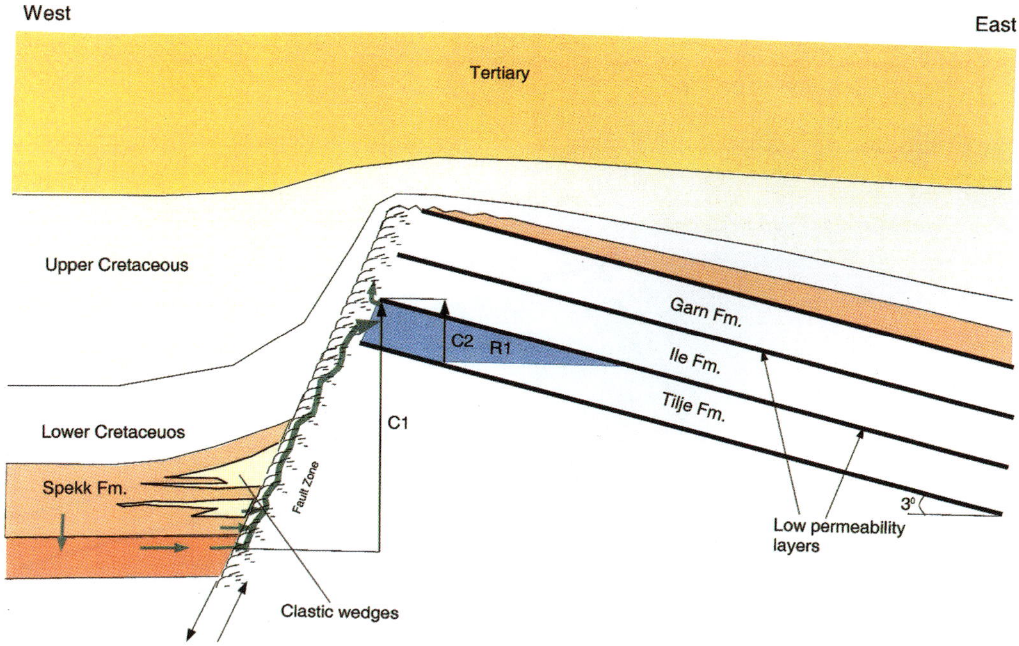


**Fig. 31.** (**a**) Filling a fault block through a fault. A petroleum column is building up in carrier sands towards a fault zone. As the petroleum head builds up, a slug of oil is injected into the fault. Petroleum bleeds off into the deepest reservoir strata (R1) until equilibrium is established between the height of the slug in the fault zone (C1) and the petroleum head in the reservoir strata (C2).

and Smørbukk Sør is likely to have occurred as shown in Figure 27, and accordingly Smørbukk Sør, which hardly contains oil inclusions (cf. Fig. 19) filled only recently.

If initial filling occurred into Smørbukk from the western side through the fault (Fig. 31a,b), we may assume that individual petroleum pulses first filled the lower reservoir units in Tilje (Fig. 31a). Expulsion of HC from the Spekk into deeper Jurassic sands is likely to have occurred either directly downwards or by faults pseudo-laterally into conducting Jurassic sands. Migration from the shales into the fault zone is facilitated by siltstones and sands in the shale in the form of turbidites or clastic wedges (cf. Leythaeuser *et al.* 1982). In the North Sea, the Hugin Formation represents such local erosional products from the rotated fault blocks and consists generally of interfingering sands into shales such as described for the Brae Field in the North Sea (Mackenzie *et al.* 1987), causing effective expulsions from the SRs. Similar settings are likely off Mid-Norway and might apply to the Smørbukk case (cf. Fig. 31a,b).

Filling of the reservoir sections from the fault would occur until there is equilibrium between the pressure head of the Tilje petroleum charge in the reservoir and the head of the petroleum in the 'feeder' fault zone. This zone would in this model be slightly overpressured due to the petroleum column (Fig. 31a). Progressive diagenesis on the fault will reduce the ability for fluid flow, and the same burial event causes more mature petroleum with higher GOR to migrate towards and into the deeper parts of the fault zone. A petroleum head may thus build up with sufficient buoyancy to breach the fault and to cause filling of adjacent shallower hydrostatic formations, i.e. the Ile and Garn Formations (Fig. 31b).

This pattern of repeated filling events from a progressively maturing source rock basin, structural trap tilting, and migration up the fault zone, coupled with phase fractionation and separation of petroleum into oil and gas phases in the reservoir intervals (Fig. 32), which were subsequently progressively sealed off due to increased quartz cementation in response to increased burial, may explain the complex patterns of variable GOR, API and carbon isotope values observed in Smørbukk (cf. Fig. 14).

The final and most recent complication to the filling history is represented by migration from

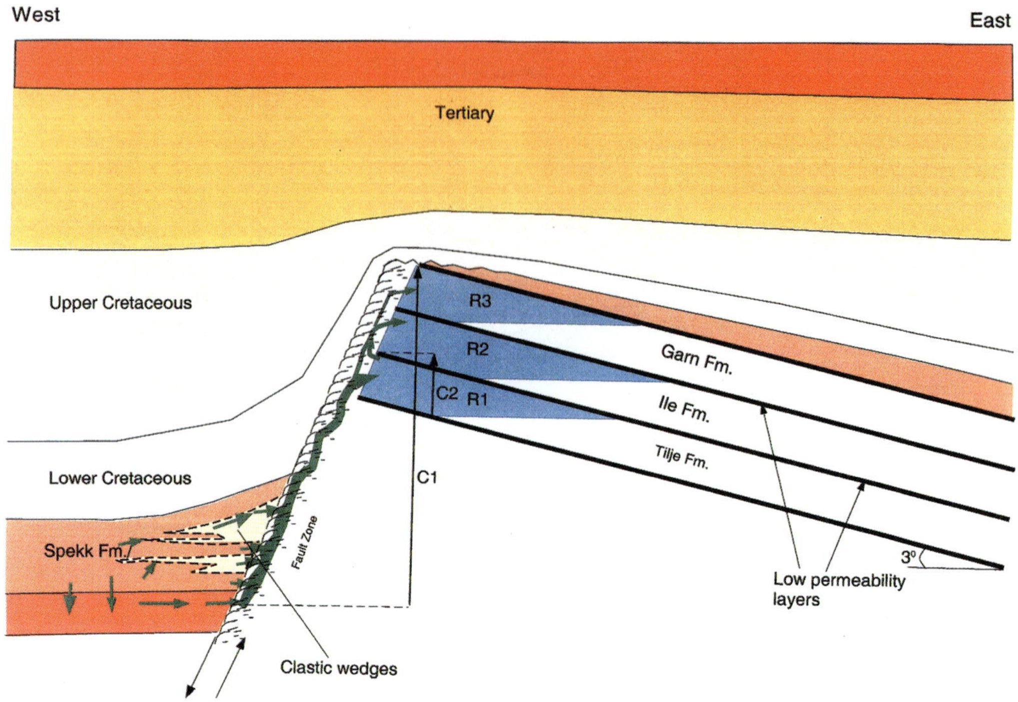


**Fig. 31.** (**b**) Diagenesis may partly heal the fault zone during progressive burial and the same burial matures the source rock causing another slug of oil to build up and force its way, not into R1 which is not hydrostatic, but into the next hydrostatic reservoir unit (R2). Individual reservoir units will contain petroleum of varied maturity, GOR and API.

the area which matured recently between Smørbukk and Smørbukk Sør and dysmigration from Garn into the Lange Formation.

This example illustrates that faults cannot be treated simplistically as statically sealing or closed. Hippler (1997), who studied faults in the North Sea, suggested that faults, which are able to seal a petroleum column, might actually be able to allow hydrocarbon movement within the fault zone over geological time.

It may seem contradictory at first that a fault may at the same time be able to act as a seal and as a conduit for oil. Still, it is clear that the vertical 'head' of petroleum in the fault zone may be hundreds of metres, i.e. large compared with the vertical relief of the trap *sensu stricto*.

Thus, the complex rheological characteristics of a fault zone suggests that it might periodically be able to communicate petroleum up and into reservoir units in a 'step-wise' fashion as illustrated in Figure 31a,b.

The complex macro and micro geometrical layout of faults and diagenetic reactions are highly variable and it follows that the flow properties of faults, changing with $P$, $T$ and time, must be understood in relation to the fault-zone properties (Hippler 1997).

Knipe *et al.* (1991) and Knipe (1992) have provided ample evidence to illustrate the interplay between tectonics, fault microstructure, diagenesis and fluid flow. Knipe (1993) concluded that there is a need for focusing on better models for describing how the nature of the fault changes during its 'active life', i.e. how progressively decreasing displacement (death) of faults influences the rate of sealing mechanisms. We have not studied the fault itself, *sensu stricto*, but we have utilized petroleum geochemical tools to inferfault zone behaviour over time, with respect to flow properties and overpressure build-up.

It is clear that faults and fault deformation products like clay smear, grain reorganization and, in general, mineral cementation act together in different fault settings to influence fluid flow. Secondary migration of oil in faults zones may appear as a plausible way of moving hydrocarbons from deep source rock to shallower reservoirs (cf. Leonard 1984; Tissot & Welte 1984; Hunt 1996). Faults as a means of focusing

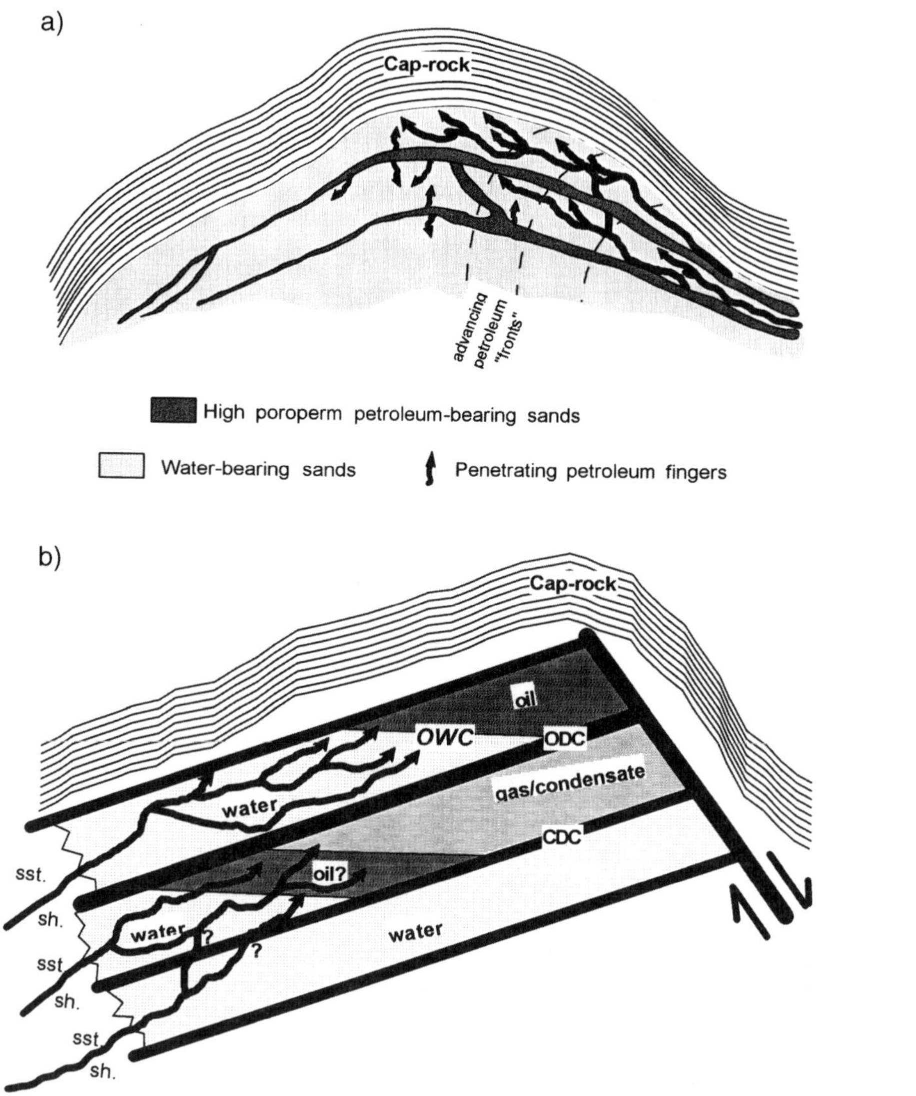


**Fig. 32.** Filling of simple dome-shaped structures (a) results in more homogeneous petroleum charges than filling through the fault zone (Figs 31a,b). Up-flank filling of rotated fault block (b) would be expected to result in more uniform GORs, APIs and maturities than filling through the fault zone (Figs 31a,b).

fluid flow were discussed by Sibson (1981, 1987, 1990) and Carter *et al.* (1990).

It is important to realize that small faults which may be important with respect to fluid flow properties can be in some cases easily overlooked in cores or thin sections (Sverdrup & Bjørlykke 1992). It is generally assumed that in the Halten Terrace most of the faulting occurred pene-contemporaneously with the rifting in the Upper Jurassic. Most of the faults seen on seismic section only extend into the base Cretaceous or into the lower Cretaceous section. It is therefore tempting to conclude that faulting therefore occurred when the sediments were in most case buried to less than 1 km. The shale and mudstones would then have been soft clays and loose sands. Such shear zones formed at moderate depth are characterized by interlocking grains producing a strain hardening without much grain crushing because of the moderate horizontal stress (Sverdrup & Bjørlykke 1992, 1997).

During progressive burial, sediments will be normally consolidated and ductile and open fractures are not likely to form (Bjørlykke & Heg 1997). Still, there is no *a priori* foundation to conclude that all tectonic activity ceased because no fault planes are seismically discernible in the generally ductile lower Cretaceous section overlying the rotated Jurassic fault blocks. On the contrary, faults in tectonically active regions are often reactivated and pulse-like petroleum migration might be associated with tectonic events. Such migration could very well occur in parts of the fault zone (cf. Teige & Hermanrud 2004), and we feel convinced that there may be such relationships between tectonic events, inter-seismic healing of the fault zone (cf. Muhuri *et al.* 2003), and pulse-like migration events such as petroleum columns, which build up over time in response to SR maturation, and which are suddenly set in motion.

Some would argue that such fault zones are unlikely conduits for fluid flow (petroleum migration) along the fault plane because the permeability along the fault plane is nearly always lower than in the matrix next to the fault. These faults may, however, not be impermeable for flow across the faults at moderate depth (<2 km) prior to cementation.

At greater depth, however, quartz cementation may destroy the remaining porosity along the fault zones and make them nearly impermeable, but fault planes are not mathematical planes, rather real complex physical phenomena which are likely zones of renewed seismic activity. When this occurs, the fault may open to migrating petroleum. These fault zones have highly variable rheological characteristics and are not planes but rather 10–30 m wide, to be seismically recognizable (depending on depth), and even if cementation eventually seals a (at one time open) fault brecciated zone during the inter-seismic period, new pathways for petroleum will inevitably form if fault reactivation occurs. We believe petroleum migration to be associated with these tectonic events, not only across faults, but also along the fault zones.

### *Palaeo-leakage from Smørbukk and the time for leakage and overpressure development in Halten Vest*

If a structure such as that shown in Figure 7, with a uniform caprock, is to leak, the leakage will occur at the maximum pressure differential, i.e. between the deepest oil–water contact (OWC) or gas–water contact (GOC) and the apex of the structure. These strata comprise the Garn Formation in Smørbukk, which today contain producible petroleum only in well 6506/11-2 in the southern part of the field. This may be surprising as this formation elsewhere in Haltenbanken, together with the Ile, Tile and the Åre Formations, makes up the main reservoir units. In several of the Garn units in Smørbukk there are residual oil saturations today, indicating clearly that these sands held petroleum at an earlier time.

We assume that this evidence reflects that petroleum was massively lost from these Garn sands through the top of the Smørbukk structure. This leakage may indeed account for the petroleum in the Cretaceous Lysing and Lange Formation strata above Smørbukk (cf. Fig. 14).

The fact that the deeper Garn horizons in the off-crest well 6506/11-2 hold petroleum today, and not the crestal wells 6506/12-1, is circumstantial evidence that palaeo-leakage from Garn sands occurred with the present Smørbukk structural configuration, i.e. after the tilting episode discussed above.

It is, in our opinion, based on the geochemistry of the undersaturated oil in the Cretaceous Lange Formation strata above Smørbukk, a likely assumption that parts of the leaked oil from a palaeo-accumulation in Garn in Smørbukk migrated vertically by fracture-induced mechanisms into the Cretaceous strata above (cf. Fig. 7).

Geochemical data cannot prove this linkage, but there is certainly no geochemical evidence excluding filling of the Cretaceous strata via fractures from below. The oil and condensate in the Cretaceous strata are similar in facies and maturity to the varieties of oil found in the Jurassic field below, but are clearly phase fractionated, evidence indicative of vertical migration possibly via self-propagating open fractures (cf. Karlsen *et al.* 1995; Angard 1996; Steinhoff 1996; Dale 1997). In particular, there is no evidence for oleanane in the oil, which otherwise is found *in situ* in Cretaceous siltstones at several locations throughout the Viking and Central Graben and also in offshore Mid-Norway.

Vik *et al.* (1991) suggested that hydraulic fracturing caused gas to escape from reservoirs in the Halten Vest region, and used the coinciding overlap between overpressured regions and the regions where the Åre Formation coals today are overmature, to tentatively indicate this assumed Åre derived gas to be responsible for the overpressure development. Vik *et al.* (1991) did not have data available to indicate that several of these structures they report to be dry and 'ideally located with respect to mature source rocks and migration' actually have held oil earlier.

This serves to illustrate the importance of detailed core analysis and geochemistry to decipher the status of drilled structures, i.e. other data points than well tests and modelling are needed to construct a model for palaeo- and recent petroleum migration in a basin. The fact that the structures earlier contained a fluid phase (not, as assumed, only gas) places a totally different perspective on the timing of oil migration relative to caprock formation, structural development, and field filling in Halten Vest.

Some structures, like the palaeo-petroleum accumulation 6506/12-4 and 6506/11-1 in Halten Vest, contain sands with two types of petroleum inclusion. One type is a greenish-yellowish inclusion population with low GOR oil (10–20% gas bubbles), and the other type is bluish with large gas bubbles (70–80%), signifying a high GOR petroleum (cf. Fig. 17). This suggests to us that the petroleum in the trap changed dramatically in GOR before caprock failure occurred. As the GOR of petroleums is also a function of maturity (cf. England and Mackenzie 1989), this could reflect a significant increase in the maturity of the expelling source rock and hence a significant time period for petroleum accumulation in the traps.

The fact that the overpressured regions coincide with the areas where the Åre Formation is over-mature simply reflects the deep burial in these regions with associated diagenesis on fault zones. Thus deep burial, not gas generation from Åre, *sensu-stricto*, resulted in dysmigration via a loss of communication and pressure release in the general updip direction eastward.

Today, the shallow petroleum-containing Cretaceous sands (Lysing and Lange Formation) at *ca.* 3380 m in the Smørbukk well 6506/11-2 are overpressured, whilst the Jurassic Smørbukk reservoirs at depths of between 4380 and 4700 m are normally pressured (excluding the effect of the HC columns).

It is thus a likely interpretation that this migration event from Garn in Smørbukk into the Cretaceous strata occurred *before* the onset of overpressure development.

The undersaturated oils in the Cretaceous sands generally occur at depths *ca.* 600–900 m shallower than the shallowest portion of the Jurassic reservoir strata, and are found in overpressured strata. The overpressure today at the Lysing level above Smørbukk is *ca.* 140 bar, and above Smørbukk in well 6506/11-2 the saturation pressure is 434 bar at a reservoir pressure of 630 bar. Above Smørbukk Sør the GOR of the Lysing oil is in the range 160–200 $m^3/m^3$ at depths of *ca.* 3160–3170 m. In the Lysing Formation above Smørbukk the GOR is in the range 170 $m^3/m^3$ at 3380 m, whilst in the deeper Lange Formation at *ca.* 4020 m the GOR is 513 $m^3/m^3$ (cf. Fig. 14).

None of these Cretaceous petroleums show signs of being biodegraded. A lack of biodegradation could be significant circumstantial evidence to suggest that these oils were emplaced when these sands were buried to depths deeper than corresponding to *ca.* 70–80 °C. This corresponds to *ca.* 10 Ma bp for the Lysing Formation (Fig. 9), i.e. a time at which we know that oil existed in the Jurassic strata in Smørbukk, 6506/11-1, and in 6506/12-4. The temperature interval of 70–80 °C is commonly used as a biodegradation-seizure temperature limit for significant aerobe biodegradation of oil accumulations (cf. Connan 1984; Ahsan *et al.* 1997).

Today, these Cretaceous aged reservoirs are overpressured and it is hence not a likely assumption that oil could have migrated recently from the presently normally-pressured Smørbukk Sør field into the overpressured Cretaceous strata above.

Rather, it is clear that progressive maturation of the Spekk Formation source rocks in the drainage areas of the Smørbukk Sør and the 6406/3-1 structures, which lead to their filling, occurred recently, during the past 10 Ma, before overpressure development in the Cretaceous strata above. We believe that the filling of these structures and the Cretaceous sands above Smørbukk Sør are time-related phenomena. The petroleum in the Cretaceous strata above Smørbukk may be related to this same time event, and may not necessarily have anything to do with the advent of local pressure build-up in the region, as we do not see hydrocarbons in the Cretaceous strata above the Jurassic 6506/12-4 structure. We prefer to interpret the hydrocarbons in the Cretaceous sands above Smørbukk as most likely resulting from catastrophic open-fracture-style dysmigration from the Garn Formation in Smørbukk, i.e. related to a local phenomenon of too excessive petroleum column build-up (cf. Scenario-1 in Fig. 7).

This combined evidence points to a time-transgressive upwards-directed overpressure development in Halten Vest, starting at Jurassic levels and progressively including shallower parts of the sedimentary prism, until today including also the Lysing and Lange Formation sands. The mechanism of this development is accelerated burial and diagenesis, resulting in shut-off of pressure communication towards the east. This all points to a semi-coinciding overpressure development and caprock failure during the past 5 Ma.

## HC loss from traps—mechanisms

We have discussed above some geochemical evidence indicative of a 'new stepwise' up-fault filling model for traps with multi-stacked pay zones. The model for dysmigration, due to a too extensive petroleum column in combination with progressive overpressure development at Jurassic levels, is more traditional and corresponds to a modified Sales model (cf. Sales 1993, 1997). What we try to do in a novel manner is to link these geochemical, diagenetic and structural evidence and models together to produce a semi-time-resolved scenario for the sequence of events including initial long range filling across the now sealing fault, dysmigration and later filling from the local kitchen, and to explain the origin and emplacement of the petroleum in the Cretaceous section when these were at normal pressure (Scenario-1, Fig. 7).

Based upon the comparison of petroleum inclusions between Smørbukk, Smørbukk Sør and 6406/3-1 (Figs 1a, 18 & 19), it is fairly safe to suggest that the latter structure also received hydrocarbons at the same time as did Smørbukk Sør, i.e. based upon basin modelling, during the very last part of a period starting *ca.* 10 Ma bp.

Thus, we have reached the conclusion that the 6406/3-1 structure did not fill and drain in the distant past; rather it filled pene-contemporaneously with Smørbukk Sør, i.e. at least 50 Ma after initial filling of Smørbukk and the Halten Vest structures penetrated by wells 6506/11-1 and 6506/12-4, and it drained, geologically speaking, very recently.

We suggested above that all the dry Halten Vest structures that previously held oil and leaked did so during the past 3 Ma and before general overpressure build-up in the Cretaceous strata, which must have been even more recent.

Still, Smørbukk, which we advocate had experienced an event of leakage, remains oil filled today throughout, apart from the Garn Formation. Furthermore, Smørbukk Sør is oil-filled despite evidence for leakage into the Cretaceous strata above. The nearby Kristin field is overpressured but is sufficiently deep as to avoid the fracture gradient, and contains a condensate as does the normally pressured Lavrans (Fig. 5), which all, accordingly 'fall into' Scenario-1 in Fig. 7.

Thin, eroded caprocks may allow steady state gas diffusive leakage, preserving oil in place (cf. Fig. 7) whilst thick tight caprocks can hold back gas, resulting in traps filled mainly with gas and/or thin oil legs (Sales 1993, 1997). It is obvious that diffusive loss of gas will cause the remaining reservoired oils to become undersaturated with respect to methane; the Draugen, Heidrun, Norne and Svale are in our opinion relevant examples as are, from other regions, Snorre, Oseberg, Gullfaks, Statfjord and Ekofisk. Other examples may include the Njord oil field with its thin partly eroded silty caprock. In Njord the capillary entry pressures and pressure distributions in the reservoir and in the caprock suggest that a membrane seal failure mechanism controlled hydrocarbon columns (Lilleng & Gundesø 1997), and this mechanism may also apply to Svale and Falk (Scenario-2 in Fig. 7).

Such selective gas leakage may represent an alternative mechanism for explaining oil fields containing undersaturated oil, as opposed to assuming that the field was filled at shallow levels and then buried so that the reservoir pressures exceed the saturation pressures. This method was utilized by Karlsen *et al.* (1993) on the Ula field in the Central Graben to tentatively advocate a Neogene or younger filling event from the Kimmeridge-aged Mandal Formation

Besides subsequent burial, it is possible that biodegradation and methane consumption are capable of reducing GOR as associated carbon dioxide and aldehydes/ketones/organo-acids (which may be intermediates from the oxidation events) are highly water soluble and readily lost from the trap by diffusion, even through interconnected films of water in less than perfectly tight caprocks, or from the OWC transition zone.

Examples of traps with thin caprocks and low GOR oil in Haltenbanken include Draugen (GOR = 18–38 $m^3/m^3$ at 1.6 km) and Heidrun (GOR = 48–90 $m^3/m^3$ at 2.4 km, see (Karlsen *et al.* 1995). The caprock on oil-filled Norne (GORs mostly 170–220 $m^3/m^3$ at 2.9 km) is thin as is also the caprock over Tyrihans Nord and Trestakk which hold oils with GORs of 148 $m^3/m^3$ at *ca.* 3.7 km, and 200 $m^3/m^3$ at 3.9 km, respectively, whilst Tyrihans Sør with its thicker caprock holds oil with a GOR of 870 $m^3/m^3$ at 3.7 km (cf. Scenario-2, Fig. 7).

It is thus clear that caprock thickness, which relates linearly to oil column retention capacity (Sales 1997) or rather caprock tightness (which is also affected by mineralogy/diagenesis), is an important parameter which must be risked in petroleum exploration and GOR predictions. This is especially important in basins where volumes of generated and expelled oil are likely to exceed the trap volumes, i.e. in 'overfilled' basins like generally on the NOCS (cf. Sales 1997).

It is usually estimated that only a tiny fraction of expelled petroleum reaches commercial sized accumulations (cf. England 1994). Significant loss of asphaltenes and high boiling compounds will occur during migration to shallower levels (lower *P* and *T*). Exsolved gas is also lost during migration. The petroleum reaching the shallower reservoirs is therefore compositionally strongly fractionated (cf. England *et al.* 1987 and Figs 29, 30). On this basis it may be advocated that it is of greater importance to model the in-reservoir processes which affect GOR rather than to try to estimate the GOR of expelled petroleum, most of which does not reach the trap. This modelling is usually based upon pyrolysis experiments of source rocks with their intrinsic experimental limitations. It is therefore possibly less relevant to model GOR of expelled petroleum, than to model caprock quality and selective gas loss through the caprock. It follows from this that PVT modelling of the migrating HC phases are possibly of secondary importance for trap systems sealed by gas loosing caprocks (Sales type 3 traps like Ekofisk, Snorre, Statfjord, Oseberg and Gullfaks, Scenario-2 in Fig. 7).

The ability of certain tight shales to sieve petroleums on a molecular size basis is considerable. Numerous examples of such fractionative behaviour were described by Karlsen *et al.* (1997, 1988), i.e. more effective loss of hydrocarbons with small kinetic diameters like methane and ethane (4.4 Å) compared with propane (with a length of 6.5 Å and a width of 4.9Å), not to mention cyclohexane with a diameter of 6.0 Å (Breck 1974). Fractionative depletion trends, from a Permian Ravnefjeld Formation shale, towards local fractures and the surface were evident for all molecules apart from methane, which even in this shale showed the ability to move effectively through the whole rock profile, equalizing concentrations, irrespective of local fractures (Karlsen 1987; Karlsen *et al.* 1988).

We suggest that Kristin today has a very thick and strong caprock, making this an example of a type 1 trap according to Sales (1997). Yet evidence exists in Kristin to suggest a palaeo-leakage event that may have occurred when the caprock did not have sufficient strength. This evidence includes organic coatings (cf. Storvoll 1999; Storvoll *et al.* 2002) and biodegraded hydrocarbons extracted from inclusions in Kristin. This suggests clearly that Kristin contained black oil when the trap was as shallow as to have been biodegraded. We suggest that the caprock quality of Kristin, and similarly on Smørbukk, improved over time as diagenesis and compaction proceeded (Scenario-1 in Fig. 7).

It may appear puzzling that there is a clear gas chimney above Smørbukk, which hold condensates, oil and oil-condensates, and also above Lavrans. We believe that the gas may have originated from palaeo-leakage events as discussed above and, as such, reflect the diffusive halo-effect around induced palaeo-fractures. If this is correct, a lack of gas chimneys as e.g. above Kristin may not necessarily prove an absence of palaeo-, or recent leakage from a structure. It is plausible that a recent focused-fracturing event with bleed-off of petroleum into shallow hydrostatic strata would not produce an immediately recognizable gas chimney. These examples illustrate the necessity for considering caprock strength and quality as time dependent/transgressive parameters, and that geochemical tools may help in recognizing evidence of leakage.

## Refilling of leaked traps

We have shown above evidence to suggest that the ratio between filled and dry structures in the hydrostatic regions of Haltenbanken is of the same order of magnitude as the ratio of 'always dry' to palaeo-filled traps in Halten Vest. This suggests that overpressure in itself had no relevance for the initial filling of these traps. We have also documented that some of these traps in Halten Vest were filled initially with oil pene-contemporaneously to Smørbukk. It follows that leakage from the presently dry structures in Halten Vest must be a recent phenomenon caused by the accelerated burial during the past 10 Ma; modelling suggests that overpressure did build up in Haltenbanken during the past 3 Ma (Vik *et al.* 1991; Lilleng & Gundesø 1997). Evidence from inclusions supports these modelling results and we infer that leakage from the traps also occurred during the last 3 Ma. Our suggestion is that the leakage most likely occurred after establishment of overpressure in the Cretaceous strata in Halten Vest, but before overpressure was established in the Cretaceous strata above Smørbukk and Smørbukk Sør.

Why did these drained structures not refill later with hydrocarbons after leakage? It follows logically that structures which earlier received hydrocarbons, like the 6505/11-1, 6506/12-4 and 6406/3-1 structures, have communicated with a generative source rock basin where the Spekk Formation was mature. Hence, palaeo-migration conduits towards the traps existed for these structures. Is this migration avenue still active or was it shut off, e.g. by structural movement or sealing faults? Or could it be that the leakage was so recent that the traps did not have sufficient time to refill?

As the basins are not uplifted, we must assume that hydrocarbons are still entering the migration avenues that at some time in the past fed the traps. If the migration avenues are still in existence, it would then follow that the flux out of the traps would have to be identical to the influx, i.e. that the structures no longer have closure, but rather constitute part of the migration avenue. Modelling of cycling rates of reservoir filling, drainage and recharge in overpressured regimes by Ortoleva (1994) indicates that such cycles may repeat in less than 100 000 years. Roberts and Nunn (1995) modelled the expulsion through faults and suggested these to last on the order of 100 years, whilst it would take on the order of 10 000 to 50 000 years to build up pressures allowing caprock seal breaching.

We believe that it is likely that a failed caprock will remain close to fracturing conditions (cf. Figs 4 & 7) in a subsiding basin and that these reservoirs for that reason are unable to build up noticeable new petroleum columns. Our few data points on trap pressures in Halten Vest do suggest this to be the case (Fig. 4).

The status of the faults in such a situation, however, is not a simple matter, as structural elements like faults in sedimentary basins have physical and chemical attributes that influence fluid dynamics, and which are best understood in the context of sedimentary poroperm and diagenetic mineral reaction kinetics and thermodynamics. Faults appear as geometrical phenomena, seen as lines on seismic sections or planes in 3D modelling but are in reality complex.

Sverdrup & Bjørlykke (1997) conclude that brittle deformation may occasionally generate faults which may remain open to fluid flow along the fault planes for some time, whilst ductile deformation and abundant clay smear as found in the fault-zone of clay-rich sandstones generally will have the opposite effect. Most of the observed carbonate cements in faults in the North Sea were found to have an *in-situ* rather than autochthonous origin by fault parallel fluid flow.

Still, the very fact that the leaked structures discussed in this paper do not contain free producible hydrocarbons could indicate either a very recent event of leakage (not sufficient time for the fault to heal and the trap to receive new petroleum charges) or that, as suggested above, the incoming petroleum moves through the trap. Alternatively, it could suggest that the hydrocarbons move in an entirely different direction, but we do not favour this interpretation in simple subsiding basins.

Indeed all our evidence from studying the filling histories of fields (cf. Karlsen & Larter 1989, 1991; Horstad *et al.* 1990; Karlsen *et al.* 1993, 1995; Skålnes 1993; Skålnes *et al.* 1993; Horstad 1995; Larter & Aplin 1995; Angard 1996; Steinhoff 1996; Bhullar *et al.* 1998, 1999a,b; Sletten 2003; Winterstad 2003; Xu 2003) indicate that structures which in the distant past (e.g. 70–50 Ma bp), were linked to establish oil wetting migration systems, will continue to utilize reservoir-proximal parts of this system even when other more shallowly buried source rocks become mature.

It is likely that this very model can account also for increased secondary migration efficiencies (cf. Figs 10a,b & Paper I). It is likely in Haltenbanken that early migration from the Åre Formation may have acted in this manner, saturating migration avenues, increasing the efficiency of secondary migration once the Spekk Formation becomes mature. The structure 25/2-15R2 in Rind in the Viking Graben (cf. Bhullar *et al.* 1998, 1999a,b) is overpressured and dry today, but contains significant residual oil saturations, which, together with the petroleum inclusions, prove that this structure held oil prior to 10 Ma bp. The question remains open as to why it did not refill. Perhaps the reservoir pressure remained too close to the fracture pressure. The overpressure in this case suggests that the 'pressure valve' (cf. Fig. 6) occurs farther updip, in Cretaceous or Tertiary strata, and this may be the location of the lost reservoir charge.

With respect to the time for the leakage event from 6506/11-1, 6506/12-4 and 6406/3-1, it is relevant to consider the overpressure development at Jurassic and also Cretaceous levels. Above, we have presented evidence to suggest that caprock failure in the dry structures is related to accelerated burial in the past 10 Ma and that faults disrupted regional pressure communication at Jurassic levels, resulting in isolated pressure blocks which then resulted in the accumulations experiencing caprock failure. This may have occurred by the pressure differential developing across the Spekk/Melke caprock boundary or also due to more mature higher GOR petroleums (with lower densities) flowing into the traps. The governing mechanism and main factor here is still a lack of pressure release in an easterly direction and we believe this to be related to diagenesis on the earlier described SSW–NNE main fault in this region (Figs 1a & 2).

There exist explorational interests in the regions west of Haltenbanken which also concern the potential existence of Cretaceous-aged source rocks. We have only found oleanane in siltstone/shale intervals, which in all likelihood represents *in-situ* generated hydrocarbons. The Nyk and Gjallar structures hold only dry gas that could have originated from deep overmature Jurassic-aged source rocks. However, an origin from Cretaceous strata is also plausible. Mature gas with an associated oil rim exists in the Ormen Lange dome and tentative maturity modelling suggests that a Cretaceous play scenario is possibly for this structure, as Jurassic strata would have been overmature before caprock integrity was established. The 6507/2-2 OST 2 petroleum west of Alve is most certainly of Cretaceous age (Xu, 2003).

## Conclusions

The occurrence and geochemistry of fluid inclusions in authigenic mineral cements and chemical characterization of bitumen and oil were used to investigate several dry and filled structures at Haltenbanken. Lack of fluorescent petroleum inclusions and core extracts suggests that the 6406/6-1 and 6407/8-1 structures never contained liquid petroleum, nor free HC gas. In contrast most of the dry structures in Halten Vest originally contained oil, e.g. 6506/11-1, 6506/11-3, 6506/12-4 and 6406/3-1. The core extracts, gas and biomarkers from inclusions indicate strongly that the Halten Vest and Smørbukk region were at one time one unified migration system. The inclusion gas composition and lean core extracts suggest that the 6407/5-1 structure west of Mikkel may have contained a condensate.

These data are used to provide a model for hydrocarbon sourcing and migration in the Smørbukk and Halten Vest region.

Evidence suggests that the main fault zone west of Smørbukk, which today forms a pressure 'seal' by impeding fluid flow, was open for petroleum migration at an earlier time, as initially tentatively suggested by Olstad *et al.* (1997).

Three out of four structures in the so-called 'Halten Vest High Pressure Dry-Hole Region' west of Smørbukk were found to have contained petroleum of the same organic facies as found today in Smørbukk, and these structures filled with oil from the Upper Jurassic-aged Spekk Formation at least as early as the Smørbukk Field.

Portions of this oil occur today in petroleum inclusions and mineral cements, as well as in reservoir core extracts at residual saturation. The burial history for Smørbukk was used together with the known temperatures of formation of inclusions in quartz to infer the time when petroleum occurred in Smørbukk and, by analogy, in the Halten Vest High Pressure Dry-Hole Region. The derived time is between 70 and 50 Ma bp

The dry structures in the Halten Vest region experienced caprock leakage by fracturing upon receiving high GOR petroleum from the Spekk Formation. Petroleum inclusions in these traps are of two types, one with small gas bubbles (standard black oil type with low to medium GOR) and one type in which the much larger gas bubbles reflect that the traps received petroleum with higher GOR before leakage occurred. The genetically associated development of extensive low-density petroleum columns, and generation of pressure 'cells' due to accelerated diagenesis on fault zones, caused leakage from the traps.

The regional pressure build-up thus occurred after the time at which the fault zone west of Smørbukk transformed from open to 'sealing' with respect to petroleum movement.

The same abundance of fluorescent petroleum inclusions in the Halten Vest dry wells as was found in Smørbukk implies, according to our model, that the time during which the presently dry Halten Vest structures actually held oil is roughly similar to the time petroleum has been present in Smørbukk. In contrast the Smørbukk Sør and the 6406/3-1 structures filled during the past 10 Ma from local basins.

The very few petroleum inclusions found in Smørbukk Sør and 6406/3-1 implies that there is no geological evidence to suggest filling of either Smørbukk or these structures from the area between Smørbukk and Smørbukk Sør earlier than about 10 Ma bp

Geochemical evidence suggests that caprock leakage in Halten Vest must have occurred, in geological terms, very recently, and after the build up of overpressure which took place during the past 3 Ma due to the accelerated burial (cf. Vik *et al.* 1991) and in response to accelerated diagenesis on deep faults.

Sandstone porosity loss versus depth in Haltenbanken is similar to trends seen for the North Sea, especially between 1.4 and 2.6 km (Bjørlykke *et al.* 1989). This could suggest that overpressure does not influence the porosity depth trends in these basins. It is possible, accordingly, that the lack of correlation between overpressure and shale porosity reported at Haltenbanken by Hermanrud *et al.* (1998) is a manifestation of 'disequilibrium' in the sense that the shales did not have time to respond chemically to the overpressure development. This suggests that mechanical models for compaction cannot adequately explain porosity loss in deep shales and that a time/temperature element must be included to account for chemical compaction.

The fact that the shales have not had time to re-equilibrate to the new pressure regime must imply that overpressure development in Halten Vest is very recent, underlined by the fact that the strata today are at maximum burial depth and that relevant mineral reactions are exponential in temperature.

Diagenesis on the fault planes is believed to be the main cause of overpressure development in Halten Vest, as the region was effectively sealed off from pressure leakage towards the east.

This led to seal failure in the Halten Vest region whilst in the shallower and normally pressured Halten region continuous strata allow pressure bleed-off towards the east. This acts to preserve petroleum in traps.

We have documented the close genetic relationship between the bitumen and petroleums in presently filled and drained traps and ascribe the minor variations that exist, not to source rocks other than the Spekk, but rather to east-to-west lateral systematic variations in terms of proximal to distal source rock organic-facies development of the Spekk Formation source rock.

Hence, our geochemical data from the Smørbukk giant oil field, the Smørbukk Sør and the Heidrun oil fields all point to the same marine source rock, but with local facies variation. Fields are sourced mainly with lateral migration distances on the order of inter-field distances, i.e. *ca.* 15–20 km. More extensive migration is likely only for the Draugen and the Midgard Fields.

We exclude, in contrast to earlier published models by other workers, any quantitatively important contributions of light hydrocarbons ($C_5$–$C_8$) and heavier hydrocarbons ($C_{15+}$) from the coal-bearing Åre Formation to the presently reservoired hydrocarbons. The Åre coals have clearly expelled vast quantities of methane. This methane would have been expected to be isotopically heavy. We have not seen any evidence for isotopically heavy, non-biodegraded methane, ethane or propane except in one proximal trap offshore Mid-Norway. This suggest that gas from the Åre coals must generally have passed through the structures before the development of tight caprocks. Some structures like Midgard show tentative evidence that they may have contained contributions from the Åre Formation at an earlier time. The free HC charges are, however, long since lost from the traps (underlining their dynamic nature).

Expelled condensate from the Åre Formation at an earlier time may have 'paved the way' for higher migration efficiencies for Spekk-derived oil. This would have occurred because early Åre-derived petroleum would have increased HC saturations in laterally continuous carriers as long range HCs condense in the decreasing *P* and *T* field during migration. A likely candidate for Åre-derived condensate is the Idun trap which is highly untypical for this region in terms of its organic geochemistry.

The high oil generative capability of the Spekk Formation and the arrival of the expelled products to the traps so much later than those from the Åre Formation means that the geochemical signatures from the Spekk Formation have mostly obliterated geochemical signatures from Åre.

The generative capability of the Spekk Formation in deep basinal depressions is likely to be much higher than often indicated from published models, as traps are generally overfilled with type II derived petroleum whilst models often indicate Spekk incapable of even filling discovered traps. Both TOC and HI values would tend to be higher in sub-basinal depressions than elsewhere due to co-occurring increased trapping of low density OM in depocentres and enhanced preservation potential. Expulsion from these top quality SR intervals may occur shallower than often implied in modelling studies. Condensates in Jurassic traps of the region are, apart from the 6407/4-1 and the Smørbukk and Kristin condensates, all evaporative condensates that owe their existence to phase separation rather than to high maturity or terrestrial kerogen quality.

Traps in the western Haltenbanken region like Lavrans, Kristin, Trestakk and Smørbukk contain mainly petroleum of high maturity, yet in the compartmentalized Smørbukk field some medium maturity bitumen remains in particular reservoir sections. This is most likely due to strong compartmentalization caused primarily by the lithological heterogeneities and secondarily by diagenetic poroperm modifications.

Deeper fields in the western regions of Haltenbanken, e.g. Trestakk and Kristin and also 6407/4-1, contain only mature petroleum, i.e. there is a lack of medium to low maturity petroleum in the fields which would reflect accumulation of 'full spectrum oils' (cf. Cornford *et al.* 1986). This suggests that even these structures have experienced leakage. Accordingly, the caprocks on these structures must have 'welded' and the structures could refill with more mature petroleum, as the Spekk Formation was still not exhausted in their drainage areas.

These cases may serve to underline the dynamic nature of petroleum entrapment.

There is no evidence for oleanane in any fluid inclusions or in any core extracts, suggesting that expelling Cretaceous source rocks have not contributed to hydrocarbons in the studied traps. However, Cretaceous strata seem clearly to be the source rock for some of the shallow high GOR traps in the deep water regions and may in our perspective represent an attractive gas play scenario for Ormen Lange and are certainly the source of the 6507/2-2 DST.2 petroleum.

We infer that several oil accumulations in Haltenbanken are generally in a state of 'steady gas depletion', e.g. Draugen, Njord, Tyrihans Nord, Trestakk, Heidrun, Norne, Svale and Falk, with gas escaping through thinned, partly eroded caprocks. This produces *in-situ* low GOR or undersaturated oils in the traps. The undersaturated oils in the Cretaceous sands above Smørbukk and Smørbukk Sør traps originated most likely (based on their geochemical characteristics) directly from the traps below by fracture induced dysmigration with gas-components entering shallower strata like the Lysing Formation. In the case of the oil above Smørbukk is it likely that the oil escaped via palaeo-caprock leakage from the now, mostly, drained Garn Formation at the Jurassic level.

Fields with thicker tighter caprocks with perhaps a higher content of carbonate in the shales are likely to be able to hold significantly higher GOR fluids in place before fill–spill or eventual caprock leakage occurs. The pressure valve mechanism and the delicate relationship between fracture pressure and trap pressure (which will vary according to reservoir depth, caprock strength and hydrocarbon column) are preserving hydrocarbons in place in some of these deeper structures. Examples of this category may include Smørbukk, Kristin, Lavrans and 6407/4-1, whilst shallower overpressured fields without a pressure vent mechanism will eventually destructively leak (cf. Halten Vest).

Seismic brightening, which occurs over large regions in shallow Lysing Formation strata, might have originated either due to *in-situ* gas generation in the Cretaceous strata or, less likely, in a regional perspective, from dysmigration from deeper traps and SRs, albeit this latter mechanism certainly is likely above leaking traps. Investigations of carbon isotope profiles of the gas (canned cuttings) will possibly resolve the issue of the origin of the gas responsible for the regional seismic brightening.

The data and interpretations presented here are based on several projects undertaken during the past decade. Several students participated during this time in projects financed by Statoil and we commend the company and the former Cand. Scient students Kristian Angard, Inga Steinhoff, Kristin Dale, Camilla K. M. Winterstad, Bowan Xu and Elisabeth B. Sletten. We are also obliged to the former Saga Petroleum company which financed the Cand. Scient project of Vidar Storvoll on Kristin and Lavrans and to Enterprise Oil Norway Ltd. for continued financial support and project cooperation during the last four years before the company was taken over by Shell. This project could not have been undertaken were it not for the oil and core bank of the Norwegian Petroleum Directorate. The views presented reflect those of the authors and not the institutions/companies mentioned above. Elements of this paper were originally presented in 1999 at the AAPG Hedberg meeting, 9–13 May, Colorado Springs, USA, and the first presentation of the model with migration across the fault west of Smørbukk was given at the NPF meeting January 1996 in Trondheim, Norway. Modelling of burial and overpressure build up was performed by Jon Erik Skeie as part of his Dr.scient project on overpressure development in this region. ChevronTexaco Norge AS and RWE Dea Norge AS are acknowledged for financing the Dr.scient project of Jon Erik Skeie. Reviewing of an early version of this manuscript by Michael Lines and an anonymous reviewer, plus the final review by William A. England is greatly appreciated.

## Appendix

### *Analytical methods*

All examined core samples were extracted with dichloromethane:methanol (93:7 vol%) and bitumen and coextractives were analysed using the Iatroscan TLC-FID method (Karlsen and Larter 1991) to determine the amount and percentage composition in terms of saturated hydrocarbons, aromatic hydrocarbons and polar compounds. This type of screening may help to differentiate between migrated hydrocarbons, *in-situ* generated hydrocarbons and also diesel drilling fluids (Karlsen & Larter 1991) and has recently become a standard tool in reservoir geochemical studies (Larter & Aplin 1995).

From selected homogeneous well-sorted reservoir sandstones, individual grains were produced by manual crushing and sieving according to the procedures outlined in Karlsen *et al.* (1993). Extensive cleansing of crushed core material using inorganic oxidizing fluids is done to ensure that organic material is removed from the sand grains. This is done to ensure that optical properties of the sand grains are well suited for microscopic observations of petroleum inclusions, and to reduce and remove if possible organic material external to the sand grains. This, if successful, will allow bulk crushing of grains and extraction of gas and oil range hydrocarbons from inclusions (cf. Karlsen *et al.* 1993).

The core sandstone samples were crushed and extracted with dichloromethane : methanol (DCM : MeOH) 93 : 7 vol% for total extract analysis. Some samples selected for geochemical analysis of the petroleum inclusions were carefully disintegrated further by manual crushing to produce fractions of individual grains. Samples were carefully examined optically throughout the disintegration procedure to ensure that single grains were being produced. These disintegrated sands were then extracted by DCM : MeOH (93 : 7 vol%) to remove bitumen from the sand grains and from associated clay and silt. This extract was used for Iatroscan TLC-FID, GC-FID and GC-MS analysis.

Three procedures were adopted for cleaning the disintegrated sand grains. First, the chromic acid treatment (6 h) followed by washing with tap water removed any clay and most organic matter from the grains (Karlsen *et al.* 1993). Second, the less viscous oxidizing agent, hydrogen peroxide (cf. George *et al.* 1997), removed residual or non-extractable organic matter from fissures and cracks. $H_2O_2$ also acts to destroy iron containing oxides, hydroxides and biotite. Finally, the remaining grains were extracted by Soxtec to remove any possible bitumen from the grain surfaces and fissures opened after the first two treatments. We believe that combined acid and $H_2O_2$ treatment helps to oxidize more effectively organic matter residing in fissures and cracks than can be done by either method alone. This probably has to do with modifications in wettability of the mineral matrix during this alternative procedure and also their different ability to destroy minerals, e.g. carbonate destruction by the chromic acid.

No attempt was done here to separate gravimetrically K-feldspar from plagioclase or quartz using the polytungstate solution used by Karlsen *et al.* (1993). Hence our crushing procedure will potentially liberate gas and biomarkers from inclusions in these three mineral groups. It is still clear that in most of these samples the bulk of K-feldspar is dissolved due to the temperature in the reservoirs exceeding the temperature at which the kaolinite to illite transition becomes intensive, i.e. at about 120–140 °C (cf. Bjørlykke *et al.* 1980, 1983) this process requires potassium taken from K-feldspar. Abundant secondary porosity, testifying to this process, is observed in all sands at depth exceeding about 4000 m. It is thus likely that petroleum inclusions which may have existed in K-feldspars were not analysed in this study. If some still existed, metastably or in some of the shallower samples, then it is likely that these inclusions were grossly outnumbered by inclusions in plagioclase and quartz.

The inclusion gases ($C_1$–$C_4$) were obtained using a steel chamber crusher with a brass cylinder and a septum through which the gas sample was withdrawn via a gas-tight syringe (Karlsen *et al.* 1993). The sample was then analysed on a Varian 3500 GC-FID (nitrogen carrier gas) using a 30 m DB-1 column with 0.32 mm ID and a film thickness of 5.0 μm). The higher boiling *n*-alkanes and biomarkers from fluid inclusions were liberated by manual crushing using an agate mortar and pestle as described in Karlsen *et al.* (1993). This crushing was done under DCM to reduce the loss of light-boiling hydrocarbons.

Whole core extracts were analysed using the Iatroscan TLC-FID instrument and methods described in Karlsen & Larter (1991) in order to determine their total composition of saturated hydrocarbons, aromatic hydrocarbons and polar compounds. The *n*-alkanes and isoprenoids in the whole oils and core extracts were analysed on a Varian 3400 GC-FID equipped with a 30 m DB-1 column with 0.25 mm ID and 0.25 μm film thickness, using nitrogen as the carrier gas. The temperature programme used was 60 °C for core extracts and 40 °C for whole oils, held for 2 min, ramped at 4.5 °C/min to 310 °C and held for 15 min. During runs, a house standard was run every tenth sample. A Fisons MD 800 gas chromatography–mass spectrometry (GC–MS) with a 60 m DB-5 ms column (0.25 mm ID and 0.25 μm film thickness) and helium as the carrier gas was used for biomarker analysis. The temperature programme used for both biomarkers and aromatic hydrocarbon was 80 °C held for 10 min, 10 °C/min to 180 °C and 1.7 °C/min to 310 °C, held for 30 min. The instrument was operated in selected ion mode (SIM) with most of the polar fraction and the *n*-alkanes having been removed by molecular sieving using silicalite from the core extracts, whole oils and petroleum inclusion extracts before injection. Ions with $m/z = 191$, 218, 217, 231, 253, 198, 192 and 178 were monitored. 50 μm thin-sections made from cleansed sand grains impregnated with non-fluorescing epoxy and polished on both sides, were used for microthermometry and mapping of fluorescent inclusions. Also used were free sand grains on stray slides. Observations of water and fluorescent inclusions were made using a Nikon Microphot-SA equipped with Nikon 10×, 40×, 80× and 100× long working distance objectives. Microthermometry was done using a Linpam THMG 600 heating/freezing stage.

### *Inclusion formation—themes of dispute*

Formation of water filled inclusions is well known and documented but some elements of the formation of HC inclusions are in part less well understood. Whilst inclusion formation is readily rationalized for minerals with oil wetting characteristics, e.g. carbonates or evaporates, the formation is somewhat more difficult to rationalize in water wetting minerals like quartz. Quartz and plagioclase are well suited for determining the time of entrapment by the use of microthermometry (cf. Karlsen *et al.* 1993; Roedder 1984) of oil in reservoirs as these minerals are generally less likely to dissolve or experience recrystallization in the temperature range experienced by a progressively buried HC trap.

In contrast, inclusions form more readily in carbonates and evaporites but these have an intrinsic tendency to experience recrystallization or neomorphism. Leaking inclusions, precipitation of minerals inside the inclusion volume, and also necking-down phenomena may alter dramatically the internal pressure and volume ratios in the inclusions from the time of its initial entrapment, making microthermometric observations more difficult to translate directly into trapping temperature (cf. Roedder 1984; Nedkvitne *et al.* 1993). This temperature is needed to decipher the time for entrapment by the use of the burial history for the trap (Karlsen *et al.* 1993). This type of application of inclusions as 'tape recorders' for reservoir filling is in carbonates best done with the help of detailed cathodoluminesence or backscatter techniques so that the various diagenetic events can be traced in detail.

It is for these reasons that the use of clastic minerals like quartz and feldspars, with their simpler diagenesis, are attractive as 'tape recorders' for deciphering both the type of oil, i.e. source facies and maturity, API and GOR, and also when this oil existed in the trap (cf. Karlsen *et al.* 1993).

Formation of organic inclusions, i.e. petroleum inclusions as opposed to water inclusions in hydrophilic minerals, poses several interesting questions as to how one may get the oil into the necessary intimate contact with the liphophobe water wetted mineral.

The existence of fluid inclusions in quartz (Reese 1898) and, especially, entrapment of organic matter in such fluid inclusions in mineral phases in the form of porphyrins and other fluorescing organic matter, is known and documented since Haberland (1939) and Neumann & Rosenquist (1940).

Forming petroleum inclusions in quartz is conceptualized as incorporating movement of ions along water films on the grains and these ions result from pressure solution (cf. Boer 1974; Bjørlykke *et al.* 1989; Walderhaug *et al.* 2000). This occurs at the interface between mineral and petroleum, possibly utilizing adsorbed water, from sites of pressure solution to sites of precipitation (cf. Riecke 1894; Weyl 1959; Boer 1977; Hutcheon 1983). The increased pressure on the point of contact between the quartz grains is likely, as for all solid phases, to increase molecular interaction and to decrease the stability of the crystal lattice. Petroleum inclusions are likely to contain, in addition to the HCs (liquid and gas), also water lining the inner part of the inclusion.

In the case of petroleum inclusions the amount of water is most likely so minute as to not be readily observed during microscopy and one may speculate as to why more three-phase inclusions (containing gas, petroleum and water) are not observed. It seems from microscope observations tempting to suggest that the inclusion mechanism is to some extent 'binary', i.e. water either with or without HCs in solution is entrapped as inclusions, or pure petroleum inclusions form. In the case of the petroleum inclusions is it likely that the bulk of the water which is not lining the inside of the inclusion is simply partly dissolved in the oil or exists as part of the gaseous phase in the gas bubble under ambient conditions. It would take much more water to make the phase a discrete microscopically observable one and such quantities seem empirically unlikely.

As the source of quartz in this case is internal and the migration of ions is driven by diffusion, cementation can proceed even under the conditions of high oil saturation for clastic water wetting minerals. This is why we do not believe there to be a need for 'pulses of water and oil' to move through the reservoir in order to have the necessary conditions for inclusion growth, nor is the inclusion growth likely to occur more rapidly close to the OWC where capillary rise of water into oil may occur.

Perhaps the strongest set of evidence to suggest that petroleum inclusion formation neither requires, nor is promoted by, alternating pulses of oil and water, nor the mixed wettability characteristics close to the OWC, is the fact that most reservoirs we have examined show clear cut depth profiles with increasing petroleum inclusions with decreasing (cf. Fig. 18) depth. As inclusions take a finite time to form (cf. insert in Fig. 20) and most traps have a geometric configuration of a dome or triangle (Fig. 32), it is clear that during filling the OWC is moving downwards with exponentially reduced rate as the trap fills (provided influx of petroleum remains constant). This is a simply a consequence of geometrics and, if mixed wettability in the transition zone within the OWC was important to petroleum inclusion, formation traps should show highest petroleum inclusion occurrences towards the present OWC. They generally do not.

In most of our reservoir studies on clastic reservoirs we come across cases with generally more petroleum inclusions having formed in the upper and most permeable and porous layers of the reservoirs, and this may reflect that oil has occurred for more time in the upper part of the reservoir, giving inclusions more time to form than deeper down. Still, it could also simply be that the higher pressures in the oil column in the top of the reservoir as compared with farther down forces the oil into more intimate contact with the cementing quartz grains and that this results in more entrapment of oil as inclusions higher up in the oil column.

Nedkvitne *et al.* (1993) calculated, based on mercury injection capillary pressure data, the petroleum column height needed for oil to penetrate into openings and cavities in water wetting minerals. An oil column of 5 m is sufficient to force oil into 5 μm cavities which are typically observed at the interface between the detrital and the authigenic mineral as along the 'dust rim' on quartz (cf. Fig. 20). From this it is gathered that the increased invasion of oil into smaller porosities which exist in the top of an oil column may provide the mechanism we believe to be necessary for getting oil into the intimate contact with the minerals needed to produce more effective inclusion formation. It was concluded by Nedkvitne *et al.* (1993) that this was the reason why there tend to be more petroleum inclusions towards the crests of traps.

A column height of 5 m is not large by any standards and it follows that petroleum inclusion formation most likely is initiated early in the trap filling process, provided the reservoir temperature is sufficiently high (see below), e.g. already at the stage when individual petroleum fingers still exist in individual sediment surface units in the reservoir (cf. Fig. 3). Closer to the OWC inclusion formation in lipophobe minerals may to a larger extent be controlled by the existence of sites with mixed wettability characteristics, i.e. solid organic matter on grains.

Nedkvitne *et al.* (1993) were the first to present down-hole logs of the abundance of fluorescent inclusions in traps and to use these to argue that the number of inclusions represents the time-resolved evolution of the oil column in reservoirs. Similar profiles, based not on relative but rather absolute numbers of fluorescent inclusions, were used by Oxteby *et al.* (1995) who also discussed, on the basis of a larger database, the reasons for small scale variations in inclusion abundance in the same lithologies evaluated by Nedkvitne *et al.* (1993).

The principle of decreasing amounts of fluorescent inclusions with increasing depth within oil columns in relatively homogenous clastic lithologies might have a bearing on the rate of inclusion formation and also trap filling. England and Mackenzie (1989) suggest that trap filling may typically occur within a 10 Ma time perspective. This time estimate may reflect both progressive SR burial and the dynamic fill-depletion mechanism of reservoirs more than actual filling rates, if the petroleum flow is sufficiently focused. Walderhaug (1994) suggested that it takes on the order of 40 Ma to form a 4 μm inclusion in overgrowths in quartz at 80 °C. At the higher temperatures of 100, 120 and 140 °C the time estimates shrink to about 17, 6 and 2 Ma, respectively. These formation rates compare favourably with the reservoir filling rates, meaning that inclusion formation in quartz may act as a 'tape recorder' with reasonable resolution to track the progressive lowering of the OWC in a trap which is progressively filling.

It is not clear, however, whether draining a trap, which will result in decreased oil saturation, results in a complete halt of petroleum inclusion formation. This is likely, however, as any reduction in oil saturation due to caprock leakage from a filled situation with, e.g. 95–80% oil saturation towards 50–40%, as is found in drained reservoirs (cf. Bhullar *et al.* 1999a), is likely to result in oil forming non-continuous isolated oil globules in the centre of pores in high porosity sandstone porosity units (cf. 'The Augustson Method', Augustson 1992; Bhullar *et al.* 1999a). This situation is not ideally suited for inclusion formation as the oil is no longer forced into grain contact when the pressure in the oil phase is relaxed.

It is well established from case histories (e.g. Skålnes 1993; Johansen 1997; and Fig. 17) that two and sometime three different types of petroleum inclusion can conveniently be detected and classified in individual thick section samples representing individual sandstone units (see also this principle illustrated in Fig. 20). These inclusions are discernible from their physical characteristics, i.e. their fluorescent colours and the relative size of the gas bubble as compared with inclusion volume. These inclusions are normally interpreted as representing different generations of petroleum occurring in the trap at different times. A plausible scenario would be to assume that the inclusions with the smaller gas bubbles and lower API represent low- to medium-mature oil with low to medium GOR, and that the inclusions with large gas bubbles and higher API (cf. Bodnar 1990) represent a more mature condensate type of included petroleum. A trap with two or more such discernible inclusion populations may reflect that the trap received petroleum from a continuously subsiding and maturing SR system. The trap, 'seeing down into' the generative basin during its burial, thus evolved from containing black oil to holding a condensate and as such represents a 'tape recorder' of the SR evolution history.

Still, it is difficult to foresee that inclusions can be used effectively to trace out much more detailed maturity traits in traps simply because the formation process in any given mineral is a continuous process and as oil dissolves oil, i.e. mixing results for biomarkers and other maturity and facies parameters. It is for these reasons that Karlsen *et al.* (1993) resorted to utilizing the quartz-plagioclase and the K-feldspar system, which are minerals with quite different diagenetic temperature responses, to get a more time resolved picture of the HC phases in the trap investigated.

We conclude from this that thc typically observed 'down-well' decrease in the abundance of inclusions in homogeneous clastic lithologies results from potentially two effects. First, progressively lower pressures exist in the oil column as the OWC is approached. This pressure is the main driving force for getting oil into the required intimate contact with cavities in the water wetting mineral grains, which will form petroleum inclusions. Second, as inclusions take on the order of single digit millions of years to form at typical reservoir temperatures in a trap taking 10 Ma to fill (England & Mackenzie 1989), it is likely that inclusions in the top of the oil column have more time to form than farther down in the reservoir. Inclusion formation is then sufficiently rapid as to act as a 'tape recorder' for the rate of trap filling.

It follows from the same argument that within individual upward-coarsening lithologies, the highest inclusion abundances should generally be observed in the upper part of the unit.

Furthermore, it can be argued that there is probably little to be gained by future geochemical analysis of HCs from individual inclusions. This is because the rate of fluid inclusion formation in any give type of mineral, e.g. quartz, despite occurring over temperature intervals of typically 80–170 °C (Walderhaug *et al.* 2000), is slow compared with the short time between individual pulses of petroleum arriving in the trap, which may range on the order of 10 000 to 100 000 years rather than millions.

Organic compounds in any given inclusion are therefore likely to contain a mixture or a mean geochemical signal resulting from the oils that existed in the trap in the past. This will also be the case if a more mature high GOR petroleum replaces less mature black oil. However, the gas ($C_1$–$C_4$) and the medium range parameters, e.g. methyldibenzothiophenes and phenathrenes, pristane and phytane, are in this case more likely to represent the later phase in comparison with the biomarkers, which will be slanted towards the maturity signature of the black oil. This is because the biomarker content in the high GOR petroleum or condensate will be orders of magnitude lower than in a low maturity black oil. Similarly, the distribution of the gaseous, light and medium HCs will be dominated by that in the high GOR petroleum.

Adding to this the simple volumetric consequences of mixing medium volumes of oil with a huge, in part gaseous HC system, in a trap at say 3–2 km, the original oil can be 'smeared' out over a huge reservoir cross-section, forming only a relatively thin phase fractionated oil leg in the lower reservoir units, or the oil may be displaced out of the structure altogether (cf. Gussov 1954, 1968; Sales 1993, 1997).

It is clear that the entrapment of petroleum in inclusions is likely to result in gross compositional fractionation, i.e. resulting in the entrapped petroleum becoming enriched in asphaltenes and resins, and such has been observed (Karlsen *et al.* 1993; Wilhelms *et al.* 1996). However, there is no reason to assume, nor observations to suggest, that the fractionation affects the typically employed molecular ratios between isomers of biomarkers and related compounds.

Wilhelms *et al.* (1996) used sequential extraction of core extracts believed to represent different generations of reservoir bitumen to investigate the palaeo-hydrocarbon composition in traps and found that extracts representing the most strongly adsorbed hydrocarbon phase approached the geochemical signature of petroleum inclusions.

Meaningful organic analysis of fluid inclusions requires thorough examination of the disintegrated sands, several steps of cleansing, and extraction with organic solvents, before final microscopic examination. Certain sands may contain fluorescent inclusions, yet be unsuited for biomarker inclusion analysis due to low effectiveness of the cleansing procedure. Such sands may still be useful for microscopic examinations of fluorescent colours as an indicator of API gravity, and measurement of the relative size of the gas bubbles as an indicator of the GOR. Furthermore, gas inclusion analysis and measurement of homogenization temperatures will provide useful information even on such samples.

Hindrances to precipitation of quartz and hence inclusion formation is represented by clay coatings on minerals (cf. Tada & Siever 1989; Nedkvitne *et al.* 1993; Ehrenberg *et al.* 1995; Storvoll 1999; Storvoll *et al.* 2002) and also extreme organic coatings (cf. Skålnes 1993; Storvoll 1999; Storvoll *et al.* 2002). In such extreme cases precipitation of quartz is clearly kinetically hindered due to a lack of nucleation sites (cf. Bjørkum 1996) and inclusion formation rate is reduced or halted. When quartz grains are more or less totally covered in asphaltenes, as in some of the sediment surfaces in the Hild Field discussed by Skålnes (1993), is it also likely that the pressure solution of quartz is kinetically hindered.

Formation of 'clean' fractures represents an extremely attractive site for entrapment of petroleum inclusions as the surfaces are clean, i.e. without impurities on the quartz surface, poisoning minerals or organic matter, and inclusion formation proceeds rapidly.

Along the undulating topography of natural clastic detrital quartz grains an abundance of nucleation sites with mixed wettability characteristics will normally exist during initial diagenesis (Fig. 20). It is likely that organic dust particles as in the 'dust rim' (*sic.*) of clastic grains will increase oil wetting and potentially increase the chance for entrapment of oil. Often, minor amounts of solid organic matter are observed in authigenic petroleum inclusions and this may reflect that, in some cases, solid organic matter may play a role in inclusion entrapment. Competitive mineral growth from these nucleation sites may trap petroleum and water by a pincer movement, as the authigenic mineral phases enclose the detrital grain (Fig. 20).

The homogenization temperatures of the 1080 inclusions (water and petroleum) studied by Walderhaug *et al.* (2000) from offshore Norway reflect that the first inclusions closed at about 80 °C, and tailed off from a maximum of 100 °C towards higher temperatures of 150–170 °C. This may seem at first to conflict with the fact that quartz cementation is exponential in temperature. Stabilization of grain frameworks could result in reduced quartz being available from pressure solution, but more likely the reduced number of inclusions formed at high temperatures simply reflects a reduction in potential sites for inclusion formation.

We believe petroleum inclusion formation to be motivated or helped by trace amounts of organic matter or oil wetting clays like chlorite occurring on the detrital surface of quartz grains during the early phase of diagenesis, i.e. from temperatures of around 80–90 to *ca.* 100 °C. High numbers of later formed inclusions will occur mainly in fractures. In albitized plagioclase, petroleum inclusions are *also* observed *within* the boundary of the detrital grain (cf. Nedkvitne *et al.* 1993). This occurrence is related to partial dissolution of detrital plagioclase grains and subsequent polynucleative growth of diagenetic albite at many sites, leading to abundant entrapment of petroleum within the intragranular porosity of the skeletal morphology of the partly dissolved plagioclase grain.

The authigenic phase may enclose voids or inclusions at the interstices of several overgrowths (Hutcheon 1983). Overgrowths in crystallographic continuity with the host may grow from several nucleation sites, and thus form larger overgrowths in overall crystallographic continuity with the detrital grain. Thus, no crystal boundary need be present at the inclusions. As the water film is last to disappear during mineral growth, it will remain until sealing of an inclusion is complete, leaving residual water mainly as adsorbed water lining the inside of the inclusion (Kvenvolden & Roedder 1971). The occurrence of petroleum inclusions in the authigenic minerals is evidence that diagenesis proceeds even after petroleum emplacement, although the actual diagenetic reaction rate and the range of petroleum saturation remain obscure.

Whilst certain aspects of the exact formation mechanism for these petroleum inclusions is unknown (cf. Roedder 1984; Nedkvitne *et al.* 1993), the data presented here, as well as unpublished work from several petroleum accumulations in Jurassic sands from the Viking Graben and the Haltenbanken area (Norwegian Shelf), suggest that diagenesis proceeds, utilizing residual water films, even under petroleum saturations corresponding to those found today in these petroleum-filled accumulations (up to 80% hydrocarbon saturation).

Petroleum inclusions formed in quartz are either closely associated with the dust rims of the detrital grains (primary inclusions), or they occur in fractures (secondary inclusions). In palgioclase the host of petroleum inclusions is typically found within the skeletal morphology of partly albitized grains (Karlsen *et al.* 1993; Nedkvitne *et al.* 1993). In all cases is it clear that the GOR of the petroleum inclusions in quartz and feldspars will reflect the petroleum present during the time the inclusion actually closes, whilst the biomarker content and the total composition of $C_{10+}$ hydrocarbons and polar compounds is likely to reflect a 'mean' of the reservoired petroleum composition during the entire period since invasion of oil into the strata.

Several studies have described the molecular composition of hydrocarbon inclusions (Murray 1957; Kvenvolden & Roedder 1971). Application of fluid inclusions to constrain diagenesis was demonstrated by Burruss (1981) and Etminan & Hoffmann (1989), the latter utilizing biomarker distributions in the inclusions. Burruss *et al.* (1985) used the occurrence of fluorescing petroleum inclusions within authigenic carbonate phases of reservoir rocks, dated by cross-cutting relationships and stylolitization, together with the burial history to decipher events concerning petroleum migration within the Cretaceous reservoir limestones from Oman and the United Arabian Emirates. The use of microthermometry (cf. Roedder 1984) on fluid inclusions in diagenetic minerals may allow more accurate dating of the petroleum entrapped during diagenetic growth (Pagel *et al.* 1986) and is especially valuable when combined with data on the petroleum composition of inclusions (Horsfield & McLimans 1984; McLimans 1987; Karlsen *et al.* 1993; Nedkvitne *et al.* 1993; George *et al.* 1997, 1998; Scotchman *et al.* 1998). Karlsen *et al.* (1993) presented the first isotope determination of hydrocarbon gas in diagenetic inclusions and used biomarker distributions from inclusions in K-feldspars, and quartz plus plagioclase, to determine that the Ula field in the Central Graben, Offshore Norway, contained oil 75–45 Ma bp and that the Kimmeridge derived oil currently filling the trap arrived during the last 5 Ma.

In this concept one relies on knowing the formation temperature for the inclusions, and this temperature is then transformed into time using burial history and geothermal gradients (cf. Fig. 20).

Application of automated instrumentation using vacuum-treated washed core or cutting samples, evolved from an original Amoco 1994 patented mass spectrometer set-up, allows high sample throughput and helps to identify hydrocarbon zones from cuttings (cf. Hall *et al.* 1997; Hadley *et al.* 1998; Barclay *et al.* 2000). This approach may be most helpful as a screening tool for inclusions and the very high sample throughput is an asset of the method. However, the methodology of direct injection of desorbed compounds from inclusions into a mass spectrometer, without chromatographic separation, does not allow conventional identification of specific biomarkers and facies and maturity parameters. It is not clear, furthermore, how one may guarantee elimination of signals from adsorbed petroleum compounds in the mineral matrix, e.g. clays (SRs) by the standard treatment. This means that compounds other than from inclusions *sensu stricto* cannot be avoided and it is recommended that the procedure be used in combination with microscopy observations of individual inclusions, and extensive treatment to cleanse the grains (see below).

We also suggest application of GC-FID analysis of gas, long chained alkanes and biomarkers to realize the full potential of inclusions to reservoir studies as these may represent 'hermetically sealed' testers of palaeo-petroleum. Furthermore, only the use of fluid inclusions in combination with standard geochemical core bitumen quantification, compositional analysis and GC-FID plus GC-MS characterization, can fully realize the potential of these methods for deciphering movement of OWCs during time and determining the time for field filling.

## Limitations of the approach and new methods

One of the main uncertainties related to the application of petroleum inclusions as indicators of earlier reservoir crude oil concerns the possibility that the growing authigenic phase has trapped not only oil but also immature kerogen which may mature *in-situ* in the progressively buried structure. This material could potentially contribute to the oil inclusion biomarker composition, making it less mature.

The presence of such material in inclusions, in addition to the gross compositional fractionation (relative proportions of saturated HCs to aromatic HCs and polars) which occurs during inclusion entrapment (cf. Karlsen *et al.* 1993), poses a definite problem to detailed PVT modelling of petroleum inclusions as attempted by, for example, Munz (2001) and Aplin *et al.* (1999). Microscopic investigations will often reveal solid organic matter in inclusions and it is clear that detailed microscopic examinations will allow one to avoid such inclusions. The approaches of Munz (2001) and Aplin *et al.* (1999) are intriguing, the latter using confocal laser scanning microscopy in an attempt to better quantify inclusion volumes and the ratio of gas to liquid phase. Still, the problem with the internally adsorbed water layer means that apparently pure HC inclusions in water wetting minerals are in fact all three-phase inclusions, of which one phase is indeed inorganic and non-observable in the microscope. It is obvious that this, in addition to the fractionation of the oil which influences the gross compositional make-up of the oil during inclusion entrapment, and which makes the included oil (IP) richer in polar compounds as compared with the free oil (FP) (cf. Karlsen *et al.* 1993), would be expected to cause problems to detailed PVT estimates. This is because it is not at all clear if one may use inferred PVT characteristics from the GC-FID alkane envelope of the entrapped oil in the inclusions, nor the PVT characteristics of the presently-reservoired oil as analogues for the real PVT properties of the petroleum occurring in the inclusions.

We have often observed that certain sands may contain abundant fluorescent material in cracks and fissures that can be almost inaccessible to chromic acid and hydrogen peroxide. Organic material can also in some lithologies be intimately associated with in-organic grain coatings. Such material can clearly contaminate 'inclusion extracts', and we go to great lengths to reduce the probability of incorporating such samples in biomarker examination of inclusion extracts. Jones & Macleod (2000) presented an elegant 'spiking' method in which they add to their rock samples an organic marker and follow the degree to which that organic marker is removed by their treatments, designed to rid the rock of all external organic material. This method can clearly help in a number of cases and represents an improvement. However, it is difficult to see if it can effectively handle 'contaminates' in the form of oil trapped in open or semi-open fractures and in intimate association with mineral coatings. This is because the spiking agent will not necessarily be as strongly adsorbed or absorbed as the bitumen under progressive diagenesis and under the longer time and higher temperatures of the reservoir conditions.

Recent advances in isotope analysis include the ability to analyse isotopically smaller proportions of individual compounds by the use of more sensitive online isotope ratio mass spectrometers (cf. Merritt & Hayes 1994; Merritt *et al.* 1994) and stable carbon isotope determination of individual *n*-alkanes, isoprenoids and biomarkers in petroleums is now a standard tool. Despite early work on collection of chromatographically separated individual *n*-alkanes followed by combustion and isotope determination (Welte 1969) it was for a number of years impossible to analyse isotopically long chained alkanes with the same comfort and ease as the gaseous compounds methane, ethane, propane and the butanes, using traditionally packed columns. Karlsen *et al.* (1993) were the first to present carbon isotope determination of gaseous HCs from diagenetic inclusions in separated mineral fractions from reservoir sediment surfaces. They used packed columns, separate ovens for the individual gaseous species and crushing of large quantities of minerals to get sufficient gas for accurate and precise measurements. The problems associated with online analysis of HCs directly from a capillary column, using only one single oven for combustion of HCs to carbon dioxide, includes how to avoid fractionation going from the overpressured capillary column system to the vacuum system of the mass spectrometer. Memory effects associated with passing all the different HC species through the same oven also had to be overcome. Today, online-gas chromatography–isotope ratio mass spectrometry (GC-IRMS) has become a standard tool for carbon isotope determination on individual GC resolvable compounds like *n*-alkanes, isoprenoids and biomarkers and we use the method also for carbon isotope determination of gas from inclusions. This allows gas from inclusions to be readily compared with free production gas or dissolved gas in petroleums using the standard set of characterization diagrams for gases (cf. Bernard, *et al.* 1978; James 1983; Schoell 1983; Whiticar 1994) and the more specialized and theoretically based approaches introduced by Galimov (1973, 1985), Sunberg & Bennet (1981), and more recently Chung *et al.* (1988).

Still, it is not yet possible to accomplish stable hydrogen isotope determination on HCs using IRMS as a more complex step involving reduction of water from the combustion process is necessary to liberate hydrogen.

## References

AHSAN, A. 1993. Petroleum biodegradation in the tertiary reservoirs of the north sea. Cand. Scient. Thesis in Geology, Department of Geology, University of Oslo, Norway.

AHSAN, A., KARLSEN, D. A. & PATIENCE, R. L. 1997. Petroleum biodegradation in the Tertiary reservoirs of the North Sea. *Marine and Petroleum Geology*, **14**, 55–64.

AHSAN, S. A., KARLSEN, D. A., MITCHELL, A. W. & ROTHWELL, N. 1998. Inter and intrafield hydrocarbon compositional variation in the Ula and the Gyda Fields (Central Graben–North Sea)—Implications for understanding the controls on hydrocarbon distribution within and between these fields. *Organic Geochemistry*, **29**, 429–448.

ALEXANDER, R., KAGI, R. I. & WOODHOUSE, G. W. 1981. Variation in the ratio of isomeric butanes with sediment temperature in the Carnavon Basin of Western Australia. *In*: BJORØY, M. (ed.) *Advances in Organic Geochemistry 1981*, Wiley, 76–79.

ANDREYEV, P. F., BOGOMOLOV, A. I., DOMRYANSKII, A. F. & KARTSEV, A. A. 1968. Transformation of petroleum in nature. *International Series Mon. Earth Science*, **29**, 466.

ANGARD, K. A. 1996. Reservoir geochemistry of the Smørbukk, Smørbukk Sør, and the Heidrun fields—Haltenbanken, Norwegian offshore continental shelf. Master's dissertation, University of Oslo, Oslo.

ASTAFEV, V. P., MAKHMUDOV, A. KH. & ZHELONKIN, A. I. 1973. Significance of isobutene/butane ratios in hydrocarbon gases in searching for petroleum (according to Baltic and Belorussian data). Tr. Vses. Neft. Nauchno-Issled. *Geologorazed. Inst.*, **338**, 35–38.

AUGUSTSON, J. H. 1992. A method on classification of oil traps based on heavy oil content in cores with relevance to filling and drainage of Barents Sea oil-bearing structures. *In*: VORREN, T. O. *ET AL.* (eds) *Arctic Geology and Petroleum Potential.* Norwegian Petroleum Society (NPF) Special Publication, **2**, 691–702.

BAILEY, N. J. L., BURWOOD, R. & HARRIMAN, G. E. 1990. Application of pyrolysate carbon isotope and biomarker technology to organofacies definition and oil correlation problems in North Sea Basins. *Organic Geochemistry*, **16**, 1157–1172.

BARCLAY, S. A., WORDEN, R. H., PARNELL, J., HALL, D. L. & STERNER, S. M. 2000. Assessment of fluid contacts and compartmentalization in sandstone reservoirs using fluid inclusions: an example from the Magnus Oil Field, North Sea. *American Association of Petroleum Geologists Bulletin*, **84**, 489–504.

BELETSKAYA, S. N. 1978. Differentiation of disseminated organic matter in the process of primary migration. *In*: NAVLIKINJ, V. D. & ALIEVE, M. M. (eds) *Fundamental Aspects of the Genesis and Accumulation of Oil and Gas.* Nauka, Moscow, 52–70.

BELETSKAYA, S. N. & SYROVA, G. M. 1972. A study of the distributions of the disseminated bitumens in pore systems of rocks related to the assessment of the state of migration processes. A comparative study of the chloroform extracts from non-crushed and crushed rocks. *Geol. Nefti i Gaza*, **16**, 44–52.

BERNARD, B. B., BROOKS, J. M. & SACKETT, W. M. 1978. Light hydrocarbons in recent Texas continental shelf and slope sediments. *Journal of Geophysical research*, **83**, 289–291.

BERTRAND. P., PITTION, J. L. & BERNAUD, C. 1986. Fluorescence of sedimentary organic matter in relation to its chemical composition. Advances in organic geochemistry 1985. *Organic Geochemistry*, **10**, 641–647.

BHULLAR, A. G., KARLSEN, D. A., HOLM, K., BACKER-OWE, K. & LE TRAN, K. 1998. Petroleum geochemistry of the Frøy Field and Rind Discovery, Norwegian Continental Shelf—Implications for reservoir characterization, compartmentalization and basin scale hydrocarbon migration patterns in the region. *In*: HORSFIELD, B., RADKE, M., SCHAEFER, R. G. & WILKES, H. (eds) Advances in Organic Geochemistry. *Organic Geochemistry*, **29**(1–3), 735–768.

BHULLAR, A. G., KARLSEN, D. A., LACHARPAGNE, J. C. & HOLM, K. 1999a. Reservoir screening using Iatroscan TLC-FID and identification of palaeo-oil zones, oil water contacts, tar-mats and residual oil saturations in the Frøy and Rind petroleum accumulations. *Journal of Petroleum Science and Engineering*, **23**, 41–63.

BHULLAR, A. G., KARLSEN, D. A., BACKER-OWE, K., SELAND, R. T. & LE TRAN, K. 1999b. Dating reservoir filling—A case history from the North Sea. *Marine and Petroleum Geology*, **16**, 581–603.

BHULLAR, A. G., KARLSEN, D. A., BACKER-OWE, K., LE TRAN, K., SKÅLNES, E., BERCHELMANN, H. H. & KITTELSEN, J. E. 2000. Reservoir characterization by a combined micro-extraction—micro thin-layer chromatography (Iatroscan) method, A calibration study with examples from the Norwegian North Sea. *Journal of Petroleum Geology*, **23**, 221–244.

BJØRKUM, P. A. 1996. How important is pressure in causing dissolution of quartz in sandstones? *Journal of Sedimentary Research*, **66**, 147–154.

BJØRLYKKE, K. 1980. Clastic diagenesis and basin evolution. *Revista del Instituto de Investigaciones Geologicas Universidad Barcelona*, **34**, 21–44.

BJØRLYKKE, K. 1983. Diagenetic reactions in sandstones. *In*: PARKER, A. & SELLWOOD, B. W. (eds) *Sediment Diagenesis.* NATO Advanced Study, Reidel, 169–213.

BJØRLYKKE, K. & HEG, K. 1997. Effects of burial diagenesis on stresses, compaction and fluid flow in sedimentary basins. *Marine and Petroleum Geology*, **14**, 267–276.

BJØRLYKKE, K., AAGAARD, P., DYPVIK, H., HASTINGS, D. S. & HARPER, A. S. 1986. Diagenesis and reservoir properties of Jurassic sandstones from the Haltenbanken area, offshore mid Norway. *In*: SPENCER, A. M. (ed.) *Habitat of Hydrocarbons on the Norwegian Continental Shelf.* Norwegian Petroleum Society, 275–286.

BJØRLYKKE, K., RAMM, M. & SAIGAL, G. C. 1989. Sandstone diagenesis and porosity modification during basin evolution. *Geologische Rundschau*, **78**, 243–268.

BLYSTAD, P., BREKKE, H., FÆRSETH, R. B., LARSEN, B. T., SKOGSEID, J. & TØRUDBAKKEN, B. 1995. Structural elements of the Norwegian continental shelf, Part II. The Norwegian Sea Region. *NPD Bulletin*, **8**, The Norwegian Petroleum Directorate, Stavanger, Norway.

BODNAR, R. J. 1990. Petroleum migration in the Miocene Monterey Formation, California, USA. Constraints from fluid-inclusion studies. *Mineralogical Magazine*, **54**, 295–304.

BØEN, F. S., EGGEN, S. & VOLLSET, J. 1984. Structures and basins of the margin from 62 to 69°N and their development. *In*: SPENCER, A. M., JOHNSEN, S. O., MOERK, A., NYSAETHER, E., SONGSTAD, P. & SPINNANGR, A. (eds) *Petroleum Geology of the North European Margin*. Norwegian Petroleum Society, 253–270.

BOER, DE, R. B. 1974. Thermodynamic and experimental aspects of pressure solution. International symposium on water-rock interaction; abstracts. *Geol. Surv. Prague*, Chechoslovakia, p. 29

BOER, DE, R. B. 1977. Pressure solution; theory and experiments. *Tectonophysics*, **39**, 287–301.

BRECK, D. W. 1974. *Zeolite Molecular Sieves—Structure, Chemistry, and Use*. Wiley, London.

BUGGE, T., KNARUD, R. & MØRK, A. 1984. Bedrock geology on the Mid-Norwegian Continental Shelf. *In*: SPENCER, A. M., JOHNSEN, S. O., MOERK, A., NYSAETHER, E., SONGSTAD, P. & SPINNANGR, A. (eds) *Petroleum Geology of the North European Margin*. Norwegian Petroleum Society, 271–284.

BUGGE, T., RINGÅS, J. E., LEITH, D. A., MANGERUD, G., WEISS, H. M. & LEITH, T. L. 2002. Upper Permian as a new play Model on the mid-Norwegian continental shelf: Investigated by shallow stratigraphic drilling. *American Association of Petroleum Geologists Bulletin*, **86**, 107–127.

BUKOVICS, C. & ZIEGLER, P. A. 1985. Tectonic development of the Mid-Norway continental margin. *Marine and Petroleum Geology*, **2**, 2–22.

BURRUSS, R. C., CERCONE, K. R. & HARRIS, P. M. 1985. Timing of hydrocarbon migration, Evidence from fluid inclusions in calcite cements, tectonic and burial history. *In*: SCHNEIERMANN, N. & HARRIS, P. M. (eds) *Carbonate Cements*. SEPM Special Publication, **26**, 277–289.

CAMPBELL, C. J. & ORMAASEN, E. 1987. The discovery of oil and gas in Norway, a historical synopsis. *In*: SPENCER, A. (ed.) *Geology of the Norwegian Oil and Gas Fields*. Graham & Trotman, 1–37.

CARTER, K. E., & DWORKIN, S. I. 1990. Channelized fluid flow through shear zones during fluid enhanced dynamic recrystallization. *Geology, Geological Society of America*, **18**, 720–723.

CHEN, W., GHAITH, A., PARK, A. & ORTOLEVA, P. 1990. Diagenesis through coupled processes: Modeling approach, self-organisation and Implications for exploration. *In*: MESHRI. I. D. & OTTOLEVA, P. J. (eds) *Prediction of Reservoir Quality through Chemical Modeling*. American Association of Petroleum Geologists Memoir, **49**, 103–130.

CHUNG, H. M., GORMLY, J. R. & SQUIRES, R. M. 1988. Origin of gaseous hydrocarbons in subsurface environments: theoretical considerations of carbon isotope distribution. *Chemical Geology*, **71**, 97–103.

COHEN, M. J. & DUNN, M. E. 1987. The hydrocarbon habitat of the Haltenbanken-Trænabanken area offshore Mid-Norway. *In*: BROOKS, J. & GLENNIE, K. (eds) *Petroleum Geology of North West Europe*. Graham & Trotman, 1091–1104.

CONNAN, J. 1984. Biodegradation of crude oils in reservoirs. *In*: BROOKS, J. & WELTE, D. H. (eds) *Advances in Petroleum Geochemistry*, **1**, 299–335.

CONNAN, J. & CASSOU, A. M. 1980. Properties of gases and petroleum liquids derived from terrestrial kerogen at various maturation levels. *Geochimica et Cosmochimica Acta*, **44**, 1–23.

COOLES, G. P., MACKENZIE, A. S. & QUIGLEY, T. M. 1986. Calculation of petroleum masses generated and expelled from source rocks. *Organic Geochemistry*, **10**, 235–245.

CORNFORD, C., NEEDHAM, C. E. J. & DE WALQUE, L. 1986. Geochemical habitat of North Sea oils. *In*: SPENCER, A. M. (ed.) *Habitat of Hydrocarbons on the Norwegian Continental Shelf, Proceedings of an International Conference*. Graham & Trotman, 39–54.

DAHLGREN, S., JAMTVEIT, B., KARLSEN, D. A. & KJÆRNET, T. 1998. Thermal effects of basaltic sill emplacement in source rocks for maturation of hydrocarbon generation. *In*: *60th EAGE Conference and Technical Exhibition, Germany, 8–12 June 1998, Leipzig*. European Association of Geoscientists & Engineers, extended abstract book, **2**, 4–34.

DALE, K. 1997. *Use of light hydrocarbons for classification of oils and condensates in the Haltenbanken area*. Master's dissertation, University of Oslo, Oslo.

DURAND, B. 1988. Understanding of HC migration in sedimentary basins (present state of knowledge). *In*: MATTAVELLI, L. & NOVELLI, L. (eds) Advances in Organic Geochemistry. *Organic Geochemistry*, **13**, 445–459.

DURAND, B., & ESPITALIÉ, J. 1972. Formation et evolution des hydrocarbures de C1 a C15 et des gaz permanents dans les' argiles du toarcien du bassin de Paris. *In*: GAERTNER, H. R. V. & WEHNER, H. (eds) *Advances in Organic Geochemistry 1971*, Pergamon, 455–468.

DURAND, B., HUC, A. Y. & OUDIN, L. 1987. Oil saturation and primary migration, observations in shales and coals from the Kerbau wells, Mahakam Delta, Indonesia. *In*: DOLOGEZ, B. (ed.) *Migration of HC in sedimentary basins*. Editions Technip, 173–195.

EHRENBERG, S. N. 1990. Relationship between diagenesis and reservoir quality in sandstones of the Garn Formation, Haltenbanken, Mid-Norwegian Continental Shelf. *American Association of Petroleum Geologists Bulletin*, **74**, 1538–1558.

EHRENBERG, S. N. 1991. Kaolinized, potassium-leached zones at the contacts of the Garn Formation, Haltenbanken, mid-Norwegian continental shelf. *Marine Petrol. Geol.* **8**, 250–269.

EHRENBERG, S. N., GJERSTAD, H. M. & HADLER-JACOBSEN, F. 1992. Smørbukk Field—A gas condensate fault trap in the Haltenbanken Province, offshore Mid-Norway. *In*: HALBOUTY, M. T. (ed.) *Giant Oil and Gas Fields of the Decade 1978–1988*. American Association of Petroleum Geologists Memoir, **54**, 323–348.

EHRENBERG, S. N., SKJERVRAK, I. & GILJE, A. E. 1995. Asphaltene-rich residues in sandstone reservoirs of Haltenbanken province, mid-Norwegian continental shelf. *Marine and Petroleum Geology*, **12**, 53–69.

ELVSBORG, A., HAGEVANG, T. & THRONDSEN, T. 1985. Origin of the gas-condensate of the Midgard Field at Haltenbanken. *In*: THOMAS, B. M., DORE, A. G., EGGEN, S. S., HOME, P. V. & LARSEN, R. M. (eds) *Petroleum Geochemistry in Exploration of the Norwegian Shelf*. Graham & Trotman, 213–219.

EMERY, D. & ROBINSON, A. 1993. *Inorganic Geochemistry. Applications to Petroleum Geology*. Oxford, Blackwell Scientific Publications.

ENGLAND, W. A. 1994. Secondary Migration and Accumulation of Hydrocarbons. *In*: MAGOON, L. B. & DOW, W. G. (eds) *The Petroleum System—From Source to Trap*. American Association of Petroleum Geologists Memoir, **60**, 211–217.

ENGLAND, W. A. & MACKENZIE, A. S. 1989. Some aspects of the organic geochemistry of petroleum fluids. *Geologisch Rundschau*, **78**, 291–303.

ENGLAND, W. A., MACKENZIE, A. S., MANN, D. M. & QUIGLEY, T. M. 1987. The movement and entrapment of petroleum fluids in the subsurface. *Journal of the Geological Society, London*, **144**, 327–347.

ESPACH, R. H. & FRY, J. 1951. Variable characteristics of the oil in the Tensleep sandstone reservoir, Elk Basin Field, Wyoming and Montana. *Petroleum Transactions, American Institute of Mechanical Engineers*, **192**, 75–83.

ESPITALIÉ, J., LAFARGUE, E. & EGGEN, S. 1991. Petroleum potential of terrestrial and marine organic matter in Jurassic sequences of the northern North Sea and Offshore mid-Norway. *In*: SPENCER, A. M. (ed.) *Generation, Accumulation and Production of Europe's Hydrocarbons*. Special Publication of the European Association of Petroleum Geoscientists, **1**, Oxford University Press, 49–63.

ETMINAN, H. & HOFFMANN, C. F. 1989. Biomarkers in fluid inclusions, A new tool in constraining source regimes and its implications for the genesis of Mississippi Valley-type deposits. *Geology*, **17**, 19–22.

FORBES, P. L., UNGERER, P. M., KUHFUSS, A. B., RIIS, F. & EGGEN, S. 1991. Compositional modelling of petroleum generation and expulsion, trial application to local mass balance in the Smørbukk Sør Field, Haltenbanken Area, Norway. *American Association of Petroleum Geologists Bulletin*, **75**, 873–893.

GABRIELSEN, R. H. & DORÉ, A. G. 1995. History of tectonic models on the Norwegian continental shelf. *In*: HANSLIEN, S. (ed.) *Petroleum Exploration and Exploitation in Norway*. Norwegian Petroleum Society, Special Publication, **4**. Elsevier, Amsterdam, 333–368.

GABRIELSEN, R. H., FAERSETH, R., HAMAR, G. & RØNNEVIK, H. 1984. Nomenclature of the main structural features of the Norwegian Continental Shelf north of the 62nd parallel. *In*: SPENCER, A. M. (ed.) *Petroleum Geology of the North European Margin*. Graham & Trotman, 41–60.

GALIMOV, E. M. 1973. *Carbon isotopes in oil and gas geology*. Nedra, Moscow (in Russian). English translation, Washington, NASA TT F–682, Washington, D. C. 1975.

GALIMOV, E. M. 1985. *The Biological Fractionation of Isotopes*. Academic Press, Orlando.

GEORGE, S. C. *ET AL*. 1997. Geochemical comparison of oil-bearing fluid inclusions and produced oil from the Toro sandstone, Papua New Guinea. *Organic Geochemistry*, **26**, 155–173.

GEORGE, S. C., LISK, M., SUMMONS, E. & QUEZADA, R. A. 1998. Constraining the oil charge history of the South Pepper oilfield from the analysis of oil-bearing fluid inclusions. *Organic Geochemistry*, **2**, 631–648.

GIVEN, P. H., MARZEC, A., BARTON, W. A., LYNCH, L. J. & GERSTEIN, B. C. 1986. The concept of a mobile phase within the macromolecular network of coals, a debate. *Fuel*, **56**, 155–163.

GOWERES, M. B. & LUNDE, G. 1984. Geological history of Trænabanken. *In*: SPENCER, A. M. (ed.) *Petroleum Geology of the North European Margin*. Graham & Trotman, 237–252.

GUSSOV, W. C. 1954. Differential entrapment of oil and gas, a fundamental principle. *American Association of Petroleum Geologists Bulletin*, **38**, 816–853.

HABERLAND H. 1939. Über den nachweis von porphyrin und anderen organischen substanzen im kalkspat und aragonit. *Naturwissenschaften*, **27**, 613–614.

HADLEY, S., HALL, D., STERNER, M. & SHENTWU, W. 1998. Hydrocarbon pay delineation and production characterization with fluid inclusions. Examples from East Coast Canada and Western Canada sedimentary basin. *In Situ* (Newsletter of the Canadian Well Logging Society), **1**, 2–4.

HALL, D., SHENTWU, W., STERNER, S. M. & WAGNER, P. 1997. Using fluid inclusions to explore for oil and gas. *Hart's Petroleum Engineer International* (**November 1997**), 29–34.

HANBABA, P., JÜNTGEN, H. & PETERS, W. 1968. Nichtisotherme instationäre Messung der aktivierten Diffusion von Gasen in Festkorpern am Beispiel der Steinkohle. *Bericht der Bunsengesellschaft für Physikalische Chemie*, **72**, 554–562.

HERMANRUD, C., WENSAAS, L., TEIGE, G. M. G., NORDGÅRD BOLÅS, H. M., HANSEN, S. & VIK, E. 1998. Shale porosities from well logs on Haltenbanken (Offshoree Mid-Norway) show no influence of overpressuring. *In*: LAW, B. E., ULMISHEK, G. F. & SLAVIN, V. I. (eds) *Abnormal Pressures in Hydrocarbon Environments*. American Association of Petroleum Geologists Memoir, **70**, 65–85.

HEUM, O. R., DALLAND, A. & MEISINGSET, K. K. 1986. Habitat of hydrocarbons at Haltenbanken (PVT-modelling as predictive tool in hydrocarbon exploration). *In*: SPENCER, A. M. (ed.) *Habitat of Hydrocarbons in the Norwegian Continental Shelf*. Graham & Trotman, 259–274.

HIPPLER, S. J. 1997. Microstructures and Diagenesis in North Sea Fault Zones, Implications for Fault-Seal Potential and Fault-Migration Rates. *In*: SURDAM, R. C. (ed.) *Seals, Traps and the Petroleum System*. American Association of Petroleum Geologists Memoir, **67**, 85–101.

HODGSON, N. A., FARNSWORTH, J. & FRASER, A. J. 1992. Salt-related Tectonics, sedimentation and hydrocarbon palys in the Central Graben, North Sea, UKCS. *In*: HARDMAN, R. F. P. (ed.) *Exploration Britain*. Geological Society, London, Special Publications, **67**, 31–63.

HOLBA, A. G. *ET AL.* 1998a. Application of 24-norcholestanes for constraining source age of petroleum. *Organic Geochemistry*, **29**(5–7), 1269–1283.

HOLBA, A. G. *ET AL.* 1998b. 24-norcholestanes as age-sensitive molecular fossils. *Geology*, **26**(9) 783–786.

HOLLANDER, N. B. 1984. Geohistory and hydrocarbon evolution of the Haltenbanken area. *In*: SPENCER, A. M. (ed.) *Petroleum Geology of the North European Margin*. Graham & Trotman, 383–388.

HOLM, G. M. 1996. The Central Graben: a dynamic overpressure system. *In*: GLENNIE, K. & HURST, A. (eds) *NW Europe's Hydrocarbon Industry*. Geological Society, London, 107–122.

HORSFIELD, B. & MCLIMANS, R. K. 1984. Geothermometry and geochemistry of aqueous and oil-bearing fluid inclusions from Fateh Field, Dubai. *Organic Geochemistry*, **6**, 733–740.

HORSTAD, I. 1995. Petroleum migration, filling and biological degradation in Mesozoic reservoirs in the Northern North Sea. Ph.D. dissertation, University of Oslo, Oslo.

HORSTAD, I. *ET AL.* 1990. Degradation and maturity controls on oil field petroleum column heterogeneity in the Gullfaks field, Norwegian North Sea. *Organic Geochemistry*, **16**, 497–510.

HUC, A. Y., IRWIN, H. & SCHOELL, M. 1985. Organic matter quality changes in an Upper Jurassic shales sequence from the Viking Graben. *In*: THOMAS, B. M., DORÉ, A. G., EGGEN, S. S., HOME, P. V. & LARSEN, R. M. (eds) *Petroleum Geochemistry in Exploration of the Norwegian Shelf*. Graham & Trotman, 179–183.

HUGHES, W. B., HOLBA, A. G. & DZOU, L. I. P. 1995. The ratios of dibenzothiophene to phenathrene and pristine to phytane as indicators of depositional environment and lithology of petroleum source rocks. *Geochimica et Cosmochimica Acta*, **59**, 3581–3598.

HUNT, J. M. 1996. *Petroleum geochemistry and geology*. W. H. Freeman and Company.

HUTCHEON, I. 1983. Aspects of the diagenesis of coarse-grained siliciclastic rocks. *Geoscience Canada*, **10**(1), 4–14.

INAN, S., YALCIN, M. N. & MANN, U. 1998. Expulsion of oil from petroleum source rocks, inferences from pyrolysis of samples of unconventional grain size. *In*: HORSFIELD, B., RADKE, M., SCHAEFER, R. G. & WILKES, H. (eds) Advances in Organic Geochemistry 1997. *Organic Geochemistry*, **29**, 45–61.

ISAKSEN, G. H., CURRY, D. J., YEAKEL, J. D. & JENSSEN, A. I. 1998. Controls on the oil and gas potential of humic coals. *In*: HORSFIELD, B., RADKE, M., SCHAEFER, R. G. & WILKES, H. (eds) Advances in Organic Geochemistry 1997. *Organic Geochemistry*, **29**, 23–44.

JACKSON, J. S. & HASTINGS, D. S. 1986. The role of salt movement in the tectonic history of Haltenbanken and Trænabanken and its relationship to structural style. *In*: SPENCER, A. M. (ed.) *Habitat of Hydrocarbons in the Norwegian Continental Shelf*. Graham & Trotman, 241–257.

JAMES, A. T. 1983. Correlation of natural gas by se of carbon isotope distribution between hydrocarbon components. *AAPG Bulletin*, **67**, 1176–1191.

JOHANSEN, M. 1997. Well 7219/9-1 in the Ringvassøy—Loppa High Fault Complex, Hydrocarbon History, Filling and Dismigration. Master's dissertation, University of Oslo, Oslo.

JONES, D. M. & MACLEOD, G. 2000. Molecular analysis of petroleum in fluid inclusions; a practical methodology. *Organic Geochemistry*, **31**, 1163–1173.

KARLSEN, D. A. 1987. Light hydrocarbon redistribution in shallow core from the Upper Ravnefjeld Formation on the Wegener Halvø, East Greenland. Master's dissertation, University of Oslo, Oslo.

KARLSEN, D. A. & LARTER, S. R. 1989. A rapid correlation method for petroleum population mapping within individual petroleum reservoirs—applications to petroleum reservoir description. *In*: HARESNAPE, J. (ed.) *Correlation in Hydrocarbon Exploration*. Graham & Trotman, 77–85.

KARLSEN, D. A. & LARTER, S. R. 1991. Analysis of petroleum fractions by TLC-FID. Applications to petroleum reservoir description. *Organic Geochemistry*, **17**, 603–617.

KARLSEN, D. A., LEYTHAEUSER, D. & SCHAEFER, R. G. 1988. Light hydrocarbon redistribution in shallow core from the Ravnefjeld Formation on the Wegener Halvø, East Greenland. *In*: MATTAVELLI, L. & NOVELLI, L. (eds) Advances in Organic Geochemistry 1987. *Organic Geochemistry*, **13**, 393–398.

KARLSEN, D. A., NEDKVITNE, T., LARTER, S. R. & BJØRLYKKE, K. 1993. Hydrocarbon composition of authigenic inclusions, Application to elucidation of petroleum reservoir filling history. *Geochemica et Cosmochimica Acta*, **57**, 3641–3659.

KARLSEN, D. A. *ET AL.* 1995. Petroleum geochemistry of the Haltenbanken, Norwegian continental shelf. *In*: CUBITT, J. M. & ENGLAND, W. A. (eds) *The Geochemistry of Reservoirs*. Geological Society, London, Special Publications, **86**, 203–256.

KNIPE, R. J. 1992. Faulting processes and fault seal. *In*: LARSEN, R. M. (ed.) *Structural and Tectonic Modelling and its Application to Petroleum Geology*. Norges Petroleumsforening, 325–342.

KNIPE, R. J. 1993. The influence of fault zone processes and diagenesis on fluid flow. *In*: HORBURY, A. D. & ROBINSON, A. G. (eds) *American Association of Petroleum Geologists Studies in Geology*, **36**, 135–150.

KNIPE, R. J., AGAR, S. M. & PRIOR, D. J. 1991. The microstructural evolution of fluid flow paths in semi-lithified sediments from subduction complexes. *Philosophical Transactions of the Royal Society of London*, **A13**, 261–273.

KROOSS, B. 1985. Experimentelle Untersuchung der Diffusion niedrigmolekularer Kohlenwasserstoffe in wassergesättigten Sedimenten. Ph.D. dissertation, Technische Hochschule Aachen, Aachen.

KROSS, B. 1988. Experimental investigations of the molecular migration of C1–C6 hydrocarbons. Kinetics of hydrocarbon release from source rocks. *In*: MATTAVELLI, L. & NOVELLI, L. (eds) Advances in Organic Geochemistry 1987. *Organic Geochemistry*, **13**, 513–523.

KVALHEIM, O. M., CHRISTY, A. A., TELNÆS, N. & BJØRSETH, A. 1987. Maturity determination of organic matter in coals using methylphenanthrene distribution. *Geochimica et Cosmochimica Acta*, **51**, 1883–1888.

KVENVOLDEN, K. A. & ROEDDER, E. 1971. Fluid inclusions in quartz crystals from south-west Africa. *Geochimica et Cosmochimica Acta*, **35**, 1209–1229.

LARTER, S. R. & APLIN, A. C. 1995. Reservoir geochemistry, methods, applications and opportunities. *In*: CUBITT, J. M. & ENGLAND, W. A. (eds) *The Geochemistry of Reservoirs*. Geological Society, London, Special Publications, **86**, 5–32.

LARTER, S. R. & MILLS, N. 1991. Phase-controlled molecular fractionations in migrating petroleum charges. *In*: ENGLAND, W. A. & FLEET, A. J. (eds) *Petroleum Migration*. Geological Society, London, Special Publications, **59**, 137–147.

LEADHOLM, R. H., HO, T. T. Y. & SHAHAI, S. K. 1985. Heat flow, geothermal gradients and maturation modelling on the Norwegian continental shelf using computer methods. *In*: THOMAS, B. M., DORÉ, A. G., EGGEN, S. S., HOME, P. V. & LARSEN, R. M. (eds) *Petroleum Geochemistry in Exploration of the Norwegian Shelf*. Graham & Trotman, 131–143.

LEHNER, F. K., MARSAL, D., HERMANS, L. & VAN KUYK, A. 1987. A model of secondary hydrocarbon migration as a buoyancy-driven separate phase flow. *In*: DOLIGEZ, B. (ed.) *Migration of Hydrocarbons in Sedimentary Basins*. 2nd IFP Exploration Research Conference, Carcans, France, Editions Technip, Paris, 457–471.

LEONARD, R. C. 1984. Generation and migration of hydrocarbons on Southern Norwegian Shelf. Abstract, *American Association of Petroleum Geologists Bulletin*, **68**, 796.

LEONARD, R. C. 1993. Distribution of sub-surface pressure in the Norwegian Central Graben and applications for exploration. *In*: PARKER, J. R, (ed.) *Petroleum Geology of the Northwest Europe*. Proceedings of the 4th Conference, Geological Society, London, 1295–1303.

LEWAN, M. D. 1994. Assessing natural oil exulsion from source rocks by laboratory pyrolysis. *In*: MAGOON, J. B. & DOW, W. G. (eds) *The Petroleum system—from source to trap*. American Association of Petroleum Geologists Memoir, **60**, 201–210.

LEYTHAEUSER, D., SCHAEFER, R. G. & YÜKLER, A. 1982. Role of diffusion in primary migration of hydrocarbons. *American Association of Petroleum Geologists Bulletin*, **66**, 408–429.

LILLENG, T. & GUNDESØ, R. 1997. The Njord Field, a dynamic hydrocarbon trap. *In*: MØLLER-PEDERSEN, P. & KOESTLER, A. G. (eds) *Hydrocarbon Seals—Importance for Exploration and Production*. Norwegian Petroleum Society (NPF) Special Publication, **7**, 201–217.

MACKENZIE, A. S. 1984. Application of biological markers in petroleum geochemistry. *In*: BROOKS, J. & WELTE, D. (eds) *Advances in Petroleum Geochemistry*, Vol. 1. Academic Press, 115–214.

MACKENZIE, A. S., MAXWELL, J. R., COLEMAN, M. L. & DEEGAN, C. E. 1984. Biological marker and isotope studies of North Sea crude oils and sediments. *In*: *Proceedings of the Eleventh World Petroleum Congress*, Wiley & Sons, New York, **2**, 45–56.

MACKENZIE, A. S., PATIENCE, R. L., MAXWELL, J. R., VANDENBROUCKE, M. & DURAND, B. 1980. Molecular parameters of maturation in the Toarcian shales, Paris Basin, France. I. Changes in configuration of acyclic isoprenoid alkanes, steranes, and triterpanes. *Geochimica et Cosmochimica Acta*, **44**, 1709–1721.

MACKENZIE, A. S., PRICE, I., LEYTHAEUSER, D., MÜLLER, P., RADKE, M. & SCHAEFER, R. G. 1987. The expulsion of petroleum from Kimmeridge clay source-rocks in the area of the Brae field, UK continental shelf. *In*: BROOKS, J. & GLENNIE, K. (eds) *Petroleum Geology of North Western Europe*. Graham & Trotman, 865–877.

MACKENZIE, A. S., RULLKÖTTER, J., WELTE, D. H. & MANKIEWICS, P. 1985. Reconstruction of oil formation and accumulation in North Slope, Alaska, using quantitative gas chromatography-mass spectrometry. *In*: MAGOON, L. B. & CLAYPOOL, G. E. (eds) *Alaska North Slope Oil/Rock Correlation Study*. American Association of Petroleum Geologists Studies in Geology, **20**, 319–377.

MCLIMANS, R. K. 1987. The application of fluid inclusions to migration of oil and diagenesis in petroleum reservoirs. *Applied Geochemistry*, **2**, 585–603.

MELLO, M. R., GAGLIANONE, P. C., BRASSELL, S. C. & MAXWELL, J. R. 1988. Geochemical and biological marker assessment of depositional environments using Brazilian offshore oils. *Marine and Petroleum Geology*, **5**, 205–224.

MELLO, M. R., KOUTSOUKOS, E. A. M., MOHRIAK, W. U. & BACOCCOLI, G. 1994. *In*: MAGOON, L. B. & DOW, W. G. (eds) *The Petroleum System—from Source to Trap*. American Association of Petroleum Geologists Memoir, **60**, 499–512.

MO, E. S. *ET AL*. 1989. A dynamic deterministic model of hydrocarbon generation in the Midgard Field drainage area offshore Mid-Norway. *Geologische Rundschau*, **78**, 305–317.

MUHURI, S. K., DEWERS, T. A. & SCOTT JR, T. E. 2003. Interseismic fault strengthening and earthquake-slip instability: Friction or cohesion? *Geology*, **31**, 881–884.

MURRAY, R. C. 1957. Hydrocarbon fluid inclusions in quartz. *American Association of Petroleum Geologists Bulletin*, **41**, 950–956.

NEDKVITNE, T., KARLSEN, D. A., BJØRLYKKE, K. & LARTER, S. R. 1993. The relationship between diagenetic evolution and petroleum emplacement in the Ula Field, North Sea. *Marine and Petroleum Geology*, **10**, 255–270.

NEUMAN, H. & ROSENQUIST, I. TH. 1940. On red, fluorescent calcite from the Fen area near Ulefoss. *Norsk Geologisk Tidsskrift*, **20**, 267–268.

NORDGÅRD BOLÅS, H. M. & HERMANRUD, C. 2003. Hydrocarbon leakage processes and trap retention capacities offshore Norway. *Petroleum Geoscience*, **9**, 321–332.

ODDEN, W. 2000a. A study of light hydrocarbons ($C_4$–$C_{13}$) in source rocks and petroleum fluids. Dissertation for the Degree Doctor Scientiarum, Department of Geology, University of Bergen, Norway.

ODDEN, W. 2000b. A study of the composition of light hydrocarbons from pyrolysis of source rock samples. *Organic Geochemistry*, **31**, 211–229.

ODDEN, W., PATIENCE, R. L. & GRAAS, G. W. 1998. Application of light hydrocarbons ($C_4$–$C_{13}$) to oil/source rock correlations. A study of light hydrocarbon compositions of source rocks and test fluids from offshore Mid-Norway. *Organic Geochemistry*, **28**, 823–847.

OLSTAD, R. 1997. Migration and trapping of fluids in sedimentary basins. Ph.D. dissertation, University of Oslo, Oslo.

OLSTAD, R., BJØRLYKKE, K. & KARLSEN, D. A. 1997. Pore water flow and migration in the Smørbukk field area, offshore mid-Norway. *In*: MØLLER-PEDERSEN, P. & KOESTLER, A. G. (eds) *Hydrocarbon Seals—Importance for Exploration and Production*. Norwegian Petroleum Society (NPF) Special Publication, **7**, 201–217.

ORTOLEVA, P. J. 1994. *Basin Compartments and Seals*. American Association of Petroleum Geologists Memoir, **61**, 477.

OXTOBY, N. H., MITCHELL, A. W. & GLUYAS, J. G. 1995. The filling and emptying of the Ula Oilfield; fluid inclusion constraints. *In*: CUBITT, J. M. & ENGLAND, W. A. (eds) *The Geochemistry of Reservoirs*, Geological Society, London, Special Publications, **86**, 141–157.

PAGEL, M., WALGENWITZ, F. & DUBESSY, J. 1986. Fluid inclusions in oil and gas-bearing sedimentary formations. *In*: BURRUS, J. (ed.) *Thermal Modelling in Sedimentary Basins*. Collection Colloques et Séminaires, **44**, 565–583.

PATIENCE, R. 2003. Where did all the coal gas go? *Organic Geochemistry*, **34**, 375–387.

PEPPER, A. S. 1991. Estimating the petroleum expulsion behaviour of source rocks, a novel quantitative approach. *In*: ENGLAND, W. A. & FLEET, A. J. (eds) *Petroleum Migration*. Geological Society, London, Special Publications, **59**, 9–31.

PEPPER, A. S. & CORVI, P. J. 1995. Simple kinetic models for petroleum formation. *Marine and Petroleum Geology*, **12**(4), 417–452.

PETERS, W. & JÜNTGEN, H. 1968. Einfluss der Hohlraum-Structur auf Gas/Feststoff-reaktionen an Steinkohle. *Chem. Ing. Tech.*, **40**, 1039–1044.

PRICE, L. C. & CLAYTON, J. L. 1992. Extraction of whole versus ground source rocks, Fundamental petroleum geochemical implications including oil-source rock correlation. *Geochimica et Cosmochimica Acta*, **56**, 1213–1222.

RADKE, M., WELTE, D. H. & WILLSCH, H. 1982a. Geochemical study on a well in the Western Canada Basin: relation of the aromatic distribution pattern to maturity of organic matter. *Geochim. Cosmochim. Acta*, **46**, 1–10.

RADKE, M., WILLSCH, H., LEYTHAEUSER, D. & TEICHMULLER, M. 1982b.Aromatic components of coal: relation of distribution pattern to rank. *Geochim. Cosmochim. Acta*, **46**, 1831–1841.

RADKE, M. 1988. Application of aromatic compounds as maturity indicators in source rocks and crude oils. *Marine and Petroleum Geology*, **5**, 224–236.

REESE, C. L. 1898. Petroleum inclusion in quartz crystals. *Journal of the American Chemical Society*, **20**, 795–797.

REZNIKOV, A. N. 1969. On the geochemical significance of the ratio of n-butane and isobutane concentrations in petroleum gases. *Geol. Neft. Gaza*, **4**, 43–47.

RIECKE, E. 1894. Über das gleichwicht zwischen einem festen, homogenen deformierten körper und einer flüssigen phase. Insbesondere über die Depression des Schmelzpunktes durch einseitige Spannung. *Nachr. Kgl. Ges. Wiss. Göttingen, Mathm.+ Naturwiss.*, **4**, 278–284.

ROBERTS, S. J. & NUNN, J. A. 1995. Episodic fluid expulsion from geopressured sediments. *Marine and Petroleum Geology*, **12**, 195–204.

ROEDDER, E. 1984. Fluid inclusions. *In*: RIBBE, P. H. (ed.) *Reviews in Mineralogy*, Vol. 12. Mineral Society of America, Washington, DC Book Crafters, Inc.

RØNNEVIK, H. C. 2000. The exploration experience from Midgard to Kristin, Norwegian Sea. *In*: OFSTAD, K., KITTILSEN, J. E. & ALEXANDER-MARRACK, P. (eds) *Improving the Exploration Process by Learning from the Past*. Norwegian Petroleum Society (NPF) Special Publication, **9**, Elsevier Science B. V., Amsterdam, 113–129.

SAJGÓ, C. S., MAXWELL, J. R. & MACKENZIE, A. S. 1983. Evaluation of fractionation effects during the early stages of primary migration. *Organic Geochemistry*, **5**, 65–73.

SALES, J. K. 1993. Closure vs seal capacity—a fundamental control on the distribution of oil and gas. *In*: DORÉ, A. G., AUGUSTSON, J. H., HERMANRUD, C., STEWARD, D. J. & OYVIND, S. (eds) *Basin Modelling: Advances and Application*. Norwegian Petroleum Society (NPF) Special Publication, **3**, Elsevier, Amsterdam, 399–414.

SALES, J. K. 1997. Seal strength vs. trap closure—a fundamental control on the distribution of oil and gas. *In*: SURDAM, R. C. (ed.) *Seals, Traps and the Petroleum System*. American Association of Petroleum Geologists Memoir, **67**, 57–83.

SANDVIK, E. I., YOUNG, W. A. & CURRY, D. J. 1991. Expulsion from hydrocarbon sources, the role of organic adsorption. *In*: ECKARDT, C. B., MAXWELL, J. R., LARTER, S. R. & MANNING, D. A. C. (eds) Advances in Organic Geochemistry 1991. *Organic Geochemistry*, **19**, 77–87.

SCHAEFER, R. G. & LEYTHAEUSER, D. 1984. $C_2$–$C_8$ Hydrocarbons in sediments from Deep Sea Drilling Project Leg 75, Holes 530A, Angola Basin, and 532 Walvis Ridge. *In*: HAY, W. W. ET AL. (eds) *Initial Reports of the Deep Sea Drilling Project*. US Government Printing Office, Washington, **75**, 1055–1067

SCHOELL, M. 1983. Genetic characterisation of natural gases. *American Association of Petroleum Geologists Bulletin*, **67**, 2225–2238.

SCHOU, L., EGGEN, S. & SCHOELL, M. 1985. Oil–oil and oil–source rock correlation, Northern North Sea. *In*: THOMAS, B. M., DORÉ, A. G., EGGEN, S. S., HOME, P. V. & LARSEN, R. M. (eds) *Petroleum Geochemistry in Exploration of the Norwegian Shelf*. Graham & Trotman, 101–117.

SCHOWALTER, T. 1979. Mechanisms of secondary hydrocarbon migration and entrapment. *American Association of Petroleum Geologists Bulletin*, **63**, 723–760.

SCOTCHMAN, I. C., GRIFFITH, C. E., HOLMES, A. J. & JONES, D. M. 1998. The Jurassic petroleum system north and west of Britain: a geochemical oil-source correlation study. *Org. Geochem.* **29**, 671–700.

SCOTCHMAN, I. C. 1991. Kerogen facies and maturity of the Kimmeridge Clay Formation in the southern and eastern England. *Marine and Petroleum Geology*, **8**, 278–295.

SEIFERT, W. K. & MOLDOWAN, J. M. 1978. Applications of steranes, terpanes and monoaromatics to the maturation, migration and source of crude oils. *Cosmochimica Acta*, **42**, 77–92.

SEIFERT, W. K. & MOLDOWAN, J. M. 1981. Paleoreconstruction by biological markers. *Geochim. Cosmochim. Acta*, **45**, 783–794.

SELLWOOD, B. W., WILKES, M. & JAMES, B. 1993. Hydrocarbon inclusions in late calcite cements: migration indicators in the Great Oolite Group, Weald Basin, S. England. *Sedimentary Geology*, **84**, 51–55.

SHANMUGAM, G. 1985. Significance of coniferous rain forests and related organic matter in generating commercial quantities of oil, Gippsland Basin, Australia. *American Association of Petroleum Geologists Bulletin*, **69**, 1241–1254.

SIBSON, R. H. 1981. Fluid flow accompanying faulting, Field evidences and models. *In*: SIMPSON, D. W. & RICHARDS, P. G. (eds) *Earthquake Prediction. An International Review*. American Geophysical Union, Maurice Ewing Series, **4**, 593–603.

SIBSON, R. H. 1987. Earthquake rupturing as a hydrothermal mineralizing agent. *Geology*, **15**, 701–704.

SIBSON, R. H. 1990. Conditions of fault-valve behaviour. *In*: KNIPE, R. J. & RUTTER, E. H. (eds) *Deformation mechanisms, rheology and tectonics*. Geological Society, London, Special Publications, **54**, 15–28.

SILVERMAN, S. R. 1965. Migration and segregation of oil and gas. *In*: YOUNG, A. & GALLEY, J. E. (eds) *Fluids in Subsurface Environments*. American Association of Petroleum Geologists Memoir, **4**, 53–65.

SKÅLNES, E. 1993. Petroleum geochemistry and filling history of the Hild Field, Norwegian Continental Shelf. Master's dissertation, University of Oslo, Oslo.

SKÅLNES, E., PATIENCE, R., BJØRLYKKE, K. & KARLSEN, D. A. 1993. Petroleum geochemistry and filling history of the Hild Field, Norwegian Continental Shelf. *In*: *Advances in Organic Geochemistry*. Falch, Oslo, Extended Abstracts, 51–55.

SLETTEN, E. B. 2003. A comparison of petroleum from reservoirs and petroleum inclusions in authigenic mineral cements—Haltenbanken. Cand. Scient. thesis, Department of Geology, University of Oslo.

STAINFORTH, J. G. & REINDERS, J. E. A. 1990. Primary migration of hydrocarbons by diffusion through organic matter networks, and its effect on oil and gas generation. *Organic Geochemistry*, **16**, 61–74.

STEINHOFF, I. 1996. Aspects of reservoir geochemistry for the Smørbukk field, the Smørbukk Sør field and the Heidrun field in the Haltenbanken area on the Norwegian Continental Shelf. Master's dissertation, University of Köln, Köln, 126.

STORVOLL, V. 1999. Diagenesis and reservoir quality in the Garn Fm. in the Kristin and Lavrans Fields, Haltenbanken. Master's dissertation, University of Oslo, Oslo, 160 (in Norwegian).

STORVOLL,V., BJØRLYKKE, K., KARLSEN, D. & SAIGAL, G. 2002. Porosity preservation in reservoir sandstones due to grain-coating illite: a study of the Jurassic Garn Formation from the Kristin and Lavaran fields, offshore Mid-Norway. *Marine and Petroleum Geology*, **19**, 767–781.

SUNBERG, K. R. & BENNET, C. R. 1981. Carbon isotope paleothermometry of natural gas. *In*: BJORØY, M. (ed.) *Advances in Organic Geochemsitry 1980*, Wiley, 769–774.

SURLYK, F., PIASECKI, S., ROLLE, F., STEMMERIK, L., THOMSEN, E. & WRANG, P. 1984. The Permian base of East Greenland. *In*: *Petroleum Geology of the North European Margin*. Norwegian Petroleum Society, Graham & Trotman, 303–315.

SVERDRUP, E. & BJØRLYKKE, K. 1992. Small faults from Spitsbergen and Haltenbanken. A study of diagenetic and deformational structures, and their relation to fluid flow. *In*: LARSEN, M., BREKKE, H., LAREN, B. T. & THALLER, E. (eds) *Structural and Tectonic Modelling and its Application to Petroleum Geology*. Norwegian Petroleum Society (NPF), Proceedings of NPF Workshop, 18–20 October 1989, Stavanger, Norway, Elsevier, 507–517.

SVERDRUP, E. & BJØRLYKKE, K. 1997. Fault properties and development of cemented fault zones in sedimentary basins. Field examples and predictive models. *In*: MØLLER-PEDERSEN, P. & KOESTLER, A. G. (eds) *Hydrocarbon Seals. Importance for Exploration and Production*. Norwegian Petroleum Society (NPF), Special Publication, **7**, 91–106.

TADA, R. & SIEVER, R. 1989. Pressure solution during diagenesis. *Annual Review of Earth and Planetary Sciences*, **17**, 89–118.

TEIGE, G. M. & HERMANRUD, C. 2004. Seismic characteristics of fluid leakage from underfilled and overpressured Jurassic fault trap in the Norwegian North Sea. *Petroleum Geoscience*, **10**, 35–42.

THOMPSON, K. F. M. 1987. Fractionated aromatic petroleums and the generation of gas-condensates. *Organic Geochemistry*, **11**, 573–590.

THOMPSON, K. F. M. 1988. Gas-condensate migration and oil fractionation in deltaic systems. *Marine and Petroleum Geology*, **5**, 237–246.

THOMPSON, K. F. M., KENNICUTT, M. C. & BROOKS, J. M. 1990. Classification of offshore Gulf of Mexico oils and gas condensates. *American Association of Petroleum Geologists Bulletin*, **74**, 187–198.

TISSOT, B. P. & WELTE, D. H. 1984. *Petroleum Formation and Occurrence*. 2nd edition, Springer, Berlin, 699.

VAN KREVELEN, D. W. 1961. Geochemistry of Coal. *In*: BERGER, I. A. (ed.) *Organic Geochemistry*. Macmillan, New York, 183–247.

VIK, E., HEUM, R. & AMALIKSEN, K. G. 1991. Leakage from deep reservoirs, possible mechanisms and relationship to shallow gas in the Haltenbanken area, mid-Norwegian Shelf. *In*: ENGLAND, W. A. & FLEET, A. J. (eds) *Petroleum Migration*. Geological Society, London, Special Publications, **59**, 273

WALDERHAUG, O. 1990. A fluid inclusion study of quartz-cemented sandstones from offshore Mid-Norway—possible evidence for continued quartz cementation during oil emplacement. *Journal of Sedimentary Petrology*, **60**, 203–210.

WALDERHAUG, O. 1994. Precipitation rates for quartz cement in sandstones determined by fluid inclusion microthermometry and temperature history modelling. *Journal of Sedimentary Petrology*, Research, **A64**, 324–333.

WALDERHAUG, O. 1995. Evidences for resetting of fluid inclusion temperatures from quartz cements in oilfields, Discussion. *Marine and Petroleum Geology*, **12**, 559–561.

WALDERHAUG, O. 2000. Modeling Quartz cementation and porosity in Middle Jurassic Brent Group sandstones of the Kvitebjørn Field, Northern North Sea. *American Association of Petroleum Geologists Bulletin*, **84**, 1325–1339.

WALDERHAUG, O., LANDER, R. H., BJØRKUM, P. A., OELKERS, E. H., BJØRLYKKE, K. & NADEAU, P. H. 2000. Modelling quartz cementation and porosity in reservoir sandstones. Examples from the Norwegian continental shelf. *Special Publications of the International Association of Sedimentologists*, **29**, 39–49.

WENSAAS, L. T, SHAW, H. F., GIBBONS, K., AAGAARD, P. & DYPVIK, H. 1994. Nature and causes of overpressure in the mudrocks of the Gullfaks Area, North Sea. *Clay Minerals*, **29**, 439–449.

WEYL, P. K. 1959. Pressure solution and the force of crystallization—a phenomenological theory. *Jour. Geol. Phys. Res.*, **64**, 2001–2025.

WHITICAR, M. J. 1994. Correlation of natural gases with their sources. *In*: MAGOON, L. B. & DOW, W. G. (eds) *The Petroleum System—From Source to Trap*. American Association of Petroleum Geologists Memoir, **60**, 261–283.

WHITLEY, P. K. 1992. The geology of the Heidrun, a giant oil and gas field on the Mid-Norwegian Shelf. *In*: HALBOUTY, M. T. (ed.) *Giant Oil and Gas Fields of the Decade 1978–1988*. American Association of Petroleum Geologists Memoir, **54**, 383–406.

WILHELMS, A., HORSTAD, I. & KARLSEN, D. 1996. Sequential extraction—a useful tool for reservoir geochemistry? *Organic Geochemistry*, **24**, 1157–1172.

WILHELMS, A., LARTER, S. R., HEAD. I., FARRIMOND, P., DI PRIMIO, R. & ZWACH, C. 2000. Biodegradation of oil in uplifted basins prevented by deep-burial sterilization. *Nature*, **411**, 1034–1036.

WILKINS, R. W. T. & GEORGE, S. C. 2002. Coal as a source rock for oil: a review. *International Journal of Coal Geology*, **50**, 317–361.

WINTERSTAD, C. K. 2003. Variasjoner i petroleumssammensetningen til brønn 6507/5–4 på Skarvfeltet på Dønnaterassen, Haltenbanken. Diskusjon av in situ organisk materiale i den jurassiske og den kritassiske fellegeometrien. Sammenligning av ekstrakter fra Skarv 5–4 med oljer/kondensater fra 5–1, 5–2 og 5–4 brønnene. Cand. Scient. thesis, Department of Geology, University of Oslo, 142.

XU, B. 2003. Biodegradation and accumulation of oils and condensates in the Dønna Region—Norwegian Offshore Continental Shelf (NOCS), Cand. Scient. thesis, Department of Geology, University of Oslo.

# A mass balance approach for assessing basin-centred gas prospects: integrating reservoir engineering, geochemistry and petrophysics

J. A. RUSHING,[1] A. CHAOUCHE[2] & K. E. NEWSHAM[3]

[1]*Anadarko Petroleum Corp., 1201 Lake Robbins Drive, The Woodlands, Texas 77038, USA (e-mail: jay_rushing@anadarko.com)*
[2]*Anadarko Petroleum Corp., 1201 Lake Robbins Drive, The Woodlands, Texas 77038, USA (e-mail: ahmed_chaouche@anadarko.com)*
[3]*Apache Corp. 2000 Post OakBoulevard, Houston, Texas 77056, USA (e-mail: kent.newsham@apachecorp.com)*

**Abstract:** This paper presents a mass balance technique for evaluating the resource-in-place potential of basin-centred gas exploration prospects. As the name implies, the methodology is derived from the conservation of mass principle. Unlike conventional reservoir engineering techniques, the mass balance approach directly integrates dynamic elements of the total petroleum system with all static reservoir engineering, geological and petrophysical data to estimate the resource-in-place. This prospect evaluation technique not only provides an independent estimate of hydrocarbon resource-in-place for comparison with that computed from the conventional reservoir engineering approach, but also gives estimates of hydrocarbons lost from the petroleum system. Accordingly, this technique reduces the inherent uncertainty associated with conventional methods and provides more realistic estimates of resource-in-place.

The fundamental question asked when evaluating an exploration prospect is 'how much hydrocarbon resource is in place?' To answer this question, the conventional reservoir engineering approach is to compute the probable reservoir hydrocarbon pore volume, i.e. the size and extent of the hydrocarbon 'container', from estimates of reservoir rock and fluid properties. Uncertainties in the geological model, source rock characteristics, trapping mechanisms and seal rock effectiveness, and the presence of sufficient reservoir rock for commercial hydrocarbon production, are often represented as a single-valued probability of success.

The probability of finding a commercially viable reservoir is usually computed as the product of all probabilities of success for each geological, petrophysical, geochemical and reservoir engineering input parameter. Assigning a probability of success is often a qualitative procedure with little or no direct integration of the dynamic processes associated with each of the total petroleum system components. Consequently, the single-valued assessment technique can be considered a static process.

Although inclusion of a probability of success in prospect evaluation does indirectly capture some elements of the total petroleum system, the lack of a consistent methodology to integrate dynamic petroleum system components directly with all reservoir engineering, geological and petrophysical data may still cause significant resource assessment errors. Further, the key to successful exploration and development of a basin-centred gas system (BCGS) is early identification of the system and its attributes. As a result, resource assessment errors may be exacerbated when exploring for these unconventional gas systems since their unique attributes are often difficult to discern from a conventional system.

To address these problems, this paper presents a mass balance technique specifically developed for evaluating the resource-in-place potential of basin-centred gas prospects. This paper begins with a general overview of basin-centred gas systems (BCGS), including a summary of common attributes identified from a literature survey. A derivation of the mass balance technique and an explanation of its elements follow this summary. Application of the technique is illustrated with an example from the Bossier tight gas sand play in the East Texas Basin located in eastern Texas, USA. Uncertainty in resource-in-place estimates are quantified by incorporating a Monte Carlo simulation technique with the mass balance computations.

## Overview of basin-centred gas systems: conventional wisdom

Neither the concept nor the characteristics of a basin-centred gas system (BCGS) are widely

*From*: CUBITT, J. M., ENGLAND, W. A. & LARTER, S. (eds) 2004. *Understanding Petroleum Reservoirs: towards an Integrated Reservoir Engineering and Geochemical Approach*. Geological Society, London, Special Publications, **237**, 373–390. 0305-8719/04/$15.00 

understood within the petroleum industry. This lack of understanding is the primary reason why no single comprehensive definition has been put forward in the literature. The most concise definition to date is attributed to Law (2000) who describes a BCGS as 'an abnormally-pressured, gas-saturated accumulation in a low permeability reservoir lacking a downdip water contact'.

Historically, a BCGS has been defined in terms of the fundamental petroleum system elements and processes associated with development and formation of this resource. Similar to conventional hydrocarbon reservoirs, the BCGS process requirements generally include deposition of the reservoir rock, hydrocarbon generation, migration and entrapment by geological structural elements and/or seal capacity genesis.

Unlike conventional hydrocarbon reservoirs, however, a BCGS develops as a result of unique temporal and spatial interactions among these various petroleum system elements. The dynamic processes responsible for fluid distribution in a conventional gas accumulation are characteristically in static capillary equilibrium, whereas most BCGSs are not currently in an equilibrium state. This disequilibrium state is most often a result of ongoing hydrocarbon generation.

General BCGS attributes are summarized in the next few sections. Note that these characteristics were taken from a literature survey but do not necessarily represent the views of the authors.

## *Geological properties*

BCGS reservoirs documented in the literature may either be single, isolated sand accumulations or vertically stacked sands that are several thousands of feet thick (Law 1984a,b, 1985, 1995). These reservoirs are most often composed of a sandstone lithology and are Cretaceous age or older (Masters 1979, 1984, 2000). Note that the absence of any documented carbonate BCGS reservoirs probably does not preclude their existence.

Unlike conventional oil and gas reservoirs, structure and stratigraphy often play a secondary role in the accumulation of hydrocarbons within a BCGS, primarily because of the dynamics associated with the abnormal pressure generation (Law 1984a,b, 1985, 1995; Masters 1979, 1984; Nuccio *et al.* 1992). BCGSs are often not in static capillary equilibrium which is usually manifested by capillary pressure transitions that are more a function of two-phase capillary seals than from structural closure (Iverson 1994; Surdam 1997, Surdam *et al.* 1997). The top of a BCGS may be defined by the top of abnormal pressure and is typically observed to cut across both structure and stratigraphy (Masters 1979, 1984).

## *Source rock characteristics*

Many of the source rock characteristics observed in BCGSs are also common to conventional hydrocarbon systems; however, development of a BCGS is a manifestation of unique temporal and spatial interaction among the various petroleum system elements. Law further classifies two types of BCGS—direct and indirect—on the basis of their source rock characteristics (Law 1984a,b, 1985, 2000). An indirect BCGS is generally more prone to generate significant liquid hydrocarbons initially, is more likely to be under-pressured, and metagenesis of liquid hydrocarbons to gas is the primary source of pore pressure generation.

Conversely, a direct BCGS is prone to produce less liquid hydrocarbons initially, will generate more gas than liquid, may be either under- or over-pressured, and gas generation is the primary source of pore pressure. In addition, the gas generation is of thermogenic origin. Reservoirs in both systems are frequently proximal to the source rock, so gas migration distances are shorter and less tortuous than in a conventional hydrocarbon system.

## *Reservoir pressure and temperature*

Hydrocarbon generation is the primary source of abnormal pressure conditions in a BCGS. Compaction disequilibrium and chemical compaction effects, when present, are often secondary. Abnormal pressures in BCGSs have been extensively documented in the literature (Ortolevac 1994; Surdam 1997). Abnormal pressure can occur as either over- or under-pressured gradients (Law 1984a,b, 1985, 1994; Masters 1979, 1984; Ryder *et al.* 1996). Furthermore, the creation of an abnormally-pressured compartment is a dynamic, rate-dependent process that balances the rate of hydrocarbon generation and volume expansion from liquid hydrocarbon metagenesis with limited but intermittent fluid leakage when seal capacity limits are exceeded. In a BCGS that is not in static equilibrium, the pressure generation process may be cyclical, depending on the genesis of seal characteristics.

Elevated or abnormal temperature gradients are another characteristic of BCGSs. Geothermal gradients of 1.5 °F per 100 ft have been documented in BCGSs (Ryder *et al.* 1996) resulting in formation temperatures ranging from 150 °F to more than 400 °F in deep reservoirs. These abnormal temperature gradients primarily affect the type of BCGS present and the associated hydrocarbon genesis.

### *Rock and fluid properties*

Most BCGSs exhibit low porosity and permeability (Law 1984a,b, 1985, 1994; Masters 1979, 1984; Ryder *et al.* 1996). Effective porosity is typically less than 15%, with many reservoirs having average porosities less than 10%. Similarly, effective hydrocarbon permeability is often less than 1 md. Consequently, many BCGSs in the USA are categorized as 'tight gas sands'.

Formation waters are often very saline, often containing more than 200 000 ppm total dissolved solids (Ryder *et al.* 1996). In addition, the *in-situ* connate water is most often present as an irreducible (non-mobile) water phase (Law 1984a,b, 1985; Ryder *et al.* 1996). The hydrocarbon properties vary depending on the type of BCGS, the source rock characteristics, and the effects of burial history on temperature and pressure gradients.

### *Producing characteristics*

BCGSs normally have large in-place hydrocarbon volumes but very low recovery factors (Masters 1979, 1984, 2001; Nuccio *et al.* 1992). Unlike conventional hydrocarbon systems, gas accumulations are sometimes downdip from normally-pressured, water-bearing reservoirs (Masters 1979, 1984, 2001; Nuccio *et al.* 1992). Furthermore, gas accumulations often do not have a downdip water contact (Law 1984a,b, 1985; Masters 1979, 1984, 2001).

Average produced water yields range from 1 to 20 bbl $H_2O$/MMscf gas (Ryder *et al.* 1996). Because of the presence of an irreducible (non-mobile) water phase, much, if not all, of this produced water can be attributed to surface condensation of water vapour that is initially dissolved in the hydrocarbon gas at reservoir conditions.

## Description of the mass balance approach

The mass balance technique for evaluating the resource-in-place of a BCGS is derived from the inviolable conservation of mass principle, i.e. mass can neither be created nor destroyed. The process model incorporating the mass balance technique, when presented in terms of the conservation of mass principle, considers the following basic elements:

hydrocarbon mass remaining in the reservoir

= hydrocarbon mass entering the reservoir

− hydrocarbon mass leaving the reservoir.

Descriptions and data requirements for each element of the mass balance model are discussed in the next sections. Additionally, the working equations required to compute the resource-in-place for a BCGS are derived and discussed.

### *Hydrocarbon mass entering the reservoir*

The hydrocarbon mass entering the reservoir is a function of the source rock and its hydrocarbon generation and expulsion capacity. This element essentially characterizes the petroleum source rock at both initial and current conditions; quantifies its hydrocarbon generation capacity, including source rock diagenesis, catagenesis and metagenesis; identifies the type and quantity of hydrocarbons generated; and describes the hydrocarbon migration paths to the reservoir. Data requirements typically include:

- *Total organic carbon* (TOC): The TOC defines the quantity of organic matter, including both bitumen and kerogen, of the source rock at current conditions.
- *Vitrinite reflectance* (Ro): Ro defines the source rock thermal maturity level.
- *Kerogen content*: The type and quantity defines the quality of the source rock material and correlates to the hydrocarbon generation efficiency. The kerogen content also provides indications of source rock maturity and residual hydrocarbon generation capacity.
- *Source rock gas content*: This property essentially defines the quantity of gas remaining in the source rock at current conditions. It is estimated from gas desorption tests. Conversely, this measurement can also help estimate the gas expulsion efficiency.
- *Hydrocarbon migration paths*: As discussed in the previous section, an important attribute of many basin-centred gas systems is the proximity of the source rock to the reservoir. Consequently, the migration pathway is less critical to the formation of a BCGS reservoir than in conventional petroleum systems.

### *Hydrocarbon mass leaving the reservoir*

This element of the mass balance model refers to the efficiency with which the generated hydrocarbons migrate to and are trapped within the reservoir. Components of this element include the trapping mechanisms identified from the geological model as well as characterization of the reservoir seals. As discussed previously, structure often plays little or no role in the accumulation of hydrocarbons in a BCGS, so evaluation of the seals is especially critical to an accurate evaluation. Key components and data requirements include:

- *Type of seal*: The type of reservoir seal depends on lithology, intrinsic properties, type and severity of diagenesis, and timing of burial history relative to hydrocarbon generation and migration. A seal is usually defined as a rock that completely prevents fluid flow over all time periods, while a barrier is a type of seal that prevents fluid flow during the reservoir's producing life, but not necessarily over geological time. A baffle is a rock that may impede but not completely stop fluid flow. Seals, barriers and baffles can be found externally or as part of the internal structure of a reservoir.
- *Seal properties*: Rock seal properties are usually described in terms of their capillary pressure characteristics, primarily wettability, entry and displacement pressures, and irreducible wetting phase saturation. Wettability defines which fluids will preferentially occupy the smallest rock pores. Entry pressure is the capillary pressure at which the non-wetting phase first displaces the wetting phase, while displacement pressure is the capillary pressure at which the non-wetting phase first forms a continuous network within the pore structure. The irreducible wetting phase saturation describes the initial connate fluid saturation at the top of the capillary column.
- *Rock diagenesis processes*: The primary diagenetic processes associated with seal creation are mechanical and chemical compaction. Mechanical compaction results from an increase in overburden with burial, resulting in grain dislocation, re-orientation, and repacking. Chemical compaction includes mineral cementation, mineral and grain dissolution, and clay genesis. The net result of diagenesis is to modify the initial pore structure and geometry, typically causing a reduction in rock flow and storage capacities.
- *Temporal aspects of seal generation*: The seal capacity of both shales and reservoir rock will evolve over geological time with changes in pressure and temperature conditions associated with the regional burial history. Much of this change in seal capacity is associated with rock diagenesis and the reduction in seal permeability and porosity caused by compaction.

### *Hydrocarbon mass remaining in the reservoir*

The hydrocarbon mass remaining in the reservoir is calculated as the arithmetic difference between the hydrocarbon masses entering and leaving the reservoir. The hydrocarbon mass remaining in the reservoir reflects not only the source rock characteristics and hydrocarbon generation efficiency, but also the tortuosity of the hydrocarbon migration paths and the sealing efficiency of the reservoir rock.

Note that this mass is computed independently from and without consideration of the reservoir rock and fluid properties. Consequently, this hydrocarbon mass should provide an excellent validation of that computed from the conventional reservoir engineering approach. Moreover, the difference between the hydrocarbon mass remaining in the reservoir and the mass balance and conventional reservoir engineering techniques provides an estimate of the volume of generated hydrocarbon that has escaped from the reservoir because of insufficient seals or traps.

### *Derivation of the mass balance equations*

To facilitate computations, the mass balance model equations derived in this section are presented in terms of hydrocarbon volume rather than mass. Further, this derivation is more appropriate and applicable to a direct rather than an indirect BCGS (Law 1984a,b, 1985, 2000) as defined in the previous section. A schematic diagram of the mass balance process model is shown in Figure 1.

The conservation of mass principle, when presented in terms of hydrocarbon volumes, may be written as

$$V_{\text{remaining}} = V_{\text{entering}} - V_{\text{leaving}} \qquad (1)$$

where $V_{\text{remaining}}$ is the gas volume remaining in the reservoir, $V_{\text{entering}}$ is the gas volume entering the reservoir, and $V_{\text{leaving}}$ is the gas volume leaving the reservoir.

Note that the gas volume entering the reservoir represents a fraction of the total gas volume that has been generated and subsequently expelled

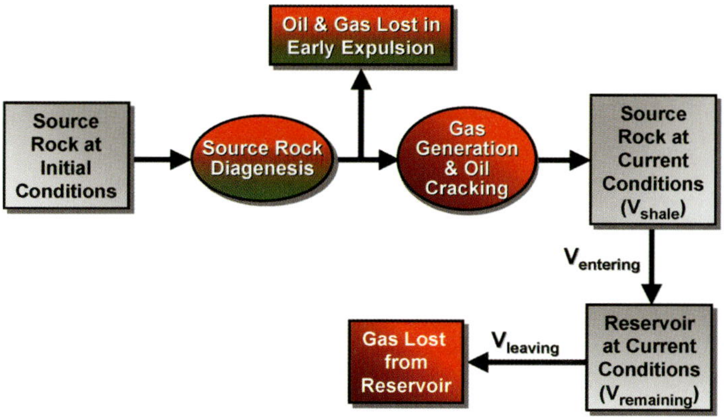


**Fig. 1.** Schematic diagram illustrating elements of the mass balance process model for computing gas volumes in a BCGS.

from the source rock. In equation form, this volume becomes

$$V_{entering} = V_{expelled} = E_E V_{generated} \quad (2)$$

where $E_E$ is defined as the source rock expulsion efficiency. $V_{generated}$, which is the total gas volume generated from the source rock, is computed from Shmoker (1994) and Tissot & Welter (1984),

$$V_{generated} = \frac{43.56(h_{shale})(\Delta HI)(\text{TOC})(\rho_{shale})(1 - E_L)}{\rho_{oil}} \quad (3)$$

where $V_{generated}$ is the gas volume generated from source rock (MMscf/acre), $h_{shale}$ is the net shale (source rock) thickness (ft), $\Delta HI$ is the hydrogen index differential $= HI_{original} - HI_{current}$ (mg HC/g TOC), TOC is the original total organic carbon (fraction), $\rho_{shale}$ is the bulk shale (source rock) density (g/cm$^3$), $\rho_{oil}$ is the density of liquid hydrocarbon, (g/cm$^3$), and $E_L$ is the fraction of liquid hydrocarbons lost from the system.

Equation (3) also assumes that a portion of the liquid hydrocarbon initially generated during source rock catagenesis is expelled and lost from the system prior to entering the gas generation window. The fraction of liquid hydrocarbon volume remaining in the source rock is represented by $(1 - E_L)$. Further, it is assumed that this lost liquid hydrocarbon volume is not available to generate gas hydrocarbons from cracking. Consequently, this hydrocarbon is not considered in the final calculations.

The value of the source rock's hydrogen index ($HI$) at initial or original conditions is estimated from values of the current hydrogen index using the following equation (Waples 2000):

$$HI_{original} = \frac{1000(HI_{current})(1 + ST)}{1000(1 - TR) + (TR)(HI_{current})(1 + ST)} \quad (4)$$

where $HI_{original}$ is the source rock's hydrogen index at original conditions (mg HC/g TOC), $HI_{current}$ is the source rock's hydrogen index at current conditions (mg HC/g TOC), $ST$ is the saturation threshold expressed as fraction of the kerogen present (fraction), and $TR$ is the transformation ratio (fraction).

Since only a fraction of the gas migrating to or entering the reservoir is trapped, then the volume of gas leaving the reservoir can be written in terms of the gas volume generated as follows:

$$V_{leaving} = (1 - E_T)V_{entering} = (1 - E_T)E_E V_{generated} \quad (5)$$

where $V_{leaving}$ is the gas volume leaving the reservoir (MMscf/acre), $E_T$ is the reservoir trapping efficiency (fraction), and $E_E$ is the source rock expulsion efficiency (fraction).

Finally, the gas volume remaining in the reservoir can now be computed by combining equations (1), (2) and (5) as follows:

$$V_{remaining} = E_T E_E V_{generated}. \quad (6)$$

Note that the source rock's gas expulsion efficiency is very difficult to quantify directly from source rock measurements. However, this parameter may be estimated from a mass balance calculation on the source rock. The gas volume remaining in the source rock or shale at present-day conditions may be estimated using gas content values measured from desorption tests. The gas content or shale gas volume, $V_{\text{shale}}$, is computed from

$$V_{\text{shale}} = 0.001359 h_{\text{shale}} C_{\text{adsorbed}} \rho_{\text{shale}} \quad (7)$$

where $V_{\text{shale}}$ is the generated gas volume remaining in the source rock (MMscf/acre), $h_{\text{shale}}$ is the net shale thickness (ft), $C_{\text{adsorbed}}$ is the adsorbed gas content from desorption measurements (scf gas per ton shale), and $\rho_{\text{shale}}$ is the bulk shale density (g/cm$^3$).

Note that equation (7) assumes all gas is currently present in an adsorbed state. If desired, however, equation (7) can be modified to account for gas volumes stored in a more conventional pore system.

From additional mass balance considerations, the total volume of gas generated by the source rock equals the sum of gas volumes expelled and remaining in the shale, or

$$V_{\text{generated}} = V_{\text{expelled}} + V_{\text{shale}}. \quad (8)$$

The gas volume remaining in the source rock can be related mathematically to the generated gas volume and source rock expulsion efficiency by combining equations (2) and (8) as follows:

$$V_{\text{generated}} = \frac{V_{\text{shale}}}{1 - E_{\text{E}}}. \quad (9)$$

Further, combining equations (7) and (9) yields an equation to compute gas generated from the source rock in terms of the gas desorption parameters and expulsion efficiency as

$$V_{\text{generated}} = \frac{0.001359 h_{\text{shale}} C_{\text{adsorbed}} \rho_{\text{shale}}}{1 - E_{\text{E}}}. \quad (10)$$

Finally, equating equations (3) and (10) yields an equation to calculate the gas expulsion efficiency directly as

$$E_{\text{E}} = 1 - \frac{0.0000312 \rho_{\text{oil}} C_{\text{adsorbed}}}{(\Delta HI)(\text{TOC})(1 - E_{\text{L}})}. \quad (11)$$

The equation to compute the gas volume remaining in the reservoir can be developed by substituting equation (3) into equation (6) to yield

$$V_{\text{remaining}} = \frac{43.56(h_{\text{shale}})(\Delta HI)(\text{TOC}) \dfrac{(\rho_{\text{shale}})(1 - E_{\text{L}}) E_{\text{E}} E_{\text{T}}}{}}{\rho_{\text{oil}}} \quad (12)$$

where the value of gas expulsion efficiency, $E_{\text{E}}$, is computed with equation (11).

Note that equation (12) also requires an estimate of reservoir trapping efficiency. As discussed previously, seal effectiveness is generally determined qualitatively from knowledge of the intrinsic rock properties, but it is very difficult to quantify this parameter. Alternatively, an equation to compute the gas volume expelled from the source rock may be developed by combing equations (2) and (3),

$$V_{\text{expelled}} = \frac{43.56(h_{\text{shale}})(\Delta HI)(\text{TOC}) \dfrac{(\rho_{\text{shale}})(1 - E_{\text{L}}) E_{\text{E}}}{}}{\rho_{\text{oil}}}. \quad (13)$$

The reservoir trapping efficiency may then be estimated by taking the ratio of the remaining gas volume computed using the conventional reservoir engineering approach with the gas volume expelled from the source rock (equation (13)). This calculation is illustrated in a subsequent section.

### *Conventional reservoir engineering technique*

For comparison with the mass balance approach, remaining gas volumes may also be computed using the conventional reservoir engineering approach (Lee & Wattenbarger 1996). The gas volume remaining in the reservoir is defined by the physical dimensions of the reservoir pore volume, *in-situ* gas and water properties, and the pressure and temperature conditions. The gas volume remaining in the reservoir is calculated as

$$V_{\text{RE}} = 43.56(h_{\text{gross}}) \frac{(h_{\text{net}}/h_{\text{gross}})\phi(1 - S_{\text{w}})}{B_{\text{g}}} \quad (14)$$

where $V_{\text{RE}}$ is the gas volume remaining in reservoir (MMscf/acre), $h_{\text{gross}}$ is the gross sand thickness (ft), $h_{\text{net}}/h_{\text{gross}}$ is the net-to-gross sand thickness ratio (fraction), $\phi$ is the effective sand porosity (fraction), $S_{\text{w}}$ is the *in-situ* connate water saturation (fraction), and $B_{\text{g}}$ is the gas formation volume, converts gas volume at reservoir conditions to surface conditions (reservoir ft$^3$ of gas per standard ft$^3$ of gas).

## Overview of the Bossier tight gas sand play

Application of the mass balance approach is illustrated with a case study from the Bossier tight gas sand play. Over the past four years, this play has been one of the most active exploration and development areas in the USA. Reservoir rock and fluid properties, as well as source rock attributes, determined from a very rigorous and comprehensive description and characterization programme, suggest the Bossier sands are part of a basin-centred gas system (Emme & Stancil 2002; Newsham & Rushing 2002).

### *Regional geology and depositional environment*

The Bossier sands, which are Upper Jurassic in age, were deposited in the East Texas Basin located in northeastern Texas. This sedimentary basin is a deep, elongated trough structure with shelf-slope systems on the basin flanks. Several major tectonic features, including the Sabine Uplift to the east, the Mexia–Talco Fault System to the north and west, and the Angelina–Caldwell Flexure to the south, bound the basin. The East Texas Basin is also characterized by major salt features within its interior (Montgomery 2000). Current geological models suggest a relationship between generation of the major fault systems, salt deformation and migration, basin subsidence, and sediment deposition during middle to late Mesozoic. The significant salt structures also appear to control the distribution of sediments within the basin interior.

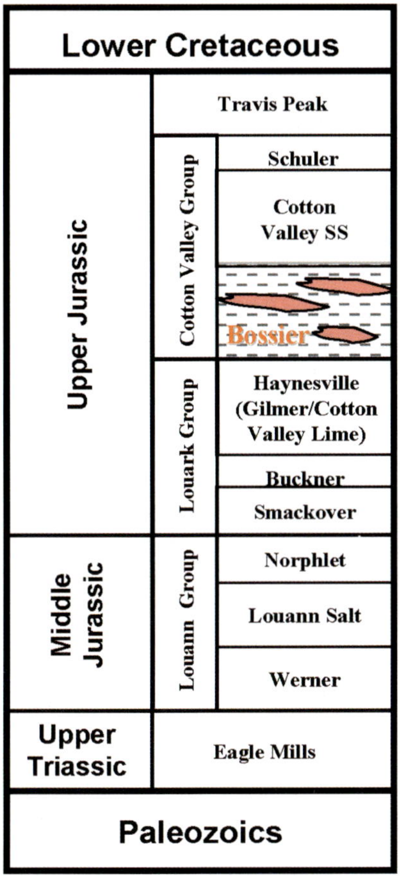


**Fig. 2.** Stratigraphic column for the Bossier Sands, East Texas Basin (Montgomery 2000).

A typical stratigraphic column for the East Texas Basin (Fig. 2) shows the Bossier sands are part of the Upper Jurassic Cotton Valley Group and are overlain by the Cotton Valley Sandstones (Montgomery 2000). The Bossier interval is a thick, lithologically complex black to grey-black shale interbedded with fine-grained argillaceous sandstone. Similarly, the Cotton Valley Sandstones are composed of interbedded shale and quartz-rich sandstone layers. The Cotton Valley Group is underlain regionally by the Upper Jurassic Louark Group, which includes other hydrocarbon-bearing formations such as the Smackover carbonates and Haynesville/Cotton Valley Limestones. Overlying the Cotton Valley Group is the regionally productive Lower Cretaceous Travis Peak formation.

As illustrated by the generalized regional dip-section in Figure 3, Bossier deposition represents cycles of sand progradation into the basin and onto organic-rich mud, succeeded by marine transgression. Much of the Bossier interval downdip appears to be time equivalent to the Cotton Valley Sandstones updip and represents pro-delta/delta-front material related to Cotton Valley deltaic systems (Montgomery 2000). The Bossier sands probably originate from the north and west and were transported downslope by slumping, debris flow and turbidity currents. Significant Bossier sand thickness is located in topographic lows created by a combination of faulting, subsidence and salt movement within the basin.

### *Regional overview of the Bossier shales*

The Bossier shale is not only the basal formation for much of the Cotton Valley Group, but is also interbedded with isolated sands throughout most of eastern Texas and northwestern Louisiana. These shales are typically black, organic-rich, calcareous, fossiliferous, marine deposits which are the primary source rock for much of the entire Upper Jurassic and Lower Cretaceous vertical section. Thinner shale intervals in the western part of the basin thicken to the east correlative to general depositional patterns in the East Texas Basin. Source rock quality is generally poorer in the western and northwestern parts of the basin, but improves significantly easterly towards the basin centre. From west to east, the near-shore sand and low-stand fan deposits are usually completely encased in the marine Bossier shale.

*Source rock richness.* Current or present-day source rock richness, as indicated by total

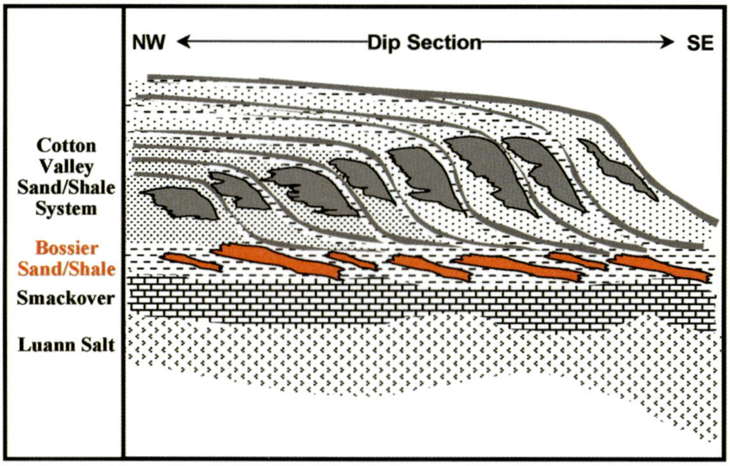


**Fig. 3.** Schematic diagram of the Bossier sands showing high-stand progradational, deltaic depositional model.

organic carbon (TOC), was measured on cuttings samples collected from 15 deep wells completed in the deep Bossier sands. These wells represent an area of several hundred square miles in the East Texas Basin. Cuttings samples were generally obtained over shales in the entire Cotton Valley Group interval. The highest TOC was recorded in the organic-rich black Bossier shale located at the base of the Cotton Valley sequence. TOC values ranged from about 0.25% in the organic-poor northwestern part of the East Texas Basin to more than 5% in the more organic-rich distal shales of the deeper parts of the basin (Fig. 4). This range reflects the heterogeneous nature of source rock quality. The average current TOC from 739 samples is about 1%.

*Kerogen type, content and maturity.* The source rock quality and hydrocarbon generation

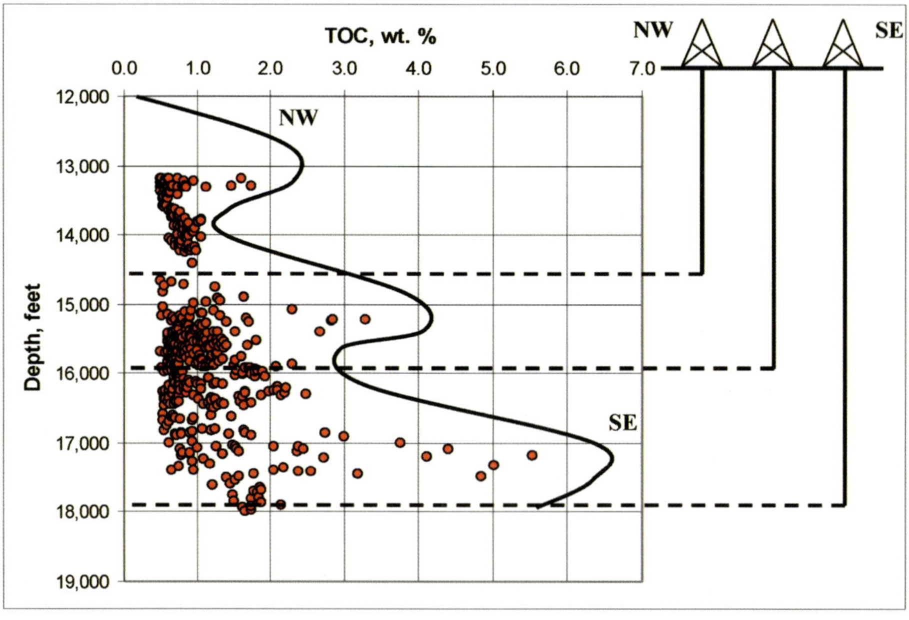


**Fig. 4.** Vertical distribution of present-day Bossier shale TOC values taken from wells located in the northwestern, central and southeastern areas of the East Texas Basin. Note the TOC distribution mimics the fluvial-deltaic progradation of sediments NW–SE.

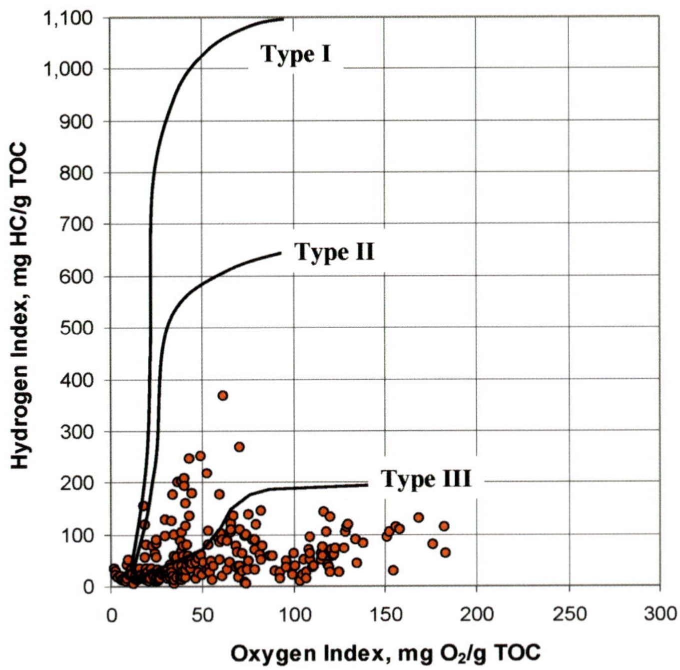


**Fig. 5.** Present-day kerogen type and quality of Bossier shales in the East Texas Basin range from type II to type III.

efficiency were estimated from measurements of kerogen type and content. The Rock-Eval pyrolysis technique (Espitalié *et al.* 1977, 1985) was performed on several hundred shale samples taken from the same fifteen wells used to measure TOC. Figure 5 is a plot of hydrogen index (*HI*) versus oxygen index illustrating the present-day or current Bossier shale kerogen type. Similarly, a plot of kerogen yield (S2) from the rock pyrolysis against TOC (Fig. 6) indicates the same type and quality of kerogen currently present in the Bossier shales.

As determined from the Rock-Eval measurements, the average current hydrogen index value is about 90 mg HC/g TOC, which suggests the Bossier shale is mature but still has capacity to generate some gas. Figure 7, which is plot of *HI* against $T_{max}$, shows the majority of data have values of $T_{max}$ greater than 430 °C, also indicating Bossier shale is mature but has some gas generative capacity remaining. In fact, the average residual kerogen is around 50 mg HC/g TOC.

The Bossier shale ranges from either mixed type II and type III kerogen in the northwestern part of the basin to pure marine (type II) kerogen in the east, thus corroborating the depositional environment of the Bossier shales. Most of the HI values are less than 150–200 mg HC/g TOC, indicating the source rock is currently in the gas generation window.

*Source rock maturity.* In a study of the diagenesis and burial history of the Lower Cretaceous Travis Peak Formation in eastern Texas, Dutton (1987) reported measured vitrinite reflectance values for Travis Peak shales generally ranged from 1.0 to 1.2%. Westcott & Hood (1994) studied regional source rock maturity trends in the East Texas Basin and published thermal maturity and geothermal gradient data indicating vitrinite reflectance data for the Bossier shales ranging from 1.2 to 1.6%. Further, source rock maturity was measured on cuttings samples collected from 15 deep Bossier wells and, as indicated by Figure 8, the vitrinite reflectance data ranged from 0.9% at 10 000 ft to 2.0% at 14 000 ft.

The geothermal gradient linearity together with measured regional vitrinite reflectance values suggest a normal burial history uninterrupted by thermal events. Small variations in the geothermal gradient trend have been attributed to localized effects of salt movement. This burial history was validated from a proprietary basin modelling study, which indicates the Bossier shale reached the oil generation window during early Cretaceous period. At the end of early Palaeocene,

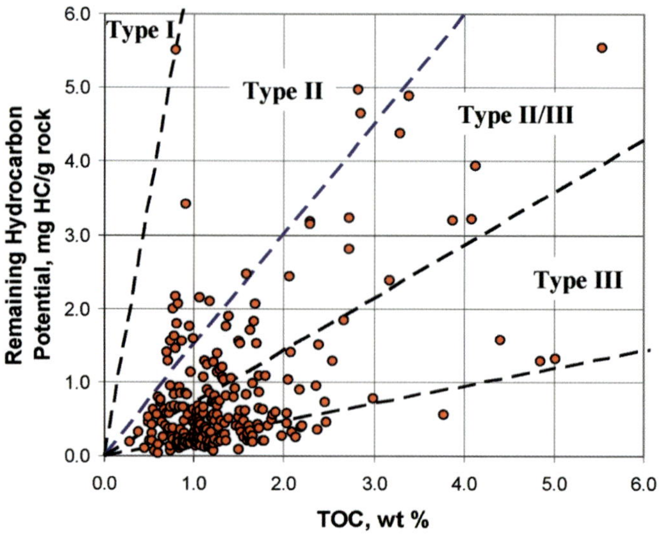


**Fig. 6.** Present-day kerogen content and quality of Bossier shales in the East Texas Basin range from type II to type III.

most of the kerogen had been transformed, and the centre of the basin reached the gas generation window. The data also suggest that gas generation is still ongoing.

*Hydrocarbon and pressure generation.* The proprietary basin modelling study also indicates the Bossier shales reached the oil generation window during early Cretaceous time. Continued burial and the ensuing pressure and temperature increases caused source rock metagenesis and the associated increase in reservoir pressure caused by the liquid hydrocarbon cracking to gas. The Bossier shales reached the gas generation window during early Palaeocene, thus generating significant gas volumes.

Evidence of source rock metagenesis or oil cracking has been indicated by the presence of pyrobitumen observed in micrographic thin sections of Bossier sands (Fig. 9). Gas compositional analysis suggests hydrocarbon liquid

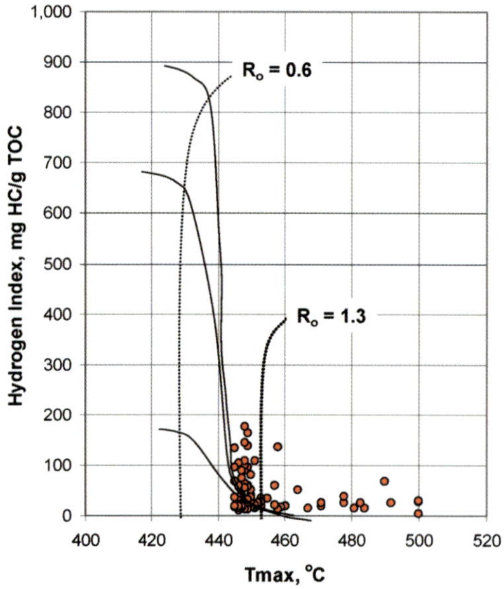


**Fig. 7.** Present-day kerogen type and related maturity of Bossier shales in the East Texas Basin.

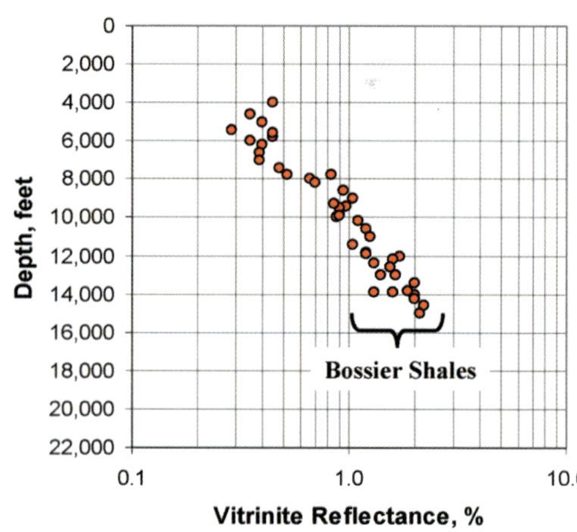


**Fig. 8.** Variation of vitrinite reflectance with depth in the East Texas Basin.

**Fig. 9.** Bossier sand thin sections showing pyrobitumen (dark mass), which is evidence of liquid hydrocarbon cracking to gas.

cracking to gas is continuing today. This phenomenon is validated by the large ratio of iso-butane to normal-butane measured from oil and gas samples obtained from producing wells.

The Bossier Sands are abnormally overpressured throughout the entire East Texas Basin. Pressure gradients range from 0.50 to 0.55 psi/ft in the upper Cotton Valley Sands, 0.60 to 0.65 psi/ft in the upper Bossier Sands, and 0.70 to 0.90 psi/ft in the lower Bossier Sands. A pressure transition exists throughout the Bossier interval with the pressure gradient increasing with depth in a basin-ward direction. This over-pressured system has been confirmed using static bottom-hole pressure measurements.

Bossier sands in this area and throughout the East Texas Basin also exhibit abnormally high temperature gradients. Bottom-hole temperatures range from 280 to 325 °F at depths ranging from 12 500 to 13 500 ft. This corresponds to temperature gradients of 2.2 to 2.4 °F per 100 ft of depth.

## Dew/Mimms Creek Fields: an analogue for petrophysical and reservoir engineering data

The fields chosen as the petrophysical and reservoir engineering analogue for exploration prospects in the Bossier tight gas sand play are the Dew/Mimms Creek Fields. Note that, rather than using these fields as the geological analogue, specific geological models were developed for each exploration prospect, especially in the deeper parts of the basin.

Located in Freestone County in the eastern portion of Texas, these fields currently have more than 300 wells completed in and producing from the low-permeability Bossier sands. These fields have also been the subject of extensive reservoir engineering and petrophysical studies over the past four years, so a vast database was available as input for the mass balance model. A comprehensive Bossier sand description programme, including more than 1000 ft of whole core obtained from three wells in the Dew/Mimms Creek Fields, provided the petrophysical data base for the Bossier exploration programme. In order to evaluate the sedimentary structure and depositional environment, the entire Bossier section, including both reservoir and non-reservoir rock, was cored. This section summarizes the rock and fluid properties of Bossier sands in the Dew/Mimms Creek Fields.

### *Produced fluid properties*

Most wells completed in the Dew/Mimms Creek Fields produce from two to four sands shown in the type log in Figure 10. These wells produce a dry to slightly wet gas with specific gravities ranging from 0.58 to 0.62. Although most wells produce no liquid hydrocarbons, some wells

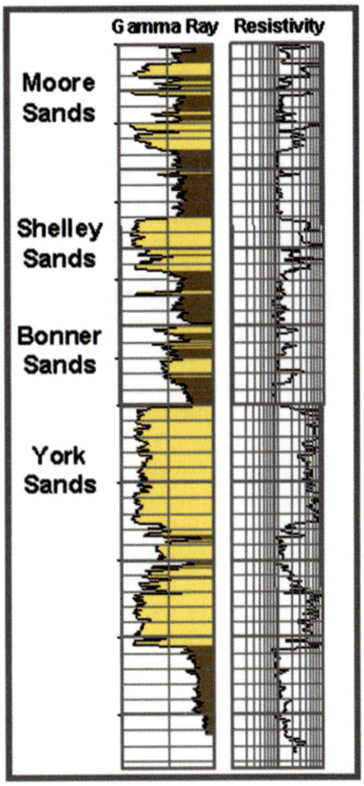


**Fig. 10.** Type log for the Bossier sands in the Dew/Mimms Creek Fields, Freestone County, Texas.

produce condensates with rates averaging one to three STB/MMscf over the life of the well. Gas composition averages 94 mol% methane and 2 mol% ethane. The remaining hydrocarbon mixture includes fractional percentages of propane through hexane with no heptanes or heavier hydrocarbon molecules. Non-hydrocarbon components include 2–3 mol% $CO_2$, 0.2–0.5 mol% $N_2$, and no $H_2S$.

The Bossier sands produce some water, with rates ranging from 1 to 5 bbl/day for the life of the well. It is believed that no mobile liquid water phase exists in the reservoir. The source of long-term water production is probably water vapour that is originally dissolved in the hydrocarbon gas phase at reservoir conditions but condenses at the surface. This observation seems to also be confirmed by the low salinity, ranging from 40 000 to 60 000 ppm NaCl, of the produced waters. These salinities are significantly lower than the *in-situ* connate waters which often have salinities exceeding 200 000 ppm (Newsham & Rushing 2002).

### *In-situ water saturation and salinity*

One of the most critical properties controlling the gas-in-place volume as well as gas productivity is the *in-situ* connate water saturation. Because of their low permeability and porosity, quantification of water saturation is especially important in tight gas sands since very small volumes of water can dramatically and adversely affect rock flow and storage properties. Quantification of water saturation was also important for evaluation of Bossier exploration prospects. Accordingly, a comprehensive program was initiated to measure and verify water saturations in the Dew/Mimms Creek Fields.

*In situ* connate water saturations were measured from several hundred core samples obtained with an oil-based mud programme (Newsham & Rushing 2002). Water saturation in the reservoir-quality rock ranged from as low as 5% in the most permeable rock to as high as 50% in the lower-quality reservoir rock. Water saturations generally exceeded 60% in the non-reservoir rock types. Core-based vertical distributions of water saturation were also validated using both log- and capillary-based calculations.

Salinity of the *in situ* connate water was also measured using both micro-thermometry fluid inclusion and commutation (i.e. residual salt) analyses. The results suggested the range of salinity in the Bossier sand connate waters is 180 000 to more than 350 000 ppm. It is believed these abnormally high salinities and ultra-low water saturations are probably a result of a water vaporization process caused by movement of dry hydrocarbon gas through the system (Newsham & Rushing 2002).

### *Petrographic rock types*

Petrographic rock types were identified based on the constituent mineral distribution, clay composition and clay habitat determined from thin section descriptions, x-ray diffraction (XRD) analysis, and scanning electron microscopy (SEM) imaging. Bossier sand petrographic rock types include clean sandstone, argillaceous/weakly laminated sandstone, dolomitic sandstone, and argillaceous/burrowed siltstone (Newsham & Rushing 2001). The Folk (Folk 1974) classification ranges from sub-arkose to sub-litharentite to lithic wackes. Framework grains average 84% quartz, 6.5% feldspar and 9.5% rock fragments.

Inter-granular constituents are primarily quartz overgrowths, diagenetic clay in the sands, detrital clay found in sand and silt, dolomitic cement, and local pyrite. The clay fraction is dominantly grain coating chlorite and illite. Texturally, the Bossier sands have a narrow range of grain size, typically from upper very fine to fine. The sands are medium to well sorted, while the silts are typically poorly sorted. Sand grain shape is sub-angular to well rounded.

Bossier sands also exhibit a significant diagenetic overprint, including mechanical compaction, cementation from quartz overgrowths, grain-coating/pore lining clay development, and grain dissolution. Figure 11a is an SEM photograph showing the presence of illite and chlorites and their effect on the rock pore structure, while Figure 11b is a thin section micro-photograph of the same rock illustrating the presence of quartz overgrowths.

### *Hydraulic rock types*

Several Bossier sand hydraulic rock types have also been identified (Newsham & Rushing 2001). The hydraulic rock type classification provides a physical measure of a rock's flow and storage properties at current conditions. When described on the basis of the dominant pore throat diameter determined from high-pressure mercury capillary pressure data, distinct groupings of rocks having similar flow and storage properties, i.e. hydraulic rock types, are

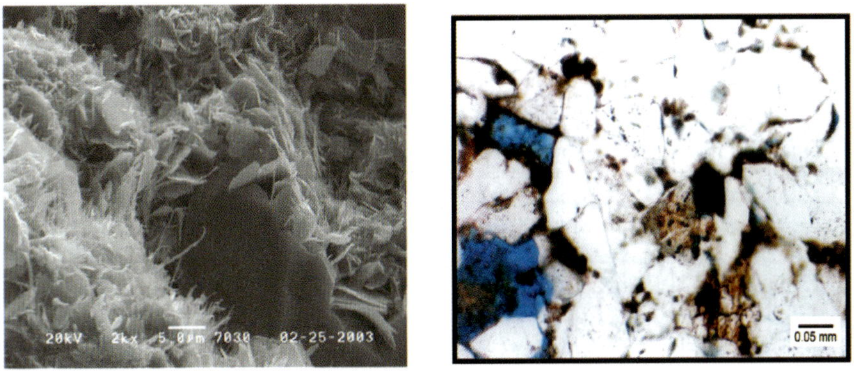


**Fig. 11.** (**a**) SEM micro-photograph showing illite and chlorite clays in the Bossier tight gas sands. (**b**) Thin section micro-photograph showing quartz overgrowths in the Bossier tight gas sands.

observed. Figure 12 is an example of an incremental mercury intrusion plot used to identify rock types (Pittman 1992; Gunter *et al.* 1997; Washburn 1921; Swanson 1979; Thompson *et al.* 1987).

Figure 13 shows the general region of each rock type in porosity–permeability space. Hydraulic rock types 1, 2A, 2B and 3A are all considered to be reservoir-quality rock. Because of the low permeability, high initial water saturation and significant degree of vertical heterogeneity, rock types 3B and 4 are non-reservoir rocks and probably act as flow baffles, barriers and seals. Effective porosity in the reservoir rocks varies from 4 to 15%, while absolute permeability of reservoir rock ranges from 0.001 to 1 md. Non-reservoir and seal rocks have permeability values lower than 0.001 md.

## Application of the mass balance technique

This section illustrates application of the mass balance technique for two exploration prospects in the East Texas Basin. Gas volumes were computed using both the conventional reservoir engineering and mass balance approaches. Furthermore, uncertainty was incorporated into the calculations with a Monte Carlo simulation technique that generates probabilistic distributions of gas volumes.

### *Overview of the Monte Carlo simulation technique*

Monte Carlo simulation is a statistics-based analysis tool that yields probability versus value

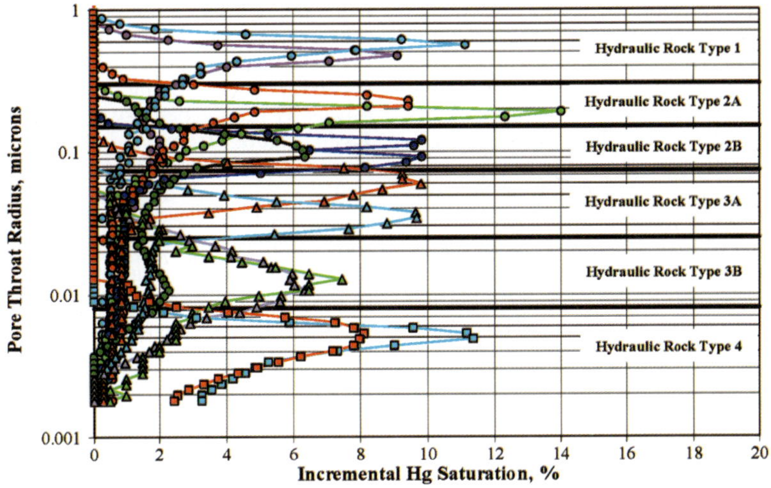


**Fig. 12.** Incremental mercury intrusion plot used to identify hydraulic rock types for the Bossier sands, Dew/Mimms Creek Fields.

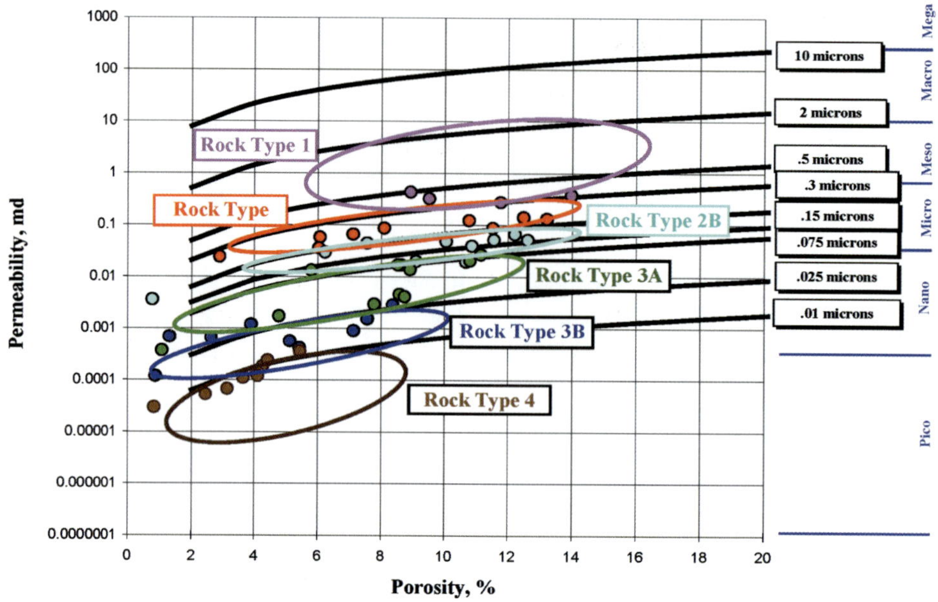


**Fig. 13.** Range of effective porosity and absolute permeability for Bossier sand hydraulic rock types, Dew/Mimms Creek Fields (Winland 1972).

relationships for any parameter of interest. These probability relationships provide indications of the probability or likelihood of some occurrence of that key parameter. The term Monte Carlo dates to the 1940s and refers to the code name assigned to the Manhattan project (Halton 1970; Murtha 1994). The name is also an inference to the European gambling mecca where games of chance are commonly employed.

The Monte Carlo simulation technique begins with a model, i.e. one or more equations, together with the underlying assumptions and logic relating the input parameters in the equation. Each of the input parameters is considered to be an independent variable whose probability of occurrence is defined by a cumulative probability curve. A trial consists of randomly selecting, according to the probability function, one value for each input parameter and calculating the output defined by the model equation. A particular combination of the input parameters represents a single realization, while a simulation represents a succession of hundreds or even thousands of repeated trials or realizations. The resulting realizations are used to develop a cumulative distribution of the output that defines the likelihood of occurrence for the key parameter.

For this study, Monte Carlo simulation was used to capture uncertainty in the evaluation of several basin-centred gas exploration prospects. Specifically, this technique was used to estimate the range of gas-in-place volumes—i.e. the key parameter—and their probability of occurrence. Equations (12) and (14) are the models defining the gas volumes computed from the mass balance and conventional reservoir engineering techniques, respectively.

As discussed previously, the primary analogue and data sources were the Dew/Mimms Creek Fields. Recall that this Bossier tight gas sand field was chosen as the petrophysical and reservoir engineering analogue for the exploration prospects, while specific geological models were used for each of the exploration prospects, especially in deeper parts of the East Texas Basin. Data from the Dew/Mimms Creek Fields were supplemented with some geological and geochemical data from 15 wells in the deeper parts of the basin. When sufficient data were available, unique probability functions (e.g. normal, log-normal etc.) were fit or matched to the data. However, in some cases, very simple functions, such as triangular distributions, were applied, particularly when few data were available. Except for those parameters exhibiting a correlation or dependency, each parameter was assumed to be an independent variable.

For this paper, the distributions were generated using 10 000 realizations. The final results are presented as $P_{10}$, $P_{50}$ and $P_{90}$ gas volumes, where the $P_{10}$ and $P_{90}$ are the low and high values, respectively. The $P_{50}$ represents the most likely value. Note that the definitions used in this paper suggest the actual gas volume has a 10% chance of being less than the $P_{10}$ value or a 10% chance of being greater than the $P_{90}$ value.

## *Summary of results*

Evaluations of two Bossier sand exploration prospects are presented to illustrate the calculation techniques. Although these two prospects were both located in the deeper parts of the basin, several hundred miles separate them. Exploration prospect number one is situated closer to the shelf area, while exploration prospect number two is located closer to the centre of the East Texas Basin. Histograms of $P_{10}$, $P_{50}$ and $P_{90}$ values for gas volumes computed using both mass balance and conventional reservoir engineering approaches are shown for comparison.

Rather than estimating reservoir trapping efficiency from seal capacity measurements, gas volumes expelled from the source rock (equation (13)) were compared with gas volumes computed from conventional reservoir engineering approach (equation (14)). These comparisons provide an indication of the range of possible trapping efficiencies in each prospect area. Additionally, the distribution of source rock expulsion efficiency was computed using a Monte Carlo simulation technique and equation (11). Finally, to eliminate the effect of prospect area on the calculations, gas volumes are essentially normalized by computing the volumes in units of millions of standard cubic feet of gas per prospect acre.

The evaluation of exploration prospect number one is shown in Figure 14. Histograms of the $P_{10}$, $P_{50}$ and $P_{90}$ gas volumes computed with the conventional reservoir engineering method are shown in solid blue, red and green colours, respectively. Furthermore, the $P_{10}$, $P_{50}$ and $P_{90}$ gas volumes expelled from the source rock computed with the mass balance approach are illustrated in chequered blue, red and green colours, respectively. Also shown is the reservoir trapping efficiency computed as the ratio of gas volumes from the conventional reservoir engineering and mass balance approaches. These values range from 39% for the $P_{10}$ case to 19% for the $P_{90}$ case.

Similar results are shown in Figure 15 for exploration prospect number two. Note that the second prospect not only has larger gas-in-place volumes, but also has greater reservoir trapping efficiencies than the first prospect. The larger gas volumes computed from the conventional reservoir engineering method are indicative of the higher pressures in the deeper parts of the East Texas Basin. This comparison also suggests more of the gas generated from the Bossier shale has escaped from the first prospect, thus migrating up-section to other shallower reservoirs. In fact, gas isotope analysis from several shallower zones in prospect area one indicates gas generated from the Bossier shale has mixed with gas from more proximal but less organic-rich shale, thus confirming the observation that gas generated from the Bossier shale has migrated up the column.

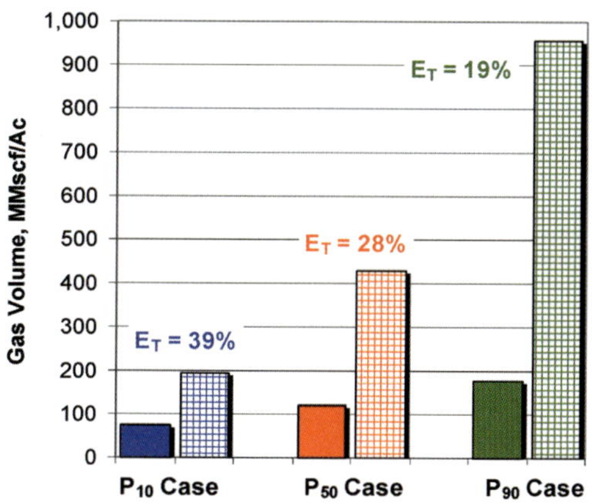


**Fig. 14.** Histograms of the $P_{10}$, $P_{50}$ and $P_{90}$ gas volumes computed with the conventional reservoir engineering (equation (14)) and mass balance (equation (13)) techniques for exploration prospect number one.

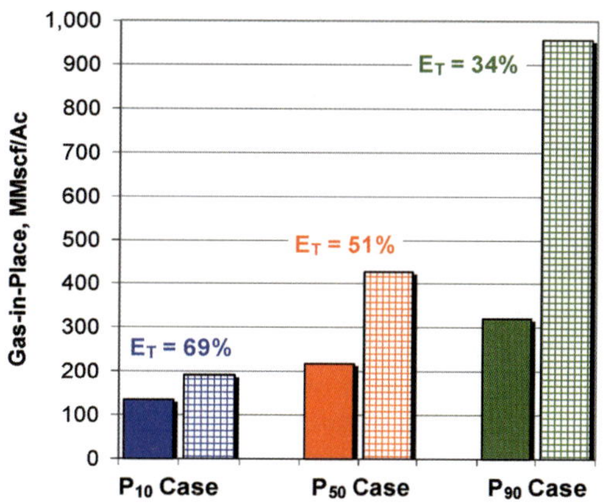


**Fig. 15.** Histograms of the $P_{10}$, $P_{50}$ and $P_{90}$ gas volumes computed with the conventional reservoir engineering (equation (14)) and mass balance (equation (13)) techniques for exploration prospect number two.

## Summary and conclusions

This paper presents a mass balance approach that directly integrates dynamic elements of the total petroleum system with all static reservoir engineering, geochemical and petrophysical components. This prospect evaluation technique provides an independent estimate of hydrocarbon resource-in-place for comparison with that computed from the conventional reservoir engineering approach. In addition, the mass balance approach may be used to quantify the resource lost from the system and quantify the reservoir trapping efficiency. Although presented in the context of a basin-centred gas accumulation, the methodology with some modifications is also applicable to the evaluation of any conventional oil or gas exploration prospect.

## Nomenclature

$B_g$ = gas formation volume, converts gas volume at reservoir conditions to surface conditions, reservoir $ft^3$ of gas per standard $ft^3$ of gas.
$C_{adsorbed}$ = adsorbed gas content from desorption measurements, scf gas per ton shale.
$E_E$ = source rock expulsion efficiency, fraction.
$E_L$ = fraction of liquid hydrocarbons lost from source rock during early hydrocarbon generation, fraction.
$E_T$ = reservoir trapping efficiency, fraction.
$HI_{original}$ = source rock's hydrogen index at original conditions, mg HC/g TOC.
$HI_{current}$ = source rock's hydrogen index at current conditions, mg HC/g TOC.
$\Delta HI$ = hydrogen index differential ($HI_{original} - HI_{current}$, mg HC/g TOC.
$h_{gross}$ = gross sand thickness, ft.
$h_{net}$ = net sand thickness ratio, fraction.
$h_{net}/h_{gross}$ = net-to-gross sand thickness ratio, fraction.
$h_{shale}$ = net shale thickness, ft.
$P_{10}$ = value or occurrence at the 10th percentile or the value corresponding to 0.10 on the cumulative-probability distribution curve.
$P_{50}$ = value or occurrence at the 50th percentile or the value corresponding to 0.50 on the cumulative-probability distribution curve.
$P_{90}$ = value or occurrence at the 90th percentile or the value corresponding to 0.90 on the cumulative-probability distribution curve.
Ro = vitrinite reflectance, percent.
ST = saturation threshold expressed as fraction of the kerogen present, fraction.
$S_w$ = *in-situ* connate water saturation, fraction.
TOC = original total organic carbon, fraction.
$T_{max}$ = maximum temperature reached by present day kerogen simulation.
TR = transformation ratio, fraction.
$V_{entering}$ = gas volume entering the reservoir from source rock.
$V_{expelled}$ = gas volume expelled from the source rock.
$V_{generated}$ = gas volume generated from source rock.
$V_{leaving}$ = gas volume leaving the reservoir.
$V_{remaining}$ = gas volume remaining in the reservoir as computed using the mass balance method.
$V_{RE}$ = gas volume remaining in the reservoir as computed using the conventional reservoir engineering method.
$V_{shale}$ = gas volume generated and remaining in the source rock.
$\phi$ = effective sand porosity, fraction.
$\rho_{oil}$ = density of liquid hydrocarbon, $g/cm^3$.
$\rho_{shale}$ = bulk shale density, $g/cm^3$.

## Abbreviations

BCGS = basin-centred gas system.
bbl = barrels of water or liquid hydrocarbon.
ft = ft.
g = grams.
$g/cm^3$ = grams per cubic centimetre.
mg = milligrams.
md = milli-darcys.
Mscf = thousands of standard cubic feet of gas.
MMscf = millions of standard cubic feet of gas.
MMscf/day = millions of standard cubic feet of gas per day.
ppm = parts per million.
scf = standard cubic feet of gas.
STB = stock tank barrels.
STB/day = stock tank barrels per day.
TDS = total dissolved solids.

°C = degrees Celsius.
°F = degrees Fahrenheit.

The authors would like to thank the management at Anadarko Petroleum Corp. for permission to publish this paper. We would also like to acknowledge all staff geologists, geophysicists, geochemists and reservoir engineers at Anadarko who contributed to the development of this methodology.

## References

DUTTON, S. P. 1987. *Diagenesis and Burial History of the Lower Cretaceous Travis Peak Formation, East Texas*. Report of Investigations No. **164**, Bureau of Economic Geology, The University of Texas at Austin.

EMME, J. J. & STANCIL, R. W. 2002. Anadarko's Bossier gas play—a sleeping giant in a mature basin,' Abstract, *American Association of Petroleum Geologists Bulletin*, **86**(13), Supplement.

ESPITALIÉ, J. *ET AL.* 1977. Source rock characterization method for petroleum exploration. *In*: *Proceedings, 9th Offshore Technology Conference*. OTS, Houston, Texas, 439–442.

ESPITALIÉ, J. *ET AL.* 1985. La pyrolyse Rock-Eval et ses applications. *Revue de l'Institut Français du Pétrole*, **40**, 563–579, 755–784; **41**, 73–89.

FOLK, R. L. 1974. *Petrology of Sedimentary Rocks*. Hemphill Publishing Co., Austin, Texas.

GUNTER, G. W. *ET AL.* 1997. Early determination of reservoir flow units using an integrated petrophysical model. Paper SPE **38679** presented at the SPE Annual Technical Conference and Exhibition, San Antonio, Texas (5–8 October).

HALTON, J. H. 1970. A retrospective and prospective survey of the Monte Carlo method. *SIAM Reviews*, **12**(1).

IVERSON, W. P. *ET AL.* 1994. Pressure seal permeability and two-phase flow. *American Association of Petroleum Geologists, Memoir*, **61**, 313–320.

LAW, B. E. 1984a. Relationships of source rocks, thermal maturity, and overpressuring to gas generation and occurrence in low-permeability Upper Cretaceous and lower Tertiary rocks, Greater Green River Basin, Wyoming, Colorado, and Utah. *In*: WOODWARD, J., MEISSNER, F. F. & CLAYTON, J. L. (eds) *Hydrocarbon Source Rocks of the Greater Rocky Mountain Region*. Rocky Mountain Association of Geologists, 469–490.

LAW, B. E. (ed.) 1984b. *Geological characteristics of low-permeability Upper Cretaceous and lower Tertiary rocks in the Pinedale anticline area, Sublette County, Wyoming*. U.S. Geological Survey Open-File Report **84**, 753.

LAW, B. E. 1995. Columbia basin-centred gas play (0503) in Tennyson, M.E., Eastern Oregon Washington Province (005). *In*: GAUTIER, D. L., DOLTON, G. L., TAKAHASHI, K. L. & VARNES, K. (eds) *1995 National Assessment in the United States of Oil and Gas Resources: Results, Methodology, and Supporting Data*. U.S. Geological Survey Data Series 30.

LAW, B. E. 2000. What is a basin-centered gas system? *In*: *2000 Basin-Center Gas Symposium*. Rocky Mountain Association of Geologists, Denver, Colorado (6 October).

LAW, B. E. & DICKINSON, W. W. 1985. A conceptual model for the origin of abnormally pressured gas accumulations in low-permeability reservoirs. *American Association of Petroleum Geologists Bulletin*, **69**, 1295–1304.

LEE, W. J. & WATTENBARGER, R. A. 1996. *Gas Reservoir Engineering*. Society of Petroleum Engineers, Textbook Series, **5**, 230–232.

MASTERS, J. A. 1979. Deep basin gas trap, western Canada. *American Association of Petroleum Geologists Bulletin*, **63**, 152–181.

MASTERS, J. A. (ed.) 1984. Elmworth—Case study of a deep basin gas field. *American Association of Petroleum Geologists Memoir*, **38**.

MASTERS, J. A. 2000. New game. *In*: 2000 *Basin-Center Gas Symposium*. Rocky Mountain Association of Geologists, Denver, Colorado (6 October).

MONTGOMERY, S. 2000. East Texas Basin Bossier gas play. *In*: ANDERSON, J. (ed.) *Petroleum Frontiers*. IHS Energy Group, Englewood, Colorado, 1–60.

MURTHA, J. A. 1994. Incorporating historical data in Monte Carlo simulation. *SPE Computer Applications*, 11.

NEWSHAM, K. E. & RUSHING J. A. 2001. An integrated work-flow model to characterize unconventional gas resources. Part I. Geological assessment and petrophysical evaluation. Paper SPE **71351** presented at the SPE Annual Technical Conference and Exhibition, New Orleans, Louisiana (1–3 October).

NEWSHAM, K. E. & RUSHING, J. A. 2002. Laboratory and field observations of an apparent sub capillary-equilibrium water saturation distribution in a tight gas sand reservoir. *In*: *Proceedings, 2002 Gas Technology Symposium* Society of Petroleum Engineers.

NUCCIO, V. F., SCHMOKER, J. W. & FOUCH, T. D. 1992. Thermal maturity, porosity, and lithofacies relationships applied to gas generation and production in Cretaceous and Tertiary low-permeability (tight) sandstones, Uinta Basin, Utah. *In*: FOUCH, T. D., NUCCIO, V. F. & CHIDSEY, T. C. JR. (eds) *Hydrocarbon and Mineral Resources of the Uinta Basin, Utah and Colorado*. Utah Geological Association Guidebook, **20**, 77–94.

ORTOLEVAC, P. J. (ed.) 1994. Basin compartments and seals. *American Association of Petroleum Geologists Memoir*, **61**.

PITTMAN, E. D. 1992. Relationship of porosity and permeability to various parameters derived from mercury injection-capillary pressure curves for sandstone. *American Association of Petroleum Geologists Memoir*, **76**(2), 191–198.

RYDER, R. T. *ET AL.* 1996. Continuous-type (basin-centred) gas accumulation in the Lower Silurian 'Clinton' sands, Medina Group, and Tuscarora Sandstone in the Appalachian Basin. *American Association of Petroleum Geologists Bulletin*, **80**, 1531.

SHMOKER, J. W. 1994. Volumetric calculation of hydrocarbon generated. *American Association of Petroleum Geologists Memoir*, **60**, 323–326.

SURDAM, R. C. (ed.) 1997. Seals, traps, and the petroleum system. *American Association of Petroleum Geologists Memoir*, **67**.

SURDAM, R. C. ET AL. 1997. Anomalously pressured gas compartments in Cretaceous rocks of the Laramide Basins of Wyoming: a new class of hydrocarbon accumulation. *American Association of Petroleum Geologists Memoir*, **67**.

SWANSON, B. F. 1979. A simple correlation between permeabilities and mercury capillary pressure. Paper SPE **8234** presented at the SPE Annual Technical Conference and Exhibition, Las Vegas, Nevada (23–26 September).

THOMPSON, A. H. ET AL. 1987. Estimation of absolute permeability from capillary pressure measurements. Paper SPE **16794** presented at the SPE Annual Technical Conference and Exhibition, Dallas, Texas (27–30 September).

TISSOT, B. P. & WELTE, D. H. 1984. *Petroleum Formation and Occurrence*. Springer, New York.

WAPLES, D. 2000. *Basic Geochemistry*. Anadarko Petroleum Corp, short course, Houston, Texas.

WASHBURN, E. W. 1921. Note on a method of determining the distribution of pore sizes in a porous material. *Proceedings of the National Academy of Science*, **7**, 15–116.

WESTCOTT, W. A. & HOOD, W. C. 1994. Hydrocarbon generation and migration routes in East Texas Basin. *American Association of Petroleum Geologists Bulletin*, **78**(2), 287–307.

WINLAND, H. D. 1972. Internal document from Amoco Production Company.

# Index

**Note:** Page numbers in **bold** refer to tables; those in *italics* refer to figures (while some are included within page ranges).